T0298323

Random matrix theory is at the intersection of linear algebra, probability theory and integrable systems, and has a wide range of applications in physics, engineering, multivariate statistics and beyond. This volume is based on a Fall 2010 MSRI program which generated the solution of long standing questions on universalities of Wigner matrices and beta-ensembles, and opened new research directions especially in relation to the KPZ universality class of interacting particle systems, and low rank perturbations. The book contains review articles and research contributions on all these topics, in addition to other core aspects of random matrix theory such as integrability and free probability theory. It will give both established and new researchers insights into the most recent advances in field and the connections among many subfields.

Mathematical Sciences Research Institute
Publications

65

Random Matrix Theory, Interacting Particle Systems, and Integrable Systems

Mathematical Sciences Research Institute Publications

Random Matrix Theory, Interacting Particle Systems, and Integrable Systems

Edited by

Percy Deift

Courant Institute

Peter Forrester

University of Melbourne

Percy Deift
Courant Institute
deift@cims.nyu.edu

Peter Forrester
University of Melbourne
p.forrester@ms.unimelb.edu.au

Silvio Levy (Series Editor)
Mathematical Sciences Research Institute
levy@msri.org

The Mathematical Sciences Research Institute wishes to acknowledge support by the National Science Foundation and the *Pacific Journal of Mathematics* for the publication of this series.

CAMBRIDGE
UNIVERSITY PRESS

32 Avenue of the Americas, New York NY 10013-2473, USA

Cambridge University Press is part of the University of Cambridge.

It furthers the University's mission by disseminating knowledge in the pursuit of education, learning and research at the highest international levels of excellence.

www.cambridge.org
Information on this title: www.cambridge.org/9781107079922

© Mathematical Sciences Research Institute 2014

First published 2014

A catalogue record for this publication is available from the British Library

Library of Congress Cataloguing in Publication data
  Random matrix theory, interacting particle systems, and integrable systems / edited by Percy Deift, Courant Institute, Peter Forrester, University of Melbourne.
    pages   cm. – (Mathematical Sciences Research Institute publications)
  "Mathematical Sciences Research Institute."
  Includes bibliographical references and index.
  ISBN 978-1-107-07992-2 (hardback)
  1. Random matrices.   I. Deift, Percy, 1945– editor.   II. Forrester, Peter (Peter John) editor.   III. Mathematical Sciences Research Institute (Berkeley, Calif.)
  QA196.5.R37   2014
  512.9´434–dc23   2014043446

ISBN 978-1-107-07992-2 Hardback

Random Matrices
MSRI Publications
Volume **65**, 2014

Contents

Random Matrices
MSRI Publications
Volume 65, 2014

Preface

In the spring of 1999, MSRI hosted a very successful and influential one-semester program on random matrix theory (RMT) and its applications. At the workshops during the semester, there was a sense of excitement as brand new and very recent results were reported. The goal of the 2010 Program has been to showcase the many remarkable developments that have taken place since 1999 and to spur further developments in RMT and related areas of interacting particle systems (IPS) and integrable systems (IS) as well as to highlight various applications of RMT.

One of the outputs of the 1999 program was volume 40 in the MSRI Publications series, entitled "Random matrix models and their applications". Looking back on this publication today, it is clear that this volume gave a representative snapshot of topics that were occupying the attention of researchers in the field then. Moreover, the papers — consisting of a mix of research articles and reviews — provide a conveniently bundled resource for researchers in the field to this day.

Since 1999 random matrix theory has captured the imagine of a whole new generation of researchers, and through a collective effort some outstanding questions have been settled, and new highly promising research areas initiated. One example of the former is work on universality questions for Wigner matrices, where the task is to show that for large dimension a symmetric matrix with independent entries of mean zero and standard deviation 1 has the same statistical properties as in the case of standard Gaussian entries. Another is universality questions for β ensembles, where one wants to show that the statistical properties are independent of the one-body potential. New research areas include the KPZ equation and related growth processes, which has led to the precise experimental realization of some random matrix distributions, and also to quite spectacular theoretical advances relating to a rigorous understanding of the replica trick via so called Macdonald processes; analytic predictions of the β-generalization of the so-called Dyson constant in the asymptotic expansion of spacing distributions in β-ensembles; and stochastic differential equations and PDEs for eigenvalue distributions in the case of a low rank perturbation leading to eigenvalue separation.

A core aim of the 2010 semester was to spur further developments in RMT and the related areas of interacting particle systems and integrable systems. It is our hope that this new MSRI Publications volume based on the 2010 semester will lend weight to this cause. Each author was a participant of the semester. Articles on all topics nominated above relating to solutions of outstanding questions and new research areas can be found: universality for Wigner matrices (Tao and Vu), universality for β ensembles (Borodin, Shcherbina); KPZ equation (Quastel, Sasamoto, Spohn, Takeuchi), Macdonald process (O'Connell), Dyson constant and asymptotics of spacing distributions (Forrester), low rank perturbations (Baik and Wang, Maida). One should also highlight the work on RMT and numerical algorithms by Pfrang, Deift and Menon, which is in the spirit of one of the very early uses of random matrices by von Neumann and co-workers at the dawn of the computer era, and the extensive review of free probability theory by Novak, the latter being a write up of a series of lectures he delivered during the semester.

Percy Deift (Courant Institute)
Peter Forrester (University of Melbourne)

Random Matrices
MSRI Publications
Volume **65**, 2014

Universality conjecture for all Airy, sine and Bessel kernels in the complex plane

GERNOT AKEMANN AND MICHAEL J. PHILLIPS

We address the question of how the celebrated universality of local correlations for the real eigenvalues of Hermitian random matrices of size $N \times N$ can be extended to complex eigenvalues in the case of random matrices without symmetry. Depending on the location in the spectrum, particular large-N limits (the so-called weakly non-Hermitian limits) lead to one-parameter deformations of the Airy, sine and Bessel kernels into the complex plane. This makes their universality highly suggestive for all symmetry classes. We compare all the known limiting real kernels and their deformations into the complex plane for all three Dyson indices $\beta = 1, 2, 4$, corresponding to real, complex and quaternion real matrix elements. This includes new results for Airy kernels in the complex plane for $\beta = 1, 4$. For the Gaussian ensembles of elliptic Ginibre and non-Hermitian Wishart matrices we give all kernels for finite N, built from orthogonal and skew-orthogonal polynomials in the complex plane. Finally we comment on how much is known to date regarding the universality of these kernels in the complex plane, and discuss some open problems.

1. Introduction

The topic of universality in Hermitian random matrix theory (RMT) has attracted a lot of attention in the mathematics community recently, particularly in the context of matrices with elements that are independent random variables, as reviewed in [Tao and Vu 2012; Erdős and Yau 2012]. The question that one tries to answer is this: under what conditions are the statistics of eigenvalues of $N \times N$ matrices with independent Gaussian variables the same (for large matrices) as for more general RMT where matrix elements may become coupled? This has been answered under very general assumptions, and we refer to some recent reviews on invariant [Kuijlaars 2011; Deift and Gioev 2009] and noninvariant [Tao and Vu 2012; Erdős and Yau 2012] ensembles.

In this short note we would like to advocate the idea that non-Hermitian RMT with eigenvalues in the complex plane also warrants the investigation of universality. Apart from the interest in its own right, these models have important applications in physics and other sciences (see, e.g., [Akemann et al. 2011a]). We will focus here on RMT that is close to Hermitian, a regime that is particularly important for applications in quantum chaotic scattering (see [Fyodorov and Sommers 2003] for a review) and quantum chromodynamics (QCD), for example. In the latter case, the non-Hermiticity may arise from describing the effect of quark chemical potential (as reviewed in [Verbaarschot 2011; Akemann 2007]), or from finite lattice spacing effects of the Wilson–Dirac operator (see [Damgaard et al. 2010; Akemann et al. 2011b] as well as [Kieburg 2012] for the solution of this non-Hermitian RMT).

Being a system of N coupled eigenvalues, Hermitian RMT already offers a rich variety of large-N limits, where one has to distinguish the bulk and (soft) edge of the spectrum for Wigner–Dyson (WD) ensembles, and in addition the origin (hard edge) for Wishart–Laguerre (WL, or chiral) RMT. Not surprisingly complex eigenvalues offer even more possibilities. The limit we will investigate is known as the weakly non-Hermitian regime; it *connects* Hermitian and (strongly) non-Hermitian RMT, and was first introduced in [Fyodorov et al. 1997a; 1997b] in the bulk of the spectrum. For strong non-Hermiticity — which includes the well-known circular law and the corresponding universality results — we refer to [Khoruzhenko and Sommers 2011] and references therein, although the picture there is also far from being complete; an important breakthrough was published recently in [Tao and Vu 2014].

In the next section we give a brief list of the six non-Hermitian WD and WL ensembles, and indicate where they were first solved in the weak limit. There are three principal reasons why we believe that universality may hold. First, in some cases two different Gaussian RMT both give the same answers. Second, there are heuristic arguments available for i.i.d. matrix elements using supersymmetry [Fyodorov et al. 1998], as well as for invariant non-Gaussian ensembles using large-N factorisation and orthogonal polynomials (OP) [Akemann 2002]. Third, the resulting limiting kernels of (skew-) OP look very similar to the corresponding kernels of real eigenvalues, being merely one-parameter deformations of them. One of the main goals of this paper is to illustrate this fact. For this purpose we give a complete list of all the known Airy, sine, and Bessel kernels for real eigenvalues, side-by-side with their deformed kernels in the complex plane, where some of our results are new.

2. Random matrices and their limiting kernels

In this section we briefly introduce the Gaussian random matrix ensembles that we consider, and give a list of the limiting kernels they lead to, for both real and complex eigenvalues. For simplicity we have restricted ourselves to Gaussian ensembles in the Hermitian cases, in order to highlight the parallels to their non-Hermitian counterparts.

We begin with the classical WD and Ginibre ensembles in Section 2.1, displaying Airy (Section 2.2) and sine (Section 2.3) behaviour at the (soft) edge and in the bulk of the spectrum respectively, as well as their deformations. We then introduce the WL ensembles and their non-Hermitian counterparts in Section 2.4, in order to access the Bessel behaviour (Section 2.5) at the origin (or hard edge). The corresponding orthogonal and skew-orthogonal Hermite and Laguerre polynomials are given in Appendix A, and precise statements of the limits that lead to the microscopic kernels can be found in Appendix B.

2.1. Gaussian ensembles with eigenvalues on $\mathbb{R}$ and $\mathbb{C}$. The three classical Gaussian Wigner–Dyson ensembles (the GOE, GUE and GSE) are defined as [Mehta 2004]

$$
\begin{aligned}
\mathcal{Z}_N^{\text{G}\beta\text{E}} &= \int dH \exp[-\beta \operatorname{Tr} H^2/4] \\
&= c_{N,\beta} \prod_{j=1}^{N} \int_{\mathbb{R}} dx_j \, w_\beta(x_j) \, |\Delta_N(\{x\})|^\beta.
\end{aligned}
\tag{2.1}
$$

The random matrix elements H_{kl} are real, complex, or quaternion real numbers for $\beta = 1, 2, 4$ respectively, with the condition that the $N \times N$ matrix H (N is taken to be even for simplicity) is real symmetric, complex Hermitian or complex Hermitian and self-dual for $\beta = 1, 2, 4$. In the first equation we integrate over all independent matrix elements denoted by dH. The Gaussian weight completely factorises and thus the independent elements are normal random variables; for $\beta = 1$, for example, the real elements are distributed $\mathcal{N}(0, 1)$ for off-diagonal elements, and $\mathcal{N}(0, 2)$ for diagonal elements.

In the second equality of (2.1), we diagonalised the matrix

$$
H = U \operatorname{diag}(x_1, \ldots, x_N) \, U^{-1},
$$

where U is an orthogonal, unitary or unitary-symplectic matrix for $\beta = 1, 2, 4$. The integral over U factorises and leads to the known constants $c_{N,\beta}$. We obtain a Gaussian weight $w_\beta(x)$ and the Vandermonde determinant $\Delta_N(\{x\})$ from the Jacobian of the diagonalisation,

$$
w_\beta(x) = \exp[-\beta x^2/4], \quad \Delta_N(\{x\}) = \prod_{1 \leq l < k \leq N} (x_k - x_l).
\tag{2.2}
$$

The integrand on the right-hand side of (2.1) times $c_{N,\beta}/\mathcal{Z}_N^{\text{G}\beta\text{E}}$ defines the normalised joint probability distribution function (jpdf) of all eigenvalues. The k-point correlation function R_k^β, which is proportional to the jpdf integrated over

$N - k$ eigenvalues, can be expressed through a single kernel $K_N^{\beta=2}$ of orthogonal polynomials (OP) for $\beta = 2$, or through a 2×2 matrix-valued kernel involving skew-OP for $\beta = 1, 4$:

$$
\begin{aligned}
R_k^{\beta=2}(x_1, \ldots, x_k) &= \det_{i,j=1,\ldots,k} [K_N^{\beta=2}(x_i, x_j)], \\
R_k^{\beta=1,4}(x_1, \ldots, x_k) &= \Pf_{i,j=1,\ldots,k} \left[\begin{pmatrix} K_N^{\beta=1,4}(x_i, x_j) & -G_N^{\beta=1,4}(x_i, x_j) \\ G_N^{\beta=1,4}(x_j, x_i) & -W_N^{\beta=1,4}(x_i, x_j) \end{pmatrix} \right].
\end{aligned}
\tag{2.3}
$$

The matrix kernel elements K_N and W_N are not independent of G_N but are related by differentiation and integration, respectively. These relations will be given later for the limiting kernels.

The three parameter-dependent Ginibre (i.e., elliptic or Ginibre–Girko) ensembles, denoted by GinOE, GinUE, and GinSE, can be written as

$$
\begin{aligned}
\mathscr{L}_N^{\mathrm{Gin}\beta\mathrm{E}}(\tau) &= \int dJ \exp\left[\frac{-\gamma_\beta}{1 - \tau^2} \mathrm{Tr}\left(JJ^\dagger - \frac{\tau}{2}(J^2 + J^{\dagger 2}) \right) \right] \\
&= \int dH_1 \, dH_2 \exp\left[-\frac{\gamma_\beta \, \mathrm{Tr}\, H_1^2}{1 + \tau} - \frac{\gamma_\beta \, \mathrm{Tr}\, H_2^2}{1 - \tau} \right],
\end{aligned}
\tag{2.4}
$$

with $\tau \in [0, 1)$. We use the parametrisation of [Khoruzhenko and Sommers 2011], with $\gamma_{\beta=2} = 1$ and $\gamma_{\beta=1,4} = \frac{1}{2}$. The matrix elements of J are of the same types as for H for all three values of β, but without any further symmetry constraint. Decomposing $J = H_1 + i H_2$ into its Hermitian and anti-Hermitian parts, these ensembles can be viewed as Gaussian two-matrix models. For $\tau = 0$ (maximal non-Hermiticity) the distribution for all matrix elements again factorises. In the opposite, that is, Hermitian, limit ($\tau \to 1$), the parameter-dependent Ginibre ensembles become the Wigner–Dyson ensembles. The jpdf of complex (and real) eigenvalues can be computed by transforming J into the following form, $J = U(Z + T)U^{-1}$. For $\beta = 2$ this is the Schur decomposition, with $Z = \mathrm{diag}(z_1, \ldots, z_N)$ containing the complex eigenvalues, and T being upper triangular:[1]

$$
\begin{aligned}
\mathscr{L}_N^{\mathrm{GinUE}}(\tau) &= c_{N,\mathbb{C}}^{\beta=2} \prod_{j=1}^N \int_{\mathbb{C}} d^2 z_j \, w_{\beta=2}^{\mathbb{C}}(z_j) \, |\Delta_N(\{z\})|^2, \\
w_{\beta=2}^{\mathbb{C}}(z) &= \exp\left[\frac{-1}{1 - \tau^2} \left(|z|^2 - \frac{\tau}{2}(z^2 + z^{*2}) \right) \right].
\end{aligned}
\tag{2.5}
$$

For $\beta = 1, 4$ we follow [Khoruzhenko and Sommers 2011] where the two ensembles have been cast into a unifying framework. For simplicity we choose

---

[1] The resulting jpdf of complex eigenvalues for normal matrices with $T \equiv 0$ at $\beta = 2$ is the same.

N to be even. Here the matrix Z can be chosen to be 2×2 block diagonal and T to be upper block triangular. The calculation of the jpdf reduces to a 2×2 calculation, yielding

$$\mathcal{L}_N^{\mathrm{GinO/SE}}(\tau) = c_{N,\mathbb{C}}^{\beta=1,4} \prod_{j=1}^{N} \int_{\mathbb{C}} d^2 z_j \prod_{k=1}^{N/2} \mathcal{F}_{\beta=1,4}^{\mathbb{C}}(z_{2k-1}, z_{2k}) \, \Delta_N(\{z\}), \qquad (2.6)$$

where we have introduced an antisymmetric bivariate weight function. For $\beta = 1$, this is given by

$$\mathcal{F}_{\beta=1}^{\mathbb{C}}(z_1, z_2) = w_{\beta=1}^{\mathbb{C}}(z_1) w_{\beta=1}^{\mathbb{C}}(z_2)$$
$$\times \left(2i \delta^2(z_1 - z_2^*) \operatorname{sign}(y_1) + \delta^1(y_1) \delta^1(y_2) \operatorname{sign}(x_2 - x_1) \right),$$

$$(w_{\beta=1}^{\mathbb{C}}(z))^2 = \operatorname{erfc}\left(\frac{|z - z^*|}{\sqrt{2(1-\tau^2)}} \right) \exp\left[\frac{-1}{2(1+\tau)}(z^2 + z^{*2}) \right], \qquad (2.7)$$

and for $\beta = 4$ by

$$\mathcal{F}_{\beta=4}^{\mathbb{C}}(z_1, z_2) = w_{\beta=4}^{\mathbb{C}}(z_1) w_{\beta=4}^{\mathbb{C}}(z_2) (z_1 - z_2) \delta(z_1 - z_2^*),$$
$$(w_{\beta=4}^{\mathbb{C}}(z))^2 = w_{\beta=2}^{\mathbb{C}}(z). \qquad (2.8)$$

For $\beta = 1$, it should be noted that the integrand in (2.6) is not always positive, and so a symmetrisation must be applied when determining the correlation functions below.[2] For $\beta = 4$, the parameter N in (2.6) should — in our convention — be taken to be the size of the complex-valued matrix that is equivalent to the original quaternion real matrix.

The correlation functions can be written in a similar form as for the real eigenvalues

$$R_{k,\mathbb{C}}^{\beta=2}(z_1, \ldots, z_k) = \det_{i,j=1,\ldots,k} [K_{N,\mathbb{C}}^{\beta=2}(z_i, z_j^*)],$$

$$R_{k,\mathbb{C}}^{\beta=1,4}(z_1, \ldots, z_k) = \Pf_{i,j=1,\ldots,k} \left[\begin{pmatrix} K_{N,\mathbb{C}}^{\beta=1,4}(z_i, z_j) & -G_{N,\mathbb{C}}^{\beta=1,4}(z_i, z_j) \\ G_{N,\mathbb{C}}^{\beta=1,4}(z_j, z_i) & -W_{N,\mathbb{C}}^{\beta=1,4}(z_i, z_j) \end{pmatrix} \right], \qquad (2.9)$$

where the elements of the matrix kernels are related through

$$G_{N,\mathbb{C}}^{\beta=1,4}(z_i, z_j) = -\int_{\mathbb{C}} d^2 z \, K_{N,\mathbb{C}}^{\beta=1,4}(z_i, z) \mathcal{F}_{\beta=1,4}^{\mathbb{C}}(z, z_j),$$

$$W_{N,\mathbb{C}}^{\beta=1,4}(z_i, z_j) = \int_{\mathbb{C}^2} d^2 z \, d^2 z' \, \mathcal{F}_{\beta=1,4}^{\mathbb{C}}(z_i, z) K_{N,\mathbb{C}}^{\beta=1,4}(z, z') \mathcal{F}_{\beta=1,4}^{\mathbb{C}}(z', z_j) \qquad (2.10)$$
$$- \mathcal{F}_{\beta=1,4}^{\mathbb{C}}(z_i, z_j).$$

---

[2]It is, however, possible to write the partition function $\mathcal{L}_N^{\mathrm{GinOE}}$ as an integral over a true (i.e., positive) jpdf, by, for example, appropriately ordering the eigenvalues; however, such a representation is technically more difficult to work with.

The kernels $K_{N,\mathbb{C}}^{\beta}(z, z')$ are given explicitly in Appendix A.

For $\beta = 1$, we can write

$$G_{N,\mathbb{C}}^{\beta=1}(z_1, z_2) = \delta^1(y_2) G_{N,\mathbb{C},\text{real}}^{\beta=1}(z_1, x_2) + G_{N,\mathbb{C},\text{com}}^{\beta=1}(z_1, z_2),$$

$$\begin{aligned}
W_{N,\mathbb{C}}^{\beta=1}(z_1, z_2) &= \delta^1(y_1)\delta^1(y_2) W_{N,\mathbb{C},\text{real,real}}^{\beta=1}(x_1, x_2) \\
&\quad + \delta^1(y_1) W_{N,\mathbb{C},\text{real,com}}^{\beta=1}(x_1, z_2) + \delta^1(y_2) W_{N,\mathbb{C},\text{com,real}}^{\beta=1}(z_1, x_2) \\
&\quad + W_{N,\mathbb{C},\text{com,com}}^{\beta=1}(z_1, z_2) - \mathscr{F}_{\beta=1}^{\mathbb{C}}(z_1, z_2),
\end{aligned} \tag{2.11}$$

whereas, for $\beta = 4$, (2.8) implies the following relations:

$$\begin{aligned}
G_{N,\mathbb{C}}^{\beta=4}(z_1, z_2) &= (z_2 - z_2^*) w_{\beta=2}^{\mathbb{C}}(z_2) K_{N,\mathbb{C}}^{\beta=4}(z_1, z_2^*), \\
W_{N,\mathbb{C}}^{\beta=4}(z_1, z_2) &= -(z_1 - z_1^*)(z_2 - z_2^*) w_{\beta=2}^{\mathbb{C}}(z_1) w_{\beta=2}^{\mathbb{C}}(z_2) K_{N,\mathbb{C}}^{\beta=4}(z_1^*, z_2^*),
\end{aligned} \tag{2.12}$$

where in the final expression we have dropped the term representing the perfect correlation between an eigenvalue z and its complex conjugate z^*. For this reason, for $\beta = 4$ we will only give one of the matrix kernel elements in the following.

Note that $\beta = 1$ is special as the eigenvalues of a real asymmetric matrix are either real or come in complex conjugate pairs. Therefore we will have to distinguish kernels (and k-point densities) of real, complex or mixed arguments.

In order to specify the limiting kernels we first need the behaviour of the mean (or macroscopic) spectral density. At large N, and for all three values of β, the (real) eigenvalues in the Hermitian ensembles are predominantly concentrated within the Wigner semicircle $\rho_{\text{sc}}(x) = (2\pi N)^{-1}\sqrt{4N - x^2}$ on $[-2\sqrt{N}, 2\sqrt{N}]$, whereas in the non-Hermitian ensemble, the complex eigenvalues lie mostly within an ellipse with half-axes of lengths $(1 + \tau)\sqrt{N}$ and $(1 - \tau)\sqrt{N}$, with constant density $\rho_{\text{el}}(z) = (N\pi(1 - \tau^2))^{-1}$. Depending on where (and how) we magnify the spectrum locally, we obtain different asymptotic Airy or sine kernels for each $\beta = 1, 2, 4$. In the following we will give all of the known real kernels; see [Kuijlaars 2011], for example, for a complete list and references, together with their deformations into the complex plane. For the Bessel kernels which will be introduced later we need to consider different matrix ensembles, see Section 2.4 below.

2.2. Limiting Airy kernels on $\mathbb{R}$ and $\mathbb{C}$. When appropriately zooming into the "square root" edge of the semicircle, the three well-known Airy kernels (matrix-valued for $\beta = 1, 4$) are obtained for real eigenvalues. For complex eigenvalues we have to consider the vicinity of the eigenvalues on a thin ellipse which have the largest real parts, and where the weakly non-Hermitian limit introduced in [Bender 2010] is defined such that

$$\sigma = N^{\frac{1}{6}}\sqrt{1 - \tau} \tag{2.13}$$

remains fixed (see Appendix B for the precise details of the scaling of the eigenvalues). This leads to one-parameter deformations of the Airy kernels in the complex plane. Whilst the results for $\beta = 2$ are already known [Bender 2010; Akemann and Bender 2010], our results for $\beta = 1, 4$, stated below, are new [Akemann and Phillips 2014]:

$\beta = 2$:

$$K_{\text{Ai}}^{\beta=2}(x_1, x_2) = \frac{\text{Ai}(x_1)\,\text{Ai}'(x_2) - \text{Ai}'(x_1)\,\text{Ai}(x_2)}{x_1 - x_2}$$

$$= \int_0^\infty dt\,\text{Ai}(x_1 + t)\,\text{Ai}(x_2 + t), \tag{2.14}$$

$$K_{\text{Ai},\mathbb{C}}^{\beta=2}(z_1, z_2) = \frac{1}{\sigma\sqrt{\pi}}\exp\left(-\frac{y_1^2 + y_2^2}{2\sigma^2} + \frac{\sigma^6}{6} + \frac{\sigma^2(z_1 + z_2)}{2}\right)$$

$$\times \int_0^\infty dt\,e^{\sigma^2 t}\,\text{Ai}\left(z_1 + t + \frac{\sigma^4}{4}\right)\text{Ai}\left(z_2 + t + \frac{\sigma^4}{4}\right). \tag{2.15}$$

In the Hermitian limit $\sigma \to 0$ we obtain

$$K_{\text{Ai},\mathbb{C}}^{\beta=2}(z_1, z_2) \to \sqrt{\delta^1(y_1)\delta^1(y_2)}\,K_{\text{Ai}}^{\beta=2}(x_1, x_2),$$

with the factor in front of the integral in (2.15) projecting the imaginary parts of the eigenvalues to zero. For the integral itself—which is obtained from the limit of the sum of the OP on $\mathbb{C}$ given in (A.4)—the deformation in σ is very smooth. The same deformed Airy kernel can be obtained from the corresponding WL ensemble equation (2.29) [Akemann and Bender 2010] with kernel equation (A.5), and is thus universal.

$\beta = 4$:

$$G_{\text{Ai}}^{\beta=4}(x_1, x_2) = -\tfrac{1}{2}K_{\text{Ai}}^{\beta=2}(x_1, x_2) + \tfrac{1}{4}\text{Ai}(x_1)\int_{x_2}^\infty dt\,\text{Ai}(t),$$

$$K_{\text{Ai}}^{\beta=4}(x_1, x_2) = \frac{\partial}{\partial x_2}G_{\text{Ai}}^{\beta=4}(x_1, x_2),$$

$$W_{\text{Ai}}^{\beta=4}(x_1, x_2) = -\int_{x_1}^\infty ds\,G_{\text{Ai}}^{\beta=4}(s, x_2) \tag{2.16}$$

$$= -\tfrac{1}{4}\int_0^\infty ds\int_0^s dt\big(\text{Ai}(x_2 + t)\,\text{Ai}(x_1 + s) - \text{Ai}(x_2 + s)\,\text{Ai}(x_1 + t)\big),$$

$$G_{Ai,\mathbb{C}}^{\beta=4}(z_1, z_2) = \frac{iy_2}{4\sigma^3\sqrt{\pi}} \exp\left(-\frac{y_1^2 + y_2^2}{2\sigma^2} + \frac{\sigma^6}{6} + \frac{\sigma^2(z_1 + z_2^*)}{2}\right)$$

$$\times \int_0^\infty ds \int_0^s dt \, e^{\frac{1}{2}\sigma^2(s+t)}$$

$$\times \left(\text{Ai}\left(z_2^* + s + \frac{\sigma^4}{4}\right)\text{Ai}\left(z_1 + t + \frac{\sigma^4}{4}\right) - (z_1 \leftrightarrow z_2^*)\right). \quad (2.17)$$

The integral in (2.17) which is also present in the other two kernel elements — see (2.12) — clearly reduces to that in (2.16) in the Hermitian limit, whereas the prefactors provide the appropriate Dirac delta functions. When analysing the Hermitian limit in detail, the real kernel elements $G_{Ai}^{\beta=4}$ and $K_{Ai}^{\beta=4}$ follow from a Taylor expansion of $W_{Ai,\mathbb{C}}^{\beta=4}$; see [Akemann and Basile 2007] for a discussion of the analogous Hermitian limit of the Bessel kernel.

$\beta = 1$:

$$G_{Ai}^{\beta=1}(x_1, x_2) = -\int_0^\infty dt \, \text{Ai}(x_1 + t)\,\text{Ai}(x_2 + t) - \frac{1}{2}\,\text{Ai}(x_1)\left(1 - \int_{x_2}^\infty dt \, \text{Ai}(t)\right),$$

$$K_{Ai}^{\beta=1}(x_1, x_2) = \frac{\partial}{\partial x_2} G_{Ai}^{\beta=1}(x_1, x_2),$$

$$W_{Ai}^{\beta=1}(x_1, x_2) = -\int_{x_1}^\infty ds \, G_{Ai}^{\beta=1}(s, x_2) - \frac{1}{2}\int_{x_1}^{x_2} dt \, \text{Ai}(t)$$

$$+ \frac{1}{2}\int_{x_1}^\infty ds \, \text{Ai}(s) \int_{x_2}^\infty dt \, \text{Ai}(t) - \frac{1}{2}\,\text{sign}(x_1 - x_2), \quad (2.18)$$

$$G_{Ai,\mathbb{C},\text{real}}^{\beta=1}(x_1, x_2) = -\exp\left(\frac{\sigma^6}{6} + \frac{\sigma^2(x_1 + x_2)}{2}\right)$$

$$\times \int_0^\infty dt \, e^{\sigma^2 t}\,\text{Ai}\left(x_1 + t + \frac{\sigma^4}{4}\right)\text{Ai}\left(x_2 + t + \frac{\sigma^4}{4}\right)$$

$$- \frac{1}{2}\exp\left(\frac{\sigma^6}{12} + \frac{\sigma^2 x_1}{2}\right)\text{Ai}\left(x_1 + \frac{\sigma^4}{4}\right)$$

$$\times \left(1 - e^{\sigma^6/12}\int_{x_2}^\infty dt \, e^{\sigma^2 t/2}\,\text{Ai}\left(t + \frac{\sigma^4}{4}\right)\right),$$

$$G_{Ai,\mathbb{C},\text{com}}^{\beta=1}(z_1, z_2) = -\frac{i}{2\sigma^2}\,\text{sign}(y_2)(z_1 - z_2^*)\exp\left(\frac{\sigma^6}{6} + \frac{\sigma^2(x_1 + x_2)}{2}\right)$$

$$\times \sqrt{\text{erfc}(|y_1|/\sigma)\,\text{erfc}(|y_2|/\sigma)}$$

$$\times \int_0^\infty dt\,(e^{\sigma^2 t} - 1)\,\text{Ai}\left(z_1 + t + \frac{\sigma^4}{4}\right)\text{Ai}\left(z_2^* + t + \frac{\sigma^4}{4}\right),$$

$$K_{\text{Ai},\mathbb{C}}^{\beta=1}(z_1, z_2) = \frac{i}{2} \operatorname{sign}(y_2) G_{\text{Ai},\mathbb{C},\text{com}}^{\beta=1}(z_1, z_2^*),$$

$$\begin{aligned}
W_{\text{Ai},\mathbb{C}}^{\beta=1}(z_1, z_2) = {} & \left(-2A(x_1, x_2) + B(x_1)B(x_2) + B(x_2) - B(x_1)\right)\delta^1(y_1)\delta^1(y_2) \\
& + 2i \operatorname{sign}(y_2) G_{\text{Ai},\mathbb{C},\text{real}}^{\beta=1}(z_2^*, x_1)\delta^1(y_1) \\
& - 2i \operatorname{sign}(y_1) G_{\text{Ai},\mathbb{C},\text{real}}^{\beta=1}(z_1^*, x_2)\delta^1(y_2) \\
& - 2i \operatorname{sign}(y_1) G_{\text{Ai},\mathbb{C},\text{com}}^{\beta=1}(z_1^*, z_2) \\
& - 2i\delta^2(z_1 - z_2^*)\operatorname{sign}(y_1) - \delta^1(y_1)\delta^1(y_2)\operatorname{sign}(x_2 - x_1),
\end{aligned}$$

$$\begin{aligned}
A(x_1, x_2) = {} & \exp\left(\frac{\sigma^6}{6} + \frac{\sigma^2(x_1 + x_2)}{2}\right) \\
& \times \int_0^\infty ds \int_0^s dt\, e^{\frac{1}{2}\sigma^2(s+t)} \operatorname{Ai}\left(x_1 + s + \frac{\sigma^4}{4}\right) \operatorname{Ai}\left(x_2 + t + \frac{\sigma^4}{4}\right),
\end{aligned}$$

$$B(x) = \exp\left(\frac{\sigma^6}{12} + \frac{\sigma^2 x}{2}\right) \int_0^\infty dt\, e^{\frac{1}{2}\sigma^2 t} \operatorname{Ai}\left(x + t + \frac{\sigma^4}{4}\right). \tag{2.19}$$

Clearly $G_{\text{Ai},\mathbb{C},\text{real}}^{\beta=1}(x_1, x_2) \to G_{\text{Ai}}^{\beta=1}(x_1, x_2)$ as $\sigma \to 0$, whereas the complex part vanishes in this Hermitian limit: $G_{\text{Ai},\mathbb{C},\text{com}}^{\beta=1}(z_1, z_2) \to 0$. We have also explicitly verified the corresponding limits for $K_{\text{Ai},\mathbb{C}}^{\beta=1}(z_1, z_2)$ and $W_{\text{Ai},\mathbb{C}}^{\beta=1}(z_1, z_2)$.

2.3. Limiting sine kernels on $\mathbb{R}$ and $\mathbb{C}$.

For real eigenvalues the sine kernels are obtained by zooming into the bulk of the spectrum, sufficiently far away from the edges. The weakly non-Hermitian limit of the complex eigenvalues introduced in [Fyodorov et al. 1997a; 1997b] is taken such that

$$\sigma = N^{1/2}\sqrt{1-\tau} \tag{2.20}$$

remains finite (see Appendix B for further details). In this limit the macroscopic support of the spectral density on an ellipse shrinks to the semicircle distribution on the real axis, whereas microscopically we still have correlations of the eigenvalues in the complex plane.

The list of the known one-parameter deformations of the sine kernels for $\beta = 2$ is as follows:

$\beta = 2$:

$$K_{\sin}^{\beta=2}(x_1, x_2) = \frac{\sin(x_1 - x_2)}{\pi(x_1 - x_2)} = \frac{1}{\pi}\int_0^1 dt\, \cos[(x_1 - x_2)t], \tag{2.21}$$

$$K_{\sin,\mathbb{C}}^{\beta=2}(z_1, z_2) = \frac{1}{\sigma\pi^{3/2}} e^{-(y_1^2 + y_2^2)/(2\sigma^2)} \int_0^1 dt\, e^{-\sigma^2 t^2} \cos[(z_1 - z_2)t]. \tag{2.22}$$

The corresponding spectral density of complex eigenvalues was first derived in [Fyodorov et al. 1997a] using supersymmetry, and the kernel with all correlation functions in [Fyodorov et al. 1997b; 1998] using OP; see (A.4). In the Hermitian limit $\sigma \to 0$, we have

$$K_{\sin,\mathbb{C}}^{\beta=2}(z_1, z_2) \to \sqrt{\delta^1(y_1)\delta^1(y_2)}\, K_{\sin}^{\beta=2}(x_1, x_2).$$

In [Fyodorov et al. 1998] it was shown using supersymmetric techniques that the same result holds for the microscopic density of random matrices with i.i.d. matrix elements for $\beta = 1, 2$. Further arguments in favour of universality were added in [Akemann 2002] for the kernel for $\beta = 2$ using large-N factorisation and asymptotic OP. The universal parameter is the mean macroscopic spectral density $\rho(x_0)$.

$\beta = 4$:

$$G_{\sin}^{\beta=4}(x_1, x_2) = -\frac{\sin[2(x_1 - x_2)]}{2\pi(x_1 - x_2)},$$

$$K_{\sin}^{\beta=4}(x_1, x_2) = \frac{\partial}{\partial x_1} G_{\sin}^{\beta=4}(x_1, x_2),$$

$$W_{\sin}^{\beta=4}(x_1, x_2) = \int_0^{x_1-x_2} dt\, G_{\sin}^{\beta=4}(t, 0) = \frac{1}{2\pi}\int_0^1 \frac{dt}{t} \sin[2(x_1 - x_2)t], \quad (2.23)$$

$$G_{\sin,\mathbb{C}}^{\beta=4}(z_1, z_2) = \frac{i2\sqrt{2}\, y_2}{\pi^{3/2}\sigma^3} e^{-2y_2^2/\sigma^2} \int_0^1 \frac{dt}{t} e^{-2\sigma^2 t^2} \sin[2(z_1 - z_2^*)t]. \quad (2.24)$$

The corresponding spectral density of complex eigenvalues was derived in [Kolesnikov and Efetov 1999] using supersymmetry, and the kernel with all correlation functions was derived in [Kanzieper 2002] using skew-OP leading to (A.9).

$\beta = 1$:

$$G_{\sin}^{\beta=1}(x_1, x_2) = -K_{\sin}^{\beta=2}(x_1, x_2),$$

$$K_{\sin}^{\beta=1}(x_1, x_2) = \frac{\partial}{\partial x_1} G_{\sin}^{\beta=1}(x_1, x_2) = \frac{1}{\pi}\int_0^1 dt\, t \sin[(x_2 - x_1)t],$$

$$W_{\sin}^{\beta=1}(x_1, x_2) = \int_0^{x_1-x_2} dt\, G_{\sin}^{\beta=1}(t, 0) + \tfrac{1}{2}\operatorname{sign}(x_1 - x_2), \quad (2.25)$$

$$G_{\sin,\mathbb{C},\text{real}}^{\beta=1}(z_1, x_2) = -\frac{1}{\pi}\int_0^1 dt\, e^{-\sigma^2 t^2} \cos[(z_1 - x_2)t],$$

$$G_{\sin,\mathbb{C},\text{com}}^{\beta=1}(z_1, z_2) = -2i\operatorname{sign}(y_2)\operatorname{erfc}\left|\frac{y_1}{\sigma}\right| K_{\sin,\mathbb{C}}^{\beta=1}(z_1, z_2^*),$$

$$K^{\beta=1}_{\sin,\mathbb{C}}(z_1, z_2) = \frac{1}{\pi} \int_0^1 dt \, t \, e^{-\sigma^2 t^2} \sin[(z_2 - z_1)t],$$

$$W^{\beta=1}_{\sin,\mathbb{C}}(z_1, z_2) = -\left(\frac{1}{\pi} \int_0^1 \frac{dt}{t} e^{-\sigma^2 t^2} \sin[(x_2 - x_1)t]\right) \delta^1(y_1) \delta^1(y_2) \quad (2.26)$$

$$- 2i \, \text{sign}(y_2) \, \text{erfc}\left|\frac{y_2}{\sigma}\right| G^{\beta=1}_{\sin,\mathbb{C},\text{real}}(z_2^*, x_1) \delta^1(y_1)$$

$$+ 2i \, \text{sign}(y_1) \, \text{erfc}\left|\frac{y_1}{\sigma}\right| G^{\beta=1}_{\sin,\mathbb{C},\text{real}}(z_1^*, x_2) \delta^1(y_2)$$

$$+ 4 \, \text{sign}(y_1) \, \text{sign}(y_2) \, \text{erfc}\left|\frac{y_1}{\sigma}\right| \text{erfc}\left|\frac{y_2}{\sigma}\right| K^{\beta=1}_{\sin,\mathbb{C}}(z_1^*, z_2^*)$$

$$- \sqrt{\text{erfc}\left|\frac{y_1}{\sigma}\right| \text{erfc}\left|\frac{y_2}{\sigma}\right|} \left(2i \, \delta^2(z_1 - z_2^*) \, \text{sign}(y_1)\right.$$

$$\left. + \delta^1(y_1) \delta^1(y_2) \, \text{sign}(x_2 - x_1)\right).$$

The kernel elements $G^{\beta=1}_{\sin,\mathbb{C}}(z_1, z_2)$ and $K^{\beta=1}_{\sin,\mathbb{C}}(z_1, z_2)$ were derived in [Forrester and Nagao 2008] using skew-OP; see also (A.11). The same resulting spectral densities of complex and real eigenvalues were derived previously in [Efetov 1997a; 1997b] using a sigma-model calculation, which again indicates universality. It can easily be verified that the Hermitian limit $\sigma \to 0$ of $G^{\beta=1}_{\sin,\mathbb{C},\text{real}}(x_1, x_2)$ is indeed $G^{\beta=1}_{\sin}(x_1, x_2)$, and that $G^{\beta=1}_{\sin,\mathbb{C},\text{com}}(z_1, z_2)$ vanishes in this limit.

2.4. Wishart–Laguerre ensembles with eigenvalues on $\mathbb{R}$ and $\mathbb{C}$. In order to be able to access the Bessel kernels for real and complex eigenvalues as well, we briefly introduce the Wishart–Laguerre (or chiral) ensembles (LβE) and their non-Hermitian counterparts ($\mathbb{C}$LβE). We begin with the former which are defined as

$$\mathcal{Z}^{L\beta E}_N = \int dW \, \exp[-\beta \, \text{Tr} \, WW^\dagger/2]$$

$$= c_{N,\beta,\nu} \prod_{j=1}^N \int_{\mathbb{R}_+} dx_j \, w^\nu_\beta(x_j) \, |\Delta_N(\{x\})|^\beta. \quad (2.27)$$

The elements of the rectangular $N \times (N + \nu)$ matrix W are again real, complex, or quaternion real for $\beta = 1, 2, 4$, without further symmetry constraints. The integration denoted by dW runs over all the independent matrix elements. Because we want to access the so-called hard edge of the spectrum we will only consider fixed $\nu = O(1)$ in the following. The distribution of the (positive definite) eigenvalues x_j of $WW^\dagger$ in the Wishart picture (or equivalently the distribution of the singular values of W in the Dirac picture used in QCD) is of the same form as (2.1), but with different weight functions

$$w^\nu_\beta(x) = x^{\frac{1}{2}\beta(\nu+1)-1} \exp[-\beta x/2], \quad (2.28)$$

that now depend on β in a nontrivial way. Consequently the k-point correlation functions take the same form as in (2.3), with the corresponding kernels.

In analogy to the Ginibre ensembles we define a parameter-dependent family of non-Hermitian Wishart–Laguerre (also called complex chiral) ensembles as the following two-matrix model

$$\mathscr{Z}_N^{\mathrm{CL}\beta\mathrm{E}}(\tau)$$
$$= \int dW\, dV \exp\left[-\frac{1}{1-\tau}\mathrm{Tr}\big(WW^\dagger + V^\dagger V - \tau(WV + V^\dagger W^\dagger)\big)\right], \quad (2.29)$$

with W and $V^\dagger$ being two rectangular $N \times (N+\nu)$ matrices. Here we follow the notation of [Akemann and Bender 2010]. This two-matrix model was first introduced and solved for $\beta = 1$ [Akemann et al. 2009; 2010b], $\beta = 2$ [Osborn 2004] and $\beta = 4$ in [Akemann 2005]. For $\tau = 0$ the jpdf of all the matrix elements again factorises, and in the opposite limit we have

$$\mathscr{Z}_N^{\mathrm{CL}\beta\mathrm{E}}(\tau) \to \mathscr{Z}_N^{\mathrm{L}\beta\mathrm{E}}$$

as $\tau \to 1$. Here we are seeking the complex (and real) eigenvalues of the product matrix WV (W and V are the off-diagonal blocks of the Dirac matrix that we diagonalise). Its jpdf takes the same form as in (2.5) and (2.6), but with different weight functions that are no longer Gaussian

$$w_{\beta=2}^{\nu,\mathbb{C}}(z) = |z|^\nu \exp\left[\frac{\tau(z+z^*)}{1-\tau^2}\right] K_\nu\left(\frac{2|z|}{1-\tau^2}\right),$$
$$w_{\beta=4}^{\nu,\mathbb{C}}(z) = \sqrt{w_{\beta=2}^{2\nu,\mathbb{C}}(z)}. \quad (2.30)$$

The function $K_\nu(z)$ here is the modified Bessel function. For $\beta = 4$ the antisymmetric weight function is defined as in (2.8). For $\beta = 1$ we explicitly specify two functions in the antisymmetric weight function

$$\mathscr{F}_{\beta=1}^{\nu,\mathbb{C}}(z_1, z_2) = i g_\nu(z_1, z_2)\,\mathrm{sign}(y_1)\,\delta^2(z_1 - z_2^*)$$
$$+ \tfrac{1}{2} h_\nu(x_1) h_\nu(x_2)\delta(y_1)\delta(y_2)\,\mathrm{sign}(x_2 - x_1),$$

$$h_\nu(x) = 2|x|^{\frac{\nu}{2}} \exp\left[\frac{\tau x}{1-\tau^2}\right] K_{\frac{\nu}{2}}\left(\frac{|x|}{1-\tau^2}\right), \quad (2.31)$$

$$g_\nu(z_1, z_2) = 2|z_1 z_2|^{\frac{\nu}{2}} \exp\left[\frac{\tau(z_1 + z_2)}{1-\tau^2}\right]$$
$$\times \int_0^\infty \frac{dt}{t} \exp\left[-\frac{(z_1^2 + z_2^2)t}{(1-\tau^2)^2} - \frac{1}{4t}\right]$$
$$\times K_{\frac{\nu}{2}}\left(\frac{2z_1 z_2 t}{(1-\tau^2)^2}\right) \mathrm{erfc}\left(\frac{|z_2 - z_1|\sqrt{t}}{1-\tau^2}\right),$$

which are related by $g_\nu(z, z^*) \to h_\nu(x)^2$ as $y \to 0$. We give the corresponding kernels in Appendix A.

In the large-N limit (with $\nu = O(1)$ fixed), for all three β the real positive Wishart eigenvalues of $WW^\dagger$ are concentrated on the interval $(0, 4N]$, with a density $\rho(x) = (2\pi N)^{-1}\sqrt{(4N-x)/x}$. This is a special case of the Marchenko–Pastur density. After mapping to Dirac eigenvalues $\lambda = \sqrt{x}$, this becomes the same semicircle distribution as for WD, but with eigenvalues coming in $\pm\lambda$ pairs. The density of the Wishart eigenvalues has a singularity at the origin; however, after mapping to the Dirac picture, we obtain a macroscopic density function that is flat on an ellipse, just as in the Ginibre ensemble[3].

We now give a list of all the known Bessel kernels. For real eigenvalues we follow [Deift et al. 2007; Nagao and Forrester 1995] where a most comprehensive list and references can be found. In some cases the parallel between kernels of real and complex eigenvalues is more transparent after using some identities for Bessel functions.

2.5. Limiting Bessel kernels on $\mathbb{R}$ and $\mathbb{C}$. The hard-edge limit is defined by zooming into the origin (see Appendix B), where for the complex eigenvalues we have to keep $\sigma = \sqrt{N(1-\tau)}$ fixed as in the weakly non-Hermitian bulk limit equation (2.20). The corresponding limiting kernels are given as follows:

$\beta = 2$:

$$K_{\mathrm{Bes}}^{\beta=2}(x_1, x_2) = \frac{J_\nu(\sqrt{x_1})\sqrt{x_2}\, J_{\nu-1}(\sqrt{x_2}) - (x_1 \leftrightarrow x_2)}{2(x_1 - x_2)}$$

$$= \frac{1}{2}\int_0^1 dt\, t\, J_\nu(\sqrt{x_1}\, t) J_\nu(\sqrt{x_2}\, t), \tag{2.32}$$

$$K_{\mathrm{Bes},\mathbb{C}}^{\beta=2}(z_1, z_2) = \frac{1}{8\pi\sigma^2} K_\nu\left(\frac{|z_1|}{4\sigma^2}\right)^{1/2} K_\nu\left(\frac{|z_2|}{4\sigma^2}\right)^{1/2} \exp\left(\frac{x_1 + x_2}{8\sigma^2}\right)$$

$$\times \int_0^1 dt\, t\, e^{-2\sigma^2 t^2} J_\nu(t\sqrt{z_1}) J_\nu(t\sqrt{z_2}). \tag{2.33}$$

It can be shown that

$$K_{\mathrm{Bes},\mathbb{C}}^{\beta=2}(z_1, z_2) \to \sqrt{\delta^1(y_1)\delta^1(y_2)\Theta(x_1)\Theta(x_2)}\, K_{\mathrm{Bes}}^{\beta=2}(x_1, x_2)$$

as $\sigma \to 0$, where $\Theta(x)$ is the Heaviside step function. The kernel of complex eigenvalues was derived in [Osborn 2004]. The same density following from this kernel was obtained from a different Gaussian non-Hermitian one-matrix model [Splittorff and Verbaarschot 2004] using replicas, and is in that sense universal.

---

[3]Note that the ν exact eigenvalues do not contribute to the macroscopic spectral density.

$\beta = 4$:

$$K_{\text{Bes}}^{\beta=4}(x_1, x_2) = -\frac{\partial}{\partial x_2}G_{\text{Bes}}^{\beta=4}(x_1, x_2),$$

$$G_{\text{Bes}}^{\beta=4}(x_1, x_2) = -2\sqrt{x_1}\int_0^1 dt \int_0^1 ds\, s^2\big(J_{2\nu}(2\sqrt{x_1}\, st)J_{2\nu+1}(2\sqrt{x_2}\, s) \\ - t\,J_{2\nu}(2\sqrt{x_1}\, s)J_{2\nu+1}(2\sqrt{x_2}\, st)\big),$$

$$W_{\text{Bes}}^{\beta=4}(x_1, x_2) = \sqrt{x_1 x_2}\int_0^1 dt\int_0^1 ds\, s\big(J_{2\nu}(2\sqrt{x_1}\, st)J_{2\nu}(2\sqrt{x_2}\, s) - (x_1 \leftrightarrow x_2)\big)$$

$$= \int_{x_1}^{x_2} dx'\, G_{\text{Bes}}^{\beta=4}(x_1, x'), \tag{2.34}$$

$$K_{\text{Bes},\mathbb{C}}^{\beta=4}(z_1, z_2) = \frac{1}{\sigma^4}\int_0^1 dt\int_0^1 ds\, s\, e^{-2\sigma^2 s^2(1+t^2)} \\ \times \big(J_{2\nu}(2\sqrt{z_1}\, st)J_{2\nu}(2\sqrt{z_2}\, s) - (z_1 \leftrightarrow z_2)\big). \tag{2.35}$$

The complex kernel was first derived in [Akemann 2005], whereas the matching in the Hermitian limit — which can best be seen when comparing the kernels $K_{\text{Bes}}^{\beta=4}$ and $K_{\text{Bes},(\mathbb{C})}^{\beta=4}$ — is discussed in detail in [Akemann and Basile 2007].

$\beta = 1$:

$$K_{\text{Bes}}^{\beta=1}(x_1, x_2) = \frac{-1}{8\sqrt{x_1 x_2}}\int_0^1 ds\, s^2\left(\sqrt{x_1}\, J_{\nu+1}(s\sqrt{x_1})J_\nu(s\sqrt{x_2}) - (x_1 \leftrightarrow x_2)\right)$$

$$= -\frac{\partial}{\partial x_2}G_{\text{Bes}}^{\beta=1}(x_1, x_2),$$

$$G_{\text{Bes}}^{\beta=1}(x_1, x_2) = -\frac{1}{2}\int_0^1 dt\, t\, J_{\nu-1}(\sqrt{x_1}\, t)J_{\nu-1}(\sqrt{x_2}\, t) \\ - \frac{1}{4\sqrt{x_1}}J_\nu(\sqrt{x_1})\int_{\sqrt{x_2}}^\infty ds\, J_{\nu-2}(s),$$

$$W_{\text{Bes}}^{\beta=1}(x_1, x_2) = -\int_{x_1}^{x_2} ds\, G_{\text{Bes}}^{\beta=1}(s, x_2) - \frac{1}{2}\,\text{sign}(x_1 - x_2), \tag{2.36}$$

$$K_{\text{Bes},\mathbb{C}}^{\beta=1}(z_1, z_2) = \frac{1}{256\pi\sigma^2}\int_0^1 ds\, s^2\, e^{-2\sigma^2 s^2} \\ \times \left(\sqrt{z_1}\, J_{\nu+1}(s\sqrt{z_1})J_\nu(s\sqrt{z_2}) - (z_1 \leftrightarrow z_2)\right),$$

$$G_{\text{Bes},\mathbb{C},\text{com}}^{\beta=1}(z_1, z_2) = -2i\,\text{sign}(y_2)\, e^{x_2/(4\sigma^2)}\int_0^\infty \frac{dt}{t}\exp\left(-\frac{t(z_2^2 + z_2^{*2})}{64\sigma^4} - \frac{1}{4t}\right) \\ \times K_{\frac{\nu}{2}}\left(\frac{t}{32\sigma^4}|z_2|^2\right)\text{erfc}\left(\frac{\sqrt{t}\,|y_2|}{4\sigma^2}\right)K_{\text{Bes},\mathbb{C}}^{\beta=1}(z_1, z_2^*),$$

$$G^{\beta=1}_{\text{Bes},\mathbb{C},\text{real}}(x_1, x_2) \tag{2.37}$$

$$= -\frac{2\, e^{x_2/(8\sigma^2)}\, K_{\frac{\nu}{2}}\left(|x_2|/8\sigma^2\right)}{[\text{sign}(x_2)]^{\nu/2}}$$

$$\times \left\{ \left((-i)^\nu \int_{-\infty}^0 dy + \frac{2}{[\text{sign}(x_2)]^{\frac{\nu}{2}}} \int_0^{x_2} dy \right) K^{\beta=1}_{\text{Bes},\mathbb{C}}(x_1, y) 2\, e^{y/(8\sigma^2)} K_{\frac{\nu}{2}}\left(\frac{|y|}{8\sigma^2} \right) \right.$$

$$- \frac{1}{32\sqrt{\pi}} \left[-\frac{1}{\sigma} e^{-\sigma^2} J_\nu(\sqrt{x_1}) \right.$$

$$+ \frac{2\sigma^\nu}{\Gamma\left(\frac{\nu+1}{2}\right)} \int_0^1 ds\, e^{-\sigma^2 s^2} s^{\nu+2}$$

$$\times \left(\frac{\sqrt{x_1}}{2} E_{\frac{1-\nu}{2}}(\sigma^2 s^2)\, J_{\nu+1}(s\sqrt{x_1}) \right.$$

$$\left.\left.\left. - \sigma^2 s \left(E_{\frac{-1-\nu}{2}}(\sigma^2 s^2) - E_{\frac{1-\nu}{2}}(\sigma^2 s^2) \right) J_\nu(s\sqrt{x_1}) \right) \right] \right\}.$$

In the final equation above,

$$E_n(x) = \int_1^\infty dt\, \frac{e^{-xt}}{t^n} \tag{2.38}$$

is the exponential integral. The kernels $K^{\beta=1}_{\text{Bes},\mathbb{C}}(z_1, z_2)$ and $G^{\beta=1}_{\text{Bes},\mathbb{C},\text{com}}(z_1, z_2)$ were derived in [Akemann et al. 2010b], and $G^{\beta=1}_{\text{Bes},\mathbb{C},\text{real}}(x_1, x_2)$ in [Akemann et al. 2011c; Phillips 2011]; the latter case is somewhat subtle, since we cannot simply commute the (weakly non-Hermitian) large-N limit operation with the integral that appears in the finite-N expression, to give an integral over the limit of the integrand. We also note that there is numerical evidence from studying some examples of non-Gaussian RMT that the density of the real eigenvalues resulting from $G^{\beta=1}_{\text{Bes},\mathbb{C},\text{real}}(x_1, x_2)$ and the corresponding distribution of the smallest eigenvalues may be universal [Phillips 2011].

The Hermitian limit is much more involved here; but compare $K^{\beta=1}_{\text{Bes},\mathbb{C}}(z_1, z_2)$ and $K^{\beta=1}_{\text{Bes}}(z_1, z_2)$ — in particular, as $N \to \infty$ and $\sigma \to 0$, we find that

$$G^{\beta=1}_{\text{Bes},\mathbb{C},\text{com}}(z_1, z_2) \to 0 \quad \text{and} \quad G^{\beta=1}_{\text{Bes},\mathbb{C},\text{real}}(x_1, x_2) \to G^{\beta=1}_{\text{Bes}}(x_1, x_2),$$

and we refer to [Akemann et al. 2011c; Phillips 2011] for more details.

3. Discussion and open problems

In this short article we have collected together all the known kernels for RMT with real eigenvalues along with all the known and new kernels for RMT with complex eigenvalues at weak non-Hermiticity. This comprises the Airy, sine and Bessel kernels of the three Wigner–Dyson and the three Wishart–Laguerre ensembles, as well as their non-Hermitian counterparts. In order to highlight the nature of this deformation we have used real integral representations for the kernels of real eigenvalues, rather than the asymptotic forms resulting from the Christoffel–Darboux identity for $\beta = 2$ or from the rewriting à la Tracy–Widom for $\beta = 1, 4$. The extra exponential factor (and shift for the Airy case) in the integral representation of the kernels on $\mathbb{C}$ is a very smooth deformation. This makes it very plausible that the universality which is very well studied for real eigenvalues extends to the weakly non-Hermitian limit for all ensembles, beyond what is already known for $\beta = 2$. The universality of the factor in front of the integral which contains special functions such as the complementary error function or modified Bessel function will be more difficult to establish. However, the presence of these factors is crucial when taking the Hermitian limit, in projecting the imaginary parts of the eigenvalues to zero.

Whilst we have already mentioned what is known about universality in the weak limit so far, let us give some more open problems. To date, a mathematically rigorous derivation of most of the limiting kernels on $\mathbb{C}$ is lacking, apart from the complex Airy kernel for $\beta = 2$ [Bender 2010]. There is no doubt that the kernels we have listed and which have been derived using different techniques such as asymptotic OP, supersymmetry or replicas are correct. This is based not only on numerical evidence but also, and more importantly, on a comparison with complex eigenvalue spectra in physics; see, for example, [Verbaarschot 2011; Akemann 2007] and references therein, where the complex Bessel kernels for $\beta = 2$ and 4 were successfully compared with complex spectra from QCD and QCD-like theories. Because the latter are field theories and not Gaussian RMT this gives a further indication that universality holds in this regime. Preliminary numerical investigations with non-Gaussian, non-Hermitian RMT appear to confirm this [Phillips 2011] for $\beta = 1$.

A much more challenging problem will be to show the universality of these kernels on $\mathbb{C}$, either by going to non-Gaussian potentials of polynomial or harmonic form, or by considering non-Hermitian Wigner matrix ensembles, with elements being independent random variables.

A further reason why we believe that this universality question is important is that some of the kernels on $\mathbb{C}$ reappear in the same integral form (with real arguments) when looking at symmetry transitions between two different *Hermitian* RMTs, say from one GUE to another GUE, in a corresponding "weak" limit. Their eigenvalue correlations are also called parametric. For $\beta = 2$ this fact can be observed for the Bessel, sine [Akemann et al. 2007; Forrester et al. 1999] and Airy [Macêdo 1994; Forrester et al. 1999] kernels.

Appendix A: finite-N (skew-) orthogonal polynomials and kernels on $\mathbb{C}$

In this appendix we specify the orthogonal polynomials (OP) and skew-OP as well as their (skew-) symmetric scalar products. These may be used to construct the kernels — which we will list for all the above matrix ensembles with complex eigenvalues — in terms of which all k-point correlation functions can be expressed; see (2.3) and (2.9). We also highlight the relations between expectation values of characteristic polynomials on the one hand and (skew-) OP and their kernels on the other, valid on both $\mathbb{R}$ and $\mathbb{C}$.

Starting with $\beta = 2$ we define the monic OP on $\mathbb{R}$ (and $\mathbb{C}$) by

$$\int_{\mathbb{R}(\mathbb{C})} d^{(2)}z \, w_{\beta=2}^{(v,\mathbb{C})}(z) P_k(z) P_l(z)^{(*)} = h_{k(v,\mathbb{C})}^{\beta=2} \delta_{kl}, \tag{A.1}$$

with squared norms $h_{k(v,\mathbb{C})}^{\beta=2}$. The symbol v labels the rectangular $N \times (N + v)$ matrices of the Wishart–Laguerre ensembles considered in Section 2.4. Because all moments exist in our examples these OP can be constructed via the Gram–Schmidt procedure. Alternatively, they can be written as

$$P_k(z) = \langle \det[z - H] \rangle_k = \frac{1}{\mathscr{L}_k^{\text{GUE}}} \int dH \det[z - H] \exp[- \operatorname{Tr} H^2/2], \tag{A.2}$$

and similarly for the GinUE and $(\mathbb{C})$LUE, replacing the $k \times k$ matrix H inside the determinant with J, or with the Wishart matrices $WW^\dagger$ and WV, respectively. In fact, this relation holds for general weight functions. For the Gaussian ensembles we obtain Hermite, and for the WL ensembles Laguerre polynomials on $\mathbb{R}$ and on $\mathbb{C}$. The corresponding kernels are then obtained by summing over the *normalised* OP (multiplied by the weights). Most conveniently, a second relation to characteristic polynomials exists [Akemann and Vernizzi 2003],

$$K_{N,\mathbb{C}}^{\beta=2}(u, v) = w_{\beta=2}^{\mathbb{C}}(u)^{\frac{1}{2}} w_{\beta=2}^{\mathbb{C}}(v)^{\frac{1}{2}} \frac{1}{h_{N-1,\mathbb{C}}^{\beta=2}} \langle \det[u - J] \det[v - J^\dagger] \rangle_{N-1}, \tag{A.3}$$

which we state here for the GinUE. Correspondingly it holds for the GUE and $\beta = 2$ WL ensembles, and, indeed, for arbitrary weights. We can now give the two $\beta = 2$ kernels in the complex plane, following [Fyodorov et al. 1998] and [Osborn 2004] respectively:

$\underline{\beta = 2}$:

$$K_{N,\mathbb{C}}^{\beta=2}(u, v) = w_{\beta=2}^{\mathbb{C}}(u)^{\frac{1}{2}} w_{\beta=2}^{\mathbb{C}}(v)^{\frac{1}{2}} \frac{1}{\pi \sqrt{1-\tau^2}} \sum_{j=0}^{N-1} \frac{\tau^j}{2^j j!} H_j\left(\frac{u}{\sqrt{2\tau}}\right) H_j\left(\frac{v}{\sqrt{2\tau}}\right),$$

(A.4)

$$K_{N,\nu,\mathbb{C}}^{\beta=2}(u, v) = w_{\beta=2}^{\nu,\mathbb{C}}(u)^{\frac{1}{2}} w_{\beta=2}^{\nu,\mathbb{C}}(v)^{\frac{1}{2}} \frac{2}{\pi(1-\tau^2)} \sum_{j=0}^{N-1} \frac{\tau^{2j} j!}{(j+\nu)!} L_j^\nu\left(\frac{u}{\tau}\right) L_j^\nu\left(\frac{v}{\tau}\right).$$

(A.5)

For the skew-OP related to $\beta = 1, 4$, we have to distinguish between the skew products for complex and real eigenvalues. Because the latter are very well known (see, e.g., [Mehta 2004]) we will focus on the former, which can be written in a unified way [Akemann et al. 2010a] as

$$\int_{\mathbb{C}^2} d^2 z_1 d^2 z_2 \, \mathcal{F}_{\beta=1,4}^{(\nu)\mathbb{C}}(z_1, z_2) \det\begin{bmatrix} Q_{2k}^{\beta=1,4}(z_1) & Q_{2l+1}^{\beta=1,4}(z_1) \\ Q_{2k}^{\beta=1,4}(z_2) & Q_{2l+1}^{\beta=1,4}(z_2) \end{bmatrix} = h_{k,(\nu)\mathbb{C}}^{\beta=1,4} \delta_{kl},$$

(A.6)

for skew-OP of even-odd degree, and which is vanishing for even-even and odd-odd degree. Here and in the following we again choose N even. Once more, the skew-OP satisfying this can be written as follows for $\beta = 4$ [Kanzieper 2002] and $\beta = 1$ [Akemann et al. 2010a]:

$$Q_{2k}^{\beta=1,4}(z) = \langle \det[z - J] \rangle_{2k}, \quad Q_{2k+1}^{\beta=1,4}(z) = \langle \det[z - J](z + c + \text{Tr } J) \rangle_{2k}. \quad \text{(A.7)}$$

Note that, for $\beta = 4$, the matrix J here should be taken as the complex-valued matrix of size $2k \times 2k$. The odd skew-OP in (A.7) are defined only up to a constant c times the even skew-OP. The same relation holds for real eigenvalues and arbitrary weights, see [Akemann et al. 2010a] for references. Moreover, the antisymmetric kernel matrix element $K_{N,\mathbb{C}}^{\beta=1,4}$ (sometimes called the prekernel) enjoys a similar relation to that in (A.3), as was observed for $\beta = 4$ [Akemann and Basile 2007] and $\beta = 1$ [Akemann et al. 2009]:

$$K_{N,\mathbb{C}}^{\beta=1,4}(u, v) = (u - v) \frac{1}{h_{\frac{N}{2}-1,\mathbb{C}}^{\beta=1,4}} \langle \det[u - J] \det[v - J^\dagger] \rangle_{N-2}. \quad \text{(A.8)}$$

An analogous relation holds for the Wishart–Laguerre ensembles (see, e.g., in [Akemann et al. 2010a]). We list the corresponding kernel matrix elements. From [Kanzieper 2002] and [Akemann 2005], respectively, we have:

$\underline{\beta = 4:}$

$$K_{N,\mathbb{C}}^{\beta=4}(u, v) = \frac{1}{\pi(1 - \tau)\sqrt{1 - \tau^2}} \sum_{k=0}^{N/2-1} \sum_{l=0}^{k} \frac{1}{(2k + 1)!!(2l)!!} \left(\frac{\tau}{2}\right)^{k+l+\frac{1}{2}}$$

$$\times \left(H_{2k+1}\left(\frac{u}{\sqrt{2\tau}}\right)H_{2l}\left(\frac{v}{\sqrt{2\tau}}\right) - (u \leftrightarrow v)\right), \tag{A.9}$$

$$K_{N,v,\mathbb{C}}^{\beta=4}(u, v) = -\frac{2}{\pi(1 - \tau^2)^2} \sum_{k=0}^{N/2-1} \sum_{j=0}^{k} \frac{2^{2k-2j}k!(k + v)!(2j)!}{(2k + 2v + 1)!j!(j + v)!} \tau^{2k+2j+1}$$

$$\times \left(L_{2k+1}^{2v}\left(\frac{u}{\tau}\right)L_{2j}^{2v}\left(\frac{v}{\tau}\right) - (u \leftrightarrow v)\right), \tag{A.10}$$

recalling that N here is the size of the complex-valued matrix that is equivalent to the original quaternion real matrix. From [Forrester and Nagao 2008] and [Akemann et al. 2009] we have

$\underline{\beta = 1:}$

$$K_{N,\mathbb{C}}^{\beta=1}(u, v) = \frac{1}{2\sqrt{2\pi}(1 + \tau)} \sum_{l=0}^{N-2} \frac{1}{l!}$$

$$\times \left(\frac{\tau}{2}\right)^{l+\frac{1}{2}} \left(H_{l+1}\left(\frac{u}{\sqrt{2\tau}}\right)H_l\left(\frac{v}{\sqrt{2\tau}}\right) - (u \leftrightarrow v)\right), \tag{A.11}$$

$$K_{N,v,\mathbb{C}}^{\beta=1}(u, v) = -\frac{1}{8\pi(1 - \tau^2)} \sum_{l=0}^{N-2} \frac{(l + 1)!}{(l + v)!} \tau^{2l+1}$$

$$\times \left(L_{l+1}^{v}\left(\frac{u}{\tau}\right)L_l^{v}\left(\frac{v}{\tau}\right) - (u \leftrightarrow v)\right). \tag{A.12}$$

The other elements of the matrix-valued kernel follow by integration.

Appendix B: large-N limits at weak non-Hermiticity

In this appendix we will specify the different large-N limits that lead to the limiting kernels listed in Section 2. Let us emphasise that these are not all of the possible large-N limits of the above finite-N kernels that one can take. We will give only those limits where $(1 - \tau)N^\delta = \sigma$ is kept fixed for some $\delta > 0$, limits where the degree of non-Hermiticity is weak. The reason is that it is only these particular limiting kernels that relate closely to the known universal kernels on $\mathbb{R}$. However, many of the results at strong non-Hermiticity (i.e., where τ is N-independent) can be recovered from the weak limit by taking $\sigma \to \infty$ and rescaling the complex eigenvalues accordingly.

Soft edge limit. We consider fluctuations around the right end-point of the long half-axis of the supporting ellipse, to obtain from (A.4) [Bender 2010]

$$z = (1+\tau)\sqrt{N} + \frac{X}{N^{1/6}} + i\frac{Y}{N^{1/6}}, \quad \sigma = N^{1/6}\sqrt{1-\tau}, \quad (B.1)$$

$$K_{Ai,\mathbb{C}}^{\beta=2}(X_1 + iY_1, X_2 + iY_2) \equiv \lim_{\substack{N\to\infty \\ \tau\to 1}} \frac{1}{N^{1/3}} K_{N,\mathbb{C}}^{\beta=2}(z_1, z_2). \quad (B.2)$$

The same limit applies to the WL kernel on $\mathbb{C}$ given in (A.5) [Akemann and Bender 2010]. By symmetry we expect the same limiting behaviour around the left end-point $-(1+\tau)\sqrt{N}$ as well as for $\nu = O(N)$. The limiting kernels for $\beta = 1, 4$ are defined in the same way. Note that the eigenvalues in the bulk of the spectrum are at strong non-Hermiticity in this limit, since $\sigma_{sine} = N^{1/3}\sigma_{Airy} \to \infty$ as $N \to \infty$. In fact, in order to reach weak non-Hermiticity in the bulk we need to consider the following scaling limit.

Bulk limit. Without loss of generality we consider fluctuations around the origin, being representative of the Gaussian ensembles equation (2.4). On rescaling, we obtain from (A.4) [Fyodorov et al. 1997a; 1997b]

$$z = \frac{X}{N^{1/2}} + i\frac{Y}{N^{1/2}}, \quad \sigma = N^{1/2}\sqrt{1-\tau}, \quad (B.3)$$

$$K_{sin,\mathbb{C}}^{\beta=2}(X_1 + iY_1, X_2 + iY_2) \equiv \lim_{\substack{N\to\infty \\ \tau\to 1}} \frac{1}{N} K_{N,\mathbb{C}}^{\beta=2}(z_1, z_2), \quad (B.4)$$

and likewise for $\beta = 1, 4$. The macroscopic spectral density collapses onto the real axis, and becomes the semicircle for our Gaussian ensembles. The functions $R_{k,\mathbb{C}}$ given by the determinant or Pfaffian of the rescaled kernels describe the microscopic correlations in the complex plane.

If we magnify around any other point $|x_0| < 2\sqrt{N}$ inside the bulk, then we rescale the fluctuations $z - x_0$ as in (B.3). The correlations are then universal when measured in units of the local mean density $\pi\rho_{sc}(x_0)$ [Fyodorov et al. 1998; Akemann 2002].

Hard edge limit. Whilst we expect that in the bulk of the spectrum the WL and Gaussian ensembles (Equation (2.29) and (2.4) respectively) show the same behaviour, the origin is singled out in the latter case. The rescaling here is given by [Osborn 2004]

$$z = \frac{X}{4N} + i\frac{Y}{4N}, \quad \sigma = N^{1/2}\sqrt{1-\tau}, \quad (B.5)$$

$$K_{Bes,\mathbb{C}}^{\beta=2}(X_1 + iY_1, X_2 + iY_2) \equiv \lim_{\substack{N\to\infty \\ \tau\to 1}} \frac{1}{(4N)^2} K_{N,\nu,\mathbb{C}}^{\beta=2}(z_1, z_2), \quad (B.6)$$

with the Laguerre polynomials in (A.5) displaying a Bessel function asymptotic.

In all three scaling limits the asymptotic kernels are obtained by replacing the sums with integrals (the Christoffel–Darboux identity does not hold for OP in the complex plane), and the Hermite and Laguerre polynomials by their corresponding Plancherel–Rotach asymptotics (proven for real arguments) in the corresponding region.

An additional problem arises from the integrations with the antisymmetric weight function $\mathscr{F}$ used to obtain the limiting kernel elements G and W. For $\beta = 1$ these integrals are not absolutely convergent, and hence the limit $N \to \infty$ and the integration cannot be interchanged. For a detailed discussion we refer to [Akemann et al. 2011c; Phillips 2011].

Acknowledgements

The organisers and participants of the workshop "Random matrix theory and its applications" at MSRI Berkeley, 13–17 September 2010, are thanked for many inspiring talks and discussions.

References

[Akemann 2002] G. Akemann, "Microscopic universality of complex matrix model correlation functions at weak non-Hermiticity", *Phys. Lett. B* **547**:1-2 (2002), 100–108.

[Akemann 2005] G. Akemann, "The complex Laguerre symplectic ensemble of non-Hermitian matrices", *Nuclear Phys. B* **730**:3 (2005), 253–299.

[Akemann 2007] G. Akemann, "Matrix models and QCD with chemical potential", *Internat. J. Modern Phys. A* **22**:6 (2007), 1077–1122.

[Akemann and Basile 2007] G. Akemann and F. Basile, "Massive partition functions and complex eigenvalue correlations in matrix models with symplectic symmetry", *Nuclear Phys. B* **766**:1-3 (2007), 150–177.

[Akemann and Bender 2010] G. Akemann and M. Bender, "Interpolation between Airy and Poisson statistics for unitary chiral non-Hermitian random matrix ensembles", *J. Math. Phys.* **51**:10 (2010), 103524.

[Akemann and Phillips 2014] G. Akemann and M. J. Phillips, "The interpolating Airy kernels for the $\beta = 1$ and $\beta = 4$ elliptic Ginibre ensembles", *J. Stat. Phys.* **155**:3 (2014), 421–465.

[Akemann and Vernizzi 2003] G. Akemann and G. Vernizzi, "Characteristic polynomials of complex random matrix models", *Nuclear Phys. B* **660**:3 (2003), 532–556.

[Akemann et al. 2007] G. Akemann, P. H. Damgaard, J. C. Osborn, and K. Splittorff, "A new chiral two-matrix theory for Dirac spectra with imaginary chemical potential", *Nuclear Phys. B* **766**:1-3 (2007), 34–67. Erratum: "A new chiral two-matrix theory for Dirac spectra with imaginary chemical potential", *Nuclear Phys. B*, **800** (2008), pp. 406-407.

[Akemann et al. 2009] G. Akemann, M. J. Phillips, and H.-J. Sommers, "Characteristic polynomials in real Ginibre ensembles", *J. Phys. A* **42**:1 (2009), 012001.

[Akemann et al. 2010a] G. Akemann, M. Kieburg, and M. J. Phillips, "Skew-orthogonal Laguerre polynomials for chiral real asymmetric random matrices", *J. Phys. A* **43**:37 (2010), 375207.

[Akemann et al. 2010b] G. Akemann, M. J. Phillips, and H.-J. Sommers, "The chiral Gaussian two-matrix ensemble of real asymmetric matrices", *J. Phys. A* **43**:8 (2010), 085211.

[Akemann et al. 2011a] G. Akemann, J. Baik, and P. Di Francesco (editors), *The Oxford handbook of random matrix theory*, Oxford University Press, Oxford, 2011.

[Akemann et al. 2011b] G. Akemann, P. H. Damgaard, K. Splittorff, and J. J. M. Verbaarschot, "Spectrum of the Wilson Dirac operator at finite lattice spacings", *Phys. Rev. D* **83** (2011), 085014.

[Akemann et al. 2011c] G. Akemann, T. Kanazawa, M. J. Phillips, and T. Wettig, "Random matrix theory of unquenched two-colour QCD with nonzero chemical potential", *J. High Energy Phys.* **2011**:3 (2011), #066.

[Bender 2010] M. Bender, "Edge scaling limits for a family of non-Hermitian random matrix ensembles", *Probab. Theory Related Fields* **147**:1-2 (2010), 241–271.

[Damgaard et al. 2010] P. H. Damgaard, K. Splittorff, and J. J. M. Verbaarschot, "Microscopic spectrum of the Wilson Dirac operator", *Phys. Rev. Lett.* **105**:16 (2010), 162002.

[Deift and Gioev 2009] P. Deift and D. Gioev, *Random matrix theory: invariant ensembles and universality*, Courant Lecture Notes in Mathematics **18**, Courant Institute of Mathematical Sciences, New York, 2009.

[Deift et al. 2007] P. Deift, D. Gioev, T. Kriecherbauer, and M. Vanlessen, "Universality for orthogonal and symplectic Laguerre-type ensembles", *J. Stat. Phys.* **129**:5-6 (2007), 949–1053.

[Efetov 1997a] K. B. Efetov, "Directed Quantum Chaos", *Phys. Rev. Lett.* **79** (Jul 1997), 491–494.

[Efetov 1997b] K. B. Efetov, "Quantum disordered systems with a direction", *Phys. Rev. B* **56** (Oct 1997), 9630–9648.

[Erdős and Yau 2012] L. Erdős and H.-T. Yau, "Universality of local spectral statistics of random matrices", *Bull. Amer. Math. Soc. (N.S.)* **49**:3 (2012), 377–414.

[Forrester and Nagao 2008] P. J. Forrester and T. Nagao, "Skew orthogonal polynomials and the partly symmetric real Ginibre ensemble", *J. Phys. A* **41**:37 (2008), 375003.

[Forrester et al. 1999] P. J. Forrester, T. Nagao, and G. Honner, "Correlations for the orthogonal-unitary and symplectic-unitary transitions at the hard and soft edges", *Nuclear Phys. B* **553**:3 (1999), 601–643.

[Fyodorov and Sommers 2003] Y. V. Fyodorov and H.-J. Sommers, "Random matrices close to Hermitian or unitary: overview of methods and results", *J. Phys. A* **36**:12 (2003), 3303–3347.

[Fyodorov et al. 1997a] Y. V. Fyodorov, B. A. Khoruzhenko, and H.-J. Sommers, "Almost-Hermitian random matrices: eigenvalue density in the complex plane", *Phys. Lett. A* **226**:1-2 (1997), 46–52.

[Fyodorov et al. 1997b] Y. V. Fyodorov, B. A. Khoruzhenko, and H.-J. Sommers, "Almost Hermitian Random Matrices: Crossover from Wigner-Dyson to Ginibre Eigenvalue Statistics", *Phys. Rev. Lett.* **79** (Jul 1997), 557–560.

[Fyodorov et al. 1998] Y. V. Fyodorov, H.-J. Sommers, and B. A. Khoruzhenko, "Universality in the random matrix spectra in the regime of weak non-Hermiticity", *Ann. Inst. H. Poincaré Phys. Théor.* **68**:4 (1998), 449–489.

[Kanzieper 2002] E. Kanzieper, "Eigenvalue correlations in non-Hermitean symplectic random matrices", *J. Phys. A* **35**:31 (2002), 6631–6644.

[Khoruzhenko and Sommers 2011] B. A. Khoruzhenko and H.-J. Sommers, "Non-Hermitian ensembles", pp. 376–397 in *The Oxford handbook of random matrix theory*, edited by G. Akemann et al., Oxford Univ. Press, Oxford, 2011.

[Kieburg 2012] M. Kieburg, "Mixing of orthogonal and skew-orthogonal polynomials and its relation to Wilson RMT", *J. Phys. A* **45**:20 (2012), 205203.

[Kolesnikov and Efetov 1999] A. V. Kolesnikov and K. B. Efetov, "Distribution of complex eigenvalues for symplectic ensembles of non-Hermitian matrices", *Wav. Rand. Media* **9** (1999), 71.

[Kuijlaars 2011] A. B. J. Kuijlaars, "Universality", pp. 103–134 in *The Oxford handbook of random matrix theory*, edited by G. Akemann et al., Oxford Univ. Press, Oxford, 2011.

[Macêdo 1994] A. M. S. Macêdo, "Universal parametric correlations at the soft edge of the spectrum of random matrix ensembles", *EPL* **26**:9 (1994), 641.

[Mehta 2004] M. L. Mehta, *Random matrices*, 3rd ed., Pure and Applied Mathematics **142**, Academic Press/Elsevier, Amsterdam, 2004.

[Nagao and Forrester 1995] T. Nagao and P. J. Forrester, "Asymptotic correlations at the spectrum edge of random matrices", *Nuclear Phys. B* **435**:3 (1995), 401–420.

[Osborn 2004] J. C. Osborn, "Universal Results from an alternate random-matrix model for QCD with a baryon chemical potential", *Phys. Rev. Lett.* **93** (Nov 2004), 222001.

[Phillips 2011] M. J. Phillips, *A random matrix model for two-colour QCD at non-zero quark density*, Ph.D. thesis, Brunel University, 2011, http://bura.brunel.ac.uk/bitstream/2438/5084/1/FulltextThesis.pdf.

[Splittorff and Verbaarschot 2004] K. Splittorff and J. J. M. Verbaarschot, "Factorization of correlation functions and the replica limit of the Toda lattice equation", *Nuclear Phys. B* **683**:3 (2004), 467–507.

[Tao and Vu 2012] T. Tao and V. Vu, "Random matrices: the universality phenomenon for Wigner ensembles", 2012. arXiv 1202.0068

[Tao and Vu 2014] T. Tao and V. Vu, "Random matrices: universality of local spectral statistics of non-Hermitian matrices", 2014, http://goo.gl/bppHGP. To appear in *Annals of Probability*. arXiv 1206.1893

[Verbaarschot 2011] J. J. M. Verbaarschot, "Quantum chromodynamics", pp. 661–682 in *The Oxford handbook of random matrix theory*, edited by G. Akemann et al., Oxford Univ. Press, Oxford, 2011.

akemann@physik.uni-bielefeld.de   *Department of Physics, Bielefeld University,*
Postfach 100131, D-D-33501 Bielefeld, Germany

michael.phillips@qmul.ac.uk   *School of Mathematical Sciences,*
Queen Mary University of London, Mile End Road,
London, E1 4NS, United Kingdom

Random Matrices
MSRI Publications
Volume **65**, 2014

On a relationship between high rank cases and rank one cases of Hermitian random matrix models with external source

JINHO BAIK AND DONG WANG

We prove an identity on Hermitian random matrix models with external source relating the high rank cases to the rank 1 cases. This identity was proved and used in a previous paper of ours to study the asymptotics of the top eigenvalues. In this paper, we give an alternative, more conceptual proof of this identity based on a connection between the Hermitian matrix models with external source and the discrete KP hierarchy. This connection is obtained using the vertex operator method of Adler and van Moerbeke. The desired identity then follows from the Fay-like identity of the discrete KP τ vector.

1. Introduction

The subject of this paper is an identity between a Hermitian random matrix model with external source of rank m and m such models with external source of rank 1. This identity allows us to reduce the asymptotic study of "spiked Hermitian random matrix models" of rank higher than 1 to that of the models of rank 1. This reduction formula was used in [Baik and Wang 2013] to evaluate the limiting fluctuations of the top eigenvalue(s) of spiked models of arbitrary fixed rank with a general class of potentials. In [Baik and Wang 2013] we gave a direct proof of this identity using the formula of [Baik 2009] on the Christoffel–Darboux kernel. Here we give a different, more conceptual proof using the relation between the random matrix model with external source and discrete KP hierarchy. We show how the general results of [Adler and van Moerbeke 1999] on the construction of solutions of discrete KP hierarchy can be used on the partition functions of the Hermitian matrix model with external source.

We now introduce the model. Let $W(x)$ be a piecewise continuous function on $\mathbb{R}$. We assume that $W(x)$ is nonnegative, has infinite support, and vanishes sufficiently fast as $|x| \to \infty$ so that (3) converges. Let A be a $d \times d$ Hermitian

The work of Jinho Baik was supported in part by NSF grant DMS1068646.

matrix with eigenvalues $a_1, \ldots, a_d$. We call A the external source matrix and $W(x)$ the weight function. Consider the following measure on the set $\mathcal{H}_d$ of $d \times d$ Hermitian matrix M:

$$P(M)\,dM := \frac{1}{Z_d(a_1, \ldots, a_d)}\det(W(M))e^{\operatorname{Tr}(AM)}\,dM, \qquad (1)$$

where $W(M)$ is defined in terms of a functional calculus and

$$Z_d(a_1, \ldots, a_d) := \int_{M \in \mathcal{H}_d} \det(W(M))e^{\operatorname{Tr}(AM)}\,dM \qquad (2)$$

is the partition function. Note that the partition function depends only on the eigenvalues of A, but not its eigenvectors. It is also symmetric in $a_1, \ldots, a_d$. The Harish-Chandra–Itzykson–Zuber integral formula [Harish-Chandra 1957; Itzykson and Zuber 1980] implies that

$$Z_d(a_1, \ldots, a_d)$$
$$= \frac{C_d}{\displaystyle\prod_{1 \le j < k \le d}(a_k - a_j)} \int_{\mathbb{R}^d} \det[e^{a_j \lambda_k}]_{j,k=1}^d \prod_{1 \le j < k \le d}(\lambda_k - \lambda_j) \prod_{j=1}^d W(\lambda_j)\,d\lambda_j \qquad (3)$$

for a constant C_d that depends only on d. There is a similar formula for the density function of the eigenvalues.

If some of the eigenvalues of A are zero, we use the following short-hand: For $\boldsymbol{m} \le d$,

$$Z_d(a_1, \ldots, a_{\boldsymbol{m}}) := Z_d(a_1, \ldots, a_{\boldsymbol{m}}, \underbrace{0, \ldots, 0}_{d-\boldsymbol{m}}),$$
$$Z_d := Z_d(\underbrace{0, \ldots, 0}_{d}). \qquad (4)$$

The main theorem of this paper is about the following expectations. For $E \subset \mathbb{R}$, $s \in \mathbb{C}$, and $\boldsymbol{m} \le d$, we define

$$\mathfrak{E}_d(a_1, \ldots, a_{\boldsymbol{m}}; E; s) := \mathbb{E}\left[\prod_{j=1}^d (1 - s\chi_E(\lambda_j))\right]$$
$$= \int_{\mathcal{H}_d} \prod_{j=1}^d (1 - s\chi_E(\lambda_j))P(M)\,dM, \qquad (5)$$

where $\lambda_1, \ldots, \lambda_d$ are eigenvalues of M and the expectation is with respect to the measure (1) when the eigenvalues of the external source matrix A are

$$a_1, \ldots, a_{\boldsymbol{m}}, \underbrace{0, \ldots, 0}_{d-\boldsymbol{m}}.$$

We note that the last integral is the partition function with new weight

$$(1 - s\chi_E(x))W(x)$$

divided by the original partition function with the weight $W(x)$. This observation will be used in the proof later. When $s = 1$ the above expectation is a gap probability. Set

$$\overline{\mathcal{E}}_d(a_1, \ldots, a_m; E; s) := \frac{\mathcal{E}_d(a_1, \ldots, a_m; E; s)}{\mathcal{E}_d(E; s)}. \tag{6}$$

Let $p_j(x)$ be orthonormal polynomials with respect to $W(x)\,dx$. For a real number a, define the constant

$$\Gamma_j(a) := \int_{\mathbb{R}} p_j(s)e^{as}W(s)\,ds. \tag{7}$$

The main result is this:

Theorem 1.1. *We have*

$$\overline{\mathcal{E}}_d(a_1, \ldots, a_m; E; s) = \frac{\det\left[\Gamma_{d-j}(a_k)\overline{\mathcal{E}}_{d-j+1}(a_k; E; s)\right]_{j,k=1}^m}{\det[\Gamma_{d-j}(a_k)]_{j,k=1}^m}. \tag{8}$$

if $a_1, \ldots, a_m$ are distinct and nonzero.

Since both sides of (8) are analytic in a_j, the above identity still holds when some of a_j are the same or equal to zero if we interpret the right-hand side using l'Hôpital's rule.

The term $\overline{\mathcal{E}}_{d-j+1}(a_k; E; s)$ on the right-hand side of (8) is given by (6) with the rank $m = 1$ and the only nonzero eigenvalue of the external source A being a_k, and the dimension is changed to $d - j + 1$. Hence the identity (8) relates the rank m case to m rank 1 cases.

In [Baik and Wang 2011], the asymptotics $\overline{\mathcal{E}}_{d-j+1}(a_k; E; s)$, the rank 1 cases, were obtained. The quantities $\Gamma_{d-j}(a_k)$ were also analyzed asymptotically in the same paper as an intermediate step. In [Baik and Wang 2013] the higher rank cases were then analyzed asymptotically using the identity (8). The main result was that when the potential, assuming that it is real analytic, is convex to the right of the right-end point of the support of the equilibrium measure, the phase transition behavior of the fluctuations of the top eigenvalues is same as in the Gaussian unitary ensemble. Otherwise, new types of jump transitions are possible to occur. A characterization of possible new transitions was also obtained.

2. Proof

Set $\Delta(x) := \prod_{1 \leq j < k \leq d}(x_k - x_j) = \det(x_j^{k-1})_{j,k=1}^d$, for $x = (x_1, \ldots, x_d)$. Recall that, setting $a = (a_1, \ldots, a_d)$ and $\lambda = (\lambda_1, \ldots, \lambda_d)$,

$$\frac{\det[e^{a_j \lambda_k}]_{j,k=1}^d}{\Delta(a)\Delta(\lambda)} = \sum_{\ell(\kappa) \leq d} \frac{s_\kappa(\lambda)s_\kappa(a)}{\prod_{q=1}^d (\kappa_q + d - q)!}, \tag{9}$$

where the sum is over all partitions κ with at most d parts. Here s_κ denotes the Schur polynomial and $\ell(\kappa)$ denotes the number of parts of partition κ. This identity can be proved as follows. First, from Andréief's identity [Andréief 1886] (equivalently the Cauchy–Binnet formula),

$$\det[e^{a_j \lambda_k}]_{j,k=1}^d = \det\left[\sum_{n=0}^\infty \frac{a_j^n \lambda_k^n}{n!}\right] = \frac{1}{d!} \sum_{n_1, \ldots, n_d = 0}^\infty \frac{\det[a_j^{n_q}] \det[\lambda_j^{n_q}]}{\prod_{q=1}^d n_q!}. \tag{10}$$

Since the summand is symmetric in the n_q and equals to zero when two of the summation indices are the same, we can replace $(1/d!) \sum_{n_1, \ldots, n_d = 0}^\infty$ by $\sum_{0 \leq n_d < \cdots < n_2 < n_1}$.

Now set $\kappa_q := n_q - d + q$. Then the summation condition becomes $\kappa_1 \geq \cdots \geq \kappa_d \geq 0$, that is, all partitions with at most d parts. The identity (9) follows by recalling the classical definition of the Schur polynomial

$$s_\kappa(a) = \frac{\det(a_j^{\kappa_q + d - q})}{\Delta(\lambda)}.$$

Inserting (9) into (3), we obtain the Schur polynomial expansion of the partition function:

$$Z_d(a_1, \ldots, a_d) = C_d \sum_{\ell(\kappa) \leq d} G_\kappa \frac{s_\kappa(a)}{\prod_{q=1}^d (\kappa_q + d - q)!} \tag{11}$$

where

$$G_\kappa := \int_{\mathbb{R}^d} s_\kappa(\lambda) \Delta(\lambda)^2 \prod_{j=1}^d W(\lambda_j) \, d\lambda_j. \tag{12}$$

Using the classical definition of the Schur function, the determinantal form of $\Delta(\lambda)$, and the Andréief's identity, we obtain

$$G_\kappa = d! \cdot \det[M_{\kappa_p + d - p + q - 1}]_{p,q=1}^d, \qquad M_j := \int_\mathbb{R} x^j W(x) \, dx. \tag{13}$$

Here M_j are the moments of the measure $W(x) \, dx$.

We insert (13) into (11) and use the Jacobi–Trudi identity,

$$s_\kappa(a) = \det[h_{\kappa_p - p + q}(a)]_{p,q=1}^d,$$

where $h_j(a)$ denotes the complete symmetric function. The sum is over the partitions $\kappa = (\kappa_1, \ldots, \kappa_d)$ where $\kappa_1 \geq \cdots \geq \kappa_d \geq 0$. We set $j_p := \kappa_p + d - p$. Then $j_1 > \cdots > j_d \geq 0$. Since the summand is symmetric in the j_p and vanishes when two indices are the same, we arrive at the formula

$$Z_d(a_1, \ldots, a_d) = C_d \sum_{j_1, \ldots, j_d = 0}^{\infty} \frac{\det[M_{j_p + q - 1}]_{p,q=1}^d \det[h_{j_p - d + q}(a)]_{p,q=1}^d}{\prod_{q=1}^d j_q!}.$$

(14)

In the below, the notation $t = (t_1, t_2, \ldots)$ denotes a sequence of variables. We also use the notation $[c] = (c, \frac{1}{2}c^2, \frac{1}{3}c^3, \ldots)$ for the evaluation of t by powers of c. The notation $[c_1] + [c_2] + \cdots + [c_m]$ stands for the evaluation of t obtained by substituting $t_j = \sum_{i=1}^m c_i^j / j$.

Following [Kac and Raina 1987, Definition 6.1], we define the so-called "elementary Schur polynomials" $h_j(t)$ by the generating function

$$\sum_{j=-\infty}^{\infty} h_j(t) w^j = e^{\sum_{j=1}^\infty t_j w^j}.$$

(15)

If $t = [a_1] + [a_2] + \cdots + [a_d]$ (i.e., $t_j = \sum_{i=1}^d a_i^j / j$), $h_j(t)$ is the complete symmetric function in $a_1, \ldots, a_d$, which we denoted earlier by $h_j(a)$. This abuse of notations is unfortunate but in the below we only use the definition $h(t)$ given in (15).

Now define the formal power series in $t = (t_1, t_2, \ldots)$

$$\hat{Z}_d(t) := \frac{1}{d!\hat{C}_d} \sum_{j_1, \ldots, j_d = 0}^{\infty} \frac{\det[M_{j_p + q - 1}]_{p,q=1}^d \det[h_{j_p - d + q}(t)]_{p,q=1}^d}{\prod_{k=1}^d j_k!},$$

(16)

where $\hat{C}_d := \frac{1}{d! C_d}$. This definition is equivalent to [Wang 2009, (26)]. Then

$$Z_d(a_1, \ldots, a_d) = \hat{Z}_d([a_1] + \cdots + [a_d]).$$

(17)

Setting some of the parameters to zero, we also have for $m \leq d$

$$Z_d(a_1, \ldots, a_m) = \hat{Z}_d([a_1] + \cdots + [a_m]).$$

(18)

We now show that $\hat{Z}_d(t)$ solves the discrete KP hierarchy following the vertex operator construction of the general solutions due to Adler and van Moerbeke. We then show that a general property of the vertex operator solution implies an identity of which the identity (8) is a special case. Note that the definition (16) does not require the assumption that the weight function $W(x)$ is nonnegative. In the next two subsections we drop this assumption.

Discrete KP τ vectors.

Proposition 2.1. *Let $\hat{Z}_d(t)$ be defined in (16) and set $\hat{Z}_0(t) := 1$. Then the sequence*

$$\left(\ldots, \hat{Z}_2(-t), \hat{Z}_1(-t), \hat{Z}_0(t), \hat{Z}_1(t), \hat{Z}_2(t), \ldots\right) \tag{19}$$

constitutes a discrete KP τ vector.

Discrete KP τ vectors are solutions to a system of differential-difference equations in the discrete KP hierarchy. In [Adler and van Moerbeke 1999] several characterizations of discrete KP τ vectors were established. Here we use two of them: the vertex operator characterization (see Proposition 2.2 below) and the Hirota bilinear identity characterization (see (34) below).

Remark. Each component of a discrete KP τ vector is a KP τ function. The fact that $\hat{Z}_d(t)$ is a KP τ function for each $d \in \mathbb{N}$ was proved in [Wang 2009]. In [Harnad and Orlov 2014] it was proved further that for each $d \in \mathbb{N}$, $\hat{Z}_d(t)$ is a so-called 1-KP-Toda τ function and moreover these 1-KP-Toda τ functions are derived from the same Grassmannian structure. It should also be possible to prove the above proposition from this fact.

The vertex operator is a differential operator defined by

$$X(t, z) := \exp\left(\sum_{k=1}^{\infty} t_k z^k\right) \exp\left(-\sum_{k=1}^{\infty} \frac{z^{-k}}{k} \frac{\partial}{\partial t_k}\right); \tag{20}$$

see [Adler and van Moerbeke 1999, (0.22)]. The vertex operator acts on a formal power series $f(t)$ as

$$X(t, z)f(t) = \exp\left(\sum_{k=1}^{\infty} t_k z^k\right) f(t - [z^{-1}])$$

$$= \left(\sum_{k=-\infty}^{\infty} h_k(t) z^k\right) f(t - [z^{-1}]). \tag{21}$$

One can construct a discrete KP τ vector from one KP τ function and a sequence of measures by applying the vertex operator repeatedly:

Proposition 2.2 [Adler and van Moerbeke 1999, Theorem 0.3]. *Let $\tau(t)$ be a KP τ function. Let $\left(\ldots, v_{-1}(z)\,dz, v_0(z)\,dz, v_1(z)\,dz, \ldots\right)$ be a sequence of arbitrary measures. Then the infinite sequence $(\ldots, \tau_{-1}(t), \tau_0(t), \tau_1(t), \ldots)$ defined by $\tau_0(t) := \tau(t)$ and*

$$\tau_d(t) := \left(\int X(t,z) v_{d-1}(z)\, dz \right) \cdots \left(\int X(t,z) v_0(z)\, dz \right) \tau(t), \qquad (22)$$

$$\tau_{-d}(t) := \left(\int X(-t,z) v_{-d}(z)\, dz \right) \cdots \left(\int X(-t,z) v_{-1}(z)\, dz \right) \tau(t), \quad (23)$$

for $d \in \mathbb{N}$, forms a discrete KP τ vector.

Remark. The original Theorem 0.3 of [Adler and van Moerbeke 1999] assumes that the measures $v_k(z)\, dz$ are defined on $\mathbb{R}$. However, it is easy to check that the proof to Proposition 2.2 in [Adler and van Moerbeke 1999] holds almost verbatim if we change $\mathbb{R}$ into $\mathbb{C}$ (or more restrictively the unit circle $\{z \in \mathbb{C} \mid |z| = 1\}$ that we are going to use in this section).

Proof of Proposition 2.1. We set $\tau(t) = \hat{Z}_0(t) := 1$. This is trivially a KP τ function. Hence Proposition 2.1 is proved if we can construct a sequence of measures $(\ldots, v_{-1}(z)\, dz,\ v_0(z)\, dz,\ v_1(z)\, dz, \ldots)$ such that $\hat{Z}_d(t)$ equals the right-hand side of (22) (resp. (23)) with $\tau(t) = 1$ for $d > 0$ (resp. $d < 0$).

Define the measure on the circle $\{z \in \mathbb{C} : |z| = 1\}$ as

$$v_d(z)\, dz := \frac{(-1)^d \hat{C}_d}{2\pi i \hat{C}_{d+1}} \sum_{j=0}^{\infty} \frac{M_{j+d}}{j!} z^{d-j-1}\, dz, \quad d = 0, 1, 2, \ldots, \qquad (24)$$

where $\hat{C}_0 := 1$. We also define

$$v_d(z) := v_{-d-1}(z) \quad d = -1, -2, \ldots. \qquad (25)$$

We now show that these measures satisfy the desired property.

For $d = 0$, since $\hat{Z}_0(t) = 1$, (21) implies $X(t,z)\hat{Z}_0(t) = \sum_{k=-\infty}^{\infty} h_k(t) z^k$. Hence we find from a direct evaluation of the integral using the Cauchy integral formula (the integral is over the unit circle) that

$$\oint X(t,z)\hat{Z}_0(t) v_0(z)\, dz = \frac{1}{\hat{C}_1} \sum_{j=0}^{\infty} \frac{h_j(t) M_j}{j!} = \hat{Z}_1(t) \qquad (26)$$

from the definition (16).

We now consider $d > 0$. From (21),

$$X(t,z)\hat{Z}_d(t) = \frac{1}{d!\,\hat{C}_d} \left(\sum_{k=-\infty}^{\infty} h_k(t) z^k \right)$$

$$\times \sum_{j_1,\ldots,j_d=0}^{\infty} \frac{\det[M_{j_p+q-1}]_{p,q=1}^d\ \det[h_{j_p-d+q}(t-[z^{-1}])]_{p,q=1}^d}{\prod_{k=1}^d j_k!}. \qquad (27)$$

Note that from (15) we have

$$\sum_{j=-\infty}^{\infty} h_j(t - [z^{-1}])w^j = \exp\left(\sum_{j=1}^{\infty}(t_j - \frac{z^{-j}}{j})w^j\right)$$

$$= \left(1 - \frac{w}{z}\right)\sum_{j=-\infty}^{\infty} h_j(t)w^j.$$

Comparing the coefficients of w^j, we find that

$$h_j(t - [z^{-1}]) = h_j(t) - z^{-1}h_{j-1}(t). \tag{28}$$

By (28), we have

$$\det[h_{j_p-d+q}(t - [z^{-1}])]_{p,q=1}^d = \sum_{l=0}^{d} z^{-l} \det[h_{j_p-d+q-H[l-q]}(t)]_{p,q=1}^d, \tag{29}$$

where H is the discrete form of the Heaviside function such that $H[n] = 0$ for $n < 0$ and $H[n] = 1$ for $n \geq 0$. Substituting (29) into (27), we have, after the change of variable $j_0 = k - l + d$, that

$$X(t,z)\hat{Z}_d(t) = \frac{1}{d!\,\hat{C}_d} \sum_{l=0}^{d}\left(\sum_{j_0=0}^{\infty} h_{j_0-d+l}(t)z^{j_0-d}\right)$$

$$\times \sum_{j_1,\ldots,j_d=0}^{\infty} \frac{\det[M_{j_p+q-1}]_{p,q=1}^d \det[h_{j_p-d+q-H[l-q]}(t)]_{p,q=1}^d}{\prod_{k=1}^{d} j_k!}$$

We can reexpress this as follows, using that the right-hand side of (30) is symmetric in $j_0, j_1, \ldots, j_d$:

$$X(t,z)\hat{Z}_d(t)$$

$$= \frac{(-1)^d}{(d+1)!\,\hat{C}_d} \sum_{j_0,j_1,\ldots,j_d=0}^{\infty} \prod_{k=0}^{n} \frac{1}{j_k!}$$

$$\times \begin{vmatrix} M_{j_0} & \cdots & M_{j_0+d-1} & j_0!\,z^{j_0-d} \\ M_{j_1} & \cdots & M_{j_1+d-1} & j_1!\,z^{j_1-d} \\ \vdots & \ddots & \vdots & \vdots \\ M_{j_d} & \cdots & M_{j_d+d-1} & j_d!\,z^{j_d-d} \end{vmatrix} \begin{vmatrix} h_{j_0-d}(t) & h_{j_0-d+1}(t) & \cdots & h_{j_0}(t) \\ h_{j_1-d}(t) & h_{j_1-d+1}(t) & \cdots & h_{j_1}(t) \\ \vdots & \vdots & \ddots & \vdots \\ h_{j_d-d}(t) & h_{j_d-d+1}(t) & \cdots & h_{j_d}(t) \end{vmatrix}. \tag{30}$$

Note that in here there is one more summation index j_0 than in (27) and the determinants are of $d + 1$ by $d + 1$ matrices. Then from (30) and (16), and noting that the variable z appears only in the last column of the first matrix in

(30), we can check directly using the Cauchy integral formula that

$$\oint X(t,z)\hat{Z}_d(t)v_d(z)\,dz = \hat{Z}_{d+1}(t), \quad d > 0. \tag{31}$$

Successive applications of the relation (31) imply that, for all $d > 0$,

$$\hat{Z}_d(t) = \left(\int X(t,z)v_{d-1}(z)\,dz\right)\cdots\left(\int X(t,z)v_0(z)\,dz\right)\hat{Z}_0(t), \tag{32}$$

which is same as (22).

Finally, (25) and (32) imply that

$$\hat{Z}_d(-t) = \left(\int X(-t,z)v_{-d}(z)\,dz\right)\cdots\left(\int X(-t,z)v_{-1}(z)\,dz\right)\hat{Z}_0(t). \tag{33}$$

This is same as (23). Hence the proposition is proved. $\qquad\qquad\qquad\square$

Fay-like identity. An importance property of discrete KP τ vector is that its components satisfy a Hirota bilinear identity (see [Adler and van Moerbeke 1999, Theorem 0.2(iii)]). (Adler and van Moerbeke, moreover, showed that the Hirota bilinear identity actually characterizes the discrete KP τ vector.) For the discrete KP τ vector (19) in our situation, replacing the notations τ_n and τ_m for components of a general KP τ vector in [Adler and van Moerbeke 1999, (0.18)] into our specific $\hat{Z}_{d_1}$ and $\hat{Z}_{d_2}$, we find that the Hirota bilinear identity becomes

$$\frac{1}{2\pi i}\oint_{z=\infty}\hat{Z}_{d_1}(\tilde{t} - [z^{-1}])\hat{Z}_{d_2+1}(t + [z^{-1}])$$

$$\times \exp\left(\sum_{j=1}^{\infty}(\tilde{t}_j - t_j)z^j\right)z^{d_1-d_2-1}\,dz = 0, \tag{34}$$

for all $d_1 > d_2 \geq 0$. Here the formal integral of a formal Laurent series is defined by

$$\frac{1}{2\pi i}\oint_{z=\infty}\left(\sum_{j=-\infty}^{\infty}a_jz^j\right)dz = a_{-1}. \tag{35}$$

We now show that this Hirota bilinear identity implies a Fay-like identity (41). Such a derivation of a Fay-like identity from the Hirota bilinear identity was obtained in the Toda lattice hierarchy by [Teo 2006] and we adapt this approach.

We take the special choices $d_1 = d$, $d_2 = d - 2$ and $\tilde{t} = t + [a] + [b]$ in (34). Then the factor $\exp\left(\sum_{j=1}^{\infty}(\tilde{t}_j - t_j)z^j\right)z^{d_1-d_2-1}$ equals

$$z\exp\left(\sum_{j=1}^{\infty}\left(\frac{a^j}{j} + \frac{b^j}{j}\right)z^j\right). \tag{36}$$

After rewriting the sum in the exponential as $-\log(1-az)-\log(1-bz)$, and using the simple identity

$$\frac{abz}{(1-az)(1-bz)} = \frac{1}{(a-b)z}\left(\frac{b}{1-az} - \frac{a}{1-bz}\right) + \frac{1}{z},$$

we find that (36) equals

$$\frac{1}{a(a-b)z}\sum_{j=0}^{\infty}a^j z^j - \frac{1}{b(a-b)z}\sum_{j=0}^{\infty}b^j z^j + \frac{1}{abz}.$$

Using this, (34) implies that

$$\frac{a}{2\pi i}\oint_{z=\infty} Q(z^{-1})\left(\sum_{j=0}^{\infty}b^j z^j\right)\frac{dz}{z} - \frac{b}{2\pi i}\oint_{z=\infty} Q(z^{-1})\left(\sum_{j=0}^{\infty}a^j z^j\right)\frac{dz}{z}$$

$$= \frac{a-b}{2\pi i}\oint_{z=\infty} Q(z^{-1})\frac{dz}{z}, \quad (37)$$

where Q is defined by

$$Q(w) := \hat{Z}_d(t + [a] + [b] - [w])\hat{Z}_{d-1}(t + [w]). \quad (38)$$

Observe that the Laurent series of $Q(w)$ consists only of nonnegative powers of w. Hence

$$Q(w) = \sum_{n=0}^{\infty} q_n w^n, \quad (39)$$

for some $q_0, q_1, \ldots$. Thus, from (35) we have

$$\frac{1}{2\pi i}\oint_{z=\infty} Q(z^{-1})\left(\sum_{j=0}^{\infty}a^j z^j\right)\frac{dz}{z} = \sum_{n=0}^{\infty}q_n a^n = Q(a)$$

$$= \hat{Z}_d(t + [a])\hat{Z}_{d-1}(t + [b]). \quad (40)$$

Similar evaluations of the other integrals of (37) imply the following Fay-like identity:

$$a\hat{Z}_d(t + [a])\hat{Z}_{d-1}(t + [b]) - b\hat{Z}_d(t + [b])\hat{Z}_{d-1}(t + [a])$$

$$= (a-b)\hat{Z}_d(t + [a] + [b])\hat{Z}_{d-1}(t). \quad (41)$$

In the remainder of this section we use identity (41) to prove the following:

Proposition 2.3. *For any $d \geq m \geq 1$ and $a_1, \ldots, a_m \in \mathbb{C}$,*

$$\frac{\hat{Z}_d(t + [a_1] + \cdots + [a_m])}{\hat{Z}_d(t)}$$

$$= \frac{1}{\Delta_m(a_1, \ldots, a_m)} \det\left[a_k^{m-j} \frac{\hat{Z}_{d+1-j}(t + [a_k])}{\hat{Z}_{d+1-j}(t)} \right]_{j,k=1}^m, \quad (42)$$

where $\Delta_m(a_1, \ldots, a_m) := \prod_{1 \leq j < k \leq m} (a_j - a_k).$

Proof. After dividing the identity (41) by $\hat{Z}_d(t)\hat{Z}_{d-1}(t)$, we obtain

$$\frac{\hat{Z}_d(t + [a] + [b])}{\hat{Z}_d(t)} = \frac{1}{a - b} \det\begin{bmatrix} a\dfrac{\hat{Z}_d(t + [a])}{\hat{Z}_d(t)} & b\dfrac{\hat{Z}_d(t + [b])}{\hat{Z}_d(t)} \\ \dfrac{\hat{Z}_{d-1}(t + [a])}{\hat{Z}_{d-1}(t)} & \dfrac{\hat{Z}_{d-1}(t + [b])}{\hat{Z}_{d-1}(t)} \end{bmatrix}. \quad (43)$$

This is the identity (42) when $m = 2$ for all $d \geq 2$.

We now prove the general case using an induction in m. Suppose that (42) holds with $m \leq m - 1$ for all $d \geq m - 1$ and $a_1, \ldots, a_{m-1} \in \mathbb{C}$. We are to prove that it holds with $m = m$ for all $d \geq m$ and $a_1, \ldots, a_m \in \mathbb{C}$. For this purpose, we set $a = a_1$, $b = a_m$, and $t \mapsto t + [a_2] + \cdots + [a_{m-1}]$ in (43). After pulling out the denominators of the entries of the determinant outside, we obtain

$$(a_1 - a_m)\hat{Z}_d(t + [a_1] + \cdots + [a_m])\hat{Z}_{d-1}(t + [a_2] + \cdots + [a_{m-1}])$$

$$= \det\begin{bmatrix} a_1\hat{Z}_d(t + [a_1] + \cdots + [a_{m-1}]) & a_m\hat{Z}_d(t + [a_2] + \cdots + [a_m]) \\ \hat{Z}_{d-1}(t + [a_1] + \cdots + [a_{m-1}]) & \hat{Z}_{d-1}(t + [a_2] + \cdots + [a_m]) \end{bmatrix}. \quad (44)$$

Let us call the entries of the last determinant A_{ij}, $i, j = 1, 2$. First we consider A_{ij} on the first row. From the induction hypothesis,

$$\frac{A_{11}}{\hat{Z}_d(t)} = \frac{a_1}{\Delta_{m-1}(a_1, \ldots, a_{m-1})} \det\left[a_k^{m-1-j} \frac{\hat{Z}_{d+1-j}(t + [a_k])}{\hat{Z}_{d+1-j}(t)} \right]_{j,k=1}^{m-1}. \quad (45)$$

If we multiply $a_2 \cdots a_{m-1}$ on both sides and bring the factor $a_1 \cdots a_{m-1}$ inside the determinant, we find that

$$a_2 \cdots a_{m-1} \frac{A_{11}}{\hat{Z}_d(t)} = \frac{1}{\Delta_{m-1}(a_1, \ldots, a_{m-1})} \det[B_{jk}]_{1 \leq j \leq m-1, \, 1 \leq k \leq m-1}, \quad (46)$$

where

$$B_{jk} := a_k^{m-j} \frac{\hat{Z}_{d+1-j}(t + [a_k])}{\hat{Z}_{d+1-j}(t)}. \quad (47)$$

Note that the power of a_k has changed to $m - j$, from $m - 1 - j$ in (45). Similarly, we find that

$$a_2 \cdots a_{m-1} \frac{A_{12}}{\hat{Z}_d(t)} = \frac{1}{\Delta_{m-1}(a_2, \ldots, a_m)} \det[B_{jk}]_{1 \le j \le m-1, 2 \le k \le m}, \quad (48)$$

with the same definition (47) of B_{jk}. Note the difference of the indices of the determinant from (46).

Now we consider A_{ij} in the second row. The induction hypothesis implies that

$$\frac{A_{21}}{\hat{Z}_{d-1}(t)} = \frac{1}{\Delta_{m-1}(a_1, \ldots, a_{m-1})} \det\left[a_k^{m-1-j} \frac{\hat{Z}_{d-j}(t + [a_k])}{\hat{Z}_{d-j}(t)}\right]_{j,k=1}^{m-1}. \quad (49)$$

Note that d is changed to $d - 1$ in the determinant from (46). If we shift the index j by $j - 1$ in the determinant, we can write the above as

$$\frac{A_{21}}{\hat{Z}_{d-1}(t)} = \frac{1}{\Delta_{m-1}(a_1, \ldots, a_{m-1})} \det[B_{jk}]_{2 \le j \le m, 1 \le k \le m-1}. \quad (50)$$

Similarly,

$$\frac{A_{22}}{\hat{Z}_{d-1}(t)} = \frac{1}{\Delta_{m-1}(a_2, \ldots, a_m)} \det[B_{jk}]_{2 \le j \le m, 2 \le k \le m}. \quad (51)$$

Consider the matrix $\mathcal{B}$ of size m whose entries are B_{jk}, $j, k = 1, \ldots, m$. Let $\mathcal{B}_a^b$ denote the matrix of size $m - 1$ obtained from $\mathcal{B}$ by deleting the row a and the column b. The determinants in (46), (48), (50), and (51) are those of the matrices $\mathcal{B}_m^m$, $\mathcal{B}_m^1$, $\mathcal{B}_1^m$, and $\mathcal{B}_1^1$, respectively. Hence we find that

$$\frac{a_2 \cdots a_{m-1} \Delta_{m-1}(a_1, \ldots, a_{m-1}) \Delta_{m-1}(a_2, \ldots, a_m)}{\hat{Z}_{d-1}(t) \hat{Z}_d(t)} \det\begin{bmatrix} A_{11} & A_{12} \\ A_{21} & A_{22} \end{bmatrix}$$

$$= \det[\mathcal{B}_m^m] \det[\mathcal{B}_1^1] - \det[\mathcal{B}_1^m] \det[\mathcal{B}_m^1]. \quad (52)$$

Now the Desnanot–Jacobi identity, often attributed to Charles Ludwig Dodgson, aka Lewis Carroll (see, e.g., [Krattenthaler 1999, Proposition 10] and references therein) implies that the above equals $\det[\mathcal{B}] \det\left[\mathcal{B}_{1,m}^{1,m}\right]$ where $\mathcal{B}_{1,m}^{1,m}$ is the matrix of size $m - 2$ obtained by deleting the rows $1, m$ and the columns $1, m$ from $\mathcal{B}$. The determinant $\det\left[\mathcal{B}\right]$ is precisely the determinant in (42) with $\boldsymbol{m} = m$. On the other hand,

$$\frac{\det\left[\mathcal{B}_{1,m}^{1,m}\right]}{a_2 \cdots a_{m-1}} = \det\left[a_k^{m-1-j} \frac{\hat{Z}_{d+1-j}(t + [a_k])}{\hat{Z}_{d+1-j}(t)}\right]_{j,k=2}^{m-1}$$

$$= \det\left[a_{k+1}^{m-2-j} \frac{\hat{Z}_{d-j}(t + [a_{k+1}])}{\hat{Z}_{d-j}(t)}\right]_{j,k=1}^{m-2}. \quad (53)$$

The last determinant is precisely the determinant in (42) with $m = m - 1$, d replaced by $d - 1$, and the complex numbers given by $a_2, \ldots, a_{m-1}$. The induction hypothesis implies the identity

$$\frac{\det\left[\mathscr{B}_{1,m}^{1,m}\right]}{a_2 \ldots a_{m-1}} = \Delta_{m-2}(a_2, \ldots, a_{m-1}) \frac{\widehat{Z}_{d-1}(t + [a_2] + \cdots + [a_{m-1}])}{\widehat{Z}_{d-1}(t)}. \quad (54)$$

Combining (44), (52), and (54), and noting that

$$\frac{\Delta_{m-2}(a_2, \ldots, a_{m-1})}{(a_1 - a_m)\Delta_{m-1}(a_1, \ldots, a_{m-1})\Delta_{m-1}(a_2, \ldots, a_m)} = \frac{1}{\Delta_m(a_1, \ldots, a_m)}, \quad (55)$$

we obtain (42) with $m = m$. Hence the induction step is established and the proposition is proved. $\qquad \square$

Completion of proof of Theorem 1.1. For any subset $E \in \mathbb{R}$ and $s \in \mathbb{C}$, consider the new weight function $W_{E,s}(x) := W(x)(1 - s\chi_E(x))$. Let $Z_d^{E,s}(a_1, \ldots, a_d)$ be the partition function (2) with W replaced by $W_{E,s}$. We also use a similar short-hand notation as (4). Then

$$\mathfrak{E}_d(a_1, \ldots, a_m; E; s) = \frac{Z_d^{E,s}(a_1, \ldots, a_m)}{Z_d(a_1, \ldots, a_m)}. \quad (56)$$

Taking $t = (0, 0, \ldots)$ in (42) and recalling (17), we find

$$\frac{Z_d(a_1, \ldots, a_m)}{Z_d} = \frac{1}{\Delta_m(a_1, \ldots, a_m)} \det\left[a_k^{m-j} \frac{Z_{d+1-j}(a_k)}{Z_{d+1-j}}\right]_{j,k=1}^m. \quad (57)$$

Note that this holds for any weight function W. We substitute $W \mapsto W_{E,s}$ in (57) and divide this identity by (57) with W. From this we obtain

$$\bar{\mathfrak{E}}_d(a_1, \ldots, a_m; E; s) = \frac{\det\left[a_k^{m-j} Z_{d+1-j}(a_k)\bar{\mathfrak{E}}_{d+1-j}(a_k; E; s)\right]_{j,k=1}^m}{\det\left[a_k^{m-j} Z_{d+1-j}(a_k)\right]_{j,k=1}^m}. \quad (58)$$

We now consider the terms $a_k^{m-j} Z_{d+1-j}(a_k)$. For any dimension l, if $a_l = a$ and $a_1 = \cdots = a_{l-1} = 0$, applying l'Hôpital's rule to (3), we have a formula of the partition function $Z_l(a)$:

$$Z_l(a) = \frac{C_l}{a^{l-1} \prod_{j=0}^{l-2} j!} \int_{\mathbb{R}^l} \det[V]\det[\lambda_i^{j-1}] \prod_{j=1}^l W(\lambda_j)\, d\lambda_j, \quad (59)$$

where $V = (V_{ij})_{i,j=1}^l$, with $V_{ij} = \lambda_i^{j-1}$ for $j = 1, \ldots, l-1$ and $V_{il} = e^{a\lambda_i}$. Let p_j be orthonormal polynomials with respect $W(x)\, dx$. By using elementary row operations, we get

$$Z_l(a) = \frac{C_l'}{a^{l-1}} \int_{\mathbb{R}^l} \det[\tilde{V}] \det[p_{j-1}(\lambda_i)] \prod_{j=1}^{l} W(\lambda_j) \, d\lambda_j, \tag{60}$$

where $\tilde{V} = (\tilde{V}_{ij})_{i,j=1}^l$, with $\tilde{V}_{ij} = p_{j-1}(\lambda_i)$ for $j = 1, \ldots, l-1$ and $\tilde{V}_{il} = e^{a\lambda_i}$, and C_l' is a new constant which depends only on l and W. Using the Andréief's formula and the fact that p_j are orthonormal polynomials, we obtain

$$Z_l(a) = \frac{l! \, C_l'}{a^{l-1}} \int_{\mathbb{R}} e^{a\lambda} p_{l-1}(\lambda) W(\lambda) \, d\lambda = l! \, C_k' a^{-l+1} \Gamma_{l-1}(a). \tag{61}$$

Inserting this into (58), we obtain Theorem 1.1.   □

References

[Adler and van Moerbeke 1999] M. Adler and P. van Moerbeke, "Vertex operator solutions to the discrete KP-hierarchy", *Comm. Math. Phys.* **203**:1 (1999), 185–210.

[Andréief 1886] C. Andréief, "Note sur une relation entre les intégrales définines des produits des fonctions", *Mém. Soc. Sci. Phys. Nat. Bordeaux, Sér. 3* **2** (1886), 1–14.

[Baik 2009] J. Baik, "On the Christoffel–Darboux kernel for random Hermitian matrices with external source", *Comput. Methods Funct. Theory* **9**:2 (2009), 455–471.

[Baik and Wang 2011] J. Baik and D. Wang, "On the largest eigenvalue of a Hermitian random matrix model with spiked external source, I: Rank 1 case", *Int. Math. Res. Not.* **2011**:22 (2011), 5164–5240.

[Baik and Wang 2013] J. Baik and D. Wang, "On the largest eigenvalue of a Hermitian random matrix model with spiked external source, II: Higher rank cases", *Int. Math. Res. Not.* **2013**:14 (2013), 3304–3370.

[Harish-Chandra 1957] Harish-Chandra, "Differential operators on a semisimple Lie algebra", *Amer. J. Math.* **79** (1957), 87–120.

[Harnad and Orlov 2014] J. Harnad and A. Y. Orlov, "Convolution symmetries of integrable hierarchies, matrix models and τ-functions", pp. 247–275 in *Random matrix theory, interacting particle systems, and integrable systems*, Mathematical Sciences Research Institute Publications **65**, Cambridge University Press, New York, 2014.

[Itzykson and Zuber 1980] C. Itzykson and J. B. Zuber, "The planar approximation, II", *J. Math. Phys.* **21**:3 (1980), 411–421.

[Kac and Raina 1987] V. G. Kac and A. K. Raina, *Bombay lectures on highest weight representations of infinite-dimensional Lie algebras*, Advanced Series in Mathematical Physics **2**, World Scientific, Teaneck, NJ, 1987.

[Krattenthaler 1999] C. Krattenthaler, "Advanced determinant calculus", *Sém. Lothar. Combin.* **42** (1999), Art. B42q, 67 pp.

[Teo 2006] L.-P. Teo, "Fay-like identities of the Toda lattice hierarchy and its dispersionless limit", *Rev. Math. Phys.* **18**:10 (2006), 1055–1073.

[Wang 2009] D. Wang, "Random matrices with external source and KP τ functions", *J. Math. Phys.* **50**:7 (2009), 073506, 10.

baik@umich.edu   *Department of Mathematics, University of Michigan,*
                                  Ann Arbor, MI 48109, United States

matwd@nus.edu.sg   *Department of Mathematics,*
                                  National University of Singapore, Singapore 119076

Random Matrices
MSRI Publications
Volume **65**, 2014

Riemann–Hilbert approach
to the six-vertex model

PAVEL BLEHER AND KARL LIECHTY

The six-vertex model, or the square ice model, with domain wall boundary conditions (DWBC) has been introduced and solved for finite n by Korepin and Izergin. The solution is based on the Yang–Baxter equations and it represents the free energy in terms of an $n \times n$ Hankel determinant. Paul Zinn-Justin observed that the Izergin–Korepin formula can be expressed in terms of the partition function of a random matrix model with a nonpolynomial interaction. We use this observation to obtain the large n asymptotics of the six-vertex model with DWBC. The solution is based on the Riemann–Hilbert approach. In this paper we review asymptotic results obtained in different regions of the phase diagram.

1. Six-vertex model

The *six-vertex model*, or the model of *two-dimensional ice*, is stated on a square lattice with arrows on edges. The arrows obey the rule that at every vertex there are two arrows pointing in and two arrows pointing out. This rule is sometimes called the ice-rule. There are only six possible configurations of arrows at each vertex, hence the name of the model; see Figure 1.

We will consider the *domain wall boundary conditions* (DWBC), in which the arrows on the upper and lower boundaries point into the square, and the ones on the left and right boundaries point out. One possible configuration with DWBC on the 4×4 lattice is shown on Figure 2.

The name of the *square ice* comes from the two-dimensional arrangement of water molecules, H_2O, with oxygen atoms at the vertices of the lattice and one hydrogen atom between each pair of adjacent oxygen atoms. We place an arrow in the direction from a hydrogen atom toward an oxygen atom if there is a bond between them. Thus, as we already noticed before, there are two in-bound and two out-bound arrows at each vertex.

Bleher is supported in part by the National Science Foundation (NSF) grant DMS-0969254.
Keywords: none.

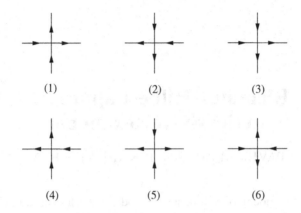

Figure 1. The six arrow configurations allowed at a vertex.

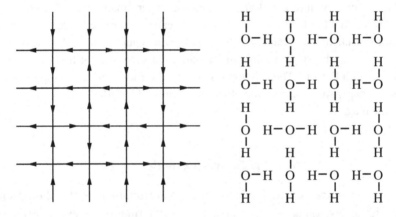

Figure 2. An example of a 4×4 configuration (left) and the corresponding ice crystal (right).

For each possible vertex state we assign a *weight* w_i, $i = 1, \ldots, 6$, and define, as usual, the *partition function*, as a sum over all possible arrow configurations of the product of the vertex weights,

$$
Z_n = \sum_{\substack{\text{arrow} \\ \text{configurations } \sigma}} w(\sigma), \quad w(\sigma) = \prod_{x \in V_n} w_{\sigma(x)} = \prod_{i=1}^{6} w_i^{N_i(\sigma)}, \qquad (1\text{-}1)
$$

where V_n is the $n \times n$ set of vertices, $\sigma(x) \in \{1, \ldots, 6\}$ is the vertex configuration of σ at vertex x according to Figure 1, and $N_i(\sigma)$ is the number of vertices of type i in the configuration σ. The sum is taken over all possible configurations obeying the given boundary condition. The *Gibbs measure* is defined then as

$$
\mu_n(\sigma) = \frac{w(\sigma)}{Z_n}. \qquad (1\text{-}2)
$$

Our main goal is to obtain the *large n asymptotics* of the partition function Z_n.

In general, the six-vertex model has *six parameters*: the weights w_i. However, by using some conservation laws we can reduce these to only *two parameters*. Any fixed boundary conditions impose some conservation laws on the six-vertex model. In the case of DWBC, they are

$$N_1(\sigma) = N_2(\sigma), \quad N_3(\sigma) = N_4(\sigma), \quad N_5(\sigma) = N_6(\sigma) + n. \qquad (1\text{-}3)$$

This allows us to reduce to the case

$$w_1 = w_2 \equiv a, \quad w_3 = w_4 \equiv b, \quad w_5 = w_6 \equiv c. \qquad (1\text{-}4)$$

Then by using the identity,

$$Z_n(a, a, b, b, c, c) = c^{n^2} Z_n\left(\frac{a}{c}, \frac{a}{c}, \frac{b}{c}, \frac{b}{c}, 1, 1\right), \qquad (1\text{-}5)$$

we can reduce to the two parameters, a/c and b/c. For details on how we make this reduction, see, e.g., [Allison and Reshetikhin 2005; Ferrari and Spohn 2006; Bleher and Liechty 2009a].

2. Phase diagram of the six-vertex model

Introduce the parameter

$$\Delta = \frac{a^2 + b^2 - c^2}{2ab}. \qquad (2\text{-}1)$$

The *phase diagram* of the six-vertex model consists of the following three regions: the *ferroelectric phase region*, $\Delta > 1$; the *antiferroelectric phase region*, $\Delta < -1$; and, the *disordered phase region*, $-1 < \Delta < 1$ (see, e.g., [Lieb and Wu 1972]). In these three regions we parametrize the weights in the standard way: in the ferroelectric phase region,

$$a = \sinh(t - \gamma), \quad b = \sinh(t + \gamma), \quad c = \sinh(2|\gamma|), \quad 0 < |\gamma| < t; \quad (2\text{-}2)$$

in the antiferroelectric phase region,

$$a = \sinh(\gamma - t), \quad b = \sinh(\gamma + t), \quad c = \sinh(2\gamma), \quad |t| < \gamma; \qquad (2\text{-}3)$$

and in the disordered phase region,

$$a = \sin(\gamma - t), \quad b = \sin(\gamma + t), \quad c = \sin(2\gamma), \quad |t| < \gamma. \qquad (2\text{-}4)$$

The phase diagram of the model is shown on Figure 3.

The phase diagram and the Bethe-ansatz solution of the *six-vertex model for periodic and antiperiodic boundary conditions* are thoroughly discussed in [Lieb 1967a; 1967b; 1967c; 1967d; Lieb and Wu 1972; Sutherland 1967; Baxter 1989;

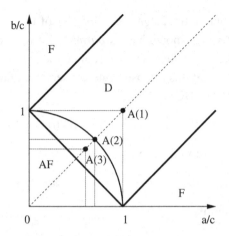

Figure 3. Phase diagram of the model. F, AF and D mark the ferro-electric, antiferroelectric, and disordered phases. The circular arc corresponds to the so-called "free fermion" line, where $\Delta = 0$, and the three dots correspond to 1-, 2-, and 3-enumeration of alternating sign matrices.

Batchelor et al. 1995]. See also [Wu and Lin 1975], in which the Pfaffian solution for the six-vertex model with periodic boundary conditions is obtained on the free fermion line, $\Delta = 0$.

3. Izergin–Korepin determinantal formula

The *six-vertex model with DWBC* was introduced by Korepin [1982], who derived an important recursion relation for the partition function of the model. This led to a beautiful *determinantal formula* of Izergin [1987] for the partition function of the six-vertex model with DWBC, known as the *Izergin–Korepin formula*. A detailed proof of this formula and its generalizations are given in the paper of Izergin, Coker and Korepin [Izergin et al. 1992]. When the weights are parametrized according to (2-4), the Izergin–Korepin formula is

$$Z_n = \frac{(ab)^{n^2}}{\left(\prod_{k=0}^{n-1} k!\right)^2} \tau_n, \tag{3-1}$$

where τ_n is the Hankel determinant

$$\tau_n = \det\left(\frac{d^{i+k-2}\phi}{dt^{i+k-2}}\right)_{1\le i,k\le n}, \tag{3-2}$$

and

$$\phi(t) = \frac{c}{ab}. \tag{3-3}$$

Observe that a, b, c have different parametrizations (2-2)–(2-4) in different phase regions. An elegant derivation of the Izergin–Korepin determinantal formula from the *Yang–Baxter equations* is given in [Korepin and Zinn-Justin 2000; Kuperberg 1996].

One of the applications of the determinantal formula is that it implies that the partition function τ_n solves the *Toda equation*,

$$\tau_N \tau_n'' - {\tau_n'}^2 = \tau_{n+1}\tau_{n-1}, \quad n \geq 1, \quad (') = \frac{\partial}{\partial t}; \tag{3-4}$$

cf. [Sogo 1993]. This was used by Korepin and Zinn-Justin [2000] to derive the free energy of the six-vertex model with DWBC, assuming some ansatz on the behavior of subdominant terms in the large n asymptotics of the free energy.

4. The six-vertex model with DWBC and a random matrix model

Another application of the Izergin–Korepin determinantal formula is that τ_n can be expressed in terms of a partition function of a *random matrix model*. The relation to the random matrix model was obtained and used in [Zinn-Justin 2000]. It can be derived as follows. Consider first the disordered phase region.

Disordered phase region. For the evaluation of the Hankel determinant (3-2), it is convenient to use an integral representation of the function

$$\phi(t) = \frac{\sin 2\gamma}{\sin(\gamma - t)\sin(\gamma + t)}; \tag{4-1}$$

namely, to write it in the form of the Laplace transform,

$$\phi(t) = \int_{-\infty}^{\infty} e^{t\lambda} m(\lambda)\, d\lambda, \tag{4-2}$$

where

$$m(\lambda) = \frac{\sinh \frac{\lambda}{2}(\pi - 2\gamma)}{\sinh \frac{\lambda}{2}\pi}. \tag{4-3}$$

Then

$$\frac{d^i \phi}{dt^i} = \int_{-\infty}^{\infty} \lambda^i e^{t\lambda} m(\lambda)\, d\lambda, \tag{4-4}$$

and by substituting this into the Hankel determinant, (3-2), we obtain

$$\tau_n = \int \prod_{i=1}^{n} [e^{t\lambda_i} m(\lambda_i)\, d\lambda_i] \det(\lambda_i^{i+k-2})_{1 \leq i,k \leq n}$$

$$= \int \prod_{i=1}^{n} [e^{t\lambda_i} m(\lambda_i)\, d\lambda_i] \det(\lambda_i^{k-1})_{1 \leq i,k \leq n} \prod_{i=1}^{n} \lambda_i^{i-1}. \tag{4-5}$$

Consider any permutation $\sigma \in S_n$ of variables λ_i. From the last equation we have that

$$\tau_n = \int \prod_{i=1}^{n} [e^{t\lambda_i} m(\lambda_i) \, d\lambda_i](-1)^{\sigma} \det(\lambda_i^{k-1})_{1 \leq i,k \leq n} \prod_{i=1}^{n} \lambda_{\sigma(i)}^{i-1}. \tag{4-6}$$

By summing over $\sigma \in S_n$, we obtain that

$$\tau_n = \frac{1}{n!} \int \prod_{i=1}^{n} [e^{t\lambda_i} m(\lambda_i) d\lambda_i] \Delta(\lambda)^2, \tag{4-7}$$

where $\Delta(\lambda)$ is the Vandermonde determinant,

$$\Delta(\lambda) = \det(\lambda_i^{k-1})_{1 \leq i,k \leq n} = \prod_{i < k} (\lambda_k - \lambda_i). \tag{4-8}$$

Equation (4-7) expresses τ_n in terms of a matrix model integral. Namely, if $m(x) = e^{-V(x)}$, then

$$\tau_n = \frac{\prod_{n=0}^{n-1} n!}{\pi^{n(n-1)/2}} \int dM e^{\operatorname{Tr}[tM - V(M)]}, \tag{4-9}$$

where the integration is over the space of $n \times n$ Hermitian matrices. The matrix model integral can be solved, furthermore, in terms of *orthogonal polynomials*.

Introduce monic polynomials $P_k(x) = x^k + \cdots$ orthogonal on the line with respect to the weight

$$w(x) = e^{tx} m(x), \tag{4-10}$$

so that

$$\int_{-\infty}^{\infty} P_j(x) P_k(x) e^{tx} m(x) \, dx = h_k \delta_{nm}. \tag{4-11}$$

Then it follows from (4-7) that

$$\tau_n = \prod_{k=0}^{n-1} h_k. \tag{4-12}$$

The orthogonal polynomials satisfy the three term recurrence relation,

$$x P_k(x) = P_{k+1}(x) + Q_k P_k(x) + R_k P_{k-1}(x), \tag{4-13}$$

where R_k can be found as

$$R_k = \frac{h_k}{h_{k-1}}; \tag{4-14}$$

see, e.g., [Szegő 1975]. This gives

$$h_k = h_0 \prod_{j=1}^{k} R_j, \tag{4-15}$$

where

$$h_0 = \int_{-\infty}^{\infty} e^{tx} m(x) \, dx = \frac{\sin(2\gamma)}{\sin(\gamma + t) \sin(\gamma - t)}. \tag{4-16}$$

By substituting (4-15) into (4-12), we obtain that

$$\tau_n = h_0^n \prod_{k=1}^{n-1} R_k^{n-k}. \tag{4-17}$$

Colomo and Pronko identified the orthogonal polynomials $\{P_k\}$ at some points on the phase diagram with classical orthogonal polynomials, see [Colomo and Pronko 2003; 2004; 2005; 2006]. They showed that on the free fermion line, $\{P_k\}$ are the Meixner–Pollaczek polynomials, at the ice point $(1, 1)$ they are the continuous Hahn polynomials, and at the point $(\frac{1}{\sqrt{3}}, \frac{1}{\sqrt{3}})$, $\{P_k\}$ can be expressed in terms of the continuous dual Hahn polynomials. The ice point $(1, 1)$ and the point $(\frac{1}{\sqrt{3}}, \frac{1}{\sqrt{3}})$ correspond to the 1- and 3-enumerations, respectively, of alternating sign matrices The 2-enumeration of ASMs corresponds to the point on the free fermion line at which $a = b$. The 1-, 2-, and 3-enumerations of ASMs are marked A(1), A(2), and A(3), respectively, in Figure 3, and the full free fermion line is marked there as well. In all these cases the normalizing constants h_k are known explicitly, and formula (4-12) can be used to find the asymptotic behavior of τ_n as $n \to \infty$. At all other points on the phase diagrams, no reduction to classical orthogonal polynomials is known.

Ferroelectric phase. In the ferroelectric phase, the parameters a, b, and c are parametrized by (2-2). We consider the case $\gamma > 0$, which corresponds to the region $b > a + c$ in the phase diagram. The case $\gamma < 0$ is similar, and a and b should be exchanged in that case. The function ϕ is the Laplace transform of a discrete measure supported on the positive integers:

$$\phi(t) = \frac{\sinh(2\gamma)}{\sinh(t + \gamma) \sinh(t - \gamma)} = 4 \sum_{l=1}^{\infty} e^{-2tl} \sinh(2\gamma l). \tag{4-18}$$

Then, similar to (4-7), we find that

$$\tau_n = \frac{2^{n^2}}{n!} \sum_{l_1, \ldots, l_n = 1}^{\infty} \Delta(l_i)^2 \prod_{i=1}^{n} [2e^{-2tl_i} \sinh(2\gamma l_i)]. \tag{4-19}$$

This is the partition function for a discrete version of a Hermitian random matrix model, often called a *discrete orthogonal polynomial ensemble* (DOPE), and can also be solved in terms of orthogonal polynomials. The appropriate polynomials in this case are the monic polynomials $P_n(l) = l^n + \cdots$ with the orthogonality

$$\sum_{l=1}^{\infty} P_j(l) P_k(l) w(l) = h_k \delta_{jk},$$

$$w(l) = 2e^{-2tl} \sinh(2\gamma l) = e^{-2tl+2\gamma l} - e^{-2tl-2\gamma l}. \tag{4-20}$$

Then it follows from (4-19) that

$$\tau_n = 2^{n^2} \prod_{k=0}^{n-1} h_k. \tag{4-21}$$

Critical line between disordered and ferroelectric phase. When the parameters a, b, and c are such that $b - a = c$ (so $\Delta = 1$ in (2-1)), the Izergin–Korepin formula is not directly applicable. However, we may consider a limiting case of the orthogonal polynomial formula (4-21). On the critical line

$$\frac{b}{c} - \frac{a}{c} = 1, \tag{4-22}$$

we fix a point,

$$\frac{a}{c} = \frac{\alpha - 1}{2}, \quad \frac{b}{c} = \frac{\alpha + 1}{2}; \quad \alpha > 1, \tag{4-23}$$

and consider the partition function

$$Z_n = Z_n\left(\frac{\alpha-1}{2}, \frac{\alpha-1}{2}, \frac{\alpha+1}{2}, \frac{\alpha+1}{2}, 1, 1\right). \tag{4-24}$$

Consider the limit of (4-21) as

$$t, \gamma \to +0, \quad \frac{t}{\gamma} \to \alpha. \tag{4-25}$$

Observe that in this limit,

$$\frac{a}{c} = \frac{\sinh(t-\gamma)}{\sinh(2\gamma)} \to \frac{\alpha-1}{2}, \quad \frac{b}{c} = \frac{\sinh(t+\gamma)}{\sinh(2\gamma)} \to \frac{\alpha+1}{2}. \tag{4-26}$$

By (1-5), (3-1), and (4-12), we have

$$Z_n\left(\frac{a}{c}, \frac{a}{c}, \frac{b}{c}, \frac{b}{c}, 1, 1\right) = \left(\frac{2\sinh(t-\gamma)\sinh(t+\gamma)}{\sinh(2\gamma)}\right)^{n^2} \prod_{k=0}^{n-1} \frac{h_k}{(k!)^2}. \tag{4-27}$$

To deal with limit (4-25) we need to rescale the orthogonal polynomials $P_k(l)$. Introduce the rescaled variable

$$x = 2tl - 2\gamma l, \tag{4-28}$$

and the rescaled limiting weight,

$$w_\alpha(x) = \lim_{\substack{t,\gamma \to +0 \\ t/\gamma \to \alpha}} (e^{-2tl+2\gamma l} - e^{-2tl-2\gamma l}) = e^{-x} - e^{-rx}, \qquad (4\text{-}29)$$

$$r = \frac{\alpha+1}{\alpha-1} > 1.$$

Consider monic orthogonal polynomials $P_j(x;\alpha)$ satisfying the orthogonality condition,

$$\int_0^\infty P_j(x;\alpha) P_k(x;\alpha) w_\alpha(x)\, dx = h_{k,\alpha} \delta_{jk}. \qquad (4\text{-}30)$$

To find a relation between $P_k(l)$ and $P_k(x;\alpha)$, introduce the monic polynomials

$$\tilde{P}_k(x) = \delta^k P_k(x/\delta), \qquad (4\text{-}31)$$

where

$$\delta = 2t - 2\gamma, \qquad (4\text{-}32)$$

and rewrite orthogonality condition (4-11) in the form

$$\sum_{l=1}^\infty \tilde{P}_j(l\delta) \tilde{P}_k(l\delta) w_\alpha(l\delta)\delta = \delta^{2k+1} h_k \delta_{jk}, \qquad (4\text{-}33)$$

which is a Riemann sum for the integral in orthogonality condition (4-30). Therefore,

$$\lim_{\substack{t,\gamma \to +0 \\ t/\gamma \to \alpha}} \tilde{P}_k(x) = P_k(x;\alpha), \qquad \lim_{\substack{t,\gamma \to +0 \\ t/\gamma \to \alpha}} \delta^{2k+1} h_k = h_{k,\alpha}. \qquad (4\text{-}34)$$

Thus, if we rewrite formula (4-27) as

$$Z_n\left(\frac{a}{c},\frac{a}{c},\frac{b}{c},\frac{b}{c},1,1\right) = \left(\frac{2\sinh(t-\gamma)\sinh(t+\gamma)}{\sinh(2\gamma)\delta}\right)^{n^2} \prod_{k=0}^{n-1} \frac{\delta^{2k+1} h_k}{(k!)^2}, \qquad (4\text{-}35)$$

we can take limit (4-25). In the limit we obtain that

$$Z_n = Z_n\left(\frac{\alpha-1}{2},\frac{\alpha-1}{2},\frac{\alpha+1}{2},\frac{\alpha+1}{2},1,1\right) = \left(\frac{\alpha+1}{2}\right)^{n^2} \prod_{k=0}^{n-1} \frac{h_{k,\alpha}}{(k!)^2}. \qquad (4\text{-}36)$$

Antiferroelectric phase. In the antiferroelectric phase, the parameters a, b, and c are parametrized by (2-3), and the function

$$\phi(t) = \frac{\sinh(2\gamma)}{\sinh(\gamma-t)\sinh(\gamma+t)}, \qquad |t| < \gamma, \qquad (4\text{-}37)$$

is the Laplace transform of a discrete measure supported on the integers:

$$\phi(t) = \frac{\sinh(2\gamma)}{\sinh(\gamma - t)\sinh(\gamma + t)} = 2\sum_{l=-\infty}^{\infty} e^{2tl-2\gamma|l|}. \tag{4-38}$$

Then

$$\tau_n = \frac{2^{n^2}}{n!}\sum_{l_1,\ldots,l_n=-\infty}^{\infty}\Delta(l)^2\prod_{i=1}^{n}e^{2tl_i-2\gamma|l_i|}. \tag{4-39}$$

This is again the partition function of a DOPE, and we introduce the discrete monic polynomials $P_n(l) = l^n + \cdots$ via the orthogonality condition

$$\sum_{l=-\infty}^{\infty} P_j(l)P_k(l)w(l) = h_k\delta_{jk}, \quad w(l) = e^{2tl-2\gamma|l|}. \tag{4-40}$$

Then it follows from (4-39) that

$$\tau_n = 2^{n^2}\prod_{k=0}^{n-1} h_k. \tag{4-41}$$

Critical line between the antiferroelectric and disordered phases. When the parameters a, b, and c are such that $a + b = c$, (so $\Delta = -1$ in (2-1)), the Izergin–Korepin formula is not directly applicable, and we must consider a limiting case of the orthogonal polynomial formula (4-41). On the critical line

$$\frac{a}{c} + \frac{b}{c} = 1, \tag{4-42}$$

we fix a point,

$$\frac{a}{c} = \frac{1-\alpha}{2}, \quad \frac{b}{c} = \frac{1+\alpha}{2}, \quad -1 < \alpha < 1, \tag{4-43}$$

and consider the partition function

$$Z_n = Z_n\left(\frac{1-\alpha}{2}, \frac{1-\alpha}{2}, \frac{1+\alpha}{2}, \frac{1+\alpha}{2}, 1, 1\right). \tag{4-44}$$

This corresponds to taking a limit of the Izergin–Korepin formula in the antiferroelectric phase as $t, \gamma \to 0$, and $t/\gamma = \alpha$. Introduce the rescaled variable

$$x = -2tl + 2\gamma l, \tag{4-45}$$

and the rescaled limiting weight,

$$w_\alpha(x) = \lim_{t,\gamma\to+0,\, \frac{t}{\gamma}\to\alpha} e^{2tl-2\gamma|l|} = \begin{cases} e^{-x}, & x \geq 0, \\ e^{rx}, & x < 0, \end{cases} \tag{4-46}$$

where

$$r = \frac{1+\alpha}{1-\alpha} > 0. \tag{4-47}$$

Consider monic orthogonal polynomials $P_j(x; \alpha)$ satisfying the orthogonality condition,

$$\int_{\mathbb{R}} P_j(x; \alpha) P_k(x; \alpha) w_\alpha(x)\, dx = h_{k,\alpha} \delta_{jk}, \tag{4-48}$$

which can be obtained from the polynomials (4-40) by taking the appropriate scaling limit as $t, \gamma \to 0$, and $t/\gamma = \alpha$. Similar to (4-36), we obtain

$$Z_n = Z_n\left(\frac{\alpha-1}{2}, \frac{\alpha-1}{2}, \frac{\alpha+1}{2}, \frac{\alpha+1}{2}, 1, 1\right) = \left(\frac{1+\alpha}{2}\right)^{n^2} \prod_{k=0}^{n-1} \frac{h_{k,\alpha}}{(k!)^2}. \tag{4-49}$$

5. Large n asymptotics of Z_n

The asymptotic evaluation of Z_n in the different regions of the phase diagram thus reduces to asymptotic evaluation of different systems of orthogonal polynomials. In general, this may be done by formulating the orthogonal polynomials as the solution to a 2×2 matrix valued Riemann–Hilbert problem as in [Fokas et al. 1992]. One may then perform the steepest descent analysis of [Deift and Zhou 1993]. In the case that the weight of orthogonality is a continuous one on $\mathbb{R}$, this analysis was performed for weights of the form $\exp(-nV(x))$ for a very general class of analytic potential functions $V(x)$ in [Deift et al. 1999]. The analysis was adapted to the case that the orthogonality is with respect to a discrete measure in [Baik et al. 2007; Bleher and Liechty 2011]. The steepest descent analysis yields the following results in the different regions of the phase diagram.

Disordered phase.

Theorem 5.1 [Bleher and Fokin 2006]. *Let the weights a, b, and c, in the six-vertex model with DWBC be parametrized as in (2-4). Then, as $n \to \infty$, the partition function Z_n has the asymptotic expansion*

$$Z_n = C n^\kappa F^{n^2} (1 + O(n^{-\varepsilon})), \quad \varepsilon > 0, \tag{5-1}$$

where

$$F = \frac{\pi ab}{2\gamma \cos \frac{\pi t}{2\gamma}}, \quad \kappa = \frac{1}{12} - \frac{2\gamma^2}{3\pi(\pi - 2\gamma)}, \tag{5-2}$$

and $C > 0$ is a constant.

This proves the conjecture of Zinn-Justin, and it gives the exact value of the exponent κ. Let us remark that the presence of the power-like factor n^κ in the asymptotic expansion of Z_n in (5-1) is rather unusual from the point of view of random matrix models. Also, in the one-cut case the usual large n asymptotics of $\log Z_n$ in a noncritical random matrix model is the so called "topological expansion", which gives $(\log Z_n)/n^2$ as an asymptotic series in powers of $1/n^2$ (see, e.g., [Ercolani and McLaughlin 2003; Bleher and Its 2005]). In this case the asymptotic expansion of $\log Z_n$ includes the term $\kappa \log n$.

It is noteworthy that, as shown in [Bogoliubov et al. 2002], asymptotic formula (5-1) remains valid on the borderline between the disordered and antiferroelectric phases. In this case $\kappa = \frac{1}{12}$, which corresponds to $\gamma = 0$. In [Bleher and Bothner 2012] the constant C in the asymptotic expansion (5-1) is calculated on the borderline between the disordered and antiferroelectric phases, up to a universal constant factor, which is still unknown. Also, it is shown in [Bleher and Bothner 2012] that the error term $O(n^{-\varepsilon})$ in (5-1) can be replaced by $O(n^{-1})$. The calculations of [Bleher and Bothner 2012] can be extended to the whole disordered region, where they give an explicit dependence of the constant C on the parameter t in parametrization (2-4), and improve the error term in (5-1) to $O(n^{-1})$.

Ferroelectric phase. We have obtained the large n asymptotics of Z_n in the ferroelectric phase, $\Delta > 1$ [Bleher and Liechty 2009a], and also on the critical line between the ferroelectric and disordered phases, $\Delta = 1$ [2009b]. In the ferroelectric phase we use parametrization (2-2) for a, b and c. The large n asymptotics of Z_n in the ferroelectric phase is given by the following theorem.

Theorem 5.2 [Bleher and Liechty 2009a]. *Let the weights a, b, and c in the six-vertex model with DWBC be parametrized as in (2-2) with $t > \gamma > 0$. For any $\varepsilon > 0$, as $n \to \infty$,*

$$Z_n = CG^n F^{n^2}\left(1 + O(e^{-n^{1-\varepsilon}})\right), \qquad (5\text{-}3)$$

where $C = 1 - e^{-4\gamma}$, $G = e^{\gamma - t}$ and $F = b$.

On the critical line between the ferroelectric and disordered phases we use the parametrization $b = a + 1$, $c = 1$. The main result here is the following asymptotic formula for Z_n.

Theorem 5.3 [Bleher and Liechty 2009b]. *As $n \to \infty$,*

$$Z_n = Cn^\kappa G^{\sqrt{n}} F^{n^2}[1 + O(n^{-1/2})], \qquad (5\text{-}4)$$

where $C > 0$,

$$\kappa = \tfrac{1}{4}, \quad G = \exp\left[-\zeta\left(\tfrac{3}{2}\right)\sqrt{\tfrac{a}{\pi}}\right], \qquad (5\text{-}5)$$

and

$$F = b. \qquad (5\text{-}6)$$

Notice that in both Theorems 5.2 and 5.3, the limiting free energy F is the weight b. The ground state in this phase is unique and is achieved when there is exactly one c-type vertex in each row and column, and the rest of the vertices are of type b. That is, the diagonal consists of type 5 vertices while above the diagonal all vertices are type 3 and below all vertices are type 4. The weight of the ground state is $b^{n^2}(c/b)^n$, and thus the free energy in the ferroelectric phase is completely determined by the ground state. This is a reflection of the fact that local fluctuations from the ground state can take place only in a thin neighborhood of the diagonal. The conservation laws (1-3) forbid local fluctuations away from the diagonal.

Antiferroelectric phase. The large n asymptotics in the antiferroelectric phase were obtained nonrigorously in [Zinn-Justin 2000], and rigorously, using the Riemann–Hilbert method, in [Bleher and Liechty 2010]. They are given in the following theorem. In this theorem ϑ_1 and ϑ_4 are the Jacobi theta functions with elliptic nome $q = e^{-\pi^2/2\gamma}$ (see, e.g., [Whittaker and Watson 1996]), and the phase ω is given as

$$\omega = \frac{\pi}{2}\left(1 + \frac{t}{\gamma}\right). \tag{5-7}$$

Theorem 5.4 [Bleher and Liechty 2010]. *Let the weights a, b, and c in the six-vertex model with DWBC be parametrized as in (2-2). As $n \to \infty$,*

$$Z_n = C\vartheta_4(n\omega)F^{n^2}(1 + O(n^{-1})), \tag{5-8}$$

where $C > 0$ is a constant, and

$$F = \frac{\pi a b \vartheta_1'(0)}{2\gamma \vartheta_1(\omega)}. \tag{5-9}$$

In contrast to the disordered phase, note the lack of a power like term. In contrast to the ferroelectric phase, notice that the free energy depends transcendentally on the weight of the ground state configuration. Only in the limit as $\gamma \to \infty$, which can be regarded as the low temperature limit, does the weight of the ground state become dominant. For a discussion of this limit, see [Zinn-Justin 2000].

6. The Riemann–Hilbert approach

All the above asymptotic results are obtained in the Riemann–Hilbert approach, but the concrete asymptotic analysis of the Riemann–Hilbert problem is quite different in the different phase regions. Let us discuss it.

Disordered phase region. To apply the Riemann–Hilbert approach, we introduce a rescaled weight as

$$w_n(x) = w\left(\frac{nx}{\gamma}\right). \tag{6-1}$$

It can be written as

$$w_n(x) = e^{-nV_n(x)}, \tag{6-2}$$

where

$$V_n(x) = -\zeta x - \frac{1}{n}\ln\frac{\sinh\left(n\left(\frac{\pi}{2\gamma} - 1\right)x\right)}{\sinh\frac{n\pi x}{2\gamma}}, \qquad \zeta = \frac{t}{\gamma}. \tag{6-3}$$

The external potential $V_n(x)$ is real analytic for any finite n, but it has logarithmic singularities on the imaginary axis, which accumulate to the origin as $n \to \infty$. In fact, the limiting external potential,

$$\lim_{n\to\infty} V_n(x) = V(x) = -\zeta x + |x|, \tag{6-4}$$

is not analytic at $x = 0$. The Riemann–Hilbert approach developed in [Bleher and Fokin 2006] is based on an opening of lenses whose boundary approaches the origin as $n \to \infty$. This turns out to be possible due to the fact that the density of the equilibrium measure $\rho_n(x)$ for the external potential $V_n(x)$ diverges logarithmically at the origin as $n \to \infty$, and as a result, the jump matrix on the boundary of the lenses converges to the unit matrix (for details, see [ibid.]). The calculation of subdominant asymptotic terms in the partition function as $n \to \infty$ is the central difficult part of the work [ibid.], and it is done by an asymptotic analysis of the solution to the Riemann–Hilbert problem near the turning points and near the origin.

Ferroelectric phase region. In the ferroelectric region, the measure of orthogonality is a discrete one on $\mathbb{N}$. To apply the Riemann–Hilbert approach to discrete orthogonal polynomials, we need to rescale both the weight and the lattice that supports the measure so that the mesh of the lattice goes to zero as $n \to \infty$. Introduce the rescaled lattice and weight

$$L_n = \left(\frac{2t}{n}\right)\mathbb{N}, \qquad w_n(x) = e^{-nx(1-\zeta)}(1 - e^{-4nx}) = e^{-nV_n(x)}, \tag{6-5}$$

where

$$V_n(x) = x(1-\zeta) - \frac{1}{n}\log(1 - e^{-2nx\zeta}), \qquad 0 < \zeta = \frac{\gamma}{t} < 1. \tag{6-6}$$

Then the orthogonality condition (4-20) can be written as

$$\sum_{x\in L_n} P_j\left(\frac{nx}{2t}\right) P_k\left(\frac{nx}{2t}\right) w_n(x) = h_k\delta_{jk}. \tag{6-7}$$

Notice that, as $n \to \infty$, $V_n(x)$ has the limit

$$\lim_{n\to\infty} V_n(x) = x(1-\zeta), \qquad (6\text{-}8)$$

which would indicate that, in the large n limit, the polynomials (4-20) behave as polynomials orthogonal on $\mathbb{N}$ with a simple exponential weight. These polynomials are a special case of the classical *Meixner polynomials*, and there are exact formulae for their recurrence coefficients (see, e.g., [Koekoek et al. 2010]). The monic Meixner polynomials which concern us are defined from the orthogonality condition

$$\sum_{l=1}^{\infty} Q_j(l) Q_k(l) q^l = h_k^Q \delta_{jk}, \qquad q = e^{2\gamma-2t}, \qquad (6\text{-}9)$$

and the normalizing constants are given exactly as

$$h_k^Q = \frac{(k!)^2 q^{k+1}}{(1-q)^{2k+1}}. \qquad (6\text{-}10)$$

Up to the constant factor, Theorem 5.2 can therefore be proven by showing that h_k and h_k^Q are asymptotically close as $k \to \infty$. More precisely, it is shown in [Bleher and Liechty 2009a] that as $k \to \infty$, for any $\varepsilon > 0$,

$$h_k = h_k^Q(1 + O(e^{-k^{1-\varepsilon}})). \qquad (6\text{-}11)$$

Antiferroelectric region. In the antiferroelectric region, the orthogonal polynomials are with respect to a discrete weight, and we rescale the weight in (4-40) and the integer lattice as

$$L_n = \left(\frac{2\gamma}{n}\right)\mathbb{Z}, \quad w_n(x) = e^{-nV(x)}, \quad V(x) = |x| - \zeta x, \quad \zeta = \frac{t}{\gamma} < 1, \quad (6\text{-}12)$$

so that the orthogonality condition (4-40) can be written as

$$\sum_{x \in L_n} P_j\left(\frac{nx}{2\gamma}\right) P_k\left(\frac{nx}{2\gamma}\right) w_n(x) = h_k \delta_{jk}. \qquad (6\text{-}13)$$

The mesh of the lattice L_n is $2\gamma/n$, which places an upper constraint on the equilibrium measure, which is the limiting distribution of zeroes of the orthogonal polynomials. This upper constraint is realized. The equilibrium measure, which has density $\rho(x)$, is supported on a single interval $[\alpha, \beta]$, but within that interval is an interval $[\alpha', \beta']$ on which $\rho(x) \equiv 1/2\gamma$. This interval is called the *saturated region*, and it separates the single band of support $[\alpha, \beta]$ into the two bands of analyticity $[\alpha, \alpha']$ and $[\beta', \beta]$. Thus in effect we have a "two-cut" situation, which is the source of the quasiperiodic factor $\vartheta_4(n\omega)$ in Theorem 5.4.

In principle, a problem could come from the fact that the potential $V(x)$ is not analytic at the origin. However, it turns out that this point of nonanalyticity is always in the saturated region and therefore does not present a problem in the steepest descent analysis.

As previously noted, there is no power-like term in the asymptotic formula for Z_n in the antiferroelectric phase. The Riemann–Hilbert approach to orthogonal polynomials generally gives an expansion of the normalizing constants h_n in inverse powers of n. In the two-cut case, the coefficients in this expansion may be quasiperiodic functions of n. For the orthogonal polynomials (4-40) it is a tedious calculation involving the Jacobi theta functions to show that the term of order n^{-1} vanishes in the expansion of h_n, which then implies the absence of the power-like term in Z_n.

References

[Allison and Reshetikhin 2005] D. Allison and N. Reshetikhin, "Numerical study of the 6-vertex model with domain wall boundary conditions", *Ann. Inst. Fourier (Grenoble)* **55**:6 (2005), 1847–1869.

[Baik et al. 2007] J. Baik, T. Kriecherbauer, K. T.-R. McLaughlin, and P. D. Miller, *Discrete orthogonal polynomials: asymptotics and applications*, Annals of Mathematics Studies **164**, Princeton University Press, 2007.

[Batchelor et al. 1995] M. T. Batchelor, R. J. Baxter, M. J. O'Rourke, and C. M. Yung, "Exact solution and interfacial tension of the six-vertex model with anti-periodic boundary conditions", *J. Phys. A* **28**:10 (1995), 2759–2770.

[Baxter 1989] R. J. Baxter, *Exactly solved models in statistical mechanics*, reprint of the 1982 ed., Academic Press, London, 1989.

[Bleher and Bothner 2012] P. Bleher and T. Bothner, "Exact solution of the six-vertex model with domain wall boundary conditions: critical line between disordered and antiferroelectric phases", *Random Matrices Theory Appl.* **1**:4 (2012), 1250012, 43.

[Bleher and Fokin 2006] P. M. Bleher and V. V. Fokin, "Exact solution of the six-vertex model with domain wall boundary conditions: disordered phase", *Comm. Math. Phys.* **268**:1 (2006), 223–284.

[Bleher and Its 2005] P. M. Bleher and A. R. Its, "Asymptotics of the partition function of a random matrix model", *Ann. Inst. Fourier (Grenoble)* **55**:6 (2005), 1943–2000.

[Bleher and Liechty 2009a] P. Bleher and K. Liechty, "Exact solution of the six-vertex model with domain wall boundary conditions: ferroelectric phase", *Comm. Math. Phys.* **286**:2 (2009), 777–801.

[Bleher and Liechty 2009b] P. Bleher and K. Liechty, "Exact solution of the six-vertex model with domain wall boundary conditions: critical line between ferroelectric and disordered phases", *J. Stat. Phys.* **134**:3 (2009), 463–485.

[Bleher and Liechty 2010] P. Bleher and K. Liechty, "Exact solution of the six-vertex model with domain wall boundary conditions: antiferroelectric phase", *Comm. Pure Appl. Math.* **63**:6 (2010), 779–829.

[Bleher and Liechty 2011] P. Bleher and K. Liechty, "Uniform asymptotics for discrete orthogonal polynomials with respect to varying exponential weights on a regular infinite lattice", *Int. Math. Res. Not.* **2011** (2011), 342–386.

[Bogoliubov et al. 2002] N. M. Bogoliubov, A. V. Kitaev, and M. B. Zvonarev, "Boundary polarization in the six-vertex model", *Phys. Rev. E* (3) **65**:2 (2002), 026126, 4.

[Colomo and Pronko 2003] F. Colomo and A. G. Pronko, "On some representations of the six vertex model partition function", *Phys. Lett. A* **315**:3-4 (2003), 231–236.

[Colomo and Pronko 2004] F. Colomo and A. G. Pronko, "On the partition function of the six-vertex model with domain wall boundary conditions", *J. Phys. A* **37**:6 (2004), 1987–2002.

[Colomo and Pronko 2005] F. Colomo and A. G. Pronko, "Square ice, alternating sign matrices, and classical orthogonal polynomials", *J. Stat. Mech. Theory Exp.* 1 (2005), 005, 33 pp.

[Colomo and Pronko 2006] F. Colomo and A. G. Pronko, "The role of orthogonal polynomials in the six-vertex model and its combinatorial applications", *J. Phys. A* **39**:28 (2006), 9015–9033.

[Deift and Zhou 1993] P. Deift and X. Zhou, "A steepest descent method for oscillatory Riemann–Hilbert problems: asymptotics for the MKdV equation", *Ann. of Math.* (2) **137**:2 (1993), 295–368.

[Deift et al. 1999] P. Deift, T. Kriecherbauer, K. T.-R. McLaughlin, S. Venakides, and X. Zhou, "Uniform asymptotics for polynomials orthogonal with respect to varying exponential weights and applications to universality questions in random matrix theory", *Comm. Pure Appl. Math.* **52**:11 (1999), 1335–1425.

[Ercolani and McLaughlin 2003] N. M. Ercolani and K. D. T.-R. McLaughlin, "Asymptotics of the partition function for random matrices via Riemann–Hilbert techniques and applications to graphical enumeration", *Int. Math. Res. Not.* **2003** (2003), 755–820.

[Ferrari and Spohn 2006] P. L. Ferrari and H. Spohn, "Domino tilings and the six-vertex model at its free-fermion point", *J. Phys. A* **39**:33 (2006), 10297–10306.

[Fokas et al. 1992] A. S. Fokas, A. R. It·s, and A. V. Kitaev, "The isomonodromy approach to matrix models in 2D quantum gravity", *Comm. Math. Phys.* **147**:2 (1992), 395–430.

[Izergin 1987] A. G. Izergin, "Partition function of a six-vertex model in a finite volume", *Dokl. Akad. Nauk SSSR* **297**:2 (1987), 331–333. In Russian; translated in *Soviet Phys. Dokl.* **32**:11 (1987), 878–879.

[Izergin et al. 1992] A. G. Izergin, D. A. Coker, and V. E. Korepin, "Determinant formula for the six-vertex model", *J. Phys. A* **25**:16 (1992), 4315–4334.

[Koekoek et al. 2010] R. Koekoek, P. A. Lesky, and R. F. Swarttouw, *Hypergeometric orthogonal polynomials and their q-analogues*, Springer, Berlin, 2010.

[Korepin 1982] V. E. Korepin, "Calculation of norms of Bethe wave functions", *Comm. Math. Phys.* **86**:3 (1982), 391–418.

[Korepin and Zinn-Justin 2000] V. Korepin and P. Zinn-Justin, "Thermodynamic limit of the six-vertex model with domain wall boundary conditions", *J. Phys. A* **33**:40 (2000), 7053–7066.

[Kuperberg 1996] G. Kuperberg, "Another proof of the alternating-sign matrix conjecture", *Internat. Math. Res. Notices* **3** (1996), 139–150.

[Lieb 1967a] E. H. Lieb, "Exact solution of the problem of the entropy of two-dimensional ice", *Phys. Rev. Lett.* **18** (Apr 1967), 692–694.

[Lieb 1967b] E. H. Lieb, "Exact solution of the F model of an antiferroelectric", *Phys. Rev. Lett.* **18** (1967), 1046–1048.

[Lieb 1967c] E. H. Lieb, "Exact solution of the two-dimensional Slater KDP Model of a ferroelectric", *Phys. Rev. Lett.* **19** (Jul 1967), 108–110.

[Lieb 1967d] E. H. Lieb, "Residual entropy of square ice", *Phys. Rev.* **162** (Oct 1967), 162–172.

[Lieb and Wu 1972] E. H. Lieb and F. Y. Wu, "Two dimensional ferroelectric models", pp. 331–490 in *Phase transitions and critical phenomena*, vol. 1, edited by C. Domb and M. Green, Academic Press, 1972.

[Sogo 1993] K. Sogo, "Toda molecule equation and quotient-difference method", *J. Phys. Soc. Japan* **62**:4 (1993), 1081–1084.

[Sutherland 1967] B. Sutherland, "Exact solution of a two-dimensional model for hydrogen-bonded crystals", *Phys. Rev. Lett.* **19** (Jul 1967), 103–104.

[Szegő 1975] G. Szegő, *Orthogonal polynomials*, 4th ed., Ame. Math. Soc., Colloquium Publications **23**, American Mathematical Society, Providence, R.I., 1975.

[Whittaker and Watson 1996] E. T. Whittaker and G. N. Watson, *A course of modern analysis*, reprint of the 4th (1927) edition ed., Cambridge University Press, 1996.

[Wu and Lin 1975] F. Y. Wu and K. Y. Lin, "Staggered ice-rule vertex model: the Pfaffian solution", *Phys. Rev. B* **12** (Jul 1975), 419–428.

[Zinn-Justin 2000] P. Zinn-Justin, "Six-vertex model with domain wall boundary conditions and one-matrix model", *Phys. Rev. E* (3) **62**:3, part A (2000), 3411–3418.

bleher@math.iupui.edu   *Department of Mathematical Sciences, Indiana University–Purdue University, 402 North Blackford Street, Indianapolis, IN 46202, United States*

kliechty@depaul.edu   *Department of Mathematical Sciences, DePaul University, Chicago, IL 60614, United States*

Random Matrices
MSRI Publications
Volume **65**, 2014

CLT for spectra of submatrices of Wigner random matrices, II: Stochastic evolution

ALEXEI BORODIN

We show that the global fluctuations of spectra of GOE and GUE matrices and their principal submatrices executing Dyson's Brownian motion are Gaussian in the limit of large matrix dimensions. For nested submatrices one obtains a limiting three-dimensional generalized Gaussian process; its restrictions to two-dimensional sections that are monotone in matrix sizes and time moments coincide with the two-dimensional Gaussian free field with zero boundary conditions. The proof is by moment convergence, and it extends to more general Wigner matrices and their stochastic evolution.

Introduction. The fact that the global spectral fluctuations of a GOE or a GUE random matrix evolving under Dyson's Brownian motion, are asymptotically Gaussian is well-known; see [Anderson et al. 2010, Section 4.3.3] and references therein, and also [Spohn 1998] for a general β analog. On the other hand, it was shown in [Borodin 2014] that the global fluctuations of spectra of various principal submatrices of a single GOE or GUE matrix are also Gaussian. The goal of this article is to put these two statements together.

We prove the asymptotic Gaussian behavior for submatrices of a class of stochastically evolving Wigner random matrices that includes Dyson's Brownian motion for GOE and GUE. The proof is by the method of moments, and the argument is slightly more general than the one presented in [Anderson et al. 2010] for a single Wigner matrix.

We also compute the resulting covariance kernel explicitly. In the case of nesting submatrices, it represents a three-dimensional generalized Gaussian process, where one dimension comes from the position of the spectral variable, the second dimension reflects the size of the submatrix, and the third dimension is the time variable. When restricted to the two-dimensional sections that are monotone in matrix size and time variables, it reproduces the two-dimensional Gaussian free field (GFF) with zero boundary conditions.

In the case of GUE, the appearance of GFF on monotone sections could have been predicted from the determinantal structure of the correlation functions

[Ferrari and Frings 2010; Adler et al. 2010b], and from the analysis of [Borodin and Ferrari 2014] that showed how such a structure leads to GFF covariances in the global asymptotic regime. However, the complete three-dimensional covariance structure seems to be inaccessible via that approach for example because the spectra of the full set of submatrices evolve in a non-Markovian way [Adler et al. 2010a].

Wigner matrices. Let $\{Z_{ij}(t)\}_{j>i\geq 1, t\in\mathbb{R}}$ and $\{Y_i(t)\}_{i\geq 1, t\in\mathbb{R}}$ be two families of independent identically distributed real-valued stochastic (not necessarily Markov) processes with zero mean such that for any $k \geq 1$,

$$\max_{t\in\mathbb{R}}(\mathbb{E}|Z_{12}(t)|^k, \mathbb{E}|Y_1(t)|^k) < \infty.$$

Set $c(s, t) = \frac{1}{2}\mathbb{E}Y_1(s)Y_1(t)$ and assume that

$$c(s, t) \geq 0, \qquad\qquad c(t, t) \equiv 1,$$
$$\mathbb{E}Z_{12}(s)Z_{12}(t) \equiv c(s, t), \quad \mathbb{E}Z_{12}^2(s)Z_{12}^2(t) \equiv 2c(s, t)^2 + 1.$$

Note that by Cauchy's inequality $c(s, t) \leq \sqrt{c(s, s)c(t, t)} = 1$. We say that a function $c(s, t)$ is admissible if it arises in this way.

One possibility for the above relations to be satisfied is to take all $\{2^{-\frac{1}{2}}Y_i(t)\}$ and $\{Z_{ij}(t)\}$ to be independent standard Ornstein–Uhlenbeck processes on $\mathbb{R}$; then $c(s, t) = \exp(-|s - t|)$. We will refer to this possibility as to *Gaussian specialization*.

Define a *real symmetric time-dependent Wigner matrix* $X(t) = [X(i, j \mid t)]_{i,j\geq 1}$ by

$$X(i, j \mid t) = X(j, i \mid t) = \begin{cases} Z_{ij}(t), & i < j, \\ Y_i, & i = j. \end{cases}$$

A Hermitian variation of the same definition is as follows: Let $\{Z_{ij}\}_{j>i\geq 1}$ now be complex-valued (i.i.d., zero-mean) stochastic processes with the same uniform bound on all moments. Denote $c(s, t) = \mathbb{E}Y_1(s)Y_1(t)$ and assume that

$$c(s, t) \geq 0, \quad c(t, t) \equiv 1,$$
$$\mathbb{E}Z_{12}(s)Z_{12}(t) \equiv 0, \quad \mathbb{E}Z_{12}(s)\overline{Z_{12}(t)} \equiv c(s, t),$$
$$\mathbb{E}|Z_{12}(s)|^2|Z_{12}(t)|^2 \equiv c(s, t)^2 + 1.$$

We will also say that a function $c(s, t)$ is admissible if it arises in this way.

There is also a Gaussian specialization that corresponds to $\{Y_i(t)\}$ and

$$\{2^{\frac{1}{2}}\Re Z_{ij}(t)\}, \quad \{2^{\frac{1}{2}}\Im Z_{ij}(t)\}$$

being independent standard Ornstein–Uhlenbeck processes on $\mathbb{R}$; in that case $c(s, t) = \exp(-|s - t|)$.

Define a *Hermitian time-dependent Wigner matrix* $X(t) = [X(i, j \mid t)]_{i,j \geq 1}$ by

$$X(i, j \mid t) = \overline{X(j, i \mid t)} = \begin{cases} Z_{ij}(t), & i < j, \\ Y_i, & i = j. \end{cases}$$

Under the Gaussian specializations, the matrix stochastic processes defined above are called *Dyson's Brownian motions*. Traditionally one distinguishes the two cases by a parameter β that takes value 1 in the real symmetric case and value 2 in the Hermitian case. The random matrices arising at a single time moment are said to belong to the Gaussian orthogonal ensemble (GOE) in the $\beta = 1$ case, and Gaussian unitary ensemble (GUE) in the $\beta = 2$ case.

The height function. For any finite set $B \subset \{1, 2, \dots\}$ denote by X_B the $|B| \times |B|$ submatrix of a matrix X formed by the intersections of the rows and columns of X marked by elements of B.

The *height function* H associated to a time-dependent Wigner matrix X is a random integer-valued function on $\mathbb{R} \times \mathbb{R}_{\geq 1} \times \mathbb{R}$ defined by

$$H(x, y, t) = \sqrt{\frac{\beta\pi}{2}} \{ \text{the number of eigenvalues of } X_{\{1,2,\dots,[y]\}}(t) \text{ that are } \geq x \}.$$

More generally, let $A = \{a_n\}_{n \geq 1}$ be an arbitrary sequence of pairwise distinct natural numbers. Then we define the height function H_A via

$$H_A(x, y) = \sqrt{\frac{\beta\pi}{2}} \{ \text{the number of eigenvalues of } X_{\{a_1,\dots,a_{[y]}\}}(t) \text{ that are } \geq x \}.$$

The first definition corresponds to $A = \mathbb{N}$.

The convenience of the constant prefactor $\sqrt{\beta\pi/2}$ will be evident shortly.

A three-dimensional Gaussian field. Let $c(s, t)$ be an admissible function as defined above. Set $\mathbb{H} = \{z \in \mathbb{C} \mid \Im z > 0\}$ and introduce a function

$$C : (\mathbb{H} \times \mathbb{R}) \times (\mathbb{H} \times \mathbb{R}) \to \mathbb{R} \cup \{+\infty\}$$

via

$$C(z, s; w, t) = \frac{1}{2\pi} \ln \left| \frac{c(s, t) \min(|z|^2, |w|^2) - zw}{c(s, t) \min(|z|^2, |w|^2) - z\overline{w}} \right|$$

$$= \begin{cases} -\dfrac{1}{2\pi} \ln \left| \dfrac{c(s, t)z - w}{c(s, t)z - \overline{w}} \right|, & |z| \leq |w|, \\[4mm] -\dfrac{1}{2\pi} \ln \left| \dfrac{c(s, t)w - z}{c(s, t)w - \overline{z}} \right|, & |z| > |w|. \end{cases}$$

It is easy to see that for any (s, t) with $c(s, t) < 1$, $C(\cdot, s; \cdot, t)$ is a continuous function on $\mathbb{H} \times \mathbb{H}$. Note also that if $c(s, t) = 1$ then

$$C(z, s; w, t) = -\frac{1}{2\pi} \ln \left| \frac{z - w}{z - \bar{w}} \right|$$

is the Green function for the Laplace operator on $\mathbb{H}$ with Dirichlet boundary conditions. Viewed as a function in (z, w), it represents the covariance for the two-dimensional Gaussian free field on $\mathbb{H}$ with zero boundary conditions.

Proposition 1. *For any admissible function $c(s, t)$ as above, there exists a generalized Gaussian process on $\mathbb{H} \times \mathbb{R}$ with the covariance kernel $C(z, s; w, t)$ as above. More exactly, for any finite family of test functions $f_m(z) \in C_0(\mathbb{H} \times \mathbb{R})$ the covariance matrix*

$$\text{cov}(f_k, f_l) = \int_{\mathbb{H} \times \mathbb{R}} \int_{\mathbb{H} \times \mathbb{R}} f_k(z, s) f_l(w, t) C(z, s; w, t) \, dz \, d\bar{z} \, ds \, dw \, d\bar{w} \, dt,$$

$$k, l = 1, \ldots, M,$$

is positive definite.

Denote the resulting generalized Gaussian process by $\mathcal{G}_{c(s,t)}$.

A proof of Proposition 1 will be given later.

Complex structure. Let A be a sequence of pairwise distinct integers. The height function $H_A(x, y, t)$ (or $H(x, y, t) = H_\mathbb{N}(x, y, t)$) is naturally defined on $\mathbb{R} \times \mathbb{R}_{\geq 1} \times \mathbb{R}$. Having the large parameter L, we would like to scale $(x, y) \mapsto (L^{-\frac{1}{2}}x, L^{-1}y)$, which lands us in $\mathbb{R} \times \mathbb{R}_{>0} \times \mathbb{R}$.

Wigner's semicircle law implies that for any $t \in \mathbb{R}$, with $L \gg 1$, $x \sim L^{\frac{1}{2}}$, $y \sim L$, after rescaling with overwhelming probability the eigenvalues (or, equivalently, the places of growth of the height function in x-direction) are concentrated in the domain

$$\{(x, y) \in \mathbb{R} \times \mathbb{R}_{>0} \mid -2\sqrt{y} \leq x \leq 2\sqrt{y}\}.$$

Let us identify the interior of this domain with $\mathbb{H}$ via the map

$$\Omega : (x, y) \mapsto \frac{x}{2} + i\sqrt{y - \left(\frac{x}{2}\right)^2}.$$

Its inverse has the form

$$\Omega^{-1}(z) = (x(z), y(z)) = (2\Re(z), |z|^2).$$

Note that this map sends the boundary of the domain to the real line.

Thanks to Ω we can now speak of the height function H_A as being defined on $\mathbb{H} \times \mathbb{R}$; we will use the notation

$$H_A^\Omega(z; t) = H_A(L^{\frac{1}{2}}x(z), Ly(z), t), \quad z \in \mathbb{H}.$$

Note that we have incorporated rescaling in this definition.

Main result. Let X be a (real symmetric or Hermitian) time-dependent Wigner matrix. We argue that the centralized random height function

$$H^{\Omega}(z; t) - \mathbb{E}H^{\Omega}(z; t), \quad z \in \mathbb{H}, \ t \in \mathbb{R},$$

viewed as distribution, converges as $L \to \infty$ to the generalized Gaussian process $\mathcal{G}_{c(s,t)}$ with $c(s, t) = \frac{1}{2}\beta \, \mathbb{E}Y_1(s)Y_1(t)$.

One needs to verify the convergence on a suitable set of test functions. The exact statement that we prove is the following.

Theorem 2. *Pick $\tau \in \mathbb{R}$, $y > 0$, and $k \in \mathbb{Z}_{\geq 0}$. Define a moment of the random height function by*

$$M_{\tau,y,k} = \int_{-\infty}^{+\infty} x^k \big(H(L^{\frac{1}{2}}x, Ly, \tau) - \mathbb{E}H(L^{\frac{1}{2}}x, Ly, \tau)\big) \, dx.$$

Then as $L \to \infty$, these moments converge, in the sense of finite dimensional distributions, to the moments of $\mathcal{G}_{c(s,t)}$ defined as

$$\mathcal{M}_{\tau,y,k} = \int_{\substack{z \in \mathbb{H} \\ |z|^2 = y}} (x(z))^k \, \mathcal{G}_{c(s,t)}(z; \tau) \, \frac{dx(z)}{dz} \, dz.$$

Monotone sections as two-dimensional Gaussian free fields. Consider a time-dependent Wigner matrix and assume that the function $c(s, t) = \frac{1}{2}\beta \, \mathbb{E}Y_1(s)Y_1(t)$ is continuous and that it has the following monotonicity property: For any $s \in \mathbb{R}$, $c(s, t)$ is strictly increasing in $t \in (-\infty, s]$ and it is strictly decreasing in $t \in [s, +\infty)$. In other words, as time distance between matrices grows, the correlation decays. Further, assume that $c(s, t) \neq 0$ for any $s, t \in \mathbb{R}$.

Let $\phi : \mathbb{R} \to \mathbb{R}_{>0}$ and $\psi : \mathbb{R} \to \mathbb{R}$ be a continuous nonincreasing and a continuous nondecreasing functions, and assume that for at least one of these functions the monotonicity is strict.

Our goal is to consider the joint fluctuations of spectra of matrices

$$X_{\{1,2,\dots,[L\phi(t)]\}}(\psi(t)), \quad t \in \mathbb{R}, \tag{1}$$

where $L \gg 1$ is a large parameter. By Wigner's semicircle law, the spectrum of such a matrix scaled by $L^{\frac{1}{2}}$ is concentrated on $[-2\sqrt{\phi(t)}, 2\sqrt{\phi(t)}]$.

The two extreme cases are $\phi(t) \equiv$ const (the size of the matrices is fixed and the time is moving) and $\psi(t) \equiv$ const (the time moment is fixed and the size of the matrices is changing).

Let us choose a reference time moment $t_0 \in \mathbb{R}$ and introduce a map

$$\Xi : \big\{(x, t) \in \mathbb{R} \times \mathbb{R} \mid -2\sqrt{\phi(t)} < x < 2\sqrt{\phi(t)}\big\} \to \mathbb{H}$$

as

$$\Xi(x, t) = \begin{cases} c(\psi(t_0), \psi(t))\left(\dfrac{x}{2} + i\sqrt{\phi(t) - \left(\dfrac{x}{2}\right)^2}\right), & t \geq t_0, \\[4mm] \dfrac{1}{c(\psi(t), \psi(t_0))}\left(\dfrac{x}{2} + i\sqrt{\phi(t) - \left(\dfrac{x}{2}\right)^2}\right), & t < t_0. \end{cases}$$

The continuity and monotonicity assumptions on c, ϕ, and ψ are needed for Ξ to be a bijection. Hence, its inverse is correctly defined, denote it as $\Xi^{-1}(\zeta) = (x(\zeta), t(\zeta))$.

We can now view the height function H for matrices (1) as a function on $\mathbb{H}$ via

$$H^{\Xi}(\zeta) = H\left(L^{\frac{1}{2}}x(\zeta), L\phi(t(\zeta)), \psi(t(\zeta))\right).$$

Our main result implies that the centralized height function

$$H^{\Xi}(\zeta) - \mathbb{E}H^{\Xi}(\zeta), \quad \zeta \in \mathbb{H},$$

viewed as a distribution, converges as $L \to \infty$ to the Gaussian free field on $\mathbb{H}$ in the sense of Theorem 2.

Moments as traces. Let us rescale the variable $x = L^{-\frac{1}{2}}u$ in the definition of $M_{\tau, y, k}$ and then integrate by parts. Since the derivative of the height function $H(u, [Ly], t)$ in u is

$$\frac{d}{du}H(u, [Ly], t) = -\sqrt{\frac{\beta\pi}{2}} \sum_{s=1}^{[Ly]} \delta(u - \lambda_s),$$

where $\{\lambda_s\}_{1 \leq s \leq [Ly]}$ are the eigenvalues of $X_{\{1,\ldots,[Ly]\}}(t)$, we obtain

$$M_{\tau, y, k} = L^{-\frac{k+1}{2}} \sqrt{\frac{\beta\pi}{2}} \left(\sum_{s=1}^{[Ly]} \frac{\lambda_s^{k+1}}{k+1} - \mathbb{E}\sum_{s=1}^{[Ly]} \frac{\lambda_s^{k+1}}{k+1}\right)$$

$$= \frac{L^{-\frac{k+1}{2}}}{k+1} \sqrt{\frac{\beta\pi}{2}} \left(\text{Tr}(X_{\{1,\ldots,[Ly]\}}^{k+1}(t)) - \mathbb{E}\,\text{Tr}(X_{\{1,\ldots,[Ly]\}}^{k+1}(t))\right).$$

We can now reformulate the statement of Theorem 2 as follows.

Theorem 2′. *Let $X(t)$ be a time-dependent (real symmetric or Hermitian) Wigner matrix with*

$$c(s, t) = \frac{\beta}{2}\, \mathbb{E}Y_1(s)Y_1(t).$$

Let $k_1, \ldots, k_m \geq 1$ be integers, $t_1, \ldots, t_m$ real numbers, and $y_1, \ldots, y_m$ positive real numbers. The m-dimensional random vector

$$\left(L^{-\frac{k_p}{2}}\left(\text{Tr}(X_{\{1,\ldots,[Ly_p]\}}^{k_p}(t_p)) - \mathbb{E}\,\text{Tr}(X_{\{1,\ldots,[Ly_p]\}}^{k_p}(t_p))\right)\right)_{p=1}^{m}$$

converges (in distribution and with all moments) to the zero mean m-dimensional Gaussian random vector $(\xi_p)_{p=1}^m$ with covariance

$$\mathbb{E}\xi_p\xi_q = \frac{2k_p k_q}{\beta\pi} \oint_{\substack{|z|^2=b_p \\ \Im z>0}} \oint_{\substack{|w|^2=b_q \\ \Im w>0}} (x(z))^{k_p-1}(x(w))^{k_q-1}$$

$$\times \frac{1}{2\pi} \ln\left|\frac{c(t_p,t_q)\min(y_p,y_q)-zw}{c(t_p,t_q)\min(y_p,y_q)-z\bar{w}}\right| \frac{dx(z)}{dz}\frac{dx(w)}{dw} dz\,dw.$$

More general submatrices. In the spirit of [Borodin 2014], we will actually prove a more general claim that involves arbitrary sequences of symmetric submatrices of the Wigner matrix that are sufficiently well-behaved. The exact statement is as follows.

Theorem 2″. *Let $X(t)$ be a time-dependent (real symmetric or Hermitian) Wigner matrix with*

$$c(s,t) = \frac{\beta}{2}\, \mathbb{E}Y_1(s)Y_1(t).$$

Let $k_1,\ldots,k_m \geq 1$ be integers, $t_1,\ldots,t_m \in \mathbb{R}$, and let $B_1,\ldots,B_m$ be subsets of $\mathbb{N}$ dependent on the large parameter L so that there exist limits

$$b_p = \lim_{L\to\infty} \frac{|B_p|}{L} > 0, \quad b_{pq} = \lim_{L\to\infty} \frac{|B_p \cap B_q|}{L}, \qquad p,q = 1,\ldots,m.$$

Then the m-dimensional random vector

$$\left(L^{-\frac{k_p}{2}}\left(\mathrm{Tr}(X_{B_p}^{k_p}(t_p)) - \mathbb{E}\,\mathrm{Tr}(X_{B_p}^{k_p}(t_p))\right)\right)_{p=1}^m \tag{2}$$

converges (in distribution and with all moments) to the zero mean m-dimensional Gaussian random variable $(\xi_p)_{p=1}^m$ with covariance

$$\mathbb{E}\xi_p\xi_q = \frac{2k_p k_q}{\beta\pi} \oint_{\substack{|z|^2=b_p \\ \Im z>0}} \oint_{\substack{|w|^2=b_q \\ \Im w>0}} (x(z))^{k_p-1}(x(w))^{k_q-1}$$

$$\times \frac{1}{2\pi} \ln\left|\frac{c(t_p,t_q)b_{pq}-zw}{c(t_p,t_q)b_{pq}-z\bar{w}}\right| \frac{dx(z)}{dz}\frac{dx(w)}{dw} dz\,dw. \tag{3}$$

Theorem 2″ can also be viewed as the moment convergence of the centralized height function $H_A(x,y,t)$ to a limiting generalized Gaussian process but we do not give further details here. The static variant of this convergence is discussed in [Borodin 2014].

Proof of Theorem 2″. The argument closely follows that given in Section 2.1.7 of [Anderson et al. 2010] in the case of one set $B_j \equiv B$, and the proof of Theorem 2′ in [Borodin 2014] in the static case. One proves the convergence of moments, which is sufficient to also claim the convergence in distribution for Gaussian limits.

Any joint moment of the coordinates of (2) is written as a finite combination of contributions corresponding to suitably defined graphs that are in their turn associated to words. This reduction is explained in Section 2.1.7 of [Anderson et al. 2010]. The key fact in the real symmetric case is that averages of products of powers of matrix elements that involve at least one matrix element with exponent 1 vanish. The time-dependent analog of this fact is that averages of products of powers of matrix elements taken at different time moments that involve one matrix element with exponent 1 *at only one time moment* vanish. This clearly holds by independence of matrix elements and our zero mean assumption. In the static Hermitian case, one needs in addition that $\mathbb{E}Z_{12}^2 = 0$. The time-dependent analog reads $\mathbb{E}Z_{12}(s)Z_{12}(t) = 0$ for any $s, t \in \mathbb{R}$, which is one of our assumptions. This allows the exact same reduction to go through in the time-dependent setting.

The only difference of the multiset case from the one-set case is that one needs to keep track of the *alphabets* the words are built from: A word corresponding to coordinate number p of (2) would have to be built from the alphabet that coincides with the set B_p. Equivalently, the corresponding graphs will have their vertices labeled by elements of B_p.

Since all sizes $|B_p|$ have order L, and $|B_1 \cup \cdots \cup B_m| = O(L)$, and also the moments of matrix elements at all times are uniformly bounded, the estimate showing that all contributions not coming from matchings are negligible in [Anderson et al. 2010, Lemma 2.1.34] carries over without difficulty. It only remains to compute the covariance.

We start with the case of the real symmetric Wigner matrices.

In the one-set case, the limits of the variances of the coordinates of (2) are given by (2.1.44) in [Anderson et al. 2010]. It reads (with $k = k_p$ for a p between 1 and m)

$$2k^2 C_{\frac{k-1}{2}}^2 + k^2 C_{\frac{k}{2}}^2 + \sum_{r=3}^{\infty} \frac{2k^2}{r} \left(\sum_{\substack{k_i \geq 0 \\ 2\sum_{i=1}^{r} k_i = k-r}} \prod_{i=1}^{r} C_{k_i} \right)^2, \qquad (4)$$

where $\{C_k\}_{k\geq 0}$ are the Catalan numbers, and we set $C_a = 0$ if $a \notin \{0, 1, 2, \ldots\}$. The Catalan number C_k counts the number of rooted planar trees with k edges, and different terms of (4) have the following interpretation (see [Anderson et al. 2010] for detailed explanations):

• The first term comes from two trees with $(k-1)/2$ edges each that hang from a common vertex; the factor k^2 originates from choices of certain starting points on each tree united with the common vertex, and the extra 2 is actually $\mathbb{E}Y_1^2$.

• The second term comes from two trees with $k/2$ edges each that are glued along one edge. There are $k/2$ choices of this edge for each of the trees, there is an additional $2 = \mathbb{E}Z_{12}^4 - 1$, and another additional 2 responsible of the choice of the orientation of the gluing.

• The third term comes from two graphs each of which is a cycle of length r with pendant trees hanging off each of the vertices of the cycle; the total number of edges in the extra trees being $(k - r)/2$ (this must be an integer). As for the first term, there is an extra $k^2 = k \cdot k$ coming from the choice of the starting points and also an extra 2 for the choice of the gluing orientation along the cycle.

For each of the three terms the total number of vertices in the resulting graph is equal to k, and if one labels each vertex with a letter from an alphabet of cardinality $|B|$ this would yield a factor of

$$|B|(|B| - 1) \cdots (|B| - k + 1) = |B|^k + O(|B|^{k-1}).$$

Normalization by $|B|^k$ yields (4).

In the general case, in order to evaluate the covariance

$$L^{-\frac{k_p + k_q}{2}} \mathbb{E}\big[\big(\mathrm{Tr}(X_{B_p}^{k_p}(t_p)) - \mathbb{E}\,\mathrm{Tr}(X_{B_p}^{k_p}(t_p))\big)\big(\mathrm{Tr}(X_{B_q}^{k_q}(t_q)) - \mathbb{E}\,\mathrm{Tr}(X_{B_q}^{k_q}(t_q))\big)\big] \quad (5)$$

in the limit, we need to employ the same graph counting, except for the two graphs being glued now correspond to different values k_p and k_q of k, and their vertices are marked by letters of different alphabets B_p and B_q.

• The first term gives $2 k_p k_q C_{\frac{k_p - 1}{2}} C_{\frac{k_q - 1}{2}}$ for the graph counting, and an extra

$$|B_p \cap B_q| \cdot (|B_p| - 1)(|B_p| - 2) \cdots \big(|B_p| - \tfrac{1}{2}(k_p + 1)\big)$$
$$\cdot (|B_q| - 1)(|B_q| - 2) \cdots \big(|B_q| - \tfrac{1}{2}(k_q + 1)\big)$$

for the vertex labeling (the factor $|B_p \cap B_q|$ comes from the only common vertex). Moreover, $\mathbb{E}Y_1^2$ is replaced by $\mathbb{E}Y_1(t_p)Y_1(t_q) = c(t_p, t_q)$. Normalized by $L^{-\frac{k_p + k_q}{2}}$ this yields

$$2 k_p k_q C_{\frac{k_p - 1}{2}} C_{\frac{k_q - 1}{2}} c(t_p, t_q) b_{pq} b_p^{\frac{k_p - 1}{2}} b_q^{\frac{k_q - 1}{2}}.$$

• The second term has

$$k_p k_q C_{\frac{k_p}{2}} C_{\frac{k_q}{2}}$$

from the graph counting and

$$c_{pq}^2 b_p^{\frac{k_p}{2} - 1} b_q^{\frac{k_q}{2} - 1}$$

from the label counting. In addition, $\mathbb{E}Z_{12}^4 - 1$ is replaced by

$$\mathbb{E}Z_{12}^2(t_p)Z_{12}^2(t_q) - 1 = 2 c^2(t_p, t_q).$$

The total contribution is

$$k_p k_q C_{\frac{k_p}{2}} C_{\frac{k_q}{2}} \left(c(t_p, t_q) b_{pq}\right)^2 b_p^{\frac{k_p}{2}-1} b_q^{\frac{k_q}{2}-1}.$$

- For the third term in the same way we obtain

$$\sum_{r=3}^{\infty} \frac{2 k_p k_q}{r} \left(\sum_{\substack{s_i \geq 0 \\ 2\sum_{i=1}^{r} s_i = k_p - r}} \prod_{i=1}^{r} C_{s_i} \right) \left(\sum_{\substack{t_i \geq 0 \\ 2\sum_{i=1}^{r} t_i = k_q - r}} \prod_{i=1}^{r} C_{t_i} \right) (c(t_p, t_q) b_{pq})^r b_p^{\frac{k_p - r}{2}} b_q^{\frac{k_q - r}{2}}$$

where $c^r(t_p, t_q)$ appeared as $(\mathbb{E} Z_{12}(t_p) Z_{12}(t_q))^r$, which in its turn came from the edges of the r-cycle.

Thus, the asymptotic value of the covariance (5) is

$$2 k_p k_q C_{\frac{k_p - 1}{2}} C_{\frac{k_q - 1}{2}} \left(c(t_p, t_q) b_{pq}\right) b_p^{\frac{k_p - 1}{2}} b_q^{\frac{k_q - 1}{2}}$$

$$+ k_p k_q C_{\frac{k_p}{2}} C_{\frac{k_q}{2}} \left(c(t_p, t_q) b_{pq}\right)^2 b_p^{\frac{k_p}{2}-1} b_q^{\frac{k_q}{2}-1}$$

$$+ \sum_{r=3}^{\infty} \frac{2 k_p k_q}{r} \left(\sum_{\substack{s_i \geq 0 \\ 2\sum_{i=1}^{r} s_i = k_p - r}} \prod_{i=1}^{r} C_{s_i} \right) \left(\sum_{\substack{t_i \geq 0 \\ 2\sum_{i=1}^{r} t_i = k_q - r}} \prod_{i=1}^{r} C_{t_i} \right) (c(t_p, t_q) b_{pq})^r b_p^{\frac{k_p - r}{2}} b_q^{\frac{k_q - r}{2}}.$$

We now use the fact that for any $S = 0, 1, 2, \ldots,$

$$\sum_{\substack{s_i \geq 0 \\ \sum_{i=1}^{r} s_i = S}} \prod_{i=1}^{r} C_{s_i} = \binom{2S + r}{S} \frac{r}{2S + r};$$

see [Graham et al. 1989, (5.70)]. This allows us to rewrite the asymptotic covariance in terms of binomial coefficients: interdisplaylinepenalty0

$$2 \binom{k_p}{(k_p - 1)/2} \binom{k_q}{(k_q - 1)/2} (c(t_p, t_q) b_{pq}) b_p^{\frac{k_p - 1}{2}} b_q^{\frac{k_q - 1}{2}}$$

$$+ 4 \binom{k_p}{k_p/2 - 1} \binom{k_q}{k_q/2 - 1} (c(t_p, t_q) b_{pq})^2 b_p^{\frac{k_p - 2}{2}} b_q^{\frac{k_q - 2}{2}}$$

$$+ \sum_{r=3}^{\infty} 2r \binom{k_p}{(k_p - r)/2} \binom{k_q}{(k_q - r)/2} (c(t_p, t_q) b_{pq})^r b_p^{\frac{k_p - r}{2}} b_q^{\frac{k_q - r}{2}}$$

$$= \sum_{r=1}^{\infty} 2r \binom{k_p}{(k_p - r)/2} \binom{k_q}{(k_q - r)/2} (c(t_p, t_q) b_{pq})^r b_p^{\frac{k_p - r}{2}} b_q^{\frac{k_q - r}{2}}.$$

Using the binomial theorem, we can write this expression as a double contour integral

$$\frac{2}{(2\pi i)^2} \iint\limits_{K_1=|z|<|w|=K_2} \left(z+\frac{b_p}{z}\right)^{k_p} \left(w+\frac{b_q}{w}\right)^{k_q} \frac{c(t_p,t_q)b_{pq}}{b_p} \frac{dz\,dw}{\left(\dfrac{c(t_p,t_q)b_{pq}}{b_p}z-w\right)^2}, \tag{6}$$

with K_1, K_2 fixed. Note that $c(t_p,t_q)b_{pq}/b_p < 1$, so $(c(t_p,t_q)b_{pq}z/b_p - w)^{-2}$ on our domain of integration can be expanded using a derivative of the geometric series.

Consider the right-hand side of (3) and assume that $|z|^2 = b_p < b_q = |w|^2$. Observe that

$$2\ln\left|\frac{c(t_p,t_q)b_{pq}-zw}{c(t_p,t_q)b_{pq}-z\overline{w}}\right|$$

$$= -2\ln\left|\frac{\dfrac{c(t_p,t_q)b_{pq}}{b_p}\overline{z}-w}{\dfrac{c(t_p,t_q)b_{pq}}{b_p}\overline{z}-\overline{w}}\right|$$

$$= -\ln\left(\frac{c(t_p,t_q)b_{pq}}{b_p}z-w\right) + \ln\left(\frac{c(t_p,t_q)b_{pq}}{b_p}z-\overline{w}\right)$$

$$+ \ln\left(\frac{c(t_p,t_q)b_{pq}}{b_p}\overline{z}-w\right) - \ln\left(\frac{c(t_p,t_q)b_{pq}}{b_p}\overline{z}-\overline{w}\right).$$

This allows us to rewrite the right-hand side of (3) as a double contour integral over complete circles in the form

$$-\frac{k_pk_q}{2\beta\pi^2} \oint\limits_{|z|^2=b_p} \oint\limits_{|w|^2=b_q} (x(z))^{k_p-1}(x(w))^{k_q-1} \ln\left(\frac{c(t_p,t_q)b_{pq}}{b_p}z-w\right)$$

$$\cdot \frac{dx(z)}{dz}\frac{dx(w)}{dw}dz\,dw.$$

Recalling that $\beta = 1$ and noting that

$$k_p(x(z))^{k_p-1}\frac{dx(z)}{dz} = \frac{d(x(z))^{k_p}}{dz}, \quad k_q(x(w))^{k_q-1}\frac{dx(w)}{dw} = \frac{d(x(w))^{k_q}}{dw},$$

we integrate by parts in z and w and recover (6). The proof for $b_p = b_q$ is obtained by continuity of both sides, and to see that the needed identity holds for $b_p > b_q$ it suffices to observe that both sides are symmetric in p and q.

The argument in the case of Hermitian Wigner matrices is exactly the same, except in the combinatorial part for the first term the factor 2 is missing due to the change in $\mathbb{E}Y_1(s)Y_1(t)$, in the second term 2 is missing due to the change in $\mathbb{E}|Z_{12}(s)|^2|Z_{12}(t)|^2$, and in the third term 2 is missing because there is no choice in the orientation of two r-cycles that are being glued together. □

Proof of Proposition 1. We need to show that for any complex numbers $\{u_k\}_{k=1}^M$,

$$\sum_{k,l=1}^M u_k \overline{u_l} \int_{\mathbb{H}\times\mathbb{R}} \int_{\mathbb{H}\times\mathbb{R}} f_k(z,s) f_l(w,t) C(z,s;w,t)\, dz\, d\bar{z}\, ds\, dw\, d\bar{w}\, dt \geq 0.$$

We can approximate the integration over the three-dimensional domains by finite sums of one-dimensional integrals over semicircles of the form $|z| = \text{const}$, $s = \text{const}$. On each semicircle we further uniformly approximate the (continuous) integrand by a polynomial in $\mathfrak{R}(z)$. Finally, for the polynomials the nonnegativity follows from Theorem 2'. $\qquad\square$

Chebyshev polynomials. One way to describe the limiting covariance structure in the one-matrix static case is to show that traces of the Chebyshev polynomials of the matrix are asymptotically independent; see [Johansson 1998]. A similar effect takes place for time-dependent submatrices as well.

For $n = 0, 1, 2, \ldots$, let $T_n(x)$ be the n-th degree Chebyshev polynomial of the first kind:

$$T_n(x) = \cos(n \arccos x),$$

or equivalently,

$$T_n(\cos x) = \cos(nx).$$

For any $a > 0$, let $T_n^a(x) = T_n(x/a)$ be the rescaled version of T_n.

Proposition 5. *In the assumptions of Theorem 2'', for any $p, q = 1, \ldots, m$*

$$\lim_{L\to\infty} \mathbb{E}\left[\left(\text{Tr}\big(T_{k_p}^{2\sqrt{b_p L^{k_p}}}(X_{B_p}(t_p))\big) - \mathbb{E}\,\text{Tr}\big(T_{k_p}^{2\sqrt{b_p L^{k_p}}}(X_{B_p}(t_p))\big) \right) \right.$$
$$\left. \times \left(\text{Tr}\big(T_{k_q}^{2\sqrt{b_q L^{k_q}}}(X_{B_q}(t_q))\big) - \mathbb{E}\,\text{Tr}\big(T_{k_q}^{2\sqrt{b_q L^{k_q}}}(X_{B_q}(t_q))\big) \right) \right]$$
$$= \delta_{k_p k_q} \frac{k_p}{2\beta} \left(\frac{c(t_p,t_q) b_{pq}}{\sqrt{b_p b_q}} \right)^{k_p}.$$

Proof. Using (6) and assuming $b_p < b_q$ we obtain that the needed limit equals

$$\frac{2}{\beta(2\pi i)^2} \iint_{b_p=|z|<|w|=b_q} T_{k_p}(\cos\arg z) T_{k_q}(\cos\arg w) \frac{c(t_p,t_q) b_{pq}}{b_p} \frac{dz\,dw}{\left(\dfrac{c(t_p,t_q) b_{pq}}{b_p} z - w\right)^2}$$

$$= \frac{1}{2\beta(2\pi i)^2} \iint_{b_p=|z|<|w|=b_q} \left(\left(\frac{z}{\sqrt{b_p}}\right)^{k_p} + \left(\frac{\sqrt{b_p}}{z}\right)^{k_p} \right) \left(\left(\frac{w}{\sqrt{b_q}}\right)^{k_q} + \left(\frac{\sqrt{b_q}}{w}\right)^{k_q} \right)$$

$$\times \frac{c(t_p,t_q) b_{pq}}{b_p} \frac{dz\,dw}{\left(\dfrac{c(t_p,t_q) b_{pq}}{b_p} z - w\right)^2}.$$

Writing $(c(t_p, t_q)b_{pq}z/b_p - w)^{-2}$ as a series in z/w we arrive at the result. Continuity and symmetry of both sides of the limiting relation removes the assumption $b_p < b_q$. $\qquad\qquad\qquad\qquad\qquad\qquad\qquad\qquad\qquad\qquad\qquad$ $\square$

Note that in the Gaussian specialization (when $c(s, t) = \exp(-|s - t|)$) and for a single size L time-dependent Wigner matrix (i.e., $b_p = b_q = b_{pq} = 1$), the centralized traces of Chebyshev polynomials of this matrix evolve as independent Ornstein–Uhlenbeck processes with speeds equal to the degrees of the polynomials.

Acknowledgements

The author is very grateful to an anonymous referee for a number of very helpful suggestions. This work was partially supported by NSF grant DMS-1056390.

References

[Adler et al. 2010a] M. Adler, E. Nordenstam, and P. van Moerbeke, "Consecutive minors for Dyson's Brownian motions", preprint, 2010. arXiv 1007.0220

[Adler et al. 2010b] M. Adler, E. Nordenstam, and P. van Moerbeke, "The Dyson Brownian minor process", preprint, 2010. arXiv 1006.2956

[Anderson et al. 2010] G. W. Anderson, A. Guionnet, and O. Zeitouni, *An introduction to random matrices*, Cambridge Studies in Advanced Mathematics **118**, Cambridge University Press, Cambridge, 2010.

[Borodin 2014] A. Borodin, "CLT for spectra of submatrices of Wigner random matrices", *Moscow Math. J.* **14**:1 (2014), 29–38.

[Borodin and Ferrari 2014] A. Borodin and P. L. Ferrari, "Anisotropic growth of random surfaces in 2 + 1 dimensions", *Comm. Math. Phys.* **325**:2 (2014), 603–684.

[Ferrari and Frings 2010] P. L. Ferrari and R. Frings, "On the partial connection between random matrices and interacting particle systems", *J. Stat. Phys.* **141**:4 (2010), 613–637.

[Graham et al. 1989] R. L. Graham, D. E. Knuth, and O. Patashnik, *Concrete mathematics: a foundation for computer science*, Addison-Wesley, Reading, MA, 1989.

[Johansson 1998] K. Johansson, "On fluctuations of eigenvalues of random Hermitian matrices", *Duke Math. J.* **91**:1 (1998), 151–204.

[Spohn 1998] H. Spohn, "Dyson's model of interacting Brownian motions at arbitrary coupling strength", *Markov Process. Related Fields* **4**:4 (1998), 649–661.

borodin@math.mit.edu   *Department of Mathematics, Massachusetts Institute of Technology, Building E18, Room 369, 77 Massachusetts Avenue, Cambridge, 02193-4307, United States*

Random Matrices
MSRI Publications
Volume **65**, 2014

Critical asymptotic behavior for the Korteweg–de Vries equation and in random matrix theory

TOM CLAEYS AND TAMARA GRAVA

We discuss universality in random matrix theory and in the study of Hamiltonian partial differential equations. We focus on universality of critical behavior and we compare results in unitary random matrix ensembles with their counterparts for the Korteweg–de Vries equation, emphasizing the similarities between both subjects.

1. Introduction

It has been observed and conjectured that the critical behavior of solutions to Hamiltonian perturbations of hyperbolic and elliptic systems of partial differential equations near points of gradient catastrophe is asymptotically independent of the chosen initial data and independent of the chosen equation [Dubrovin 2006; Dubrovin et al. 2009]. A classical example of a Hamiltonian perturbation of a hyperbolic equation which exhibits such universal behavior, is the Korteweg–de Vries (KdV) equation

$$u_t + 6uu_x + \epsilon^2 u_{xxx} = 0, \quad \epsilon > 0. \tag{1-1}$$

If one is interested in the behavior of KdV solutions in the small dispersion limit $\epsilon \to 0$, it is natural to study first the inviscid Burgers' or Hopf equation $u_t + 6uu_x = 0$. Given smooth initial data $u(x, 0) = u_0(x)$ decaying at $\pm\infty$, the solution of this equation is, for t sufficiently small, given by the method of characteristics: we have $u(x, t) = u(\xi(x, t))$, where $\xi(x, t)$ is given as the solution to the equation

$$x = 6tu_0(\xi) + \xi. \tag{1-2}$$

It is easily derived from this implicit form of the solution that the x-derivative of $u(x, t)$ blows up at time

$$t_c = \frac{1}{\max_{\xi \in \mathbb{R}}(-6u_0'(\xi))},$$

which is called the time of gradient catastrophe. After this time, the Hopf solution $u(x, t)$ ceases to exist in the classical sense. For t slightly smaller than the critical time t_c the KdV solution starts to oscillate as shown numerically in [Grava and Klein 2007]. For $t > t_c$ the KdV solution develops a train of rapid oscillations of wavelength of order ϵ. In general, the asymptotics for the KdV solution as $\epsilon \to 0$ can be described in terms of an equilibrium problem, discovered by Lax and Levermore [1983a; 1983b; 1983c; Lax et al. 1993].

The support of the solution of the equilibrium problem, which depends on x and t, consists of a finite or infinite union of intervals [Grava 2004; Deift et al. 1998b], and the endpoints evolve according to the Whitham equations [Whitham 1974; Flaschka et al. 1980]. For $t < t_c$, the support of the equilibrium problem consists of one interval and the KdV solution as $\epsilon \to 0$ is approximated by the Hopf solution. For $t > t_c$ the support of the equilibrium problem may consists of several intervals and the KdV solution is approximated as $\epsilon \to 0$ by Riemann θ-functions [Gurevich and Pitaevskii 1973; Lax and Levermore 1983a; 1983b; 1983c; Deift et al. 1997; Venakides 1990].

The (x, t)-plane can thus be divided into different regions labeled by the number of intervals in the support of the Lax–Levermore minimization problem. Such regions are independent of ϵ and depend only on the initial data. Those regions are separated by a collection of breaking curves where the number of intervals in the support changes. We will review recently obtained results concerning the asymptotic behavior of KdV solutions near curves separating a one-interval region from a two-interval region. The two interval region corresponds to the solution of KdV being approximated as $\epsilon \to 0$ by the Jacobi elliptic function, the one interval region corresponds to the solution of KdV being approximated by the Hopf solution (1-2).

On the space of $n \times n$ Hermitian matrices, one can define unitary invariant probability measures of the form

$$\frac{1}{\tilde{Z}_n} \exp(-N \operatorname{Tr} V(M)) \, dM, \quad dM = \prod_{i=1}^{n} dM_{ii} \prod_{i<j} d\operatorname{Re} M_{ij} \, d\operatorname{Im} M_{ij}, \quad (1\text{-}3)$$

where $\tilde{Z}_n = \tilde{Z}_n(N)$ is a normalization constant which depends on the integer N and V is a real polynomial of even degree with positive leading coefficient. The eigenvalues of random matrices in such a unitary ensemble follow a determinantal point process defined by

$$\frac{1}{Z_n} \prod_{i<j} (\lambda_i - \lambda_j)^2 \prod_{i=1}^{n} e^{-NV(\lambda_i)} \, d\lambda_i, \quad (1\text{-}4)$$

with correlation kernel

$$K_n(u, v) = \frac{e^{-\frac{N}{2}V(u)}e^{-\frac{N}{2}V(v)}}{u - v}\frac{\kappa_{n-1}}{\kappa_n}\big(p_n(u)p_{n-1}(v) - p_n(v)p_{n-1}(u)\big), \quad (1\text{-}5)$$

where p_k is the degree k orthonormal polynomial with respect to the weight e^{-NV} defined by

$$\int_{\mathbb{R}} p_j(s)p_k(s)e^{-NV(s)}\,ds = \delta_{jk}, \quad j, k \in \mathbb{R},$$

and $\kappa_k > 0$ is the leading coefficient of p_k. The average counting measure of the eigenvalues has a limit as $n = N \to \infty$. We will denote this limiting mean eigenvalue distribution by μ_V. For a general polynomial external field V of degree $2m$, the support of μ_V consists of a finite union of at most m intervals [Deift et al. 1998a]. If V depends on one or more parameters, the measure μ_V will in general also vary with those parameters. Critical phenomena occur when the number of intervals in the support of μ_V changes. A decrease in the number of intervals can be caused essentially by three different events:

  (i) shrinking of an interval, which disappears ultimately;

 (ii) merging of two intervals to a single interval;

(iii) simultaneous merging of two intervals and shrinking of one of them.

Near such transitions, double scaling limits of the correlation kernel are different from the usual sine or Airy kernel. At a type (i) transition, the limiting kernel is built out of Hermite polynomials [Eynard 2006; Claeys 2008; Mo 2008; Bertola and Lee 2009], at a type (ii) transition the limiting kernel is built out of functions related to the Painlevé II equation [Bleher and Its 2003; Claeys et al. 2008], and at a type (iii) transition the limiting kernel is related to the Painlevé I hierarchy [Brézin et al. 1990; Claeys and Vanlessen 2007b]. Higher order transitions, such as the simultaneous merging and/or shrinking of more than two intervals, can also take place but will not be considered here. Rather than on the limiting kernels, we will concentrate on the asymptotic behavior of the recurrence coefficients of the orthogonal polynomials, defined by the three-term recurrence relation

$$sp_n(s) = \gamma_{n+1}p_{n+1}(s) + \beta_n p_n(s) + \gamma_n p_{n-1}(s). \quad (1\text{-}6)$$

The recurrence coefficients contain information about the orthogonal polynomials and about the partition function Z_n of the determinantal point process (1-4) [Bessis et al. 1980; Bleher and Its 2005; Ercolani and McLaughlin 2003]. The large n, N asymptotics for the recurrence coefficients show remarkable similarities with the asymptotic behavior for KdV solution $u(x, t, \epsilon)$ as $\epsilon \to 0$.

2. Phase diagram for the KdV equation

We assume throughout this section that the (ϵ-independent) initial data $u_0(x)$ for the KdV equation are real analytic in a neighborhood of the real line, negative, have a single local minimum x_M for which $u_0(x_M) = -1$, and that they decay sufficiently rapidly as $x \to \infty$ in a complex neighborhood of the real line. The neighborhood of the real line where u_0 is analytic and where the decay holds should contain a sector $\{|\arg x| < \delta\} \cup \{|\arg(-x)| < \delta\}$. In addition certain generic conditions have to be valid; we refer to [Claeys and Grava 2009] for details about those. A simple example of admissible initial data is given by $u_0(x) = -\operatorname{sech}^2 x$.

2.1. Regular asymptotics for the KdV solution. Before the time of gradient catastrophe

$$t_c = \frac{1}{\max_{\xi \in \mathbb{R}}(-6u_0'(\xi))},$$

the asymptotics for the KdV solution $u(x, t, \epsilon)$ as $\epsilon \to 0$ are given by

$$u(x, t, \epsilon) = u(x, t) + \mathcal{O}(\epsilon^2),$$

where $u(x, t)$ is the solution to the Hopf equation with initial data $u_0(x)$, that is, the implicit solution $u_0(\xi(x, t))$ defined by (1-2). The leading term of the above asymptotic expansion was obtained in [Lax and Levermore 1983a; 1983b; 1983c] while the error term was obtained only recently for a larger class of equations and initial data in [Masoero and Raimondo 2013]. Such an expansion still holds true after the time of gradient catastrophe as long as x is outside the interval where the KdV solution develops oscillations. In the oscillatory region, the oscillations for some time $t > t_c$ can be approximated as $\epsilon \to 0$, by the elliptic function

$$u(x, t, \epsilon) = \beta_1 + \beta_2 + \beta_3 + 2\alpha$$

$$+ 2\epsilon^2 \frac{\partial^2}{\partial x^2} \log \vartheta \left(\frac{\sqrt{\beta_1 - \beta_3}}{2\epsilon K(s)} [x - 2t(\beta_1 + \beta_2 + \beta_3) - q]; \tau \right) + \mathcal{O}(\epsilon). \quad (2\text{-}1)$$

Here

$$\alpha = -\beta_1 + (\beta_1 - \beta_3)\frac{E(s)}{K(s)}, \qquad \tau = i\frac{K'(s)}{K(s)}, \qquad s^2 = \frac{\beta_2 - \beta_3}{\beta_1 - \beta_3}, \qquad (2\text{-}2)$$

where $K(s)$ and $E(s)$ are the complete elliptic integrals of the first and second kind, $K'(s) = K(\sqrt{1 - s^2})$, and $\vartheta(z; \tau)$ is the Jacobi elliptic theta function. In the formula (2-1) the term $\beta_1 + \beta_2 + \beta_3 + 2\alpha$ is the weak limit of the solution $u(x, t, \epsilon)$ of KdV as $\epsilon \to 0$ and it was derived in the seminal papers [Lax and Levermore 1983a; 1983b; 1983c]. The asymptotic description of the oscillations by theta-function was obtained in [Venakides 1990]. A heuristic derivation of formula (2-1) without the phase, was first obtained in [Gurevich and Pitaevskii

1973]. The phase q in the argument of the Jacobi elliptic theta function (2-1) was derived in [Deift et al. 1997]. It depends on $\beta_1, \beta_2, \beta_3$ and on the initial data and it was observed in [Grava and Klein 2007] that q satisfies a linear over-determined system of Euler–Poisson–Darboux type derived in [Tian 1994b; Gurevich et al. 1992]. The negative numbers $\beta_1 > \beta_2 > \beta_3$ depend on x and t and solve the genus one Whitham equations [Whitham 1974]. The complete solution of the Whitham equation for the class of initial data considered, was derived in [Tian 1994a].

At later times, the KdV solution can, depending on the initial data, develop multiphase oscillations which can be described in terms of higher genus Whitham equations [Flaschka et al. 1980] and in terms of Riemann θ functions [Venakides 1990; Lax and Levermore 1983a; 1983b; 1983c; Deift et al. 1997].

The parameters $\beta_1, \beta_2, \beta_3$ can be interpreted in terms of the endpoints of the support $[0, \sqrt{\beta_3 + 1}] \cup [\sqrt{\beta_2 + 1}, \sqrt{\beta_1 + 1}]$ of the minimizer of the Lax–Levermore energy functional [Lax and Levermore 1983a; 1983b; 1983c; 1993; 1997].

A transition from the elliptic asymptotic region to the Hopf region can happen in three different ways:

(i) β_1 approaches β_2 (shrinking of an interval),

(ii) β_2 approaches β_3 (merging of two intervals),

(iii) $\beta_1, \beta_2,$ and β_3 approach each other (simultaneous shrinking and merging of intervals).

The transitions (i), (ii) and (iii) will lead to an asymptotic description of the KdV solution which is similar to the asymptotic description for the recurrence coefficients of orthogonal polynomials when the number of intervals in the support of the limiting mean eigenvalue density of random matrix ensembles changes. A transition of type (iii) takes place at the point of gradient catastrophe. In the (x, t) plane after the time of gradient catastrophe, the oscillations asymptotically develop in a V-shape region that does not depend on ϵ; see Figure 1. At the left boundary (the leading edge), a transition of type (ii) takes place, and at the right boundary (the trailing edge) we have a type (i) transition. Given t sufficiently short after the time of gradient catastrophe t_c, the leading edge $x^-(t)$ is characterized by the system of equations

$$x^-(t) = 6tu(t) + f_L(u(t)), \tag{2-3}$$

$$6t + \theta(v(t); u(t)) = 0, \tag{2-4}$$

$$\partial_v \theta(v(t); u(t)) = 0, \tag{2-5}$$

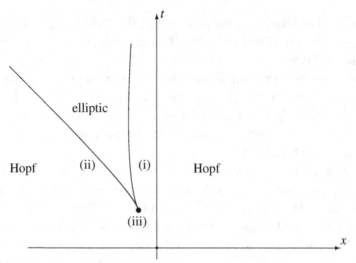

Figure 1. Sketch of the phase diagram for the equilibrium problem associated to the KdV equation. Outside the cusp-shaped region, the support of the Lax–Levermore minimizer consists of one interval; inside of two intervals. At the cusp point, we have a type (iii) transition, at the left breaking curve one of type (ii), and at the right breaking curve, one of type (i).

where $u(t) > v(t)$, $f_L(u)$ is the inverse of the decreasing part of $u_0(x)$, and θ is given by

$$\theta(\lambda; u) = \frac{1}{2\sqrt{2}} \int_{-1}^{1} \frac{f_L'\left(\frac{1+m}{2}\lambda + \frac{1-m}{2}u\right) dm}{\sqrt{1 - m}}. \tag{2-6}$$

This corresponds to the confluent case where the elliptic solution (2-1) degenerates formally to linear oscillations, namely $\beta_2 = \beta_3 = v$ and $\beta_1 = u$. The trailing edge on the other hand is characterized by

$$x^+(t) = 6tu(t) + f_L(u(t)), \tag{2-7}$$

$$6t + \theta(v(t); u(t)) = 0, \tag{2-8}$$

$$\int_{u(t)}^{v(t)} (6t + \theta(\lambda; u(t)))\sqrt{\lambda - u(t)}\, d\lambda = 0, \tag{2-9}$$

with $u(t) < v(t)$, and $\theta(\lambda; u)$ defined in (2-6). In this case we have $\beta_1 = \beta_2 = v$ and $\beta_3 = u$. In this case the solution (2-1) degenerates formally to a soliton.

2.2. *Critical asymptotics for the KdV solution.*

2.2.1. *Point of gradient catastrophe.* Near the first break-up time, the KdV solution starts developing oscillations for small ϵ. These oscillations are modeled by a Painlevé transcendent $U(X, T)$, defined as the unique real smooth solution to the fourth order ODE

$$X = TU - \left[\tfrac{1}{6}U^3 + \tfrac{1}{24}(U_X^2 + 2U\,U_{XX}) + \tfrac{1}{240}U_{XXXX}\right], \qquad (2\text{-}10)$$

with asymptotic behavior given by

$$U(X, T) = \mp(6|X|)^{1/3} \mp \tfrac{1}{3}6^{2/3}T|X|^{-1/3} + \mathcal{O}(|X|^{-1}), \quad \text{as } X \to \pm\infty, \quad (2\text{-}11)$$

for each fixed $T \in \mathbb{R}$. The existence of a pole free solution of (2-10) with asymptotic conditions (2-11) was conjectured in [Dubrovin 2006] and proved in [Claeys and Vanlessen 2007a]. Let us denote t_c for the time of gradient catastrophe, x_c for the point where the x-derivative of the Hopf solution blows up, and $u_c = u(x_c, t_c)$. We take a double scaling limit where we let $\epsilon \to 0$ and at the same time we let $x \to x_c$ and $t \to t_c$ in such a way that, for fixed $X, T \in \mathbb{R}$,

$$\lim \frac{x - x_c - 6u_c(t - t_c)}{(8k\epsilon^6)^{1/7}} = X, \quad \lim \frac{6(t - t_c)}{(4k^3\epsilon^4)^{1/7}} = T, \qquad (2\text{-}12)$$

where

$$k = -f_L'''(u_c).$$

In this double scaling limit the solution $u(x, t, \epsilon)$ of the KdV Equation (1-1) has the expansion

$$u(x, t, \epsilon) = u_c + \left(\frac{2\epsilon^2}{k^2}\right)^{1/7} U\left(\frac{x - x_c - 6u_c(t - t_c)}{(8k\epsilon^6)^{1/7}}, \frac{6(t - t_c)}{(4k^3\epsilon^4)^{1/7}}\right) + O(\epsilon^{4/7}).$$
$$(2\text{-}13)$$

The idea that the solution of KdV near the point of gradient catastrophe can be approximated by the solution of (2-10) appeared first in [Suleĭmanov 1994; 1993] and in a more general setting in [Dubrovin 2006], and was confirmed rigorously in [Claeys and Grava 2009]. In [Claeys and Grava 2012] the correction term of order $\epsilon^{4/7}$ was determined.

2.2.2. *Leading edge.* Near the leading edge, the onset of the oscillations is described by the Hastings–McLeod solution to the Painlevé II equation

$$q''(s) = sq + 2q^3(s). \qquad (2\text{-}14)$$

The Hastings–McLeod solution is characterized by the asymptotics

$$q(s) = \sqrt{-s/2}(1 + o(1)), \quad \text{as } s \to -\infty, \qquad (2\text{-}15)$$

$$q(s) = \text{Ai}(s)(1 + o(1)), \quad \text{as } s \to +\infty, \qquad (2\text{-}16)$$

where $\text{Ai}(s)$ is the Airy function. The leading edge $x^-(t)$ is, for t sufficiently short after t_c, determined by the system of equations (2-3)–(2-5). Let us consider a double scaling limit where we let $\epsilon \to 0$ and at the same time we let $x \to x^-(t)$ in such a way that

$$\lim \frac{x - x^-(t)}{\epsilon^{2/3}} = X \in \mathbb{R}, \qquad (2\text{-}17)$$

for $t > t_c$ fixed. In this double scaling limit, the solution $u(x, t, \epsilon)$ of the KdV equation with initial data u_0 has the asymptotic expansion

$$u(x, t, \epsilon) = u - \frac{4\epsilon^{1/3}}{c^{1/3}} q[s(x, t, \epsilon)] \cos \frac{\Theta(x, t)}{\epsilon} + O(\epsilon^{2/3}), \qquad (2\text{-}18)$$

where

$$\Theta(x, t) = 2\sqrt{u - v}(x - x^-) + 2 \int_v^u (f_L'(\xi) + 6t)\sqrt{\xi - v}\, d\xi, \qquad (2\text{-}19)$$

and

$$c = -\sqrt{u - v}\frac{\partial^2}{\partial v^2}\theta(v; u) > 0, \quad s(x, t, \epsilon) = -\frac{x - x^-}{c^{1/3}\sqrt{u - v}\,\epsilon^{2/3}}, \qquad (2\text{-}20)$$

with θ defined by (2-6), and q is the Hastings–McLeod solution to the Painlevé II equation. Here x^- and $v < u$ (each of them depending on t) solve the system (2-3)–(2-5). The above result was proved in [Claeys and Grava 2010a], confirming numerical results in [Grava and Klein 2007]. In [Claeys and Grava 2010a], an explicit formula for the correction term of order $\epsilon^{2/3}$ was obtained as well. We remark that a connection between leading edge asymptotics and the Painlevé II equation also appeared in [Kudashev and Suleĭmanov 1999].

2.2.3. *Trailing edge.* The trailing edge $x^+(t)$ of the oscillatory interval (i.e., the right edge of the cusp-shaped region in Figure 2) is determined by the equations (2-7)–(2-9). As $\epsilon \to 0$, we have, for fixed y and t,

$$u\left(x^+ + \frac{\epsilon \ln \epsilon}{2\sqrt{v - u}}y, t, \epsilon\right) = u + 2(v - u)\sum_{k=0}^{\infty} \operatorname{sech}^2 X_k + O(\epsilon \ln^2 \epsilon), \quad (2\text{-}21)$$

where

$$X_k = \tfrac{1}{2}\left(\tfrac{1}{2} - y + k\right)\ln \epsilon - \ln(\sqrt{2\pi} h_k) - \left(k + \tfrac{1}{2}\right)\ln \gamma,$$

$$h_k = \frac{2^{k/2}}{\pi^{1/4}\sqrt{k!}}, \quad \gamma = 4(v - u)^{5/4}\sqrt{-\partial_v \theta(v; u)}, \qquad (2\text{-}22)$$

and θ is given by (2-6) [Claeys and Grava 2010b]. It should be noted in this context that the KdV equation admits soliton solutions of the form $a \operatorname{sech}^2(bx - ct)$. This means that the last oscillations of the KdV solution resemble, at the local scale, solitons.

3. Phase diagram for unitary random matrix ensembles

3.1. Equilibrium problem. In unitary random matrix ensembles of the form (1-3), the limiting mean eigenvalue density is characterized as the equilibrium measure minimizing the logarithmic energy

$$I_V(\mu) = \iint \log \frac{1}{|s-y|} \, d\mu(s) \, d\mu(y) + \int V(s) \, d\mu(s), \qquad (3\text{-}1)$$

among all probability measures on $\mathbb{R}$. For a polynomial external field of degree $2m$, the equilibrium measure is supported on a union S_V of at most m disjoint intervals. Its density can be written in the form [Deift et al. 1998a]

$$\psi_V(s) = \prod_{j=1}^{k} \sqrt{(b_j - s)(s - a_j)} \, h(s), \quad s \in \bigcup_{j=1}^{k} [a_j, b_j], \ k \le m, \qquad (3\text{-}2)$$

where h is a polynomial of degree at most $2(m-k)$. The equilibrium measure is characterized by the variational conditions

$$2 \int \log |s-y| \, d\mu(y) - V(s) = \ell_V, \quad s \in \bigcup_{j=1}^{k} [a_j, b_j], \qquad (3\text{-}3)$$

$$2 \int \log |s-y| \, d\mu(y) - V(s) \le \ell_V, \quad s \in \mathbb{R}. \qquad (3\text{-}4)$$

The external field V is called k-cut regular if $h(s)$ in (3-2) is strictly positive on $\bigcup_{j=1}^{k} [a_j, b_j]$ and if (3-4) is strict for $s \in \mathbb{R} \setminus \bigcup_{j=1}^{k} [a_j, b_j]$. In other words, it is singular if

(i) equality in (3-4) holds at a point $s^* \in \mathbb{R} \setminus \bigcup_{j=1}^{k} [a_j, b_j]$,

(ii) $h(s^*) = 0$ with $s^* \in \bigcup_{j=1}^{k} (a_j, b_j)$,

(iii) $h(s^*) = 0$ with $s^* = a_j$ or $s^* = b_j$.

3.2. Example: quartic external field. Let us now study a two-parameter family of quartic external fields

$$V_{x,t}(s) = e^x \left[(1-t)\frac{s^2}{2} + t \left(\frac{s^4}{20} - \frac{4s^3}{15} + \frac{s^2}{5} + \frac{8}{5}s \right) \right]. \qquad (3\text{-}5)$$

For $t = 0$, we have $V_{x,0}(s) = e^x s^2/2$, which means that the random matrix ensemble is a rescaled Gaussian Unitary Ensemble. The equilibrium measure $\mu_{x,0}$ is then given by

$$d\mu_{x,0}(s) = \frac{e^x}{2\pi} \sqrt{4e^{-x} - s^2} \, ds, \quad s \in [-2e^{-x/2}, 2e^{-x/2}]. \qquad (3\text{-}6)$$

It can indeed be verified directly that this measure satisfies the variational conditions (3-3)–(3-4). For $x = 0$ and $0 < t \le 1$, one can verify that

$$d\mu_{0,t}(s) = \frac{1}{2\pi(5+\gamma^2)} \sqrt{s^2 - 4((s-2)^2+\gamma^2)}\, ds, \quad s \in [-2, 2], \gamma = \sqrt{\frac{5}{t} - 5}. \quad (3\text{-}7)$$

This shows that $V_{0,1}$ has a singular point of type (iii) at $s = 2$. On the line $t = 9$, $V_{x,9}(s)$ is symmetric around $s^* = \frac{4}{3}$. The external field is one-cut regular for $x < x^* =: -\log\frac{245}{9}$, and presumably two-cut for $x > x^*$. For $x \le x^*$, the equilibrium measure is given by

$$d\mu_{x,9}(s) = \frac{8}{\pi b^2(b^2+4C)} \sqrt{\left(s-\frac{4}{3}+b\right)\left(\frac{4}{3}+b-s\right)\left(\left(s-\frac{4}{3}\right)^2+C\right)}\, ds, \quad s \in [s^*-b, s^*+b],$$
$$(3\text{-}8)$$

where

$$b = \sqrt{\frac{140}{27} + \frac{4}{27}\sqrt{5}e^{-x}\sqrt{27e^x + 245e^{2x}}}, \quad C = \frac{e^{-x}}{36b^2}(80 - 9b^4 e^x). \quad (3\text{-}9)$$

At $x = x^*$, the equilibrium measure is given by

$$d\mu_{x^*,9}(s) = \frac{8}{\pi b^4} \sqrt{\left(s-\frac{4}{3}+b\right)\left(\frac{4}{3}+b-s\right)\left(s-\frac{4}{3}\right)^2}\, ds, \quad s \in \left[\frac{4}{3}-b, \frac{4}{3}+b\right], b = \frac{2}{3}\sqrt{35},$$

which means that there is a type (ii) singular point at $s^* = \frac{4}{3}$.

For t fixed and x sufficiently large and positive, it follows from results in [Kuijlaars and McLaughlin 2000] that the number of intervals is equal to the number of global minima of $V_{x,t}$, which is one for $t < 9$ and two for $t = 9$. For t fixed and x sufficiently large negative, one can show that the equilibrium measure is supported on a single interval. Also, for any t, when x decreases, the support of the equilibrium measure increases. This suggests that there are, as shown in Figure 2, two curves in the (x, t)-plane where $V_{x,t}$ is singular: one connecting $(0, 1)$ with $(x^*, 9)$ where a singular point of type (ii) is present, and one connecting $(0, 1)$ with $(+\infty, 9)$ where a singular point of type (i) occurs.

Remark. In [Bertola and Tovbis 2011], orthogonal polynomials with respect to complex weights of the form $e^{-nV(x)}$ were considered, with V quartic symmetric with complex-valued leading coefficient. This lead to a phase diagram which shows certain similarities with ours, but also with breaking curves of a different nature.

3.3. *Regular asymptotics.*
If V is a one-cut regular external field, the leading term of the asymptotics for the recurrence coefficients depends in a very simple way on the endpoints a and b: we have [Deift et al. 1999]

$$\gamma_n = \frac{b-a}{4} + \mathbb{O}(n^{-2}) \quad \text{as } n \to \infty \quad (3\text{-}10)$$

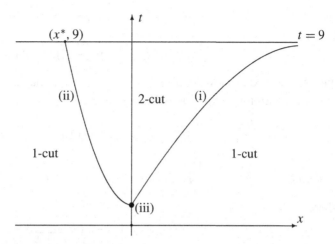

Figure 2. Sketch of the phase diagram for the equilibrium measure in external field $V_{x,t}$. The one-cut region and the two-cut region are separated by two curves, at the left curve a type (ii) singular point is present, at the right curve a type (i) singular point, and at the intersection point $(0, 1)$ there is a type (iii) singular point.

and

$$\beta_n = \frac{b+a}{2} + \mathcal{O}(n^{-2}) \quad \text{as } n \to \infty. \tag{3-11}$$

If V is a two-cut regular external field, the leading order term in the asymptotic expansion for the recurrence coefficients is still determined by the endpoints a_1, b_1, a_2, b_2, but the dependence is somewhat more complicated, and the leading term is oscillating with n. An explicit formula for the leading order asymptotics was given and proved in [Deift et al. 1999] for k-cut regular external fields V, with k arbitrary. We will not give details about those asymptotics, but we note that the expansion is of a similar nature as (2-1) in the two-cut case.

3.4. *Critical asymptotics.* We will now describe the critical asymptotics for the recurrence coefficients $\gamma_n(x, t)$ and $\beta_n(x, t)$ of the orthogonal polynomials with respect to the weight $e^{-nV_{x,t}}$. It should be noted that critical asymptotics near type (ii) and type (iii) singular points are known for more general deformations of external fields $V_{x,t}$ than only the one defined by (3-5).

3.4.1. *Singular interior points.* Assume that $V_{x^*,t^*}(s)$ is a singular external field with a singular point s^* of type (ii) (a singular interior point), and with support $[a, b]$ of the equilibrium measure. Asymptotics for the recurrence coefficients were obtained in [Bleher and Its 2003] for quartic symmetric V and in [Claeys et al. 2008] for real analytic V. Let us specialize the results to our example where $V_{x,t}$ is given by (3-5). Since V_{x^*,t^*} is quartic, this implies that ψ_{x^*,t^*} has the form

$$\psi_{x^*,t^*}(s) = C\sqrt{(s-a)(b-s)}(s-s^*)^2. \tag{3-12}$$

Then as $n \to \infty$ simultaneously with $x \to x^*$ such that $x - x^* = \mathcal{O}(n^{-2/3})$, we have the asymptotic expansions [Claeys et al. 2008]

$$\gamma_n(x,t^*) = \frac{b-a}{4} - \frac{1}{2c}q(s_{x,n})\cos(2\pi n\omega(x))n^{-1/3} + \mathcal{O}(n^{-2/3}), \tag{3-13}$$

$$\beta_n(x,t^*) = \frac{b+a}{2} + \frac{1}{c}q(s_{x,n})\sin(2\pi n\omega(x)+\theta)n^{-1/3} + \mathcal{O}(n^{-2/3}), \tag{3-14}$$

where

$$s_{x,n} = n^{2/3}(e^{x^*-x}-1)\frac{1}{c\sqrt{(s^*-a)(b-s^*)}},$$

and where c, θ, and ω are given by

$$c = \left(\frac{\pi C\sqrt{(s^*-a)(b-s^*)}}{4}\right)^{1/3}, \qquad \theta = \arcsin\frac{b+a}{b-a},$$

$$\omega(x) = \int_0^b \psi_{x^*,t^*}(s)\,ds + \mathcal{O}(n^{-2/3}) \qquad \text{as } n \to \infty.$$

An exact formula for ω can be given in terms of a modified equilibrium problem. When x approaches x^*, we observe that the recurrence coefficients develop oscillations. The envelope of the oscillations is described by the Hastings–McLeod solution q. One should compare formulas (3-13)–(3-14) with (2-13) and note that the scalings correspond after identifying ϵ with $1/n$.

3.4.2. *Singular edge points.* Asymptotics for the recurrence coefficients for general one-cut external fields V with a singular endpoint were obtained in [Claeys and Vanlessen 2007b]. Let V_0 be an external field such that the equilibrium measure is supported on $[a, b]$ and such that the density ψ_0 behaves like $\psi_0(s) \sim c(b-s)^{5/2}$ as $s \to b$, $c \neq 0$. Double scaling asymptotics were obtained for external fields of the form $V_0 + SV_1 + TV_2$ with real $S, T \to 0$, where V_1 is arbitrary and V_2 satisfies the condition

$$\int_a^b \sqrt{\frac{s-a}{b-s}}V_2'(s)\,ds = 0.$$

We can write $V_{x,t}$ in the form

$$V_{x,t}(s) = V_{0,1}(s) + (e^x - 1)V_{0,1}(s) + e^x(t-1)(V_{0,1}(s) - V_{0,0}(s)). \tag{3-15}$$

Since

$$\int_{-2}^2 \sqrt{\frac{s+2}{2-s}}(V_{0,1}'(s) - V_{0,0}'(s))\,ds = 0,$$

we can apply the results of [Claeys and Vanlessen 2007b]. In the double scaling limit where $n \to \infty$ and simultaneously $x \to 0$, $t \to 1$ in such a way that $\lim n^{6/7}(e^x - 1)$ and $\lim n^{4/7} e^x (t - 1)$ exist, we have

$$\gamma_n(x, t) = 1 + \frac{1}{2c} U\big(c_1 n^{6/7}(e^x - 1), c_2 n^{4/7} e^x (t-1)\big) n^{-2/7} + \mathbb{O}(n^{-4/7}), \quad (3\text{-}16)$$

$$\beta_n(x, t) = \frac{1}{c} U\big(c_1 n^{6/7}(e^x - 1), c_2 n^{4/7} e^x (t-1)\big) n^{-2/7} + \mathbb{O}(n^{-4/7}). \quad (3\text{-}17)$$

The constants c, c_1, c_2 are given by

$$c = 6^{2/7} > 0, \quad c_1 = \frac{1}{2\pi c^{1/2}} \int_{-2}^{2} \sqrt{\frac{u-2}{2-u}} V'_{0,1}(u)\, du = 6^{-1/7},$$

$$c_2 = \frac{1}{4\pi i c^{3/2}} \int_{\gamma} \sqrt{\frac{2+u}{(2-u)^3}} (V'_{0,1}(u) - V'_{0,0}(u))\, du = 2 \cdot 6^{-3/7},$$

where γ is a counterclockwise oriented contour encircling $[-2, 2]$.

Remark. Applying the results from [Claeys and Vanlessen 2007b] directly, one has an error term $\mathbb{O}(n^{-3/7})$ in (3-16) and (3-17), but going through the calculations, it can be verified that the error term is actually $\mathbb{O}(n^{-4/7})$. The analogy between (3-16)–(3-17) and (2-13) is obvious.

3.4.3. *Singular exterior points.* Asymptotics for the recurrence coefficients in the vicinity of a singular exterior point have not appeared in the literature to the best of our knowledge. Asymptotics for orthogonal polynomials associated to an external field V with a singular exterior point and for the correlation kernel (1-5) have been studied in [Claeys 2008; Bertola and Lee 2009; Mo 2008] using the Riemann–Hilbert approach. We are convinced that the same analysis can be used, with some additional effort, to compute asymptotics for the recurrence coefficients. If V_{x^*, t^*} is an external field with a singular exterior point, the analogy with the KdV asymptotics suggests asymptotic expansions of the form

$$\gamma_n\left(x^* - y \frac{\ln n}{c_0 n}, t\right) = \frac{b(x^*, t) - a(x^*, t)}{4} + c_1 \sum_{k=0}^{\infty} \operatorname{sech}^2 X_k + \mathbb{O}(n^{-1} \ln^2 n) \tag{3-18}$$

$$\beta_n\left(x^* - y \frac{\ln n}{c_0 n}, t\right) = \frac{b(x^*, t) + a(x^*, t)}{2} + c_1 \sum_{k=0}^{\infty} \operatorname{sech}^2 X_k + \mathbb{O}(n^{-1} \ln^2 n), \tag{3-19}$$

as $n \to \infty$, where

$$X_k = -c_2(y, k) \ln n + c_3(k).$$

4. The problem of matching

Asymptotic expansions for KdV solutions are known in the regular regions and in critical regions, but we do not have uniform asymptotics for $u(x, t, \epsilon)$ in x and t. Indeed, the critical asymptotics are only valid in shrinking neighborhoods of the breaking curves: a neighborhood of size $\mathcal{O}(\epsilon^{2/3})$ near the leading edge, a neighborhood of size $\mathcal{O}(\epsilon \ln \epsilon)$ near the trailing edge, and a neighborhood of size $\mathcal{O}(\epsilon^{4/7})$ at the point of gradient catastrophe. On the other hand, the regular asymptotics are only proved to hold uniformly for x and t at a fixed distance away from the breaking curves. However one can see easily that (2-13) and (2-21) match formally with the regular asymptotics for x close to the breaking curves but outside the cusp-shaped region. Indeed for (2-13) this follows from the decay of the Hastings–McLeod solution q at $+\infty$. When x is close to the boundary but inside the cusp-shaped region, the situation is more complicated. One can hope that the regular asymptotics can be improved in such a way that they hold also when x, t approach a breaking curve sufficiently slowly when ϵ tends to 0, and that the critical asymptotics can be improved to hold in a slightly bigger neighborhood of the breaking curves. It would be of interest to see if such an approach could provide uniform asymptotics for the KdV solution as $\epsilon \to 0$.

The problem of obtaining uniform asymptotics in x and t for the recurrence coefficients $\gamma_n(x, t)$ and $\beta_n(x, t)$ may seem an artificial one at first sight, since one is often interested in a random matrix with a fixed external field V instead of letting V vary. However, it becomes more relevant when studying the partition function

$$Z_n = \int_{\mathbb{R}^n} \prod_{i<j} (\lambda_i - \lambda_j)^2 \prod_{i=1}^{n} e^{-nV(\lambda_i)} \, d\lambda_i.$$

It is well-known that

$$Z_n = n! \prod_{j=1}^{n-1} \kappa_j^{-2},$$

where κ_j is the leading coefficient of the normalized orthogonal polynomial p_j with respect to the weight e^{-nV}. A consequence of this formula is that, if one lets V vary with a parameter τ in a convenient way, it is possible to derive various identities for τ-derivatives of $\ln Z_n$ in terms of the recurrence coefficients $\gamma_k(\tau)$ and $\beta_k(\tau)$ for k large [Bleher and Its 2005; Ercolani and McLaughlin 2003]. A possible strategy to obtain asymptotics for the partition function, is to let the τ-dependence be such that V interpolates between the Gaussian $V(z; \tau_0) = \frac{z^2}{2}$ and $V(z; \tau_1) = V(z)$. Integrating the differential identity then requires asymptotics for the Gaussian partition function (which are known) and uniform asymptotics for the recurrence coefficients $\gamma_n(\tau)$ and $\beta_n(\tau)$ over the whole range $[\tau_0, \tau_1]$.

Depending on the chosen deformation, this could require uniform asymptotics for the recurrence coefficients near a singular point of type (i), (ii), or (iii). The results presented in the previous section do not provide sufficiently detailed asymptotics for the recurrence coefficients: they are not uniform near the breaking curves. For example near a critical point of type (ii), formulas (3-13)–(3-14) are only valid for $x - x^* = \mathbb{O}(n^{-2/3})$ as $n \to \infty$, whereas the asymptotic formula in the two-cut region is valid only at a fixed distance away from a critical point.

5. The Toda lattice and KdV

It is well-known that recurrence coefficients for orthogonal polynomials follow the time flows of the Toda hierarchy. In this section, following [Dubrovin 2009; Dubrovin et al. 2013] we will formally derive the KdV equation as a scaling limit of the continuum limit of the Toda lattice. This gives a heuristic argument why asymptotics for KdV and the recurrence coefficients show similarities.

The Toda lattice is a Hamiltonian system described by the equations

$$\frac{du_n}{dt} = v_n - v_{n-1}, \qquad \frac{dv_n}{dt} = e^{u_{n+1}} - e^{u_n}, \qquad n \in \mathbb{Z}. \tag{5-1}$$

The Toda lattice is a prototypical example of a completely integrable system [Flaschka 1974]. Let

$$V(\xi) = V_0(\xi) + \sum_{j=1}^{2d} t_j \xi^j, \qquad t_{2d} > 0, \tag{5-2}$$

where $V_0(\xi)$ is a fixed polynomial of even degree with positive leading coefficient, and let p_j be the orthogonal polynomials defined by

$$\int_{-\infty}^{\infty} p_n(\xi) p_m(\xi) e^{-\frac{1}{\epsilon} V(\xi)} \, d\xi = \delta_{nm}, \tag{5-3}$$

where $\epsilon = \frac{1}{N}$ is a small positive parameter. As mentioned before, the polynomials $p_n(\xi)$ satisfy a three term recurrence relation of the form (1-6).

The recurrence coefficients γ_n and β_n in (1-6) evolve with respect to the times t_k defined in (5-2) according to the equations [Eynard 2001; Douglas and Shenker 1990; Fokas et al. 1992; Bertola et al. 2003]

$$\epsilon \frac{\partial \gamma_n}{\partial t_k} = \frac{\gamma_n}{2} \left([Q^k]_{n-1,n-1} - [Q^k]_{nn} \right), \tag{5-4}$$

$$\epsilon \frac{\partial \beta_n}{\partial t_k} = \gamma_n [Q^k]_{n,n-1} - \gamma_{n+1} [Q^k]_{n+1,n}, \tag{5-5}$$

where $[Q^k]_{n,m}$ denotes the n, m-th element of the matrix Q^k and Q is the tridiagonal matrix

$$Q = \begin{pmatrix} \beta_0 & \gamma_1 & 0 & 0 & 0 & \cdots \\ \gamma_1 & \beta_1 & \gamma_2 & 0 & 0 & \cdots \\ 0 & \gamma_2 & \beta_2 & \gamma_3 & 0 & \cdots \\ 0 & 0 & \gamma_3 & \beta_3 & \gamma_4 & \cdots \\ 0 & 0 & 0 & \gamma_4 & \beta_4 & \cdots \\ \vdots & \vdots & \vdots & \vdots & \vdots & \ddots \end{pmatrix}. \tag{5-6}$$

The equations (5-4)–(5-5) are the Toda lattice hierarchy in the Flaschka variables [Flaschka 1974]. In particular the first flow of the hierarchy takes the form

$$\epsilon \frac{\partial \gamma_n}{\partial t_1} = \frac{\gamma_n}{2}(\beta_{n-1} - \beta_n),$$

$$\epsilon \frac{\partial \beta_n}{\partial t_1} = \gamma_n^2 - \gamma_{n+1}^2. \tag{5-7}$$

These equations correspond to the Toda lattice (5-1) by identifying $t_1 = t$, $\beta_n = -v_n$ and

$$u_n = \log \gamma_n^2. \tag{5-8}$$

In addition to the Toda equations, the recurrence coefficients for the orthogonal polynomials satisfy a constraint that is given by the discrete string equation which takes the form [Fokas et al. 1992]

$$\gamma_n [V'(Q)]_{n,n-1} = n\epsilon, \quad [V'(Q)]_{n,n} = 0. \tag{5-9}$$

For example, choosing $V_0(\xi) = \frac{1}{2}\xi^2$ one obtains

$$\beta_n(t=0) = 0, \quad \gamma_n^2(t=0) = n\epsilon, \quad t = (t_1, t_2, \ldots, t_{2d}). \tag{5-10}$$

To obtain the continuum limit of the Toda lattice, let us assume that $u(x)$ and $v(x)$ are smooth functions that interpolate the sequences u_n, v_n in the following way: $u(\epsilon n) = u_n$ and $v(\epsilon n) = v_n$ for some small $\epsilon > 0$, $n > 0$, $x = \epsilon n$. Then the Toda lattice (5-1) reduces to an evolutionary PDE of the form [Eguchi and Yang 1994; Deift and McLaughlin 1998]

$$u_t = \frac{1}{\epsilon}[v(x) - v(x - \epsilon)] = v_x - \frac{1}{2}\epsilon v_{xx} + O(\epsilon^2),$$

$$v_t = \frac{1}{\epsilon}[e^{u(x+\epsilon)} - e^{u(x)}] = e^u u_x + \frac{1}{2}\epsilon(e^u)_{xx} + O(\epsilon^2). \tag{5-11}$$

In order to write the continuum limit of the Toda lattice in a canonical Hamiltonian form, following [Dubrovin and Zhang 2001], we introduce $w(x)$ by

$$w(x) = \epsilon \partial_x [1 - e^{-\epsilon \partial_x}]^{-1} u(x) = u + \frac{\epsilon}{2} u_x + \frac{\epsilon^2}{12} u_{xx} + \cdots \qquad (5\text{-}12)$$

In the coordinates v, w the continuum limit of the Toda lattice equations takes the form

$$w_t = v_x,$$
$$v_t = e^w \left(w_x + \frac{\epsilon^2}{24} (2w_{xxx} + 4w_x w_{xx} + w_x^3) \right) + O(\epsilon^4), \qquad (5\text{-}13)$$

with the corresponding Hamiltonian given by

$$H = \int \left(\frac{v^2}{2} + e^w - \frac{\epsilon^2}{24} e^w w_x^2 + \cdots \right) dx$$

and Poisson bracket $\{v(x), w(y)\} = \delta'(x - y)$ where $\delta(x)$ is the Dirac δ function. We remark that in these coordinates the continuum limit of the Toda equation contains only even terms in ϵ. For $\epsilon = 0$, (5-13) reduces to

$$w_t = v_x, \qquad v_t = e^w w_x. \qquad (5\text{-}14)$$

The solution of equations (5-14) can be obtained by the method of characteristics. The initial data relevant to us should satisfy the continuum limit of the string equation (5-9) for $t = 0$. The Riemann invariants of (5-14) are

$$r_\pm = v \pm 2e^{w/2},$$

so that (5-14) takes the form

$$\frac{\partial}{\partial t} r_\pm + \lambda_\pm \frac{\partial}{\partial x} r_\pm = 0, \qquad \lambda_\pm = \mp e^{w/2} = \mp \frac{r_+ - r_-}{4}.$$

The generic solution of (5-14) can be written in the form [Tsarëv 1990; Whitham 1974]

$$x = \lambda_\pm t + f_\pm(r_+, r_-), \qquad (5\text{-}15)$$

where $f_\pm(r_+, r_-)$ are two functions that satisfy the equations [Tsarëv 1990]

$$\frac{\partial}{\partial r_-} f_+ = \frac{\partial \lambda_+}{\partial r_-} \frac{f_+ - f_-}{\lambda_+ - \lambda_-} = -\frac{f_+ - f_-}{2(r_+ - r_-)} = \frac{\partial}{\partial r_+} f_-. \qquad (5\text{-}16)$$

From (5-16) one can conclude that there exists a function $f = f(r_+, r_-)$ such that

$$f_\pm = \frac{\partial f}{\partial r_\pm}.$$

The explicit dependence of f for a certain class of initial data can be found in [Deift and McLaughlin 1998]. To obtain f in the random matrix case we impose that the equations (5-15) are consistent with the continuum limit of the discrete string equation (5-9) for $t_1 = t \geq 0$ and $t_j = 0$ for $j > 1$. At the leading order in ϵ the string equation (5-9) in the Riemann invariants $r_\pm = -\beta \pm 2\gamma$ gives after straightforward but long calculations, the following expression for the function $f(r_+, r_-)$:

$$f(r_+, r_-) = -\text{Res}_{\xi=\infty}\left[V_0'(\xi)\sqrt{(\xi - r_+)(\xi - r_-)}\, d\xi \right]. \qquad (5\text{-}17)$$

Remark. The equations (5-15) with f given in (5-17), coincide with the equations that define the support of the equilibrium measure for the variational problem

$$\inf_{\int_{\mathbb{R}} d\nu(\xi)=1}\left[\int_{\mathbb{R}}\int_{\mathbb{R}} \log\frac{1}{|\xi - \eta|}\, d\nu(\xi)d\nu(\eta) + \frac{1}{x}\int_{\mathbb{R}} V(\xi)d\nu(\xi) \right]$$

in the case where the equilibrium measure is supported on one interval. The Riemann invariants r_+ and r_- can thus be interpreted as the end-points of the support of the equilibrium measure.

In what follows, we are going to show that the solution of the Equation (5-13) in the vicinity of a singular point of type (iii) reduces to the KdV equation, in agreement with [Dubrovin 2009]. First we consider the solution of the hodograph Equation (5-15) near a singular point of type (iii); namely, let (x_c, t_c) be a point of gradient catastrophe for the Riemann invariant r_+, which means that $\partial_x r_+$ goes to infinity at the critical point (x_c, t_c). We define $r_\pm(x_c, t_c) = r_\pm^c$. Such a critical point is characterized by the conditions

$$\lambda_{+,+}^c t_c + f_{+,+}^c = 0, \quad \lambda_{+,++}^c t_c + f_{+,++}^c = 0,$$

and the critical point is generic if

$$\lambda_{+,+++}^c t_c + f_{+,+++}^c \neq 0, \quad \lambda_{-,-}^c t_c + f_{-,-}^c \neq 0,$$

where we have used the notation $\lambda_{-,-}^c = \dfrac{\partial}{\partial r_-}\lambda_-(r_+ = r_+^c, r_- = r_-^c)$ and consistently for the other terms.

Expanding in power series (5-15) near (x_c, t_c) and using (5-16) after the rescalings

$$\begin{aligned}
&x_- = k^{-2/3}(x - x_c - \lambda_-^c(t - t_c)), \quad x_+ = k^{-1}(x - x_c - \lambda_+^c(t - t_c)) \\
&\bar{r}_- = k^{-2/3}(r_- - r_-^c), \qquad\qquad\qquad \bar{r}_+ = k^{-1/3}(r_+ - r_+^c),
\end{aligned} \qquad (5\text{-}18)$$

one obtains, letting $k \to 0$,

$$x_- = c_1 \bar{r}_-, \quad x_+ = c_2 x_- \bar{r}_+ + c_3 \bar{r}_+^3, \tag{5-19}$$

where

$$c_1 = (f_{-,-}^c + \lambda_{-,-}^c t_c), \quad c_2 = \frac{\lambda_{+,+}^c}{\lambda_+^c - \lambda_-^c}, \quad c_3 = \frac{1}{6}(\lambda_{+,+++}^c t_c + f_{+,+++}^c). \tag{5-20}$$

We observe that (5-19) describes a Withney singularity in the neighborhood of $(0, 0)$ [Dubrovin 2009]. Performing the same rescalings (5-18) to the equations (5-14) and letting $k \to 0$ one obtains

$$\frac{\partial \bar{r}_-}{\partial x_+} = 0, \quad \frac{\partial \bar{r}_+}{\partial x_-} + c_2 \bar{r}_+ \frac{\partial \bar{r}_+}{\partial x_+} = 0,$$

with c_2 as in (5-20). Clearly (5-19) represents a solution of the above equations with singularity in $(x_+ = 0, x_- = 0)$ and at $\bar{r}_\pm = 0$. The next step is to perform the rescaling (5-18) to (5-13) and letting $\epsilon \to k^{7/6} \epsilon$. One obtains, in the limit $k \to 0$,

$$\bar{r}_- = \frac{x_-}{c_1} + c_4 \epsilon^2 \frac{\partial^2}{\partial x_+^2} \bar{r}_+, \quad c_4 = \frac{r_+^c - r_-^c}{192(\lambda_-^c - \lambda_+^c)} = \frac{1}{96},$$

$$\frac{\partial \bar{r}_+}{\partial x_-} + c_2 \bar{r}_+ \frac{\partial \bar{r}_+}{\partial x_+} + c_4 \epsilon^2 \frac{\partial^3}{\partial x_+^3} \bar{r}_+ = 0. \tag{5-21}$$

The first of the above equations has been obtained after integration with respect to x_+ using (5-19). The second one is the KdV equation for r_+ with time variable x_- and space variable x_+. Such derivation has been obtained in a more general setting in [Dubrovin et al. 2013]. On the formal level, the above calculations explain why the asymptotic behavior of the solution of the continuum limit of Toda lattice and in particular of the recurrence coefficients of orthogonal polynomials near the point of gradient catastrophe is of a similar nature as the KdV case. However, a rigorous proof of the generic behavior of the solution of the continuum limit of Toda lattice near the point of gradient catastrophe cannot be derived from the KdV case but a separate proof is needed.

Acknowledgements

The authors acknowledge support by ERC Advanced Grant FroMPDE. Claeys was also supported by FNRS, by the Belgian Interuniversity Attraction Pole P06/02, P07/18 and by the European Research Council under the European Union's Seventh Framework Programme (FP/2007/2013)/ERC Grant Agreement no. 307074. Grava also acknowledges Italian PRIN Research Project Geometric Methods in the Theory of Nonlinear Waves and their Applications.

References

[Bertola and Lee 2009] M. Bertola and S. Y. Lee, "First colonization of a spectral outpost in random matrix theory", *Constr. Approx.* **30**:2 (2009), 225–263.

[Bertola and Tovbis 2011] M. Bertola and A. Tovbis, "Asymptotics of orthogonal polynomials with complex varying quartic weight: global structure, critical point behaviour and the first Painlevé equation", 2011. arXiv 1108.0321

[Bertola et al. 2003] M. Bertola, B. Eynard, and J. Harnad, "Partition functions for matrix models and isomonodromic tau functions", *J. Phys. A* **36**:12 (2003), 3067–3083.

[Bessis et al. 1980] D. Bessis, C. Itzykson, and J. B. Zuber, "Quantum field theory techniques in graphical enumeration", *Adv. in Appl. Math.* **1**:2 (1980), 109–157.

[Bleher and Its 2003] P. Bleher and A. Its, "Double scaling limit in the random matrix model: the Riemann–Hilbert approach", *Comm. Pure Appl. Math.* **56**:4 (2003), 433–516.

[Bleher and Its 2005] P. M. Bleher and A. R. Its, "Asymptotics of the partition function of a random matrix model", *Ann. Inst. Fourier (Grenoble)* **55**:6 (2005), 1943–2000.

[Brézin et al. 1990] É. Brézin, E. Marinari, and G. Parisi, "A nonperturbative ambiguity free solution of a string model", *Phys. Lett. B* **242**:1 (1990), 35–38.

[Claeys 2008] T. Claeys, "Birth of a cut in unitary random matrix ensembles", *Int. Math. Res. Not.* **2008**:6 (2008), Art. ID rnm166, 40.

[Claeys and Grava 2009] T. Claeys and T. Grava, "Universality of the break-up profile for the KdV equation in the small dispersion limit using the Riemann–Hilbert approach", *Comm. Math. Phys.* **286**:3 (2009), 979–1009.

[Claeys and Grava 2010a] T. Claeys and T. Grava, "Painlevé II asymptotics near the leading edge of the oscillatory zone for the Korteweg–de Vries equation in the small-dispersion limit", *Comm. Pure Appl. Math.* **63**:2 (2010), 203–232.

[Claeys and Grava 2010b] T. Claeys and T. Grava, "Solitonic asymptotics for the Korteweg–de Vries equation in the small dispersion limit", *SIAM J. Math. Anal.* **42**:5 (2010), 2132–2154.

[Claeys and Grava 2012] T. Claeys and T. Grava, "The KdV hierarchy: universality and a Painlevé transcendent", *Int. Math. Res. Not.* **2012**:22 (2012), 5063–5099.

[Claeys and Vanlessen 2007a] T. Claeys and M. Vanlessen, "The existence of a real pole-free solution of the fourth order analogue of the Painlevé I equation", *Nonlinearity* **20**:5 (2007), 1163–1184.

[Claeys and Vanlessen 2007b] T. Claeys and M. Vanlessen, "Universality of a double scaling limit near singular edge points in random matrix models", *Comm. Math. Phys.* **273**:2 (2007), 499–532.

[Claeys et al. 2008] T. Claeys, A. B. J. Kuijlaars, and M. Vanlessen, "Multi-critical unitary random matrix ensembles and the general Painlevé II equation", *Ann. of Math.* (2) **168**:2 (2008), 601–641.

[Deift and McLaughlin 1998] P. Deift and K. T.-R. McLaughlin, *A continuum limit of the Toda lattice*, vol. 131, Memoirs of the American Mathematical Society **624**, 1998.

[Deift et al. 1997] P. Deift, S. Venakides, and X. Zhou, "New results in small dispersion KdV by an extension of the steepest descent method for Riemann–Hilbert problems", *Internat. Math. Res. Notices* 6 (1997), 286–299.

[Deift et al. 1998a] P. Deift, T. Kriecherbauer, and K. T.-R. McLaughlin, "New results on the equilibrium measure for logarithmic potentials in the presence of an external field", *J. Approx. Theory* **95**:3 (1998), 388–475.

[Deift et al. 1998b] P. Deift, T. Kriecherbauer, K. T.-R. McLaughlin, S. Venakides, and X. Zhou, "Uniform asymptotics for orthogonal polynomials", pp. 491–501 in *Proceedings of the International Congress of Mathematicians* (Berlin, 1998), vol. III, 1998.

[Deift et al. 1999] P. Deift, T. Kriecherbauer, K. T.-R. McLaughlin, S. Venakides, and X. Zhou, "Uniform asymptotics for polynomials orthogonal with respect to varying exponential weights and applications to universality questions in random matrix theory", *Comm. Pure Appl. Math.* **52**:11 (1999), 1335–1425.

[Douglas and Shenker 1990] M. R. Douglas and S. H. Shenker, "Strings in less than one dimension", *Nuclear Phys. B* **335**:3 (1990), 635–654.

[Dubrovin 2006] B. Dubrovin, "On Hamiltonian perturbations of hyperbolic systems of conservation laws, II: universality of critical behaviour", *Comm. Math. Phys.* **267**:1 (2006), 117–139.

[Dubrovin 2009] B. Dubrovin, "Hamiltonian perturbations of hyperbolic PDEs: from classification results to the properties of solutions", pp. 231–276 in *New trends in mathematical physics: selected contributions of the XVth international congress on mathematical physics*, edited by V. Sidoravicius, Springer, Houten, 2009.

[Dubrovin and Zhang 2001] B. Dubrovin and Y. Zhang, "Normal forms of hierarchies of integrable PDEs, Frobenius manifolds and Gromov–Witten invariants", 2001. arXiv math/0108160

[Dubrovin et al. 2009] B. Dubrovin, T. Grava, and C. Klein, "On universality of critical behavior in the focusing nonlinear Schrödinger equation, elliptic umbilic catastrophe and the tritronquée solution to the Painlevé-I equation", *J. Nonlinear Sci.* **19**:1 (2009), 57–94.

[Dubrovin et al. 2013] B. Dubrovin, T. Grava, C. Klein, and A. Moro, "On critical behaviour in systems of Hamiltonian partial differential equations", 2013. arXiv 1311.7166

[Eguchi and Yang 1994] T. Eguchi and S.-K. Yang, "The topological CP^1 model and the large-N matrix integral", *Modern Phys. Lett. A* **9**:31 (1994), 2893–2902.

[Ercolani and McLaughlin 2003] N. M. Ercolani and K. D. T.-R. McLaughlin, "Asymptotics of the partition function for random matrices via Riemann–Hilbert techniques and applications to graphical enumeration", *Int. Math. Res. Not.* **2003**:14 (2003), 755–820.

[Eynard 2001] B. Eynard, "A concise expression for the ODE's of orthogonal polynomials", 2001. arXiv math-ph/0109018

[Eynard 2006] B. Eynard, "Universal distribution of random matrix eigenvalues near the 'birth of a cut' transition", *J. Stat. Mech. Theory Exp.* **7** (2006), P07005.

[Flaschka 1974] H. Flaschka, "The Toda lattice, I: existence of integrals", *Phys. Rev. B* (3) **9** (1974), 1924–1925.

[Flaschka et al. 1980] H. Flaschka, M. G. Forest, and D. W. McLaughlin, "Multiphase averaging and the inverse spectral solution of the Korteweg-de Vries equation", *Comm. Pure Appl. Math.* **33**:6 (1980), 739–784.

[Fokas et al. 1992] A. S. Fokas, A. R. It·s, and A. V. Kitaev, "The isomonodromy approach to matrix models in 2D quantum gravity", *Comm. Math. Phys.* **147**:2 (1992), 395–430.

[Grava 2004] T. Grava, "Whitham equations, Bergmann kernel and Lax–Levermore minimizer", *Acta Appl. Math.* **82**:1 (2004), 1–86.

[Grava and Klein 2007] T. Grava and C. Klein, "Numerical solution of the small dispersion limit of Korteweg-de Vries and Whitham equations", *Comm. Pure Appl. Math.* **60**:11 (2007), 1623–1664.

[Gurevich and Pitaevskii 1973] A. V. Gurevich and L. P. Pitaevskii, "Non stationary structure of a collisionless shock waves", *JEPT Letters* **17** (1973), 193–195.

[Gurevich et al. 1992] A. V. Gurevich, A. L. Krylov, N. G. Mazur, and G. A. Èl, "Evolution of a localized perturbation in Korteweg-de Vries hydrodynamics", *Dokl. Akad. Nauk* **323**:5 (1992), 876–879. In Russian; translated in *Soviet Phys. Dokl.* **37**:4 (1992), 198–200.

[Kudashev and Suleĭmanov 1999] V. R. Kudashev and B. I. Suleĭmanov, "Small-amplitude dispersion oscillations on the background of the nonlinear geometric optics approximation",

Teoret. Mat. Fiz. **118**:3 (1999), 413–422. In Russian; translated in *Theoret. and Math. Phys.* **118**:3 (1999), 325–332.

[Kuijlaars and McLaughlin 2000] A. B. J. Kuijlaars and K. T.-R. McLaughlin, "Generic behavior of the density of states in random matrix theory and equilibrium problems in the presence of real analytic external fields", *Comm. Pure Appl. Math.* **53**:6 (2000), 736–785.

[Lax and Levermore 1983a] P. D. Lax and C. D. Levermore, "The small dispersion limit of the Korteweg-de Vries equation, I", *Comm. Pure Appl. Math.* **36**:5 (1983), 253–290.

[Lax and Levermore 1983b] P. D. Lax and C. D. Levermore, "The small dispersion limit of the Korteweg-de Vries equation, II", *Comm. Pure Appl. Math.* **36**:5 (1983), 571–593.

[Lax and Levermore 1983c] P. D. Lax and C. D. Levermore, "The small dispersion limit of the Korteweg-de Vries equation, III", *Comm. Pure Appl. Math.* **36**:5 (1983), 809–830.

[Lax et al. 1993] P. D. Lax, C. D. Levermore, and S. Venakides, "The generation and propagation of oscillations in dispersive initial value problems and their limiting behavior", pp. 205–241 in *Important developments in soliton theory*, edited by A. S. Fokas and V. E. Zakharov, Springer, Berlin, 1993.

[Masoero and Raimondo 2013] D. Masoero and A. Raimondo, "Semiclassical limit for generalized KdV equations before the gradient catastrophe", *Lett. Math. Phys.* **103**:5 (2013), 559–583.

[Mo 2008] M. Y. Mo, "The Riemann–Hilbert approach to double scaling limit of random matrix eigenvalues near the "birth of a cut" transition", *Int. Math. Res. Not.* **2008**:13 (2008), Art. ID rnn042, 51.

[Suleĭmanov 1993] B. I. Suleĭmanov, "Solution of the Korteweg-de Vries equation which arises near the breaking point in problems with a slight dispersion", *Pis'ma Zh. Èksper. Teoret. Fiz.* **58**:11 (1993), 906–910. In Russian; translated in *JETP Lett.* **58**:11 (1993), 849–854.

[Suleĭmanov 1994] B. I. Suleĭmanov, "Onset of nondissipative shock waves and the "nonperturbative" quantum theory of gravitation", *Zh. Èksper. Teoret. Fiz.* **105**:5 (1994), 1089–1097. In Russian; translated in *J. Experiment. Theoret. Phys.* **78**:5 (1994), 583–587.

[Tian 1994a] F. R. Tian, "The initial value problem for the Whitham averaged system", *Comm. Math. Phys.* **166**:1 (1994), 79–115.

[Tian 1994b] F. R. Tian, "The Whitham-type equations and linear overdetermined systems of Euler–Poisson–Darboux type", *Duke Math. J.* **74**:1 (1994), 203–221.

[Tsarëv 1990] S. P. Tsarëv, "The geometry of Hamiltonian systems of hydrodynamic type: the generalized hodograph method", *Izv. Akad. Nauk SSSR Ser. Mat.* **54**:5 (1990), 1048–1068. In Russian; translated in *Math. USSR Izv.* **37** (1991), 397–419.

[Venakides 1990] S. Venakides, "The Korteweg–de Vries equation with small dispersion: higher order Lax–Levermore theory", *Comm. Pure Appl. Math.* **43**:3 (1990), 335–361.

[Whitham 1974] G. B. Whitham, *Linear and nonlinear waves*, Wiley, New York, 1974.

tom.claeys@uclouvain.be   *Institut de Recherche en Mathématique et Physique,*
                                 Université Catholique de Louvain, Chemin du Cyclotron 2,
                                 B-1348 Louvain-La-Neuve, Belgium

tg13161@bristol.ac.uk   *Department of Mathematics, University of Bristol,*
                                 University Walk, Clifton, Bristol BS8 1TW, United Kingdom

= grava@sissa.it   *Mathematics Area, Scuola Internazionale Superiore di Studi*
                                 Avanzati (SISSA), via Bonomea 265, 34100 Trieste, Italy

Random Matrices
MSRI Publications
Volume **65**, 2014

On the asymptotics of a Toeplitz determinant with singularities

PERCY DEIFT, ALEXANDER ITS AND IGOR KRASOVSKY

We provide an alternative proof of the classical single-term asymptotics for Toeplitz determinants whose symbols possess Fisher–Hartwig singularities. We also relax the smoothness conditions on the regular part of the symbols and obtain an estimate for the error term in the asymptotics. Our proof is based on the Riemann–Hilbert analysis of the related systems of orthogonal polynomials and on differential identities for Toeplitz determinants. The result discussed in this paper is crucial for the proof of the asymptotics in the general case of Fisher–Hartwig's singularities and extensions to Hankel and Toeplitz+Hankel determinants.

1. Introduction

Let $f(z)$ be a complex-valued function integrable over the unit circle. Denote its Fourier coefficients

$$f_j = \frac{1}{2\pi} \int_0^{2\pi} f(e^{i\theta}) e^{-ij\theta} d\theta, \quad j = 0, \pm 1, \pm 2, \ldots$$

We are interested in the n-dimensional Toeplitz determinant with symbol $f(z)$,

$$D_n(f(z)) = \det(f_{j-k})_{j,k=0}^{n-1}, \quad n \geq 1, \tag{1-1}$$

where $f(e^{i\theta})$ has a fixed number of Fisher–Hartwig singularities [Fisher and Hartwig 1968; Lenard 1964; 1972], i.e., f has the following form on the unit circle:

$$f(z) = e^{V(z)} z^{\sum_{j=0}^m \beta_j} \prod_{j=0}^m |z - z_j|^{2\alpha_j} g_{z_j, \beta_j}(z) z_j^{-\beta_j},$$

$$z = e^{i\theta}, \quad \theta \in [0, 2\pi), \tag{1-2}$$

for some $m = 0, 1, \ldots$, where

$$z_j = e^{i\theta_j}, \quad j = 0, \ldots, m, \quad 0 = \theta_0 < \theta_1 < \cdots < \theta_m < 2\pi, \tag{1-3}$$

$$g_{z_j, \beta_j}(z) := g_{\beta_j}(z) = \begin{cases} e^{i\pi\beta_j} & \text{if } 0 \leq \arg z < \theta_j, \\ e^{-i\pi\beta_j} & \text{if } \theta_j \leq \arg z < 2\pi, \end{cases} \tag{1-4}$$

$$\Re\alpha_j > -\tfrac{1}{2}, \quad \beta_j \in \mathbb{C}, \quad j = 0, \ldots, m, \tag{1-5}$$

and $V(e^{i\theta})$ is a sufficiently smooth function on the unit circle (see below). The condition on the α_j insures integrability. Note that a single Fisher–Hartwig singularity at z_j consists of a root-type singularity

$$|z - z_j|^{2\alpha_j} = \left| 2 \sin \frac{\theta - \theta_j}{2} \right|^{2\alpha_j} \tag{1-6}$$

and a jump $e^{i\pi\beta} \to e^{-i\pi\beta}$. We assume that z_j, $j = 1, \ldots, m$, are genuine singular points, i.e., either $\alpha_j \neq 0$ or $\beta_j \neq 0$. However, we always include $z_0 = 1$ explicitly in (1-2), even when $\alpha_0 = \beta_0 = 0$: this convention was adopted in [Deift et al. 2011] in order to facilitate the application of our Toeplitz methods to Hankel determinants. Note that $g_{\beta_0}(z) = e^{-i\pi\beta_0}$. Observe that for each $j \neq 0$, $z^{\beta_j} g_{\beta_j}(z)$ is continuous at $z = 1$, and so for each j each "beta" singularity produces a jump only at the point z_j. The factors $z_j^{-\beta_j}$ are singled out to simplify comparisons with the existing literature. Indeed, (1-2) with the notation $b(\theta) = e^{V(e^{i\theta})}$ is exactly the symbol considered in [Fisher and Hartwig 1968; Basor 1978; 1979; Böttcher and Silbermann 1981; 1985; 1986; Ehrhardt and Silbermann 1997; Ehrhardt 2001; Widom 1973]. However, we write the symbol in a form with $z^{\sum_{j=0}^m \beta_j}$ factored out. The representation (1-2) is more natural for our analysis.

On the unit circle, $V(z)$ is represented by its Fourier expansion:

$$V(z) = \sum_{k=-\infty}^{\infty} V_k z^k, \quad V_k = \frac{1}{2\pi} \int_0^{2\pi} V(e^{i\theta}) e^{-ki\theta} d\theta. \tag{1-7}$$

The canonical Wiener–Hopf factorization of $e^{V(z)}$ is given by

$$e^{V(z)} = b_+(z) e^{V_0} b_-(z), \quad b_+(z) = e^{\sum_{k=1}^{\infty} V_k z^k}, \quad b_-(z) = e^{\sum_{k=-\infty}^{-1} V_k z^k}. \tag{1-8}$$

Define a seminorm

$$\||\beta|\| = \max_{j,k} |\Re\beta_j - \Re\beta_k|, \tag{1-9}$$

where the indices $j, k = 0$ are omitted if $z = 1$ is not a singular point, i.e., if $\alpha_0 = \beta_0 = 0$. If $m = 0$, set $\||\beta|\| = 0$.

In this paper we consider the asymptotics of $D_n(f)$, $n \to \infty$, in the case $\||\beta|\| < 1$. The asymptotic behavior of $D_n(f)$ has been studied by many authors (see [Deift et al. 2011; Ehrhardt 2001] for a review). An expansion of $D_n(f)$ for the case $V \in C^\infty$, $\||\beta|\| < 1$, was obtained by Ehrhardt in [2001]. The aim of this paper is to provide an alternative proof of this result based on differential identities for $D_n(f)$ and a Riemann–Hilbert-problem analysis of the corresponding system of orthogonal polynomials. We also obtain estimates for the error term and extend the validity of the result to less smooth V. The behavior of $D_n(f)$ as $n \to \infty$ for $|||\beta||| \geq 1$ was conjectured by Basor and Tracy [1991], and their conjecture was eventually proved in [Deift et al. 2011]. It turns out that the case $|||\beta||| < 1$ plays a crucial role in our proof of the conjecture for $|||\beta||| \geq 1$.

Theorem 1.1. *Let $f(e^{i\theta})$ be defined in (1-2), $|||\beta||| < 1$, $\Re\alpha_j > -\frac{1}{2}$, $\alpha_j \pm \beta_j \neq -1, -2, \ldots$ for $j, k = 0, 1, \ldots, m$, and let $V(z)$ satisfy the condition (1-11), (1-12) below. Then, as $n \to \infty$,*

$$D_n(f) = \exp\left[nV_0 + \sum_{k=1}^{\infty} kV_kV_{-k}\right]\prod_{j=0}^{m} b_+(z_j)^{-\alpha_j+\beta_j} b_-(z_j)^{-\alpha_j-\beta_j}$$

$$\times n^{\sum_{j=0}^{m}(\alpha_j^2-\beta_j^2)} \prod_{0 \leq j < k \leq m} |z_j - z_k|^{2(\beta_j\beta_k - \alpha_j\alpha_k)} \left(\frac{z_k}{z_j e^{i\pi}}\right)^{\alpha_j\beta_k - \alpha_k\beta_j}$$

$$\times \prod_{j=0}^{m} \frac{G(1+\alpha_j+\beta_j)G(1+\alpha_j-\beta_j)}{G(1+2\alpha_j)}(1+o(1)), \quad (1\text{-}10)$$

where $G(x)$ is Barnes' G-function. The double product over $j < k$ is set to 1 if $m = 0$.

Remark 1.2. As indicated above, this result was first obtained by Ehrhardt in the case $V \in C^{\infty}$. We prove the theorem for $V(z)$ satisfying the smoothness condition

$$\sum_{k=-\infty}^{\infty} |k|^s |V_k| < \infty, \quad (1\text{-}11)$$

where

$$s > \frac{1 + \sum_{j=0}^{m}\left[(\Im\alpha_j)^2 + (\Re\beta_j)^2\right]}{1 - |||\beta|||}. \quad (1\text{-}12)$$

Remark 1.3. In the case of a single singularity, i.e., when $m = 1$ and $\alpha_0 = \beta_0 = 0$, or when $m = 0$, the seminorm $|||\beta||| = 0$, and the theorem implies that the asymptotic form (1-10) holds for all

$$\Re\alpha_m > -\frac{1}{2}, \quad \beta_m \in \mathbb{C}, \quad \alpha_m \pm \beta_m \neq -1, -2, \ldots. \quad (1\text{-}13)$$

In fact, if there is only one singularity, say $m = 0$, and $V \equiv 0$, an explicit formula is known for $D_n(f)$ for any n in terms of the G-functions:

$$D_n(f) = \frac{G(1+\alpha_0+\beta_0)G(1+\alpha_0-\beta_0)}{G(1+2\alpha_0)} \frac{G(n+1)G(n+1+2\alpha_0)}{G(n+1+\alpha_0+\beta_0)G(n+1+\alpha_0-\beta_0)},$$

$$\Re\alpha_0 > -\frac{1}{2}, \quad \alpha_0 \pm \beta_0 \neq -1, -2, \ldots, \quad n \geq 1, \quad (1\text{-}14)$$

and (1-10) can then be read off from the known asymptotics of the G-function (see, e.g., [Barnes 1900]). The formula (1-14) was found by Böttcher and Silbermann

[1985]. However, it also follows from a more general one for a Selberg-type integral which was discovered by Selberg himself (see [Forrester and Warnaar 2008]) a few years earlier.

Remark 1.4. Assume that the function $V(z)$ is sufficiently smooth, i.e., such that s in (1-11) is, in addition to satisfying (1-12), sufficiently large in comparison with $\||\beta\||$. Then we show that the error term $o(1) = O(n^{\||\beta\||-1})$ in (1-10). In particular, the error term $o(1) = O(n^{\||\beta\||-1})$ if $V(z)$ is analytic in a neighborhood of the unit circle. Moreover, for analytic $V(z)$, our methods would allow us to calculate, in principle, the full asymptotic expansion rather than just the leading term presented in (1-10). Various regularity properties of the expansion (uniformity, differentiability) in compact sets of parameters satisfying $\Re \alpha_j > -\frac{1}{2}$, $\||\beta\|| < 1$, $\alpha_j \pm \beta_j \neq -1, -2, \dots, \theta_j \neq \theta_k$, are easy to deduce from our analysis.

Remark 1.5. Since $G(-k) = 0$, $k = 0, 1, \dots$, the formula (1-10) no longer represents the leading asymptotics if $\alpha_j + \beta_j$ or $\alpha_j - \beta_j$ is a negative integer for some j. Although our method applies, we do not address these cases in this paper. It is simply a matter of going deeper in the asymptotic expansion for $D_n(f)$. It can happen that $D_n(f)$ vanishes to all orders (cf. discussion of the Ising model at temperatures above the critical temperature in [Deift et al. 2012]), and $e^{-nV_0} D_n(f)$ is exponentially decreasing.

We prove Theorem 1.1 in the following way. We begin by deriving differential identities for the logarithm of $D_n(f)$ in Section 3, in the spirit of [Deift 1999; Krasovsky 2004; 2007; Its and Krasovsky 2008; Deift et al. 2007; 2008], utilizing the polynomials orthogonal with respect to the weight $f(z)$ on the unit circle. Then, assuming that $V(z)$ is analytic in a neighborhood of the unit circle, we analyze in Section 4 the asymptotics of these polynomials using Riemann–Hilbert/steepest-descent methods as in [Deift et al. 2011]. This gives in turn the asymptotics of the differential identities from which the formula (1-10) follows in the $V \equiv 0$ case by integration with respect to α_j, β_j in Section 5A. However, the error term that results is of order $n^{2\||\beta\||-1} \ln n$ (see (5-35)), which is asymptotically small only for $\||\beta\|| < \frac{1}{2}$, rather than in the full range $\||\beta\|| < 1$. To prove (1-10) for all $\||\beta\|| < 1$, we need a finer analysis of cancellations in the Riemann–Hilbert problem as $n \to \infty$. We carry this out in Section 5B and reduce the leading order terms in $D_n(f)$ to a telescopic form (see (5-65)), which leads to a uniform bound on $D_n(f) n^{\sum_{j=0}^{m}(\beta_j^2 - \alpha_j^2)}$ for large n which is valid for all $\||\beta\|| < 1$ and away from the points $\alpha_j \pm \beta_j = -1, -2, \dots$. We then apply Vitali's theorem together with the previous result for $\||\beta\|| < \frac{1}{2}$. This proves Theorem 1.1 in the $V \equiv 0$ case as desired, with the error term of order $n^{\||\beta\||-1}$. In Sections 5C and 5D, we then extend the result to the case of analytic $V \not\equiv 0$ by applying another differential identity from Section 3.

Ehrhardt proves (1-10) using a "localization" or "separation" technique, introduced by Basor [1979], where the effect of adding in Fisher–Hartwig singularities one at a time is controlled. One may also view our approach as a "separation" technique, but in contrast to [Basor 1979; Ehrhardt 2001], we add in the Fisher–Hartwig singularities, as well as the regular term $e^{V(z)}$, in a continuous fashion.

Remark 1.6. An alternative approach to proving Theorem 1.1 in the V-analytic case is to apply, ab initio, the finer analysis of Section 5B to the orthogonal polynomials which appear in the differential identities. This approach is more direct, but is considerably more involved technically. The analysis is resolved, as above, by reduction of the problem to an appropriate telescopic form.

Finally in Section 5E, we extend our result to the case when $V(z)$ is not analytic and only satisfies the smoothness condition (1-11), (1-12). We approximate such $V(z)$ by trigonometric polynomials $V^{(n)}(z) = \sum_{k=-p}^{p} V_k z^k$ with an appropriate $p = p(n)$ and modify the Riemann–Hilbert analysis accordingly. This produces asymptotics of a Toeplitz determinant in whose symbol $f^{(n)}(z)$ the function $V(z)$ is replaced by $V^{(n)}(z)$. We then use the Heine representation of Toeplitz determinants by multiple integrals to show that $D_n(f^{(n)})$ approximates $D_n(f)$ as $n \to \infty$, sufficiently strongly to conclude Theorem 1.1 in the general case.

2. Riemann–Hilbert problem

In this section we formulate a Riemann–Hilbert problem (RHP) for the polynomials orthogonal on the unit circle (which, oriented in the positive direction, we denote C) with weight $f(z)$ given by (1-2). We use this RHP in Section 4 to find the asymptotics of the polynomials in the case of analytic $V(z)$. Suppose that all $D_k(f) \neq 0$, $k = k_0, k_0 + 1 \ldots$, for some sufficiently large k_0 (see discussion below). Then the polynomials $\phi_k(z) = \chi_k z^k + \cdots$, $\hat{\phi}_k(z) = \chi_k z^k + \cdots$ of degree k, $k = k_0, k_0 + 1, \ldots$, with nonzero leading coefficients χ_k, satisfying the orthogonality conditions

$$\frac{1}{2\pi} \int_0^{2\pi} \phi_k(z) z^{-j} f(z)\, d\theta = \chi_k^{-1} \delta_{jk}, \quad \frac{1}{2\pi} \int_0^{2\pi} \hat{\phi}_k(z^{-1}) z^j f(z)\, d\theta = \chi_k^{-1} \delta_{jk},$$

$$z = e^{i\theta}, \quad j = 0, 1, \ldots, k, \quad (2\text{-}1)$$

exist and are given by the following expressions:

$$\phi_k(z) = \frac{1}{\sqrt{D_k D_{k+1}}} \begin{vmatrix} f_{00} & f_{01} & \cdots & f_{0k} \\ f_{10} & f_{11} & \cdots & f_{1k} \\ \vdots & \vdots & & \vdots \\ f_{k-10} & f_{k-11} & \cdots & f_{k-1k} \\ 1 & z & \cdots & z^k \end{vmatrix}, \quad (2\text{-}2)$$

$$\hat{\phi}_k(z^{-1}) = \frac{1}{\sqrt{D_k D_{k+1}}} \begin{vmatrix} f_{00} & f_{01} & \cdots & f_{0k-1} & 1 \\ f_{10} & f_{11} & \cdots & f_{1k-1} & z^{-1} \\ \vdots & \vdots & & \vdots & \vdots \\ f_{k0} & f_{k1} & \cdots & f_{kk-1} & z^{-k} \end{vmatrix}, \tag{2-3}$$

where

$$f_{st} = \frac{1}{2\pi} \int_0^{2\pi} f(z) z^{-(s-t)} d\theta, \quad s, t = 0, 1, \ldots, k.$$

We obviously have

$$\chi_k = \sqrt{\frac{D_k}{D_{k+1}}}. \tag{2-4}$$

Consider the matrix-valued function $Y^{(k)}(z) = Y(z)$, $k \geq k_0$, given by

$$Y^{(k)}(z) := \begin{pmatrix} \chi_k^{-1} \phi_k(z) & \chi_k^{-1} \int_C \frac{\phi_k(\xi)}{\xi - z} \frac{f(\xi) d\xi}{2\pi i \xi^k} \\ -\chi_{k-1} z^{k-1} \hat{\phi}_{k-1}(z^{-1}) & -\chi_{k-1} \int_C \frac{\hat{\phi}_{k-1}(\xi^{-1})}{\xi - z} \frac{f(\xi) d\xi}{2\pi i \xi} \end{pmatrix}. \tag{2-5}$$

It is easy to verify that $Y(z)$ solves the following Riemann–Hilbert problem:

(a) $Y(z)$ is analytic for $z \in \mathbb{C} \setminus C$.

(b) Let $z \in C \setminus \bigcup_{j=0}^m z_j$. Y has continuous boundary values $Y_+(z)$ as z approaches the unit circle from the inside, and $Y_-(z)$, from the outside, related by the jump condition

$$Y_+(z) = Y_-(z) \begin{pmatrix} 1 & z^{-k} f(z) \\ 0 & 1 \end{pmatrix}, \quad z \in C \setminus \bigcup_{j=0}^m z_j. \tag{2-6}$$

(c) $Y(z)$ has the following asymptotic behavior at infinity:

$$Y(z) = \left(I + O\left(\frac{1}{z}\right) \right) \begin{pmatrix} z^k & 0 \\ 0 & z^{-k} \end{pmatrix} \quad \text{as } z \to \infty. \tag{2-7}$$

(d) As $z \to z_j$, $j = 0, 1, \ldots, m$, $z \in \mathbb{C} \setminus C$,

$$Y(z) = \begin{pmatrix} O(1) & O(1) + O(|z - z_j|^{2\alpha_j}) \\ O(1) & O(1) + O(|z - z_j|^{2\alpha_j}) \end{pmatrix} \quad \text{if } \alpha_j \neq 0, \tag{2-8}$$

and

$$Y(z) = \begin{pmatrix} O(1) & O(\ln|z - z_j|) \\ O(1) & O(\ln|z - z_j|) \end{pmatrix} \quad \text{if } \alpha_j = 0, \beta_j \neq 0. \tag{2-9}$$

(Here and below $O(a)$ stands for $O(|a|)$.) If $\alpha_0 = \beta_0 = 0$, $Y(z)$ is bounded at $z = 1$.

A general fact that orthogonal polynomials can be so represented as a solution of a Riemann–Hilbert problem was noticed in [Fokas et al. 1992] for polynomials on the line and extended to polynomials on the circle in [Baik et al. 1999]. This fact is important because it turns out that the RHP can be efficiently analyzed for large k by a steepest-descent type method found in [Deift and Zhou 1993] and developed further in many subsequent works. Thus, we first find the solution to the problem (a)–(d) for large k (applying this method) and then interpret it as the asymptotics of the orthogonal polynomials by (2-5).

Recall the Heine representation for a Toeplitz determinant:

$$D_n(f) = \frac{1}{(2\pi)^n n!} \int_0^{2\pi} \cdots \int_0^{2\pi} \prod_{1 \le j < k \le n} |e^{i\phi_j} - e^{i\phi_k}|^2 \prod_{j=1}^n f(e^{i\phi_j}) d\phi_j. \quad (2\text{-}10)$$

If $f(z)$ is positive on the unit circle, it follows from (2-10) that $D_k(f) > 0$, $k = 1, 2, \ldots$, i.e., $k_0 = 1$. In the general case, let Λ be a compact subset in the subset $\|\|\beta\|\| < 1$, $\alpha_j \pm \beta_j \ne -1, -2, \ldots$ of the parameter space $\mathcal{P} = \{(\alpha_0, \beta_0, \ldots, \alpha_m, \beta_m) : \alpha_j, \beta_j \in \mathbb{C}, \Re \alpha_j > -\frac{1}{2}\}$. We will show in Section 4 that the RHP (a)–(d) is solvable in Λ, in particular χ_k are finite and nonzero, for all sufficiently large k ($k \ge k_0(\Lambda)$). Let Ω_{k_0} be the set of parameters in $\mathcal{P}$ such that $D_k(f) = 0$ for some $k = 1, 2, \ldots, k_0 - 1$. We will then have $D_k(f) \ne 0$, $k = 1, 2, \ldots$ for all points in $\Lambda \setminus \Omega_{k_0}$. Note that $D_n(f)$ depends analytically on α_j, β_j in $\mathcal{P}$: this is true, in particular, on (the interior of) $\Lambda \setminus \Omega_{k_0}$.

The solution to the RHP (a)–(d) is unique. Note first that $\det Y(z) = 1$. Indeed, from the conditions on $Y(z)$, $\det Y(z)$ is analytic across the unit circle, has all singularities removable, and tends to 1 as $z \to \infty$. It is then identically 1 by Liouville's theorem. Now if there is another solution $\tilde{Y}(z)$, we easily obtain by Liouville's theorem that $\tilde{Y}(z) Y(z)^{-1} \equiv 1$.

3. Differential identities

In this section we derive expressions for the derivative $(\partial/\partial \gamma) \ln D_n(f(z))$, where either $\gamma = \alpha_j$ or $\gamma = \beta_j$, $j = 0, 1, \ldots, m$, in terms of the matrix elements of (2-5). These will be exact differential identities valid for all $n = 1, 2, \ldots$ (see Proposition 3.1 below), provided all the $D_n(f) \ne 0$. We will use these expressions in Section 5A to obtain the asymptotics (1-10) in the case $V \equiv 0$, $\|\|\beta\|\| < \frac{1}{2}$ (improved to $\|\|\beta\|\| < 1$ in Section 5B). Furthermore, in this section we will derive a differential identity (see Proposition 3.3 below) which will enable us in Sections 5C and 5D to extend the results to analytic $V \ne 0$ (improved to sufficiently smooth V in Section 5E).

Set $D_0 \equiv 1$, $\phi_0(z) \equiv \hat{\phi}_0(z) \equiv 1$, and suppose that $D_n(f) \neq 0$ for all $n = 1, 2, \ldots$. Then the orthogonal polynomials (2-2), (2-3) exist and are analytic in the α_j and β_j for all $k = 1, 2, \ldots$. Moreover, (2-4) implies that

$$D_n(f(z)) = \prod_{j=0}^{n-1} \chi_j^{-2}. \tag{3-1}$$

Note that by orthogonality,

$$\frac{1}{2\pi} \int_0^{2\pi} \frac{\partial \phi_j(z)}{\partial \gamma} \hat{\phi}_j(z^{-1}) f(z) \, d\theta$$

$$= \frac{1}{2\pi} \int_0^{2\pi} \left(\frac{\partial \chi_j}{\partial \gamma} z^j + \text{polynomial of degree } j-1 \right) \hat{\phi}_j(z^{-1}) f(z) \, d\theta$$

$$= \frac{1}{\chi_j} \frac{\partial \chi_j}{\partial \gamma}. \tag{3-2}$$

Similarly,

$$\frac{1}{2\pi} \int_0^{2\pi} \phi_j(z) \frac{\partial \hat{\phi}_j(z^{-1})}{\partial \gamma} f(z) \, d\theta = \frac{1}{\chi_j} \frac{\partial \chi_j}{\partial \gamma}. \tag{3-3}$$

Therefore, using (3-1), we obtain

$$\frac{\partial}{\partial \gamma} \ln D_n(f(z)) = \frac{\partial}{\partial \gamma} \ln \prod_{j=0}^{n-1} \chi_j^{-2} = -2 \sum_{j=0}^{n-1} \frac{\frac{\partial \chi_j}{\partial \gamma}}{\chi_j}$$

$$= -\frac{1}{2\pi} \int_0^{2\pi} \frac{\partial}{\partial \gamma} \left(\sum_{j=0}^{n-1} \phi_j(z) \hat{\phi}_j(z^{-1}) \right) f(z) \, d\theta. \tag{3-4}$$

Using here the Christoffel–Darboux identity (see Lemma 2.3 of [Deift et al. 2011], for example),

$$\sum_{k=0}^{n-1} \hat{\phi}_k(z^{-1}) \phi_k(z) = -n \phi_n(z) \hat{\phi}_n(z^{-1}) + z \left(\hat{\phi}_n(z^{-1}) \frac{d}{dz} \phi_n(z) - \phi_n(z) \frac{d}{dz} \hat{\phi}_n(z^{-1}) \right),$$

and then orthogonality, we can write

$$\frac{\partial}{\partial \gamma} \ln D_n(f(z))$$

$$= 2n \frac{\frac{\partial \chi_n}{\partial \gamma}}{\chi_n} + \frac{1}{2\pi} \int_0^{2\pi} \frac{\partial}{\partial \gamma} \left(\phi_n(z) \frac{d \hat{\phi}_n(z^{-1})}{dz} - \hat{\phi}_n(z^{-1}) \frac{d \phi_n(z)}{dz} \right) z f(z) \, d\theta. \tag{3-5}$$

Writing out the derivative with respect to γ in the integral and using orthogonality, we obtain

$$\frac{\partial}{\partial \gamma} \ln D_n(f(z)) = I_1 - I_2, \tag{3-6}$$

where

$$
\begin{aligned}
I_1 &= \frac{1}{2\pi i} \int_0^{2\pi} \frac{\partial \phi_n(z)}{\partial \gamma} \frac{\partial \hat{\phi}_n(z^{-1})}{\partial \theta} f(z)\, d\theta, \\
I_2 &= \frac{1}{2\pi i} \int_0^{2\pi} \frac{\partial \phi_n(z)}{\partial \theta} \frac{\partial \hat{\phi}_n(z^{-1})}{\partial \gamma} f(z)\, d\theta.
\end{aligned}
\tag{3-7}
$$

It turns out that the particular structure of Fisher–Hartwig singularities allows us to reduce (3-6) to a local formula, i.e., to replace the integrals by the polynomials (and their Cauchy transforms) evaluated only at several points (compare [Its and Krasovsky 2008; Krasovsky 2007]).

Let us encircle each of the points z_j by a sufficiently small disc,

$$U_{z_j} = \left\{ z : |z - z_j| < \varepsilon \right\}. \tag{3-8}$$

Denote

$$C_\varepsilon = \bigcup_{j=0}^{m} \left(U_{z_j} \cap C \right). \tag{3-9}$$

We now integrate I_1 by parts. First assume that $V(z) \equiv 0$. Then, using the expression

$$\frac{\partial f(z)}{\partial \theta} = \sum_{j=0}^{m} \left(\alpha_j \cot \frac{\theta - \theta_j}{2} + i\beta_j \right) f(z) = \left(\sum_{j=0}^{m} \alpha_j \frac{z + z_j}{z - z_j} + \beta_j \right) i f(z),$$

we obtain

$$
\begin{aligned}
I_1 = {}& -\chi_n^{-1} \frac{\partial \chi_n}{\partial \gamma} \left(n + \sum_{j=0}^{m} \beta_j \right) \\
& - \lim_{\varepsilon \to 0} \Bigg[\frac{1}{2\pi} \int_{C \backslash C_\varepsilon} \frac{\partial \phi_n(z)}{\partial \gamma} \hat{\phi}_n(z^{-1}) \left(\sum_{j=0}^{m} \alpha_j \frac{z + z_j}{z - z_j} \right) f(z)\, d\theta \\
& - \frac{1}{2\pi i} \sum_{j=0}^{m} \frac{\partial \phi_n(z_j)}{\partial \gamma} \hat{\phi}_n(z_j^{-1}) (f(z_j e^{-i\varepsilon}) - f(z_j e^{i\varepsilon})) \Bigg], \tag{3-10}
\end{aligned}
$$

where the integration is over $C \setminus C_\varepsilon$ in the positive direction around the unit circle.

Note that by adding and subtracting $\dfrac{\partial \phi_n(z_j)}{\partial \gamma}$, and by using orthogonality we can write

$$
\int_{C \backslash C_\varepsilon} \frac{\partial \phi_n(z)}{\partial \gamma} \hat{\phi}_n(z^{-1}) \frac{z+z_j}{z-z_j} f(z)\, d\theta =
$$

$$
\int_{C \backslash C_\varepsilon} \hat{\phi}_n(z^{-1}) \frac{\partial \phi_n(z)/\partial \gamma - \partial \phi_n(z_j)/\partial \gamma}{z - z_j}(z + z_j) f(z)\, d\theta
$$

$$
+ \frac{\partial \phi_n(z_j)}{\partial \gamma} \int_{C \backslash C_\varepsilon} \hat{\phi}_n(z^{-1}) \frac{2 z_j}{z - z_j} f(z)\, d\theta + O(\varepsilon^{2\Re \alpha_j + 1}). \quad (3\text{-}11)
$$

Obviously, the fraction in the first integral on the right-hand side is a polynomial in z of degree $n - 1$ with leading coefficient $\partial \chi_n / \partial \gamma$. Therefore, the integral equals $2\pi (\partial \chi_n / \partial \gamma)/\chi_n$ up to $O(\varepsilon^{2\Re \alpha_j + 1})$. The second integral can be written in terms of the element Y_{22} of (2-5) for $z \to z_j$. Let us estimate therefore the following expression for $\alpha_j \neq 0$:

$$
\int_C \hat{\phi}_n(s^{-1}) \frac{2 z_j f(s)}{s - z} \frac{ds}{is}
$$

$$
- \lim_{\varepsilon \to 0} \left[\int_{C \backslash C_\varepsilon} \hat{\phi}_n(s^{-1}) \frac{2 z_j f(s)}{s - z_j} \frac{ds}{is} - \frac{1}{i \alpha_j} \hat{\phi}_n(z_j^{-1})(f(z_j e^{-i\varepsilon}) - f(z_j e^{i\varepsilon})) \right],
$$

$$
z \to z_j, \quad |z| > 1. \quad (3\text{-}12)
$$

This difference tends to zero as $z \to z_j$, for $\Re \alpha_j > 0$. When $\Re \alpha_j < 0$, it is a growing function as $z \to z_j$, and when $\Re \alpha_j = 0$, $\Im \alpha_j \neq 0$, an oscillating one. The analysis is similar to that of Section 3 in [Krasovsky 2007]. For future use, we now fix an analytical continuation of the absolute value, namely, write for z on the unit circle,

$$
|z - z_j|^{\alpha_j} = (z - z_j)^{\alpha_j/2} (z^{-1} - z_j^{-1})^{\alpha_j/2} = \frac{(z - z_j)^{\alpha_j}}{(z z_j e^{i\ell_j})^{\alpha_j/2}}, \quad z = e^{i\theta}, \quad (3\text{-}13)
$$

where ℓ_j is found from the condition that the argument of the above function is zero on the unit circle. Let us fix the cut of $(z - z_j)^{\alpha_j}$ going along the line $\theta = \theta_j$ from z_j to infinity. Fix the branch by the condition that on the line going from z_j to the right parallel to the real axis, $\arg(z - z_j) = 2\pi$. For $z^{\alpha_j/2}$ in the denominator, $0 < \arg z < 2\pi$. If $z_0 = 1$ let $0 < \arg(z - 1) < 2\pi$. (This choice will enable us to use the standard asymptotics for a confluent hypergeometric function in the RH analysis in Section 4 below.) Then, a simple consideration of triangles shows that

$$
\ell_j = \begin{cases} 3\pi & \text{if } 0 < \theta < \theta_j, \\ \pi & \text{if } \theta_j < \theta < 2\pi. \end{cases} \quad (3\text{-}14)
$$

Thus (3-13) is continued analytically to neighborhoods of the arcs $0 < \theta < \theta_j$ and $\theta_j < \theta < 2\pi$. We now analyze (3-12) in the same way that equation (26) was analyzed in [Krasovsky 2007]. For this analysis, however, we will need two other choices of the function $(z - z_j)^{2\alpha_j}$: one choice with the cut going a short distance clockwise along the unit circle C from z_j, and another, with the cut going a short distance anticlockwise along C from z_j. Let c_j and d_j be some points on C between z_j and the neighboring singularity in the clockwise and anticlockwise directions, respectively. In a neighborhood of z_j, let $g(s)$ be defined by

$$\frac{\hat{\phi}_n(s^{-1})f(s)}{is} = |s - z_j|^{2\alpha_j} g(s).$$

We then obtain as in [Krasovsky 2007] for the part of (3-12) on the arc (c_j, d_j):

$$\int_{c_j}^{d_j} \frac{|s - z_j|^{2\alpha_j}}{s - z} g(s)\, ds$$

$$- \lim_{\varepsilon \to 0}\left[\left(\int_{c_j}^{z_j e^{-i\varepsilon}} + \int_{z_j e^{i\varepsilon}}^{d_j}\right)\frac{|s - z_j|^{2\alpha_j}}{s - z_j} g(s)\, ds - \frac{\varepsilon^{2\alpha_j}}{2\alpha_j}(g(z_j e^{-i\varepsilon}) - g(z_j e^{i\varepsilon}))\right]$$

$$= \lim_{\varepsilon \to 0} \frac{\pi (z - z_j)^{2\alpha_j}}{(zz_j)^{\alpha_j}\sin(2\pi\alpha_j)}(e^{2\pi i\alpha_j - i\alpha\ell_R} g(z_j e^{-i\varepsilon}) - e^{-2\pi i\alpha_j - i\alpha\ell_L} g(z_j e^{i\varepsilon}))$$

$$+ \alpha_j^{-1} O(z - z_j), \quad z \to z_j, \quad |z| > 1, \quad (3\text{-}15)$$

for $\alpha_j \neq 0, \frac{1}{2}, 1, \frac{3}{2}, \ldots$ (when $\alpha_j = \frac{1}{2}, 1, \frac{3}{2}, \ldots$ one obtains terms involving $(z - z_j)^k \ln(z - z_j)$ vanishing as $z \to z_j$). Here the constants ℓ_R, ℓ_L depend on the choice of a branch for $(z - z_j)^{2\alpha}$ (whose cut is, recall, along the circle) and their values will not be important below.

Introduce a "regularized" version of the integral in a neighborhood of z_j:

$$\int_{c_j}^{d_j^{(r)}} \frac{|s - z_j|^{2\alpha_j}}{s - z} g(s)\, ds \equiv \int_{c_j}^{d_j} \frac{|s - z_j|^{2\alpha_j}}{s - z} g(s)\, ds$$

$$- \lim_{\varepsilon \to 0} \frac{\pi (z - z_j)^{2\alpha_j}}{(zz_j)^{\alpha_j}\sin(2\pi\alpha_j)}(e^{2\pi i\alpha_j - i\alpha\ell_R} g(z_j e^{-i\varepsilon}) - e^{-2\pi i\alpha_j - i\alpha\ell_L} g(z_j e^{i\varepsilon})),$$

for z in a complex neighborhood of z_j and $-\frac{1}{2} < \Re\alpha_j \leq 0$, $\alpha_j \neq 0$. If $\Re\alpha_j > 0$, we set the "regularized" integral equal to the integral itself.

Denote by $\widetilde{Y}$ the matrix (2-5), in which the integrals of the second column are replaced by their "regularized" values in a neighborhood of each z_j.

Then, collecting our observations together, we can write (3-10) in the form

$$I_1 = -\chi_n^{-1}\frac{\partial \chi_n}{\partial \gamma}\left(n + \sum_{j=0}^{m}(\alpha_j + \beta_j)\right) + \sum_{j=0}^{m} G_j^+(z_j), \quad (3\text{-}16)$$

where

$$
G_j^+(z_j) = \begin{cases} \dfrac{2\alpha_j z_j}{\chi_n} \dfrac{\partial}{\partial\gamma}\left(\chi_n Y_{11}^{(n)}(z_j)\right)\widetilde{Y}_{22}^{(n+1)}(z_j) & \text{if } \alpha_j \neq 0, \\[12pt] \dfrac{1}{2\pi i}\dfrac{\partial\phi_n(z_j)}{\partial\gamma}\hat{\phi}_n(z_j^{-1})\Delta f(z_j) & \text{if } \alpha_j = 0, \end{cases}
\tag{3-17}
$$

with

$$
\Delta f(z_j) = \lim_{\varepsilon\to 0}\left(f(z_j e^{-i\varepsilon}) - f(z_j e^{i\varepsilon})\right).
\tag{3-18}
$$

A similar analysis yields

$$
I_2 = \chi_n^{-1}\frac{\partial\chi_n}{\partial\gamma}\left(n + \sum_{j=0}^{m}(\alpha_j - \beta_j)\right) + \sum_{j=0}^{m} G_j^-(z_j),
$$

where

$$
G_j^-(z_j) = \begin{cases} 2\chi_n\alpha_j\dfrac{\partial}{\partial\gamma}\left(\chi_n^{-1}Y_{21}^{(n+1)}(z_j)\right)\widetilde{Y}_{12}^{(n)}(z_j) & \text{if } \alpha_j \neq 0, \\[12pt] \dfrac{1}{2\pi i}\dfrac{\partial\hat{\phi}_n(z_j^{-1})}{\partial\gamma}\phi_n(z_j)\Delta f(z_j) & \text{if } \alpha_j = 0. \end{cases}
\tag{3-19}
$$

Substituting these results into (3-6) we obtain:

Proposition 3.1. *Let* $V(z) \equiv 0$. *Let* $\gamma = \alpha_k$ *or* $\gamma = \beta_k$, $k = 0, 1, \ldots, m$, *and* $D_n(f(z)) \neq 0$ *for all* n. *Then for any* $n = 1, 2, \ldots,$

$$
\frac{\partial}{\partial\gamma}\ln D_n(f(z)) = -2\chi_n^{-1}\frac{\partial\chi_n}{\partial\gamma}\left(n + \sum_{j=0}^{m}\alpha_j\right) + \sum_{j=0}^{m}(G_j^+(z_j) - G_j^-(z_j)),
\tag{3-20}
$$

where $G_j^+(z_j)$ *and* $G_j^-(z_j)$ *are defined in* (3-17) *and* (3-19), *with reference to* (3-18).

In Section 5A we substitute the asymptotics for Y (found in Section 4) in (3-20) and, by integrating, obtain part of Theorem 1.1 for $f(z)$ with $V(z) \equiv 0$, $\|\beta\| < \frac{1}{2}$. Further analysis of Section 5B extends the result to $\|\beta\| < 1$.

Remark 3.2. The differential identities (3-20) admit an interesting interpretation in the context of the monodromy theory of the Fuchsian system of linear ODEs canonically related to the Riemann–Hilbert problem (2-6)–(2-9). We explain this connection in some detail in the Appendix. The results presented there, however, are not used in the main body of the paper.

To extend the theorem to nonzero $V(z)$ we will use another differential identity. Let us introduce a parametric family of weights and the corresponding orthogonal polynomials indexed by $t \in [0, 1]$. Namely, let

$$f(z, t) = (1 - t + te^{V(z)})e^{-V(z)}f(z). \tag{3-21}$$

Thus $f(z, 0)$ corresponds to $f(z)$ with $V = 0$, whereas $f(z, 1)$ gives the function (1-2) we are interested in.

Note that

$$\frac{\partial f(z, t)}{\partial t} = \frac{f(z, t) - f(z, 0)}{t}. \tag{3-22}$$

Set now $\gamma = t$ and replace the function $f(z)$ and the orthogonal polynomials in (3-5) by $f(z, t)$ and the polynomials orthogonal with respect to $f(z, t)$. Then the integral in the right-hand side of (3-5) can be written as follows (we assume $D_n \neq 0$ for all n):

$$\frac{\partial}{\partial t}\left[\frac{1}{2\pi}\int_0^{2\pi}\left(\phi_n(z, t)\frac{d\hat{\phi}_n(z^{-1}, t)}{dz} - \hat{\phi}_n(z^{-1}, t)\frac{d\phi_n(z, t)}{dz}\right)zf(z, t)\,d\theta\right]$$

$$- \frac{1}{2\pi}\int_0^{2\pi}\left(\phi_n(z, t)\frac{d\hat{\phi}_n(z^{-1}, t)}{dz} - \hat{\phi}_n(z^{-1}, t)\frac{d\phi_n(z, t)}{dz}\right)z\frac{f(z, t) - f(z, 0)}{t}\,d\theta$$

$$= \frac{2n}{t} + \frac{1}{2\pi t}\int_C\left(\phi_n(z, t)\frac{d\hat{\phi}_n(z^{-1}, t)}{dz} - \hat{\phi}_n(z^{-1}, t)\frac{d\phi_n(z, t)}{dz}\right)zf(z, 0)\,d\theta.$$

Therefore, we obtain

$$\frac{\partial}{\partial t}\ln D_n(f(z, t)) = 2n\left(\frac{1}{t} + \chi_n(t)^{-1}\frac{\partial\chi_n(t)}{\partial t}\right)$$

$$+ \frac{1}{2\pi t}\int_C\left(\phi_n(z, t)\frac{d\hat{\phi}_n(z^{-1}, t)}{dz} - \hat{\phi}_n(z^{-1}, t)\frac{d\phi_n(z, t)}{dz}\right)zf(z, 0)\,d\theta. \tag{3-23}$$

To write this identity in terms of the solution to the RHP (2-6)–(2-7), note first that using the recurrence relation (see, e.g., [Deift et al. 2011, Lemma 2.2])

$$\chi_n z^{-1}\hat{\phi}_n(z^{-1}) = \chi_{n+1}\hat{\phi}_{n+1}(z^{-1}) - \hat{\phi}_{n+1}(0)z^{-n-1}\phi_{n+1}(z), \tag{3-24}$$

we have

$$Y_{21}^{(n)}(z, t) = -\chi_{n-1}(t)z^{n-1}\hat{\phi}_{n-1}(z^{-1}, t)$$

$$= -\chi_n(t)z^n\hat{\phi}_n(z^{-1}, t) + \hat{\phi}_n(0, t)\phi_n(z, t). \tag{3-25}$$

Now using the orthogonality relations (2-1) and the formulae (3-2) and (3-3), we obtain from (3-23):

Proposition 3.3. *Let $f(z, t)$ be given by (3-21) and $D_n(f(z, t)) \neq 0$ for all n. Let $\phi_k(z, t)$, $\hat{\phi}_k(z, t)$, $k = 0, 1, \ldots,$ be the corresponding orthogonal polynomials.*

Then, for any $n = 1, 2, \ldots,$

$$\frac{\partial}{\partial t} \ln D_n(f(z, t))$$

$$= \frac{1}{2\pi i} \int_C z^{-n} \left(Y_{11}(z, t) \frac{\partial Y_{21}(z, t)}{\partial z} - Y_{21}(z, t) \frac{\partial Y_{11}(z, t)}{\partial z} \right) \frac{\partial f(z, t)}{\partial t} \, dz, \quad (3\text{-}26)$$

where the integration is over the unit circle and $Y(z, t) := Y^{(n)}(z, t)$.

4. Asymptotics for the Riemann–Hilbert problem

The RHP of Section 2 was solved in [Deift et al. 2011]. In this section we list the results from that paper needed for the proof of Theorem 1.1. We always assume (for the rest of the paper) that $f(z)$ is given by (1-2) and that $\alpha_j \pm \beta_j \neq -1, -2, \ldots$ for all $j = 0, 1, \ldots, m$. In this section we also assume for simplicity that $z_0 = 1$ is a singularity. However, the results trivially extend to the case $\alpha_0 = \beta_0 = 0$. In this section, we further assume that $V(z)$ is analytic in a neighborhood of the unit circle.

Let $Y(z) := Y^{(n)}(z)$. First, set

$$T(z) = Y(z) \begin{cases} z^{-n\sigma_3} & \text{if } |z| > 1 \\ I & \text{if } |z| < 1. \end{cases} \quad (4\text{-}1)$$

Now split the contour as shown in Figure 1. We call lenses the $m + 1$ regions of the complex plane containing the arcs Σ'_j of the unit circle and bounded by the curves Σ_j, Σ''_j.

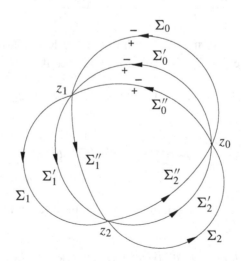

Figure 1. Contour for the S-Riemann–Hilbert problem ($m = 2$).

Set

$$
S(z) = \begin{cases}
T(z) & \text{for } z \text{ outside the lenses,} \\[2mm]
T(z)\begin{pmatrix} 1 & 0 \\ f(z)^{-1}z^{-n} & 1 \end{pmatrix} & \text{for } |z| > 1 \text{ and inside the lenses,} \\[3mm]
T(z)\begin{pmatrix} 1 & 0 \\ -f(z)^{-1}z^{n} & 1 \end{pmatrix} & \text{for } |z| < 1 \text{ and inside the lenses.}
\end{cases}
\tag{4-2}
$$

Here $f(z)$ is the analytic continuation of $f(z)$ off the unit circle into the inside of the lenses as discussed following (3-13).

The function $S(z)$ satisfies the following Riemann–Hilbert problem:

(a) $S(z)$ is analytic for $z \in \mathbb{C} \setminus \Sigma$, where $\Sigma = \bigcup_{j=0}^{m}(\Sigma_j \cup \Sigma_j' \cup \Sigma_j'')$.

(b) The boundary values of $S(z)$ are related by the jump condition

$$
S_+(z) = S_-(z)\begin{pmatrix} 1 & 0 \\ f(z)^{-1}z^{\mp n} & 1 \end{pmatrix}, \quad z \in \bigcup_{j=0}^{m}(\Sigma_j \cup \Sigma_j''),
\tag{4-3}
$$

where the minus sign in the exponent is on Σ_j, and plus on Σ_j'',

$$
S_+(z) = S_-(z)\begin{pmatrix} 0 & f(z) \\ -f(z)^{-1} & 0 \end{pmatrix}, \quad z \in \bigcup_{j=0}^{m}\Sigma_j'.
\tag{4-4}
$$

(c) $S(z) = I + O(1/z)$ as $z \to \infty$,

(d) As $z \to z_j$, $j = 0, \dots, m$, $z \in \mathbb{C} \setminus C$ outside the lenses,

$$
S(z) = \begin{pmatrix} O(1) & O(1) + O(|z - z_j|^{2\alpha_j}) \\ O(1) & O(1) + O(|z - z_j|^{2\alpha_j}) \end{pmatrix}
\tag{4-5}
$$

if $\alpha_j \neq 0$, and

$$
S(z) = \begin{pmatrix} O(1) & O(\ln|z - z_j|) \\ O(1) & O(\ln|z - z_j|) \end{pmatrix}
\tag{4-6}
$$

if $\alpha_j = 0$, $\beta_j \neq 0$. The behavior of $S(z)$ for $z \to z_j$ in other sectors is obtained from these expressions by application of the appropriate jump conditions.

We now present formulae for the parametrices which solve the model Riemann–Hilbert problems outside the neighborhoods U_{z_j} of the points z_j, and inside those neighborhoods, respectively. These parametrices match to the leading order in n on the boundaries of the neighborhoods U_{z_j}, and this matching allows us to construct the asymptotic solution to the RHP for Y.

The parametrix outside the U_{z_j} is

$$N(z) = \begin{cases} \mathscr{D}(z)^{\sigma_3} & \text{if } |z| > 1, \\ \mathscr{D}(z)^{\sigma_3} \begin{pmatrix} 0 & 1 \\ -1 & 0 \end{pmatrix} & \text{if } |z| < 1, \end{cases} \qquad z \in \mathbb{C} \setminus \bigcup_{j=0}^{m} U_{z_j}, \qquad (4\text{-}7)$$

where the Szegő function

$$\mathscr{D}(z) = \exp \frac{1}{2\pi i} \int_C \frac{\ln f(s)}{s - z} ds \qquad (4\text{-}8)$$

is analytic away from the unit circle, and we have

$$\mathscr{D}(z) = e^{V_0} b_+(z) \prod_{k=0}^{m} \left(\frac{z - z_k}{z_k e^{i\pi}} \right)^{\alpha_k + \beta_k}, \qquad |z| < 1. \qquad (4\text{-}9)$$

and

$$\mathscr{D}(z) = b_-(z)^{-1} \prod_{k=0}^{m} \left(\frac{z - z_k}{z} \right)^{-\alpha_k + \beta_k}, \qquad |z| > 1, \qquad (4\text{-}10)$$

where V_0, $b_\pm(z)$ are defined in (1-8). Note that the branch of $(z - z_k)^{\pm \alpha_k + \beta_k}$ in (4-9) and (4-10) is taken as discussed following Equation (3-13) above. In (4-10) for any k, the cut of the root $z^{-\alpha_k + \beta_k}$ is the line $\theta = \theta_k$ from $z = 0$ to infinity, and $\theta_k < \arg z < 2\pi + \theta_k$.

Inside each neighborhood U_{z_j} the parametrix is given in terms of a confluent hypergeometric function. First, set

$$\zeta = n \ln \frac{z}{z_j}, \qquad (4\text{-}11)$$

where $\ln x > 0$ for $x > 1$, and has a cut on the negative half of the real axis. Under this transformation the neighborhood U_{z_j} is mapped onto a neighborhood of zero in the ζ-plane. Note that the transformation $\zeta(z)$ is analytic, one-to-one, and it takes an arc of the unit circle to an interval of the imaginary axis. Let us now choose the exact form of the cuts Σ in U_{z_j} so that their images under the mapping $\zeta(z)$ are straight lines (Figure 2).

We add one more jump contour to Σ in U_{z_j} which is the preimage of the real line Γ_3 and Γ_7 in the ζ-plane. This is needed below because of the nonanalyticity of the function $|z - z_j|^{\alpha_j}$. Note that we can construct two different analytic continuations of this function off the unit circle to the preimages of the upper and lower half ζ-plane, respectively. Namely, let

$$h_{\alpha_j}(z) = |z - z_j|^{\alpha_j}, \qquad z = e^{i\theta} \qquad (4\text{-}12)$$

with the branches chosen as in (3-13). As remarked above, (3-13) is continued analytically to neighborhoods of the arcs $0 < \theta < \theta_j$ and $\theta_j < \theta < 2\pi$. In U_{z_j}, we

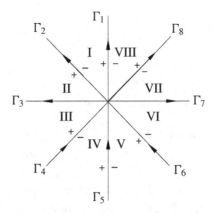

Figure 2. The auxiliary contour for the parametrix at z_j.

extend these neighborhoods to the preimages of the lower and upper half ζ-plane (intersected with $\zeta(U_{z_j})$), respectively. The cut of h_{α_j} is along the contours Γ_3 and Γ_7 in the ζ-plane.

For $z \to z_j$, $\zeta = n(z - z_j)/z_j + O((z - z_j)^2)$. We have $0 < \arg \zeta < 2\pi$, which follows from the choice of $\arg(z - z_j)$ in (3-13).

From now on we will provide the formulae for the parametrix only in the region I (see [Deift et al. 2011] for complete results). Set

$$F_j(z) = e^{V(z)/2} \prod_{k=0}^{m} \left(\frac{z}{z_k}\right)^{\beta_k/2} \prod_{k \neq j} h_{\alpha_k}(z) g_{\beta_k}(z)^{1/2} h_{\alpha_j}(z) e^{-i\pi\alpha_j},$$

$$\zeta \in I, \quad z \in U_{z_j}, \quad j \neq 0. \quad (4\text{-}13)$$

Note that this function is related to $f(z)$ as follows:

$$F_j(z)^2 = f(z) e^{-2\pi i \alpha_j} g_{\beta_j}^{-1}(z) \quad \zeta \in I. \quad (4\text{-}14)$$

The functions $g_{\beta_k}(z)$ are defined in (1-4). The formulae for $F_0(z)$ are the same, but with $g_{\beta_0}(z)$ replaced by

$$\widehat{g}_{\beta_0}(z) = \begin{cases} e^{-i\pi\beta_0} & \text{if } \arg z > 0, \\ e^{i\pi\beta_0} & \text{if } \arg z < 2, \pi, \end{cases} \quad z \in U_{z_0}. \quad (4\text{-}15)$$

We then have the following expression for the parametrix $P_j(z)$ in the region $z(I)$ of U_{z_j}:

$$P_{z_j}(z) = E(z) \Psi_j(\zeta) F_j(z)^{-\sigma_3} z^{n\sigma_3/2}, \quad \zeta \in I. \quad (4\text{-}16)$$

Here

$$E(z) = N(z) \zeta^{\beta_j \sigma_3} F_j^{\sigma_3}(z) z_j^{-n\sigma_3/2} \begin{pmatrix} e^{-i\pi(2\beta_j + \alpha_j)} & 0 \\ 0 & e^{i\pi(\beta_j + 2\alpha_j)} \end{pmatrix} \quad (4\text{-}17)$$

and

$$\Psi_j(\zeta)$$

$$= \begin{pmatrix} \zeta^{\alpha_j}\psi(\alpha_j+\beta_j,1+2\alpha_j,\zeta)e^{i\pi(2\beta_j+\alpha_j)}e^{-\zeta/2} \\ -\zeta^{-\alpha_j}\psi(1-\alpha_j+\beta_j,1-2\alpha_j,\zeta)e^{i\pi(\beta_j-3\alpha_j)}e^{-\zeta/2}\frac{\Gamma(1+\alpha_j+\beta_j)}{\Gamma(\alpha_j-\beta_j)} \\[1ex] -\zeta^{\alpha_j}\psi(1+\alpha_j-\beta_j,1+2\alpha_j,e^{-i\pi}\zeta)e^{i\pi(\beta_j+\alpha_j)}e^{\zeta/2}\frac{\Gamma(1+\alpha_j-\beta_j)}{\Gamma(\alpha_j+\beta_j)} \\ \zeta^{-\alpha_j}\psi(-\alpha_j-\beta_j,1-2\alpha_j,e^{-i\pi}\zeta)e^{-i\pi\alpha_j}e^{\zeta/2} \end{pmatrix}, \quad (4\text{-}18)$$

where $\psi(a,b,x)$ is the confluent hypergeometric function of the second kind, and $\Gamma(x)$ is Euler's Γ-function. Recall our assumption that $\alpha_j \pm \beta_j \neq -1, -2, \ldots$.

The matching condition for the parametrices P_{z_j} and N is the following for any $k = 1, 2, \ldots$:

$$P_{z_j}(z)N^{-1}(z) = I + \Delta_1(z) + \Delta_2(z) + \cdots + \Delta_k(z) + \Delta_{k+1}^{(r)}, \quad z \in \partial U_{z_j}. \quad (4\text{-}19)$$

Every $\Delta_p(z)$, $\Delta_p^{(r)}(z)$, $p = 1, 2, \ldots$, with $z \in \partial U_{z_j}$ is of the form

$$a_j^{-\sigma_3} O(n^{-p}) a_j^{\sigma_3}, \quad a_j := n^{\beta_j} z_j^{-n/2}. \quad (4\text{-}20)$$

In particular, explicitly, on the part of ∂U_{z_j} whose ζ-image is in I,

$$\Delta_1(z) = \frac{1}{\zeta}\begin{pmatrix} -(\alpha_j^2-\beta_j^2) \\ -\dfrac{\Gamma(1+\alpha_j-\beta_j)}{\Gamma(\alpha_j+\beta_j)}\left(\dfrac{\mathcal{D}(z)}{\zeta^{\beta_j}F_j(z)}\right)^{-2}z_j^{-n}e^{-i\pi(2\beta_j-\alpha_j)} \\[2ex] \dfrac{\Gamma(1+\alpha_j+\beta_j)}{\Gamma(\alpha_j-\beta_j)}\left(\dfrac{\mathcal{D}(z)}{\zeta^{\beta_j}F_j(z)}\right)^{2}z_j^{n}e^{i\pi(2\beta_j-\alpha_j)} \\ \alpha_j^2-\beta_j^2 \end{pmatrix}, \quad (4\text{-}21)$$

which extends to a meromorphic function in a neighborhood of U_{z_j} with a simple pole at $z = z_j$.

The error term $\Delta_{k+1}^{(r)}$ in (4-19) is uniform in z on ∂U_{z_j}.

At the point z_j we have

$$F_j(z) = \eta_j e^{-3i\pi\alpha_j/2} z_j^{-\alpha_j} u^{\alpha_j}(1 + O(u)), \quad u = z - z_j, \quad \zeta \in I, \quad (4\text{-}22)$$

where

$$\eta_j = e^{V(z_j)/2}\exp\left\{-\frac{i\pi}{2}\left(\sum_{k=0}^{j-1}\beta_k - \sum_{k=j+1}^{m}\beta_k\right)\right\}\prod_{k\neq j}\left(\frac{z_j}{z_k}\right)^{\beta_k/2}|z_j - z_k|^{\alpha_k}, \quad (4\text{-}23)$$

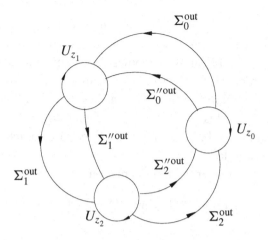

Figure 3. Contour Γ for the R and $\tilde{R}$ Riemann–Hilbert problems ($m = 2$).

and

$$\left(\frac{\mathscr{D}(z)}{\zeta^{\beta_j} F_j(z)}\right)^2 = \mu_j^2 e^{i\pi(\alpha_j - 2\beta_j)} n^{-2\beta_j}(1 + O(u)), \quad u = z - z_j, \quad \zeta \in I, \quad (4\text{-}24)$$

$$\mu_j = \left(e^{V_0}\frac{b_+(z_j)}{b_-(z_j)}\right)^{1/2} \exp\left\{-\frac{i\pi}{2}\left(\sum_{k=0}^{j-1}\alpha_k - \sum_{k=j+1}^{m}\alpha_k\right)\right\}\prod_{k\neq j}\left(\frac{z_j}{z_k}\right)^{\alpha_k/2}|z_j - z_k|^{\beta_k}. \tag{4-25}$$

The sums from 0 to -1 for $j = 0$ and from $m + 1$ to m for $j = m$ are set to zero.

4A. R-RHP. Let

$$R(z) = \begin{cases} S(z)N^{-1}(z) & \text{if } z \in U_\infty \setminus \Gamma, \text{ with } U_\infty = \mathbb{C} \setminus \bigcup_{j=0}^m U_{z_j}, \\ S(z)P_{z_j}^{-1}(z) & \text{if } z \in U_{z_j} \setminus \Gamma \text{ for } j = 0, \ldots, m. \end{cases} \tag{4-26}$$

It is easy to verify that this function has jumps only on ∂U_{z_j}, and the parts of Σ_j, Σ_j'' lying outside the neighborhoods U_{z_j} (we denote these parts without the end-points Σ^{out}, Σ''^{out}). The full contour Γ is shown in Figure 3. Away from Γ, as a standard argument shows, $R(z)$ is analytic. Moreover, we have: $R(z) = I + O(1/z)$ as $z \to \infty$.

The jumps of $R(z)$ are as follows, with $j = 0, \ldots, m$:

$$R_+(z) = R_-(z)N(z)\begin{pmatrix} 1 & 0 \\ f(z)^{-1}z^{-n} & 1 \end{pmatrix}N(z)^{-1}, \qquad z \in \Sigma_j^{\text{out}}, \tag{4-27}$$

$$R_+(z) = R_-(z)N(z)\begin{pmatrix} 1 & 0 \\ f(z)^{-1}z^n & 1 \end{pmatrix}N(z)^{-1}, \qquad z \in \Sigma_j''^{\text{out}}, \tag{4-28}$$

$$R_+(z) = R_-(z)P_{z_j}(z)N(z)^{-1}, \quad z \in \partial U_{z_j}\setminus\{\text{intersection points}\}, \tag{4-29}$$

The jump matrix on Σ^{out}, Σ''^{out} can be estimated uniformly in α_j, β_j as $I + O(\exp(-\varepsilon n))$, where ε is a positive constant. The jump matrices on ∂U_{z_j} admit a uniform expansion (4-19) in inverse powers of n conjugated by $n^{\beta_j \sigma_3} z_j^{-n\sigma_3/2}$, and (4-20) is of order $n^{2\max_j |\Re\beta_j|-P}$. To obtain the standard solution of the R-RHP in terms of a Neuman series (see, e.g., [Deift et al. 1999]) we must have $n^{2\max_j |\Re\beta_j|-1} = o(1)$, that is $\Re\beta_j \in (-\frac{1}{2}, \frac{1}{2})$ for all $j = 0, 1, \ldots, m$. However, it is possible to obtain the solution in any half-closed or open interval of length 1, i.e., for $\|\|\beta\|\| < 1$, as follows.

Let $\|\|\beta\|\| < 1$ and consider the transformation

$$\widetilde{R}(z) = n^{\omega\sigma_3} R(z) n^{-\omega\sigma_3} \quad z \in \mathbb{C} \setminus \Gamma, \tag{4-30}$$

where

$$\omega = \tfrac{1}{2}(\min_j \Re\beta_j + \max_j \Re\beta_j) \tag{4-31}$$

which "shifts" all $\Re\beta_j$ (in the conjugation n^{β_j} terms of (4-20) for the jump matrix in (4-29)) into the interval $(-\frac{1}{2}, \frac{1}{2})$. Note that $\omega = \Re\beta_{j_0}$ if only one $\Re\beta_{j_0} \neq 0$, and $\omega = 0$ if all $\Re\beta_j = 0$.

Now in the RHP for $\widetilde{R}(z)$, the condition at infinity and the uniform exponential estimate $I + O(\exp(-\varepsilon n))$ (with different ε) of the jump matrices on Σ^{out}, Σ''^{out} is preserved, while the jump matrices on ∂U_{z_j} have the form

$$I + n^{\omega\sigma_3} \Delta_1(z) n^{-\omega\sigma_3} + \cdots + n^{\omega\sigma_3} \Delta_k(z) n^{-\omega\sigma_3} + n^{\omega\sigma_3} \Delta_{k+1}^{(r)}(z) n^{-\omega\sigma_3},$$
$$z \in \partial U_{z_j}, \tag{4-32}$$

where the order of each $n^{\omega\sigma_3} \Delta_p(z) n^{-\omega\sigma_3}$, $n^{\omega\sigma_3} \Delta_p^{(r)}(z) n^{-\omega\sigma_3}$, $p = 1, 2, \ldots$, with $z \in \bigcup_{j=0}^m \partial U_{z_j}$ is

$$O(n^{2\max_j |\Re\beta_j - \omega|-P}) = O(n^{\|\|\beta\|\|-P}).$$

This implies that the standard analysis can be applied to the $\widetilde{R}$-RHP problem in the range $\Re\beta_j \in (q - \frac{1}{2}, q + \frac{1}{2})$, $j = 0, 1, \ldots, m$, for any $q \in \mathbb{R}$, and we obtain the asymptotic expansion

$$\widetilde{R}(z) = I + \sum_{p=1}^k \widetilde{R}_p(z) + \widetilde{R}_{k+1}^{(r)}(z), \quad p = 1, 2 \ldots \tag{4-33}$$

uniformly for all z and for β_j in bounded sets of the strip $q - \frac{1}{2} < \Re\beta_j < q + \frac{1}{2}$, $j = 0, 1, \ldots m$, i.e., $\|\|\beta\|\| < 1$, provided $\alpha_j \pm \beta_j$ are outside neighborhoods of the points $-1, -2, \ldots$ (compare (4-21)).

The functions $\widetilde{R}_j(z)$ are computed recursively. We will need explicit expressions only for the first two. The first one is found from the conditions that $\widetilde{R}_1(z)$

is analytic outside $\partial U = \bigcup_{j=0}^m \partial U_{z_j}$, $\tilde{R}_1(z) \to 0$ as $z \to \infty$, and

$$\tilde{R}_{1,+}(z) = \tilde{R}_{1,-}(z) + n^{\omega\sigma_3}\Delta_1(z)n^{-\omega\sigma_3}, \quad z \in \partial U. \tag{4-34}$$

The solution is easily found. First set

$$R_k(z) := n^{-\omega\sigma_3}\tilde{R}_k(z)n^{\omega\sigma_3}, \quad R_k^{(r)}(z) := n^{-\omega\sigma_3}\tilde{R}_k^{(r)}(z)n^{\omega\sigma_3}, \quad k = 1, 2, \ldots,$$

and write for R:

$$R_1(z) = \frac{1}{2\pi i}\int_{\partial U}\frac{\Delta_1(x)\,dx}{x-z}$$

$$= \begin{cases} \displaystyle\sum_{k=0}^m\frac{A_k}{z-z_k} & \text{if } z \in \mathbb{C}\setminus\bigcup_{j=0}^m U_{z_j}, \\[4mm] \displaystyle\sum_{k=0}^m\frac{A_k}{z-z_k} - \Delta_1(z) & \text{if } z \in U_{z_j}, \quad j = 0, 1, \ldots, m, \end{cases} \tag{4-35}$$

where

$$\partial U = \bigcup_{j=0}^m \partial U_{z_j}, \tag{4-36}$$

the contours in the integral are traversed in the negative direction, and the A_k are the coefficients in the Laurent expansion of $\Delta_1(z)$:

$$\Delta_1(z) = \frac{A_k}{z-z_k} + B_k + O(z-z_k), \quad z \to z_k, \quad k = 0, 1, \ldots, m. \tag{4-37}$$

The coefficients are easy to compute using (4-19) and (4-25):

$$A_k = A_k^{(n)} = \frac{z_k}{n}\begin{pmatrix} -(\alpha_k^2 - \beta_k^2) & \dfrac{\Gamma(1+\alpha_k+\beta_k)}{\Gamma(\alpha_k-\beta_k)}z_k^n\mu_k^2 n^{-2\beta_k} \\[4mm] -\dfrac{\Gamma(1+\alpha_k-\beta_k)}{\Gamma(\alpha_k+\beta_k)}z_k^{-n}\mu_k^{-2}n^{2\beta_k} & \alpha_k^2 - \beta_k^2 \end{pmatrix}. \tag{4-38}$$

The function $\tilde{R}_2$ is now found from the conditions that $\tilde{R}_2(z) \to 0$ as $z \to \infty$, is analytic outside ∂U, and

$$\tilde{R}_{2,+}(z) = \tilde{R}_{2,-}(z) + \tilde{R}_{1,-}(z)n^{\omega\sigma_3}\Delta_1(z)n^{-\omega\sigma_3} + n^{\omega\sigma_3}\Delta_2(z)n^{-\omega\sigma_3}, \quad z \in \partial U.$$

The solution to this RHP is

$$\tilde{R}_2(z) = \frac{1}{2\pi i}\int_{\partial U}\left(\tilde{R}_{1,-}(x)n^{\omega\sigma_3}\Delta_1(x)n^{-\omega\sigma_3} + n^{\omega\sigma_3}\Delta_2(x)n^{-\omega\sigma_3}\right)\frac{dx}{x-z}. \tag{4-39}$$

At the k-th step we have the RHP for $\tilde{R}_k(z)$ with the same analyticity condition and the condition at infinity, and the following jump, where $\tilde{R}_0(z) \equiv I$:

$$\tilde{R}_{k,+}(z) = \tilde{R}_{k,-}(z) + \sum_{p=1}^k \tilde{R}_{k-p,-}(z)n^{\omega\sigma_3}\Delta_p(z)n^{-\omega\sigma_3}, \quad z \in \partial U. \tag{4-40}$$

We will now discuss the way in which the general $\widetilde{R}_k(z)$ depends on n. In particular, we will discuss its order in n. First note that

$$\widetilde{R}_1(z) \sim \frac{1}{n} \begin{pmatrix} 1 & \sum_j b_j^{-2} \\ \sum_j b_j^2 & 1 \end{pmatrix}, \tag{4-41}$$

$$\widetilde{R}_2(z) \sim \frac{1}{n^2} \begin{pmatrix} 1 + \delta' n^2 & \sum_j b_j^{-2} \\ \sum_j b_j^2 & 1 + \delta' n^2 \end{pmatrix}, \tag{4-42}$$

where

$$b_j := n^{\beta_j - \omega} z_j^{-n/2}, \qquad \delta' \sim \sum_{j,k} n^{2((\beta_j - \beta_k) - 1)} \left(\frac{z_k}{z_j} \right)^n. \tag{4-43}$$

Here the notation $A \sim B$ means $A' = B'$, where X' is X in which each matrix element and each term in the sums is multiplied by a suitable constant *independent* of n. Starting with these expressions, and noting from (4-40) that

$$\widetilde{R}_k(z) \sim \sum_{p=1}^{k} \widetilde{R}_{k-p,-}(z) n^{\omega \sigma_3} \Delta_p(z) n^{-\omega \sigma_3}$$

$$\sim \widetilde{R}_{k-1,-}(z) n^{\omega \sigma_3} \Delta_1(z) n^{-\omega \sigma_3} + \frac{1}{n} \widetilde{R}_{k-1,-}(z), \tag{4-44}$$

we obtain by induction, for $p = 0, 1, \ldots$:

$$\widetilde{R}_{2p+1}(z) \sim \frac{1}{n^{2p+1}} \sum_{k=0}^{p} (\delta' n^2)^k \begin{pmatrix} 1 & \sum_j b_j^{-2} \\ \sum_j b_j^2 & 1 \end{pmatrix}, \tag{4-45}$$

$$\widetilde{R}_{2p+2}(z) \sim \frac{1}{n^{2p+2}} \sum_{k=0}^{p} (\delta' n^2)^k \begin{pmatrix} 1 + \delta' n^2 & \sum_j b_j^{-2} \\ \sum_j b_j^2 & 1 + \delta' n^2 \end{pmatrix}. \tag{4-46}$$

In particular, as $n \to \infty$, and again for $p = 0, 1, \ldots$ we have,

$$\widetilde{R}_{2p+1}(z) = \frac{\delta'^p}{n} O \begin{pmatrix} 1 & \sum_j b_j^{-2} \\ \sum_j b_j^2 & 1 \end{pmatrix}, \tag{4-47}$$

$$\widetilde{R}_{2p+2}(z) = \frac{\delta'^p}{n^2} O \begin{pmatrix} 1 + \delta' n^2 & \sum_j b_j^{-2} \\ \sum_j b_j^2 & 1 + \delta' n^2 \end{pmatrix}, \tag{4-48}$$

$$O(\delta') = O(\delta), \qquad \delta = \max_{j,k} n^{2(\Re(\beta_j - \beta_k) - 1)} = n^{2(\||\beta\|| - 1)}, \tag{4-49}$$

Here $O(A)$ represent 2×2 matrices with elements of the corresponding order. Finally, note that the error term in (4-33) is

$$\widetilde{R}_k^{(r)}(z) = O(|\widetilde{R}_k(z)| + |\widetilde{R}_{k+1}(z)|).$$

In particular, as is clear from the above, if there is only one nonzero β_{j_0}, we obtain the expansion purely in inverse integer powers of n valid in fact for all $\beta_{j_0} \in \mathbb{C}$ uniformly in bounded sets of the complex plane.

It is clear from the construction and the properties of the asymptotic series of the confluent hypergeometric function that the error terms $\widetilde{R}_k^{(r)}(z)$ are uniform for β_j in bounded subsets of the strip $q - \frac{1}{2} < \Re \beta_j < q + \frac{1}{2}$, $j = 0, 1, \ldots m$, for α_j in bounded sets of the half-plane $\Re \alpha_j > -\frac{1}{2}$, and for $\alpha_j \pm \beta_j$ away from neighborhoods of the negative integers. Moreover, the series (4-33) is differentiable in α_j, β_j.

5. Asymptotics for differential identities and integration: proof of Theorem 1.1

5A. *Pure Fisher–Hartwig singularities: the case* $\||\beta\|| < \frac{1}{2}$. First, we will prove the theorem for $V(z) \equiv 0$ and $\||\beta\|| = \max_{j,k} |\Re \beta_j - \Re \beta_k| < \frac{1}{2}$. The proof is based on the differential identity (3-20). First, we show that (3-20) has the following asymptotic form.

Proposition 5.1. *Let* $(\alpha_0, \beta_0, \ldots, \alpha_m, \beta_m)$ *be in a compact subset, denote it* Λ, *belonging to the subset* $\||\beta\|| < 1$, $\alpha_j \pm \beta_j \neq -1, -2, \ldots$ *of the parameter space* $\mathcal{P} = \{(\alpha_0, \beta_0, \ldots, \alpha_m, \beta_m) : \alpha_j, \beta_j \in \mathbb{C}, \Re \alpha_j > -\frac{1}{2}\}$ *and including the point* $\alpha_j = \beta_j = 0$, $j = 0, 1, \ldots, m$. *Let* $\beta_j = 0$ *if* $\alpha_j = 0$, $j = 0, 1, \ldots, m$, $\delta = n^{2(\||\beta\||-1)}$. *Then for* $n \to \infty$, *and* $\nu = 0, 1, \ldots, m$,

$$\frac{\partial}{\partial \alpha_\nu} \ln D_n(f(z)) = 2\alpha_\nu + (\alpha_\nu + \beta_\nu)\left[\frac{\partial}{\partial \alpha_\nu} \ln \frac{\Gamma(1 + \alpha_\nu + \beta_\nu)}{\Gamma(1 + 2\alpha_\nu)} + \ln n\right]$$

$$+ (\alpha_\nu - \beta_\nu)\left[\frac{\partial}{\partial \alpha_\nu} \ln \frac{\Gamma(1 + \alpha_\nu - \beta_\nu)}{\Gamma(1 + 2\alpha_\nu)} + \ln n\right]$$

$$- \sum_{j \neq \nu}\left[(\alpha_j + \beta_j) \ln \frac{z_j - z_\nu}{z_j} + (\alpha_j - \beta_j) \ln \frac{z_j - z_\nu}{z_\nu e^{i\pi}}\right]$$

$$+ 2\pi i \sum_{j=0}^{\nu-1}(\alpha_j + \beta_j) + O(n^{-1} \ln n) + O(\delta n \ln n), \quad (5\text{-}1)$$

$$\frac{\partial}{\partial \beta_\nu} \ln D_n(f(z)) = -2\beta_\nu + (\alpha_\nu + \beta_\nu)\left[\frac{\partial}{\partial \beta_\nu} \ln \Gamma(1 + \alpha_\nu + \beta_\nu) - \ln n\right]$$

$$+ (\alpha_\nu - \beta_\nu)\left[\frac{\partial}{\partial \beta_\nu} \ln \Gamma(1 + \alpha_\nu - \beta_\nu) + \ln n\right]$$

$$+ \sum_{j \neq \nu}\left[(\alpha_j + \beta_j) \ln \frac{z_j - z_\nu}{z_j} - (\alpha_j - \beta_j) \ln \frac{z_j - z_\nu}{z_\nu e^{i\pi}}\right]$$

$$- 2\pi i \sum_{j=0}^{\nu-1}(\alpha_j + \beta_j) + O(n^{-1} \ln n) + O(\delta n \ln n). \quad (5\text{-}2)$$

Remark 5.2. The error term $O(\delta n \ln n)$ is $o(1)$ uniformly in Λ if $\|\beta\| \leq \frac{1}{2} - \varepsilon$, $\varepsilon > 0$. In fact, the estimate for the error term can be considerably improved: see next section.

Remark 5.3. The case $\beta_j = 0$ if $\alpha_j = 0$ is all we need below. After the proof of Proposition 5.1, we will integrate the identity (5-1) to obtain the asymptotics of D_n for all $\alpha_j \neq 0$, $\beta_j = 0$. We then integrate the identity (5-2) and obtain the asymptotics of D_n for all $\alpha_j \neq 0$, $\beta_j \neq 0$. This gives the general result for $V \equiv 0$, $\|\beta\| < \frac{1}{2}$, since we can set any $\alpha_j = 0$ using the uniformity of the asymptotic expansion in the α_j.

Proof. Assume that for all j, $\beta_j = 0$ if $\alpha_j = 0$, and $D_k(f) \neq 0$, $k = 1, 2, \ldots$. Then we can rewrite (3-20) in the form

$$\frac{\partial}{\partial \gamma} \ln D_n(f(z))$$

$$= -2 \frac{\frac{\partial \chi_n}{\partial \gamma}}{\chi_n} \left(n + \sum_{j=0}^{m} \alpha_j \left\{ 1 - \widetilde{Y}_{12}^{(n)}(z_j) Y_{21}^{(n+1)}(z_j) - z_j Y_{11}^{(n)}(z_j) \widetilde{Y}_{22}^{(n+1)}(z_j) \right\} \right)$$

$$- 2 \sum_{j=0}^{m} \alpha_j \left(\widetilde{Y}_{12}^{(n)}(z_j) \frac{\partial}{\partial \gamma} Y_{21}^{(n+1)}(z_j) - z_j \frac{\partial}{\partial \gamma} Y_{11}^{(n)}(z_j) \widetilde{Y}_{22}^{(n+1)}(z_j) \right), \quad (5\text{-}3)$$

where $\gamma = \alpha_j$ or $\gamma = \beta_j$. We now estimate the right-hand side of this identity as $n \to \infty$. The asymptotics of χ_n were found in [Deift et al. 2011, Theorem 1.8]. We need these asymptotics here in the case $V \equiv 0$:

$$\chi_{n-1}^2 = 1 - \frac{1}{n} \sum_{k=0}^{m} (\alpha_k^2 - \beta_k^2)$$

$$+ \sum_{j=0}^{m} \sum_{k \neq j} \frac{z_k}{z_j - z_k} \left(\frac{z_j}{z_k} \right)^n n^{2(\beta_k - \beta_j - 1)} \frac{v_j}{v_k} \frac{\Gamma(1 + \alpha_j + \beta_j) \Gamma(1 + \alpha_k - \beta_k)}{\Gamma(\alpha_j - \beta_j) \Gamma(\alpha_k + \beta_k)}$$

$$+ O(\delta^2) + O(\delta/n), \quad \delta = n^{2(\|\beta\| - 1)}, \quad n \to \infty, \quad (5\text{-}4)$$

where

$$v_j = \exp\left\{ -i\pi \left(\sum_{p=0}^{j-1} \alpha_p - \sum_{p=j+1}^{m} \alpha_p \right) \right\} \prod_{p \neq j} \left(\frac{z_j}{z_p} \right)^{\alpha_p} |z_j - z_p|^{2\beta_p}. \quad (5\text{-}5)$$

The asymptotics of $\widetilde{Y}^{(n)}(z_j)$ were also found in [Deift et al. 2011, (7.11)–(7.21)]. Namely,

$$\widetilde{Y}^{(n)}(z_j) = (I + r_j^{(n)}) L_j^{(n)}, \quad (5\text{-}6)$$

where

$$L_j^{(n)} = \begin{pmatrix} M_{21}\mu_j\eta_j^{-1}n^{\alpha_j-\beta_j}z_j^n & M_{22}\mu_j\eta_j n^{-\alpha_j-\beta_j} \\ -M_{11}\mu_j^{-1}\eta_j^{-1}n^{\alpha_j+\beta_j} & -M_{12}\mu_j^{-1}\eta_j n^{-\alpha_j+\beta_j}z_j^{-n} \end{pmatrix},$$

and $r_j = R_1^{(r)}(z_j)$, the parameters η_j, μ_j are given by (4-23), (4-25), and

$$M = \begin{pmatrix} \dfrac{\Gamma(1+\alpha_j-\beta_j)}{\Gamma(1+2\alpha_j)} & -\dfrac{\Gamma(2\alpha_j)}{\Gamma(\alpha_j+\beta_j)} \\ \dfrac{\Gamma(1+\alpha_j+\beta_j)}{\Gamma(1+2\alpha_j)} & \dfrac{\Gamma(2\alpha_j)}{\Gamma(\alpha_j-\beta_j)} \end{pmatrix}.$$

Note that the matrix $L_j^{(n)}$ has the structure

$$L_j^{(n)} = n^{-\beta_j\sigma_3}\hat{L}_j^{(n)}n^{\alpha_j\sigma_3}, \tag{5-7}$$

where $\hat{L}$ depends on n only via the oscillatory terms z_j^n.

Let us now obtain the asymptotics for the following combination appearing in (5-3):

$$\frac{\partial}{\partial\gamma}Y_{11}^{(n)}(z_j)\widetilde{Y}_{22}^{(n+1)}(z_j)$$

$$= \frac{\partial}{\partial\gamma}\left((1+r_{11}^{(n)})L_{11}^{(n)}+r_{12}^{(n)}L_{21}^{(n)}\right)\left(r_{21}^{(n+1)}L_{12}^{(n+1)}+(1+r_{22}^{(n+1)})L_{22}^{(n+1)}\right)$$

$$= \left(L_{11}^{(n)}L_{22}^{(n+1)}(1+r_{22}^{(n+1)})+L_{11}^{(n)}L_{12}^{(n+1)}r_{21}^{(n+1)}\right)\left[(1+r_{11}^{(n)})\frac{\partial}{\partial\gamma}\ln L_{11}^{(n)}+\frac{\partial}{\partial\gamma}r_{11}^{(n)}\right]$$

$$+\left(L_{21}^{(n)}L_{22}^{(n+1)}(1+r_{22}^{(n+1)})+L_{21}^{(n)}L_{12}^{(n+1)}r_{21}^{(n+1)}\right)\left[r_{12}^{(n)}\frac{\partial}{\partial\gamma}\ln L_{21}^{(n)}+\frac{\partial}{\partial\gamma}r_{12}^{(n)}\right].$$

We omit the lower index j of r and L for simplicity of notation.

Now the explicit formula for L and the estimates for $\widetilde{R}(z)$ imply that

$$L_{11}^{(n)}L_{22}^{(n+1)} = O(1), \qquad L_{21}^{(n)}L_{12}^{(n+1)} = O(1), \tag{5-8}$$

$$L_{11}^{(n)}L_{12}^{(n+1)}r_{21}^{(n+1)} = O(n\delta), \qquad L_{21}^{(n)}L_{22}^{(n+1)}r_{12}^{(n)} = O(n\delta), \tag{5-9}$$

$$r_{21}^{(n+1)}r_{12}^{(n)} = O(\delta), \tag{5-10}$$

$$\frac{\partial}{\partial\gamma}\ln L^{(n)} = O(\ln n), \qquad \frac{\partial}{\partial\gamma}r = O(r)\ln n, \tag{5-11}$$

where as before

$$\delta = n^{2(\|\!\|\beta\|\!\|-1)}.$$

Therefore,

$$\frac{\partial}{\partial \gamma} Y_{11}^{(n)}(z_j) \tilde{Y}_{22}^{(n+1)}(z_j) = L_{11}^{(n)} L_{22}^{(n+1)} \frac{\partial}{\partial \gamma} \ln L_{11}^{(n)} + O\left(\frac{\ln n}{n}\right) + O(\delta n \ln n)$$

$$= \frac{\alpha_j + \beta_j}{2\alpha_j z_j} \frac{\partial}{\partial \gamma} \ln L_{11}^{(n)} + O\left(\frac{\ln n}{n}\right) + O(\delta n \ln n). \quad (5\text{-}12)$$

Similarly, we obtain

$$\tilde{Y}_{12}^{(n)}(z_j) \frac{\partial}{\partial \gamma} Y_{21}^{(n+1)}(z_j) = -\frac{\alpha_j - \beta_j}{2\alpha_j} \frac{\partial}{\partial \gamma} \ln L_{21}^{(n+1)} + O\left(\frac{\ln n}{n}\right) + O(\delta n \ln n),$$

$$(5\text{-}13)$$

and furthermore,

$$\tilde{Y}_{12}^{(n)}(z_j) Y_{21}^{(n+1)}(z_j) = O(\delta n) + O(1),$$

$$Y_{11}^{(n)}(z_j) \tilde{Y}_{22}^{(n+1)}(z_j) = O(\delta n) + O(1). \quad (5\text{-}14)$$

Note that because of the special structure of (5-7), the quantity n^{α_j} does not appear in any of the products (5-12)–(5-14). Substituting (5-12)–(5-14) into (5-3) and using the asymptotics (5-4), we obtain

$$\frac{\partial}{\partial \gamma} \ln D_n(f(z)) =$$

$$\frac{\partial}{\partial \gamma} \left[\sum_{j=0}^{m} (\alpha_j^2 - \beta_j^2) \right] + \sum_{j=0}^{m} \left[(\alpha_j + \beta_j) \frac{\partial}{\partial \gamma} \ln L_{j,11}^{(n)} + (\alpha_j - \beta_j) \frac{\partial}{\partial \gamma} \ln L_{j,21}^{(n+1)} \right]$$

$$+ O\left(\frac{\ln n}{n}\right) + O(\delta n \ln n). \quad (5\text{-}15)$$

Let us calculate the logarithmic derivatives appearing in (5-15). From (5-6), (4-25), and (4-23) it is easy to obtain for the derivatives with respect to α_ν, $\nu = 0, 1, \ldots, m$:

$$\frac{\partial}{\partial \alpha_\nu} \ln L_{\nu,11}^{(n)} = \frac{\partial}{\partial \alpha_\nu} \ln \frac{\Gamma(1 + \alpha_\nu + \beta_\nu)}{\Gamma(1 + 2\alpha_j)} + \ln n, \quad (5\text{-}16)$$

$$\frac{\partial}{\partial \alpha_\nu} \ln L_{j,11}^{(n)} = \begin{cases} -\ln \dfrac{z_j - z_\nu}{z_j} + 2\pi i & \text{if } j < \nu, \\[2mm] -\ln \dfrac{z_j - z_\nu}{z_j} & \text{if } j > \nu, \end{cases} \quad (5\text{-}17)$$

$$\frac{\partial}{\partial \alpha_\nu} \ln L_{\nu,21}^{(n+1)} = \frac{\partial}{\partial \alpha_\nu} \ln \frac{\Gamma(1 + \alpha_\nu - \beta_\nu)}{\Gamma(1 + 2\alpha_j)} + \ln n, \quad (5\text{-}18)$$

$$\frac{\partial}{\partial \alpha_\nu} \ln L_{j,21}^{(n+1)} = -\ln \frac{z_j - z_\nu}{z_\nu e^{i\pi}}, \quad j \neq \nu. \quad (5\text{-}19)$$

Similarly, we obtain for the derivatives with respect to β_ν:

$$\frac{\partial}{\partial \beta_\nu} \ln L_{\nu,11}^{(n)} = \frac{\partial}{\partial \beta_\nu} \ln \Gamma(1 + \alpha_\nu + \beta_\nu) - \ln n, \tag{5-20}$$

$$\frac{\partial}{\partial \beta_\nu} \ln L_{j,11}^{(n)} = \begin{cases} \ln \dfrac{z_j - z_\nu}{z_j} - 2\pi i & \text{if } j < \nu, \\ \ln \dfrac{z_j - z_\nu}{z_j} & \text{if } j > \nu, \end{cases} \tag{5-21}$$

$$\frac{\partial}{\partial \beta_\nu} \ln L_{\nu,21}^{(n+1)} = \frac{\partial}{\partial \beta_\nu} \ln \Gamma(1 + \alpha_\nu - \beta_\nu) + \ln n, \tag{5-22}$$

$$\frac{\partial}{\partial \beta_\nu} \ln L_{j,21}^{(n+1)} = -\ln \frac{z_j - z_\nu}{z_\nu e^{i\pi}}, \quad j \neq \nu. \tag{5-23}$$

Combining these results with (5-15) we obtain (5-1), (5-2) on condition that $D_k(f) \neq 0$, $k = 1, 2, \ldots$, or equivalently (see Section 2), $(\alpha_0, \beta_0, \ldots, \alpha_m, \beta_m) \in \Lambda \setminus \Omega_{k_0}$. This condition can be replaced simply by $(\alpha_0, \beta_0, \ldots, \alpha_m, \beta_m) \in \Lambda$ in the following way. Let

$$\beta_0 = 0, \quad \alpha_j = \beta_j = 0, \quad j = 1, \ldots, m. \tag{5-24}$$

Let $\Omega_{k_0}(\alpha_0)$ be the subset of Ω_{k_0} with α_j, $j = 1, \ldots, m$, and β_j, $j = 0, \ldots, m$, fixed by (5-24). Since $D_k(f) \equiv D_k(f(\alpha_0; z))$ is an analytic function of α_0 and $D_k(1) \neq 0$, the set $\Omega_{k_0}(\alpha_0)$ is finite. Let us rewrite the identity (5-1) with $\nu = 0$ and assuming (5-24) in the form $H'(\alpha_0) = 0$, where $H(\alpha_0) = D_n(f(\alpha_0; z)) \exp(-\int_0^{\alpha_0} r(n, s)\, ds)$ and where $r(n, \alpha_0)$ is the right-hand side of (3-20) with $\gamma = \alpha_0$ and assuming (5-24). Since the expression (5-1) for $r(n, \alpha_0)$ holds uniformly and is continuous for $\alpha_0 \in \Lambda$ provided n is larger than some $k_0(\Lambda)$, and $D_n(f(\alpha_0; z))$ and its derivative are continuous, the function $H(\alpha_0)$ is continuously differentiable for all $n > k_0(\Lambda)$. Hence, $H'(\alpha_0) = 0$ for *all* $\alpha_0 \in \Lambda$ and $n > k_0(\Lambda)$. Taking into account that $H(0) = D_n(1) \neq 0$, we conclude that $D_n(f(\alpha_0; z))$ is nonzero, and that the identity (5-1) under (5-24) is, in fact, true for all $\alpha_0 \in \Lambda$ if n is sufficiently large (larger than $k_0(\Lambda)$). Now fix $\alpha_0 \in \Lambda$ and assume the condition $\alpha_1 = \cdots = \alpha_m = \beta_m = 0$. A similar argument as above then gives that $D_n(f(\alpha_0, \beta_0; z))$ is nonzero and the identity (5-2) with $\nu = 0$ is true for all $\beta_0 \in \Lambda$ if n is sufficiently large. Continuing this way, we complete the proof of Proposition 5.1 by induction.

Remark 5.4. A similar argument applies to the asymptotic form of the differential identity (3-26) we need in Section 5C below. We omit the discussion.   □

We will now complete the proof of Theorem 1.1 in the case $\|\|\beta\|\| < \frac{1}{2}$, $V(z) \equiv 0$ by integrating the identities of Proposition 5.1. In this case we denote

$$D_n(f(z)) \equiv D_n(\alpha_0, \ldots, \alpha_m; \beta_0, \ldots, \beta_m). \tag{5-25}$$

First, set $m = 0$ and $\beta_0 = 0$. Then (5-1) becomes

$$\frac{\partial}{\partial \alpha_0} \ln D_n(\alpha_0) = 2\alpha_0 \left(1 + \ln n + \frac{d}{d\alpha_0} \frac{\Gamma(1+\alpha_0)}{\Gamma(1+2\alpha_0)} \right) + O(n^{-1} \ln n) + O(\delta n \ln n).$$

Integrating both sides over α_0 from 0 to some α_0 and using the fact that $D_n(0) = 1$, we obtain

$$D_n(\alpha_0) = n^{\alpha_0^2} \frac{G(1+\alpha_0)^2}{G(1+2\alpha_0)}, \tag{5-26}$$

where $G(x)$ is Barnes G-function. To perform the integration we used the identity

$$\int_0^z \left(1 + \frac{d}{dx} \frac{\Gamma(1+x)}{\Gamma(1+2x)} \right) 2x \, dx = \ln \frac{G(1+z)^2}{G(1+2z)}, \tag{5-27}$$

which easily follows from the standard formula (see, e.g., [Whittaker and Watson 1927]):

$$\int_0^z \ln \Gamma(x+1) \, dx = \frac{z}{2} \ln 2\pi - \frac{z(z+1)}{2} + z \ln \Gamma(z+1) - \ln G(z+1). \tag{5-28}$$

Now set $m = 1$, α_0 fixed. Set $\beta_0 = \beta_1 = 0$. Relation (5-1) for $\nu = 1$ is then

$$\frac{\partial}{\partial \alpha_1} \ln D_n(\alpha_0, \alpha_1) = 2\alpha_1 \left(1 + \ln n + \frac{d}{d\alpha_1} \frac{\Gamma(1+\alpha_1)}{\Gamma(1+2\alpha_1)} \right)$$
$$+ \alpha_0 \ln z_0 + \alpha_0 (\ln z_1 + 3\pi i) - 2\alpha_0 \ln(z_0 - z_1) + O(n^{-1} \ln n) + O(\delta n \ln n). \tag{5-29}$$

Integrating this over α_1 from 0 to some fixed α_1 along a path lying in Λ (see Proposition 5.1) and using (5-27), we obtain

$$\ln \frac{D_n(\alpha_0, \alpha_1)}{D_n(\alpha_0, 0)} = \alpha_1^2 \ln n + 2 \ln G(1+\alpha_1) - \ln G(1+2\alpha_1) + \alpha_0 \alpha_1 \ln(z_0 z_1 e^{3\pi i})$$
$$- 2\alpha_0 \alpha_1 \ln(z_0 - z_1) + O(n^{-1} \ln n) + O(\delta n \ln n). \tag{5-30}$$

Substituting (5-26) here, we obtain

$$D_n(\alpha_0, \alpha_1)$$
$$= n^{\alpha_0^2 + \alpha_1^2} \prod_{j=0}^{1} \frac{G(1+\alpha_j)^2}{G(1+2\alpha_j)} \left[\frac{(z_0 - z_1)^2}{z_0 z_1 e^{3\pi i}} \right]^{-\alpha_0 \alpha_1} (1 + O(n^{-1} \ln n) + O(\delta n \ln n))$$
$$= n^{\alpha_0^2 + \alpha_1^2} \prod_{j=0}^{1} \frac{G(1+\alpha_j)^2}{G(1+2\alpha_j)} |z_0 - z_1|^{-\alpha_0 \alpha_1} (1 + O(n^{-1} \ln n) + O(\delta n \ln n)) \tag{5-31}$$

(to write the last equation we recall (3-13), the way the branch of $(z - z_j)^{\alpha_j}$ was fixed there, and the fact that $\arg z_0 < \arg z_1$).

Continuing this way, we finally obtain by induction for any fixed m, $\beta_j = 0$, $j = 0, 1, \ldots, m$, the asymptotic expression

$$D_n(\alpha_0, \ldots, \alpha_m) = n^{\sum_{j=0}^{m} \alpha_j^2}$$

$$\times \prod_{j=0}^{m} \frac{G(1+\alpha_j)^2}{G(1+2\alpha_j)} \prod_{0 \le j < k \le m} |z_j - z_k|^{-\alpha_j \alpha_k} (1 + O(n^{-1} \ln n) + O(\delta n \ln n)). \quad (5\text{-}32)$$

We now add in the β-singularities. We will make use of one more identity, which follows from (5-28):

$$\int_0^\beta \left((\alpha + x) \frac{d}{dx} \ln \Gamma(1 + \alpha + x) + (\alpha - x) \frac{d}{dx} \ln \Gamma(1 + \alpha - x) - 2x \right) dx$$

$$= \ln \frac{G(1 + \alpha + \beta) G(1 + \alpha - \beta)}{G(1 + \alpha)^2}. \quad (5\text{-}33)$$

First, we obtain the result for the case when $-\frac{1}{4} < \Re\beta_j < \frac{1}{4}$. This implies that the order of the error term $O(\delta n \ln n)$ remains $o(1)$ as we integrate over the β_j starting at zero. As before, we always assume integration along a path in Λ. Setting $\nu = 0$, $\beta_j = 0$, $j = 1, \ldots, m$ in (5-2), and integrating this identity over β_0 from zero to a fixed β_0, we obtain using (5-33):

$$\ln \frac{D_n(\alpha_0, \ldots, \alpha_m; \beta_0)}{D_n(\alpha_0, \ldots, \alpha_m; 0)} = -\beta_0^2 \ln n + \ln \frac{G(1 + \alpha_0 + \beta_0) G(1 + \alpha_0 - \beta_0)}{G(1 + \alpha_0)^2}$$

$$+ \beta_0 \sum_{j \neq 0} \alpha_j \ln \frac{z_0 e^{i\pi}}{z_j} + O(n^{-1} \ln n) + O(\delta n \ln n). \quad (5\text{-}34)$$

Substituting (5-32) here, we obtain

$$D_n(\alpha_0, \ldots, \alpha_m; \beta_0)$$

$$= n^{\sum_{j=0}^{m} \alpha_j^2 - \beta_0^2} \frac{G(1 + \alpha_0 + \beta_0) G(1 + \alpha_0 - \beta_0)}{G(1 + \alpha_0)^2} \prod_{j=1}^{m} \frac{G(1 + \alpha_j)^2}{G(1 + 2\alpha_j)}$$

$$\times \prod_{0 \le j < k \le m} |z_j - z_k|^{-\alpha_j \alpha_k} \prod_{j=1}^{m} \left(\frac{z_0 e^{i\pi}}{z_j} \right)^{\alpha_j \beta_0} (1 + O(n^{-1} \ln n) + O(\delta n \ln n)). \quad (5\text{-}35)$$

Next, set $\nu = 1$, $\beta_j = 0$, $j = 2, \ldots n$ in (5-2) and integrate over β_1. We obtain then the determinant $D_n(\alpha_0, \ldots, \alpha_m; \beta_0, \beta_1)$. Continuing this procedure, we finally obtain by induction at step m the asymptotics (1-10) for the case $-\frac{1}{4} < \Re\beta_j < \frac{1}{4}$ with $V \equiv 0$ and the error term $O(n^{-1} \ln n) + O(\delta n \ln n) = o(1)$.

Consider now the general case $\|\|\beta\|\| < \frac{1}{2}$. We can choose q such that all $\Re\beta_j \in (q - \frac{1}{4}, q + \frac{1}{4})$. Divide $(0, q)$ into subintervals of length less than $\frac{1}{2}$. Apply the above integration procedure to move all β_j from zero to the line where the real part of all β_j is the right end of the first subinterval. Since the length of subintervals is less than $\frac{1}{2}$, the error term $O(\delta n \ln n)$ remains $o(1)$. Recall that during the integration we avoid any points where $\alpha_j + \beta_j$ or $\alpha_j - \beta_j$ is a negative integer. Next, move the β_j to the right end of the second subinterval, and so on, until the point $\Re \beta_j = q$. From that point move the β_j as needed. We thus obtain Theorem 1.1 with $V \equiv 0$ and $|||\beta||| < \frac{1}{2}$.

5B. *Pure Fisher–Hartwig singularities: extension to $|||\beta||| < 1$.* We now show that in fact the error term in (1-10) remains $o(1)$ for the full range $|||\beta||| < 1$. First, recall the definition of $\widetilde{R}$ and ω in (4-30) and (4-31). We have $|||\beta||| = 2 \max_j (\Re \beta_j - \omega)$ and

$$-\tfrac{1}{2} < \Re \beta_j - \omega < \tfrac{1}{2}$$

for all singular points z_j. We denote $p_j = z_j$ if $\Re \beta_j - \omega > 0$, and m_+ the number of such points. Furthermore, denote $q_j = z_j$ if $\Re \beta_j - \omega < 0$, and m_- the number of such points. Finally, let $r_j = z_j$ if $\Re \beta_j - \omega = 0$.

Separating the main contributions in n (see (4-21)), we write the jump matrix for $\widetilde{R}$ on ∂U_{z_j} (cf. (4-32)) in the form

$$I + n^{\omega \sigma_3} \Delta_1(z) n^{-\omega \sigma_3} + \cdots = I + \hat{\Delta}_1(z) + \hat{D}(z) + O(n^{-1-\rho}), \quad z \in \partial U_{z_j}, \quad (5\text{-}36)$$

where $\hat{\Delta}_1(z)$ is about to be defined and

$$\rho = 1 - |||\beta|||, \tag{5-37}$$

$$\hat{D}(z) = n^{\omega \sigma_3} \Delta_1(z) n^{-\omega \sigma_3} - \hat{\Delta}_1(z), \quad z \in \partial U_{z_j}. \tag{5-38}$$

For $z \in \partial U_{p_j}$ we set

$$\hat{\Delta}_1(z) = (n^{\omega \sigma_3} \Delta_1(z) n^{-\omega \sigma_3})_{21} \sigma_- = \frac{b_j(z)}{z - p_j} \sigma_-, \quad \text{with } \sigma_- = \begin{pmatrix} 0 & 0 \\ 1 & 0 \end{pmatrix} \text{ and}$$

$$b_j(z) = -\frac{\Gamma(1+\alpha_j - \beta_j)}{\Gamma(\alpha_j + \beta_j)} \left(\frac{\mathcal{D}(z)}{\zeta^{\beta_j} F_j(z)} \right)^{-2} p_j^{-n} e^{-i\pi(2\beta_j - \alpha_j)} \frac{z - p_j}{n \ln(z/p_j)} n^{-2\omega}. \tag{5-39}$$

For $z \in \partial U_{q_j}$ we set

$$\hat{\Delta}_1(z) = (n^{\omega \sigma_3} \Delta_1(z) n^{-\omega \sigma_3})_{12} \sigma_+ = \frac{a_j(z)}{z - q_j} \sigma_+, \quad \text{with } \sigma_+ = \begin{pmatrix} 0 & 1 \\ 0 & 0 \end{pmatrix} \text{ and}$$

$$a_j(z) = \frac{\Gamma(1 + \alpha_j + \beta_j)}{\Gamma(\alpha_j - \beta_j)} \left(\frac{\mathcal{D}(z)}{\zeta^{\beta_j} F_j(z)} \right)^2 q_j^n e^{i\pi(2\beta_j - \alpha_j)} \frac{z - q_j}{n \ln(z/q_j)} n^{2\omega}. \tag{5-40}$$

Finally,

$$\hat{\Delta}_1(z) = 0 \quad \text{when } z \in \partial U_{r_j}. \tag{5-41}$$

Note that $\hat{D}(z)$ and $\hat{\Delta}_1(z)$ are meromorphic functions in a neighborhood of U_{z_j} with a simple pole at $z = z_j$. We have (see (4-24), (4-25), and recall that $V \equiv 0$)

$$b_j = b_j^{(n)} \equiv \lim_{z \to p_j} b_j(z) = -n^{2(\beta_j - \omega) - 1} p_j^{-n+1} \mu_j^{-2} \frac{\Gamma(1 + \alpha_j - \beta_j)}{\Gamma(\alpha_j + \beta_j)} = O(n^{-\rho}),$$

$$\tag{5-42}$$

$$a_j = a_j^{(n)} \equiv \lim_{z \to q_j} a_j(z) = n^{-2(\beta_j - \omega) - 1} q_j^{n+1} \mu_j^2 \frac{\Gamma(1 + \alpha_j + \beta_j)}{\Gamma(\alpha_j - \beta_j)} = O(n^{-\rho}). \tag{5-43}$$

Also note that

$$\hat{D}(z) = O(n^{-1}), \quad z \in \partial U_{z_j}. \tag{5-44}$$

The main idea which will allow us to give the required estimate for the error term in (1-10) is the following. Write $\widetilde{R}$ in the form

$$\widetilde{R}(z) = Q(z)\hat{R}(z),$$

where $\hat{R}(z)$ is the solution to the following RHP:

$$\hat{R}(z) \text{ is analytic for } z \in \mathbb{C} \setminus \bigcup_j \partial U_{z_j}. \tag{5-45}$$

$$\hat{R}(z)_+ = \hat{R}(z)_-(I + \hat{\Delta}_1) \quad \text{if } z \in \bigcup_j \partial U_{z_j}, \tag{5-46}$$

$$\hat{R}(z) = I + O(1/z) \quad \text{as } z \to \infty. \tag{5-47}$$

We solve this RHP below explicitly, however, note first that the solution exists and is unique and

$$\hat{R}(z) = I + O(n^{-\rho}) \tag{5-48}$$

uniformly in z by standard arguments. Using (5-48), the jump condition (5-46), and the nilpotency of $\hat{\Delta}_1$, we then have on ∂U_{z_j}

$$Q_+ = \widetilde{R}_+ \hat{R}_+^{-1} = \widetilde{R}_-(I + \hat{\Delta}_1(z) + \hat{D}(z) + O(n^{-1-\rho}))\hat{R}_+^{-1}$$

$$= Q_-(I + \hat{D}(z) + O(n^{-1-\rho})).$$

The jump matrix for Q on Σ^{out}, Σ''^{out} remains exponentially close to the identity. Therefore,

$$Q(z) = I + Q_1(z) + O(n^{-1-\rho}), \quad Q_1(z) = \frac{1}{2\pi i} \int_{\bigcup_j \partial U_{z_j}} \frac{\hat{D}(s)}{s - z} \, ds \tag{5-49}$$

In what follows, we will be interested in the matrix element $Y_{21}(z)$ of our original RHP in a neighborhood of $z = 0$. In this neighborhood, we have using

(5-49),

$$Y(z) = R(z)N(z) = n^{-\omega\sigma_3} Q(z) \hat{R}(z) n^{\omega\sigma_3} N(z)$$

$$= n^{-\omega\sigma_3} (I + Q_1(z) + O(n^{-1-\rho})) \hat{R}(z) n^{\omega\sigma_3} \mathcal{D}(z)^{\sigma_3} \begin{pmatrix} 0 & 1 \\ -1 & 0 \end{pmatrix}. \qquad (5\text{-}50)$$

From here, noting that $\mathcal{D}(0) = 1$ as $V(z) \equiv 0$, and using the estimates $\hat{R}(z) = I + O(n^{-\rho})$, $Q_1(0) = O(n^{-1})$, we obtain

$$\chi_{n-1}^2 = -Y_{21}(0) = (1 + Q_{1,22}(0) + O(n^{-1-\rho})) \hat{R}_{22}(0).$$

Using the expression for Q_1 from (5-49) and (5-38), we can write this equation in the form (recall that the contours ∂U_{z_j} are oriented in the negative direction):

$$\chi_{n-1}^2 = \left(1 - \frac{1}{n} \sum_{j=0}^{m} (\alpha_j^2 - \beta_j^2) + O(n^{-1-\rho})\right) \hat{R}_{22}(0). \qquad (5\text{-}51)$$

Eventually, we will use the product of these quantities over n to represent the determinant. Before doing that, we now solve the RHP for $\hat{R}$ and find $\hat{R}_{22}(0)$.

Set

$$\Phi(z) = \begin{cases} \hat{R}(z) & \text{if } z \in \mathbb{C} \setminus \bigcup_j U_{z_j}, \\ \hat{R}(z)(I + \hat{\Delta}_1(z)) & \text{if } z \in \bigcup_j U_{z_j}. \end{cases} \qquad (5\text{-}52)$$

So defined, $\Phi(z)$ is obviously a meromorphic function with simple poles at the points z_j, which tends to I at infinity. Therefore, it can be written in the form

$$\Phi(z) = I + \sum_{j=1}^{m_+} \frac{\Phi_j^+}{z - p_j} + \sum_{j=1}^{m_-} \frac{\Phi_j^-}{z - q_j} \qquad (5\text{-}53)$$

for some constant matrices $\Phi_j^\pm$.

Moreover, the function $\hat{R}(z)$, which has the expression

$$\hat{R}(z) = \Phi(z)(I + \hat{\Delta}_1)^{-1}$$

$$= \left[I + \sum_{j=1}^{m_+} \frac{\Phi_j^+}{z - p_j} + \sum_{j=1}^{m_-} \frac{\Phi_j^-}{z - q_j}\right] \left(I - \frac{b_k(z)}{z - p_k} \sigma_-\right) \qquad (5\text{-}54)$$

in U_{p_k}, $k = 1, \ldots, m_+$ (we use here the definition of $\hat{\Delta}_1$), is analytic there. Hence, the coefficients at negative powers of $z - p_k$ vanish. Equating the coefficient at $(z - p_k)^{-2}$ to zero, we obtain

$$\Phi_k^+ \sigma_- = 0, \qquad (5\text{-}55)$$

and therefore the matrix Φ_k^+ has the form

$$\Phi_k^+ = \begin{pmatrix} g_k & 0 \\ f_k & 0 \end{pmatrix} \tag{5-56}$$

for some constants g_k, f_k. The vanishing of the coefficient at $(z - p_k)^{-1}$ gives the following condition on $\Phi_j^\pm$, where we used the equation $\Phi_j^+ \sigma_- = 0$,

$$\Phi_k^+ - b_k \sum_{j=1}^{m_-} \frac{\Phi_j^- \sigma_-}{p_k - q_j} = b_k \sigma_-, \quad k = 1, \ldots, m_+, \tag{5-57}$$

with b_k given by (5-42). Similarly, using the analyticity of $\Phi(z)(I + \hat{\Delta}_1(z))^{-1}$ in U_{q_k}, we obtain for all $k = 1, \ldots, m_-$

$$\Phi_k^- = \begin{pmatrix} 0 & e_k \\ 0 & h_k \end{pmatrix}, \tag{5-58}$$

$$\Phi_k^- - a_k \sum_{j=1}^{m_+} \frac{\Phi_j^+ \sigma_+}{q_k - p_j} = a_k \sigma_+ \tag{5-59}$$

for some constants e_k, h_k. Conditions (5-59), (5-57) are equations for the constants g_k, f_k, e_k, h_k. In view of (5-51), we are interested in

$$\hat{R}_{22}(0) = \Phi_{22}(0) = 1 - \sum_{j=1}^{m_-} \frac{h_j}{q_j}, \tag{5-60}$$

where we used (5-53) to write the last equation. To calculate this quantity we first substitute Φ_j^+ from (5-57) into (5-59). Then the 2,2-element of the resulting equations for $k = 1, \ldots, m_-$ can be written as follows:

$$(I - A)h = B, \tag{5-61}$$

where h and B are m_--dimensional vectors with components h_k and

$$B_k = \sum_{j=1}^{m_+} \frac{a_k b_j}{q_k - p_j}, \quad k = 1, \ldots, m_-,$$

$A = A^{(n)}$ is an $m_- \times m_-$ matrix with matrix elements

$$A_{k,\ell} = \sum_{j=1}^{m_+} \frac{a_k b_j}{(q_k - p_j)(p_j - q_\ell)},$$

and I is the $m_- \times m_-$ identity matrix.

Define the $m_- \times m_-$ diagonal matrix Δ as follows

$$\Delta = \text{diag}\{-q_1, -q_2, \ldots, -q_{m_-}\}.$$

Then (5-61) can be written in the form

$$Tx = y, \quad T = I - \Delta^{-1} A \Delta, \quad x = \Delta^{-1} h, \quad y = \Delta^{-1} B$$

By Cramer's rule,

$$x_k = \frac{\det(T_1 \cdots y \cdots T_{m_-})}{\det T},$$

where T_j are the columns of T and y is in the place of the k-th column. We are interested in

$$1 - \sum_{j=1}^{m_-} \frac{h_j}{q_j} = 1 + \sum_{j=1}^{m_-} x_j = \frac{\det(T_1 + y, T_2 + y, \cdots, T_{m_-} + y)}{\det T}. \tag{5-62}$$

First note that

$$\det T = \det(1 - \Delta^{-1} A \Delta) = \det(I - A) = \det(I - A^{(n)}). \tag{5-63}$$

Second, a direct calculation shows that

$$(T_1 + y, T_2 + y, \cdots, T_{m_-} + y) = I - A', \quad A'_{jk} = \sum_{\ell=1}^{m_+} \frac{a'_j b'_\ell}{(q_j - p_\ell)(p_\ell - q_k)},$$

where

$$a'_j = a_j q_j^{-1}, \quad b'_\ell = b_\ell p_\ell.$$

Using the definitions (5-43), (5-42) of a_j, b_j, we note that

$$a'_j = a_j^{(n-1)} + O(n^{-1-\rho}),$$
$$b'_j = b_j^{(n-1)} + O(n^{-1-\rho}),$$

and therefore $A' = A^{(n-1)} + O(n^{-1-\rho})$. Thus we can rewrite (5-62) as

$$\hat{R}_{22}(0) = \Phi_{22}(0) = 1 - \sum_{j=1}^{m_-} \frac{h_j}{q_j} = \frac{\det(I - A^{(n-1)})}{\det(I - A^{(n)})} \left[1 + O\left(\frac{1}{n^{1+\rho}} \right) \right]. \tag{5-64}$$

Note that

$$\det(I - A^{(n)}) = 1 + O(n^{-2\rho}).$$

Recalling now (5-51) and the representation of the Toeplitz determinant $D_n(f)$ as a product of χ_k^{-2} and using (5-64), we can write, for some sufficiently large $n_0 > 0$,

$$D_n(f)$$

$$= D_{n_0}(f) \prod_{k=n_0+1}^{n} \chi_{k-1}^{-2}$$

$$= D_{n_0}(f) \prod_{k=n_0+1}^{n} \left[1 + \frac{1}{k} \sum_{j=0}^{m} (\alpha_j^2 - \beta_j^2) + O\left(\frac{1}{k^{1+\rho}}\right) \right] \frac{\det(I - A^{(k)})}{\det(I - A^{(k-1)})}$$

$$\times \left[1 + O\left(\frac{1}{k^{1+\rho}}\right) \right]$$

$$= C(n_0, \alpha_0, \ldots, \alpha_m, \beta_0, \ldots, \beta_m) n^{\sum_{j=0}^{m}(\alpha_j^2 - \beta_j^2)} \left[1 + O\left(\frac{1}{n^\rho}\right) \right], \qquad (5\text{-}65)$$

where all the error terms are uniform for α_j, β_j in compact sets, and C is a constant depending analytically on α_j, β_j, and n_0 only. Recall that in the derivation of this expression we assumed that $|||\beta||| < 1$ (and as usual $\Re \alpha_j > -\frac{1}{2}$ for all j). Under this condition $\rho > 0$ (see (5-37)), and the error term tends to zero as $n \to \infty$. In the previous section, we obtained C explicitly for β_j satisfying $|||\beta||| < \frac{1}{2}$. Obviously, if all α_j, β_j belong to fixed compact sets and the value of n_0 is fixed, the constant C in (5-65) is bounded above in absolute value by a constant independent of any α_j, β_j. By Vitali's theorem this fact implies that C can be analytically continued in β_j to the full domain $|||\beta||| < 1$ off its values on the domain $|||\beta||| < \frac{1}{2}$. Thus C is given by the same expression also for $|||\beta||| < 1$, and this concludes the proof of Theorem 1.1 for $|||\beta||| < 1$, $V(z) \equiv 0$, with the error term $o(1) = O(n^{|||\beta|||-1})$ in (1-10).

5C. *Adding special analytic $V(z)$.* In this section and in the next one, we will add the multiplicative factor $e^{V(z)}$, where V is analytic in a neighborhood of the unit circle C, to a symbol with pure Fisher–Hartwig singularities and obtain the asymptotics of the corresponding determinant. Consider the deformation of the symbol $f(z, t)$ given by (3-21) for $t \in [0, 1]$. The analysis is based on integration of the differential identity (3-26) over t. In the present section we assume that V is such that

$$1 - t + t e^{V(z)} \neq 0, \quad t \in [0, 1], \quad z \in C, \qquad (5\text{-}66)$$

and $1 - t + t e^{V(z)}$ has no winding around C for all $t \in [0, 1]$. Then the Riemann–Hilbert problem (for the polynomials orthogonal) with $f(z, t)$ has the same singularities as the problem with $f(z)$ and is solved in the same way. In the following section we remove the condition (5-66).

Let us rewrite the identity (3-26) in terms of the function $S(z) = S(z, t)$ that is the solution of the Riemann–Hilbert problem posed on the deformed contour depicted in Figure 1. From the transformation $Y \to S$ defined in (4-1), (4-2) we obtain

$$Y_{11} = z^n f^{-1} S_{12,+} + S_{11,+}, \quad Y_{21} = z^n f^{-1} S_{22,+} + S_{21,+}, \quad z \in C \equiv \Sigma'. \quad (5\text{-}67)$$

Substituting these expressions into (3-26) and taking into account that $\det S = 1$ we arrive at the formula

$$\frac{\partial}{\partial t} \ln D_n(f(z,t))$$

$$= n \int_C f^{-1} \dot{f} \frac{dz}{2\pi i z} + \int_C \left[-f^{-2} f' + 2 f^{-1} \left(S'_{22,+} S_{11,+} - S'_{12,+} S_{21,+} \right) \right] \dot{f} \frac{dz}{2\pi i}$$

$$+ \int_C \left[z^{-n} \left(S'_{21,+} S_{11,+} - S'_{11,+} S_{21,+} \right) + z^n \left(S'_{22,+} S_{12,+} - S'_{12,+} S_{22,+} \right) f^{-2} \right] \dot{f} \frac{dz}{2\pi i}, \quad (5\text{-}68)$$

where we introduced the notation $\dot{f} := \partial f / \partial t$ and $f' := \partial f / \partial z$. Using the jump relation (4-4) satisfied by the function $S(z,t)$ across the unit circle, we can rewrite (5-68) more symmetrically as

$$\frac{\partial}{\partial t} \ln D_n(f(z,t)) = n \int_C f^{-1} \dot{f} \frac{dz}{2\pi i z} + X(t), \quad (5\text{-}69)$$

with

$$X(t) := \int_C \left[S'_{22,+} S_{11,+} - S'_{12,+} S_{21,+} + S'_{11,-} S_{22,-} - S'_{21,-} S_{12,-} \right] f^{-1} \dot{f} \frac{dz}{2\pi i}$$

$$+ \int_C \left[z^{-n} \left(S'_{22,-} S_{12,-} - S'_{12,-} S_{22,-} \right) + z^n \left(S'_{22,+} S_{12,+} - S'_{12,+} S_{22,+} \right) \right] f^{-2} \dot{f} \frac{dz}{2\pi i}.$$

Analytically continuing the boundary values of $S(z,t)$ from the $+$ side of C to the $-$ side of Σ'', and from the $-$ side of C to the $+$ side of Σ, we can write the term $X(t)$ in (5-69) in the form

$$X(t) = \int_{\Sigma''} (J_- + z^n I_-) f^{-1} \dot{f} \frac{dz}{2\pi i} + \int_{\Sigma} (-J_+ + z^{-n} I_+) f^{-1} \dot{f} \frac{dz}{2\pi i}, \quad (5\text{-}70)$$

where

$$I = (S'_{22} S_{12} - S'_{12} S_{22}) f^{-1}, \quad J = S'_{22} S_{11} - S'_{12} S_{21}. \quad (5\text{-}71)$$

Denote by Σ_ε (resp., Σ''_ε) the part of Σ (resp., Σ'') that lies inside $\bigcup_{j=0}^m U_{z_j}$. Consider first

$$\int_{\Sigma''_\varepsilon} (J_- + z^n I_-) f^{-1} \dot{f} \frac{dz}{2\pi i}. \quad (5\text{-}72)$$

Using (4-3) on Σ'' and (5-71), we easily obtain that $J_- = J_+ - z^n I_+$ and $I_- = I_+$, and therefore,

$$\int_{\Sigma''_\varepsilon} (J_- + z^n I_-) f^{-1} \dot{f} \frac{dz}{2\pi i} = \int_{\Sigma''_\varepsilon} J_+ f^{-1} \dot{f} \frac{dz}{2\pi i} = \int_{C''_\varepsilon} J f^{-1} \dot{f} \frac{dz}{2\pi i}, \quad (5\text{-}73)$$

where C''_ε is the part of $\bigcup_{j=0}^{m} \partial U_{z_j}$ lying inside the unit circle from the intersection of ∂U_{z_j} with the incoming Σ'' to the intersection with the outgoing Σ'' for each j. Note that $J(z, t)$ has no nonintegrable singularity at z_j. Indeed, for z inside the smaller sector formed by Σ'' at z_j, we can write

$$S(z, t) = Y(z, t) = \hat{Y}(z, t) \begin{pmatrix} 1 & \kappa(z, t) \\ 0 & 1 \end{pmatrix}, \tag{5-74}$$

where $\kappa(z, t) = c_1(z, t)(z - z_j)^{2\alpha_j}$, if $\alpha_j \neq 0$, and $\kappa(z, t) = c_2(z, t) \ln(z - z_j)$, if $\alpha_j = 0$, $\beta_j \neq 0$, for some $c_k(z, t)$ analytic near z_j, chosen so that $\hat{Y}(z)$ is analytic in a neighborhood of z_j. Writing J in terms of the matrix elements of $\hat{Y}$, we see that the contributions of the (singular) derivative $\kappa'(z)$ cancel, and we obtain

$$J = (\hat{Y}'_{21} \hat{Y}_{11} - \hat{Y}_{21} \hat{Y}'_{11})\kappa + \hat{Y}'_{22} \hat{Y}_{11} - \hat{Y}'_{12} \hat{Y}_{21}. \tag{5-75}$$

We now analyze (5-73) asymptotically. The asymptotic expression for $S(z, t)$ inside the unit circle and outside $\bigcup_{j=0}^{m} U_{z_j}$ is given by (see (4-26), (4-7))

$$S(z, t) = R(z, t) e^{g(z,t)\sigma_3} \begin{pmatrix} 0 & 1 \\ -1 & 0 \end{pmatrix}, \tag{5-76}$$

where $g(z)$ is defined by the formula

$$\mathcal{D}(z, t) = e^{g(z,t)}. \tag{5-77}$$

We have

$$J = g' + R'_{11} R_{22} - R_{12} R'_{21} = g' + O_\varepsilon(1/n), \quad n \to \infty,$$

and we finally obtain

$$\int_{\Sigma''_\varepsilon} (J_- + z^n I_-) f^{-1} \dot{f} \frac{dz}{2\pi i} = \int_{C''_\varepsilon} J f^{-1} \dot{f} \frac{dz}{2\pi i}$$

$$= \int_{C''_\varepsilon} g'(z) f^{-1} \dot{f} \frac{dz}{2\pi i} + O_\varepsilon(1/n), \quad n \to \infty. \tag{5-78}$$

Similarly, using the jump condition for S and then the asymptotics for S outside the unit circle, we obtain

$$\int_{\Sigma_\varepsilon} (-J_+ + z^{-n} I_+) f^{-1} \dot{f} \frac{dz}{2\pi i} = -\int_{C_\varepsilon} J f^{-1} \dot{f} \frac{dz}{2\pi i}$$

$$= \int_{C_\varepsilon} g'(z) f^{-1} \dot{f} \frac{dz}{2\pi i} + O_\varepsilon(1/n), \quad n \to \infty, \tag{5-79}$$

where C_ε is the part of $\bigcup_{j=0}^{m} \partial U_{z_j}$ lying outside the unit circle from the intersection of ∂U_{z_j} with the incoming Σ to the intersection with the outgoing Σ for each j.

Returning to the integrals (5-70), we now consider the part arising from the integration over $\Sigma'' \setminus \Sigma_\varepsilon''$ and $\Sigma \setminus \Sigma_\varepsilon$. In this part the terms containing $z^{\pm n} I_\mp$ give a contribution that is exponentially small in n, while the integration of the terms with J can be replaced by the integration over $\Sigma' \setminus \Sigma_\varepsilon'$, where $\Sigma_\varepsilon' = \Sigma' \cap (\bigcup_{j=0}^m U_{z_j})$, and over parts of the boundaries ∂U_{z_j}. Thus, recalling also (5-78), (5-79), we have

$$X(t) = \int_{\Sigma' \setminus \Sigma_\varepsilon'} (J_+ - J_-) f^{-1} \dot{f} \frac{dz}{2\pi i}$$

$$+ \sum_{j=0}^m \left(\int_{\partial U_{z_j}^+} + \int_{\partial U_{z_j}^-} \right) g'(z) f^{-1} \dot{f} \frac{dz}{2\pi i} + O_\varepsilon(1/n), \quad n \to \infty, \quad (5\text{-}80)$$

where $\partial U_{z_j}^+$ (resp., $\partial U_{z_j}^-$) is the part of the boundary of U_{z_j} inside (resp., outside) the unit circle oriented from the intersection with the incoming Σ' to the intersection with the outgoing Σ'.

Note that using the same considerations as before, we can write in (5-80) $J_+ - J_- = g_+' + g_-' + O_\varepsilon(1/n)$, and therefore

$$X(t) = \int_{\Sigma_+} g_+' f^{-1} \dot{f} \frac{dz}{2\pi i} + \int_{\Sigma_-} g_-' f^{-1} \dot{f} \frac{dz}{2\pi i} + O_\varepsilon(1/n), \quad n \to \infty, \quad (5\text{-}81)$$

where the closed anticlockwise oriented contours

$$\Sigma_+ = (\Sigma' \setminus \Sigma_\varepsilon') \bigcup_{j=0}^m \partial U_{z_j}^+, \quad \Sigma_- = (\Sigma' \setminus \Sigma_\varepsilon') \bigcup_{j=0}^m \partial U_{z_j}^-. \quad (5\text{-}82)$$

(Note that one can deform Σ_+ (resp., Σ_-) to a circle around zero of radius $1 - \varepsilon$ (resp., $1 + \varepsilon$).)

By (3-21),

$$f^{-1} \dot{f} = \frac{-1 + e^{V(z)}}{1 - t + t e^{V(z)}} = \frac{\partial}{\partial t} \ln \left(1 - t + t e^{V(z)} \right). \quad (5\text{-}83)$$

Furthermore, writing g in the form

$$g(z, t) = g^{Sz}(z, t) + g^{FH}(z), \quad (5\text{-}84)$$

we have, by (5-77), (4-8), (4-9), (4-10),

$$g^{Sz}(z, t) = \int_C \frac{\ln(1 - t + t e^{V(s)})}{s - z} \frac{ds}{2\pi i},$$

$$g^{FH}(z) = \begin{cases} \displaystyle\sum_{k=1}^m (\alpha_k + \beta_k) \ln \frac{z - z_k}{z_k e^{i\pi}} & \text{if } |z| < 1, \\[2ex] \displaystyle\sum_{k=1}^m (-\alpha_k + \beta_k) \ln \frac{z - z_k}{z} & \text{if } |z| > 1. \end{cases} \quad (5\text{-}85)$$

The solution of the Riemann–Hilbert problem is uniform in $t \in [0, 1]$. From this fact and the explicit formulas (5-83), (5-84) and (5-85), we conclude, by an argument similar to the argument following (5-23) that the identity (5-69) holds for all $t \in [0, 1]$. Now integrating (5-69) from $t = 0$ to $t = 1$, we connect the Toeplitz determinant $D_n(f(z, 1))$ with the Toeplitz determinant $D_n(f(z, 0))$ that represents the "pure" Fisher–Hartwig case and whose asymptotics we evaluated in the previous section. First, using (5-83) and changing the order of integration in the first term of (5-69), we obtain

$$\int_0^1 dt \int_C f^{-1} \dot{f} \frac{dz}{2\pi i z} = \frac{1}{2\pi} \int_0^{2\pi} V(e^{i\theta})\, d\theta = V_0. \tag{5-86}$$

Furthermore, by (5-81), (5-84) and (5-85),

$$\int_0^1 dt\, X(t) = I^{Sz} + I^{FH} + O_\varepsilon(1/n), \quad n \to \infty, \tag{5-87}$$

where (compare [Deift 1999, (86), (87)])

$$I^{Sz} = \int_0^1 dt \int_C \left((g^{Sz})_+' + (g^{Sz})_-' \right) f^{-1} \dot{f} \frac{dz}{2\pi i} = \sum_{k=1}^\infty k V_k V_{-k}, \tag{5-88}$$

and

$$I^{FH} = \int_{\Sigma_+} \left(g^{FH}(z) \right)_+' V(z) \frac{dz}{2\pi i} + \int_{\Sigma_-} \left(g^{FH}(z) \right)_-' V(z) \frac{dz}{2\pi i} \tag{5-89}$$

$$= \sum_{k=0}^m \left[(\alpha_k + \beta_k) \int_{\Sigma_+} \frac{V(z)}{z - z_k} \frac{dz}{2\pi i} + (-\alpha_k + \beta_k) \int_{\Sigma_-} \left(\frac{V(z)}{z - z_k} - \frac{V(z)}{z} \right) \frac{dz}{2\pi i} \right].$$

Since

$$g^{Sz}(z, 1) = \int_C \frac{V(s)}{s - z} \frac{ds}{2\pi i} = \begin{cases} \ln b_+(z) + V_0 & \text{if } |z| < 1, \\ -\ln b_-(z) & \text{if } |z| > 1, \end{cases} \tag{5-90}$$

we obtain

$$\ln b_+(z_k) = \int_{\Sigma_-} \frac{V(z)}{z - z_k} \frac{dz}{2\pi i} - V_0, \quad \ln b_-(z_k) = -\int_{\Sigma_+} \frac{V(z)}{z - z_k} \frac{dz}{2\pi i}, \tag{5-91}$$

which finally gives

$$I^{FH} = \sum_{k=0}^m \left[-(\alpha_k + \beta_k) \ln b_-(z_k) + (-\alpha_k + \beta_k) \ln b_+(z_k) \right]. \tag{5-92}$$

Collecting (5-86), (5-87), (5-88), and (5-92), we obtain from (5-69)

$$\ln D_n(f(z, 1)) - \ln D_n(f(z, 0))$$

$$= nV_0 + \sum_{k=1}^{\infty} kV_k V_{-k} + \sum_{k=0}^{m} \left[-(\alpha_k + \beta_k) \ln b_-(z_k) + (-\alpha_k + \beta_k) \ln b_+(z_k)\right]$$

$$+ O_\varepsilon(1/n), \quad n \to \infty, \quad (5\text{-}93)$$

which, in view of the result of the previous section, concludes the proof of
Theorem 1.1 for analytic $V(z)$ satisfying the condition (5-66).

5D. *Extension to general analytic $V(z)$.* Now let $V(z)$ be any function analytic
in a neighborhood of the unite circle. Since zeros of the expression $1 - t + te^{V(z)}$,
$t \in [0, 1]$, $z \in C$, can only occur if $\Im V(z) = \pi(2k + 1)$, $k \in \mathbb{Z}$, there exists a
positive integer q such that $\frac{1}{q}V(z)$ satisfies the condition (5-66) of the previous
section, i.e.,

$$1 - t + te^{\frac{1}{q}V(z)} \neq 0 \quad \text{for all } t \in [0, 1] \text{ and } z \in C,$$

and this function has no winding around C for all $t \in [0, 1]$. Let

$$f_0(z) = f^{FH}(z), \quad f_\ell(z) = e^{\frac{1}{q}V(z)} f_{\ell-1}(z), \quad \ell = 1, \dots, q, \quad (5\text{-}94)$$

where $f^{FH}(z)$ is the symbol for the "pure" Fisher–Hartwig case. Note that

$$f(z) \equiv e^{V(z)} f^{FH}(z) = f_q(z). \quad (5\text{-}95)$$

Consider $f_\ell(z)$, $\ell = 1, \dots, q$, and introduce the deformation,

$$f_\ell(z, t) = \left(1 - t + te^{\frac{1}{q}V(z)}\right) f_{\ell-1}(z) = \left(1 - t + te^{\frac{1}{q}V(z)}\right) e^{\frac{\ell-1}{q}V(z)} f^{FH}(z). \quad (5\text{-}96)$$

All the considerations of the part of the previous section between equations
(5-67) and (5-81) go through with $f(z, t)$ replaced by $f_\ell(z, t)$ and we arrive at
the formula

$$\frac{\partial}{\partial t} \ln D_n(f_\ell(z, t)) = n \int_C f_\ell^{-1} \dot{f}_\ell \frac{dz}{2\pi i z} + X(t), \quad (5\text{-}97)$$

with

$$X(t) = \int_{\Sigma_+} g_+' f_\ell^{-1} \dot{f}_\ell \frac{dz}{2\pi i} + \int_{\Sigma_-} g_-' f_\ell^{-1} \dot{f}_\ell \frac{dz}{2\pi i} + O_\varepsilon(1/n), \, n \to \infty,$$

where we have, as before, the closed anticlockwise oriented contours

$$\Sigma_+ = (\Sigma' \setminus \Sigma_\varepsilon') \bigcup_{j=0}^{m} \partial U_{z_j}^+, \quad \Sigma_- = (\Sigma' \setminus \Sigma_\varepsilon') \bigcup_{j=0}^{m} \partial U_{z_j}^-, \quad (5\text{-}98)$$

and $g(z) := g_\ell(z)$ now corresponds to f_ℓ.

The first term in (5-97) yields (cf. (5-86))

$$\int_0^1 dt \int_C f_\ell^{-1} \dot{f}_\ell \frac{dz}{2\pi i z} = \frac{1}{2\pi q} \int_0^{2\pi} V(e^{i\theta})\, d\theta = \frac{1}{q} V_0. \qquad (5\text{-}99)$$

In order to evaluate $X(t)$, we write g in the form

$$g(z, t) = g^{Sz}(z, t) + \tilde{g}^{Sz}(z) + g^{FH}(z), \qquad (5\text{-}100)$$

where $g^{FH}(z)$ is the same as in (5-85), and

$$g^{Sz}(z, t) = \int_C \frac{\ln(1 - t + t e^{\frac{1}{q} V(s)})}{s - z} \frac{ds}{2\pi i}, \quad \tilde{g}^{Sz}(z) = \frac{\ell - 1}{q} \int_C \frac{V(s)}{s - z} \frac{ds}{2\pi i}. \qquad (5\text{-}101)$$

Then we obtain

$$\int_0^1 dt\, X(t) = I^{Sz} + \tilde{I}^{Sz} + I^{FH} + O_\varepsilon(1/n), \quad n \to \infty, \qquad (5\text{-}102)$$

where, up to the replacement $V \to V/q$, the integrals I^{Sz} and I^{FH} are the respective integrals from the previous section:

$$I^{Sz} = \frac{1}{q^2} \sum_{k=1}^{\infty} k V_k V_{-k},$$

$$I^{FH} = \frac{1}{q} \sum_{k=0}^{m} \left[-(\alpha_k + \beta_k) \ln b_-(z_k) + (-\alpha_k + \beta_k) \ln b_+(z_k) \right]. \qquad (5\text{-}103)$$

The term $\tilde{I}^{Sz}$ in (5-102) is given by the equation

$$\tilde{I}^{Sz} = \frac{1}{q} \int_C \left((\tilde{g}^{Sz}(z))'_+ + (\tilde{g}^{Sz}(z))'_- \right) V(z) \frac{dz}{2\pi i}. \qquad (5\text{-}104)$$

Note that

$$(\tilde{g}^{Sz}(z))'_+ = \frac{\ell - 1}{q} \sum_{k=1}^{\infty} k z^{k-1} V_k, \quad (\tilde{g}^{Sz}(z))'_- = \frac{\ell - 1}{q} \sum_{k=1}^{\infty} k z^{-k-1} V_{-k}.$$

Therefore, after a simple calculation we obtain

$$\tilde{I}^{Sz} = \frac{2\ell - 2}{q^2} \sum_{k=1}^{\infty} k V_k V_{-k}. \qquad (5\text{-}105)$$

Integrating (5-97) from $t = 0$ to $t = 1$ and taking into account (5-99), (5-102), (5-103), and (5-105), we obtain the following equation for the determinant $D_n(f_\ell(z))$:

$$\ln D_n(f_\ell(z)) - \ln D_n(f_{\ell-1}(z))$$

$$= \frac{1}{q} n V_0 + \frac{2\ell - 1}{q^2} \sum_{k=1}^{\infty} k V_k V_{-k}$$

$$+ \frac{1}{q} \sum_{k=0}^{m} [-(\alpha_k + \beta_k) \ln b_-(z_k) + (-\alpha_k + \beta_k) \ln b_+(z_k)]$$

$$+ O_\varepsilon(1/n), \quad n \to \infty. \quad (5\text{-}106)$$

This equation holds for any $\ell = 1, \ldots, q$. Summing up from $\ell = 1$ to $\ell = q$ we again arrive at the formula

$$\ln D_n(f(z)) - \ln D_n(f^{FH}(z))$$

$$= n V_0 + \sum_{k=1}^{\infty} k V_k V_{-k} + \sum_{k=0}^{m} [-(\alpha_k + \beta_k) \ln b_-(z_k) + (-\alpha_k + \beta_k) \ln b_+(z_k)]$$

$$+ O_\varepsilon(1/n), \quad n \to \infty, \quad (5\text{-}107)$$

which concludes the proof of Theorem 1.1 in the case of $V(z)$ analytic in a neighborhood of the unit circle.

5E. *Extension to smooth* $V(z)$. If $V(z)$ is just sufficiently smooth, in particular C^∞, on the unit circle C so that (1-11) holds for s from zero up to and including some $s \geq 0$, we can approximate $V(z)$ by trigonometric polynomials $V^{(n)}(z) = \sum_{k=-p(n)}^{p(n)} V_k z^k$, $z \in C$. First, consider the case when

$$\|\beta\| = \max_{j,k} |\Re\beta_j - \Re\beta_k| = 2 \max_j |\Re\beta_j - \omega| < 1,$$

where ω is defined by (4-31). (The indices $j, k = 0$ are omitted if $\alpha_0 = \beta_0 = 0$.) We set

$$p = [n^{1-\nu}], \quad \nu = \|\beta\| + \varepsilon_1, \quad (5\text{-}108)$$

where $\varepsilon_1 > 0$ is chosen sufficiently small so that $\nu < 1$ (square brackets denote the integer part).

First, we need to extend the RH analysis of the previous sections to symbols which depend on n, namely to the case when V in f is replaced by $V^{(n)}$. (We will denote such f by $f(z, V^{(n)})$, and the original f by $f(z, V)$.) We need to have a suitable estimate for the behavior of the error term in the asymptotics with n. For a fixed f, our analysis depended, in particular, on the fact that $f(z)^{-1} z^{-n}$ is of order $e^{-\varepsilon' n}$, $\varepsilon' > 0$, for $z \in \Sigma^{\text{out}}$ (see Section 4A), and similarly, $f(z)^{-1} z^n = O(e^{-\varepsilon' n})$ for $z \in \Sigma''^{\text{out}}$. Here the contours Σ^{out}, Σ''^{out} are outside a *fixed* neighborhood of the unit circle (outside and inside C, respectively). If V is replaced by $V^{(n)}$, let us define the curve Σ outside $\bigcup_{j=0}^{m} U_{z_j}$ by

$$z = \left(1 + \gamma \frac{\ln p}{p}\right) e^{i\theta}, \qquad \gamma > 0, \tag{5-109}$$

and Σ'' outside $\bigcup_{j=0}^{m} U_{z_j}$ by

$$z = \left(1 - \gamma \frac{\ln p}{p}\right) e^{i\theta}. \tag{5-110}$$

Inside all the sets U_{z_j}, the curves still go to z_j as discussed in Section 4. Let the radius of all U_{z_j} be $2\gamma \ln p/p$. We now fix the value of γ as follows. Using the condition (1-11) we can write (here and below c stands for various positive constants independent of n)

$$|V^{(n)}(z)| - |V_0| \leq \sum_{\substack{k=-p \\ k \neq 0}}^{p} |k^s V_k| \frac{|z|^k}{|k|^s}$$

$$< c \left(\sum_{\substack{k=-p \\ k \neq 0}}^{p} |k^s V_k|^2 \right)^{1/2} \left(\sum_{k=1}^{p} \frac{(1 \pm 3\gamma \ln p/p)^{\pm 2k}}{k^{2s}} \right)^{1/2}$$

$$< c \left(\sum_{k=1}^{p} \frac{(1 \pm 3\gamma \ln k/k)^{\pm 2k}}{k^{2s}} \right)^{1/2}$$

$$< c \left(\sum_{k=1}^{p} \frac{1}{k^{2(s-3\gamma)}} \left[1 + O\left(\frac{\ln^2 k}{k}\right) \right] \right)^{1/2}, \tag{5-111}$$

where $z \in \Sigma^{\text{out}}$, $z \in \partial U_{z_j} \cap \{|z| > 1\}$ (with the "+" sign in "$\pm$"), and $z \in \Sigma''^{\text{out}}$, $z \in \partial U_{z_j} \cap \{|z| < 1\}$ (with the "−" sign). We now set

$$3\gamma = s - (1 + \varepsilon_2)/2, \qquad \varepsilon_2 > 0, \tag{5-112}$$

and then

$$|V^{(n)}(z)| < c, \quad |b_+(z, V^{(n)})| < c, \quad |b_-(z, V^{(n)})| < c, \quad \text{for all } n \tag{5-113}$$

uniformly on Σ^{out}, Σ''^{out} and the ∂U_{z_j}, and in fact in the whole annulus

$$1 - 3\gamma \frac{\ln p}{p} < |z| < 1 + 3\gamma \frac{\ln p}{p}.$$

It is easy to adapt the considerations of the previous sections to the present case, and we again obtain the expansion (4-19) for the jump matrix of R on ∂U_{z_j}. Note that now $|\zeta(z)| = O(n^\nu \ln n)$ and $|z - z_j| = \ln n/n^{1-\nu}$ as $n \to \infty$ for $z \in \partial U_{z_j}$, and therefore using (4-19), (4-13), (3-13), (4-9) and the definition of ν in (5-108), we obtain, in particular,

$$n^{\omega\sigma_3}\Delta_1(z)n^{-\omega\sigma_3} = O\left(\frac{1}{n^{\varepsilon_1}\ln n}\right), \quad z \in \bigcup_{j=0}^{m}\partial U_{z_j}. \tag{5-114}$$

Furthermore, as follows from (5-109), (5-110), (5-113), and (4-27), (4-28), the jump matrix on Σ^{out} and Σ''^{out} is now the identity plus a function uniformly bounded in absolute value by

$$c\left(\frac{n^{1-\nu}}{\ln n}\right)^{2\max_j |\Re\beta_j|}\left(1 \pm \gamma(1-\nu)\frac{\ln n}{n^{1-\nu}}\right)^{\mp n}$$

$$< c\exp\left\{-\frac{\gamma}{2}(1-\nu)n^\nu \ln n\right\} n^{2(1-\nu)\max_j |\Re\beta_j|}, \tag{5-115}$$

where the upper sign corresponds to Σ^{out}, and the lower to Σ''^{out}.

The RH problem for $R(z)$ (see Section 4A) is therefore solvable, and we obtain $R(z)$ as a series where the first term R_1 is the same as before, and for the error term the same estimate holds for z outside a fixed neighborhood of the unit circle, e.g., for z large.

This implies that Theorem 1.1 holds for $f(z, V^{(n)})$. Note that it also holds for $|f(z, V^{(n)})|$.

We will now show that replacing $V^{(n)}$ with V in the symbol of the determinant $D_n(f(z, V^{(n)}))$ results, under a condition on s, in a small error only, so that Theorem 1.1 holds for $D_n(f(z, V))$ as well.

Using the Heine representation (2-10) for a Toeplitz determinant with (any) symbol $f(z)$, the straightforward estimate

$$b_\pm(z, V^{(n)}) = b_\pm(z, V)\left[1 + O\left(\frac{1}{n^{(1-\nu)s}}\right)\right] \quad \text{uniformly for } |z| = 1, \tag{5-116}$$

which follows from (1-11), and Theorem 1.1 applied to $D_n(|f(z, V^{(n)})|)$ and $D_n(f(z, V^{(n)}))$, we have, if $s(1-\nu) > 1$,

$$\left|D_n(f(z, V)) - D_n(f(z, V^{(n)}))\right|$$

$$< \frac{1}{(2\pi)^n n!}\int_0^{2\pi}\cdots\int_0^{2\pi}\prod_{1 \le j < k \le n}|e^{i\phi_j} - e^{i\phi_k}|^2 \prod_{j=0}^{n}|f(e^{i\phi_j}, V^{(n)})|\,d\phi_j$$
$$\times\left(\left|1 + c/n^{(1-\nu)s}\right|^n - 1\right)$$

$$< ce^{\Re V_0 n}n^{\sum_{j=0}^{m}((\Re\alpha_j)^2 + (\Im\beta_j)^2)}(e^{c/n^{(1-\nu)s-1}} - 1)$$

$$< c\left|e^{V_0 n}n^{\sum_{j=0}^{m}(\alpha_j^2 - \beta_j^2)}\right|n^{\sum_{j=0}^{m}((\Im\alpha_j)^2 + (\Re\beta_j)^2)}\frac{1}{n^{(1-\nu)s-1}}$$

$$< c\left|D_n(f(z, V^{(n)}))\right|n^{-((1-\nu)s-1-\sum_{j=0}^{m}((\Im\alpha_j)^2 + (\Re\beta_j)^2))}. \tag{5-117}$$

Therefore,

$$
D_n(f(z, V)) = D_n(f(z, V^{(n)})) \left(1 + \frac{D_n(f(z, V)) - D_n(f(z, V^{(n)}))}{D_n(f(z, V^{(n)}))}\right)
$$

$$
= D_n(f(z, V^{(n)}))(1 + o(1)), \tag{5-118}
$$

if

$$
s > \frac{1 + \sum_{j=0}^{m} ((\Im \alpha_j)^2 + (\Re \beta_j)^2)}{1 - v}. \tag{5-119}
$$

Note that this condition is consistent with (5-112) and the requirement that $\gamma > 0$. Using the expression for v in (5-108) and noting that ε_1 can be arbitrary close to zero, we replace (5-119) with (1-12). Under the condition (1-12) we then obtain the statement of the theorem for $D_n(f(z, V))$.

Appendix: the Toeplitz determinant D_n as a tau-function

Here we construct a Fuchsian system of ODE's corresponding to the Riemann–Hilbert problem of Section 2 for $V \equiv 0$. We show that the differential identities (3-20) for the Toeplitz determinant can be viewed as monodromy deformations of the tau-function associated with this Fuchsian system.

Assume the pure Fisher–Hartwig case, $V(z) \equiv 0$. Set

$$
\Phi(z) = \Lambda Y^{(n)}(z) \Lambda^{-1} \prod_{k=0}^{m} (z - z_k)^{\alpha_k \sigma_3} z^{\lambda \sigma_3}, \tag{A-1}
$$

where

$$
\Lambda = \prod_{k=0}^{m} z_k^{\frac{1}{2}(\beta_k + \alpha_k)\sigma_3},
$$

$$
\lambda = \sum_{k=0}^{m} \frac{\beta_k - \alpha_k}{2} - \frac{n}{2},
$$

and the branches of all multivalued functions are chosen as in Section 3. In terms of the function $\Phi(z)$, the Riemann–Hilbert problem (2-6)- (2-7) reads as follows:

(a) $\Phi(z)$ is analytic for $z \in \mathbb{C} \setminus (C \cup [0, 1] \cup \{\bigcup_{j=0}^{m} \Gamma_j\})$, where Γ_j is the ray $\theta = \theta_j$ from z_j to infinity. The unit circle C is oriented as before, counterclockwise, the segment $[0, 1]$ is oriented from 0 to 1, and the rays Γ_j are oriented towards infinity.

(b) The boundary values of $\Phi(z)$ are related by the jump conditions, namely

$$
\Phi_+(z) = \Phi_-(z) \begin{pmatrix} 1 & s_j \\ 0 & 1 \end{pmatrix} \tag{A-2}
$$

if $z \in C$, $\theta_j < \arg z < \theta_{j+1}$, $0 \le j \le m$, $\theta_{m+1} = 2\pi$, and otherwise

$$\Phi_+(z) = \Phi_-(z) \begin{cases} e^{-2\pi i \alpha_j \sigma_3} & \text{if } z \in \Gamma_j, \ 1 \le j \le m, \\ e^{-2\pi i(\alpha_0 + \lambda)\sigma_3} & \text{if } z \in \Gamma_0, \\ e^{-2\pi i \lambda \sigma_3} & \text{if } z \in [0,1], \end{cases}$$

where

$$s_j = \exp\left\{ -i\pi \sum_{k=0}^{j} \beta_k + i\pi \sum_{k=j+1}^{m} \beta_k - i\pi \sum_{k=0}^{j} \alpha_k - 3i\pi \sum_{k=j+1}^{m} \alpha_k \right\} \tag{A-3}$$

(for $j = m$, the second and fourth sums are absent).

(c) $\Phi(z)$ has the following asymptotic behavior at infinity:

$$\Phi(z) = \left(I + O\left(\frac{1}{z}\right) \right) z^{\left(\sum_{k=0}^{m} \frac{1}{2}(\beta_k + \alpha_k) + \frac{1}{2} n \right)\sigma_3} e^{2\pi i \sum_{k=j+1}^{m} \alpha_k \sigma_3}, \tag{A-4}$$

as $z \to \infty$ and $\theta_j < \arg z < \theta_{j+1}$, $0 \le j \le m$, $\theta_{m+1} = 2\pi$ (for $j = m$ the last factor is omitted).

(d) In the neighborhoods U_{z_j} of the points z_j, $j = 0, 1, \ldots, m$, the function $\Phi(z)$ admits the following representations, which constitute a refinement of the estimates (2-8) and (2-9).

• If $\alpha_j \ne 0$, then

$$\Phi(z) = \widetilde{\Phi}_j(z)(z - z_j)^{\alpha_j \sigma_3} C_j, \tag{A-5}$$

where $\widetilde{\Phi}_j(z)$ is holomorphic at $z = z_j$ (it is essentially the function $\widetilde{Y}(z)$ from Section 3) and the matrix C_j is given by the formula

$$C_j = \begin{pmatrix} 1 & c_j \\ 0 & 1 \end{pmatrix}, \tag{A-6}$$

with

$$c_j = s_j \begin{cases} \dfrac{1 - e^{2\pi i(\beta_j + \alpha_j)}}{1 - e^{4\pi i \alpha_j}} & \text{if } z \in U_{z_j}, \ |z| < 1, \\[2ex] \dfrac{1 - e^{2\pi i(\beta_j - \alpha_j)}}{1 - e^{4\pi i \alpha_j}} & \text{if } z \in U_{z_j}, \ |z| > 1, \ \arg z < \theta_j, \\[2ex] e^{4\pi i \alpha_j} \dfrac{1 - e^{2\pi i(\beta_j - \alpha_j)}}{1 - e^{4\pi i \alpha_j}} & \text{if } z \in U_{z_j}, \ |z| > 1, \ \arg z > \theta_j \end{cases} \tag{A-7}$$

in the case $j \ne 0$, and

$$C_0 = \begin{pmatrix} 1 & c_0 \\ 0 & 1 \end{pmatrix} \times \begin{cases} I & \text{if } \Im z > 0, \\ e^{2\pi i \kappa \sigma_3} & \text{if } \Im z < 0, \end{cases} \tag{A-8}$$

with

$$c_0 = s_0 \begin{cases} \dfrac{1-e^{2\pi i(\beta_0+\alpha_0)}}{1-e^{4\pi i\alpha_0}} & \text{if } z \in U_{z_0}, \ |z| < 1 \\[2ex] \dfrac{1-e^{2\pi i(\beta_0-\alpha_0)}}{1-e^{4\pi i\alpha_0}} & \text{if } z \in U_{z_0}, \ |z| > 1, \ \Im z < 0, \\[2ex] e^{4\pi i\alpha_0}\dfrac{1-e^{2\pi i(\beta_0-\alpha_0)}}{1-e^{4\pi i\alpha_0}} & \text{if } z \in U_{z_0}, \ |z| > 1, \ \Im z > 0, \end{cases} \tag{A-9}$$

in the case $j = 0$.

- If $\alpha_j = 0$ and $\beta_j \neq 0$, then

$$\Phi(z) = \widetilde{\Phi}_j(z) \begin{pmatrix} 1 & \frac{d_j}{2\pi i}\ln(z-z_j) \\ 0 & 1 \end{pmatrix} C_j, \quad d_j = -s_j(1 - e^{2\pi i\beta_j}), \tag{A-10}$$

where $\widetilde{\Phi}_j(z)$ is again holomorphic at $z = z_j$ and the matrix C_j this time is given by the formula

$$C_j = \begin{pmatrix} 1 & c_j \\ 0 & 1 \end{pmatrix}, \tag{A-11}$$

with

$$c_j = \begin{cases} s_{j-1} & \text{if } z \in U_{z_j}, \ |z| < 1, \\ 0 & \text{if } z \in U_{z_j}, \ |z| > 1, \ \arg z < \theta_j, \\ d_j & \text{if } z \in U_{z_j}, \ |z| > 1, \ \arg z > \theta_j, \end{cases} \tag{A-12}$$

in the case $j \neq 0$, and

$$C_0 = \begin{pmatrix} 1 & c_0 \\ 0 & 1 \end{pmatrix} \times \begin{cases} I & \text{if } \Im z > 0, \\ e^{2\pi i\lambda\sigma_3} & \text{if } \Im z < 0, \end{cases} \tag{A-13}$$

with

$$c_0 = \begin{cases} s_m e^{4\pi i\lambda} & \text{if } z \in U_{z_0}, \ |z| < 1, \\ 0 & \text{if } z \in U_{z_0}, \ |z| > 1, \ \Im z < 0, \\ d_0 & \text{if } z \in U_{z_0}, \ |z| > 1, \ \Im z > 0, \end{cases} \tag{A-14}$$

in the case $j = 0$.

(e) In a small neighborhood of $z = 0$, the function $\Phi(z)$ admits a similar representation:

$$\Phi(z) = \widetilde{\Phi}^{(0)}(z)z^{\lambda\sigma_3}, \tag{A-15}$$

where $\widetilde{\Phi}^{(0)}(z)$ is holomorphic at $z = 0$.

We also note that all the matrices above have determinants equal to 1.

A key feature of the Φ-RH problem is that all its jump matrices and the connection matrices C_j are piecewise constant in z. By standard arguments (see, e.g., [Jimbo et al. 1981; Its et al. 2001]), based on the Liouville theorem, this fact implies that the function $\Phi(z)$ satisfies a linear matrix ODE of Fuchsian type:

$$\frac{d\Phi(z)}{dz} = A(z)\Phi(z), \quad A(z) = \sum_{k=0}^{m} \frac{A_k}{z - z_k} + \frac{B}{z} \tag{A-16}$$

with

$$B = \lambda \widetilde{\Phi}^{(0)}(0)\sigma_3\big(\widetilde{\Phi}^{(0)}(0)\big)^{-1} \tag{A-17}$$

and

$$A_j = \begin{cases} \alpha_j \widetilde{\Phi}_j(z_j)\sigma_3 \widetilde{\Phi}_j^{-1}(z_j) & \text{if } \alpha_j \neq 0, \\[2mm] \dfrac{d_j}{2\pi i} \widetilde{\Phi}_j(z_j)\big(\begin{smallmatrix} 0 & 1 \\ 0 & 0 \end{smallmatrix}\big)\widetilde{\Phi}_j^{-1}(z_j) & \text{if } \alpha_j = 0, \ \beta_j \neq 0. \end{cases} \tag{A-18}$$

Moreover, since the jump matrices and the connection matrices C_j are all constant with respect to z_j, the function $\Phi(z)$ satisfies, in addition to (A-16), the equations

$$\frac{\partial \Phi(z)}{\partial z_j} = -\frac{A_j}{z - z_j}\Phi(z), \quad j = 1, \ldots, m. \tag{A-19}$$

The compatibility condition of (A-16) and (A-19) yields the following nonlinear systems of ODEs on the matrix coefficients B and A_j:

$$\frac{\partial B}{\partial z_j} = \frac{[A_j, B]}{z_j}, \quad \frac{\partial A_k}{\partial z_j} = \frac{[A_j, A_k]}{z_j - z_k}, \quad k \neq j, \tag{A-20}$$

$$\frac{\partial A_j}{\partial z_j} = -\sum_{\substack{k=0 \\ k \neq j}}^{m} \frac{[A_j, A_k]}{z_j - z_k} + \frac{[B, A_j]}{z_j}. \tag{A-21}$$

In the context of the Fuchsian system (A-16), the jump matrices and the connection matrices C_j of the Φ-Riemann–Hilbert problem form the monodromy data of the system. The fact that these data do not depend on the parameters z_j means that the functions $B = B(z_1, \ldots, z_m)$ and $A_j = A_j(z_1, \ldots, z_m)$, $j = 0, \ldots, m$ describe *isomonodromy deformations* of the system (A-16). Equations (A-20)–(A-21) are the classical Schlesinger equations.

An important role in the modern theory of isomonodromy deformations is played by the notion of a τ-function which was introduced by M. Jimbo, T. Miwa and K. Ueno in [Jimbo et al. 1981]. In the Fucshian case, the τ-function is defined as follows. Let

$$\widetilde{\omega} = \sum_{k=1}^{m} \operatorname{Res}_{z=z_k} \operatorname{trace} A(z)\frac{\partial \widetilde{\Phi}_k(z)}{\partial z}\widetilde{\Phi}_k^{-1}(z)\, dz_k$$

$$= \sum_{k=1}^{m} \operatorname{trace} A_k \frac{\partial \widetilde{\Phi}_k(z_k)}{\partial z}\widetilde{\Phi}_k^{-1}(z_k)\, dz_k. \tag{A-22}$$

As shown in [Jimbo et al. 1981], the differential form

$$\widetilde{\omega} = \widetilde{\omega}(A_0, \ldots, A_m, B; z_1, \ldots, z_m)$$

is closed on the solutions of the Schlesinger system (A-20)–(A-21). The τ-function is then defined as the exponential of the antiderivative of $\widetilde{\omega}$, i.e.,

$$\frac{\partial \ln \tau}{\partial z_k} = \text{trace } A_k \frac{\partial \widetilde{\Phi}_k(z_k)}{\partial z} \widetilde{\Phi}_k^{-1}(z_k). \tag{A-23}$$

It has been observed (see, e.g., [Its et al. 2001; Bertola 2009]) that the τ-functions evaluated on solutions of the Schlesinger equations generated by the Riemann–Hilbert problems associated with Toeplitz and Hankel determinants coincide, up to trivial factors, with the determinants themselves. In particular, for the Toeplitz determinant D_n with the pure Fisher–Hartwig symbol, $V(z) \equiv 0$, one can follow the calculations of [Its et al. 2001] and obtain that

$$\frac{\partial \ln D_n}{\partial z_k} = \text{trace } A_k \frac{\partial \widetilde{\Phi}_k(z_k)}{\partial z} \widetilde{\Phi}_k^{-1}(z_k) + 2 \sum_{\substack{j=0 \\ j \neq k}}^{m} \frac{\alpha_k \alpha_j}{z_j - z_k} - 2\lambda \frac{\alpha_k}{z_k}. \tag{A-24}$$

Therefore, in the case of the Riemann–Hilbert problem (A-2)–(A-15), the relation between the associated Toeplitz determinant and the τ-function is given by

$$D_n(f(z)) \equiv D_n(z_1, \ldots, z_m | \alpha, \beta)$$

$$= \tau(z_1, \ldots, z_m | \alpha, \beta) \prod_{j<k}(z_k - z_j)^{-2\alpha_j \alpha_k} \prod_{j=0}^{m} z_j^{-2\alpha_j \lambda}. \tag{A-25}$$

Direct substitution of (A-5), (A-10), and (A-15) into (A-16) and (A-19) yields

$$\text{trace } A_k \frac{\partial \widetilde{\Phi}_k(z_k)}{\partial z} \widetilde{\Phi}_k^{-1}(z_k)$$

$$= \sum_{\substack{j=0 \\ j \neq k}}^{m} \frac{\text{trace } A_j A_k}{z_k - z_j} + \frac{1}{z_k} \text{trace } B A_k$$

$$= \sum_{j=0}^{m} \text{trace } A_j \frac{\partial \widetilde{\Phi}_j(z_j)}{\partial z_k} \widetilde{\Phi}_j^{-1}(z_j) + \text{trace } B \frac{\partial \widetilde{\Phi}^{(0)}(0)}{\partial z_k} (\widetilde{\Phi}^{(0)}(0))^{-1}. \tag{A-26}$$

Hence (A-24) can be written as follows:

Lemma A.5. *Let the Riemann–Hilbert problem for Φ be solvable. Then for any $k = 0, 1, \ldots, m$,*

$$\frac{\partial \ln D_n}{\partial z_k} = \sum_{j=0}^{m} \text{trace } A_j \frac{\partial \widetilde{\Phi}_j(z_j)}{\partial z_k} \widetilde{\Phi}_j^{-1}(z_j) + \text{trace } B \frac{\partial \widetilde{\Phi}^{(0)}(0)}{\partial z_k} \left(\widetilde{\Phi}^{(0)}(0) \right)^{-1}$$

$$+ 2 \sum_{\substack{j=0 \\ j \neq k}}^{m} \frac{\alpha_k \alpha_j}{z_j - z_k} - 2\lambda \frac{\alpha_k}{z_k}. \quad \text{(A-27)}$$

What is of interest here is that the differential identities (3-20), which play a very important role in the main text, can be written in a matrix form closely related to (A-27). For simplicity, we present only the case $\alpha_j \neq 0$.

Lemma A.6. *Let the Riemann–Hilbert problem for* Φ *be solvable. Let* $\alpha_j \neq 0$, $j = 0, \ldots, m$. *Then, for any* $k = 0, 1, \ldots, m$,

$$\frac{\partial \ln D_n}{\partial \alpha_k} = \sum_{j=0}^{m} \text{trace } A_j \frac{\partial \widetilde{\Phi}_j(z_j)}{\partial \alpha_k} \widetilde{\Phi}_j^{-1}(z_j) + \text{trace } B \frac{\partial \widetilde{\Phi}^{(0)}(0)}{\partial \alpha_k} \left(\widetilde{\Phi}^{(0)}(0) \right)^{-1}$$

$$- 2 \sum_{\substack{j=0 \\ j \neq k}}^{m} \alpha_j \ln(z_j - z_k) - 2\lambda \ln(-z_k) + \sum_{j=0}^{m} \alpha_j \ln z_j - n \ln z_k, \quad \text{(A-28)}$$

$$\frac{\partial \ln D_n}{\partial \beta_k} = \sum_{j=0}^{m} \text{trace } A_j \frac{\partial \widetilde{\Phi}_j(z_j)}{\partial \beta_k} \widetilde{\Phi}_j^{-1}(z_j) + \text{trace } B \frac{\partial \widetilde{\Phi}^{(0)}(0)}{\partial \beta_k} \left(\widetilde{\Phi}^{(0)}(0) \right)^{-1}$$

$$- \sum_{j=0}^{m} \alpha_j \ln z_j - n \ln z_k. \quad \text{(A-29)}$$

Remark A.7. The significance of these equations is that they complement the isomonodromy deformation formula (A-27) by formulae which describe the *monodromy* deformations of the τ-function (represented by the Toeplitz determinants D_n). (Equations (A-28) and (A-29) should be compared with the general constructions of the recent paper [Bertola 2010].)

Proof. Although straightforward, it is rather tedious to derive (A-28) and (A-29) from the differential identities (3-20). There is, however, an alternative way to obtain (A-28) and (A-29) based on the direct analysis of the Riemann–Hilbert problem (2-6)–(2-7). First, we observe that the basic deformation formula for the Toeplitz determinant D_n, i.e., Equations (3-4) and (3-5), can be written, using [Deift et al. 2011, (2.4)], in the form

$$\frac{\partial}{\partial \gamma} \ln D_n = \frac{1}{2\pi i} \int_C z^{-n} \left(Y_{11}(z) \frac{dY_{21}(z)}{dz} - \frac{dY_{11}(z)}{dz} Y_{21}(z) \right) \frac{\partial f(z)}{\partial \gamma} dz. \quad \text{(A-30)}$$

Here, as in Section 3, γ is either α_k or β_k. Second, by γ–differentiating the Riemann–Hilbert problem (2-6)–(2-7) we easily obtain the following representation for the logarithmic derivative $(\partial Y(z)/\partial \gamma)Y^{-1}(z)$ of its solution (compare

[Bertola 2010, Lemma 2.1]):

$$X(z) \equiv \frac{\partial Y(z)}{\partial \gamma} Y^{-1}(z)$$

$$= \frac{1}{2\pi i} \int_C Y_-(z') \left(\begin{smallmatrix} 0 & 1 \\ 0 & 0 \end{smallmatrix}\right) Y_+^{-1}(z') \frac{\partial f(z')}{\partial \gamma} (z')^{-n} \frac{dz'}{z' - z}$$

$$= \frac{1}{2\pi i} \int_C \begin{pmatrix} -Y_{11}(z')Y_{21}(z') & Y_{11}^2(z') \\ -Y_{21}^2(z') & Y_{11}(z')Y_{21}(z') \end{pmatrix} \frac{\partial f(z')}{\partial \gamma} (z')^{-n} \frac{dz'}{z' - z}. \quad \text{(A-31)}$$

Now, from (A-1) and (A-16) we see that

$$\frac{dY_{11}(z)}{dz} = A_{11}(z)Y_{11}(z) + \Lambda_{11}^{-2} A_{12}(z) Y_{21}(z) - c(z) Y_{11}(z),$$

$$\frac{dY_{21}(z)}{dz} = \Lambda_{11}^2 A_{21}(z) Y_{11}(z) + A_{22}(z) Y_{21}(z) - c(z) Y_{21}(z),$$

where $c(z) = \sum_{k=0}^m \frac{\alpha_k}{z - z_k} + \frac{\lambda}{z}$. Further, this allows us to rewrite (A-30) in the form

$$\frac{\partial}{\partial \gamma} \ln D_n$$

$$= \frac{1}{2\pi i} \int_C z^{-n} \big(Y_{11}^2(z) \Lambda_{11}^2 A_{21}(z) - \Lambda_{11}^{-2} A_{12}(z) Y_{21}^2(z)$$

$$\qquad\qquad + Y_{11}(z) Y_{21}(z) (A_{22}(z) - A_{11}(z)) \big) \frac{\partial f(z)}{\partial \gamma} dz$$

$$= \Lambda_{11}^2 \sum_{k=0}^m A_{k,21} \int_C Y_{11}^2(z) \frac{\partial f(z)}{\partial \gamma} \frac{z^{-n} dz}{2\pi i(z - z_k)}$$

$$\quad - \Lambda_{11}^{-2} \sum_{k=0}^m A_{k,12} \int_C Y_{21}^2(z) \frac{\partial f(z)}{\partial \gamma} \frac{z^{-n} dz}{2\pi i(z - z_k)}$$

$$\quad + \sum_{k=0}^m (A_{k,22} - A_{k,11}) \int_C Y_{11}(z) Y_{21}(z) \frac{\partial f(z)}{\partial \gamma} \frac{z^{-n} dz}{2\pi i(z - z_k)}$$

$$\quad + \Lambda_{11}^2 B_{21} \int_C Y_{11}^2(z) \frac{\partial f(z)}{\partial \gamma} \frac{z^{-n} dz}{2\pi i z} - \Lambda_{11}^{-2} B_{12} \int_C Y_{21}^2(z) \frac{\partial f(z)}{\partial \gamma} \frac{z^{-n} dz}{2\pi i z}$$

$$\quad + (B_{22} - B_{11}) \int_C Y_{11}(z) Y_{21}(z) \frac{\partial f(z)}{\partial \gamma} \frac{z^{-n} dz}{2\pi i z}.$$

A comparison with (A-31) yields the following *local* representation for the γ-derivative of $\ln D_n$:

$$\frac{\partial}{\partial \gamma} \ln D_n = \Lambda_{11}^2 \sum_{k=0}^m A_{k,21} X_{12}(z_k) + \Lambda_{11}^{-2} \sum_{k=0}^m A_{k,12} X_{21}(z_k) +$$

$$+ \sum_{k=0}^{m} A_{k,11} X_{11}(z_k) + \sum_{k=0}^{m} A_{k,22} X_{22}(z_k) + \Lambda_{11}^{2} B_{21} X_{12}(0)$$

$$+ \Lambda_{11}^{-2} B_{12} X_{21}(0) + B_{11} X_{11}(0) + B_{22} X_{22}(0). \quad \text{(A-32)}$$

The last formula can be also written in the compact matrix form

$$\frac{\partial}{\partial \gamma} \ln D_n = \sum_{k=0}^{m} \text{trace } \Lambda^{-1} A_k \Lambda X(z_k) + \text{trace } \Lambda^{-1} B \Lambda X(0). \quad \text{(A-33)}$$

By evaluating (A-33), with the help of the representation (A-5), one arrives at the formulae (A-28) and (A-29). □

Acknowledgements

Deift was supported in part by NSF grants DMS 0500923 and DMS 1001886. Its was supported in part by NSF grant DMS-0701768 and EPSRC grant EP/ F014198/1. Krasovsky was supported in part by EPSRC grants EP/E022928/1 and EP/F014198/1.

References

[Baik et al. 1999] J. Baik, P. Deift, and K. Johansson, "On the distribution of the length of the longest increasing subsequence of random permutations", *J. Amer. Math. Soc.* **12**:4 (1999), 1119–1178.

[Barnes 1900] E. W. Barnes, "The theory of the *G*-function", *Quart. J. Pure and Appl. Math.* **31** (1900), 264–313.

[Basor 1978] E. L. Basor, "Asymptotic formulas for Toeplitz determinants", *Trans. Amer. Math. Soc.* **239** (1978), 33–65.

[Basor 1979] E. L. Basor, "A localization theorem for Toeplitz determinants", *Indiana Univ. Math. J.* **28**:6 (1979), 975–983.

[Basor and Tracy 1991] E. L. Basor and C. A. Tracy, "The Fisher–Hartwig conjecture and generalizations", *Phys. A* **177**:1-3 (1991), 167–173.

[Bertola 2009] M. Bertola, "Moment determinants as isomonodromic tau functions", *Nonlinearity* **22**:1 (2009), 29–50.

[Bertola 2010] M. Bertola, "The dependence on the monodromy data of the isomonodromic tau function", *Comm. Math. Phys.* **294**:2 (2010), 539–579.

[Böttcher and Silbermann 1981] A. Böttcher and B. Silbermann, "The asymptotic behavior of Toeplitz determinants for generating functions with zeros of integral orders", *Math. Nachr.* **102** (1981), 79–105.

[Böttcher and Silbermann 1985] A. Böttcher and B. Silbermann, "Toeplitz matrices and determinants with Fisher–Hartwig symbols", *J. Funct. Anal.* **63**:2 (1985), 178–214.

[Böttcher and Silbermann 1986] A. Böttcher and B. Silbermann, "Toeplitz operators and determinants generated by symbols with one Fisher–Hartwig singularity", *Math. Nachr.* **127** (1986), 95–123.

[Deift 1999] P. Deift, "Integrable operators", pp. 69–84 in *Differential operators and spectral theory*, edited by V. Buslaev et al., Amer. Math. Soc. Transl. Ser. 2 **189**, Amer. Math. Soc., Providence, RI, 1999.

[Deift and Zhou 1993] P. Deift and X. Zhou, "A steepest descent method for oscillatory Riemann–Hilbert problems: asymptotics for the MKdV equation", *Ann. of Math.* (2) **137**:2 (1993), 295–368.

[Deift et al. 1999] P. Deift, T. Kriecherbauer, K. T.-R. McLaughlin, S. Venakides, and X. Zhou, "Strong asymptotics of orthogonal polynomials with respect to exponential weights", *Comm. Pure Appl. Math.* **52**:12 (1999), 1491–1552.

[Deift et al. 2007] P. Deift, A. Its, I. Krasovsky, and X. Zhou, "The Widom–Dyson constant for the gap probability in random matrix theory", *J. Comput. Appl. Math.* **202**:1 (2007), 26–47.

[Deift et al. 2008] P. Deift, A. Its, and I. Krasovsky, "Asymptotics of the Airy-kernel determinant", *Comm. Math. Phys.* **278**:3 (2008), 643–678.

[Deift et al. 2011] P. Deift, A. Its, and I. Krasovsky, "Asymptotics of Toeplitz, Hankel, and Toeplitz+Hankel determinants with Fisher–Hartwig singularities", *Ann. of Math.* (2) **174**:2 (2011), 1243–1299.

[Deift et al. 2012] P. Deift, A. Its, and I. Krasovsky, "Toeplitz matrices and Toeplitz determinants under the impetus of the Ising model: some history and some recent results", preprint 1207.4990, 2012.

[Ehrhardt 2001] T. Ehrhardt, "A status report on the asymptotic behavior of Toeplitz determinants with Fisher–Hartwig singularities", pp. 217–241 in *Recent advances in operator theory* (Groningen, 1998), edited by A. Dijksma et al., Oper. Theory Adv. Appl. **124**, Birkhäuser, Basel, 2001.

[Ehrhardt and Silbermann 1997] T. Ehrhardt and B. Silbermann, "Toeplitz determinants with one Fisher–Hartwig singularity", *J. Funct. Anal.* **148**:1 (1997), 229–256.

[Fisher and Hartwig 1968] M. E. Fisher and R. E. Hartwig, "Toeplitz determinants: some applications, theorems, and conjectures", pp. 333–353 in *Stochastic processes in chemical physics*, edited by K. E. Shuler, Advances in Chemical Physics **15**, Wiley, 1968.

[Fokas et al. 1992] A. S. Fokas, A. R. Its, and A. V. Kitaev, "The isomonodromy approach to matrix models in 2D quantum gravity", *Comm. Math. Phys.* **147**:2 (1992), 395–430.

[Forrester and Warnaar 2008] P. J. Forrester and S. O. Warnaar, "The importance of the Selberg integral", *Bull. Amer. Math. Soc.* (*N.S.*) **45**:4 (2008), 489–534.

[Its and Krasovsky 2008] A. Its and I. Krasovsky, "Hankel determinant and orthogonal polynomials for the Gaussian weight with a jump", pp. 215–247 in *Integrable systems and random matrices*, edited by J. Baik et al., Contemp. Math. **458**, Amer. Math. Soc., Providence, RI, 2008.

[Its et al. 2001] A. R. Its, C. A. Tracy, and H. Widom, "Random words, Toeplitz determinants and integrable systems, II: advances in nonlinear mathematics and science", *Phys. D* **152/153** (2001), 199–224.

[Jimbo et al. 1981] M. Jimbo, T. Miwa, and K. Ueno, "Monodromy preserving deformation of linear ordinary differential equations with rational coefficients, I: general theory and τ-function", *Phys. D* **2**:2 (1981), 306–352.

[Krasovsky 2004] I. V. Krasovsky, "Gap probability in the spectrum of random matrices and asymptotics of polynomials orthogonal on an arc of the unit circle", *Int. Math. Res. Not.* **2004**:25 (2004), 1249–1272.

[Krasovsky 2007] I. V. Krasovsky, "Correlations of the characteristic polynomials in the Gaussian unitary ensemble or a singular Hankel determinant", *Duke Math. J.* **139**:3 (2007), 581–619.

[Lenard 1964] A. Lenard, "Momentum distribution in the ground state of the one-dimensional system of impenetrable Bosons", *J. Mathematical Phys.* **5** (1964), 930–943.

[Lenard 1972] A. Lenard, "Some remarks on large Toeplitz determinants", *Pacific J. Math.* **42** (1972), 137–145.

[Whittaker and Watson 1927] E. T. Whittaker and G. N. Watson, *A Course of Modern Analysis, an Introduction to the General Theory of Infinite Processes and of Analytic Functions*, 4th ed., Cambridge University Press, 1927. Reprinted 1969.

[Widom 1973] H. Widom, "Toeplitz determinants with singular generating functions", *Amer. J. Math.* **95** (1973), 333–383.

deift@cims.nyu.edu   *Courant Institute of Mathematical Sciences,*
                            251 Mercer Street, New York, NY 10012, United States

aits@math.iupui.edu   *Department of Mathematical Sciences,*
                            Indiana University–Purdue University Indianapolis,
                            402 N. Blackford, LD 270, Indianapolis, IN 46202,
                            United States

i.krasovsky@imperial.ac.uk *Department of Mathematics, Imperial College, South*
                            Kensington Campus, London SW7 2AZ, United Kingdom

Random Matrices
MSRI Publications
Volume **65**, 2014

Asymptotic analysis of the two-matrix model with a quartic potential

MAURICE DUITS, ARNO B. J. KUIJLAARS AND MAN YUE MO

We give a summary of the recent progress made by the authors and collaborators on the asymptotic analysis of the two-matrix model with a quartic potential. The paper also contains a list of open problems.

1. Two-matrix model: introduction

The Hermitian two-matrix model is the probability measure

$$\frac{1}{Z_n} e^{-n \operatorname{Tr}(V(M_1) + W(M_2) - \tau M_1 M_2)} \, dM_1 \, dM_2 \tag{1-1}$$

defined on pairs (M_1, M_2) of $n \times n$ Hermitian matrices. Here V and W are two polynomial potentials, $\tau \neq 0$ is a coupling constant, and

$$Z_n = \int e^{-n \operatorname{Tr}(V(M_1) + W(M_2) - \tau M_1 M_2)} \, dM_1 \, dM_2$$

is a normalization constant in order to make (1-1) a probability measure.

In recent works of the authors and collaborators [Duits et al. 2011; 2012; Duits and Kuijlaars 2009; Mo 2009] the model was studied with the aim to gain understanding in the limiting behavior of the eigenvalues of M_1 as $n \to \infty$, and to find and describe new types of critical behaviors.

The results should be compared with the well known results for the Hermitian one-matrix model

$$\frac{1}{Z_n} e^{-n \operatorname{Tr}(V(M))} \, dM, \tag{1-2}$$

which we briefly summarize here. The eigenvalues of the random matrix M

Duits is supported in part by the grant KAW 2010.0063 from the Knut and Alice Wallenberg Foundation and by the Swedish Research Council (VR) Grant 2012-3128. Kuijlaars is supported by FWO grants G.0641.11 and G.0934.13, K. U. Leuven research grants OT/08/33 and OT/12/73, and by research project MTM2011-28952-C02-01 of the Spanish Ministry of Science and Innovation. Mo is supported by the EPSRC grant EP/G019843/1.

from (1-2) have the explicit joint pdf

$$\frac{1}{\tilde{Z}_n} \prod_{j<k} (x_k - x_j)^2 \prod_{j=1}^{n} e^{-nV(x_j)},$$

which yields that the eigenvalues are a determinantal point process with correlation kernel

$$K_n(x, y) = \sqrt{e^{-nV(x)}} \sqrt{e^{-nV(y)}} \sum_{k=0}^{n-1} p_{k,n}(x) p_{k,n}(y),$$

where $(p_{k,n})_k$ is the sequence of orthonormal polynomials with respect to the weight function $e^{-nV(x)}$ on the real line. As $n \to \infty$ the empirical eigenvalue distributions have an a.s. weak limit[1]

$$\frac{1}{n} \sum_{j=1}^{n} \delta_{x_j} \to \mu^*,$$

where μ^* is a nonrandom probability measure that is characterized as the minimizer of the energy functional (Coulomb gas picture)

$$E_V(\mu) = \iint \log \frac{1}{|x - y|} \, d\mu(x) \, d\mu(y) + \int V(x) \, d\mu(x) \qquad (1\text{-}3)$$

when taken over all probability measures on the real line. For a polynomial V the minimizer μ^* is supported on a finite union of intervals [Deift et al. 1998]. In addition there is a polynomial Q of degree deg $V - 2$ such that

$$\xi(z) = V'(z) - \int \frac{d\mu_1^*(s)}{z - s}$$

is the solution of a quadratic equation

$$\xi^2 - V'(z)\xi + Q(z) = 0. \qquad (1\text{-}4)$$

From this it follows that μ_1^* has a density with respect to Lebesgue measure that is real analytic in the interior of any of the intervals and that can be written as

$$\rho(x) = \frac{d\mu^*(x)}{dx} = \frac{1}{\pi} \sqrt{q^-(x)}, \quad x \in \mathbb{R},$$

where q^- denotes the negative part of the polynomial

$$q(x) = \left(\frac{V'(x)}{2} \right)^2 - Q(x).$$

---

[1] That is, for any bounded continuous function f, we have $\lim_{n \to \infty} \frac{1}{n} \sum_{j=1}^{n} f(x_j) = \int f \, d\mu^*$ almost surely.

2. Limiting eigenvalue distribution

2.1. *Vector equilibrium problem.* Guionnet [2004] showed that the eigenvalues of the matrices M_1 and M_2 in the two-matrix model (1-1) have a limiting distribution as $n \to \infty$. The results of [Guionnet 2004] are in fact valid for a much greater class of random matrix models. The limiting distribution is characterized as the minimizer of a certain functional, which is however very different from the energy functional (1-3) for the one matrix model.

Our aim is to develop an analogue of the Coulomb gas picture for the eigenvalues of the matrices in the two-matrix model (1-1). We have been successful in doing this for the eigenvalues of M_1 in the case of even polynomial potentials V and W with W of degree 4. Thus our assumptions are:

- V is an even polynomial with positive leading coefficient.
- $W(y) = \frac{1}{4}y^4 + \frac{\alpha}{2}y^2$ with $\alpha \in \mathbb{R}$.
- $\tau > 0$ (without loss of generality).

We recall some notions from logarithmic potential theory [Saff and Totik 1997]: the mutual logarithmic energy

$$I(\mu, \nu) = \iint \log \frac{1}{|x - y|} \, d\mu(x) \, d\nu(y)$$

of two measures μ and ν, and the logarithmic energy

$$I(\mu) = I(\mu, \mu)$$

of a measure μ. Then the limiting mean distribution of the eigenvalues of M_1 is characterized by a vector equilibrium problem for three measures. This involves an energy functional

$$E(\mu_1, \mu_2, \mu_3) = I(\mu_1) + I(\mu_2) + I(\mu_3) - I(\mu_1, \mu_2) - I(\mu_2, \mu_3)$$
$$+ \int V_1(x) \, d\mu_1(x) + \int V_3(x) \, d\mu_3(x) \quad (2\text{-}1)$$

defined on three measures μ_1, μ_2, μ_3. Note that there is an attraction between the measures μ_1 and μ_2 and between the measures μ_2 and μ_3, while there is no direct interaction between the measures μ_1 and μ_3. This type of interaction is characteristic for a Nikishin system [Nikishin and Sorokin 1991].

The energy functional (2-1) depends on the external fields V_1 and V_3 that act on the measures μ_1 and μ_3 in (2-1). The vector equilibrium problem will also have an upper constraint σ_2 for the measure μ_2. These input data take a very special form that we describe next.

External field V_1. The external field that acts on μ_1 is defined by

$$V_1(x) = V(x) + \min_{s \in \mathbb{R}}(W(s) - \tau xs), \tag{2-2}$$

where we recall that $W(s) = \frac{1}{4}s^4 + \frac{\alpha}{2}s^2$. For the case $\alpha = 0$, this is simply $V_1(x) = V(x) - \frac{3}{4}|\tau x|^{4/3}$.

External field V_3. The external field that acts on the third measure is absent if $\alpha \geq 0$, i.e.,

$$V_3(x) \equiv 0 \quad \text{if } \alpha \geq 0.$$

The function $s \in \mathbb{R} \mapsto W(s) - \tau xs$ has a global minimum at $s = s_1(x)$ and this value plays a role in the definition of V_1, see (2-2). For $\alpha < 0$, and $x \in (-x^*(\alpha), x^*(\alpha))$, where

$$x^*(\alpha) = \frac{2}{\tau}\left(\frac{-\alpha}{3}\right)^{3/2}, \quad \alpha < 0,$$

the function $s \in \mathbb{R} \mapsto W(s) - \tau xs$ has another local minimum at $s = s_2(x)$, and a local maximum at $s = s_3(x)$.

Then V_3 is defined by

$$V_3(x) = (W(s_3(x)) - \tau xs_3(x)) - (W(s_2(x)) - \tau xs_2(x)) \tag{2-3}$$

if $x \in (-x^*(\alpha), x^*(\alpha))$, and $V_3(x) \equiv 0$ otherwise.

Upper constraint σ_2. The upper constraint σ_2 that acts on the second measure is the measure on the imaginary axis with the density

$$\frac{d\sigma_2(z)}{|dz|} = \frac{\tau}{\pi} \max_{s^3 + \alpha s = \tau z} \operatorname{Re} s, \quad z \in i\mathbb{R}. \tag{2-4}$$

In case $\alpha = 0$ this simplifies to

$$\frac{d\sigma_2}{|dz|} = \frac{\sqrt{3}}{2\pi}\tau^{4/3}|z|^{1/3}.$$

If $\alpha < 0$ then the density of σ_2 is positive and real analytic on the full imaginary axis. If $\alpha > 0$ then the support of σ_2 has a gap around 0:

$$\operatorname{supp}(\sigma_2) = (-i\infty, -iy^*(\alpha)] \cup [iy^*(\alpha), i\infty),$$

where

$$y^*(\alpha) = \frac{2}{\tau}\left(\frac{\alpha}{3}\right)^{3/2}, \quad \alpha > 0.$$

Theorem 1 [Duits et al. 2012, Theorem 1.1]. *There is a unique minimizer $(\mu_1^*, \mu_2^*, \mu_3^*)$ of the energy functional (2-1) subject to the following conditions, with input data V_1, V_3, and σ (as described above):*

(a) μ_1 is a measure on $\mathbb{R}$ with $\mu_1(\mathbb{R}) = 1$.

(b) μ_2 is a measure on $i\mathbb{R}$ with $\mu_2(i\mathbb{R}) = \frac{2}{3}$.

(c) μ_3 is a measure on $\mathbb{R}$ with $\mu_3(\mathbb{R}) = \frac{1}{3}$.

(d) $\mu_2 \leq \sigma_2$.

The proof of the existence of a minimizer was completed and simplified in [Hardy and Kuijlaars 2012]; see Section 4.2 below.

Now that we have existence and uniqueness, it is natural to ask about further properties of the minimizer. The three measures μ_1^*, $\sigma - \mu_2^*$ and μ_3^* are absolutely continuous with respect to the Lebesgue measure with densities that are real analytic in the interior of their supports, except possibly at the origin. Furthermore, denoting by $S(\mu)$ the support of a measure μ, we have:

- The support of μ_1^* is a finite union of bounded intervals on the real line.

- There exists $c_2 \geq 0$ such that $S(\sigma_2 - \mu_2^*) = i\mathbb{R} \setminus (-ic_2, ic_2)$, and if $c_2 > 0$ the density of $\sigma_2 - \mu_2^*$ vanishes like a square root at $\pm ic_2$.

- There exists $c_3 \geq 0$ such that $S(\mu_3^*) = \mathbb{R} \setminus (-c_3, c_3)$, and if $c_3 > 0$ the density of μ_3^* vanishes like a square root at $\pm c_3$.

In a generic situation, the density of μ_1^* is strictly positive in the interior of its support and vanishes like a square root at endpoints. In addition strict inequality holds in the variational inequality outside the support $S(\mu_1^*)$. Moreover, generically if $c_2 = 0$ the density of $\sigma - \mu_2^*$ is positive at the origin, and likewise if $c_3 = 0$ the density of μ_3^* is positive at the origin. If we are in such a generic situation, then we say that (V, W, τ) is regular. See [Duits et al. 2012, Section 1.5] for more details and a discussion on the singular situations that may occur.

Theorem 2 [Duits et al. 2012, Theorem 1.4]. *Let μ_1^* be the first component of the minimizer in Theorem 1, and assume that (V, W, τ) is regular, then as $n \to \infty$ with $n \equiv 0$ (mod 3), the mean eigenvalue distribution of M_1 convergences to μ_1^*.*

We are convinced that the theorem is also valid in the singular cases, which correspond to phase transitions in the two-matrix model. The condition that n is a multiple of three is nonessential as well. It is imposed for convenience in the analysis.

In [Duits et al. 2012] only the convergence of mean eigenvalue distributions was considered, which is a rather weak form of convergence. However, when combined with the results of [Guionnet 2004] it will actually follow that the empirical eigenvalue distributions of M_1 tend to μ_1^* almost surely.

The analysis of [Duits et al. 2012] also proves the usual universality results for local eigenvalue statistics in Hermitian matrix ensembles, given by the sine kernel in the bulk of the spectrum and by the Airy kernel at edge points. In nonregular situations one may find Pearcey and Painlevé II kernels, while in multicritical cases new kernels may appear. This was indeed proved recently in [Duits and Geudens 2013]; see Section 4.1 below.

2.2. Riemann surface. A major ingredient in the asymptotic analysis in [Duits et al. 2012] is the construction of an appropriate Riemann surface (or spectral curve), which plays a role similar to the algebraic equation (1-4) in the one-matrix model. The existence of such a Riemann surface is implied by the work of Eynard [2005] on the formal two-matrix model. Our approach is different from the one of Eynard, in that we use the vector equilibrium problem to construct the Riemann surface, and in a next step we define a meromorphic function on it.

The main point is that the supports $S(\mu_1^*)$, $S(\sigma - \mu_2^*)$ and $S(\mu_3^*)$ associated to the minimizer in Theorem 1 determine the cut structure of a Riemann surface

$$\mathcal{R} = \bigcup_{j=1}^{4} \mathcal{R}_j$$

with four sheets:

$$\mathcal{R}_1 = \overline{\mathbb{C}} \setminus S(\mu_1^*),$$
$$\mathcal{R}_2 = \mathbb{C} \setminus (S(\mu_1^*) \cup S(\sigma_2 - \mu_2^*)),$$
$$\mathcal{R}_3 = \mathbb{C} \setminus (S(\sigma_2 - \mu_2^*) \cup S(\mu_3^*)),$$
$$\mathcal{R}_4 = \mathbb{C} \setminus S(\mu_3^*).$$

The sheet $\mathcal{R}_j$ is glued to the next sheet $\mathcal{R}_{j+1}$ along the common cut in the usual crosswise manner. The meromorphic function on $\mathcal{R}$ arises in the following way:

Proposition 3 ([Duits et al. 2012, Proposition 4.8]). *The function*

$$\xi_1(z) = V'(z) - \int \frac{d\mu_1^*(x)}{z - x}, \quad z \in \mathcal{R}_1,$$

extends to a meromorphic function on the Riemann surface $\mathcal{R}$ whose only poles are at infinity. There is a pole of order $\deg V$ *at infinity on the first sheet, and a simple pole at the other point at infinity.*

The proof of Proposition 3 follows from the Euler–Lagrange variational conditions that are associated with the vector equilibrium problem. See Section 4.2 of [Duits et al. 2012] for explicit expressions for the meromorphic continuation of ξ_1 to the other sheets.

It follows from Proposition 3 that ξ_1 is one of the solutions of a quartic equation, which is the analogue of the quadratic equation (1-4) that is relevant in the one-matrix model.

3. About the proof

We describe the main tools that are used in the proof of Theorem 2.

3.1. *Biorthogonal polynomials.* We make use of the integrable structure of the two-matrix model that is described in terms of biorthogonal polynomials. In this context the biorthogonal polynomials are two sequences of monic polynomials $(p_{j,n})_j$ and $(q_{k,n})_k$ (depending on n) with $\deg p_{j,n} = j$ and $\deg q_{k,n} = k$, that satisfy

$$\int_{-\infty}^{\infty} \int_{-\infty}^{\infty} p_{j,n}(x)q_{k,n}(y)e^{-n(V(x)+W(y)-\tau xy)} \, dx \, dy = h_{k,n}\delta_{j,k};$$

see [Bertola 2011; Bertola et al. 2002; 2003; Ercolani and McLaughlin 2001; Eynard and Mehta 1998]. These polynomials uniquely exist, have real and simple zeros [Ercolani and McLaughlin 2001], and in addition the zeros of $p_{j,n}$ and $p_{j+1,n}$ interlace, as well as those of $q_{k,n}$ and $q_{k+1,n}$; see [Duits et al. 2011].

There is an explicit expression for the joint pdf of the eigenvalues of M_1 and M_2:

$$\frac{1}{(n!)^2} \det \begin{pmatrix} K_n^{(1,1)}(x_i, x_j) & K_n^{(1,2)}(x_i, y_j) \\ K_n^{(2,1)}(y_1, y_j) & K_n^{(2,2)}(y_i, y_j), \end{pmatrix} \tag{3-1}$$

with four kernels that are expressed in terms of the biorthogonal polynomials and their transformed functions

$$Q_{k,n}(x) = \int_{-\infty}^{\infty} q_{k,n}(y)e^{-n(V(x)+W(y)-\tau xy)} \, dy,$$

$$P_{j,n}(y) = \int_{-\infty}^{\infty} p_{j,n}(x)e^{-n(V(x)+W(y)-\tau xy)} \, dx,$$

as follows:

$$K_n^{(1,1)}(x_1, x_2) = \sum_{k=0}^{n-1} \frac{1}{h_{k,n}^2} p_{k,n}(x_1)Q_{k,n}(x_2),$$

$$K_n^{(1,2)}(x, y) = \sum_{k=0}^{n-1} \frac{1}{h_{k,n}^2} p_{k,n}(x)q_{k,n}(y),$$

$$K_n^{(2,1)}(y, x) = \sum_{k=0}^{n-1} \frac{1}{h_{k,n}^2} P_{k,n}(y)Q_{k,n}(x) - e^{-n(V(x)+W(y)-\tau xy)}, \tag{3-2}$$

$$K_n^{(2,2)}(y_1, y_2) = \sum_{k=0}^{n-1} \frac{1}{h_{k,n}^2} P_{k,n}(y_1)q_{k,n}(y_2).$$

The joint pdf (3-1) is determinantal, which means that eigenvalue correlation functions have determinantal expressions with the same kernels $K_n^{(i,j)}$, $i, j = 1, 2$. In particular, after averaging out the eigenvalues of M_2 we get that the eigenvalues of M_1 are a determinantal point process with kernel $K_n^{(1,1)}$.

A natural first step to compute the asymptotic behavior of the polynomials and hence the kernels, is to formulate a Riemann–Hilbert problem ($\equiv$ RH problem) for the polynomials. Several different formulations exist in the literature [Bertola et al. 2003; Ercolani and McLaughlin 2001; Kapaev 2003; Kuijlaars and McLaughlin 2005]. The analysis in [Duits and Kuijlaars 2009; Duits et al. 2012; Mo 2009] is based on the RH problem in [Kuijlaars and McLaughlin 2005] that we will discuss in the next subsection.

3.2. Riemann–Hilbert problem. It turns out that the kernel (3-2) has a special structure which relates it to multiple orthogonal polynomials and the eigenvalues of M_1 (after averaging over M_2) are an example of a multiple orthogonal polynomial ensemble [Kuijlaars 2010]. This is due to the following observation of Kuijlaars and McLaughlin [2005].

Proposition 4. *Suppose W is a polynomial of degree $r + 1$, and let*

$$w_{k,n}(x) = \int_{-\infty}^{\infty} y^k e^{-n(V(x)+W(y)-\tau xy)} \, dy, \quad k = 0, \ldots, r-1.$$

Then the biorthogonal polynomial $p_{j,n}$ satisfies

$$\int_{-\infty}^{\infty} p_{j,n}(x) x^l w_{k,n}(x) \, dx = 0, \quad l = 0, \ldots, \left\lceil \frac{j-k}{r} \right\rceil - 1, \quad (3\text{-}3)$$

for $k = 0, \ldots, r-1$.

The conditions (3-3) are known as multiple orthogonality conditions [Aptekarev 1998], and they characterize the biorthogonal polynomials.

The advantage of the formulation as multiple orthogonality is that these polynomials are characterized by a RH problem of size $(r + 1) \times (r + 1)$, [Van Assche et al. 2001], which we state here for the case $r = 3$ and for $j = n$ with n a multiple of three. Then the RH problem has size 4×4 and it asks for a 4×4 matrix valued function Y on $\mathbb{C} \setminus \mathbb{R}$ satisfying these conditions:

(a) $Y : \mathbb{C} \setminus \mathbb{R} \to \mathbb{C}^{4 \times 4}$ is analytic.

(b) For $x \in \mathbb{R}$,

$$Y_+(x) = Y_-(x) \begin{pmatrix} 1 & w_{0,n}(x) & w_{1,n}(x) & w_{2,n}(x) \\ 0 & 1 & 0 & 0 \\ 0 & 0 & 1 & 0 \\ 0 & 0 & 0 & 1 \end{pmatrix},$$

where $Y_+(x)$ $(Y_-(x))$ denotes the limiting value of $Y(z)$ as $z \to x$ from the upper (lower) half-plane.

(c) As $z \to \infty$,

$$Y(z) = \left(I_4 + O\left(\frac{1}{z}\right)\right) \begin{pmatrix} z^n & 0 & 0 & 0 \\ 0 & z^{-n/3} & 0 & 0 \\ 0 & 0 & z^{-n/3} & 0 \\ 0 & 0 & 0 & z^{-n/3} \end{pmatrix}.$$

The RH problem has a unique solution, given by

$$Y = \begin{pmatrix} p_{n,n} & C(p_{n,n}w_{0,n}) & C(p_{n,n}w_{1,n}) & C(p_{n,n}w_{2,n}) \\ p_{n,n}^{(0)} & C(p_{n,n}^{(0)}w_{0,n}) & C(p_{n,n}^{(0)}w_{1,n}) & C(p_{n,n}^{(0)}w_{2,n}) \\ p_{n,n}^{(1)} & C(p_{n,n}^{(1)}w_{0,n}) & C(p_{n,n}^{(1)}w_{1,n}) & C(p_{n,n}^{(1)}w_{2,n}) \\ p_{n,n}^{(2)} & C(p_{n,n}^{(2)}w_{0,n}) & C(p_{n,n}^{(2)}w_{1,n}) & C(p_{n,n}^{(2)}w_{2,n}) \end{pmatrix},$$

where $p_{n,n}$ is the n-th degree biorthogonal polynomial, $p_{n,n}^{(0)}$, $p_{n,n}^{(1)}$, $p_{n,n}^{(2)}$ are three polynomials of degree $\leq n - 1$ that satisfy certain multiple orthogonal conditions and Cf is the Cauchy transform

$$Cf(z) = \frac{1}{2\pi i} \int_{-\infty}^{\infty} \frac{f(x)}{x - z}\, dx.$$

By using the Christoffel–Darboux formula for multiple orthogonal polynomials [Bleher and Kuijlaars 2004; Daems and Kuijlaars 2004] the correlation kernel $K_n^{(1,1)}$ for the eigenvalues of M_1 can be expressed in terms of the solution of the RH problem:

$$K_n^{(1,1)}(x, y) = \begin{pmatrix} 0 & w_{0,n}(y) & w_{1,n}(y) & w_{2,n}(y) \end{pmatrix} \frac{Y_+^{-1}(y)Y_+(x)}{2\pi i (x - y)} \begin{pmatrix} 1 \\ 0 \\ 0 \\ 0 \end{pmatrix}. \quad (3\text{-}4)$$

Multiple orthogonal polynomials and RH problems are also used for random matrices with external source [Bleher et al. 2011; Bleher and Kuijlaars 2004] and models of nonintersecting paths [Kuijlaars et al. 2009]. In these cases, correlation kernels for the relevant statistical quantities are also expressed in terms of the corresponding RH problem through (3-4).

3.3. Steepest descent analysis.

The remaining part of the proof of Theorem 2 is an asymptotic analysis of the RH problem via an extension of the Deift–Zhou steepest descent method [Deift et al. 1999; Deift and Zhou 1993]. The vector equilibrium problem and the Riemann surface play a crucial role in the transformations in this analysis. For the precise transformations and the many details that are involved we refer the reader to [Duits et al. 2012]. Following the effect of the transformations on the kernel (3-4), one finds that

$$\lim_{n\to\infty} \frac{1}{n} K_n^{(1,1)}(x, x) = \frac{d\mu_1^*(x)}{dx},$$

which is what is needed to establish the theorem.

A somewhat similar steepest descent analysis is done in [Bleher et al. 2011] for a random matrix model with external source, where vector equilibrium problems and Riemann surfaces also play an important role.

4. Further developments

4.1. *Critical behavior in the quadratic/quartic model.* For the case $V(x)=\frac{1}{2}x^2$ the spectral curve can be computed and a classification of all possible cases can be made explicitly.

Case I: $0 \in S(\mu_1^*) \cap S(\mu_3^*)$ and $0 \notin S(\sigma_2 - \mu_2^*)$.

Case II: $0 \in S(\mu_3^*)$ and $0 \notin S(\mu_1^*) \cup S(\sigma_2 - \mu_2^*)$.

Case III: $0 \in S(\sigma_2 - \mu_2^*)$ and $0 \notin S(\mu_1^*) \cup S(\mu_3^*)$.

Case IV: $0 \in S(\mu_1^*)$ and $0 \notin S(\sigma_2 - \mu_2^*) \cup S(\mu_3^*)$.

Phase transitions between the regular cases represent the critical cases.

The quadratic/quartic model depends on two parameters, namely the coupling constant τ and the number α in the quartic potential $W(y) = y^4/4 + \alpha y^2/2$. Figure 1 (taken from [Duits et al. 2011]) shows the phase diagram in the α-τ plane. Critical behavior takes place on the curves $\tau^2 = \alpha + 2$ and $\alpha\tau^2 = -1$.

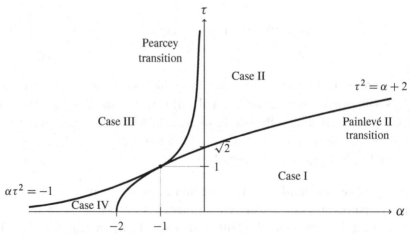

Figure 1. Phase diagram for the quadratic case $V(x) = \frac{1}{2}x^2$.

On the parabola $\tau^2 = \alpha + 2$ a gap appears around 0 in the support of either μ_1^* (if one moves from Case I to Case II) or μ_3^* (if from Case I to Case IV). This is a transition of Painlevé II type which also appears in the opening of gaps in one-matrix models [Bleher and Its 2003; Claeys and Kuijlaars 2006]. On the curve $\alpha\tau^2 = -1$ a gap appears in the support of either μ_1^* (if one moves from Case IV to Case III) or μ_3^* (if one moves from Case II to Case III), while simultaneously the gap in the support of $\sigma_2 - \mu_2^*$ closes. This is a transition of Pearcey type, which was observed before in the random matrix model with external source and in the model of nonintersecting Brownian motions [Bleher and Kuijlaars 2007; Brézin and Hikami 1998; Tracy and Widom 2006].

The phase diagram has a very special point $\alpha = -1$, $\tau = 1$ which is on both critical curves, and where all four regular cases come together. For these special values, the density of μ_1^* vanishes like a square root at the origin, which is an interior point of $S(\mu_1^*)$. The local analysis at this point was done very recently by Duits and Geudens [2013]. They found that in the asymptotic limit, the local eigenvalue correlation kernels around 0 are closely related to the limiting kernels that describe the tacnode behavior for nonintersecting Brownian motions [Adler et al. 2013; Delvaux et al. 2011; Johansson 2013]. More precisely, the kernels can be expressed in terms of an extension of the same 4×4 RH problem in [Delvaux et al. 2011]. However, they are constructed in a different way out of this 4×4 RH problem and, as a result, these kernels are not the same.

4.2. Vector equilibrium problems. The analysis in [Duits et al. 2012] of the vector equilibrium problem was not fully complete, since the lower semicontinuity of the energy functional (2-1) was implicitly assumed but not established in [Duits et al. 2012].

In [Beckermann et al. 2013; Hardy and Kuijlaars 2012] the vector equilibrium problem was studied in a more systematic way, in the more general context of an energy functional for n measures

$$E(\mu_1, \ldots, \mu_n) = \sum_{i=1}^{n}\sum_{j=1}^{n} c_{ij} I(\mu_i, \mu_j) + \sum_{j=1}^{n} \int V_j(x)\, d\mu_j(x), \qquad (4\text{-}1)$$

where $C = (c_{ij})_{i,j=1}^{n}$ is a real symmetric positive definite matrix (in [Beckermann et al. 2013] also semidefinite interaction matrices are considered). The external fields $V_j : \Sigma_j \to \mathbb{R} \cup \{\infty\}$ are lower semicontinuous with domains Σ_j that are closed subsets of $\mathbb{C}$. Let $m_1, \ldots, m_n$ be given positive numbers and assume that for every $i = 1, \ldots, n$,

$$\liminf_{|x| \to \infty} \left(V_i(x) - \left(\sum_{j=1}^{n} c_{ij} m_j \right) \log(1 + |x|^2) \right) > -\infty.$$

Under these assumptions it is shown in [Hardy and Kuijlaars 2012] that the energy functional (4-1), restricted to the set of measures with $\mu_j(\Sigma_j) = m_j$ for $j = 1, \dots, n$,

(a) has compact sublevel sets $\{E \leq \alpha\}$ for every $\alpha \in \mathbb{R}$, (so E is in particular lower semicontinuous), and

(b) is strictly convex on the subset where it is finite.

This guarantees existence and uniqueness of a minimizer of (4-1), provided that E is not identically infinite. Existence and uniqueness of a minimizer readily extends to situations where the domain of E is further restricted by upper constraints $\mu_j \leq \sigma_j$ for $j = 1, \dots, n$, again provided that E is not identically infinite on this domain. In particular, this applies to the energy functional (2-1) for the two-matrix model with quartic potential with the constraint $\mu_2 \leq \sigma_2$ described in Section 2.

4.3. *Open problems.* Numerous intriguing questions and open problems arise out of our analysis.

(a) What is the motivation for the central vector equilibrium problem? In the one-matrix model there is a direct way to come from the joint eigenvalue probability density to the equilibrium problem. We do not have this direct link for the two-matrix model.

(b) How is the vector equilibrium problem related to the variational problem from [Guionnet 2004]?

(c) A possibly related question: is there a large deviation principle associated with the vector equilibrium problem? See, for example, [Anderson et al. 2010] for the large deviations interpretation of the equilibrium problem for the one-matrix model.

(d) Our analysis is restricted to even potentials V and W. This restriction provides a symmetry of the problem around zero, which is the reason why the second measure μ_2 in the vector equilibrium problem is supported on the imaginary axis. If we remove the symmetry then probably we would have to look for a contour that replaces the imaginary axis. It is likely that such a contour would be an S-curve in a certain external field, but at this moment we do not know how to handle this situation. See [Martínez-Finkelshtein and Rakhmanov 2011; Rakhmanov 2012] for important recent developments around S-curves for scalar equilibrium problems.

(e) Extension to higher degree W is wide open. If $\deg W = d$ then one would expect a vector equilibrium problem for $d - 1$ measures. It may be that S-curves are needed for $d \geq 6$, even in the case of even potentials.

(f) Exploration of further critical phenomena in the two-matrix model.

References

[Adler et al. 2013] M. Adler, P. L. Ferrari, and P. van Moerbeke, "Nonintersecting random walks in the neighborhood of a symmetric tacnode", *Ann. Probab.* **41**:4 (2013), 2599–2647.

[Anderson et al. 2010] G. W. Anderson, A. Guionnet, and O. Zeitouni, *An introduction to random matrices*, Cambridge Studies in Advanced Mathematics **118**, Cambridge University Press, Cambridge, 2010.

[Aptekarev 1998] A. I. Aptekarev, "Multiple orthogonal polynomials", pp. 423–447 in *Proceedings of the VIIIth Symposium on Orthogonal Polynomials and Their Applications* (Seville, 1997), vol. 99, edited by A. J. Duran, 1998.

[Beckermann et al. 2013] B. Beckermann, V. Kalyagin, A. C. Matos, and F. Wielonsky, "Equilibrium problems for vector potentials with semidefinite interaction matrices and constrained masses", *Constr. Approx.* **37**:1 (2013), 101–134.

[Bertola 2011] M. Bertola, "Two-matrix models and biorthogonal polynomials", pp. 310–328 in *The Oxford handbook of random matrix theory*, edited by G. Akemann et al., Oxford Univ. Press, Oxford, 2011.

[Bertola et al. 2002] M. Bertola, B. Eynard, and J. Harnad, "Duality, biorthogonal polynomials and multi-matrix models", *Comm. Math. Phys.* **229**:1 (2002), 73–120.

[Bertola et al. 2003] M. Bertola, B. Eynard, and J. Harnad, "Differential systems for biorthogonal polynomials appearing in 2-matrix models and the associated Riemann–Hilbert problem", *Comm. Math. Phys.* **243**:2 (2003), 193–240.

[Bleher and Its 2003] P. Bleher and A. Its, "Double scaling limit in the random matrix model: the Riemann–Hilbert approach", *Comm. Pure Appl. Math.* **56**:4 (2003), 433–516.

[Bleher and Kuijlaars 2004] P. M. Bleher and A. B. J. Kuijlaars, "Random matrices with external source and multiple orthogonal polynomials", *Int. Math. Res. Not.* **2004**:3 (2004), 109–129.

[Bleher and Kuijlaars 2007] P. M. Bleher and A. B. J. Kuijlaars, "Large n limit of Gaussian random matrices with external source, III: double scaling limit", *Comm. Math. Phys.* **270**:2 (2007), 481–517.

[Bleher et al. 2011] P. Bleher, S. Delvaux, and A. B. J. Kuijlaars, "Random matrix model with external source and a constrained vector equilibrium problem", *Comm. Pure Appl. Math.* **64**:1 (2011), 116–160.

[Brézin and Hikami 1998] E. Brézin and S. Hikami, "Universal singularity at the closure of a gap in a random matrix theory", *Phys. Rev. E* (3) **57**:4 (1998), 4140–4149.

[Claeys and Kuijlaars 2006] T. Claeys and A. B. J. Kuijlaars, "Universality of the double scaling limit in random matrix models", *Comm. Pure Appl. Math.* **59**:11 (2006), 1573–1603.

[Daems and Kuijlaars 2004] E. Daems and A. B. J. Kuijlaars, "A Christoffel–Darboux formula for multiple orthogonal polynomials", *J. Approx. Theory* **130**:2 (2004), 190–202.

[Deift and Zhou 1993] P. Deift and X. Zhou, "A steepest descent method for oscillatory Riemann–Hilbert problems: asymptotics for the MKdV equation", *Ann. of Math.* (2) **137**:2 (1993), 295–368.

[Deift et al. 1998] P. Deift, T. Kriecherbauer, and K. T.-R. McLaughlin, "New results on the equilibrium measure for logarithmic potentials in the presence of an external field", *J. Approx. Theory* **95**:3 (1998), 388–475.

[Deift et al. 1999] P. Deift, T. Kriecherbauer, K. T.-R. McLaughlin, S. Venakides, and X. Zhou, "Uniform asymptotics for polynomials orthogonal with respect to varying exponential weights and applications to universality questions in random matrix theory", *Comm. Pure Appl. Math.* **52**:11 (1999), 1335–1425.

[Delvaux et al. 2011] S. Delvaux, A. B. J. Kuijlaars, and L. Zhang, "Critical behavior of noninter-secting Brownian motions at a tacnode", *Comm. Pure Appl. Math.* **64**:10 (2011), 1305–1383.

[Duits and Geudens 2013] M. Duits and D. Geudens, "A critical phenomenon in the two-matrix model in the quartic/quadratic case", *Duke Math. J.* **162**:8 (2013), 1383–1462.

[Duits and Kuijlaars 2009] M. Duits and A. B. J. Kuijlaars, "Universality in the two-matrix model: a Riemann–Hilbert steepest-descent analysis", *Comm. Pure Appl. Math.* **62**:8 (2009), 1076–1153.

[Duits et al. 2011] M. Duits, D. Geudens, and A. B. J. Kuijlaars, "A vector equilibrium problem for the two-matrix model in the quartic/quadratic case", *Nonlinearity* **24**:3 (2011), 951–993.

[Duits et al. 2012] M. Duits, A. B. J. Kuijlaars, and M. Y. Mo, *The Hermitian two matrix model with an even quartic potential*, Mem. Amer. Math. Soc. **1022**, 2012.

[Ercolani and McLaughlin 2001] N. M. Ercolani and K. T.-R. McLaughlin, "Asymptotics and integrable structures for biorthogonal polynomials associated to a random two-matrix model: advances in nonlinear mathematics and science", *Phys. D* **152/153** (2001), 232–268.

[Eynard 2005] B. Eynard, *Le modèle à deux matrices, polynômes biorthogonaux, problème de Riemann–Hilbert*, Thèse d'habilitation à diriger des recherches, Université Paris 7 Denis Diderot, Paris, 2005. arXiv math-ph/0504034

[Eynard and Mehta 1998] B. Eynard and M. L. Mehta, "Matrices coupled in a chain, I: eigenvalue correlations", *J. Phys. A* **31**:19 (1998), 4449–4456.

[Guionnet 2004] A. Guionnet, "First order asymptotics of matrix integrals; a rigorous approach towards the understanding of matrix models", *Comm. Math. Phys.* **244**:3 (2004), 527–569.

[Hardy and Kuijlaars 2012] A. Hardy and A. B. J. Kuijlaars, "Weakly admissible vector equilib-rium problems", *J. Approx. Theory* **164**:6 (2012), 854–868.

[Johansson 2013] K. Johansson, "Non-colliding Brownian motions and the extended tacnode process", *Comm. Math. Phys.* **319**:1 (2013), 231–267.

[Kapaev 2003] A. A. Kapaev, "Riemann–Hilbert problem for bi-orthogonal polynomials", *J. Phys. A* **36**:16 (2003), 4629–4640.

[Kuijlaars 2010] A. B. J. Kuijlaars, "Multiple orthogonal polynomial ensembles", pp. 155–176 in *Recent trends in orthogonal polynomials and approximation theory*, edited by F. M. Jorge Arvesú and A. Martínez-Finkelshtein, Contemp. Math. **507**, Amer. Math. Soc., Providence, RI, 2010.

[Kuijlaars and McLaughlin 2005] A. B. J. Kuijlaars and K. T.-R. McLaughlin, "A Riemann–Hilbert problem for biorthogonal polynomials", *J. Comput. Appl. Math.* **178**:1-2 (2005), 313–320.

[Kuijlaars et al. 2009] A. B. J. Kuijlaars, A. Martínez-Finkelshtein, and F. Wielonsky, "Non-intersecting squared Bessel paths and multiple orthogonal polynomials for modified Bessel weights", *Comm. Math. Phys.* **286**:1 (2009), 217–275.

[Martínez-Finkelshtein and Rakhmanov 2011] A. Martínez-Finkelshtein and E. A. Rakhmanov, "Critical measures, quadratic differentials, and weak limits of zeros of Stieltjes polynomials", *Comm. Math. Phys.* **302**:1 (2011), 53–111.

[Mo 2009] M. Y. Mo, "Universality in the two matrix model with a monomial quartic and a general even polynomial potential", *Comm. Math. Phys.* **291**:3 (2009), 863–894.

[Nikishin and Sorokin 1991] E. M. Nikishin and V. N. Sorokin, *Rational approximations and orthogonality*, Translations of Mathematical Monographs **92**, American Mathematical Society, Providence, RI, 1991.

[Rakhmanov 2012] E. A. Rakhmanov, "Orthogonal polynomials and S-curves", pp. 195–239 in *Recent advances in orthogonal polynomials, special functions, and their applications*, edited by J. Arvesú and G. L. Lagomasino, Contemp. Math. **578**, Amer. Math. Soc., Providence, RI, 2012.

[Saff and Totik 1997] E. B. Saff and V. Totik, *Logarithmic potentials with external fields*, Grundlehren der Math. Wissensch. **316**, Springer, Berlin, 1997.

[Tracy and Widom 2006] C. A. Tracy and H. Widom, "The Pearcey process", *Comm. Math. Phys.* **263**:2 (2006), 381–400.

[Van Assche et al. 2001] W. Van Assche, J. S. Geronimo, and A. B. J. Kuijlaars, "Riemann–Hilbert problems for multiple orthogonal polynomials", pp. 23–59 in *Special functions 2000: current perspective and future directions* (Tempe, AZ), edited by J. Bustoz et al., NATO Sci. Ser. II Math. Phys. Chem. **30**, Kluwer, Dordrecht, 2001.

duits@kth.se   *Department of Mathematics,*
Royal Institute of Technology (KTH), Lindstedtsvagen 25,
SE-SE-10044 Stockholm, Sweden

arno.kuijlaars@wis.kuleuven.be   *Department of Mathematics,*
Katholieke Universiteit Leuven, 3001 Leuven, Belgium

Department of Mathematics, University of Bristol, University Walk, Bristol, BS8 1TW, United Kingdom

Random Matrices
MSRI Publications
Volume **65**, 2014

Conservation laws of random matrix theory

NICHOLAS M. ERCOLANI

This paper presents an overview of the derivation and significance of recently derived conservation laws for the matrix moments of Hermitian random matrices with dominant exponential weights that may be either even or odd. This is based on a detailed asymptotic analysis of the partition function for these unitary ensembles and their scaling limits. As a particular application we derive closed form expressions for the coefficients of the genus expansion for the associated free energy in a particular class of dominant even weights. These coefficients are generating functions for enumerating g-maps, related to graphical combinatorics on Riemann surfaces. This generalizes and resolves a 30+ year old conjecture in the physics literature related to quantum gravity.

1. Introduction

We present an overview of some recent developments in the application of random matrix analysis to the *topological combinatorics* of surfaces. Such applications have a long history about which we should say a few words at the outset. The combinatorial objects of interest here are *maps*. A map is an embedding of a graph into a compact, oriented and connected surface X with the requirement that the complement of the graph in X should be a disjoint union of simply connected open sets. If the genus of X is g, this object is referred to as a *g-map*. The notion of g-maps was introduced by Tutte [1968] and his collaborators in the 1960s as part of their investigations of the four color conjecture.

In the early 1980s Bessis, Itzykson and Zuber, a group of physicists studying 't Hooft's diagrammatic approaches to large-N expansions in quantum field theory, discovered a profound connection between the problem of enumerating g-maps and random matrix theory [Bessis et al. 1980]. That seminal work was the basis for bringing asymptotic analytical methods into the study of maps and other related combinatorial problems.

Subsequently other physicists [Douglas and Shenker 1990; Gross and Migdal 1990] realized that the matrix model diagrammatics described by Bessis et al. provide a natural means for discretizing the Einstein–Hilbert action in two dimensions. From that and a formal double scaling limit, they were able to put

Ercolani was supported by NSF grant DMS-0808059.

forward a candidate for so-called *2D quantum gravity*. This generated a great
deal of interest in the emerging field of string theory. We refer to [Di Francesco
et al. 1995] for a systematic review of this activity and to [Mariño 2005] for a
description of more recent developments related to topological string theory.

All of these applications were based on the postulated existence of a $1/n^2$
asymptotic expansion of the free energy associated to the random matrix partition
function, where n denotes the size of the matrix, as n becomes large. The
combinatorial significance of this expansion is that the coefficient of $1/n^{2g}$
should be the generating function for the enumeration of g-maps (ordered by
the cardinality of the map's vertices). In [Ercolani and McLaughlin 2003] the
existence of this asymptotic expansion and several of its important analytical
properties were rigorously established. This analysis was based on a Riemann–
Hilbert problem originally introduced by Fokas, Its and Kitaev [Fokas et al.
1992] to study the 2D gravity problem.

The aim of this paper is to outline how the results of [Ercolani and McLaughlin
2003] and its sequel [Ercolani et al. 2008] have been used to gain new insights
into the map enumeration problem. In particular, we will be able to prove and
significantly extend a conjecture made in [Bessis et al. 1980] about the closed
form structure of the generating functions for map enumeration.

Over time combinatorialists have made novel use of many tools from analysis
including contour integrals and differential equations. In this work we also
introduce nonlinear partial differential equations, in particular a hierarchy of
conservation laws reminiscent of *Burgers equation* (see (4-53)) and the *shallow
water wave equations* [Whitham 1974] (see (6-80)). This appears to make
contact with the class of *differential posets* introduced by Stanley [1988] (see
Remark 4.8).

2. Background

The general class of matrix ensembles we analyze has probability measures of
the form

$$d\mu_{t_j} = \frac{1}{Z^{(n)}(g_s, t_j)} \exp\left\{-\frac{1}{g_s} \operatorname{Tr}[V_j(M, t_j)]\right\} dM, \qquad (2\text{-}1)$$

where

$$V_j(\lambda; t_j) = \frac{1}{2}\lambda^2 + \frac{t_j}{j}\lambda^j, \qquad (2\text{-}2)$$

defined on the space $\mathcal{H}_n$ of $n \times n$ Hermitian matrices, M, and where g_s is
a positive parameter, referred to as the *string coefficient*. The normalization
factor $Z^{(n)}(g_s, t_j)$, which serves to make μ_t a probability measure, is called the
partition function of this unitary ensemble.

Remark 2.1. In previous treatments [Ercolani and McLaughlin 2003; Ercolani et al. 2008; Ercolani 2011; Ercolani and Pierce 2012], we have used the parameter $1/N$ instead of g_s. This was in keeping with notational usages in some areas of random matrix theory; however, since here we are trying to make a connection to some applications in quantum gravity, we have adopted the notation traditionally used in that context. This also is why we have scaled the time parameter t_j by $1/j$ in the first three sections of this paper.

For general polynomial weights V it is possible to establish the following fundamental asymptotic expansion [Ercolani and McLaughlin 2003; Ercolani et al. 2008] of the logarithm of the *free energy* associated to the partition function. More precisely, those papers consider weights of the form

$$V(\lambda) = \frac{1}{2}\lambda^2 + \sum_{\ell=1}^{j} \frac{t_\ell}{\ell}\lambda^\ell, \tag{2-3}$$

with j even.

We introduce a renormalized partition function, which we refer to as a *tau function* representation:

$$\tau_{n,g_s}^2(\vec{t}) = \frac{Z^{(n)}(g_s, \vec{t})}{Z^{(n)}(g_s, 0)}, \tag{2-4}$$

where $\vec{t} = (t_1, \dots t_j) \in \mathbb{R}^j$. The principal object of interest is the *large n* asymptotic expansion of this representation for which one has the result [Ercolani and McLaughlin 2003; Ercolani et al. 2008]

$$\log \tau_{n,g_s}^2(\vec{t})$$
$$= n^2 e_0(x,\vec{t}) + e_1(x,\vec{t}) + \frac{1}{n^2}e_2(x,\vec{t}) + \cdots + \frac{1}{n^{2g-2}}e_g(x,\vec{t}) + \cdots \tag{2-5}$$

as $n \to \infty$ while $g_s \to 0$ with $x = ng_s$, called the 't Hooft parameter, held fixed. Moreover, for

$$\mathcal{T} = (1 - \epsilon, 1 + \epsilon) \times \left(\{ |\vec{t}| < \delta \} \cap \{ t_j > 0 \} \right)$$

and for some $\epsilon > 0$, $\delta > 0$, we have:

(i) The expansion is uniformly valid on compact subsets of $\mathcal{T}$.

(ii) $e_g(x, \vec{t})$ extends to be complex analytic in

$$\mathcal{T}^{\mathbb{C}} = \{ (x, \vec{t}) \in \mathbb{C}^{j+1} \mid |x - 1| < \epsilon, |\vec{t}| < \delta \}.$$

(iii) The expansion may be differentiated term by term in $(x, \vec{t})$ with uniform error estimates as in (i).

The meaning of (i) is that for each g there is a constant, K_g, depending only on $\mathcal{T}$ and g such that

$$\left| \log \tau_{n,g_s}^2(\vec{t}) - n^2 e_0(x,\vec{t}) - \cdots - \frac{1}{n^{2g-2}} e_g(x,\vec{t}) \right| \leq \frac{K_g}{n^{2g}}$$

for $(x,\vec{t})$ in a compact subset of $\mathcal{T}$. The estimates referred to in (iii) have a similar form with τ_{n,g_s}^2 and $e_j(x,\vec{t})$ replaced by their mixed derivatives (the same derivatives in each term) and with a possibly different set of constants K_g.

Remark 2.2. These results were extended to the case where j is odd in [Ercolani and Pierce 2012].

To explain the topological significance of the $e_g(x,\vec{t})$ as generating functions, we begin with a precise definition of the objects they enumerate. A *map* Σ on a compact, oriented and connected surface X is a pair $\Sigma = (K(\Sigma), [\imath])$ where

- $K(\Sigma)$ is a connected 1-complex;
- $\imath$ is an embedding of $K(\Sigma)$ into X;
- the complement of $K(\Sigma)$ in X is a disjoint union of open cells (faces);
- the complement of the vertices in $K(\Sigma)$ is a disjoint union of open segments (edges).

Two such maps, Σ_1 and Σ_2, are isomorphic if there is an orientation-preserving homeomorphism of X to itself which maps the associated embedding of $K(\Sigma_1)$ homeomorphically to that of $K(\Sigma_2)$. When the genus of X is g one refers to the map as a *g-map*. What Bessis et al. [1980] effectively showed was that the partial derivatives of $e_g(1,\vec{t})$ evaluated at $\vec{t} = 0$ "count" a geometric quotient of a certain class of *labeled g-maps*.

As a means to reduce from enumerating these labeled g-maps to enumerating g-maps, it is natural to try taking a geometric quotient by a "relabeling group" more properly referred to as a *cartographic group* [Bauer and Itzykson 1996].

This labeling has two parts. First, the vertices of the same valence, ℓ, have an order labeling $1, \ldots, n_\ell$. Second, at each vertex one of the edges is distinguished. Given that X is oriented, this second labeling gives a unique ordering of the edges around each vertex. The fact that the coefficients of the free energy expansion (2-5) enumerate this class of labeled g-maps is a consequence of statements (i)–(iii) on the previous page, which enable one to evaluate a mixed partial derivative of e_g in terms of the *Gaussian unitary ensemble* (GUE) where the calculation of correlation functions of matrix coefficients all reduce to calculating just the quadratic correlation functions. (A precise description of how this enumeration works may be found in [Ercolani and McLaughlin 2003]).

To help fix these ideas we consider the case of a j-regular g-map (i.e., every vertex has the same valence, j) of size m (i.e., the map has m vertices) which is the main interest of this paper. The cartographic group in this case is generated by the symmetric group S_m which permutes the vertex labels and m factors of the cyclic group C_j, which rotates the distinguished edge at a given vertex in the direction of the holomorphic (counterclockwise) orientation on X. The order of the cartographic group here is the same as that of the product of its factors which is $m! j^m$. On the other hand the generating function for g-maps in this setting is given by

$$e_g(t_j) = e_g\left(x = 1, \vec{t} = (0, \ldots, 0, t_j)\right) \tag{2-6}$$

$$= \sum_{m \geq 1} \frac{1}{m! j^m} (-t_j)^m \kappa_j^{(g)}(m), \tag{2-7}$$

where $\kappa_j^{(g)}(m)$ is the number of labeled j-regular g-maps on m vertices. The fractional factor in the sum perfectly cancels the order of the cartographic group, making this series appear to indeed be the ordinary generating function for pure g-maps. However, for some g-maps the cartographic action may have nontrivial isotropy and this can create an "over-cancellation" of the labeling. This happens when a particular relabeling of a given map can be transformed back to the original labeling by a diffeomorphism of the underlying Riemann surface X. In this event the two labelings are indistinguishable and the diffeomorphism induces an automorphism of the underlying map. In addition, the element of the cartographic group giving rise to this situation is an element of the isotropy group of the given map. Hence, as a generating function for the geometric quotient, (2-6) is expressible as

$$e_g(t_j) = \sum_{g\text{-maps} \Sigma} \frac{1}{|\operatorname{Aut}(\Sigma)|} (-t_j)^{m(\Sigma)}, \tag{2-8}$$

$$E_g(x, t_j) = e_g(x, \vec{t} = (0, \ldots, 0, t_j)) \tag{2-9}$$

$$= \sum_{g\text{-maps} \Sigma} \frac{1}{|\operatorname{Aut}(\Sigma)|} (-t_j)^{m(\Sigma)} x^{f(\Sigma)} = x^{2-2g} e_g(x^{j/2-1} t_j),$$

where $m(\Sigma)$ is the number of vertices of Σ, $f(\Sigma)$ is the number of faces of Σ and $\operatorname{Aut}(\Sigma)$ is the automorphism group of the map Σ. We have included the x-dependent form, (2-9), of e_g since that will play an important role later on and also to observe that this is in fact a *bivariate* generating function for enumerating g-maps with a fixed number of vertices and faces. Moreover, in this j-regular setting, one sees that the bivariate function is self-similar. This is a direct consequence of Euler's relation:

$$2 - 2g = \#\,\text{vertices} - \#\,\text{edges} + \#\,\text{faces} = m(\Sigma) - \frac{j}{2}m(\Sigma) + f(\Sigma),$$

$$t_j^{m(\Sigma)} x^{f(\Sigma)} = x^{2-2g} (t_j x^{j/2-1})^{m(\Sigma)}. \tag{2-10}$$

The presence of geometric factors such as $1/|\text{Aut}(\Sigma)|$ is not uncommon in enumerative graph theory, a classical example being that of Erdős–Rényi graphs [Janson et al. 1993]. In the quantum gravity setting these factors also have a natural interpretation in terms of the discretization of the reduction to conformal structures via a quotient of metrics by the action of the diffeomorphism group. We refer to [Di Francesco et al. 1995; Bauer and Itzykson 1996] for further details on this attractive set of ideas.

In [Bessis et al. 1980], e_0, e_1 and e_2 were explicitly computed for the case of valence $j = 4$. We quote, from the same paper, the following conjecture (some notation has been changed to be consistent with ours):

It would of course be very interesting to obtain $e_g(t_4)$ in closed form for any value of g. The method of this paper enabled us to do so up to $g = 2$, but works in the general case, although it requires an increasing amount of work. We conjecture a general expression of the form

$$e_g = \frac{(1 - z_0)^{2g-1}}{(2 - z_0)^{5(g-1)}} P^{(g)}(z_0), \quad g \geq 2,$$

with $P^{(g)}$ a polynomial in z_0, the degree of which could be obtained by a careful analysis of the above procedure.

Here $z_0 = z_0(t_4)$ is equal, up to a scaling, to the generating function for the Catalan numbers; below it will signify $z_0(t_{2\nu})$, which is similarly related to the generating function for the higher Catalan numbers (4-62).

The main purpose of this paper is to show how this conjecture can be verified and significantly extended. In particular, we will show, for the case of even valence, $j = 2\nu$:

Theorem 2.3. *For $g \geq 2$ and for $\nu \geq 2$,*

$$e_g(z_0) = C^{(g)} + \frac{c_0^{(g)}(\nu)}{(\nu - (\nu-1)z_0)^{2g-2}} + \cdots + \frac{c_{3g-3}^{(g)}(\nu)}{(\nu - (\nu-1)z_0)^{5g-5}} \tag{2-11}$$

$$= \frac{(z_0 - 1)^r Q_{5g-5-r}(z_0)}{(\nu - (\nu-1)z_0)^{5g-5}}, \tag{2-12}$$

$$r = \max\left\{1, \left\lfloor \frac{2g-1}{\nu-1} \right\rfloor\right\}, \tag{2-13}$$

The top coefficient and the constant term are respectively given by

$$c_{3g-3}^{(g)}(v) = \frac{1}{(5g-5)(5g-3)v^2} a_{3g-1}^{(g)}(v) \neq 0, \tag{2-14}$$

$$C^{(g)} = -2(2g-3)! \left[\frac{1}{(2g+2)!} - \frac{1}{(2g)!\,12} \right.$$
$$\left. + \frac{(1-\delta_{2,g})}{(2g-1)!} \sum_{k=2}^{g-1} \frac{(2-2k)_{2g-2k+2}}{(2g-2k+2)!} C^{(k)} \right], \tag{2-15}$$

where $a_{3g-1}^{(g)}(v)$ (see Theorem 4.7) is proportional to the g-th coefficient in the asymptotic expansion at infinity of the v-th equation in the Painlevé I hierarchy [Ercolani 2011] and $(r)_m = r(r-1)\ldots(r-m+1)$. (Explicit expressions for e_1 and e_0 are given in (5-70) and (5-71).)

Our methods can be extended to the case of j odd and the derivation of the analogue to Theorem 2.3 is in progress (see Section 6).

The route to getting these results passes through nonlinear PDE, in particular a class of nonlinear evolution equations known as conservation laws which come from studying scaling limits of the recursion operators for orthogonal polynomials whose weights match those of the matrix models.

This appeal to orthogonal polynomials also motivated the approaches of [Bessis et al. 1980; Douglas and Shenker 1990]. However to give a rigorous *and* effective treatment to the problem of finding closed form expressions for the coefficients of the asymptotic free energy, (2-5), requires essential use of Riemann–Hilbert analysis on the Riemann–Hilbert problem for orthogonal polynomials that was introduced in [Fokas et al. 1992]. Though we will not review this analysis here, we will state the consequences of it needed for our applications and reference their sources.

In Section 3 we present the necessary background on orthogonal polynomials and introduce the main equations governing their recurrence operators: the *difference string equations* and the *Toda lattice equations*. In Section 4 we describe how (2-5) can be used to derive and solve (in the case of even valence) the continuum limits of these equations which relates to the nonlinear evolution equations alluded to earlier. In Section 5 we outline the proof Theorem 2.3 and in Section 6 we describe the extension of this program to the case of odd valence and briefly mention what has been accomplished in that case thus far. This will also help to illuminate the full picture behind the idea of conservation laws for random matrices.

Over the years there have been a number of efforts to systematically address the question of graphical enumeration on Riemann surfaces by studying the resolvent of the random matrix and associated Schwinger–Dyson equations [Ambjørn et al. 1993; Chekhov and Eynard 2006; Eynard 2004; Eynard and Orantin 2009]. These approaches have led to interesting direct extensions of the equations that Tutte originally introduced to study maps. They also have particular relevance for the topics described in the concluding remark Section 7.2. Our methods take a different approach, based on orthogonal polynomials, but the development of a unified perspective that incorporates the Toda, string and Schwinger–Dyson equations would undoubtedly yield valuable insights.

3. The role of orthogonal polynomials and their asymptotics

Let us recall the classical relation between orthogonal polynomials and the space of square-integrable functions on the real line, $\mathbb{R}$, with respect to exponentially weighted measures. In particular, we want to focus attention on weights that correspond to the random matrix weights $V(\lambda)$, (2-2), with j even. (Recently this relation has been extended to the cases of j odd [Bleher and Deaño 2012; Ercolani and Pierce 2012], with the orthogonal polynomials generalized to the class of so-called *non-Hermitian* orthogonal polynomials; however, for this exposition we will stick primarily with the even case.) To that end we consider the Hilbert space $H = L^2\big(\mathbb{R}, e^{-g_s^{-1}V(\lambda)}\big)$ of weighted square integrable functions. This space has a natural polynomial basis, $\{\pi_n(\lambda)\}$, determined by the conditions that

$$\pi_n(\lambda) = \lambda^n + \text{ lower order terms},$$
$$\int \pi_n(\lambda)\pi_m(\lambda)e^{-g_s^{-1}V(\lambda)}d\lambda = 0 \quad \text{for } n \neq m.$$

For the construction of this basis and related details we refer the reader to [Deift 1999].

With respect to this basis, the operator of multiplication by λ is representable as a semiinfinite tridiagonal matrix:

$$\mathscr{L} = \begin{pmatrix} a_0 & 1 & & \\ b_1^2 & a_1 & 1 & \\ & b_2^2 & a_2 & \ddots \\ & & & \ddots & \ddots \end{pmatrix}. \tag{3-16}$$

We commonly refer to $\mathscr{L}$ as the *recursion operator* for the orthogonal polynomials and to its entries as *recursion coefficients*. (When V is an even potential, it follows from symmetry that $a_j = 0$ for all j.) We remark that often a basis of orthonormal, rather than monic orthogonal, polynomials is used to make this representation. In that case the analogue of (3-16) is a symmetric tridiagonal matrix. As long as the coefficients $\{b_n\}$ do not vanish, these two matrix representations can be related through conjugation by a semiinfinite diagonal matrix of the form $\mathrm{diag}\left(1, b_1^{-1}, (b_1 b_2)^{-1}, (b_1 b_2 b_3)^{-1}, \dots\right)$.

Similarly, the operator of differentiation with respect to λ, which is densely defined on H, has a semiinfinite matrix representation, $\mathscr{D}$, that can be expressed in terms of $\mathscr{L}$ as

$$\mathscr{D} = \frac{1}{g_s}(\mathscr{L} + t\mathscr{L}^{j-1})_-, \tag{3-17}$$

where the minus subscript denotes projection onto the strictly lower part of the matrix.

From the canonical (Heisenberg) relation on H, one sees that

$$[\partial_\lambda, \lambda] = 1,$$

where here λ in the bracket and 1 on the right-hand side are regarded as multiplication operators. With respect to the basis of orthogonal polynomials this may be reexpressed as

$$\left[\mathscr{L}, (\mathscr{L} + t\mathscr{L}^{j-1})_-\right] = g_s I. \tag{3-18}$$

The relations implicit in (3-18) have been referred to as *string equations* in the physics literature. In fact the relations that one has, row by row, in (3-18) are actually successive differences of consecutive string equations in the usual sense. However, by continuing back to the first row one may recursively decouple these differences to get the usual equations. To make this distinction clear we will refer to the row by row equations that one has directly from (3-18) as *difference string equations*.

$\mathscr{L}$ depends smoothly on the coupling parameter t_j in the weight $V(\lambda)$; see (2-2). The explicit dependence can be determined from the fact that multiplication by λ commutes with differentiation by t_j. This yields our second fundamental relation on the recurrence coefficients,

$$g_s \frac{\partial}{\partial t_j}\mathscr{L} = [(\mathscr{L}^j)_-, \mathscr{L}], \tag{3-19}$$

which is equivalent to the j-th equation of the semiinfinite Toda lattice hierarchy. The Toda equations for $j = 1$ are

$$-g_s \frac{da_{n,g_s}}{dt_1} = b_{n+1,g_s}^2 - b_{n,g_s}^2, \tag{3-20}$$

$$-g_s \frac{db_{n,g_s}^2}{dt_1} = b_{n,g_s}^2 (a_{n,g_s} - a_{n-1,g_s}). \tag{3-21}$$

Hirota equations. One may apply standard methods of orthogonal polynomial theory [Szegő 1939] to deduce the existence of a semiinfinite lower unipotent matrix A such that

$$\mathcal{L} = A^{-1}\epsilon A,$$

where

$$\epsilon = \begin{pmatrix} 0 & 1 & & & \\ & 0 & 0 & 1 & \\ & & 0 & 0 & \ddots \\ & & & \ddots & \ddots \end{pmatrix}.$$

(For a description of the construction of such a unipotent matrix we refer to Proposition 1 of [Ercolani and McLaughlin 2001].)

This is related to the Hankel matrix

$$\mathcal{H} = \begin{pmatrix} m_0 & m_1 & m_2 & \cdots \\ m_1 & m_2 & m_3 & \cdots \\ m_2 & m_3 & m_4 & \cdots \\ \vdots & \vdots & \vdots & \ddots \end{pmatrix},$$

where

$$m_k = \int_{\mathbb{R}} \lambda^k e^{-g_s^{-1}V(\lambda)} \, d\lambda$$

is the k-th moment of the measure, by

$$ADA^\dagger = \mathcal{H},$$
$$D = \text{diag}\{d_0, d_1 \ldots\},$$

with

$$d_n = \frac{\det \mathcal{H}_{n+1}}{\det \mathcal{H}_n},$$

where $\mathcal{H}_n$ denotes the $n \times n$ principal submatrix of $\mathcal{H}$ whose determinant may be expressed as (see [Szegő 1939]) as

$$\det \mathcal{H}_n = n!\hat{Z}^{(n)}(t_1, t_{2v}),$$

$$\hat{Z}^{(n)}(t_1, t_{2v}) = \int_{\mathbb{R}} \cdots \int_{\mathbb{R}} \exp\left\{-g_s^{-2}\left[g_s \sum_{m=1}^{n} V(\lambda_m; t_1, t_{2v})\right.\right.$$
$$\left.\left. - g_s^2 \sum_{m \neq \ell} \log |\lambda_m - \lambda_\ell|\right]\right\} d^n\lambda, \quad (3\text{-}22)$$

where $V(\lambda; t_1, t_{2v+1}) = \frac{1}{2}\lambda^2 + t_1\lambda + \frac{t_{2v}}{2v}\lambda^{2v}$. We set $\det \mathcal{H}_0 = 1$.

Remark 3.1. One sometimes needs to extend the domain of the tau functions to include other parameters, such as t_1, as we have done here. Doing this presents no difficulties in the prior constructions.

The diagonal elements may in fact be expressed as $d_n = \dfrac{\tau_{n+1,g_s}^2}{\tau_{n,g_s}^2} d_n(0)$, where

$$\tau_{n,g_s}^2 = \frac{\hat{Z}^{(n)}(t_1, t_{2v})}{\hat{Z}^{(n)}(0,0)} \tag{3-23}$$

$$= \frac{Z^{(n)}(t_1, t_{2v})}{Z^{(n)}(0,0)}, \tag{3-24}$$

which agrees with the definition of the tau function given in (2-4). The second equality follows by reducing the unitarily invariant matrix integrals in (3-24) to their diagonalizations, which yields (3-23) [Ercolani and McLaughlin 2003]. Tracing through these connections, from $\mathcal{L}$ to D, one may derive the fundamental identity relating the random matrix partition function to the recurrence coefficients:

$$b_{n,g_s}^2 = \frac{d_n}{d_{n-1}} = \frac{\tau_{n+1,g_s}^2 \tau_{n-1,g_s}^2}{\tau_{n,g_s}^4} b_{n,g_s}^2(0), \tag{3-25}$$

which is the basis for our analysis of continuum limits in the next section. (Note that $b_{0,g_s}^2(0) = 0$ and therefore $b_{0,g_s}^2 \equiv 0$.) We will also need a differential version of this relation:

Lemma 3.2 (Hirota).

$$a_{n,g_s} = -g_s \frac{\partial}{\partial t_1} \log\left[\frac{\tau_{n+1,g_s}^2}{\tau_{n,g_s}^2}\right] = -g_s \frac{\partial}{\partial t_1} \log\left[\frac{Z^{(n+1)}(t_1, t_{2v})}{Z^{(n)}(t_1, t_{2v})}\right], \tag{3-26}$$

$$b_{n,g_s}^2 = g_s^2 \frac{\partial^2}{\partial t_1^2} \log \tau_{n,g_s}^2 = g_s^2 \frac{\partial^2}{\partial t_1^2} \log Z^{(n)}(t_1, t_{2v}). \tag{3-27}$$

(A derivation of this lemma may be found in [Bleher and Its 2005].) It follows from (3-27), (2-5) and (2-9) that:

Corollary 3.3.

$$b_{n,g_s}^2(x; t_{2v}) = x\left(z_0(x; t_{2v}) + \cdots + \frac{1}{n^{2g}} z_g(x; t_{2v}) + \cdots\right), \tag{3-28}$$

$$z_g(x; t_{2v}) = \frac{d^2}{dt_1^2} e_g(x; t_1, t_{2v})|_{t_1=0} \tag{3-29}$$

$$= x^{1-2g} z_g(x^{v-1} t_{2v}) \tag{3-30}$$

is a uniformly valid asymptotic expansion (in the sense of (iii) on page 165).

Path weights and recurrence coefficients. In order to effectively utilize the relations (3-18)–(3-19) it will be essential to keep track of how the matrix entries of powers of the recurrence operator, $\mathcal{L}^j$, depend on the original recurrence coefficients. That is best done via the combinatorics of weighted walks on the index lattice of the orthogonal polynomials. For the case of even potentials, the relevant walks are *Dyck paths* which are walks, P, on $\mathbb{Z}$ which, at each step, can either increase by 1 or decrease by 1. Set

$$\mathcal{P}^j(m_1, m_2) = \text{the set of all Dyck paths of length } j \text{ from } m_1 \text{ to } m_2. \quad (3\text{-}31)$$

Then step weights, path weights and the (m_1, m_2)-entry of $\mathcal{L}^j$ are, respectively, given by

$$\omega(p) = \begin{cases} 1 & \text{if the } p\text{-th step moves from } n \text{ to } n+1 \text{ on the lattice,} \\ b_n^2 & \text{if the } p\text{-th step moves from } n \text{ to } n-1, \end{cases}$$

$$\omega(P) = \prod_{\text{steps } p \in P} \omega(p),$$

$$\mathcal{L}_{m_1, m_2}^j = \sum_{P \in \mathcal{P}^j(m_1, m_2)} \omega(P). \quad (3\text{-}32)$$

Dyck representation of the difference string equations. The *difference string equations* are given (for the 2ν-valent case) by (3-18):

$$[\mathcal{L}, (\mathcal{L} + t\mathcal{L}^{2\nu-1})_-] = g_s I. \quad (3\text{-}33)$$

By parity considerations, when the potential V is even, the only nontautological equations come from the diagonal entries of (3-34): *the (n, n) entry* gives

$$g_s = (\mathcal{L} + t\mathcal{L}^{2\nu-1})_{n+1,n} - (\mathcal{L} + t\mathcal{L}^{2\nu-1})_{n,n-1}. \quad (3\text{-}34)$$

In terms of Dyck paths this becomes

$$\frac{x}{n} = b_{n+1}^2 - b_n^2 + t \sum_{P \in \mathcal{P}^{2\nu-1}(1,0)} \left(\prod_{m=1}^{\nu} b_{n+\ell_m(P)+1}^2 - \prod_{m=1}^{\nu} b_{n+\ell_m(P)}^2 \right), \quad (3\text{-}35)$$

where $\ell_m(P)$ denotes the lattice location of the path P after the m-th downstep and we have used the relation $x = n g_s$ on the left-hand side of the equation.

We illustrate this more concretely for the case of $j = 2\nu = 4$. Referring to (3-31), the relevant path classes here are

$$\mathcal{P}^1(1, 0) = \text{a descent by one step,}$$

$$\mathcal{P}^3(1, 0) = \text{paths with exactly one upstep and two downsteps (Figure 1).}$$

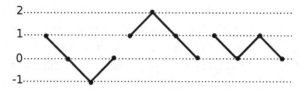

Figure 1. Elements of $\mathcal{P}^3(1,0)$.

Note that the structure of the path classes does not actually depend upon n. This is a reflection of the underlying spatial homogeneity of these equations. Thus, for the purpose of describing the path classes, one can translate n to 0.

Now applying (3-32) the difference string equation becomes, for $n > 0$,

$$\frac{1}{n} = (b_{n+1}^2 - b_n^2) + t\{b_{n+1}^2(b_n^2 + b_{n+1}^2 + b_{n+2}^2) - b_n^2(b_{n-1}^2 + b_n^2 + b_{n+1}^2)\},$$

where, for this example, we have set the parameter x equal to 1.

Dyck representation of the Toda equations. We now pass to a more explicit form of the *Toda equation* (3-19) in the case $j = 2\nu$:

$$-\frac{x}{2\nu n}\frac{db_n^2}{dt_{2\nu}} = (\mathscr{L}^{2\nu})_{n+1,n-1} - (\mathscr{L}^{2\nu})_{n,n-2}$$

$$= \sum_{P \in \mathcal{P}^{2\nu}(2,0)}\left(\prod_{m=1}^{\nu+1} b_{n+\ell_m(P)+1}^2 - \prod_{m=1}^{\nu+1} b_{n+\ell_m(P)}^2\right).$$

Once again we illustrate these equations in the tetravalent case ($\nu = 2$). The relevant path class is

$$\mathcal{P}^4(1,-1) = \text{paths with exactly one upstep and three downsteps.}$$

Applying (3-32), the tetravalent Toda equations become

$$-\frac{1}{4n}\frac{db_n^2}{dt} = b_{n+1}^2 b_n^2(b_{n-1}^2 + b_n^2 + b_{n+1}^2 + b_{n+2}^2)$$

$$-b_n^2 b_{n-1}^2(b_{n-2}^2 + b_{n-1}^2 + b_n^2 + b_{n+1}^2), \quad (3\text{-}36)$$

where we have again used the relation $x = ng_s$ and then set the parameter $x = 1$.

4. Continuum limits

The continuum limits of the difference string and Toda equations will be described in terms of certain scalings of the independent variables, both discrete and continuous. As indicated at the outset, the positive parameter g_s sets the scale for the potential in the random matrix partition function and is taken to be small.

The discrete variable n labels the lattice *position* on $\mathbb{Z}^{\geq 0}$ that marks, for instance, the n-th orthogonal polynomial and recurrence coefficients. We also always take n to be large and in fact to be of the same order as $1/g_s$; that is, as n and g_s tend to ∞ and 0 respectively, they do so in such a way that their product

$$x \doteq g_s n \tag{4-37}$$

remains fixed at a value close to 1.

In addition to the *global* or *absolute* lattice variable n, we also introduce a *local* or *relative* lattice variable denoted by k. It varies over integers but will always be taken to be small in comparison to n and independent of n. The Dyck lattice paths naturally introduce the composite discrete variable $n + k$ into the formulation of the difference string and Toda equations which we think of as a small discrete variation around a large value of n. The spatial homogeneity of those equations manifests itself in their all having the same form, independent of what n is, while k in those equations varies over $\{-\nu-1, \ldots, -1, 0, 1, \ldots, \nu+1\}$, the *bandwidth* of the (2ν)-th Toda/difference string equations. Taking $\nu + 1 \ll n$ will insure the necessary separation of scales between k and n. We define

$$w \doteq (n + k)g_s \tag{4-38}$$

$$= x + g_s k = x\left(1 + \frac{k}{n}\right) \tag{4-39}$$

as a *spatial* variation close to x which will serve as a continuous analogue of the lattice location along a Dyck path relative to the starting location of the path.

We also introduce the self-similar scalings

$$s_1 \doteq x^{-\frac{1}{2}}t_1, \tag{4-40}$$

$$s_{2\nu} \doteq x^{\nu-1}2\nu t_{2\nu}, \tag{4-41}$$

$$\widetilde{w} \doteq \left(1 + \frac{k}{n}\right), \tag{4-42}$$

that are natural given (2-10). In terms of these scalings, (3-28) may be rewritten [Ercolani et al. 2008] as

$$b_{n,g_s}^2(s_{2\nu}) = x\left(xz_0(s_{2\nu}) + \cdots + \frac{x^{1-2g}}{n^{2g}}z_g(s_{2\nu}) + \cdots\right), \tag{4-43}$$

$$z_g(s_{2\nu}) = \frac{d^2}{ds_1^2}e_g(s_1, s_{2\nu})|_{s_1=0}, \tag{4-44}$$

and

$$b_{n+k,g_s}^2(s_{2v}) = x\left(f_0(s_{2v}, w) + \cdots + \frac{1}{n^{2g}}f_g(s_{2v}, w) + \cdots\right), \qquad (4\text{-}45)$$

$$f_g(s_{2v}, w) = w^{1-2g}z_g(s_{2v}\tilde{w}^{v-1}). \qquad (4\text{-}46)$$

Remark 4.1. The variables s_j as defined above differ slightly from their usage in related works [Ercolani et al. 2008; Ercolani 2011] where $s_j = -\alpha_j t_j$ for appropriate parameters $\alpha_j > 0$.

We also introduce a shorthand notation to denote the expansion of the coefficients of $f(s_1, s_{2v+1}, w)$ around $w = x$.

Definition 4.2. For $w = x + g_s k$,

$$f(s_1, s_{2v}, w) = \sum_{j=0}^{\infty} \frac{f_{w^{(j)}}|_{w=x}}{j!}\left(\frac{kx}{n}\right)^j, \qquad (4\text{-}47)$$

where the subscript $w^{(j)}$ denotes the operation of taking the j-th derivative with respect to w of each coefficient of f:

$$f_{w^{(j)}} = \sum_{g\geq 0} \frac{\partial^j}{\partial w^j} f_g(s_1, s_{2v}, w)\frac{1}{n^{2g}}.$$

As valid asymptotic expansions, these representations denote the asymptotic series whose successive terms are gotten by collecting all terms with a common power of $1/n$ in (4-47).

In what follows we will frequently abuse notation and drop the evaluation at $w = x$. In particular, we will write:

$$b_{n+k,g_s}^2 = \sum_{j=0}^{\infty} \frac{f_{w^{(j)}}}{j!}(g_s k)^j = \sum_{j=0}^{\infty} \frac{1}{j!} \sum_{g\geq 0} \frac{\partial^j}{\partial w^j} f_g(s_{2v}, w)\frac{1}{n^{2g}}(g_s k)^j.$$

$$(4\text{-}48)$$

In doing this these series must now be regarded as formal but whose orders are still defined by collecting all terms in $1/n$ and g_s of a common order. (Recall that $g_s \sim 1/n$, so that $n^{-\alpha}g_s^{\beta} = \mathcal{O}(n^{-(\alpha+\beta)})$). They will be substituted into the difference string and the Toda equations to derive the respective continuum equations. At any point in this process, if one evaluates these expressions at $w = x$ and $g_s = x/n$ one may recover valid asymptotic expansions in which the b_{n+k,g_s}^2 have their original significance as valid asymptotic expansions of the recursion coefficients.

The continuum limit of the Toda equations. One is now in a position to study the Toda lattice equations (3-36) expanded on the asymptotic series (4-48):

$$-\frac{1}{n}\frac{d}{ds}f(s,w) =$$

$$\sum_{P\in\mathscr{P}^{2v}(1,-1)}\left(\prod_{m=1}^{v+1}\sum_{j=0}^{\infty}\frac{f_{w^{(j)}}}{j!}(g_s\,(\ell_m(P)+1))^j - \prod_{m=1}^{v+1}\sum_{j=0}^{\infty}\frac{f_{w^{(j)}}}{j!}(g_s\,\ell_m(P))^j\right).$$

From now on we will take $s_1 = 0$, since its role in determining the structure of the asymptotic expansions of the b_{n+k} is now completed, and set $s_{2v} = s$.

Collecting terms in these equations order by order in orders of $1/n$ one has a hierarchy of equations that, in principle, allows one to recursively determine the coefficients of (4-43). We will refer to this hierarchy as the *continuum Toda equations*. (Note that one has such a hierarchy for each value of v.) Of course this is a standard procedure in perturbation theory. The equations we will derive are PDEs in the form of evolution equations in which w, now regarded as a continuous variable, is the independent *spatial* variable and s_{2v} is the *temporal* variable. One must still determine, at each level of the hierarchy, which solution of the PDE is the one that corresponds to the expressions given for f_g in (4-46). This amounts to a kind of solvability condition. This process was carried out fully in [Ercolani et al. 2008; Ercolani 2011]. We will now state the results of that analysis.

Theorem 4.3 [Ercolani et al. 2008]. *The continuum limit, to all orders, of the Toda Lattice equations as $n \to \infty$ is given by the following infinite order partial differential equation for $f(s,w)$:*

$$-\frac{df}{ds} = F^{(v+1)}(g_s; f, f_w, \ldots, f_{w^{(j)}}, \ldots)$$

$$\dot{=} \sum_{g\geq 0} g_s^{2g}\, F_g^{(v+1)}(f, f_w, f_{w^{(2)}}, \cdots, f_{w^{(2g+1)}})$$

$$= c_v f^v f_w + g_s^2 F_1^{(v+1)}(f, f_w, f_{ww}, f_{www}) + \cdots \tag{4-49}$$

for (s, w) near $(0, 1)$ and initial data given by $f(0, w) = w$,

$$\tag{4-50}$$

$$F_g^{(v+1)} = \sum_{\substack{\lambda\in\Lambda^{2g+1}\\ \ell(\lambda)\leq v+1}}\left(\sum_{\vec{\ell}\in\mathscr{P}^{2v}(1,-1)}\left(m_\lambda(\ell_1+1,\ldots,\ell_{v+1}+1)\right.\right.$$

$$\left.\left. -m_\lambda(\ell_1,\ldots,\ell_{v+1})\right)\right)\frac{f_{w^\lambda}}{\lambda!}, \tag{4-51}$$

where Λ^{2g+1} denotes the set of number partitions $\lambda = (\lambda_1, \lambda_2, \ldots)$ such that $\lambda_1 \geq \lambda_2 \geq \lambda_3 \geq \cdots$ and whose size $|\lambda| = \sum_i \lambda_i$ equals $2g + 1$; we call $\ell(\lambda) = \sum_j r_j(\lambda)$ (where $r_j(\lambda) = \#\{\lambda_i \mid \lambda_i = j\}$) the length of λ. The w-derivatives of f have been expressed in multiindex notation so that

$$f_{w^\lambda} = f_{w^{(\lambda_1)}} \cdots f_{w^{(\lambda_{v+1})}}.$$

Recall also that the monomial symmetric function [Macdonald 1995] *associated to the partition* λ, *appearing inside the inner sum of* (4-51), *is given by*

$$m_\lambda(x_1, \ldots, x_{\nu+1}) = \frac{1}{r(\lambda)!} \sum_{\sigma \in S_{\nu+1}} x_{\sigma(1)}^{\lambda_1} \cdots x_{\sigma(\nu+1)}^{\lambda_{\nu+1}}, \qquad (4\text{-}52)$$

where division by the factor $r(\lambda)! = \prod_j r_j(\lambda)!$ *ensures that there is no redundancy among the (monic) terms appearing in the symmetric polynomial. Finally, since a Dyck path in* $\mathscr{P}^{2\nu}(1, -1)$ *is uniquely determined by the locations* ℓ_m *of its downsteps, the inner sum of* (4-51) *is well defined.*

To help see how the expression (4-51) for the forcing term $F_g^{(\nu+1)}$ works, let us illustrate it in the simplest case where $g = 0$. Then there is only one partition, $\lambda = (1)$ in Λ^1 and

$$m_{(1)}(x_1, \ldots, x_{\nu+1}) = x_1 + \cdots + x_{\nu+1},$$
$$m_{(1)}(\ell_1 + 1, \ldots, \ell_{\nu+1} + 1) - m_{(1)}(\ell_1, \ldots, \ell_{\nu+1}) = \nu + 1.$$

Hence, $c_\nu = (\nu + 1) \binom{2\nu}{\nu+1}$.

We now introduce the global *conservation law* structure (4-53) of these continuum Toda equations.

Proposition 4.4. *Equation* (4-49) *may be rewritten as*

$$-\frac{df}{ds} = F^{(\nu+1)} = \partial_w \hat{F}^{(\nu+1)}$$

$$\doteq \partial_w \sum_{g \geq 0} g_s^{2g} \hat{F}_g^{(\nu+1)}, \qquad (4\text{-}53)$$

where

$$\hat{F}_g^{(\nu+1)}$$
$$= \sum_{\substack{\lambda \in \Lambda^{2g} \\ \ell(\lambda) \leq \nu+1}} \sum_{\tilde{\ell} \in \mathscr{P}^{2\nu}(1,-1)} \left(\sum_{\mu \prec \lambda} \frac{1}{|\lambda - \mu| + 1} \binom{\lambda}{\mu} m_\mu(\ell_1, \ldots, \ell_{\nu+1}) \right) \frac{1}{r(\lambda)!} \frac{f_{w\lambda}}{\lambda!}.$$

Here $\mu \prec \lambda$ *means* μ *is a partition such that* $\mu_1 \leq \lambda_1, \ldots, \mu_{\nu+1} \leq \lambda_{\nu+1}$ *and* $|\lambda - \mu| = \sum_i (\lambda_i - \mu_i) = |\lambda| - |\mu|$. *The combinatorial coefficient*

$$\binom{\lambda}{\mu} = \binom{\lambda_1}{\mu_1} \cdots \binom{\lambda_{\nu+1}}{\mu_{\nu+1}}$$

together with the factor $|\lambda - \mu| + 1$ *account for the multiplicities induced by* w-*differentiation (see Remark 4.8).*

It is straightforward to check the (key) second equality in (4-53) by direct differentiation. We can again take the special case of $g = 0$ as an example for which $\mu = \lambda = \varnothing$ and $m_\varnothing \equiv 1$, so that

$$\hat{F}_0^{(\nu+1)} = \binom{2\nu}{\nu+1} f^{\nu+1},$$

$$\partial_w \hat{F}_0^{(\nu+1)} = (\nu+1)\binom{2\nu}{\nu+1} f^\nu f_w = c_\nu f^\nu f_w = F_0^{(\nu+1)}.$$

One is now in a position to deduce the form of the Toda hierarchy. This is done by setting $x = 1$ so that $g_s = 1/n$. One then collects *all* terms of order n^{-2g} in the resulting expansion of (4-49) and this will be a partial differential equation in s and w that we refer to as the g-th equation in the continuum Toda hierarchy.

At leading order in the hierarchy one observes that, for general ν, the continuum Toda equation is an inviscid Burgers equation [Whitham 1974]:

$$\frac{d}{ds} f_0 = -\frac{c_\nu}{\nu+1} \partial_w (f_0)^{\nu+1}, \tag{4-54}$$

with initial data $f_0 = w$. A solution exists and is unique for sufficiently small values of s. It may be explicitly calculated by the method of characteristics, also known as the *hodograph* method in the version we now present. Consider the (hodograph) relation among the independent variables (s, w, f_0):

$$w = f_0 + c_\nu s f_0^\nu. \tag{4-55}$$

Lemma 4.5. *A local solution of* (4-54) *is implicitly defined by* (4-55).

Proof. The annihilator of the differential of (4-55),

$$(1 + \nu c_\nu s f_0^{\nu-1})\, df_0 - dw + c_\nu f_0^\nu\, ds,$$

is a two-dimensional distribution locally on the space (s, w, f_0). An initial curve over the w-axis (parametrized as the graph of a function $f_0(w)$), transverse to the locus where $1 + \nu c_\nu s f_0^{\nu-1} = 0$ locally determines a unique integral surface foliated by the integral curves of the *characteristic* vector field

$$\frac{df_0}{ds} = 0, \tag{4-56}$$

$$\frac{dw}{ds} = c_\nu f_0^\nu, \tag{4-57}$$

$$f_0(0, w) = f_0(w). \tag{4-58}$$

Equation (4-56) requires that along an integral curve of the characteristic vector field, f_0 is constant; i.e.,

$$0 = \frac{df_0}{ds}(s, w(s)) = \frac{\partial f_0}{\partial s} + \frac{\partial f_0}{\partial w}\frac{dw}{ds} = \frac{\partial f_0}{\partial s} + c_v f_0^v \frac{\partial f_0}{\partial w},$$

by (4-57), which is equivalent to (4-54). Using (4-58) to set $f_0(0, w) = w$ pins down our solution uniquely. □

Remark 4.6. We note that the numerical coefficients appearing in these Burgers equations depend only on the total number of Dyck paths in $\mathcal{P}^{2v}(1, -1)$.

From (4-55) and the self-similar form of f_0,

$$f_0(s, w) = wz_0(sw^{v-1}), \tag{4-59}$$

one finds [Ercolani et al. 2008] that

$$z_0(s) = \sum_{j \geq 0} c_v^j \zeta_j s^j, \tag{4-60}$$

where

$$c_v = 2v\binom{2v-1}{v-1} = (v+1)\binom{2v}{v+1}, \tag{4-61}$$

$$\zeta_j = \frac{1}{j}\binom{vj}{j-1} = \frac{1}{(v-1)j+1}\binom{vj}{j}. \tag{4-62}$$

When $v = 2$, ζ_j is the j-th Catalan number. For general v these are the *higher Catalan numbers* which play a role in a wide variety of enumerative combinatorial problems [Pierce 2007].

Continuum limits of the difference string equations. The *continuum difference string* hierarchies may be derived from the difference string equations (3-35) in a manner completely analogous to what was done with the Toda equations in the previous subsection.

Expanding (3-35) on the asymptotic series (4-48) one arrives at the following asymptotic equations:

$$\frac{1}{n} = \sum_{j=1}^{\infty} \frac{f_{w^{(j)}}}{j!}\left(\frac{1}{n}\right)^j$$

$$+ 2vs \sum_{P \in \mathcal{P}^{2v-1}(0,-1)}\left(\prod_{m=1}^{v}\sum_{j=0}^{\infty}\frac{f_{w^{(j)}}}{j!}\left(\frac{\ell_m(P)+1}{n}\right)^j - \prod_{m=1}^{v}\sum_{j=0}^{\infty}\frac{f_{w^{(j)}}}{j!}\left(\frac{\ell_m(P)}{n}\right)^j\right).$$

The equations at leading order, $\mathcal{O}(n^{-1})$, are

$$1 = \partial_w f_0 + 2vs \sum_{P \in \mathscr{P}^{2v-1}(1,0)} v f_0^{v-1} \partial_w f_0 = \partial_w f_0 + 2v\binom{2v-1}{v} s v f_0^{v-1} \partial_w f_0,$$

or, equivalently,

$$\partial_w (w - f_0 - c_v s f_0^v) = 0, \tag{4-63}$$

which one directly recognizes as the spatial derivative of the hodograph solution (4-55). Evaluating that solution at $w = 1$ yields

$$c_v s z_0^v + z_0 - 1 = 0, \tag{4-64}$$

which is the functional equation for the generating function of the v-th higher Catalan numbers, mentioned in the previous subsection.

The terms of the equations at $\mathbb{O}(n^{-2g-1})$ can be computed directly and are found to have the form

$$\partial_w [f_g + c_v s v f_0^{v-1} f_g] + 2s \left(c_v \partial_w \sum_{\substack{0 \leq k_j < g \\ k_1 + \cdots + k_v = g}} f_{k_1} \cdots f_{k_v} \right) + \partial_w \sum_{k=0}^{g-1} \frac{f_{kw(2g-2k)}}{(2g-2k+1)!}$$

$$+ 2vs \left(F_1^{(v)}[2g-2] + F_2^{(v)}[2g-4] + \cdots + F_g^{(v)}[0] \right) = 0, \tag{4-65}$$

where $F_g^{(v)}[2m]$ denotes the coefficient of n^{-2m} in

$$F_g^{(v)} = \sum_{\substack{\lambda \in \Lambda^{2g+1} \\ \ell(\lambda) \leq v}} \left(\sum_{\vec{\ell} \in \mathscr{P}^{2v-1}(0,-1)} \left(m_\lambda(\ell_1 + 1, \ldots, \ell_v + 1) - m_\lambda(\ell_1, \ldots, \ell_v) \right) \right) \frac{f_w^\lambda}{\lambda!}$$

$$= \partial_w \sum_{\substack{\lambda \in \Lambda^{2g} \\ \ell(\lambda) \leq v}} \sum_{\vec{\ell} \in \mathscr{P}^{2v-1}(0,-1)} \left(\sum_{\mu < \lambda} \frac{1}{|\lambda - \mu| + 1} \binom{\lambda}{\mu} m_\mu(\ell_1, \ldots, \ell_v) \right) \frac{1}{r(\lambda)!} \frac{f_w^\lambda}{\lambda!}$$

$$= \partial_w \hat{F}_g^{(v)}.$$

The above relations are derived in exactly the same manner as those which lead to Theorem 4.3 and Proposition 4.4.

As a consequence of this result one sees that the continuum difference string equation is directly integrable:

$$f_g = \frac{-2s}{1 + c_v s v f_0^{v-1}} \left\{ \left(c_v \sum_{\substack{0 \leq k_j < g \\ k_1 + \cdots + k_v = g}} f_{k_1} \cdots f_{k_v} \right) \right.$$

$$\left. + v \left(\hat{F}_1^{(v-1)}[2g-2] + \hat{F}_2^{(v-1)}[2g-4] + \cdots + \hat{F}_g^{(v-1)}[0] \right) + \frac{1}{2s} \sum_{k=0}^{g-1} \frac{f_{kw(2g-2k)}}{(2g-2k+1)!} \right\}.$$

Setting $w = 1$ and applying (4-64) to eliminate s this reduces to

$$z_g = \frac{2z_0(z_0 - 1)}{(v - (v-1)z_0)}\left\{\left(\sum_{\substack{0 \le k_j < g \\ k_1 + \cdots + k_v = g}} \frac{z_{k_1}}{z_0} \cdots \frac{z_{k_v}}{z_0}\right) + \frac{1}{2(1-z_0)}\sum_{k=0}^{g-1} \frac{f_{kw}(2g-2k)|_{w=1}}{(2g-2k+1)!}\right.$$

$$\left. + \frac{v}{c_v z_0^v}\left(\hat{F}_1^{(v-1)}[2g-2] + \hat{F}_2^{(v-1)}[2g-4] + \cdots + \hat{F}_g^{(v-1)}[0]\right)\Big|_{w=1}\right\}.$$

It is immediate from this representation that z_g is a rational function of z_0. A priori this *antiderivative* should also include a constant term (in w; it could depend on s). This would lead to a term of the form $c(s)/(v - (v-1)z_0)$. However, in [Ercolani 2011] it is shown, by an independent argument, that the pole order in z_0 at $v/(v-1)$ is always greater than one. Hence the constant of integration must be zero. With further effort this can be refined to:

Theorem 4.7 [Ercolani 2011].

$$z_g(z_0) = \frac{z_0(z_0-1)P_{3g-2}(z_0)}{(v-(v-1)z_0)^{5g-1}}$$

$$= z_0\left\{\frac{a_0^{(g)}(v)}{(v-(v-1)z_0)^{2g}} + \frac{a_1^{(g)}(v)}{(v-(v-1)z_0)^{2g+1}} + \cdots + \frac{a_{3g-1}^{(g)}(v)}{(v-(v-1)z_0)^{5g-1}}\right\},$$

where P_{3g-2} is a polynomial of degree $3g - 2$ in z_0 whose coefficients are rational functions of v over the rational numbers $\mathbb{Q}$ and $a_{3g-1}^{(g)}(v) \ne 0$.

Remark 4.8. A key element in the proof of Proposition 4.4 is the observation that differentiation with respect to w adjusts the multinomial labeling of partial derivatives in the expansion according to the edges of the Hasse–Young graph (Figure 2). This graph describes the adjacency relations between Young diagrams of differing sizes. The edges describe which partitions of size $2g + 1$ are *covered* by a given partition of size $2g$. Conversely it describes which partitions of size $2g$ cover a partition of size $2g + 1$ which in the setting described here acts as an

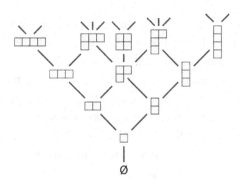

Figure 2. λ Hasse–Young graph (courtesy of D. Eppstein).

antidifferentiation operator. This kind of structure was called a differential poset by Stanley and systematically examined in [Stanley 1988].

5. Determining e_g

The spatial extension of the basic identity (3-25) reads

$$b_{n+k}^2 = \frac{\tau_{n+k+1}^2 \tau_{n+k-1}^2}{\tau_{n+k}^4} b_{n+k}^2(0); \tag{5-66}$$

we have, by taking logarithms,

$$\log \tau_{n+k+1}^2 - 2\log \tau_{n+k}^2 + \log \tau_{n+k-1}^2 = \log(b_{n+k}^2) - \log(b_{n+k}^2)(0), \tag{5-67}$$

where the initial value $b_{n+k}^2(0) = w$ is given by the recursion relations of the Hermite polynomials. As in [Ercolani et al. 2008], we can use formula (5-67) to recursively determine e_g in terms of solutions to the continuum equations. We use the asymptotic expansion of b_{n+k}^2 which has the form (4-45):

$$b_{n+k}^2 = x \sum_{g=0}^{\infty} f_g(s) n^{-2g}. \tag{5-68}$$

Note that the left-hand side of (5-67) has the form of a centered second difference: $\Delta_1 \tau_{n+k}^2 - \Delta_{-1} \tau_{n+k}^2$. It follows that this expression has an expansion for large n involving only even derivatives of the spatial variable w. We have, at order n^{-2g},

$$\frac{\partial^2}{\partial w^2} E_g(s,w) = -\sum_{\ell=1}^{g} \frac{2}{(2\ell+2)!} \frac{\partial^{2\ell+2}}{\partial w^{2\ell+2}} E_{g-\ell}(s,w)$$

$$+ \text{ the } n^{-2g} \text{ terms of } \log\left(\sum_{m=0}^{\infty} \frac{1}{n^{2m}} f_m(s)\right), \tag{5-69}$$

where $E_h(s,w) = w^{2-2h} e_h(w^{\nu-1}s)$. In [Ercolani 2011] it was shown that e_g is rational in z_0 with poles located only at $z_0 = \nu/(\nu-1)$. However we will now prove the more refined result stated in Theorem 2.3.

The proof of this theorem is by induction on g. (The base case of $g = 2$ is established by direct calculation [Ercolani et al. 2008].) We assume that (2-11) holds for all $k < g$. We state here, without proof, some straightforward lemmas and propositions describing the derivatives of (2-11) (details may be found in [Ercolani 2011] where similar lemmas are proved for the z_g).

Lemma 5.1. $(E_k)_{w^{(p)}}(s,w) = \sum_{j=0}^{p} (\nu-1)^j Q_j^{(p,k)}(\nu) w^{2-2k+(\nu-1)j-p} s^j e_k^{(j)},$

where $e_k^{(j)} = d^j e_k / d\tilde{s}^j$, $\tilde{s} = s w^{\nu-1}$,

$$Q_j^{(p,k)}(\nu) = Q_{j-1}^{(p-1,k)}(\nu) + \{(\nu-1)j - (2k-3+p)\}Q_j^{(p-1,k)}(\nu), \quad 0 < j < p,$$

$$Q_0^{(p,k)}(\nu) = (2-2k)_p, \quad p > 0,$$

$$Q_p^{(p,k)}(\nu) = 1,$$

$$Q_j^{(p,k)}(\nu) = 0, \qquad j > p \text{ or } j < 0.$$

Lemma 5.2. *For* $1 < k < g$ *and* $j > 0$,

$$e_k^{(j)} = (-1)^j c_\nu^j z_0^{j\nu+1}\left(\sum_{\ell=0}^{3k-4+j} \frac{c_\ell^{(k,j)}(\nu)}{(\nu - (\nu-1)z_0)^{2k+\ell+j-1}}\right),$$

$$c_\ell^{(k,j)}(\nu) = [(j-1)\nu - (2k+\ell+(j-3))]c_\ell^{k,j-1}(\nu)$$
$$+ \nu(2k+\ell+(j-3))c_{\ell-1}^{k,j-1}(\nu),$$

$$c_\ell^{(k,j)}(\nu) = 0 \quad \ell < 0, \ell \geq 3k-3+j,$$

$$c_\ell^{(k,0)}(\nu) = c_\ell^{(k)}(\nu).$$

Lemma 5.3. *For* $1 < k < g$ *and* $j > 0$,

$$s^j e_k^{(j)} = \frac{z_0}{(\nu-1)^j}\sum_{r=0}^{j}\sum_{m=r}^{3k-4+j+r} \frac{(-1)^{j-r}\binom{j}{r}c_{m-r}^{(k,j)}}{(\nu-(\nu-1)z_0)^{2k+m-1}},$$

where $m = \ell + r$.

Proposition 5.4. *For* $1 < k < g$ *and* $p > 0$,

$$(E_k)_{w^{(p)}}(s,1) = (2-2k)_p C^{(k)}$$

$$+ z_0 \sum_{m=0}^{3k+2p-4} \frac{\sum_{j=0}^{p} Q_j^{(p,k)}(\nu)\sum_{r=0}^{m}(-1)^{j-r}\binom{j}{r}c_{m-r}^{(k,j)}(\nu)}{(\nu-(\nu-1)z_0)^{2k+m-1}}.$$

Lemma 5.5. $\displaystyle\sum_{j=0}^{p} Q_j^{(p,k)}(\nu)\sum_{r=0}^{m}(-1)^{j-r}\binom{j}{r}c_{m-r}^{(k,j)}(\nu) = 0$,

for $m = 0, 1, \ldots, p-1$.

By this vanishing lemma, the minimal pole order of the expansion in Proposition 5.4 is $\geq 2k-2+p$. In particular the minimal pole orders coming from terms involving E_k on the right-hand side of (5-69) are all greater than $2k-2+(2g+2-2k) = 2g$.

Proposition 5.6. *The n^{-2g} term of* $\log \sum\limits_{m=0}^{\infty} \dfrac{1}{n^{2m}} f_m(s)\Big|_{w=1}$ *is given by*

$$\sum_{|\lambda|=g} \frac{(-1)^{\ell(\lambda)-1}}{\prod_{j\geq 1} r_j(\lambda)!} \prod_{j\geq 1} \left(\frac{z_j}{z_0}\right)^{r_j(\lambda)},$$

which, by Theorem 4.7, equals

$$\sum_{|\lambda|=g} \frac{(-1)^{\ell(\lambda)-1}}{\prod_{j\geq 1} r_j(\lambda)!} \prod_{j\geq 1} \left\{ \frac{a_0^{(j)}(v)}{(v-(v-1)z_0)^{2j}} + \cdots + \frac{a_{3j-1}^{(j)}(v)}{(v-(v-1)z_0)^{5j-1}} \right\}^{r_j(\lambda)}.$$

This result shows that the minimal pole order coming from the log terms in (5-69) is once again greater than $\sum_{j\geq 1} 2jr_j = 2g$.

The preceding lemmas and propositions provide explicit Laurent expansions (in z_0) for all terms on the right-hand side of (5-69), with two exceptions:

$$\frac{\partial^{2g}}{\partial w^{2g}} E_1(s,w) = \frac{\partial^{2g}}{\partial w^{2g}} e_1(w^{v-1}s), \quad e_1 = -\tfrac{1}{12}\log(v-(v-1)z_0), \quad (5\text{-}70)$$

and

$$\frac{\partial^{2g+2}}{\partial w^{2g+2}} E_0(s,w) = \frac{\partial^{2g+2}}{\partial w^{2g+2}} w^2 e_0(w^{v-1}s),$$

$$e_0 = \tfrac{1}{2}\log z_0 + \frac{(v-1)^2}{4v(v+1)}(z_0-1)\left(z_0 - \frac{3(v+1)}{v-1}\right). \quad (5\text{-}71)$$

With a small modification (5-70) may be brought into line with Proposition 5.4:

Proposition 5.7.

$(E_1)_{w^{(p)}}(s,1)$

$$= z_0 \sum_{m=0}^{2p-1} \frac{\sum_{j=1}^{p} Q_j^{(p,1)}(v) \sum_{r=0}^{m}(-1)^{j-r}\binom{j}{r} c_{m-r}^{(1,j)}(v)}{(v-(v-1)z_0)^{m+1}}$$

$$= -\frac{(p-1)!}{12}$$

$$+ \frac{1}{v-1} \sum_{m=1}^{2p} \frac{\sum_{j=1}^{p} Q_j^{(p,1)}(v) \sum_{r=0}^{m}(-1)^{j-r}\binom{j}{r}\{vc_{m-r-1}^{(1,j)}(v) - c_{m-r}^{(1,j)}(v)\}}{(v-(v-1)z_0)^{m}},$$

with $c_0^{(1,1)} = \tfrac{1}{12}(v-1)$. *All other coefficients are then specified by the corresponding recursions stated in Lemmas 5.1–5.3 with k set to 1.*

A variant of the vanishing Lemma 5.5 also holds for $(E_1)_{w^{(p)}}(s,1)$:

Lemma 5.8. $\displaystyle\sum_{j=1}^{p} Q_j^{(p,1)}(v) \sum_{r=0}^{m} (-1)^{j-r} \binom{j}{r} \left(c_{m-r}^{(1,j)}(v) - v c_{m-r-1}^{(1,j)}(v) \right) = 0,$

for $m = 0, 1, \ldots, p-1$.

It follows that the minimal pole order of the expansion in Proposition 5.7 is at least p and so the corresponding contribution to the minimal pole order of (5-69) is $\geq 2g$.

Finally we observe that for $p \geq 3$, $\dfrac{\partial^p}{\partial w^p} E_0(s, w)$ is a rational function of f_0 and its w-derivatives:

Proposition 5.9.

$$\frac{\partial^p}{\partial w^p} E_0(s, w)$$

$$= \frac{\partial^p}{\partial w^p} \left[w^2 \tfrac{1}{2} \log f_0 + \frac{(v-1)^2}{4v(v+1)} (f_0 - w) \left(f_0 - \frac{3(v+1)}{v-1} w \right) - w^2 \tfrac{1}{2} \log w \right]$$

$$= \binom{p}{2} \left(\frac{f_{0w}}{f_0} \right)_{w^{(p-3)}} + p w \left(\frac{f_{0w}}{f_0} \right)_{w^{(p-2)}} + \frac{w^2}{2} \left(\frac{f_{0w}}{f_0} \right)_{w^{(p-1)}}$$

$$+ \frac{(v-1)^2}{2v(v+1)} \left[\sum_{j=0}^{\lfloor \frac{p}{2} \rfloor} \binom{p}{j} f_{0w^{(j)}} f_{0w^{(p-j)}} - \frac{2v+1}{v-1} \left(w f_{0w^{(p)}} + p f_{0w^{(p-1)}} \right) \right]$$

$$+ (p-3)! \left(-\frac{1}{w} \right)^{p-2}. \tag{5-72}$$

Each line of the proposition can be established directly by induction, starting with the base case $p = 3$. It then follows from Proposition 3.1(iii) of [Ercolani 2011] that the minimal pole order contributed by (5-71) is $\geq 2g$.

We are now in a position to outline the

Proof of Theorem 2.3. In [Ercolani 2011, Theorem 1.3] it was shown that

$$e_g(z_0) = \frac{(z_0 - 1) q_{d(g)}(z_0)}{(v - (v-1)z_0)^{o(g)}}, \tag{5-73}$$

where $q_{d(g)}(z_0)$ denotes a polynomial of degree $d(g)$ in z_0. We first want to determine the relation between this degree and the pole order $o(g)$. To this end we observe from Propositions 5.4, 5.6, 5.7, and 5.9 that the right-hand side of (5-69), evaluated at $w = 1$, is a rational function in z_0 that approaches a finite constant value as $z_0 \to \infty$. From the form of the left-hand side of (5-69) evaluated at $w = 1$ one also sees that its asymptotic order (as $z_0 \to \infty$) is the same as that of e_g. Hence, $d(g) = o(g) - 1$ and this shows that (2-11) is valid up to the determination of the minimal and maximal pole orders at $z_0 = v/(v-1)$, to which we now turn.

In the preceding lemmas and propositions we have seen that, for all terms on the right-hand side of (5-69), the minimal pole order is $\geq 2g$. Furthermore, from these same representations together with Proposition 3.1(iii) of [Ercolani 2011] one sees that, with the possible exception of the genus 0 terms in (5-72), the maximal pole order of the terms on the right-hand side of (5-69) is $5g - 1$. The apparent maximal pole order in (5-72) is $4g + 3$, which exceeds the stated bound when $g = 2, 3$. This maximal order comes from terms containing the factor $f_{0w^{(2g+2)}}$, which are, specifically,

$$\left[\frac{1}{2} \frac{f_{0w^{(2g+2)}}}{f_0} - \frac{(2v+1)(v-1)}{2v(v+1)} f_{0w^{(2g+2)}} + \frac{(v-1)^2}{2v(v+1)} (f_0 f_{0w^{(2g+2)}}) \right]_{w=1}$$

$$= \frac{f_{0w^{(2g+2)}}}{2f_0} \bigg|_{w=1} \left[1 - \frac{(2v+1)(v-1)}{v(v+1)} z_0 + \frac{(v-1)^2}{v(v+1)} z_0^2 \right]$$

$$= \frac{f_{0w^{(2g+2)}}}{2v(v+1) f_0} \bigg|_{w=1} \left\{ (v - (v-1)z_0) + (v - (v-1)z_0)^2 \right\}.$$

Hence the maximal pole order contributed by the genus 0 terms is $\leq 4g + 2$. Indeed, one may go further with this type of analysis to show that the coefficient of order $\mathbb{O}\left((v - (v-1)z_0)^{-4g-2} \right)$ also vanishes. In establishing this the following identity, which is a direct consequence of the quadratic relation satisfied by the generating function for the Catalan numbers $\frac{1}{g+1}\binom{2g}{g}$, proves useful:

$$\frac{(2g)!}{g!} = \sum_{j=1}^{g} \binom{g+1}{j} \frac{(2j-2)!(2(g-j))!}{(j-1)!(g-j)!}.$$

It follows that the maximal pole order coming from genus 0 terms is $\leq 4g + 1$ which is $< 5g - 1$ for $g > 2$. This establishes that the pole orders on the right-hand side of (5-69) are always bounded between $2g$ and $5g - 1$. For $g > 2$, the case-by-case checking of terms on the right-hand side of (5-69) that has been carried out in this subsection shows that the maximal pole order is realized by the term in Proposition 5.6 corresponding to the partition λ of $2g$ having minimal length $(= 1)$, i.e., the partition whose Young diagram is a single row. This implies that the residue of the maximal order pole is $a_{3g-1}^{(g)}(v)$, which is nonzero by Theorem 4.7. (This also holds for $g = 2$, as can be checked by direct calculation; see, for example, [Ercolani 2011, Section 1.4.2].) Hence the maximal order pole is realized. Now, given that e_g has the form (5-73) with $d(g) = o(g) - 1$, it follows that $\partial^2 E_g(s, 1)/\partial w^2$ raises the minimum pole degree by 2 and the maximum pole degree by 4 with the coefficient at this order given by (2-14).

To establish (2-13) first note that by Euler's relation, $2 - 2g = m - vm + F$ for a g-map where m is the number of ($2v$-valent) vertices and F is the number of faces. Since $F \geq 1$, one immediately sees that the number of vertices of such a map must satisfy the inequality $m \geq \frac{2g-1}{\nu-1}$. It follows that

$$e_g^{(j)}(s=0) = 0 \quad \text{for } j \leq r = \max\left\{1, \frac{2g-1}{\nu-1}\right\}$$

(e_g must vanish at least simply at $s = 0$ since $\tau_n^2(s=0) \equiv 1$). Via Cauchy's theorem these conditions may be reexpressed as

$$0 = \frac{1}{2\pi i}\oint_{s\sim 0}\frac{e_g(s)}{s^{j+1}}\,ds = \frac{-1}{2\pi i}\oint_{z\sim 1}\frac{z^{\nu j-1}q_{5g-6}(z)}{(\nu-(\nu-1)z)^{5g-6}(z-1)^j}\,dz,$$

for $j \leq r$, where in the second line we have rewritten e_g as a rational function of z ((5-73) with $o(g) = 5g - 5$) and employed the change of variables

$$\frac{ds}{dz} = -\frac{\nu-(\nu-1)z}{z^{\nu+1}},$$

which may be deduced from the string equation (4-64). This yields a contour integral in z centered at 1. Now one can see that these vanishing conditions are satisfied if and only if $q_{5g-6}^{(j)}(z=1) = 0$ for $j \leq r$, which in turn proves (2-13).

Finally we turn to the determination of the constant $C^{(g)}$. By Proposition 5.6, contributions to the constant term of e_g come only from the first sum on the right-hand side of (5-69). The parts of this coming from $g = 0$ and $g = 1$ are, by Propositions 5.9 and 5.7, respectively,

$$-2\frac{(2g-1)!}{(2g+2)!} \quad \text{and} \quad 2\frac{(2g-1)!}{(2g)!}\frac{1}{12}.$$

At higher genus, $k < g$, the contribution to the constant term is determined by Lemma 5.1 to be

$$-2\frac{(2-2k)_{2g-2k+2}}{(2g-2k+2)!}C^{(k)}.$$

Hence, by (5-69) we have

$$(2-2g)(1-2g)C^{(g)}$$
$$= -2\frac{(2g-1)!}{(2g+2)!} + 2\frac{(2g-1)!}{(2g)!}\frac{1}{12} - 2\sum_{k=2}^{g-1}\frac{(2-2k)_{2g-2k+2}}{(2g-2k+2)!}C^{(k)},$$

from which (2-15) immediately follows. $\qquad\square$

6. The case of odd valence

In the case when j is odd in the weight (2-3) for V, there is clearly a problem in applying the method of orthogonal polynomials as it was outlined in Section 3. Very recently, however, a generalization of the *equilibrium measure* (which governs the leading order behavior of the free energy associated to (2-1)) was developed and applied to this problem [Bleher and Deaño 2012]. It is based on generalizing to a class of complex valued non-Hermitian orthogonal polynomials on a contour in the complex plane other than the real axis. These extensions were motivated by new ideas in approximation theory related to complex Gaussian quadrature of integrals with high order stationary points [Deaño et al. 2010].

But even when the issue of existence of appropriate orthogonal polynomials has been resolved, there are still a number of significant obstacles to deriving results like Theorem 2.3 that are not present when the valence j is even. For odd valence there is an additional string of recurrence coefficients, the diagonal coefficients a_n of $\mathcal{L}$, whose asymptotics need to be analyzed. This in turn requires that the lattice paths used to define and analyze the Toda and difference string equations must be generalized to the class of *Motzkin paths* which can have segments where the lattice site remains fixed rather than always taking a step (either up or down) as was the case for Dyck paths.

Nevertheless, all these constructions have been carried out in [Ercolani and Pierce 2012] to derive the hierarchies of continuum Toda and difference string equations when the valence j is odd.

The recurrence coefficients again have asymptotic expansions with continuum representations given by

$$a_{n+k,N} = h(s_1, s_{2v+1}, w) = x^{1/2} \sum_{g \geq 0} h_g(s_1, s_{2v+1}, w) n^{-g} \qquad (6\text{-}74)$$

and

$$
\begin{aligned}
&h_g(s_1, s_{2v+1}, w) \\
&= -w^{1-g} \sum_{\substack{2g_1+j=g+1 \\ g_1 \geq 0, j > 0}} \frac{1}{j!} \\
&\quad \times \frac{\partial^{j+1}}{\partial s_1 \partial \tilde{w}^j} \left[\tilde{w}^{2-2g_1} e_{g_1} \left((w\tilde{w})^{-\frac{1}{2}} s_1, (w\tilde{w})^{v-\frac{1}{2}} s_{2v+1} \right) \right]_{\tilde{w}=1}. \quad (6\text{-}75)
\end{aligned}
$$

The off-diagonal coefficients have corresponding representations which are much as they were in the even valence case,

$$b_{n+k,N}^2 = f(s_1, s_{2v+1}, w) = x \sum_{g \geq 0} f_g(s_1, s_{2v+1}, w) n^{-2g}, \quad (6\text{-}76)$$

$$f_g(s_1, s_{2v+1}, w) = w^{2-2g} \frac{\partial^2}{\partial s_1^2} e_g(w^{-1/2} s_1, w^{v-1/2} s_{2v+1}). \qquad (6\text{-}77)$$

The coefficients in these expansions have a self-similar structure given by

$$h_g(s_1, s_{2v+1}, w) = w^{\frac{1}{2}-g} u_g(s_1 w^{-1/2}, s_{2v+1} w^{v-\frac{1}{2}}), \qquad (6\text{-}78)$$

$$f_g(s_1, s_{2v+1}, w) = w^{1-2g} z_g(s_1 w^{-1/2}, s_{2v+1} w^{v-\frac{1}{2}}). \qquad (6\text{-}79)$$

At leading order the continuum Toda equations are

$$\frac{\partial}{\partial s}\begin{pmatrix} h_0 \\ f_0 \end{pmatrix} + (2\nu+1)\begin{pmatrix} B_{11} & B_{12} \\ f_0 B_{12} & B_{11} \end{pmatrix}\frac{\partial}{\partial w}\begin{pmatrix} h_0 \\ f_0 \end{pmatrix} = 0, \tag{6-80}$$

and the leading order continuum difference string equations are

$$\begin{pmatrix} 0 \\ 1 \end{pmatrix} = \begin{pmatrix} A_{11} & A_{12} \\ f_0 A_{12} & A_{11} \end{pmatrix}\frac{\partial}{\partial w}\begin{pmatrix} h_0 \\ f_0 \end{pmatrix}, \tag{6-81}$$

where the coefficients of the matrix in (6-80) are specified by

$$B_{11} = \sum_{\mu=1}^{\nu}\binom{2\nu}{2\mu-1,\, \nu-\mu,\, \nu-\mu+1}h_0^{2\mu-1}f_0^{\nu-\mu+1}, \tag{6-82}$$

$$B_{12} = \sum_{\mu=0}^{\nu}\binom{2\nu}{2\mu,\, \nu-\mu,\, \nu-\mu}h_0^{2\mu}f_0^{\nu-\mu}, \tag{6-83}$$

and those of the matrix in (6-81) by

$$A_{11}=1+(2\nu+1)s\sum_{\mu=0}^{\nu-1}\binom{2\nu}{2\mu+1,\, \nu-\mu-1,\, \nu-\mu}(\nu-\mu)h_0^{2\mu+1}f_0^{\nu-\mu-1}, \tag{6-84}$$

$$A_{12}=(2\nu+1)s\sum_{\mu=0}^{\nu-1}\binom{2\nu}{2\mu,\, \nu-\mu-1,\, \nu-\mu+1}(\nu-\mu+1)h_0^{2\mu}f_0^{\nu-\mu-1}. \tag{6-85}$$

Remark 6.1. The index μ appearing in the trinomial coefficients corresponds to the number of flat steps, $2\mu-1$ or 2μ, in the Motzkin paths giving rise to that term.

It is straightforward to see that (6-80) may be rewritten in conservation law form as

$$\frac{\partial}{\partial s}\begin{pmatrix} h_0 \\ f_0 + \frac{1}{2}h_0^2 \end{pmatrix} + \frac{\partial}{\partial w}\begin{pmatrix} \Psi_1 \\ \Psi_2 + h_0\Psi_1 \end{pmatrix} = 0, \tag{6-86}$$

where the coefficients in the flux vector are given by

$$\Psi_1 = \sum_{\mu=0}^{\nu}\binom{2\nu+1}{2\mu,\, \nu-\mu,\, \nu-\mu+1}h_0^{2\mu}f_0^{\nu-\mu+1}, \tag{6-87}$$

$$\Psi_2 = \sum_{\mu=0}^{\nu}\binom{2\nu+1}{2\mu+1,\, \nu-\mu-1,\, \nu-\mu+1}h_0^{2\mu+1}f_0^{\nu-\mu+1}. \tag{6-88}$$

Recently, we have determined that the equations (6-81) are in fact a differentiated form of the generalized hodograph solution of the conservation law (6-86).

This hodograph solution is given by

$$\Phi_1 \doteq h_0 + (2\nu + 1)sB_{12} = 0, \tag{6-89}$$
$$\Phi_2 \doteq f_0 + (2\nu + 1)sB_{11} = w. \tag{6-90}$$

Analogous to what was done in Theorem 2.3 we expect to determine closed form expressions for all the coefficients in the topological expansion with odd weights. The first few of these, for the trivalent case, are [Ercolani and Pierce 2012]

$$e_0(t_3) = \frac{1}{2}\log(z_0) + \frac{1}{12}\frac{(z_0 - 1)(z_0^2 - 6z_0 - 3)}{(z_0 + 1)},$$

$$e_1(t_3) = -\frac{1}{24}\log\left(\frac{3}{2} - \frac{z_0^2}{2}\right),$$

$$\tag{6-91}$$

$$e_2(t_3) = \frac{1}{960}\frac{(z_0^2 - 1)^3(4z_0^4 - 93z_0^2 - 261)}{(z_0^2 - 3)^5}$$

$$= \frac{1}{240} - \frac{3}{64Z} - \frac{29}{32Z^2} - \frac{191}{48Z^3} - \frac{55}{8Z^4} - \frac{21}{5Z^5}, \quad Z = z_0 - 3,$$

where z_0 is implicitly related to t_3 by the polynomial equation

$$1 = z_0^2 - 72t_3^2 z_0^3.$$

In fact, $z_0(t_3)$ is the generating function for a *fractional* generalization of the Catalan numbers. Its m-th coefficient counts the number of connected, non-crossing symmetric graphs on $2m + 1$ equidistributed vertices on the unit circle (V. U. Pierce, private communication).

7. Concluding remarks

7.1. Spectrum. Nothing has been said, in this article, about the eigenvalues of the random matrix M although this is at the heart of the Riemann–Hilbert analysis underlying all of our results. The essential link comes through the *equilibrium measure* [Ercolani and McLaughlin 2003; Ercolani et al. 2008], or density of states, for these eigenvalues. When $t_j = 0$ in (2-2), this equilibrium measure reduces to the well known Wigner semicircle law. As t_j changes this measure deforms; but, for t_j satisfying the bounds implicit in (2-5) (i), its support remains a single interval, $[\alpha, \beta]$. The *edges of the spectrum* $\alpha(t_j), \beta(t_j)$ evolve dynamically with t_j. In fact can show that α and β are the Riemann invariants of the hyperbolic system (6-80). In the case of even valence these invariants collapse to $\pm 2\sqrt{z_0}$ where z_0 is the generating function upon which all the e_g

are built, as described in Theorem 2.3. So the edge of the spectrum is indeed directly related to the genus expansion (2-5).

7.2. Random surfaces. It follows from Corollary 3.3 and the map-theoretic interpretation of e_g given by (2-8) that $z_g(t_j)$ is a generating function for enumerating j-regular g-maps with, in addition, two *legs*. A *leg* is a univalent vertex; that is, a vertex with just one adjacent edge, connected to some other vertex of the map. In particular $z_0(t)$ enumerates such maps on the Riemann sphere or what are more commonly referred to as two-legged *planar maps*. In a remarkable paper, [Schaeffer 1997], building on prior work of Cori and Vauquelin [1981], Schaeffer found a constructive correspondence between two-legged 2ν-regular planar maps and 2ν-valent *blossom trees*. A 2ν-valent blossom tree is a rooted 2ν-valent tree, with each external vertex taken to be a leaf that is colored either black or white and such that each internal (nonroot) vertex is adjacent to exactly $\nu - 1$ black leaves. This gives another interpretation of $z_0(t_{2\nu})$ as the generating function for the enumeration of blossom trees. It seems reasonable to hope that the arithmetic data implicit in the coefficients $a_m^{(g)}(\nu)$ in Theorem 4.7 (resp. $c_m^{(g)}(\nu)$ in Theorem 2.3) might provide a means, such as *sewing rules*, for constructing two-legged 2ν-regular g-maps (resp. pure 2ν-regular g-maps) from blossom trees.

In another direction Bouttier, Di Francesco and Guitter [Di Francesco 2006] have studied the combinatorics of geodesic distance for planar maps. They define the geodesic distance of a two-legged graph to be the minimum number of edges crossed by a continuous path between the two legs and study $r_d(t_{2\nu})$, the generating function for enumerating all two-legged 2ν-valent planar maps whose geodesic distance is $\leq d$. They find surprising and elegant closed form expressions for the $r_d(t_{2\nu})$. The statistics of planar maps is a natural stepping off point for the study of random surfaces. There has been a lot of recent activity in this direction by Le Gall and his collaborators related to the work of Schaeffer and Bouttier et al. See, for example, [Le Gall 2010].

7.3. Enumerative geometry of moduli spaces. A different (from matrix models) representation of 2D quantum gravity may be given in terms of intersection theory on the moduli space of stable curves (Riemann surfaces), $\overline{\mathcal{M}}_{g,n}$ and from this alternate perspective Witten conjectured that a generating function for the intersection numbers of tautological bundles on $\overline{\mathcal{M}}_{g,n}$ should be "given by" a double-scaling limit of the differentiated free energy (3-28) for matrix models. (A precise description of this double-scaling limit may be found in [Ercolani 2011].) He further conjectured that this intersection theoretic generating function should, with respect to appropriate choices of parameter variables, satisfy the Korteweg–de Vries (KdV) equation. Subsequently, Kontsevich [1992] was able to outline a proof of Witten's KdV conjecture based on a combinatorial model of intersection theory on $\overline{\mathcal{M}}_{g,n}$. This model expresses tautological intersections in terms of sums over trivalent graphs on a genus g Riemann surface. He was then able to recast this sum in terms of a special matrix integral involving cubic weights on which the proof of Witten's KdV conjecture is based. For a readable overview on the above circle of ideas we refer to [Okounkov and Pandharipande 2009].

However, the first Witten conjecture, on relating the intersection-theoretic free energy to the matrix model free energy (see (2-5)), remains open. With the results described in this paper it may now be possible to determine if, and in precisely what sense, this conjecture might be true and to see if this leads to connections between the KdV equation in the Witten–Kontsevich model and the conservation laws given by (6-80). In addition, given the recent results on matrix models with odd dominant weights [Bleher and Deaño 2012; Ercolani and Pierce 2012], it may now be possible to give a rigorous treatment of Kontsevich's matrix integral which, up to now, has been formal.

More recently there have been other, perhaps more natural, approaches to the proof of Witten's KdV conjecture [Kazarian and Lando 2007; Mirzakhani 2007; Goulden et al. 2009], in terms of coverings of the Riemann sphere and *Hurwitz numbers* for which the generating functions specified in Theorem 2.3 should also have a natural interpretation.

7.4. Analytical deformations and critical parameters. In [Ercolani 2011] it was observed that the equilibrium measure (see concluding remark Section 7.1) for the weight V, with $j = 2\nu$ in (2-2), may be reexpressed as

$$\mu_{V_t}(\lambda) = z_0 \mu_0(\lambda) + (1 - z_0)\mu_\infty(\lambda), \qquad (7\text{-}92)$$

where μ_0 is the equilibrium measure for $V = 1/2\lambda^2$ (the *semicircle* law) and μ_∞ is the equilibrium measure for $V = \lambda^{2\nu}$; that is, the general measure is a linear combination, over z_0, of two extremal monomial equilibrium measures. For $z_0 \in [0, 1]$ (which corresponds to $t_{2\nu} \in [0, \infty]$), this combination is convex and (7-92) is indeed a measure with a single interval of support in $\lambda \in \mathbb{R}$. This may be analytically continued to a complex z_0 neighborhood of $[0, 1]$ so that (7-92) remains a positive measure along an appropriate connected contour ("single interval") in the complex λ-plane. For $\nu = 2$ this continuation may be made up to a boundary curve in the complex z_0-plane passing through $z_0 = 2$ (with a corresponding image in the complex t_4 plane). (For a related result see [Bertola and Tovbis 2011].) This should be extendable for general ν. The mechanism for carrying out this continuation is to regard (7-92) as a z_0-parametrized family of holomorphic quadratic differentials. The candidate for the measure's support is then an appropriate bounded real trajectory of the quadratic differential. Outside the boundary curve, the Riemann–Hilbert analysis used in this paper may be analytically deformed and our results extended. The boundary may be regarded as a curve of critical parameters for this deformation. This curve is precisely the locus where the Riemann invariants, that determine the edge of the spectrum (as described in concluding remark 7.1) exhibit a shock.

This scenario is reminiscent of that for the small $\hbar$-limit of the nonlinear Schrödinger equation [Jin et al. 1999; Kamvissis et al. 2003] in which the analogue of our boundary curve is the envelope of *dispersive shocks*. In that setting it is the Zakharov–Shabat inverse scattering problem that shows one how to pass through the dispersive shocks and describe a continuation of measure-valued solutions with so-called *multigap* support. It is our expectation that coupling gravity to an appropriate conformal field theory (to thus arrive at a bona fide string theory) [Mariño 2005] will play a similar role in our setting to determine a unique continuation through the boundary curve of critical parameters to a unique equilibrium measure with multicut support. We also hope that this will help bring powerful methods from the study of dispersive limits of nonlinear PDE into the realm of random matrix theory.

Acknowledgement

The author wishes to thank MSRI for its hospitality and the organizers for the excellent Fall 2010 program on random matrix theory. Most of the new results described here had their inception during that happy period.

References

[Ambjørn et al. 1993] J. Ambjørn, L. Chekhov, C. F. Kristjansen, and Y. Makeenko, "Matrix model calculations beyond the spherical limit", *Nuclear Phys. B* **404**:1-2 (1993), 127–172.

[Bauer and Itzykson 1996] M. Bauer and C. Itzykson, "Triangulations", *Discrete Math.* **156**:1-3 (1996), 29–81.

[Bertola and Tovbis 2011] M. Bertola and A. Tovbis, "Asymptotics of orthogonal polynomials with complex varying quartic weight: global structure, critical point behaviour and the first Painlevé equation", 2011. arXiv 1108.0321

[Bessis et al. 1980] D. Bessis, C. Itzykson, and J. B. Zuber, "Quantum field theory techniques in graphical enumeration", *Adv. in Appl. Math.* **1**:2 (1980), 109–157.

[Bleher and Deaño 2012] P. Bleher and A. Deaño, "Topological expansion in the cubic random matrix model", *Int. Math. Res. Not.* (2012), 2699–2755.

[Bleher and Its 2005] P. M. Bleher and A. R. Its, "Asymptotics of the partition function of a random matrix model", *Ann. Inst. Fourier* (*Grenoble*) **55**:6 (2005), 1943–2000.

[Chekhov and Eynard 2006] L. Chekhov and B. Eynard, "Hermitian matrix model free energy: Feynman graph technique for all genera", *J. High Energy Phys.* **2006**:3 (2006), #014.

[Cori and Vauquelin 1981] R. Cori and B. Vauquelin, "Planar maps are well labeled trees", *Canad. J. Math.* **33**:5 (1981), 1023–1042.

[Deaño et al. 2010] A. Deaño, D. Huybrechs, and A. B. J. Kuijlaars, "Asymptotic zero distribution of complex orthogonal polynomials associated with Gaussian quadrature", *J. Approx. Theory* **162**:12 (2010), 2202–2224.

[Deift 1999] P. A. Deift, *Orthogonal polynomials and random matrices: a Riemann–Hilbert approach*, Courant Lecture Notes in Mathematics 3, New York University, Courant Institute of Mathematical Sciences, AMS, New York / Providence, RI, 1999.

[Di Francesco 2006] P. Di Francesco, "2D quantum gravity, matrix models and graph combinatorics", pp. 33–88 in *Applications of random matrices in physics*, edited by E. Brézin et al., NATO Sci. Ser. II Math. Phys. Chem. **221**, Springer, Dordrecht, 2006.

[Di Francesco et al. 1995] P. Di Francesco, P. Ginsparg, and J. Zinn-Justin, "2D gravity and random matrices", *Phys. Rep.* **254**:1-2 (1995), 133.

[Douglas and Shenker 1990] M. R. Douglas and S. H. Shenker, "Strings in less than one dimension", *Nuclear Phys. B* **335**:3 (1990), 635–654.

[Ercolani 2011] N. M. Ercolani, "Caustics, counting maps and semi-classical asymptotics", *Nonlinearity* **24**:2 (2011), 481–526.

[Ercolani and McLaughlin 2001] N. M. Ercolani and K. T.-R. McLaughlin, "Asymptotics and integrable structures for biorthogonal polynomials associated to a random two-matrix model: advances in nonlinear mathematics and science", *Phys. D* **152/153** (2001), 232–268.

[Ercolani and McLaughlin 2003] N. M. Ercolani and K. D. T.-R. McLaughlin, "Asymptotics of the partition function for random matrices via Riemann–Hilbert techniques and applications to graphical enumeration", *Int. Math. Res. Not.* **2003**:14 (2003), 755–820.

[Ercolani and Pierce 2012] N. M. Ercolani and V. U. Pierce, "The continuum limit of Toda lattices for random matrices with odd weights", *Commun. Math. Sci.* **10**:1 (2012), 267–305.

[Ercolani et al. 2008] N. M. Ercolani, K. D. T.-R. McLaughlin, and V. U. Pierce, "Random matrices, graphical enumeration and the continuum limit of Toda lattices", *Comm. Math. Phys.* **278**:1 (2008), 31–81.

[Eynard 2004] B. Eynard, "Topological expansion for the 1-Hermitian matrix model correlation functions", *J. High Energy Phys.* 11 (2004), 031.

[Eynard and Orantin 2009] B. Eynard and N. Orantin, "Topological recursion in enumerative geometry and random matrices", *J. Phys. A* **42**:29 (2009), 293001, 117.

[Fokas et al. 1992] A. S. Fokas, A. R. Its, and A. V. Kitaev, "The isomonodromy approach to matrix models in 2D quantum gravity", *Comm. Math. Phys.* **147**:2 (1992), 395–430.

[Goulden et al. 2009] I. P. Goulden, D. M. Jackson, and R. Vakil, "A short proof of the λ_g-conjecture without Gromov-Witten theory: Hurwitz theory and the moduli of curves", *J. Reine Angew. Math.* **637** (2009), 175–191.

[Gross and Migdal 1990] D. J. Gross and A. A. Migdal, "A nonperturbative treatment of two-dimensional quantum gravity", *Nuclear Phys. B* **340**:2-3 (1990), 333–365.

[Janson et al. 1993] S. Janson, D. E. Knuth, T. Łuczak, and B. Pittel, "The birth of the giant component", *Random Structures Algorithms* **4**:3 (1993), 231–358.

[Jin et al. 1999] S. Jin, C. D. Levermore, and D. W. McLaughlin, "The semiclassical limit of the defocusing NLS hierarchy", *Comm. Pure Appl. Math.* **52**:5 (1999), 613–654.

[Kamvissis et al. 2003] S. Kamvissis, K. D. T.-R. McLaughlin, and P. D. Miller, *Semiclassical soliton ensembles for the focusing nonlinear Schrödinger equation*, Annals of Mathematics Studies **154**, Princeton Univ. Press, Princeton, NJ, 2003.

[Kazarian and Lando 2007] M. E. Kazarian and S. K. Lando, "An algebro-geometric proof of Witten's conjecture", *J. Amer. Math. Soc.* **20**:4 (2007), 1079–1089.

[Kontsevich 1992] M. Kontsevich, "Intersection theory on the moduli space of curves and the matrix Airy function", *Comm. Math. Phys.* **147**:1 (1992), 1–23.

[Le Gall 2010] J.-F. Le Gall, "Geodesics in large planar maps and in the Brownian map", *Acta Math.* **205**:2 (2010), 287–360.

[Macdonald 1995] I. G. Macdonald, *Symmetric functions and Hall polynomials*, 2nd ed., Oxford Univ. Press, New York, 1995.

[Mariño 2005] M. Mariño, "Les Houches lectures on matrix models and topological strings", 2005. arXiv hep-th/0410165

[Mirzakhani 2007] M. Mirzakhani, "Weil–Petersson volumes and intersection theory on the moduli space of curves", *J. Amer. Math. Soc.* **20**:1 (2007), 1–23.

[Okounkov and Pandharipande 2009] A. Okounkov and R. Pandharipande, "Gromov–Witten theory, Hurwitz numbers, and matrix models", pp. 325–414 in *Algebraic geometry, I* (Seattle, 2005), edited by D. Abramovich et al., Proc. Sympos. Pure Math. **80**, Amer. Math. Soc., Providence, RI, 2009.

[Pierce 2007] V. U. Pierce, "Combinatoric results for graphical enumeration and the higher Catalan numbers", 2007. arXiv 0703160v1

[Schaeffer 1997] G. Schaeffer, "Bijective census and random generation of Eulerian planar maps with prescribed vertex degrees", *Electron. J. Combin.* **4**:1 (1997), R20.

[Stanley 1988] R. P. Stanley, "Differential posets", *J. Amer. Math. Soc.* **1**:4 (1988), 919–961.

[Szegő 1939] G. Szegő, *Orthogonal polynomials*, AMS Colloquium Publications **23**, Amer. Math. Soc., 1939.

[Tutte 1968] W. T. Tutte, "On the enumeration of planar maps", *Bull. Amer. Math. Soc.* **74** (1968), 64–74.

[Whitham 1974] G. B. Whitham, *Linear and nonlinear waves*, Pure and Applied Mathematics, Wiley, New York, 1974.

ercolani@math.arizona.edu   *Department of Mathematics, The University of Arizona,*
                                   617 N. Santa Rita Avenue, P.O. Box 210089,
                                   Tucxon, 85721-0089, United States

Random Matrices
MSRI Publications
Volume **65**, 2014

Asymptotics of spacing distributions
50 years later

PETER J. FORRESTER

In 1962 Dyson used a physically based, macroscopic argument to deduce the first two terms of the large spacing asymptotic expansion of the gap probability for the bulk state of random matrix ensembles with symmetry parameter β. In the ensuing years, the question of asymptotic expansions of spacing distributions in random matrix theory has shown itself to have a rich mathematical content. As well as presenting the main known formulas, we give an account of the mathematical methods used for their proofs, and provide some new formulas. We also provide a high precision numerical computation of one of the spacing probabilities to illustrate the accuracy of the corresponding asymptotics.

1. Introduction

Random matrices were introduced in physics by Wigner in the 1950s; see [Porter 1965]. Wigner's original hypothesis was that the statistical properties of energy levels of complex nuclei could be reproduced by considering an ensemble of systems rather than a single system in which all interactions are completely described. This allowed for an entirely mathematical approach where statistical properties of the spectrum of an ensemble of random matrices were considered. But coming from physics, the aim was to use mathematics to compute experimentally measurable statistical quantities, and to compare against the data.

One viewpoint on a real spectrum from a random matrix is as a point process on the real line. As such, perhaps the most natural statistical characterization is that of the distribution of the eigenvalue spacing. This choice of statistic becomes even more compelling when one considers that in many cases of interest, eigenvalue spectra can be "unfolded". This means that unlike many statistical mechanical systems, the density is not an independent control variable, but rather fixes the length scale only. Unfolding then is scaling the eigenvalues in the bulk of the spectrum so that the mean density is unity. It is indeed the bulk spacing distribution for the Gaussian orthogonal ensemble of real symmetric matrices — albeit in an approximate form known as the Wigner surmise (see, e.g., [Mehta 1991]) — which was compared against the empirical spacing distribution for the energy level of highly excited nuclei (again, see [Mehta 1991], and references therein).

Fixing length scales at the edge of the spectrum is, as a practical exercise, a more difficult task. In addition to the bulk, we will have interest in the soft and hard spectrum edges when the eigenvalue spectrum exhibits a square root profile and inverse square root profile, respectively. To specify realizations of the bulk and edge regions of the eigenvalue spectrum, we recall (see, e.g., [Forrester 2010]) that the so-called classical random matrix ensembles have their eigenvalue probability density functions (PDFs) of the form

$$\frac{1}{C}\prod_{l=1}^{N} g(\lambda_l) \prod_{1\le j<k\le N} |\lambda_k - \lambda_j|^{\beta},\qquad (1\text{-}1)$$

with β corresponding to the underlying global symmetry ($\beta = 1, 2$ or 4 for invariance under orthogonal, unitary or symplectic unitary transformations, respectively); C denotes the normalization. This is extended to general $\beta > 0$, giving the β-ensembles [Dumitriu and Edelman 2002] as specified by the eigenvalue PDF (1-1), to be denoted $\mathrm{ME}_{N,\beta}(g(\lambda))$. In particular the choice $g(\lambda) = \mathrm{e}^{-\beta\lambda^2/2}$ defines the Gaussian β-ensemble and the choice $g(\lambda) = \lambda^{\beta a/2}\mathrm{e}^{-\beta\lambda/2}$, ($\lambda > 0$), defines the Laguerre β-ensemble.

The bulk state can be realized by scaling $\lambda_l \mapsto x_l/\sqrt{2N}$ in the Gaussian β-ensemble. The soft edge is realized by the scalings

$$\lambda_l \mapsto \sqrt{2N} + \frac{x_l}{\sqrt{2}N^{1/6}} \quad \text{and} \quad \lambda_l \mapsto 4N + 2\sqrt{2}x_l$$

in the Gaussian and Laguerre β-ensembles, respectively [Forrester 1993]. Only the Laguerre β-ensemble has a hard edge, as it requires the eigenvalue density to be strictly zero on one side; it is realized by the scaling $\lambda_l \mapsto x_l/(4N)$. In all cases the limit $N \to \infty$ needs to be taken after the scaling. At an edge, the spacing between consecutive eigenvalues is not the natural observable. Instead, it is most natural to measure the distribution of the largest, second largest, etc., eigenvalue (or smallest, second smallest etc.). It is well known, and easy to verify, that all these quantities can be expressed in terms of the (conditional) gap probabilities $E_\beta^{(\cdot)}(n; J)$ for there being exactly n eigenvalues in the interval J, for the scaled state $(\cdot) = $ bulk, soft or hard indexed by β. In the case of the hard edge, the probability depends on the exponent $\beta a/2$ in the Laguerre weight $\lambda^{\beta/2}\mathrm{e}^{-\beta\lambda/2}$, so we write $E_\beta^{\mathrm{hard}}(n; J; a\beta/2)$.

Our interest in this review is on the asymptotic form of spacing distributions in the bulk, and of the distribution of large and small eigenvalues at the edge. This is a topic which (in the bulk case) occupied the attention of Dyson in one of the pioneering papers on random matrix theory in the early 1960s [Dyson

1962], and is still being written on as we stand today some 50 years later. We are seeking to catalog both the results and the methods which underlie them, and also to contribute some new formulas. Section 2 deals with results founded on Dyson's heuristic physical hypothesis; these are in the form of conjectures. The various mathematical techniques which can both prove, and build on these asymptotic expressions, are covered in Section 3. A numerical illustration of the accuracy of the asymptotic form is given in Section 4, as is a discussion of asymptotic results for the gap probability in the case that each eigenvalue is independently deleted with probability $(1 - \xi)$.

2. Macroscopic heuristics

2.1. Zero eigenvalues in the gap. The eigenvalue PDF (1-1) can be interpreted as the Boltzmann factor of a classical log-gas system interacting at inverse temperature β. The particles repel via the logarithmic potential and are subject to a one body potential with Boltzmann factor $g(\lambda) = e^{-\beta V(\lambda)}$. This interpretation led Dyson [1962] to hypothesize an ansatz for the asymptotic form of the gap probability $E_\beta(0; (-\alpha, \alpha); C\beta E_N)$, where $C\beta E_N$ denotes Dyson's circular ensembles (see, e.g., [Forrester 2010, Chapter 2]) of random unitary matrices (all eigenvalues are therefore on the unit circle; the interval $(-\alpha, \alpha)$ refers to a sector of the circumference specified by its angles):

$$E_\beta(0; (-\alpha, \alpha); C\beta E_N) \underset{N \to \infty}{\sim} e^{-\beta \delta F}. \tag{2-1}$$

Here and below the symbol $\sim$ is used to denote that the right-hand side gives leading terms, up to some order to be further specified, of the asymptotic expansion of the left-hand side. In (2-1) δF is the energy cost of conditioning the equilibrium particle density so that $\rho_{(1)}(\theta) = 0$ for $\theta \in (-\alpha, \alpha)$. This energy cost consists of an electrostatic energy

$$V_1 = -\frac{1}{2} \int_0^{2\pi} \int_0^{2\pi} \left(\rho_{(1)}(\theta_1) - N/2\pi \right) \left(\rho_{(1)}(\theta_2) - N/2\pi \right) \log |e^{i\theta_1} - e^{i\theta_2}| \, d\theta_1 \, d\theta_2 \tag{2-2}$$

and an entropy term

$$V_2 = \left(\frac{1}{\beta} - \frac{1}{2} \right) \int_0^{2\pi} \rho_{(1)}(\theta) \log \frac{\rho_{(1)}(\theta)}{N/2\pi} \, d\theta. \tag{2-3}$$

The density is chosen to minimize V_1 and then V_1 and V_2 evaluated, and we have

$$\delta F = (V_1 + V_2). \tag{2-4}$$

Proposition 1 [Dyson 1962]. *With the requirements that $\rho_{(1)}(\theta) = 0$ for $\theta \in (-\alpha, \alpha)$ and $\int_0^{2\pi} \rho_{(1)}(\theta) \, d\theta = N$, V_1 is minimized by*

$$\rho_{(1)}(\theta) = \frac{N}{2\pi} \frac{\sin \theta/2}{\sqrt{\sin^2 \theta/2 - \sin^2 \alpha/2}}. \tag{2-5}$$

We then have

$$\beta V_1 = -\frac{\beta}{2} N^2 \log \cos \frac{\alpha}{2}, \quad \beta V_2 = \left(1 - \frac{\beta}{2}\right) N \log\left(\sec \frac{\alpha}{2} + \tan \frac{\alpha}{2}\right). \tag{2-6}$$

We remark that explicit calculations in [Dyson 1962] showed that requiring $\rho_{(1)}(\theta)$ to minimize $V_1 + V_2$ (rather than V_1) results in a correction to βV_2 which for large N is of order $\log(N\alpha)$, indicating that the asymptotic expansion (2-1) will not correctly give terms of this order.

Substituting (2-6) in (2-4), and substituting the result in (2-1) gives a large deviation formula, telling us (as a conjecture) the probability of there being no eigenvalues in the interval $(-\alpha, \alpha)$. This probability decays as a Gaussian in N. An $O(1)$ expression should result from choosing the excluded interval as $(-\pi s/N, \pi s/N)$, as then there are $O(1)$ eigenvalues in the gap. Replacing α by $\pi\alpha/2$ in (2-6), then taking $N \to \infty$ (this is a double scaling limit) gives the prediction

$$\lim_{N \to \infty} E_\beta(0; (-\pi s/N, \pi s/N); C\beta E_N) \sim e^{-\beta(\pi s)^2/16 + (\beta/2 - 1)\pi s/2}. \tag{2-7}$$

Dyson was well aware that the $\sim$ symbol should be interpreted as agreeing in the large s asymptotic expansion to the order given. But the left-hand side is the definition of $E_\beta^{\text{bulk}}(0; (-s/2, s/2))$, thus providing the following conjecture.

Conjecture 2 [Dyson 1962]. *We have*

$$E_\beta^{\text{bulk}}(0; (0, s)) \underset{s \to \infty}{\sim} e^{-\beta(\pi s)^2/16 + (\beta/2 - 1)\pi s/2}. \tag{2-8}$$

As remarked above, Dyson [1962] carried through the details of the minimization of $V_1 + V_2$, resulting in a logarithmic correction to the exponent of the right-hand side of (2-8): $((1 - \beta/2)^2/(2\beta)) \log s$. However, this was later put in doubt by Mehta and des Cloizeaux [1972], who, using a method based on eigenvalues (see Section 3.3 below), obtained $-\frac{1}{8}$, $-\frac{1}{4}$ and $-\frac{1}{8}$ for the prefactor of $\log s$ for $\beta = 1$, 2 and 4, respectively. Dyson himself [1976] used inverse scattering methods applied to the Fredholm determinant form of $E_1^{\text{bulk}}(0; (0, s))$ (see Section 3.1) to also give the prediction $-\frac{1}{8}$ for the prefactor in the case $\beta = 1$. In fact the correct extension of (2-8) for general β, as proved for the

Gaussian β ensemble, is [Valkó and Virág 2010]:

$$E_\beta^{\text{bulk}}(0; (0, s)) \underset{s \to \infty}{\sim} \exp\left(-\beta(\pi s)^2/16 + (\beta/2 - 1)\pi s/2\right.$$
$$\left. + \tfrac{1}{4}(\beta/2 + 2/\beta - 3)\log s + O(1)\right). \quad (2\text{-}9)$$

Its derivation will be reviewed in Section 3.2.

The ansatz (2-1) was applied to the gap probability at the hard edge of the Laguerre ensemble by Chen and Manning [1994]. They considered the probability of there being no eigenvalues in an interval $(0, t)$.

Proposition 3 [Chen and Manning 1994]. *For the Laguerre ensemble specified by (1-1) with $g(\lambda) = \lambda^a e^{-\lambda}$, with the eigenvalues constrained to the interval (t, b), with $t > 0$ given, the minimizing solution for the level density $\rho_{(1)}(x)$ is*

$$\rho_{(1)}(x) = \frac{1}{\pi\beta}\sqrt{\frac{b-x}{x-t}}\left(1 - \frac{a}{x}\sqrt{\frac{t}{b}}\right). \quad (2\text{-}10)$$

Normalization of the density requires that b is related to N by

$$N = \frac{b-t}{2\beta} + \frac{a}{\beta}\left(\sqrt{\frac{t}{b}} - 1\right). \quad (2\text{-}11)$$

Using (2-10) appropriate analogues of (2-2) and (2-3) were computed (see also [Chen and Manning 1996]), thus giving a prediction for the large N form of $E_\beta(0; (0, t); \mathrm{ME}_{N,\beta}(\lambda^\alpha e^{-\lambda}))$. This is exponentially small in N. But with $t = s/(4N)$, the number of eigenvalues in $(0, t)$ will be $O(1)$. With the resulting expression interpreted as the large s asymptotic form of $E_\beta^{\text{hard}}(0; (0, s); a)$ (s must be scaled $s \mapsto (\beta/2)^2 s$ to account for the latter being defined as the large N limit of $E_\beta(0; (0, s/(4N)); \mathrm{ME}_{N,\beta}(\lambda^a e^{-\beta\lambda/2})))$), the following conjecture was obtained.

Conjecture 4 [Chen and Manning 1994]. *We have*

$$E_\beta^{\text{hard}}(0; (0, s); a) \underset{s \to \infty}{\sim} \exp\left(-\frac{\beta s}{8} + a\sqrt{s} - \frac{a^2}{2\beta}\log s + \left(1 - \frac{\beta}{2}\right)\frac{a}{2\beta}\log s\right). \quad (2\text{-}12)$$

This asymptotic had already been proved in [Forrester 1994] for $a \in \mathbb{Z}_{\geq 0}$ and $2/\beta \in \mathbb{Z}_{>0}$ before the work of Chen and Manning [1994]. Moreover, [Forrester 1994], which was based on a-dimensional integral forms for $E_\beta^{\text{hard}}(0; (0, s); a)$, gave the explicit form of the constant term in the extension of (2-12) to next order (see Section 3.4).

The first application of the log-gas ansatz (2-1) at the soft edge was due to Dean and Majumdar [2006; 2008].

Proposition 5 [Dean and Majumdar 2006]. *Consider the Gaussian β-ensemble $\mathrm{ME}_{\beta,N}(e^{-\beta N x^2})$. Suppose the eigenvalues are confined to the interval $(-b, t)$*

where $t < 1$ and $b > 0$ is determined by charge neutrality. The corresponding density is given by

$$\rho_{(1)}(x) = \frac{2N}{\pi} \left(\frac{l-t+x}{t-x} \right)^{1/2} \left(\frac{l}{2} - x \right),$$

where $l := b+t = \frac{2}{3}(t + \sqrt{t^2+3})$.

Only the corresponding form of V_1 was computed, and this gave the large deviation formula

$$E_\beta(0; (t, \infty); \mathrm{ME}_{\beta, N}(e^{-\beta Nx^2})) \underset{N \to \infty}{\sim}$$

$$\exp\left(-\beta N^2 \left(\frac{2t^2}{3} - \frac{t^4}{27} - \frac{5}{18} t\sqrt{3+t^2} - \frac{1}{27} t^3 \sqrt{3+t^2} - \frac{1}{2} \log \frac{t+\sqrt{t^2+3}}{3} \right) \right), \quad (2\text{-}13)$$

and thus, upon the appropriate soft edge scaling $\sqrt{2N}(1-t) = -\dfrac{s}{\sqrt{2}N^{1/6}}$, the asymptotic formula

$$E_\beta^{\mathrm{soft}}(0; (s, \infty)) \underset{s \to -\infty}{\sim} e^{-\beta|s|^3/24}. \quad (2\text{-}14)$$

This latter prediction was already implied in [Forrester 1993; Tracy and Widom 1994a].

2.2. *Loop equations.*

Borot, Eynard, Majumdar and Nadal [Borot et al. 2011] gave an alternative heuristic formalism to the Dyson log-gas ansatz, for purposes of computing the soft edge gap probability. This in based on the so-called loop equations associated with the large N form of the multiple integral definition of the latter. The approach allows for the Dyson ansatz (2-1) to be extended to include higher order terms; in practice two new terms are computed — one is termed the Polyakov anomaly, and the following result is obtained.

Conjecture 6 [Borot et al. 2011]. *We have*

$$E_\beta^{\mathrm{soft}}(0; (s, \infty)) \underset{s \to -\infty}{\sim} \exp\left(-\beta \frac{|s|^3}{24} + \frac{\sqrt{2}(\beta/2-1)}{3} |s|^{3/2} \right.$$

$$\left. + \frac{\beta/2+2/\beta-3}{8} \log|s| + \log \tau_\beta^{\mathrm{soft}} + O(|s|^{-3/2}) \right), \quad (2\text{-}15)$$

where

$$\log \tau_\beta^{\mathrm{soft}} = \left(\frac{17}{8} - \frac{25}{24}(\beta/2+2/\beta) \right) \log 2 - \frac{\log 2\pi}{2} - \frac{\log \beta/2}{2} + \kappa_{\beta/2}, \quad (2\text{-}16)$$

with κ_β the constant term in the large N expansion of

$$F(N+1) := \sum_{j=1}^{N} \log \Gamma(1 + j\beta/2).$$

(Note that in [Borot et al. 2011] what we call $\beta/2$ is written as β.)

In [Borot et al. 2011], for β rational, κ_β was evaluated in terms of the Barnes G-function, while for general $\beta > 0$ it was shown that

$$\kappa_{\beta/2} = \frac{\log 2\pi}{4} + \frac{\beta}{2}\left(\frac{1}{12} - \zeta'(-1)\right) + \frac{\gamma}{6\beta}$$

$$+ \int_0^\infty \frac{1}{e^{\beta s/2} - 1}\left(\frac{s}{e^s - 1} - 1 + \frac{s}{2} - \frac{s^2}{12}\right) ds,$$

where γ denotes Euler's constant. In fact $\kappa_{\beta/2}$ can be expressed in terms of the so-called Stirling modular form $\rho_2(1, \tau)$, which from a computational viewpoint can be defined by the infinite product [Shintani 1980]

$$\rho_2(1, \tau) = (2\pi)^{3/4} \tau^{-1/4 + (\tau + 1/\tau)/12} e^{P(\tau)} \prod_{n=1}^\infty \frac{e^{Q(n\tau)}}{\Gamma(1 + n\tau)},$$

where

$$P(\tau) = -\frac{\gamma}{12\tau} - \frac{\tau}{12} + \tau\zeta'(-1), \quad Q(x) = \left(\tfrac{1}{2} + x\right)\log x - x + \log\sqrt{2\pi} + \frac{1}{12x}.$$

The quantity $\rho_2(1, \tau)$ is fundamental to the theory of the Barnes double gamma function $\Gamma_2(z; 1, \tau)$ [Barnes 1904], the latter being related to the usual gamma function through the two functional equations

$$\frac{1}{\Gamma_2(z + 1; 1, \tau)} = \frac{\tau^{z/\tau - 1/2}}{\sqrt{2\pi}} \frac{\Gamma(z/\tau)}{\Gamma_2(z; 1, \tau)},$$

$$\frac{1}{\Gamma_2(z + \tau; 1, \tau)} = \frac{1}{\sqrt{2\pi}} \frac{\Gamma(z)}{\Gamma_2(z; 1, \tau)}, \tag{2-17}$$

and furthermore is normalized by requiring $\lim_{z \to 0} z\Gamma_2(z; 1, \tau) = 1$.

Proposition 7. *Let $\tau = 2/\beta$ and specify $F(N + 1)$ and $\kappa_{\beta/2}$ as in Conjecture 6. We have*

$$F(N + 1) = (2\pi)^{N/2} \tau^{-(N^2 - N(1-\tau))/2\tau} \frac{\Gamma(N)\Gamma(1 + N/\tau)}{\Gamma_2(N; 1, \tau)}, \tag{2-18}$$

$$\kappa_{1/\tau} = -\tfrac{1}{2}\log\tau + \log 2\pi - \log\rho_2(1, \tau), \tag{2-19}$$

with the latter equation substituted into (2-16) giving

$$\log\tau_\beta^{\text{soft}} = \left(\tfrac{17}{8} - \tfrac{25}{24}(\beta/2 + 2/\beta)\right)\log 2 + \frac{\log 2\pi}{2} - \log\rho_2(1, 2/\beta). \tag{2-20}$$

Equation (2-18) has appeared in [Brini et al. 2011]; it follows immediately by characterizing $F(N+1)$ as a first order recurrence, and using (2-17). The formula for $\kappa_{1/\tau}$ then follows by extracting the term independent of N in the corresponding asymptotic expansion. Here one uses the fact that for $\log \Gamma_2(N; 1, \tau)$ this is $\log \rho_2(1, \tau)$ [Quine et al. 1993]. A consequence of (2-20) is that

$$\log \frac{\tau_{\beta/2}^{\text{soft}}}{\tau_{2/\beta}^{\text{soft}}} = -\log \frac{\rho_2(1, 2/\beta)}{\rho_2(1, \beta/2)} = \frac{\beta/2+2/\beta-3}{8}\left(\log \frac{\beta}{2}\right)^{2/3} - \frac{1}{2}\log \frac{\beta}{2}, \quad (2\text{-}21)$$

where the final equality follows from the inversion formula for the Stirling modular form [Katayama and Ohtsuki 1998, Proposition 7(iv)]. Using this in (2-15) gives that (cf. [Borot et al. 2011, Equation (6.2)])

$$E_\beta^{\text{soft}}(0; (s, \infty)) \underset{s \to -\infty}{\sim} \left(\frac{2}{\beta}\right)^{1/2} \widetilde{E}_{4/\beta}^{\text{soft}}\left(0; \left(\left(\frac{\beta}{2}\right)^{2/3} s, \infty\right)\right), \quad (2\text{-}22)$$

where $\widetilde{E}_\beta^{\text{soft}}$ refers to the right-hand side of (2-15) with $|s|^{3/2}$ replaced by $-|s|^{3/2}$.

2.3. Conditioning n eigenvalues in the gap. Dyson [1995], and independently Fogler and Shklovskii [1995], further developed the log-gas argument by the consideration of the setting that the gap $(-t, t)$ is required to contain exactly n eigenvalues, with $0 \ll n \ll t$. Moreover, a change of viewpoint was introduced: the log-gas was taken to be infinite in extent, with the bulk state characterized by a uniform density, normalized to unity. The n eigenvalues are modeled as a continuous conductive fluid occupying the interval $(-b, b) \subset (-t, t)$. The electrostatic potential in this region must therefore be equal to a constant $-v$ say, $v > 0$, with the potential in the other conducting region $\mathbb{R}\backslash(-t, t)$ taken to be zero. The explicit form of the density was determined, and this substituted in the appropriate modification of (2-2) and (2-3) gave after some calculation the simple results

$$V_1 = -\frac{nv}{2} + \frac{\pi^2}{4}(t^2 - b^2), \quad V_2 = v. \quad (2\text{-}23)$$

The end point b is determined by n via a certain elliptic integral, and similarly v in terms of an elliptic integral of modulus b/t. Expansion of these quantities for $t \to \infty$, and substitution in (2-1) provides a generalization of (2-9).

Conjecture 8 [Dyson 1995; Fogler and Shklovskii 1995]. *For $0 \ll n \ll s$ we have*

$$\log E_\beta^{\text{bulk}}(n; (0, s)) \underset{s \to \infty}{\sim} -\beta \frac{(\pi s)^2}{16} + \left(\beta n + \frac{\beta}{2} - 1\right)\frac{\pi s}{2}$$

$$+ \left\{\frac{n}{2}\left(1 - \frac{\beta}{2} - \frac{\beta n}{2}\right) + \frac{1}{4}\left(\frac{\beta}{2} + \frac{2}{\beta} - 3\right)\right\}\log s. \quad (2\text{-}24)$$

(Here we have added the $n = 0$ contribution to the term $\log s$ as implied by (2-9) — we then expect (2-24) to hold for $0 \le n \ll s$; this is not a consequence of the calculations in [Dyson 1995; Fogler and Shklovskii 1995].)

Only very recently has this infinite log-gas formalism been applied to predict the asymptotic forms of the conditioned gap probabilities at the hard and soft edges [Forrester and Witte 2012]. Since the system is (semi-) infinite, this relies on characterizing these edges in terms of the respective background densities: $\sqrt{x}/\pi$ for the soft edge, and $1/(2\pi\sqrt{x})$ for the hard edge. In both cases the coordinates are chosen so that the edge occurs at $x = 0$. It was found in [Forrester and Witte 2012] that applying the ansatz (2-1) with δF given by (2-4) in this setting to the $n = 0$ case gave results inconsistent with both (2-12) and its soft edge analogue in the second order term. Thus the ansatz (2-1) with δF given by (2-4) is incorrect in the infinite log-gas formalism applied to the hard and soft edges. On the other hand, it was observed that replacing V_2 by the potential drop v in going from the region containing the infinite mobile log-charges, to the region containing the n charges — which, according to (2-23), is an identity for the bulk — restores the correct value for these terms. Making this replacement for general n then gives the following predictions.

Conjecture 9 [Forrester and Witte 2012]. *We have, for $0 \ll n \ll |s|$ (or more strongly $0 \leq n \ll |s|$),*

$$\log E_\beta^{\mathrm{hard}}(n; (0, s); \beta a/2)$$

$$\underset{s \to \infty}{\sim} -\beta\left\{\frac{s}{8} - \sqrt{s}\left(n + \frac{a}{2}\right) + \left[\frac{n^2}{2} + \frac{na}{2} + \frac{a(a-1)}{4} + \frac{a}{2\beta}\right]\log s^{1/2}\right\} \quad (2\text{-}25)$$

and

$$\log E_\beta^{\mathrm{soft}}(n; (s, \infty)) \underset{s \to -\infty}{\sim} -\frac{\beta|s|^3}{24} + \frac{2\sqrt{2}}{3}|s|^{3/2}\left(\beta n + \frac{\beta}{2} - 1\right)$$

$$+ \left[\frac{\beta}{2}n^2 + \left(\frac{\beta}{2} - 1\right)n + \frac{1}{6}\left(1 - \frac{2}{\beta}\left(1 - \frac{\beta}{2}\right)^2\right)\right]\log|s|^{-3/4}. \quad (2\text{-}26)$$

(As for (2-24), the results coming from the log-gas calculation have, in the case of the logarithmic term, been supplemented by knowledge of the asymptotic expansion at that order for $n = 0$.)

We remark that a check on (2-24)–(2-26) is that they obey certain asymptotic functional equations, implied by exact functional equations for spacing distributions obtained in [Forrester 2009]. For example, at the hard edge one requires

$$E_\beta^{\mathrm{hard}}(n; (0, s/\tilde{s}_\beta); \beta a/2) \underset{\substack{s \to \infty \\ n \ll t}}{\sim} E_{4/\beta}^{\mathrm{hard}}\left(\tfrac{1}{2}\beta(n+1) - 1; (0, s/\tilde{s}_{4/\beta}); a - 2 + 4/\beta\right),$$

where $\tilde{s}_\beta$ is an arbitrary length scale that satisfies $\tilde{s}_{4/\beta}(\beta/2)^2 = \tilde{s}_\beta$. This is indeed a property of (2-25).

Precise asymptotic statements can also be made concerning the asymptotic form of $E_\beta^{(\cdot)}(n; J)$, for $|J| \to \infty$ and $n \approx \langle n_J \rangle$, where n_J ($\langle n_J \rangle$) denotes the number (expected number) of particles in J for the unconstrained system. Thus macroscopic heuristics applied to this linear statistic (see, e.g., [Forrester 2012, §14.5.1]) predict that $(n_J - \langle n_J \rangle)/\sqrt{\text{Var}\, n_J}$ has a Gaussian distribution with zero mean and unit variance, and so suggesting the following result.

Conjecture 10. *For* $n \approx \langle n_J \rangle$,

$$E_\beta^{(\cdot)}(n; J) \underset{|J| \to \infty}{\sim} \frac{1}{(2\pi \,\text{Var}\, n_J)^{1/2}} e^{-(n-\langle n_J \rangle)^2/2\text{Var}\, n_J}. \tag{2-27}$$

Moreover, for $(\cdot) = $ *bulk, soft and hard we have*

$$\langle n_{(0,s)} \rangle \underset{s \to \infty}{\sim} s, \quad \langle n_{(s,\infty)} \rangle \underset{s \to -\infty}{\sim} \frac{2(-s)^{3/2}}{3\pi}, \quad \langle n_{(0,s)} \rangle \underset{s \to \infty}{\sim} \frac{s^{1/2}}{\pi}, \tag{2-28}$$

and

$$\text{Var}\, n_{(0,s)} \underset{s \to \infty}{\sim} \frac{2}{\pi^2 \beta} \log s, \quad \text{Var}\, n_{(s,\infty)} \underset{s \to -\infty}{\sim} \frac{1}{\pi^2 \beta} \log |s|^{3/2},$$

$$\text{Var}\, n_{(0,s)} \underset{s \to \infty}{\sim} \frac{1}{\pi^2 \beta} \log s^{1/2}. \tag{2-29}$$

The results (2-28) are immediate consequences of the corresponding asymptotic density profiles (recall the second sentence below Conjecture 8), while (2-29) can be derived heuristically from knowledge of the asymptotic form of the two-point correlation function (see [Forrester 2010, paragraph below (14.87)]). In the case of $(\cdot) = $ bulk, (2-27), with the corresponding values of $\langle n_{(0,s)} \rangle$ and $\text{Var}\, n_J$ as implied by (2-28) and (2-29), was derived in the context of the infinite log-gas formalism by Dyson [1995] and by Fogler and Shklovskii [1995].

3. Rigorous methods

3.1. *Toeplitz/Hankel asymptotics.* It is a fundamental result in random matrix theory (see, e.g., [Forrester 2010, §9.1]) that in the scaled limit $(\cdot)$ equal to bulk, hard or soft, and $\beta = 2$ the probability of there being no eigenvalues in an interval J, may be written in terms of a determinant of a Fredholm integral operator

$$E_2^{(\cdot)}(0; J) = \det(1 - K_J^{(\cdot)}),$$

where $K_J^{(\cdot)}$ is the integral operator on the interval J with well known sine, Bessel and Airy kernels (see, e.g., [Forrester 2010] for the precise definitions). This is related to the fact that for $\beta = 2$ the gap probabilities can be written in terms of either Toeplitz or Hankel determinants. For example, the Toeplitz determinant of a function $f(\theta)$, integrable over the unit circle, is defined as

$$D_n(f) := \det\left(\frac{1}{2\pi}\int_0^{2\pi} e^{-i(j-k)\theta} f(\theta)\, d\theta\right)_{j,k=0}^{n-1}, \qquad (3\text{-}1)$$

and one has the well known formula

$$E_2(0; (-\alpha, \alpha); \mathrm{CUE}_N) = D_N(f_\alpha), \qquad f_\alpha = \begin{cases} 1, & \theta \in (\alpha, 2\pi - \alpha), \\ 0, & \text{otherwise.} \end{cases}$$

In particular $\lim_{n\to\infty} D_n(f_{2s/n}) = \det(\mathbb{I} - K_s^{\mathrm{bulk}})$, allowing for a strategy whereby the $s \to \infty$ behavior can be extracted from the asymptotics of the Toeplitz determinant. On the other hand the Toeplitz determinant has a representation in terms of quantities associated with orthonormal polynomials $\phi_k(z) = \chi_k z^k + \cdots$ with weight $f(\theta)$ on the unit circle; explicitly $D_n(f) = \prod_{k=0}^{n-1} \chi_k^{-2}$. Krasovsky [2004] used a Riemann–Hilbert formulation to compute the large n form of $\frac{d}{d\mu}\ln D_n(f_\mu)$, uniformly in μ, providing both a proof and refinement of (2-8) in the case $\beta = 2$.

Theorem 11 [Krasovsky 2004; Ehrhardt 2006]. *We have*

$$\log E_2^{\mathrm{bulk}}(0; (0, s)) = -\frac{(\pi s)^2}{8} - \tfrac{1}{4}\log\frac{\pi s}{2} + \tfrac{1}{12}\log 2 + 3\zeta'(-1) + O\left(\frac{1}{s}\right), \quad (3\text{-}2)$$

where $\zeta(z)$ is the Riemann zeta function.

We remark that up to the constant term this result, deduced by Dyson [Dyson 1976] using a scaling argument from known Toeplitz determinant asymptotics, was first rigorously proved by Deift, Its and Zhou [Deift et al. 1997]; also the proof of Ehrhardt [2006] is operator theoretic, and does not make use of orthogonal polynomials.

Analogous strategies can be used to analyze the hard and soft edges for $\beta = 2$, giving the following results, proving and extending (2-12) and (2-14), respectively.

Theorem 12. *We have*

$$\log E_2^{\mathrm{soft}}(0; (s, \infty)) = -\tfrac{1}{12}|s|^3 - \tfrac{1}{8}\log|s| + \tfrac{1}{24}\ln 2 + \zeta'(-1) + O\left(|s|^{-\frac{3}{2}}\right) \quad (3\text{-}3)$$

for $s \to -\infty$ (see [Deift et al. 2008]), and

$$\log E_2^{\mathrm{hard}}(0; (0, s); a) = -\frac{s}{4} + a\sqrt{s} - \frac{a^2}{4}\log s + \log\frac{G(1+a)}{(2\pi)^{a/2}} + O\left(s^{-\frac{1}{2}}\right) \quad (3\text{-}4)$$

for $s \to \infty$ (see [Deift et al. 2011]), where $G(x)$ denotes the Barnes G-function.

An alternative proof of (3-3) has been given by Baik, Buckingham and DiFranco [Baik et al. 2008], using the Painlevé form of $E_2^{\mathrm{soft}}(0; (s, \infty))$ [Tracy and Widom 1994a]. This method carries over to the cases $\beta = 1$ and 4, and in [Baik et al. 2008] the expansion (2-15) with

$$\log \tau_1^{\text{soft}} = -\frac{11 \log 2}{48} + \frac{\zeta'(-1)}{2},$$

$$\log \tau_4^{\text{soft}} = -\frac{37 \log 2}{48} + \frac{\zeta'(-1)}{2}, \tag{3-5}$$

was obtained. These confirm the values implied by (2-20).

With regards to (3-4), as noted above, for $a \in \mathbb{Z}_{\geq 0}$ it was first proved by Forrester [1994]. More recently a proof of (3-4) valid for $|a| < 1$ was given by Ehrhardt [2010]. Furthermore, let the next order (constant) term in the exponent of (2-12) be included by adding $\log \tau_{a,\beta}^{\text{hard}}$. We read off from (3-4) that $\tau_{a,2}^{\text{hard}} = G(1+a)/(2\pi)^{a/2}$. For $a \in \mathbb{Z}_{\geq 0}$ a multiple integral form for E_1^{hard} [Forrester and Witte 2002], and an identity [Forrester and Rains 2001] relating E_4^{hard} to E_2^{hard} and E_1^{hard} for general $a > -1$ tells us that

$$\tau_{a,1} = 2^{-a(a+1/2)} \frac{G(3/2)G(2a+2)}{G(a+3/2)G(a+2)},$$

$$\tau_{a+1,4} = 2^{-a(a+1)/4-1} \frac{\tau_{a,2}}{\tau_{(a-1)/2,1}}. \tag{3-6}$$

In the case of bulk scaling, include a constant term by adding $\log \tau_\beta^{\text{bulk}}$ to the exponent of (2-9) with s replaced by s/π (thus the bulk density is now $1/\pi$). It follows from (3-2) that $\log \tau_2^{\text{bulk}} = \frac{1}{3} \log 2 + 3\zeta'(-1)$. And interrelations between the bulk gap probability for $\beta = 1$ and 4 with $\beta = 2$ quantities give that [Basor et al. 1992]

$$\tau_1^{\text{bulk}} = 2^{5/12} e^{(3/2)\zeta'(-1)}, \qquad \tau_4^{\text{bulk}} = 2^{-29/24} e^{(3/2)\zeta'(-1)}. \tag{3-7}$$

We observe that (3-7) is consistent with a relation analogous to (2-20).

Conjecture 13. *Let $\rho_2(1, \tau)$ denote the Stirling modular form. We have*

$$\log \tau_{\beta/2}^{\text{bulk}} = \left(3 - \tfrac{4}{3}(\beta/2 + 2/\beta)\right) \log 2 + 3\left(\tfrac{1}{2} \log 2\pi - \log \rho_2(1, 2/\beta)\right), \tag{3-8}$$

and consequently

$$\log \frac{\tau_{\beta/2}^{\text{bulk}}}{\tau_{2/\beta}^{\text{bulk}}} = -3 \log \frac{\rho_2(1, 2/\beta)}{\rho_2(1, \beta/2)},$$

$$E_\beta^{\text{bulk}}(0; (0, s/\pi)) \underset{s \to \infty}{\sim} \left(\frac{2}{\beta}\right)^{3/2} \widetilde{E_{4/\beta}^{\text{bulk}}}\left(0; \left(0, \frac{\beta}{2} s/\pi\right)\right), \tag{3-9}$$

where $\widetilde{E_\beta^{\text{bulk}}}$ refers to the right-hand side of (2-9) with s replaced by $-s$ in the second term.

3.2. Stochastic differential equations. The Gaussian and Laguerre β-ensemble, defined as eigenvalue PDFs below (1-1), admit realizations as real symmetric tridiagonal matrices [Dumitriu and Edelman 2002]. In the scaled $N \to \infty$ limit, this in turn leads to explicit characterization of gap probabilities in terms of stochastic differential equations. The first result of this type was done for the soft edge, by Ramirez, Rider and Valkó [Ramírez et al. 2011a]. With N fixed, it relies on expressing the number of eigenvalues greater than μ as the number of sign changes of the shooting vector for the tridiagonal matrix. Similarly at the hard edge [Ramírez and Rider 2009]. In the bulk, the shooting eigenvector must be parametrized in terms of the corresponding Prüfer phase [Killip and Stoiciu 2009; Valkó and Virág 2009]. The following results are obtained.

Proposition 14. *Let b_t denote standard Brownian motion. At the soft edge, define a diffusion by the Ito process by* (see [Ramírez et al. 2011a]):

$$dp(t) = \frac{2}{\sqrt{\beta}} db_t + (\lambda + t - p^2(t)) \, dt, \quad p(0) = \infty;$$

at the hard edge with parameter $\beta(a+1)/2 - 1$ by (see [Ramírez et al. 2011b])

$$dp(t) = db_t + \left(\tfrac{1}{4}\beta\left(a + \tfrac{1}{2}\right) - \tfrac{1}{2}\beta\sqrt{\lambda}e^{-\beta t/8} \cosh p(t)\right) dt, \quad p(0) = \infty;$$

and in the bulk (see [Valkó and Virág 2010]) *by*

$$dp(t) = db_t + \left(\tfrac{1}{2}\tanh p(t) - \tfrac{1}{8}\beta\lambda e^{-\beta t/4} \cosh p(t)\right) dt, \quad p(0) = \infty.$$

Let $J = (0, s/2\pi)$ for $(\cdot) = bulk$, $J = (0, s)$ for $(\cdot) = hard$, and $J = (s, \infty)$ for $(\cdot) = soft$. We have

$$E_\beta^{(\cdot)}(0; J) = \Pr\left(p(t) > -\infty \text{ for all } t \in \mathbb{R}^+ \cup \{\infty\}\right). \tag{3-10}$$

The utility of these characterizations for the purpose of asymptotics is that, via the Cameron–Martin–Girsanov formula, they allow (3-10) to be rewritten as the expectation of a functional of a transformed stochastic process. In contrast to (3-10), this functional allows for a systematic, rigorous $s \to \infty$ asymptotic analysis resulting in a proof of (2-9) — giving in the process the correct form of the general $\beta > 0$, $\log s$ term, for the first time — and a proof of (2-12) for general $\beta > 0$ and $a > -1$. At the soft edge only the leading asymptotic form (2-14) has been proved using this approach [Ramírez et al. 2011a].

For the large N limit of the circular β-ensemble, the Prüfer phase has been used to prove the analogue of the Gaussian fluctuation formula (2-27), namely

$$E_\beta(n, (-\alpha, \alpha); C\beta E_N) \sim (1/(2\pi \operatorname{Var} n_{(-\alpha,\alpha)})) \exp\left(-\frac{(n - N\alpha/\pi)^2}{2\operatorname{Var} n_{(-\alpha,\alpha)}}\right),$$

where $\operatorname{Var} n_{(-\alpha,\alpha)} \sim \dfrac{1}{\pi^2\beta} \log N$ [Killip 2008].

3.3. *Fredholm determinant/eigenvalue forms for $E_\beta^{(\cdot)}(n, J)/E_\beta^{(\cdot)}(0, J)$.* With $(\cdot)$ denoting bulk, soft or hard, let $E_\beta^{(\cdot)}(J; \xi)$ be the generating function for $\{E_\beta^{(\cdot)}(n; J)\}$, so that

$$E_\beta^{(\cdot)}(J; \xi) = \sum_{n=0}^{\infty} (1 - \xi)^n E_\beta^{(\cdot)}(n; J). \tag{3-11}$$

Generalizing the Fredholm determinant expressions for $E_\beta^{(\cdot)}(0; J)$ from Section 3.1, one has that for $\beta = 2$

$$E_2^{(\cdot)}(J; \xi) = \det(1 - \xi K_J^{(\cdot)}) = \prod_{l=0}^{\infty} (1 - \xi \lambda_l), \tag{3-12}$$

where $1 > \lambda_0 > \lambda_1 > \lambda_2 > \cdots > 0$ are the eigenvalues of $K_J^{(\cdot)}$. Consequently

$$\frac{E_2^{(\cdot)}(n; J)}{E_2^{(\cdot)}(0; J)} = \sum_{0 \le j_1 < \cdots < j_n} \frac{\lambda_{j_1} \ldots \lambda_{j_n}}{(1 - \lambda_{j_1}) \ldots (1 - \lambda_{j_n})}. \tag{3-13}$$

It has been known since the work of Gaudin [1961] that associated with $K_{(0,s)}^{\text{bulk}}$ is a commuting differential operator. Furthermore, the work of Fuchs [1964] uses this, together with a WKB asymptotic analysis, to deduce the $s \to \infty$ asymptotic form of λ_j (j fixed). It was noted by Tracy and Widom [1993] that the latter implies the term with $(j_1, j_2, \ldots, j_n) = (0, 1, \ldots, n - 1)$ dominates as $t \to \infty$. These authors carried out a similar analysis in the soft and hard edge cases [Tracy and Widom 1994a; 1994b], so arriving at the following result (stated as Proposition 9.6.6 in [Forrester 2010]).

Proposition 15. *Let $G(x)$ denote the Barnes G-function. For n fixed,*

$$\frac{E_2^{\text{bulk}}(n; (0, s))}{E_2^{\text{bulk}}(0; (0, s))} \underset{s \to \infty}{\sim} G(n + 1)\pi^{-n/2}2^{-n^2-n}(\pi s)^{-n^2/2}e^{n\pi s},$$

$$\frac{E_2^{\text{soft}}(n; (0, s))}{E_2^{\text{soft}}(0; (0, s))} \underset{s \to \infty}{\sim} \frac{G(n + 1)}{\pi^{n/2}2^{(5n^2+n)/2}}(-s/2)^{-3n^2/4}\exp\left(\frac{8n}{3}\left(-\frac{s}{2}\right)^{3/2}\right), \tag{3-14}$$

$$\frac{E_2^{\text{hard}}(n; (0, s))}{E_2^{\text{hard}}(0; (0, s))} \underset{s \to \infty}{\sim} \frac{G(a + n + 1)G(n + 1)}{G(a + 1)}\pi^{-n}2^{-n(2n+2a+1)}s^{-n^2/2-an/2}e^{2n\sqrt{s}}.$$

In [Forrester 2010, § 9.6.2], as $t \to \infty$, $E_1^{\text{bulk}}(n; t)$ and $E_4^{\text{bulk}}(n; t)$ are related to $E_2^{\text{hard}}(\cdot; \cdot)$ for particular choices of the parameters. The asymptotics of the latter are known as noted in the above proposition, allowing us to extend the first result in (3-14) to $\beta = 1$ and 4 [Forrester 2010, Equations (9.100) and (9.102)].

Proposition 16. *For n fixed and $\beta = 1$ and 4 we have*

$$\frac{E_\beta^{\text{bulk}}(n; (0, s))}{E_\beta^{\text{bulk}}(0; (0, s))} = c_{\beta,n} \frac{e^{\beta n \pi s/2}}{(\pi s)^{\beta n^2/4 + (\beta/2 - 1)n/2}} \left(1 + O\left(\frac{1}{s}\right)\right), \qquad (3\text{-}15)$$

where

$$c_{1,n} = \frac{G(n/2 + 1/2)G(n/2 + 1)}{G(1/2)} \pi^{-n/2} 2^{-n(n+1)/4},$$

$$c_{4,n} = \frac{G(n + 3/2)G(n + 1)}{G(3/2)} \pi^{-n} 2^{-2n(n+1)}.$$

According to the first asymptotic formula in (3-14), (3-15) is, for a specific $c_{2,n}$, valid too for $\beta = 2$. Furthermore the functional form (3-15) for general $\beta > 0$ coincides with the log-gas prediction (2-26), and thus validates the latter for $\beta = 1, 2$ and 4, and furthermore extends it by the evaluation of $c_{\beta,n}$.

We would like to extend Proposition 16 to the soft and hard edge cases. For this, let

$$V_j^{(\cdot)} \quad \text{for } (\cdot) = \text{soft, hard}$$

and $\tilde{J} = (0, \infty)$, $(0, 1)$, respectively, denote the integral operators on $\tilde{J}$, dependent on a parameter s, with kernels $\text{Ai}(x + y + s)$ and $\frac{\sqrt{s}}{2} J_a(\sqrt{sxy})$. Write

$$E_\pm^{(\cdot)}(\xi; J) = \det(\mathbb{I} \mp \sqrt{\xi} V_j^{(\cdot)}),$$

and define

$$E_\pm^{(\cdot)}(n; J) := \frac{(-1)^n}{n!} \frac{\partial^n}{\partial \xi^n} E_\pm^{(\cdot)}(\xi; J)\big|_{\xi=1}.$$

Results contained in [Forrester 2006], and further refined in [Bornemann 2010b], tell us that for $s \to \infty$

$$E_1^{\text{soft}}(2k; (s, \infty)) = E_+^{\text{soft}}(k; (s, \infty)) + \cdots,$$

$$E_1^{\text{soft}}(2k + 1; (s, \infty)) = \tfrac{1}{2} E_-^{\text{soft}}(; (s, \infty)) + \cdots, \qquad (3\text{-}16)$$

$$E_4^{\text{soft}}(k; (s, \infty)) = E_1^{\text{soft}}(2k + 1; (2^{2/3}s, \infty)) + \cdots.$$

Here terms not written on the right-hand side are exponentially smaller (in s) than the given term. To proceed further requires a property of the eigenvalues of $V_j^{(\cdot)}$ which although supported by numerical computations, to our knowledge is yet to be proven.

Conjecture 17. *Let $v_j^{(\cdot)}(j = 0, 1, 2, \ldots)$ denote the eigenvalues of $V_j^{(\cdot)}$, ordered so that*

$$|v_0^{(\cdot)}| < |v_1^{(\cdot)}| < |v_2^{(\cdot)}| < \cdots.$$

Then $v_{2j}^{(\cdot)} > 0$ while $v_{2j+1}^{(\cdot)} < 0$ for each $j = 0, 1, 2, \ldots$.

It is well known, and easy to verify (see, e.g., [Forrester 2010, §9.6.1]) that $(v_j^{(\cdot)})^2 = \lambda_j^{(\cdot)}$, where $\{\lambda_j^{(\cdot)}\}$ are the eigenvalues of $K_j^{(\cdot)}$. This fact, together with Conjecture 17 and the analogue of (3-13) relating to $E_{\pm}^{\text{soft}}(k; (s, \infty))$, tells us that for $s \to \infty$

$$
\begin{aligned}
\frac{E_1^{\text{soft}}(2k; (s, \infty))}{E_1^{\text{soft}}(0; (s, \infty))} &= \frac{1}{(1 - \lambda_0^s)(1 - \lambda_2^s) \ldots (1 - \lambda_{2k-2}^s)} + \cdots, \\
\frac{E_1^{\text{soft}}(2k+1; (s, \infty))}{E_4^{\text{soft}}(0; (s/2^{2/3}, \infty))} &= \frac{1}{(1 - \lambda_1^s)(1 - \lambda_3^s) \ldots (1 - \lambda_{2k-1}^s)} + \cdots.
\end{aligned}
\tag{3-17}
$$

Knowledge of the explicit asymptotic form of λ_j^s from [Tracy and Widom 1994a], together with the asymptotic form of $E_1^{\text{soft}}(0; (s, \infty))/E_4^{\text{soft}}(0; (s/2^{2/3}, \infty))$ implied by (2-15) and (3-5) then allows us to extend the second result of (3-14) to $\beta = 1$ and 4.

Proposition 18 (under the assumption of Conjecture 17). *We have*

$$
\frac{E_1^{\text{soft}}(n; (s, \infty))}{E_1^{\text{hard}}(0; (s, \infty))} \underset{s \to -\infty}{\sim} \frac{G(n/2+1/2)G(n/2+1)}{\pi^{n/2} G(1/2)} 2^{-\frac{5}{8}n^2 + \frac{1}{8}n} (-s)^{-\frac{3}{8}n^2 + \frac{3}{8}n}
$$
$$
\times \exp\left(\frac{4n}{3}\left(-\frac{s}{2}\right)^{3/2}\right),
$$

$$
\frac{E_4^{\text{soft}}(n; (s/2^{2/3}, \infty))}{E_4^{\text{soft}}(0; (s/2^{2/3}, \infty))} \underset{s \to -\infty}{\sim} \sqrt{2} e^{-\frac{\sqrt{2}}{3}(-s)^{3/2}} \frac{E_1^{\text{soft}}(2n+1; (s, \infty))}{E_1^{\text{soft}}(0; (s, \infty))}.
\tag{3-18}
$$

At the hard edge, formulas structurally identical to (3-16) hold [Forrester 2006; Bornemann 2010b], with the important qualification that the additional label need to specify the hard edge gap probabilities is $(a - 1)/2$ on the left-hand side of the first two equations, and $a + 1$ on the left-hand side of the third equation; on the right-hand sides it is a, a and $a - 1$, respectively, and in the third equation s is scaled by 4 instead of $2^{2/3}$. The analogue of (3-17) then allows the analogue of Proposition 18 to be deduced.

Proposition 19 (under the assumption of Conjecture 17). *We have*

$$
\frac{E_1^{\text{hard}}(n; (0, s); (a - 1)/2)}{E_1^{\text{hard}}(0; (0, s); (a - 1)/2)} \underset{s \to \infty}{\sim} 2^{-n(n-1+a)/2} (2\pi)^{-n}
$$
$$
\times \prod_{p=1}^{2} \frac{G((n+p)/2)G((n+p+a)/2)}{G(p/2)G((p+a)/2)} s^{-(n^2+n(a-1))/4} e^{n\sqrt{s}},
$$

$$
\frac{E_4^{\text{hard}}(n; (0, s/4); a + 1)}{E_4^{\text{hard}}(0; (0, s/4); a + 1)} \underset{s \to \infty}{\sim} e^{-\sqrt{s}} s^{a/4} \frac{2^{(a+1)/2}(2\pi)^{1/2}}{\Gamma((a+1)/2)}
$$
$$
\times \frac{E_1^{\text{hard}}(2n+1; (0, s); (a - 1)/2)}{E_1^{\text{hard}}(0; (0, s); (a - 1)/2)}.
$$

As a final point in this subsection, we remark that the Gaussian fluctuation formula (2-27) can be proved for $\beta = 2$, using only the determinantal structure (3-12) and the fact that $\text{Var}\, n_J \to \infty$ [Costin and Lebowitz 1995; Soshnikov 2000].

3.4. *Hard edge: generalized hypergeometric functions.* In the case that $a \in \mathbb{Z}^+$ and general $\beta > 0$, the hard edge gap probability $E_\beta^{\text{hard}}(0; (0, s); a)$ permits evaluation in terms of a generalized hypergeometric function based on Jack polynomials $P_\kappa^{(\alpha)}(z_1, \ldots, z_N)$. The latter are labeled by a partition

$$\kappa_1 \geq \kappa_2 \geq \cdots \geq \kappa_N \geq 0$$

of nonnegative integers, and depend on the parameter α. For $\alpha = 1$ they are the Schur polynomials, while for $\alpha = 2$ they are the zonal polynomials of mathematical statistics; their precise definition can be found in, e.g., [Forrester 2010, § 12.6]. Defining $C_\kappa^{(a)}(z_1, \ldots, z_N)$ as proportional to $P_\kappa^{(\alpha)}(z_1, \ldots, z_N)$ with a specific proportionality depending on α and κ [Forrester 2010, Equation (13.1)], and the generalized Pochhammer symbol $[u]_\kappa^{(a)}$ [Forrester 2010, Equation (12.46)], the generalized hypergeometric function $_pF_q^{(\alpha)}$ is specified by (see, e.g., [Forrester 2010, § 13.1])

$$_pF_q^{(\alpha)}(a_1, \ldots, a_p; b_1, \ldots, b_q; x_1, \ldots, x_m)$$

$$:= \sum_\kappa \frac{1}{|\kappa|!} \frac{[a_1]_\kappa^{(\alpha)} \cdots [a_p]_\kappa^{(\alpha)}}{[b_1]_\kappa^{(\alpha)} \cdots [b_q]_\kappa^{(\alpha)}} C_\kappa^{(\alpha)}(x_1, \ldots, x_m). \quad (3\text{-}19)$$

Like their classical counterpart, these exhibit the confluence property

$$\lim_{a_p \to \infty} {}_pF_q^{(\alpha)}\left(a_1, \ldots, a_p; b_1, \ldots, b_q; \frac{x_1}{a_p}, \ldots, \frac{x_m}{a_p}\right)$$

$$= {}_{p-1}F_q^{(\alpha)}(a_1, \ldots, a_{p-1}; b_1, \ldots, b_q; x_1, \ldots, x_m).$$

Using this in the case $p = q = 1$, together with an integral expression for $_1F_1$ [Forrester 2010, § 13.2.5] we can readily express the conditional gap probability $E_\beta^{\text{hard}}(n; (0, s); a)$ for $a, \beta \in \mathbb{Z}_{\geq 0}$ in terms of the generalized hypergeometric function $_0F_1^{\beta/2}$, extending the $n = 0$ result of [Forrester 1994].

Proposition 20. *Let* $\beta a/2, \beta \in \mathbb{Z}_{\geq 0}$. *We have*

$$E_\beta^{\text{hard}}(n; (0, s); \beta a/2) = A_\beta(n, a) s^{n+(\beta/2)n(n+a-1)} e^{-\beta s/8}$$

$$\times \int_0^1 dy_1 \ldots \int_0^1 dy_n \prod_{j=1}^n (1 - y_j)^{\beta a/2} \prod_{1 \leq j < k \leq n} |y_k - y_j|^\beta \quad (3\text{-}20)$$

$$\times {}_0F_1^{(\beta/2)}\left(\_; a + 2n; (s/4)^{\beta a/2}, (sy_1/4)^\beta, \ldots, (sy_n/4)^\beta\right),$$

where

$$A_\beta(n, a) = \frac{2^{-2n}}{n!}\left(\frac{\beta}{2}\right)^n\left(\frac{\beta}{4}\right)^{n(a+n-1)\beta} \frac{(\Gamma(1+\beta/2))^n}{\prod_{j=0}^{2n-1}\Gamma(a\beta/2+1+j\beta/2)}, \quad (3\text{-}21)$$

and in the argument of $_0F_1^{\beta/2}$ the notation $(u)^r$ means u repeated r times. Furthermore, in the case $n = 0$, this remains valid for general $\beta > 0$.

An integral representation of $_0F_1^{(\beta/2)}$ allows for the rigorous determination of the $x \to \infty$ asymptotic expansion of $_0F_1^{(\beta/2)}(\_; c+2(m-1)/\beta; (x)^m)$ for $c, 2/\beta \in \mathbb{Z}^+$ [Forrester 1994], implying the corresponding asymptotic form of $E_\beta^{\text{hard}}(0; (0, s))$.

Proposition 21 [Forrester 1994]. *Let $2/\beta \in \mathbb{Z}^+$, and $a\beta/2 = m \in \mathbb{Z}_{\geq 0}$. For $s \to \infty$ we have*

$$E_\beta^{\text{hard}}(0; (0, s); m) = \tau_{m,\beta}\left(\frac{1}{s}\right)^{m(m+1)/2\beta-m/4}e^{-\beta s/8+ms^{\frac{1}{2}}}\left(1+O\left(\frac{1}{s^{1/2}}\right)\right), \quad (3\text{-}22)$$

where

$$\tau_{m,\beta}^{\text{hard}} = 2^{(2/\beta-1)m}\left(\frac{1}{2\pi}\right)^{m/2}\prod_{j=1}^m\Gamma(2j/\beta). \quad (3\text{-}23)$$

We see that (3-22) is in agreement with the log-gas prediction (2-12) for general $a > -1$, $\beta > 0$, and furthermore gives the explicit form of the constant in the asymptotic expansion (to use (3-23) for $m \notin \mathbb{Z}_{\geq 0}$ and check for example (3-6) requires an appropriate rewrite of the product using (2-17)).

In the case $\beta = 4$ of $_0F_1^{\beta/2}$, an integral representation not available for general β shows that, for $s \to \infty$ and $y_1, \ldots, y_n \approx 1$ (see [Muirhead 1978]),

$$_0F_1^{(\beta/2)}\left(\_; c; (s/4)^{\beta a/2}, (sy_1/4)^\beta, \ldots, (sy_n/4)^\beta\right)$$
$$= {}_0F_1^{(\beta/2)}(\_; c; (s/4)^{\beta(a+2n)/2})e^{\beta\sqrt{s}\sum_{j=1}^n(1-y_j)/2}\left(1+O\left(\frac{1}{s^{1/2}}\right)\right). \quad (3\text{-}24)$$

This, substituted in (3-20), implies, as a conjecture, the extension of the asymptotic formula (2-26) to include the constant term.

Conjecture 22. *For $\beta n \in \mathbb{Z}_{\geq 0}$, let*

$$\tau_{\beta a/2,\beta}^{\text{hard}}(n)$$
$$= \frac{2^{-(a+n)\beta n}}{n!}\left(\frac{\beta}{2}\right)^{n(a+n-1)\beta/2}\prod_{j=1}^{\beta n}\frac{\Gamma(a+2j/\beta)}{(2\pi)^{1/2}}\frac{\prod_{j=0}^{n-1}\Gamma(1+(j+1)\beta/2)}{\prod_{j=n}^{2n-1}\Gamma(1+(j+a)\beta/2)}.$$

For $s \to \infty$ we have

$$\frac{E_\beta^{\text{hard}}(n; (0, s); \beta a/2)}{E_\beta^{\text{hard}}(0; (0, s); \beta a/2)} =$$

$$\tau_{\beta a/2, \beta}^{\text{hard}}(n) \exp\left(-\beta\left\{-\sqrt{s}n + \left(\frac{n^2}{2} + \frac{na}{2}\right)\log s^{1/2}\right\}\right)\left(1 + O\left(\frac{1}{s^{1/2}}\right)\right). \quad (3\text{-}25)$$

We can check that (3-25) is consistent with the results of Proposition 19.

3.5. Approach to unity of $E_\beta^{(\cdot)}(0; J)$ for $|J| \to 0$. Generally the gap probability is given in terms of the k-point correlation functions $\{\rho_{(k)}^{(\cdot)}\}$ according to

$$E_\beta^{(\cdot)}(0; J) = 1 - \int_J \rho_{(1)}^{(\cdot)}(x)\, dx + \frac{1}{2!}\int_J \int_J \rho_{(2)}^{(\cdot)}(x, y)\, dx\, dy - \cdots.$$

Thus the leading $|J| \to 0$ asymptotic form of $E_\beta^{(\cdot)}(0; J)$ is determined by the asymptotic form of $\rho_{(1)}^{\text{hard}}(x)$ for $x \to 0$, $\rho_{(1)}^{\text{soft}}(x)$ for $x \to \infty$ and $\rho_{(2)}^{\text{bulk}}(x, y)$ for $x, y \to 0$ (for $(\cdot) = $ bulk, $\rho_{(1)}(x) = 1$ and so gives no distinguishing information). The calculation of the first and third is elementary [Forrester 1992; 1994], while direct calculation of $\rho_{(1)}^{\text{soft}}(x)$ is only known for $\beta = 1$, and β even [Desrosiers and Forrester 2006]. Collecting these together, we have the following result.

Proposition 23. *Let*

$$A_{a, \beta} = 4^{-(a+1)}(\beta/2)^{2a+1}\frac{\Gamma(1 + \beta/2)}{\Gamma(1 + a)\Gamma(1 + a + \beta/2)},$$

$$B_\beta = (\pi\beta)^\beta\frac{(\Gamma(\beta/2 + 1))^3}{\Gamma(\beta + 1)\Gamma(3\beta/2 + 1)}.$$

For $t \to 0$,

$$E_\beta^{\text{hard}}(0; (0, t); a) = 1 - A_{a, \beta}\int_0^t s^a\, ds + O(t^{a+2}),$$

$$\quad (3\text{-}26)$$

$$E_\beta^{\text{bulk}}(0; (0, t)) = 1 - t + \tfrac{1}{2}B_\beta\int_0^t \int_0^t (s_1 - s_2)^\beta\, ds_1\, ds_2 + O(t^{\beta+3}),$$

while for $t \to \infty$,

$$E_\beta^{\text{soft}}(0; (t, \infty))$$

$$= 1 - \frac{\Gamma(1 + \beta/2)}{\pi(4\beta)^{\beta/2}}\int_t^\infty \frac{e^{-2\beta X^{3/2}/3}}{X^{3\beta/4 - 1/2}}\, dX + O\left(\int_t^\infty \frac{e^{-2\beta X^{3/2}/3}}{X^{3\beta/4 + 1}}\, dX\right). \quad (3\text{-}27)$$

Two distinct derivations of (3-27) for general $\beta > 0$ are known, both involving use of a nonrigorous double scaling limit [Forrester 2012; Borot and Nadal 2011]. In [Dumaz and Virág 2011], the stochastic differential equation characterization (recall Section 3.2) is used to give a rigorous proof for general $\beta > 0$, but without determining the prefactor of the integral.

4. Other aspects

4.1. *Numerical results.* Bornemann [2010a; 2010b] has given a detailed study of the numerical analysis relating to the precise numerical evaluation of spacing distributions for $\beta = 1, 2$ and 4, working from the Fredholm determinant forms. As an end product he has provided a suite of Matlab programs implementing the theoretical procedures. The implementation in Matlab, with the arithmetic done in the hardware, means that the tails of the spacing distributions cannot be computed: their numerical values written as decimals are typically smaller than 10^{-15}, and so double precision arithmetic typically truncates significant nonzero digits, leading to unreliable results. But with there being numerous exact and conjectured results relating to the asymptotics of spacing distributions, there is much interest in implementing the theory of [Bornemann 2010a; 2010b] using an arbitrary precision package. As a start, we have done this for the Fredholm determinant form for $E_2^{\text{bulk}}(0; (0, s))$ (in fact we have modified the procedure of [Bornemann 2010a; 2010b] by using instead of Gauss–Legendre or Clenshaw–Curtis quadrature rules, the tanh-sinh quadrature rule (see, e.g., [Ye 2006])). As a result we are able to tabulate

$$r(s) = \frac{E_2^{\text{b},as}(0; (0, s))}{E_2^{\text{bulk}}(0; (0, s))}, \qquad (4\text{-}1)$$

where $E_2^{\text{b},as}(0; (0, s))$ is the asymptotic form of $E_2^{\text{bulk}}(0; (0, s))$ as given by (3-2), extended to the next two terms: $1/(8(\pi s)^2) + 5/(8(\pi s)^4)$ (these follow from the Painlevé transcendent characterization of $E_2^{\text{bulk}}(0; (0, s))$ (see, e.g., [Forrester 2010, § 9.6.7])). The values in Table 1 clearly illustrate the accuracy of the asymptotic expansion, even for relatively small values of s.

4.2. *Diluted spectra.* For a general one-dimensional point process, the generating function (3-11) can also be interpreted as the probability that there are no eigenvalues in the interval J, given that each eigenvalue has independently been deleted with probability $(1 - \xi)$. In this setting the $|J| \to \infty$ asymptotics can readily be deduced, by making use of a heuristic analysis based on (2-27) [Bohigas and Pato 2004].

Conjecture 24. *For $0 < \xi < 1$ we have*

$$E_\beta^{(\cdot)}(J; \xi) \underset{|J| \to \infty}{\sim} e^{\langle n_J \rangle \log(1-\xi)}, \qquad (4\text{-}2)$$

where $\langle n_J \rangle$ is given by (2-28) for $(\cdot) = $ bulk, soft and hard.

s	$r(s)$
1	1.0046735914726577
2	0.9998383226940526
3	0.9999753765440204
4	0.9999961026171116
5	0.9999991096965057
6	0.9999997235559452
7	0.9999998946139279
8	0.9999999537746553
9	0.9999999775313906
10	0.9999999881794448

Table 1. Tabulation of the ratio of the asymptotic to exact bulk gap probability for $\beta = 2$.

We see from (2-28) and (2-8), (2-12), (2-14) that as a function of s the decay exhibited by (4-2) is proportional to the square root of the leading decay of $E_\beta^{(\cdot)}$. A method to prove (4-2) for $\beta = 2$, making use of (3-12), has been given in [Pastur and Shcherbina 2011]. Alternatively, for this β, (4-2) can be verified by using known asymptotics of the Painlevé transcendent evaluations, as done for $(\cdot) =$ soft in [Bohigas et al. 2009].

An interesting feature of the asymptotic expansion of the relevant Painlevé transcendents with $0 < \xi < 1$ is that they contain oscillatory terms, in contrast to their asymptotic expansion with $\xi = 1$. It is indeed the case that oscillations can clearly be seen in plots of $(d/ds)E_2^{\text{soft}}((s, \infty); \xi)$ with $0 < \xi < 1$ [Bohigas et al. 2009]. Dyson [1995] has combined Coulomb gas and Painlevé theory to deduce the asymptotic form $E_2^{\text{bulk}}((0, s); \xi)$ when $\xi \to 1$ and simultaneously $s \to \infty$, which is shown to involve an elliptic theta function; for fixed ξ the asymptotic expansion of the relevant Painlevé transcendent [McCoy and Tang 1986] involves only trigonometric functions.

Acknowledgements

The assistance of Tomasz Dutka for carrying out the numerical work of Section 4.1 during the 2012 Vacation Scholarship program in the Department of Mathematics and Statistics at the University of Melbourne, and the assistance of Mark Sorrell in the preparation of the manuscript, is acknowledged. Thanks are due to the organizers of the MSRI program "Random matrices, interacting particle systems and integrable systems" for providing financial support and a stimulating environment. This research has been supported by the Australian Research Council.

References

[Baik et al. 2008] J. Baik, R. Buckingham, and J. DiFranco, "Asymptotics of Tracy–Widom distributions and the total integral of a Painlevé II function", *Comm. Math. Phys.* **280**:2 (2008), 463–497.

[Barnes 1904] E. W. Barnes, "The theory of the multiple gamma function", *Trans. Cam. Phil. Soc.* **19** (1904), 374–425.

[Basor et al. 1992] E. L. Basor, C. A. Tracy, and H. Widom, "Asymptotics of level-spacing distributions for random matrices", *Phys. Rev. Lett.* **69**:1 (1992), 5–8.

[Bohigas and Pato 2004] O. Bohigas and M. Pato, "Missing levels in correlated spectra", *Phys. Lett. B* **595** (2004), 171–176.

[Bohigas et al. 2009] O. Bohigas, J. X. de Carvalho, and M. P. Pato, "Deformations of the Tracy–Widom distribution", *Phys. Rev. E* (3) **79**:3 (2009), 031117, 6.

[Bornemann 2010a] F. Bornemann, "On the numerical evaluation of Fredholm determinants", *Math. Comp.* **79**:270 (2010), 871–915.

[Bornemann 2010b] F. Bornemann, "On the numerical evaluation of distributions in random matrix theory: a review", *Markov Process. Related Fields* **16**:4 (2010), 803–866.

[Borot and Nadal 2011] G. Borot and C. Nadal, "Right tail expansion of Tracy–Widom beta laws", *Random Matrices: Theory and Applications* **1**:3 (2011), 1250006.

[Borot et al. 2011] G. Borot, B. Eynard, S. N. Majumdar, and C. Nadal, "Large deviations of the maximal eigenvalue of random matrices", *J. Stat. Mech. Theory Exp.* 11 (2011), P11024, 56.

[Brini et al. 2011] A. Brini, M. Mariño, and S. Stevan, "The uses of the refined matrix model recursion", *J. Math. Phys.* **52**:5 (2011), 052305, 24.

[Chen and Manning 1994] Y. Chen and S. M. Manning, "Asymptotic level spacing of the Laguerre ensemble: a Coulomb fluid approach", *J. Phys. A* **27**:11 (1994), 3615–3620.

[Chen and Manning 1996] Y. Chen and S. M. Manning, "Some eigenvalue distribution functions of the Laguerre ensemble", *J. Phys. A* **29**:23 (1996), 7561–7579.

[Costin and Lebowitz 1995] O. Costin and J. L. Lebowitz, "Gaussian fluctuation in random matrices", *Phys. Rev. Lett.* **75**:1 (1995), 69–72.

[Dean and Majumdar 2006] D. S. Dean and S. N. Majumdar, "Large deviations of extreme eigenvalues of random matrices", *Phys. Rev. Lett.* **97**:16 (2006), 160201, 4.

[Dean and Majumdar 2008] D. S. Dean and S. N. Majumdar, "Extreme value statistics of eigenvalues of Gaussian random matrices", *Phys. Rev. E* (3) **77**:4 (2008), 041108, 12.

[Deift et al. 1997] P. A. Deift, A. R. Its, and X. Zhou, "A Riemann–Hilbert approach to asymptotic problems arising in the theory of random matrix models, and also in the theory of integrable statistical mechanics", *Ann. of Math.* (2) **146**:1 (1997), 149–235.

[Deift et al. 2008] P. Deift, A. Its, and I. Krasovsky, "Asymptotics of the Airy-kernel determinant", *Comm. Math. Phys.* **278**:3 (2008), 643–678.

[Deift et al. 2011] P. Deift, I. Krasovsky, and J. Vasilevska, "Asymptotics for a determinant with a confluent hypergeometric kernel", *Int. Math. Res. Not.* **2011**:9 (2011), 2117–2160.

[Desrosiers and Forrester 2006] P. Desrosiers and P. J. Forrester, "Hermite and Laguerre β-ensembles: asymptotic corrections to the eigenvalue density", *Nuclear Phys. B* **743**:3 (2006), 307–332.

[Dumaz and Virág 2011] L. Dumaz and B. Virág, "The right tail exponent of the Tracy–Widom β distribution", *Annales de l'Institut Henri Poincaré, Probabilités et Statistiques* **49**:4 (2011), 915–933.

[Dumitriu and Edelman 2002] I. Dumitriu and A. Edelman, "Matrix models for beta ensembles", *J. Math. Phys.* **43**:11 (2002), 5830–5847.

[Dyson 1962] F. J. Dyson, "Statistical theory of the energy levels of complex systems, II", *J. Mathematical Phys.* **3** (1962), 157–165.

[Dyson 1976] F. J. Dyson, "Fredholm determinants and inverse scattering problems", *Comm. Math. Phys.* **47**:2 (1976), 171–183.

[Dyson 1995] F. J. Dyson, "The Coulomb fluid and the fifth Painlevé transcendent", pp. 131–146 in *Chen Ning Yang: a great physicist of the 20th century*, edited by C. S. Liu and S.-T. Yau, Int. Press, Cambridge, MA, 1995.

[Ehrhardt 2006] T. Ehrhardt, "Dyson's constant in the asymptotics of the Fredholm determinant of the sine kernel", *Comm. Math. Phys.* **262**:2 (2006), 317–341.

[Ehrhardt 2010] T. Ehrhardt, "The asymptotics of a Bessel-kernel determinant which arises in random matrix theory", *Adv. Math.* **225**:6 (2010), 3088–3133.

[Fogler and Shklovskii 1995] M. M. Fogler and B. I. Shklovskii, "Probability of an eigenvalue number fluctuation in an interval of a random matrix spectrum", *Phys. Rev. Lett.* **74**:17 (1995), 3312–3315.

[Forrester 1992] P. J. Forrester, "Selberg correlation integrals and the $1/r^2$ quantum many-body system", *Nuclear Phys. B* **388**:3 (1992), 671–699.

[Forrester 1993] P. J. Forrester, "The spectrum edge of random matrix ensembles", *Nuclear Phys. B* **402**:3 (1993), 709–728.

[Forrester 1994] P. J. Forrester, "Exact results and universal asymptotics in the Laguerre random matrix ensemble", *J. Math. Phys.* **35**:5 (1994), 2539–2551.

[Forrester 2006] P. J. Forrester, "Hard and soft edge spacing distributions for random matrix ensembles with orthogonal and symplectic symmetry", *Nonlinearity* **19**:12 (2006), 2989–3002.

[Forrester 2009] P. J. Forrester, "A random matrix decimation procedure relating $\beta = 2/(r+1)$ to $\beta = 2(r+1)$", *Comm. Math. Phys.* **285**:2 (2009), 653–672.

[Forrester 2010] P. J. Forrester, *Log-gases and random matrices*, London Mathematical Society Monographs Series **34**, Princeton University Press, Princeton, NJ, 2010.

[Forrester 2012] P. J. Forrester, "Spectral density asymptotics for Gaussian and Laguerre β-ensembles in the exponentially small region", *J. Phys. A* **45**:7 (2012), 075206, 17.

[Forrester and Rains 2001] P. J. Forrester and E. M. Rains, "Interrelationships between orthogonal, unitary and symplectic matrix ensembles", pp. 171–207 in *Random matrix models and their applications*, edited by P. Bleher and A. Its, Math. Sci. Res. Inst. Publ. **40**, Cambridge Univ. Press, Cambridge, 2001.

[Forrester and Witte 2002] P. J. Forrester and N. S. Witte, "Application of the τ-function theory of Painlevé equations to random matrices: P_V, P_{III}, the LUE, JUE, and CUE", *Comm. Pure Appl. Math.* **55**:6 (2002), 679–727.

[Forrester and Witte 2012] P. J. Forrester and N. S. Witte, "Asymptotic forms for hard and soft edge general β conditional gap probabilities", *Nucl. Phys. B* **859** (2012), 321–340.

[Fuchs 1964] W. H. J. Fuchs, "On the eigenvalues of an integral equation arising in the theory of band-limited signals", *J. Math. Anal. Appl.* **9** (1964), 317–330.

[Gaudin 1961] M. Gaudin, "Sur la loi limite de l'espacement des valeurs propres d'une matrice aléatoire", *Nucl. Phys.* **25** (1961), 447–458.

[Katayama and Ohtsuki 1998] K. Katayama and M. Ohtsuki, "On the multiple gamma-functions", *Tokyo J. Math.* **21**:1 (1998), 159–182.

[Killip 2008] R. Killip, "Gaussian fluctuations for β ensembles", *Int. Math. Res. Not.* **2008**:8 (2008), Art. ID rnn007, 19.

[Killip and Stoiciu 2009] R. Killip and M. Stoiciu, "Eigenvalue statistics for CMV matrices: from Poisson to clock via random matrix ensembles", *Duke Math. J.* **146**:3 (2009), 361–399.

[Krasovsky 2004] I. V. Krasovsky, "Gap probability in the spectrum of random matrices and asymptotics of polynomials orthogonal on an arc of the unit circle", *Int. Math. Res. Not.* **2004**:25 (2004), 1249–1272.

[McCoy and Tang 1986] B. M. McCoy and S. Tang, "Connection formulae for Painlevé V functions", *Phys. D* **19**:1 (1986), 42–72.

[Mehta 1991] M. L. Mehta, *Random matrices*, 2nd ed., Academic Press, Boston, MA, 1991.

[Mehta and des Cloizeaux 1972] M. L. Mehta and J. des Cloizeaux, "The probabilities for several consecutive eigenvalues of a random matrix", *Indian J. Pure Appl. Math.* **3**:2 (1972), 329–351.

[Muirhead 1978] R. J. Muirhead, "Latent roots and matrix variates: a review of some asymptotic results", *Ann. Statist.* **6**:1 (1978), 5–33.

[Pastur and Shcherbina 2011] L. Pastur and M. Shcherbina, *Eigenvalue distribution of large random matrices*, Mathematical Surveys and Monographs **171**, American Mathematical Society, Providence, RI, 2011.

[Porter 1965] C. E. Porter, *Statistical theories of spectra: fluctuations*, Academic Press, New York, 1965.

[Quine et al. 1993] J. R. Quine, S. H. Heydari, and R. Y. Song, "Zeta regularized products", *Trans. Amer. Math. Soc.* **338**:1 (1993), 213–231.

[Ramírez and Rider 2009] J. A. Ramírez and B. Rider, "Diffusion at the random matrix hard edge", *Comm. Math. Phys.* **288**:3 (2009), 887–906.

[Ramírez et al. 2011a] J. A. Ramírez, B. Rider, and B. Virág, "Beta ensembles, stochastic Airy spectrum, and a diffusion", *J. Amer. Math. Soc.* **24**:4 (2011), 919–944.

[Ramírez et al. 2011b] J. A. Ramírez, B. Rider, and O. Zeitouni, "Hard edge tail asymptotics", *Electron. Commun. Probab.* **16** (2011), 741–752.

[Shintani 1980] T. Shintani, "A proof of the classical Kronecker limit formula", *Tokyo J. Math.* **3**:2 (1980), 191–199.

[Soshnikov 2000] A. Soshnikov, "Determinantal random point fields", *Uspekhi Mat. Nauk* **55**:5(335) (2000), 107–160. In Russian; translated in *Russian Math. Surveys* **55**:5 (2000), 923–975.

[Tracy and Widom 1993] C. A. Tracy and H. Widom, "Introduction to random matrices", pp. 103–130 in *Geometric and quantum aspects of integrable systems* (Scheveningen, 1992), edited by G. F. Helminck, Lecture Notes in Phys. **424**, Springer, Berlin, 1993.

[Tracy and Widom 1994a] C. A. Tracy and H. Widom, "Level-spacing distributions and the Airy kernel", *Comm. Math. Phys.* **159**:1 (1994), 151–174.

[Tracy and Widom 1994b] C. A. Tracy and H. Widom, "Level spacing distributions and the Bessel kernel", *Comm. Math. Phys.* **161**:2 (1994), 289–309.

[Valkó and Virág 2009] B. Valkó and B. Virág, "Continuum limits of random matrices and the Brownian carousel", *Invent. Math.* **177**:3 (2009), 463–508.

[Valkó and Virág 2010] B. Valkó and B. Virág, "Large gaps between random eigenvalues", *Ann. Probab.* **38**:3 (2010), 1263–1279.

[Ye 2006] L. Ye, *Numerical quadrature: theory and computation*, Master's thesis, Dalhousie University, 2006.

P.Forrester@ms.unimelb.edu.au

Department of Mathematics and Statistics, The University of Melbourne, 1-100 Grattan Street, Parkville 3010, Australia

Random Matrices
MSRI Publications
Volume **65**, 2014

Applications of random matrix theory for sensor array imaging with measurement noise

JOSSELIN GARNIER AND KNUT SØLNA

The imaging of a small target embedded in a medium is a central problem in sensor array imaging. The goal is to find a target embedded in a medium. The medium is probed by an array of sources, and the signals backscattered by the target are recorded by an array of receivers. The responses between all pairs of source and receiver are collected so that the available information takes the form of a response matrix. When the data are corrupted by additive measurement noise we show how tools of random matrix theory can help to detect, localize, and characterize the target.

1. Introduction

The imaging of a small target embedded in a medium is a central problem in wave sensor imaging [Angelsen 2000; Stergiopoulos 2001]. Sensor array imaging involves two steps. The first step is experimental, it consists in emitting waves from an array of sources and recording the backscattered signals by an array of receivers. The data set then consists of a matrix of recorded signals whose indices are the index of the source and the index of the receiver. The second step is numerical, it consists in processing the recorded data in order to estimate the quantities of interest in the medium, such as reflector locations. The main applications of sensor array imaging are medical imaging, geophysical exploration, and nondestructive testing.

Recently it has been shown that random matrix theory could be used in order to build a detection test based on the statistical properties of the singular values of the response matrix [Aubry and Derode 2009a; 2009b; 2010; Ammari et al. 2011; 2012]. This paper summarizes the results contained in [Ammari et al. 2011; 2012] and extends them into several important directions. First we address in this paper the case in which the source array and the receiver array are not coincident, and more generally the case in which the number of sources is different from the number of receivers. As a result the noise singular value distribution has the form of a deformed quarter circle and the statistics of the singular value

MSC2010: primary 78A46; secondary 15B52.

associated to the target is also affected. Second we study carefully the estimation of the noise variance of the response matrix. Different estimators are studied and an estimator that achieves an efficient trade-off between bias and variance is proposed. The use of this estimator instead of the empirical estimator used in the previous versions significantly improves the quality of the detection test based on the singular value distribution of the measured response matrix when the number of sensors is not very large. Third we propose an algorithm that can reconstruct not only the position of the target, but also its scattering amplitude. The estimator of the scattering amplitude compensates for the level repulsion of the singular value associated to the target due to the noise.

2. The response matrix

We address the case of a point reflector that can model a small dielectric anomaly in electromagnetism, a small density anomaly in acoustics, or more generally a local variation of the index of refraction in the scalar wave equation. We consider the case in which the contrast of the anomaly (its index of refraction relative to the one of the background medium) can be of order one but its diameter is assumed to be small compared to the wavelength. In such a situation it is possible to expand the solution of the wave equation around the background solution, as we explain below [Ammari and Kang 2004; Ammari et al. 2001; Ammari and Volkov 2005].

Let us consider the scalar wave equation in a d-dimensional homogeneous medium with the index of refraction n_0. The reference speed of propagation is denoted by c. We assume that the target is a small reflector or inclusion D with the index of refraction $n_{\text{ref}} \neq n_0$. The support of the inclusion is of the form $D = x_{\text{ref}} + B$, where B is a domain with small volume and x_{ref} is the location of the reflector. Therefore the scalar wave equation with the source $S(t, x)$ takes the form

$$\frac{n^2(x)}{c^2} \partial_t^2 E - \Delta_x E = S(t, x),$$

where the index of refraction is given by

$$n(x) = n_0 + (n_{\text{ref}} - n_0) 1_D(x).$$

In this paper we consider time-harmonic point sources emitting at frequency ω. For any y_n, z_m far from x_{ref} the field $\text{Re}(\hat{E}(y_n, z_m)e^{-i\omega t})$ observed at y_n when a point source emits a time-harmonic signal with frequency ω at z_m can be expanded as powers of the volume of the inclusion as

$$\hat{E}(y_n, z_m) = \hat{G}(y_n, z_m) + k_0^2 \rho_{\text{ref}} \hat{G}(y_n, x_{\text{ref}}) \hat{G}(x_{\text{ref}}, z_m) + O\big(|B|^{\frac{d+1}{d}}\big), \quad (1)$$

where $k_0 = n_0\omega/c$ is the homogeneous wavenumber, ρ_{ref} is the scattering amplitude

$$\rho_{\text{ref}} = \left(\frac{n_{\text{ref}}^2}{n_0^2} - 1\right)|B|, \tag{2}$$

and $\hat{G}(x, z)$ is the Green's function or fundamental solution of the Helmholtz equation with a point source at z:

$$\Delta_x \hat{G}(x, z) + k_0^2 \hat{G}(x, z) = -\delta(x - z). \tag{3}$$

More explicitly we have

$$\hat{G}(x, z) = \begin{cases} \dfrac{i}{4} H_0^{(1)}(k_0|x - z|) & \text{if } d = 2, \\[2mm] \dfrac{e^{ik_0|x-z|}}{4\pi|x - z|} & \text{if } d = 3, \end{cases}$$

where $H_0^{(1)}$ is the Hankel function of the first kind of order zero.

When there are M sources $(z_m)_{m=1,\ldots,M}$ and N receivers $(y_n)_{n=1,\ldots,N}$, the response matrix is the $N \times M$ matrix $A_0 = (A_{0,nm})_{n=1,\ldots,N,m=1,\ldots,M}$ defined by

$$A_{0,nm} := \hat{E}(y_n, z_m) - \hat{G}(y_n, z_m). \tag{4}$$

This matrix has rank one:

$$A_0 = \sigma_{\text{ref}} u_{\text{ref}} v_{\text{ref}}^\dagger, \tag{5}$$

where $\dagger$ stands for the conjugate transpose. The unique nonzero singular value of this matrix is

$$\sigma_{\text{ref}} = k_0^2 \rho_{\text{ref}} \left(\sum_{l=1}^N |\hat{G}(y_l, x_{\text{ref}})|^2\right)^{1/2} \left(\sum_{l=1}^M |\hat{G}(z_l, x_{\text{ref}})|^2\right)^{1/2}. \tag{6}$$

The associated left and right singular vectors u_{ref} and v_{ref} are given by

$$u_{\text{ref}} = u(x_{\text{ref}}), \qquad v_{\text{ref}} = v(x_{\text{ref}}), \tag{7}$$

where we have defined the normalized vectors of Green's functions:

$$u(x) = \left(\frac{\hat{G}(y_n, x)}{\left(\sum_{l=1}^N |\hat{G}(y_l, x)|^2\right)^{1/2}}\right)_{n=1,\ldots,N},$$

$$\tag{8}$$

$$v(x) = \left(\frac{\overline{\hat{G}(z_m, x)}}{\left(\sum_{l=1}^M |\hat{G}(z_l, x)|^2\right)^{1/2}}\right)_{m=1,\ldots,M}.$$

The matrix A_0 is the complete data set that can be collected. In practice the measured matrix is corrupted by electronic or measurement noise that has the form of an additive noise, with uncorrelated entries. The purpose of this paper is to address the classical imaging problems given the measured data set:

(1) Is there a target in the medium? This is the detection problem. In the absence of noise this question is trivial in that we can claim that there is a target buried in the medium as soon as the response matrix is not zero. In the presence of noise, it is not so obvious to answer this question since the response matrix is not zero due to additive noise even in the absence of a target. Our purpose is to build a detection test that has the maximal probability of detection for a given false alarm rate.

(2) Where is the target? This is the localization problem. Several methods can be proposed, essentially based on the back-propagation of the data set, and we will describe robust methods in the presence of noise.

(3) What are the characteristic properties of the target? This is the reconstruction problem. One may look after geometric and physical properties. In fact, in view of the expression (1)–(2), only the product of the volume of the inclusion times the contrast can be identified in the regime we address in this paper.

The paper is organized as follows. In Section 3 we explain how the data should be collected to minimize the impact of the additive noise. In Section 4 we give results about the distribution of the singular values of the response matrix, with special attention on the maximal singular value. In Section 5 we discuss how the noise level can be estimated with minimal bias and variance. In Section 6 we build a test for the detection of the target and in Section 7 we show how the position and the scattering amplitude of the target can be estimated.

3. Data acquisition

In this section we consider that there are M sources and N receivers. The measures are noisy, which means that the signal measured by a receiver is corrupted by an additive noise that can be described in terms of a circular complex Gaussian random variable with mean zero and variance σ_n^2. The recorded noises are independent from each other.

In the standard acquisition scheme, the response matrix is measured during a sequence of M experiments. In the m-th experience, $m = 1, \ldots, M$, the m-th source (located at z_m) generates a time-harmonic signal with unit amplitude and the N receivers (located at y_n, $n = 1, \ldots, N$) record the backscattered waves which means that they measure

$$A_{\text{meas},nm} = A_{0,nm} + W_{nm}, \quad n = 1, \ldots, N, \quad m = 1, \ldots, M,$$

which gives the matrix

$$A_{\text{meas}} = A_0 + W, \tag{9}$$

where A_0 is the unperturbed response matrix of rank one (4) and W_{nm} are independent complex Gaussian random variables with mean zero and variance σ_n^2.

The Hadamard technique is a noise reduction technique in the presence of additive noise that uses the structure of Hadamard matrices.

Definition 3.1. A complex Hadamard matrix H of order M is a $M \times M$ matrix whose elements are of modulus one and such that $H^\dagger H = MI$.

Complex Hadamard matrices exist for all M. For instance the Fourier matrix

$$H_{nm} = \exp\left[i 2\pi \frac{(n-1)(m-1)}{M} \right], \quad m, n = 1, \ldots, M, \tag{10}$$

is a complex Hadamard matrix. A Hadamard matrix has maximal determinant among matrices with complex entries in the closed unit disk. More exactly Hadamard [1893] proved that the determinant of any complex $M \times M$ matrix H with entries in the closed unit disk satisfies $|\det H| \leq M^{M/2}$, with equality attained by a complex Hadamard matrix.

We now describe a general multisource acquisition scheme and show the importance of Hadamard matrices to build an optimal scheme. Let H be an invertible $M \times M$ matrix with complex entries in the closed unit disk. In the multisource acquisition scheme, the response matrix is measured during a sequence of M experiments. In the m-th experience, $m = 1, \ldots, M$, all sources generate time-harmonic signals with unit amplitude or smaller, the m' source generating $H_{m'm}$ for $m' = 1, \ldots, M$. In other words, in the m-th experiment, we can use all the sources up to their maximal transmission power (which we have normalized to one) and we are free to choose their phases in order to minimize the effective noise level in the recorded data. In the m-th experiment, the N receivers record the backscattered waves, which means that they measure

$$B_{\text{meas},nm} = \sum_{m'=1}^{M} H_{m'm} A_{0,nm'} + W_{nm} = (A_0 H)_{nm} + W_{nm}, \quad n = 1, \ldots, N.$$

Collecting the recorded signals of the M experiments gives the matrix

$$B_{\text{meas}} = A_0 H + W,$$

where A_0 is the unperturbed response matrix and W_{nm} are independent complex Gaussian random variables with mean zero and variance σ_n^2. The measured response matrix A_{meas} is obtained by right multiplying the matrix B_{meas} by the matrix H^{-1}:

$$A_{\text{meas}} := B_{\text{meas}} H^{-1} = A_0 H H^{-1} + W H^{-1}, \tag{11}$$

so that we get the unperturbed matrix A_0 up to a new noise

$$A_{\text{meas}} = A_0 + \tilde{W}, \qquad \tilde{W} = W H^{-1}. \tag{12}$$

The choice of the matrix H should fulfill the property that the new noise matrix $\tilde{W}$ has independent complex entries with Gaussian statistics, mean zero, and minimal variance. We have

$$\mathbb{E}\left[\overline{\tilde{W}_{nm}} \tilde{W}_{n'm'}\right] = \sum_{q,q'=1}^{M} \overline{(H^{-1})_{qm}} (H^{-1})_{q'm'} \mathbb{E}\left[\overline{W_{nq}} W_{n'q'}\right]$$
$$= \sigma_n^2 ((H^{-1})^\dagger H^{-1})_{mm'} \mathbf{1}_n(n').$$

This shows that we look for a complex matrix H with entries in the unit disk such that $(H^{-1})^\dagger H^{-1} = cI$ with a minimal c. This is equivalent to require that H is unitary and that $|\det H|$ is maximal. Using Hadamard result we know that the maximal determinant is $M^{M/2}$ and that a complex Hadamard matrix attains the maximum. Therefore a matrix H that minimizes the noise variance should be a Hadamard matrix, such as, for instance, the Fourier matrix (10). Note that, in the case of a linear array, the use of a Fourier matrix corresponds to an illumination in the form of plane waves with regularly sampled angles.

When the multisource acquisition scheme is used with a Hadamard technique, we have $H^{-1} = \frac{1}{M} H^\dagger$ and the new noise matrix $\tilde{W}$ in (12) has independent complex entries with Gaussian statistics, mean zero, and variance σ_n^2 / M:

$$\mathbb{E}\left[\overline{\tilde{W}_{nm}} \tilde{W}_{n'm'}\right] = \frac{\sigma_n^2}{M} \mathbf{1}_m(m') \mathbf{1}_n(n'). \tag{13}$$

This gain of a factor M in the signal-to-noise ratio is called the Hadamard advantage.

4. Singular value decomposition of the response matrix

Singular values of a noisy matrix. We consider here the situation in which the measured response matrix A_{meas} consists of independent noise coefficients with mean zero and variance σ_n^2 / M and the number of receivers is larger than the number of sources $N \geq M$. As shown in the previous section, this is the case when the response matrix is acquired with the Hadamard technique and there is no target in the medium.

We denote by $\sigma_1^{(M)} \geq \sigma_2^{(M)} \geq \sigma_3^{(M)} \geq \cdots \geq \sigma_M^{(M)}$ the singular values of the response matrix A_{meas} sorted by decreasing order and by $\Lambda^{(M)}$ the corresponding integrated density of states defined by

$$\Lambda^{(M)}([\sigma_u, \sigma_v]) := \frac{1}{M} \text{Card}\{l = 1, \ldots, M, \sigma_l^{(M)} \in [\sigma_u, \sigma_v]\} \quad \text{for } \sigma_u < \sigma_v. \tag{14}$$

For large N and M with $N/M = \gamma \geq 1$ fixed we have the following results which are classical in random matrix theory [Marchenko and Pastur 1967; Johnstone 2001; Capitaine et al. 2012].

Proposition 4.1. (a) *The random measure $\Lambda^{(M)}$ almost surely converges to the deterministic absolutely continuous measure Λ with compact support:*

$$\Lambda([\sigma_u, \sigma_v]) = \int_{\sigma_u}^{\sigma_v} \frac{1}{\sigma_n} \rho_\gamma \left(\frac{\sigma}{\sigma_n} \right) d\sigma, \quad 0 \leq \sigma_u \leq \sigma_v, \tag{15}$$

where ρ_γ is the deformed quarter-circle law given by

$$\rho_\gamma(x) = \begin{cases} \dfrac{1}{\pi x} \sqrt{((\gamma^{\frac{1}{2}}+1)^2 - x^2)(x^2 - (\gamma^{\frac{1}{2}}-1)^2)} & \text{if } \gamma^{\frac{1}{2}}-1 < x \leq \gamma^{\frac{1}{2}}+1, \\ 0 & \text{otherwise.} \end{cases} \tag{16}$$

(b) *The normalized l^2-norm of the singular values satisfies*

$$M \left[\frac{1}{M} \sum_{j=1}^{M} (\sigma_j^{(M)})^2 - \gamma \sigma_n^2 \right] \xrightarrow{M \to \infty} \sqrt{\gamma} \sigma_n^2 Z_0 \quad \text{in distribution,} \tag{17}$$

where Z_0 follows a Gaussian distribution with mean zero and variance one.

(c) *The maximal singular value satisfies*

$$M^{\frac{2}{3}} [\sigma_1^{(M)} - \sigma_n(\gamma^{\frac{1}{2}}+1)] \xrightarrow{M \to \infty} \frac{\sigma_n}{2}(1 + \gamma^{-\frac{1}{2}})^{\frac{1}{3}} Z_2 \quad \text{in distribution,} \tag{18}$$

where Z_2 follows a type-2 Tracy–Widom distribution.

The type-2 Tracy–Widom distribution has the cumulative distribution function Φ_{TW2} given by

$$\Phi_{\text{TW2}}(z) = \exp\left(-\int_z^\infty (x - z)\varphi^2(x) \, dx \right), \tag{19}$$

where $\varphi(x)$ is the solution of the Painlevé equation

$$\varphi''(x) = x\varphi(x) + 2\varphi(x)^3, \quad \varphi(x) \simeq \text{Ai}(x), \ x \to \infty, \tag{20}$$

Ai being the Airy function. The expectation and variance of Z_2 are

$$\mathbb{E}[Z_2] \simeq -1.77 \quad \text{and} \quad \text{Var}(Z_2) \simeq 0.81.$$

Detailed results about the Tracy–Widom distributions can be found in [Baik et al. 2008] and their numerical evaluations are addressed in [Bornemann 2010].

Singular values of the perturbed response matrix. The measured response matrix using the Hadamard technique in the presence of a target and in the presence of measurement noise is

$$A_{\text{meas}} = A_0 + W, \tag{21}$$

where A_0 is given by (4) and W has independent random complex entries with Gaussian statistics, mean zero and variance σ_n^2/M. We consider the critical and interesting regime in which the singular values of the unperturbed matrix are of the same order as the singular values of the noise, that is to say, σ_{ref} is of the same order of magnitude as σ_n. The following proposition shows that there is a phase transition:

(1) Either the noise level σ_n is smaller than the critical value $\gamma^{-1/4}\sigma_{\text{ref}}$ and then the maximal singular value of the perturbed response matrix is a perturbation of the nonzero singular value of the unperturbed response matrix; this perturbation has Gaussian statistics with a mean of order one and a variance of the order of $1/M$.

(2) Or the noise level σ_n is larger than the critical value $\gamma^{-1/4}\sigma_{\text{ref}}$ and then the nonzero singular value of the unperturbed response matrix is buried in the deformed quarter circle distribution of the pure noise matrix and the maximal singular value of the perturbed response matrix has a behavior similar to the pure noise case (described in Proposition 4.1).

Proposition 4.2. *In the regime $M \to \infty$:*

(a) *The normalized l^2-norm of the singular values satisfies*

$$M\left[\frac{1}{M}\sum_{j=1}^{M}(\sigma_j^{(M)})^2 - \gamma\sigma_n^2\right] \overset{M\to\infty}{\longrightarrow} \sigma_{\text{ref}}^2 + \sqrt{2\gamma}\sigma_n^2 Z_0 \quad \text{in distribution,} \tag{22}$$

where Z_0 follows a Gaussian distribution with mean zero and variance one.

(b1) *If $\sigma_{\text{ref}} < \gamma^{1/4}\sigma_n$, the maximal singular value satisfies*

$$\sigma_1^{(M)} \overset{M\to\infty}{\longrightarrow} \sigma_n(\gamma^{\frac{1}{2}} + 1) \quad \text{in probability.} \tag{23}$$

More exactly,

$$M^{\frac{2}{3}}\left[\sigma_1^{(M)} - \sigma_n\left(\gamma^{\frac{1}{2}}+1\right)\right] \xrightarrow{M\to\infty} \frac{1}{2}\sigma_n\left(1+\gamma^{-\frac{1}{2}}\right)^{\frac{1}{3}} Z_2 \quad \text{in distribution,} \quad (24)$$

where Z_2 follows a type-2 Tracy–Widom distribution.

(b2) If $\sigma_{\text{ref}} > \gamma^{1/4}\sigma_n$, then the maximal singular value satisfies

$$\sigma_1^{(M)} \xrightarrow{M\to\infty} \sigma_{\text{ref}}\left(1+\gamma\frac{\sigma_n^2}{\sigma_{\text{ref}}^2}\right)^{\frac{1}{2}}\left(1+\frac{\sigma_n^2}{\sigma_{\text{ref}}^2}\right)^{\frac{1}{2}} \quad \text{in probability.} \quad (25)$$

More exactly,

$$M^{\frac{1}{2}}\left[\sigma_1^{(M)} - \sigma_{\text{ref}}\left(1+\gamma\frac{\sigma_n^2}{\sigma_{\text{ref}}^2}\right)^{\frac{1}{2}}\left(1+\frac{\sigma_n^2}{\sigma_{\text{ref}}^2}\right)^{\frac{1}{2}}\right]$$
$$\xrightarrow{M\to\infty} \frac{\sigma_n}{2}\frac{\left(1-\gamma\sigma_n^4/\sigma_{\text{ref}}^4\right)^{\frac{1}{2}}\left(2+(1+\gamma)\sigma_n^2/\sigma_{\text{ref}}^2\right)^{\frac{1}{2}}}{\left(1+\gamma\sigma_n^2/\sigma_{\text{ref}}^2\right)^{\frac{1}{2}}\left(1+\sigma_n^2/\sigma_{\text{ref}}^2\right)^{\frac{1}{2}}} Z_0, \quad (26)$$

in distribution, where Z_0 follows a Gaussian distribution with mean zero and variance one.

These results are illustrated in Figure 1. Their proofs can be obtained from the method described in [Benaych-Georges and Nadakuditi 2011]. Extensions to heteroscedastic noise can also be obtained as in [Chapon et al. 2014]. Note that formula (26) seems to predict that the standard deviation of the maximal singular value cancels when $\sigma_{\text{ref}} \searrow \gamma^{1/4}\sigma_n$, but this is true only to the order $M^{-1/2}$, and in fact it becomes of order $M^{-2/3}$ (see Figure 1). Following [Baik et al. 2005] we can anticipate that there are interpolating distributions which appear when

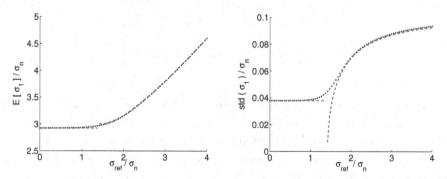

Figure 1. Mean and standard deviation of the maximal singular value. We compare the empirical means (left) and standard deviations (right) obtained from 10^4 MC simulations (blue dots) with the theoretical formulas given in Proposition 4.2 (red dashed lines). Here $N = 200$ and $M = 50$ ($\gamma = 4$).

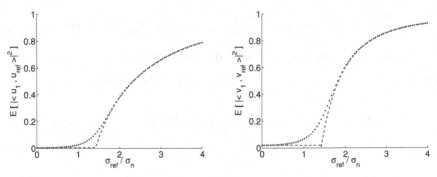

Figure 2. Means of the square angles between the perturbed and unperturbed singular vectors. We compare the empirical means obtained from 10^4 MC simulations (blue dots) with the theoretical formulas given in Proposition 4.3 (red dashed lines). Here $N = 200$ and $M = 50$ ($\gamma = 4$).

$\sigma_{\text{ref}} = \gamma^{1/4}\sigma_n + wM^{-1/3}$ for some fixed w. This problem deserves a detailed study.

Singular vectors of the perturbed response matrix. It is of interest to describe the statistical distribution of the angle between the left singular vector $u_1^{(M)}$ (resp. right singular vector $v_1^{(M)}$) of the noisy matrix A_{meas} and the left singular vector $u(x_{\text{ref}})$ (resp. right singular vector $v(x_{\text{ref}})$) of the unperturbed matrix A_0. This justifies the MUSIC-based algorithm for the target localization algorithm that we discuss in Section 7.

Proposition 4.3. *We consider the case $\sigma_{\text{ref}} > \gamma^{\frac{1}{4}}\sigma_n$. When $\gamma = N/M$ is fixed and $M \to \infty$, we have in probability*

$$\left|(u_1^{(M)})^\dagger u(x_{\text{ref}})\right|^2 \xrightarrow{M\to\infty} \frac{1 - \gamma\sigma_n^4/\sigma_{\text{ref}}^4}{1 + \gamma\sigma_n^2/\sigma_{\text{ref}}^2},$$

$$\left|(v_1^{(M)})^\dagger v(x_{\text{ref}})\right|^2 \xrightarrow{M\to\infty} \frac{1 - \gamma\sigma_n^4/\sigma_{\text{ref}}^4}{1 + \sigma_n^2/\sigma_{\text{ref}}^2}. \tag{27}$$

Proposition 4.3 shows that the first singular vectors of the perturbed matrix A_{meas} have deterministic angles with respect to the first singular vectors of the unperturbed matrix A_0 provided the first singular value emerges from the deformed quarter-circle distribution. These results are proved in [Benaych-Georges and Nadakuditi 2011] and they are illustrated in Figure 2.

5. The evaluation of the noise level

Empirical estimator. The truncated normalized l^2-norm of the singular values satisfies (22). Therefore the truncated normalized l^2-norm of the singular values satisfies

$$M\left[\frac{1}{M-(1+\gamma^{-\frac{1}{2}})^2}\sum_{j=2}^{M}(\sigma_j^{(M)})^2-\gamma\sigma_n^2\right]\overset{M\to\infty}{\longrightarrow}b_1+\sqrt{\gamma}\sigma_n^2 Z_0\quad\text{in distribution,}$$

where Z_0 follows a Gaussian distribution with mean zero and variance one, and the asymptotic bias is

$$b_1=\sigma_{\text{ref}}^2-\bar{\sigma}_1^2+\sigma_n^2(1+\gamma^{\frac{1}{2}})^2.\qquad(28)$$

Here

$$\bar{\sigma}_1=\max\left\{\sigma_{\text{ref}}\left(1+\gamma\frac{\sigma_n^2}{\sigma_{\text{ref}}^2}\right)^{\frac{1}{2}}\left(1+\frac{\sigma_n^2}{\sigma_{\text{ref}}^2}\right)^{\frac{1}{2}},\;\sigma_n(1+\gamma^{\frac{1}{2}})\right\}\qquad(29)$$

is the deterministic leading-order value of the maximal singular value as shown in Proposition 4.2. The normalization in the truncated l^2-norm has been chosen so that, in the absence of a target, the asymptotic bias is zero: $b_1\big|_{\sigma_{\text{ref}}=0}=0$. This implies that

$$\hat{\sigma}_n^e:=\gamma^{-\frac{1}{2}}\left[\frac{1}{M-(1+\gamma^{-\frac{1}{2}})^2}\sum_{j=2}^{M}(\sigma_j^{(M)})^2\right]^{\frac{1}{2}}\qquad(30)$$

is an empirical estimator of σ_n with Gaussian fluctuations of the order of M^{-1}. This estimator satisfies

$$M[\hat{\sigma}_n^e-\sigma_n]\overset{M\to\infty}{\longrightarrow}\frac{b_1}{2\gamma\sigma_n}+\frac{\sigma_n}{2\gamma^{\frac{1}{2}}}Z_0\quad\text{in distribution,}$$

and therefore

$$\hat{\sigma}_n^e=\sigma_n+o\left(\frac{1}{M^{\frac{2}{3}}}\right)\quad\text{in probability.}\qquad(31)$$

The empirical estimator is easy to compute, since it requires the evaluation of the Frobenius norm of the measured matrix A_{meas} and the maximal singular value:

$$\hat{\sigma}_n^e=\gamma^{-\frac{1}{2}}\left[\frac{\sum_{n=1}^{N}\sum_{m=1}^{M}|A_{nm}|^2-(\sigma_1^{(M)})^2}{M-(1+\gamma^{-\frac{1}{2}})^2}\right]^{\frac{1}{2}}.\qquad(32)$$

Corrected empirical estimator. It is possible to improve the quality of the estimation of the noise level and to cancel the bias of the empirical estimator. Using Proposition 4.2 we can see that the quantity

$$\hat{\sigma}_{\text{ref}}^e=\frac{\hat{\sigma}_n^e}{\sqrt{2}}\left\{\left(\frac{\sigma_1^{(M)}}{\hat{\sigma}_n^e}\right)^2-1-\gamma+\left(\left[\left(\frac{\sigma_1^{(M)}}{\hat{\sigma}_n^e}\right)^2-1-\gamma\right]^2-4\gamma\right)^{\frac{1}{2}}\right\}^{\frac{1}{2}}\qquad(33)$$

is an estimator of σ_{ref}, provided that $\sigma_{\text{ref}} > \gamma^{1/4}\sigma_n$. Therefore, when $\sigma_{\text{ref}} > \gamma^{1/4}\sigma_n$, it is possible to build an improved estimator of the noise variance by removing from the empirical estimator an estimation of the asymptotic bias which is itself based on the empirical estimator $\hat{\sigma}_n^e$. The estimator of the asymptotic bias that we propose to use is

$$\hat{b}_1^e = (\hat{\sigma}_{\text{ref}}^e)^2 - \left(\sigma_1^{(M)}\right)^2 + (\hat{\sigma}_n^e)^2(1 + \gamma^{\frac{1}{2}})^2, \tag{34}$$

and therefore we can propose the following estimator of the noise level σ_n:

$$\hat{\sigma}_n^c := \hat{\sigma}_n^e - \frac{\hat{b}_1^e}{2\gamma M \hat{\sigma}_n^e}. \tag{35}$$

This estimator satisfies

$$M[\hat{\sigma}_n^c - \sigma_n] \xrightarrow{M \to \infty} \frac{\sigma_n}{2\gamma^{\frac{1}{2}}} Z_0 \quad \text{in distribution.} \tag{36}$$

This estimator can only be used when $\hat{\sigma}_{\text{ref}}^e > \gamma^{1/4}\hat{\sigma}_n^e$ and it should then be preferred to the empirical estimator $\hat{\sigma}_n^e$.

Kolmogorov–Smirnov estimator. An alternative method to estimate σ_n is the approach outlined in [Györfi et al. 1996] and applied in [Shabalin and Nobel 2013], which consists in minimizing the Kolmogorov–Smirnov distance $\mathcal{D}(\sigma)$ between the observed sample distribution of the $M - K$ smallest singular values of the measured matrix A_{meas} and that predicted by theory, which is the deformed quarter circle distribution (16) parametrized by σ_n. Compared to [Györfi et al. 1996; Shabalin and Nobel 2013] we here introduce a cut-off parameter K that can be chosen by the user. All choices are equivalent in the asymptotic framework $M \to \infty$, but for finite M low values for K give estimators with small variances but with bias, while large values for K increase the variance but decay the bias (see Figure 3). We define the new estimator $\hat{\sigma}_n^K$ of σ_n as the parameter that minimizes the Kolmogorov–Smirnov distance. After elementary manipulations we find that the Kolmogorov–Smirnov estimator is of the form

$$\hat{\sigma}_n^K := \underset{\sigma > 0}{\operatorname{argmin}} \, \mathcal{D}_K^{(M)}(\sigma), \tag{37}$$

where $\mathcal{D}_K^{(M)}(\sigma)$ is defined by

$$\mathcal{D}_K^{(M)}(\sigma) := \max_{m=1,\dots,M-K} \left| G_\gamma\left(\frac{\sigma_{M+1-m}^{(M)}}{\sigma}\right) - \frac{m - 1/2}{M} \right| + \frac{1}{2M}, \tag{38}$$

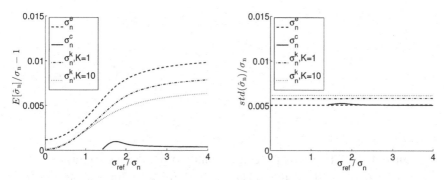

Figure 3. Relative bias (left) and standard deviations (right) of different estimators of the noise level. Here $N = 200$ and $M = 50$ ($\gamma = 4$).

G_γ is the cumulative distribution function with density (16):

$$
G_\gamma(x) = \begin{cases}
0 & \text{if } x \leq \gamma^{\frac{1}{2}} - 1, \\[2mm]
\begin{aligned}
& \frac{1}{2} + \frac{\gamma^{\frac{1}{2}}}{\pi}(1 - G(x)^2)^{\frac{1}{2}} - \frac{\gamma+1}{\pi} \arcsin G(x) \\
& - \frac{\gamma-1}{\pi} \arctan \frac{1 - (\gamma^{\frac{1}{2}} + \gamma^{-\frac{1}{2}})G(x)}{(1-G(x)^2)^{\frac{1}{2}}(\gamma^{\frac{1}{2}} - \gamma^{-\frac{1}{2}})}
\end{aligned} & \text{if } \gamma^{\frac{1}{2}} - 1 < x \leq \gamma^{\frac{1}{2}} + 1, \\[4mm]
1 & \text{if } \gamma^{\frac{1}{2}} + 1 < x,
\end{cases}
$$

with

$$
G(x) = \frac{(1+\gamma) - x^2}{2\gamma^{\frac{1}{2}}}.
$$

If $\gamma = 1$, we have

$$
G_1(x) = \begin{cases}
0 & \text{if } x \leq 0, \\[2mm]
\frac{1}{2\pi}\left(x\sqrt{4 - x^2} + 4 \arcsin \frac{x}{2}\right) & \text{if } 0 < x \leq 2, \\[2mm]
1 & \text{if } 2 < x.
\end{cases}
$$

Discussion. The three estimation methods described in the three previous subsections have been implemented and numerical results are reported in Figure 3. As predicted by the asymptotic theory, the variance of the empirical estimator is equivalent to the one of the corrected empirical estimator, and they are smaller than the ones of the Kolmogorov–Smirnov estimator. The bias of the empirical estimator is larger than the bias of the Kolmogorov–Smirnov estimator. The corrected empirical estimator has a very small bias. The variance of the

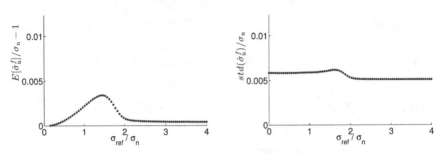

Figure 4. Relative bias (left) and standard deviations (right) of the final estimator (39) of the noise level. Here $N = 200$ and $M = 50$ ($\gamma = 4$).

Kolmogorov–Smirnov estimator increases with K, but its bias decreases with increasing K. From these observations we conclude:

• When $\hat{\sigma}^e_{\text{ref}} > \gamma^{1/4}\hat{\sigma}^e_n$, it is recommended to use the corrected empirical estimator (35). It is the one that has the minimal bias and the minimal variance amongst all the estimators studied in this paper, but it can only be applied in the regime when the singular value corresponding to the target is outside the deformed quarter-circle distribution of the noise singular values.

• When $\hat{\sigma}^e_{\text{ref}} < \gamma^{1/4}\hat{\sigma}^e_n$, it is recommended to use the Kolmogorov–Smirnov estimator (37) with $K = 1$. Although its variance is larger than the one of the empirical estimator, its bias is much smaller and, as a result, it is the one that has the minimal quadratic error (sum of the squared bias and of the variance).

To summarize, the estimator of the noise variance that we will use in the following is

$$\hat{\sigma}^f_n = \mathbf{1}_{\hat{\sigma}^e_{\text{ref}} \leq \gamma^{1/4}\hat{\sigma}^e_n} \hat{\sigma}^{K=1}_n + \mathbf{1}_{\hat{\sigma}^e_{\text{ref}} > \gamma^{1/4}\hat{\sigma}^e_n} \hat{\sigma}^c_n. \tag{39}$$

Its bias and standard deviation are plotted in Figure 4.

6. Detection test

Consider the response matrix in the presence of measurement noise:

$$A_{\text{meas}} = A_0 + W,$$

where A_0 is zero in the absence of a target and equal to (4) when there is a target. The matrix W models additive measurement noise and its complex entries are independent and identically distributed with Gaussian statistics, mean zero and variance σ^2_n/M.

The objective is to propose a detection test for the target. Since we know that the presence of a target is characterized by the existence of a significant singular value, we propose to use a test of the form $R > r$ for the alarm corresponding to the presence of a target. Here R is the quantity obtained from the measured response matrix and defined by

$$R = \frac{\sigma_1^{(M)}}{\hat{\sigma}_n}, \tag{40}$$

where $\hat{\sigma}_n$ is the known value of σ_n, if known, or the estimator (39) of σ_n. Here the threshold value r of the test has to be chosen by the user. This choice follows from Neyman–Pearson theory as we explain below. It requires the knowledge of the statistical distribution of R which we give in the following proposition, which is a corollary of Proposition 4.2 (and Slutsky's theorem).

Proposition 6.1. *In the asymptotic regime $M \gg 1$ the following statements hold.*

(a) *In the absence of a target we have up to a term of order $o(M^{-2/3})$:*

$$R \simeq 1 + \gamma^{\frac{1}{2}} + \frac{1}{2M^{\frac{2}{3}}}(1 + \gamma^{-\frac{1}{2}})^{\frac{1}{3}} Z_2, \tag{41}$$

where Z_2 follows a type-2 Tracy–Widom distribution.

(b) *In presence of a target:*

(b1) *If $\sigma_{\text{ref}} > \gamma^{1/4}\sigma_n$, then we have, up to a term of order $o(M^{-1/2})$,*

$$R \simeq \frac{\sigma_{\text{ref}}}{\sigma_n}\left(1 + \gamma \frac{\sigma_n^2}{\sigma_{\text{ref}}^2}\right)^{\frac{1}{2}}\left(1 + \frac{\sigma_n^2}{\sigma_{\text{ref}}^2}\right)^{\frac{1}{2}}$$

$$+ \frac{1}{2M^{\frac{1}{2}}}\left(\frac{\left(1 - \gamma\sigma_n^4/\sigma_{\text{ref}}^4\right)\left(2 + (1+\gamma)\sigma_n^2/\sigma_{\text{ref}}^2\right)}{\left(1 + \gamma\sigma_n^2/\sigma_{\text{ref}}^2\right)\left(1 + \sigma_n^2/\sigma_{\text{ref}}^2\right)^{\frac{1}{2}}}\right)^{\frac{1}{2}} Z_0, \tag{42}$$

where Z_0 follows a Gaussian distribution with mean zero and variance one.

(b2) *If $\sigma_{\text{ref}} < \gamma^{1/4}\sigma_n$, then we have (41).*

The data (i.e., the measured response matrix) gives the value of the ratio R. We propose to use a test of the form $R > r$ for the alarm corresponding to presence of a target. The quality of this test can be quantified by two coefficients:

(1) The false alarm rate (FAR) is the probability to sound the alarm while there is no target:

$$\text{FAR} = \mathbb{P}(R > r_\alpha \mid \text{no target}).$$

(2) The probability of detection (POD) is the probability to sound the alarm when there is a target:

$$\text{POD} = \mathbb{P}(R > r_\alpha \mid \text{target}).$$

As is well-known in statistical test theory, it is not possible to find a test that minimizes the FAR and maximizes the POD. However, by the Neyman–Pearson lemma, the decision rule of sounding the alarm if and only if $R > r_\alpha$ maximizes the POD for a given FAR α, provided the threshold is taken to be equal to

$$r_\alpha = 1 + \gamma^{\frac{1}{2}} + \frac{1}{2M^{\frac{2}{3}}}(1 + \gamma^{-\frac{1}{2}})^{\frac{1}{3}} \Phi_{TW2}^{-1}(1 - \alpha), \tag{43}$$

where Φ_{TW2} is the cumulative distribution function (19) of the Tracy–Widom distribution of type 2. The computation of the threshold r_α is easy since it depends only on the number of sensors N and M and on the FAR α. We have, for instance,

$$\Phi_{TW2}^{-1}(0.9) \simeq -0.60, \quad \Phi_{TW2}^{-1}(0.95) \simeq -0.23, \quad \Phi_{TW2}^{-1}(0.99) \simeq 0.48.$$

These values are used in the detection tests whose POD are plotted in Figure 5.

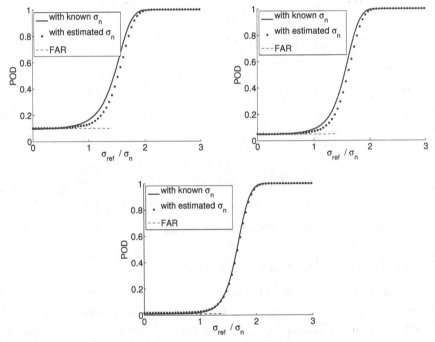

Figure 5. Probability of detection (POD) for the detection test calibrated with the threshold values r_α with $\alpha = 0.1$ (left), $\alpha = 0.05$ (right), and $\alpha = 0.01$ (bottom). Here $N = 200$ and $M = 50$. The blue solid and dotted lines correspond to the results of 10^4 MC simulations, in which the noise level is known (thick solid lines) or estimated by (39) (thick dotted lines). The dashed lines are the FAR desired in the absence of a target, which should be obtained when $\sigma_{ref} = 0$.

The POD of this optimal test (optimal amongst all tests with the FAR α) depends on the value σ_{ref} and on the noise level σ_n. The theoretical test performance improves very rapidly with M when $\sigma_{\text{ref}} > \gamma^{1/4}\sigma_n$. When $\sigma_{\text{ref}} < \gamma^{1/4}\sigma_n$, so that the target is buried in noise (more exactly, the singular value corresponding to the target is buried in the deformed quarter-circle distribution of the other singular values), then we have POD $= 1 - \Phi_{\text{TW2}}(\Phi_{\text{TW2}}^{-1}(1-\alpha)) = \alpha$.

The POD of the test (40) calibrated for different values of the FAR is plotted in Figure 5. One can observe that the calibration with r_α gives the desired FAR and that the POD rapidly goes to one when the singular value σ_{ref} of the target becomes larger than $\gamma^{1/4}\sigma_n$. Furthermore, the use of the estimator (39) for the noise level σ_n is also very efficient in that we get almost the same FAR and POD with the true value σ_n as with the estimator $\hat{\sigma}_n^f$. In Figure 6 we plot the POD obtained with other estimators of the noise level in order to confirm that the estimator $\hat{\sigma}_n^f$ defined by (39) is indeed the most appropriate.

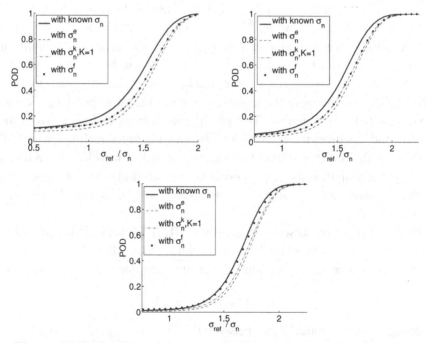

Figure 6. Probability of detection (POD) for the detection test calibrated with the threshold values r_α with $\alpha = 0.1$ (left), $\alpha = 0.05$ (right), and $\alpha = 0.01$ (bottom). Here $N = 200$ and $M = 50$. The blue lines correspond to the results of 10^4 MC simulations, in which the noise level is known (thick solid lines) or estimated by (39) (thick dotted lines) or estimated by the estimators (30) and (37) (thin dashed lines).

7. Target localization and reconstruction

In this section we would like to present simple and robust way to localize the target and reconstruct its properties once the detection test has passed. By simple we mean that we will only use the first singular value and left singular vector of the response matrix, and by robust we mean a procedure that allows for estimations with bias and variance as small as possible.

Localization. A standard imaging functional is the MUSIC functional defined by [Ammari et al. 2011]:

$$\mathcal{I}_{\text{MUSIC}}(x) := \left\| u(x) - \left(u_1^{(M)\dagger} u(x) \right) u_1^{(M)} \right\|^{-\frac{1}{2}} = \left(1 - \left| u(x)^\dagger u_1^{(M)} \right|^2 \right)^{-\frac{1}{2}},$$

where $u(x)$ is the normalized vector of Green's functions (8) and $u_1^{(M)}$ is the first left singular vector of the measured response matrix A_{meas}. The MUSIC functional is the projection of the Green's vector $u(x)$ from the receiver array to the search point x onto the noise space of the measured response matrix.

In the absence of noise the MUSIC functional presents a peak at $x = x_{\text{ref}}$. Indeed, in this case, we have $u_1^{(M)} = u(x_{\text{ref}})$ (up to a phase term) and therefore $\mathcal{I}_{\text{MUSIC}}(x)$ becomes singular at $x = x_{\text{ref}}$. Furthermore, we can quantify the accuracy of the reflector localization as follows. When the arrays surround the search region, the singular vectors $u(x)$ can be shown to be orthogonal to $u(x_{\text{ref}})$ when the distance between the search point x and the target point x_{ref} becomes larger than half a wavelength. This can be shown using Helmholtz–Kirchhoff identity and this gives the resolution of the imaging functional: one can get the position of the reflector within the accuracy of half a wavelength. When the arrays are partial, then the accuracy can be described in terms of the so-called Rayleigh resolution formulas [Elmore and Heald 1969; Garnier and Papanicolaou 2010].

In the presence of noise the peak of the MUSIC functional is affected. By Proposition 4.3, in the regime $M \gg 1$, the value of the MUSIC functional at an arbitrary point $x \neq x_{\text{ref}}$ is one while the theoretical value at $x = x_{\text{ref}}$ is given by

$$\mathcal{I}_{\text{MUSIC}}(x_{\text{ref}}) = (1 - c_u)^{-\frac{1}{2}},$$

where c_u is the theoretical angle between the first left singular vector $u(x_{\text{ref}})$ of the unperturbed matrix A_0 and the first left singular vector $u_1^{(M)}$ of the measured response matrix A_{meas}:

$$c_u = \begin{cases} \dfrac{1 - \gamma \sigma_n^4 / \sigma_{\text{ref}}^4}{1 + \gamma \sigma_n^2 / \sigma_{\text{ref}}^2} & \text{if } \sigma_{\text{ref}} > \gamma^{\frac{1}{4}} \sigma_n, \\[2mm] 0 & \text{if } \sigma_{\text{ref}} < \gamma^{\frac{1}{4}} \sigma_n. \end{cases}$$

Therefore, provided the detection test has passed, which means that the target singular value is larger than the noise singular values, the MUSIC algorithm gives a robust and simple way to estimate the position of the reflector. The estimator of x_{ref} that we propose is

$$\hat{x}_{\text{ref}} := \underset{x}{\text{argmax}} \, \mathscr{I}_{\text{MUSIC}}(x). \tag{44}$$

Note that more complex and computationally expensive algorithms (using reverse-time migration) can improve the quality of the estimation as shown in [Ammari et al. 2012].

Reconstruction. Using Proposition 4.2 we can see that the quantity

$$\hat{\sigma}_{\text{ref}} = \frac{\hat{\sigma}_n}{\sqrt{2}} \left\{ \left(\frac{\sigma_1^{(M)}}{\hat{\sigma}_n} \right)^2 - 1 - \gamma + \left(\left[\left(\frac{\sigma_1^{(M)}}{\hat{\sigma}_n} \right)^2 - 1 - \gamma \right]^2 - 4\gamma \right)^{\frac{1}{2}} \right\}^{\frac{1}{2}} \tag{45}$$

is an estimator of σ_{ref}, provided that $\sigma_{\text{ref}} > \gamma^{\frac{1}{4}} \sigma_n$. Here $\hat{\sigma}_n$ is the known value of σ_n, if known, or the estimator (39) of σ_n. In practice, if the detection test passes, then this implies that we are in this case. From (6) we can therefore estimate the scattering amplitude ρ_{ref} of the inclusion by

$$\hat{\rho}_{\text{ref}} = \frac{c_0^2}{\omega^2} \left(\sum_{n=1}^{N} |\hat{G}(\omega, \hat{x}_{\text{ref}}, y_n)|^2 \right)^{-\frac{1}{2}} \left(\sum_{m=1}^{M} |\hat{G}(\omega, \hat{x}_{\text{ref}}, z_m)|^2 \right)^{-\frac{1}{2}} \hat{\sigma}_{\text{ref}}, \tag{46}$$

with $\hat{\sigma}_{\text{ref}}$ the estimator (45) of σ_{ref} and $\hat{x}_{\text{ref}}$ is the estimator (44) of the position of the inclusion. This estimator is not biased asymptotically because it compensates for the level repulsion of the first singular value due to the noise.

Numerical simulations. We consider the following numerical set-up: the wavelength is equal to one. There is one reflector with scattering amplitude $\rho_{\text{ref}} = 1$, located at $x_{\text{ref}} = (0, 0, 50)$. We consider a linear array of $N = 200$ transducers located half a wavelength apart on the line from $(-50, 0, 0)$ to $(50, 0, 0)$. Each transducer is used as a receiver, but only one of four is used as a source (therefore, $M = 50$ and $\gamma = 4$). The noise level is $\sigma_n = \sigma_{\text{ref}}/4$ or $\sigma_{\text{ref}}/2$, where σ_{ref} is the singular value associated to the reflector (given by (6)).

We have carried out a series of 10^4 MC simulations (using the estimator (39) of σ_n). The results are reported in Figure 7 for $\sigma_n = \sigma_{\text{ref}}/4$ and in Figure 8 for $\sigma_n = \sigma_{\text{ref}}/2$:

(1) The reflector is always detected when $\sigma_n = \sigma_{\text{ref}}/4$ and it is detected with probability 97% when $\sigma_n = \sigma_{\text{ref}}/2$ (in agreement with the POD plotted in

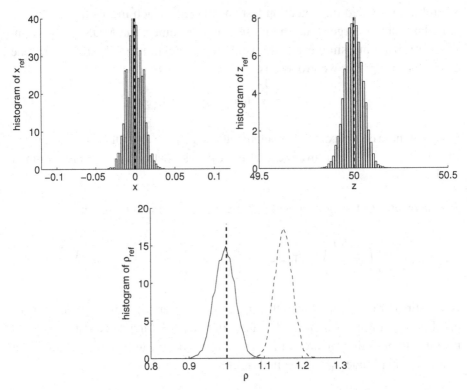

Figure 7. Top: histograms of the estimated cross-range position $\hat{x}_{ref}$ (left) and estimated range position $\hat{z}_{ref}$ (right) given by (44). Bottom: histogram of the estimated scattering amplitude $\hat{\rho}_{ref}$ given by (46) (solid lines) or $\hat{\rho}^e_{ref}$ given by (47) (dashed lines). Here $\sigma_n = \sigma_{ref}/4$.

Figure 5).

(2) The estimator $\hat{x}_{ref}$ defined by (44) of the position of the reflector has good properties. The histograms of the estimated positions $\hat{x}_{ref} = (\hat{x}_{ref}, 0, \hat{z}_{ref})$ are plotted in the top row of Figures 7 and 8.

(3) The estimator $\hat{\rho}_{ref}$ defined by (46) of the scattering amplitude has no bias because it uses the inversion formula (45) which compensates for the level repulsion of the first singular value. We plot in the bottom row of Figures 7 and 8 the histogram of the estimated scattering amplitude and we compare with the empirical estimator

$$\hat{\rho}^e_{ref} = \frac{c_0^2}{\omega^2} \left(\sum_{n=1}^{N} |\hat{G}(\omega, \hat{x}_{ref}, y_n)|^2 \right)^{-\frac{1}{2}} \left(\sum_{m=1}^{M} |\hat{G}(\omega, \hat{x}_{ref}, z_m)|^2 \right)^{-\frac{1}{2}} \sigma_1^{(M)}, \tag{47}$$

which has a large bias.

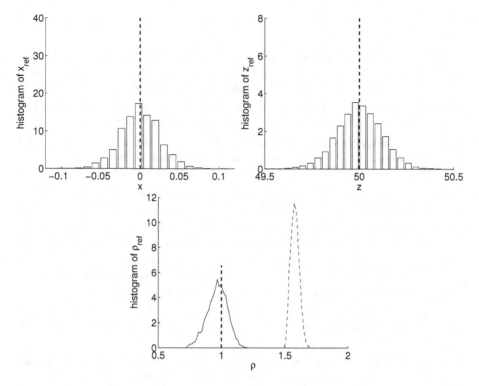

Figure 8. The same as in Figure 7, but here $\sigma_n = \sigma_{\mathrm{ref}}/2$.

8. Conclusion

In this paper we have presented a few results that show how random matrix theory can be used in sensor array imaging. It turns out that most of the needed results are already available in the literature in the case addressed in this paper, that is, when the response matrix is perturbed by an additive measurement noise. However, the most interesting questions arise in the presence of clutter noise, which is the case in which the data are corrupted by perturbations due to random heterogeneities present in the medium. In this case the random perturbation of the response matrix cannot be described in terms of an additive uncorrelated noise, but it has special correlation structure [Aubry and Derode 2009a; Fouque et al. 2007]. This case certainly deserves more attention and more work.

References

[Ammari and Kang 2004] H. Ammari and H. Kang, *Reconstruction of small inhomogeneities from boundary measurements*, Lecture Notes in Mathematics **1846**, Springer, Berlin, 2004.

[Ammari and Volkov 2005] H. Ammari and D. Volkov, "The leading-order term in the asymptotic expansion of the scattering amplitude of a collection of finite number of dielectric inhomogeneities of small diameter", *International Journal for Multiscale Computational Engineering* **3** (2005), 149–160.

[Ammari et al. 2001] H. Ammari, M. S. Vogelius, and D. Volkov, "Asymptotic formulas for perturbations in the electromagnetic fields due to the presence of inhomogeneities of small diameter, II: The full Maxwell equations", *J. Math. Pures Appl.* (9) **80**:8 (2001), 769–814.

[Ammari et al. 2011] H. Ammari, J. Garnier, H. Kang, W.-K. Park, and K. Sølna, "Imaging schemes for perfectly conducting cracks", *SIAM J. Appl. Math.* **71**:1 (2011), 68–91.

[Ammari et al. 2012] H. Ammari, J. Garnier, and K. Sølna, "A statistical approach to target detection and localization in the presence of noise", *Waves Random Complex Media* **22**:1 (2012), 40–65.

[Angelsen 2000] B. Angelsen, *Ultrasound imaging: waves, signals and signal processing*, Emantec, Trondheim, 2000.

[Aubry and Derode 2009a] A. Aubry and A. Derode, "Detection and imaging in a random medium: A matrix method to overcome multiple scattering and aberration", *J. Appl. Physics* **106** (2009), 044903.

[Aubry and Derode 2009b] A. Aubry and A. Derode, "Random matrix theory applied to acoustic backscattering and imaging in complex media", *Phys. Rev. Lett.* **102** (2009), 084301.

[Aubry and Derode 2010] A. Aubry and A. Derode, "Singular value distribution of the propagation matrix in random scattering media", *Waves in Random and Complex Media* **20**:3 (2010), 333–363.

[Baik et al. 2005] J. Baik, G. Ben Arous, and S. Péché, "Phase transition of the largest eigenvalue for nonnull complex sample covariance matrices", *Ann. Probab.* **33**:5 (2005), 1643–1697.

[Baik et al. 2008] J. Baik, R. Buckingham, and J. DiFranco, "Asymptotics of Tracy–Widom distributions and the total integral of a Painlevé II function", *Comm. Math. Phys.* **280**:2 (2008), 463–497.

[Benaych-Georges and Nadakuditi 2011] F. Benaych-Georges and R. R. Nadakuditi, "The eigenvalues and eigenvectors of finite, low rank perturbations of large random matrices", *Adv. Math.* **227**:1 (2011), 494–521.

[Bornemann 2010] F. Bornemann, "On the numerical evaluation of distributions in random matrix theory: a review", *Markov Process. Related Fields* **16**:4 (2010), 803–866.

[Capitaine et al. 2012] M. Capitaine, C. Donati-Martin, and D. Féral, "Central limit theorems for eigenvalues of deformations of Wigner matrices", *Ann. Inst. Henri Poincaré Probab. Stat.* **48**:1 (2012), 107–133.

[Chapon et al. 2014] F. Chapon, R. Couillet, W. Hachem, and X. Mestre, "The outliers among the singular values of large rectangular random matrices with additive fixed rank deformation", *Markov Processes and Related Fields* (2014). Accepted for publication. arXiv 1207.0471

[Elmore and Heald 1969] W. Elmore and M. Heald, *Physics of waves*, Dover, New York, 1969.

[Fouque et al. 2007] J.-P. Fouque, J. Garnier, G. Papanicolaou, and K. Sølna, *Wave propagation and time reversal in randomly layered media*, Stochastic Modelling and Applied Probability **56**, Springer, New York, 2007.

[Garnier and Papanicolaou 2010] J. Garnier and G. Papanicolaou, "Resolution analysis for imaging with noise", *Inverse Problems* **26**:7 (2010), 074001, 22.

[Györfi et al. 1996] L. Györfi, I. Vajda, and E. van der Meulen, "Minimum Kolmogorov distance estimates of parameters and parametrized distributions", *Metrika* **43**:3 (1996), 237–255.

[Hadamard 1893] J. Hadamard, "Résolution d'une question relative aux déterminants", *Bull. Sci. Math.* **17** (1893), 30–31.

[Johnstone 2001] I. M. Johnstone, "On the distribution of the largest eigenvalue in principal components analysis", *Ann. Statist.* **29**:2 (2001), 295–327.

[Marchenko and Pastur 1967] V. A. Marchenko and L. A. Pastur, "Distribution of eigenvalues in some sets of random matrices", *Mat. Sb. (N.S.)* **72 (114)** (1967), 507–536. In Russian. Translated in *Math. USSR Sbornik* **1** (1967), 457–483.

[Shabalin and Nobel 2013] A. A. Shabalin and A. B. Nobel, "Reconstruction of a low-rank matrix in the presence of Gaussian noise", *J. Multivariate Anal.* **118** (2013), 67–76.

[Stergiopoulos 2001] S. Stergiopoulos, *Advanced signal processing handbook: theory and implementation for radar, sonar, and medical imaging real-time systems*, CRC Press, Boca Raton, FL, 2001.

Laboratoire de Probabilités et Modèles Aléatoires & Laboratoire Jacques-Louis Lions, Université Paris VII, 75205 Paris Cedex 13, France
garnier@math.univ-paris-diderot.fr

Department of Mathematics, University of California, Irvine, CA 92697, United States
ksolna@math.uci.edu

Random Matrices
MSRI Publications
Volume **65**, 2014

Convolution symmetries of integrable hierarchies, matrix models and τ-functions

J. HARNAD AND A. YU. ORLOV

Generalized convolution symmetries of integrable hierarchies of KP and 2D-Toda type act diagonally on the Hilbert space $\mathcal{H} = L^2(S^1)$ in the standard monomial basis. The induced transformations on the Hilbert space Grassmannian $\mathrm{Gr}_{\mathcal{H}_+}(\mathcal{H})$ may be viewed as symmetries of these hierarchies, acting upon the Sato–Segal–Wilson τ-functions, and thereby generating new solutions of the hierarchies. The corresponding transformations of the associated fermionic Fock space are also diagonal in the standard orthonormal basis, labeled by integer partitions. The Plücker coordinates of the image under the Plücker map of the element $W \in \mathrm{Gr}_{\mathcal{H}_+}(\mathcal{H})$ defining the initial point under the commuting KP flows are the coefficients in the single and double Schur function expansions of the associated τ-functions. These are therefore multiplied by the eigenvalues of the convolution action in the fermionic representation. Applying such transformations to standard matrix model integrals, we obtain new matrix models of externally coupled type whose partition functions are thus also seen to be KP or 2D-Toda τ-functions. More general multiple integral representations of tau functions are similarly obtained, as well as finite determinantal expressions for them.

1. Introduction: convolution symmetries of τ-functions

Solutions of integrable hierarchies of KP and 2D-Toda type are determined by their τ-functions [Sato 1981; Sato and Sato 1983; Segal and Wilson 1985]. Infinite sequences of such KP τ-functions $\{\tau(N,t)\}_{N\in\mathbb{Z}}$, depending on the infinite set of commuting flow parameters $t = (t_1, t_2, \dots)$ and an integer lattice label N, may be associated in a standard fashion (see references just cited) to elements of a "universal phase" space, viewed as an infinite Grassmann manifold or flag manifold. These satisfy the Hirota bilinear equations of the KP hierarchy

Harnad's work is supported by the Natural Sciences and Engineering Research Council of Canada (NSERC) and the Fonds Québecois de la recherche sur la nature et les technologies (FQRNT). Orlov's work is supported by joint Russian Foundation for Basic Research (RFBR) Consortium EINSTEIN grants 06-01-92054 and 05-01-00498, and by RAS Program "Fundamental Methods in Nonlinear Physics".

and also, in certain cases (e.g., exponential flows of matrix model integrals induced by trace invariants), the equations of the Toda lattice hierarchy.

The τ-functions may be expanded as infinite series in a basis of Schur functions $s_\lambda(t)$, labelled by integer partitions $\lambda = (\lambda_1 \geq \lambda_2 \geq \cdots \geq 0)$

$$\tau(N, t) = \sum_\lambda \pi_N(\lambda) s_\lambda(t). \tag{1-1}$$

In the approach of Sato and Segal–Wilson [Sato 1981; Segal and Wilson 1985], the coefficients $\pi_N(\lambda)$ are interpreted as Plücker coordinates of the image $\mathcal{P}(W)$ of an element W of a Hilbert space Grassmannian $\mathrm{Gr}_{\mathcal{H}_+}(\mathcal{H})$ under the Plücker map

$$\mathcal{P} : \mathrm{Gr}_{\mathcal{H}_+}(\mathcal{H}) \rightarrow \mathbb{P}(\mathcal{F}) \tag{1-2}$$

into the projectivization of the semiinfinite exterior space $\mathcal{F} := \Lambda\mathcal{H}$ (the Fermionic Fock space). In [Segal and Wilson 1985], the Hilbert pace $\mathcal{H}$ is chosen as the square integrable functions $L^2(S^1)$ on the unit circle in the complex z-plane and the elements of $\mathrm{Gr}_{\mathcal{H}_+}(\mathcal{H})$ are subspaces of $\mathcal{H} = L^2(S^1)$ that are "commensurable" with the subspace $\mathcal{H}_+ \subset \mathcal{H}$ of functions admitting a holomorphic extension to the interior disk.

The image $\mathcal{P}(\mathrm{Gr}_{\mathcal{H}_+}(\mathcal{H}))$ of the Grassmannian under the Plücker map consists of all decomposable elements of $\Lambda\mathcal{H}$, which is the intersection of the infinite set of quadrics defined by the Plücker relations. The latter are equivalent to the infinite set of Hirota bilinear differential relations [Jimbo and Miwa 1983; Sato 1981; Sato and Sato 1983] for $\tau(N, t)$, which are the defining property of τ-functions. Through the Sato formula for the Baker–Akhiezer function

$$\Psi_N(z, t) = e^{\sum_{i=1}^\infty t_i z^i} \frac{\tau(N, t - [z^{-1}])}{\tau(N, t)}, \quad [z^{-1}] := (z^{-1}, 2z^{-2}, 3z^{-3}, \ldots), \tag{1-3}$$

these equations are equivalent to the KP hierarchy and their associated Lax equations.

The 2D-Toda hierarchy [Jimbo and Miwa 1983; Ueno and Takasaki 1984] can similarly be expressed in terms of τ-functions depending on N, t and a further infinite sequence of flow parameters $\tilde{t} = (\tilde{t}_1, \tilde{t}_2, \ldots)$. These admit double Schur function expansions [Takasaki 1984]

$$\tau^{(2)}(N, t, \tilde{t}) = \sum_\lambda \sum_\mu B_N(\lambda, \mu) s_\lambda(t) s_\mu(\tilde{t}), \tag{1-4}$$

in which the coefficients $B_N(\lambda, \mu)$ have a similar interpretation in terms of Plücker coordinates. They also satisfy an infinite set of bilinear differential Hirota-type relations in both sequences of flow variables and difference-differential equations relating different lattice points. For fixed N, they include the KP Hirota equations of the KP hierarchy in each of the two sets of flow variables, so we refer to them as 2D-Toda tau functions.

Starting with any given τ-function of KP-Toda or 2D-Toda type, it will be shown in the following that new τ-functions can be constructed, satisfying the same sets of bilinear relations, having the following Schur function expansions:

$$\tilde{C}_\rho(\tau)(N, t) = \sum_\lambda r_\lambda(N) \pi_N(\lambda) s_\lambda(t), \tag{1-5}$$

$$\tilde{C}^{(2)}_{\rho,\tilde\rho}(\tau^{(2)})(N, t, \tilde{t}) = \sum_\lambda \sum_\mu r_\lambda(N) B_N(\lambda, \mu) \tilde{r}_\mu(N) s_\lambda(t) s_\mu(\tilde{t}), \tag{1-6}$$

where the factors $r_\lambda(N)$, $\tilde{r}_\lambda(N)$ are defined in terms of a given pair of infinite sequences of nonvanishing constants $\{r_i\}_{i\in\mathbb{Z}}$, $\{\tilde{r}_i\}_{i\in\mathbb{Z}}$ through the formulae

$$r_\lambda(N) := c_r(N) \prod_{(i,j)\in\lambda} r_{N-i+j}, \quad \tilde{r}_\mu(N) := c_{\tilde{r}}(N) \prod_{(k,l)\in\mu} \tilde{r}_{N-k+l}. \tag{1-7}$$

Here the products are over pairs of positive integers $(i, j) \in \lambda$ and $(k, l) \in \mu$ that lie within the matrix locations represented by the Young diagrams of the partitions λ and μ, respectively,

$$c_r(N) := \prod_{i=1}^\infty \frac{\rho_{N-i}}{\rho_{-i}}, \tag{1-8}$$

and

$$r_i = \frac{\rho_i}{\rho_{i-1}}. \tag{1-9}$$

The sequence of nonvanishing parameters $\{\rho_i\}$ may be viewed as Fourier coefficients of a function $\rho(z)$ on the unit circle, or a distribution. It will be shown (Proposition 3.1) that, in terms of the elements of the subspace $W \subset L^2(S^1)$ corresponding to a point of the Grassmannian, the transformations (1-5), (1-6) mean taking a generalized convolution product with $\rho(z)$ (and similarly for $\tilde{r}_i$). These will therefore be referred to as (generalized) *convolution symmetries*.

With the usual 2D-Toda flow parameters $(t, \tilde{t})$ fixed at some specific values, such transformations extend to an infinite abelian group of commuting flows whose parameters determine the ρ_i's. This has been used to generate new classes of solutions of integrable hierarchies [Bettelheim et al. 2007; Orlov 2006; Orlov and Scherbin 2001]. In the present work, they are studied rather as individual transformations, for fixed values of the parameters ρ_i which, when applied to a given KP-Toda or 2D-Toda τ-function, produce a new one. Particular cases that implicitly use such transformations as symmetries have found applications, for example, as generating functions for topological invariants related to Riemann surfaces, such as Gromov–Witten invariants and Hurwitz numbers [Okounkov 2000; Okounkov and Pandharipande 2006].

As an immediate application, we may start with an integral over $N \times N$ Hermitian matrices:

$$Z_N(t) = \int_{M \in \mathbb{H}^{N \times N}} d\mu(M) \, e^{\mathrm{tr} \sum_{i=1}^{\infty} t_i M^i}, \qquad (1\text{-}10)$$

where $d\mu$ is a suitably defined $U(N)$ conjugation invariant measure on the space $\mathbb{H}^{N \times N}$ of Hermitian $N \times N$ matrices,. This is known to be a KP-Toda τ-function [Kharchev et al. 1991]. Applying a convolution symmetry (1-5) with $\rho(z)$ taken essentially as the exponential function e^z on the unit disc, and evaluating at flow parameter values

$$t_i = \frac{1}{i} \, \mathrm{tr}(A^i), \quad t = [A] = (t_1, t_2, \dots), \qquad (1\text{-}11)$$

for a fixed $N \times N$ Hermitian matrix A we obtain, within a constant multiplicative factor, the externally coupled matrix model integral (Proposition 4.1):

$$Z_{N,\mathrm{ext}}([A]) := \int_{M \in \mathbb{H}^{N \times N}} d\mu(M) e^{\mathrm{tr}\, AM} = \left(\prod_{i=1}^{N-1} i! \right) \tilde{C}_\rho(Z_N)([A]). \quad (1\text{-}12)$$

Such integrals arise in a number contexts, such as the Kontsevich–Witten generating function [Kontsevich 1992], the Brézin–Hikami model [Brézin and Hikami 1996; Zinn-Justin 1998; 2002] and the complex Wishart ensemble [Silverstein and Bai 1995; Wang 2009]. More general choices for the function $\rho(z)$ are shown in Proposition 4.2 to also determine KP-Toda τ-functions as externally coupled matrix integrals. It is further shown, in Proposition 4.3, that these matrix model τ-functions can be expressed as finite $N \times N$ determinants.

Similarly, Hermitian two-matrix integrals with exponential coupling of Itzykson–Zuber type [Itzykson and Zuber 1980]

$$Z_N^{(2)}(t, \tilde{t}) =$$

$$\int_{M_1 \in \mathbb{H}^{N \times N}} d\mu(M_1) \int_{M_2 \in \mathbb{H}^{N \times N}} d\tilde{\mu}(M_2) \, e^{\mathrm{tr}(\sum_{i=1}^{\infty} (t_i M_1^i + \tilde{t}_i M_2^i) + M_1 M_2)} \quad (1\text{-}13)$$

are known to be 2D-Toda τ-functions [Adler and van Moerbeke 1999; Harnad and Orlov 2002; Harnad and Orlov 2003; Orlov 2004]. Applying the convolution symmetry (1-6) to (1-13) gives an externally coupled two-matrix integral (Proposition 4.4).

$$\tilde{C}_{\rho,\tilde{\rho}}^{(2)}(Z_N^{(2)})([A], [B]) = \int_{M_1 \in \mathbb{H}^{N \times N}} d\mu(M_2) \int_{M_2 \in \mathbb{H}^{N \times N}} d\tilde{\mu}(M_2)$$

$$\times \tau_r(N, [A], [M_2]) \tau_{\tilde{r}}(N, [B], [M_2]) e^{\mathrm{tr}(M_1 M_2)}, \quad (1\text{-}14)$$

where $[A]$ and $[B]$ signify the sequences $\{\frac{1}{i}\operatorname{tr}(A^i)\}_{i\in\mathbb{N}^+}$ and $\{\frac{1}{i}\operatorname{tr}(B^i)\}_{i\in\mathbb{N}^+}$ of trace invariants for the pair of Hermitian matrices A and B and

$$\tau_r(N,[A],[M_1]) = \sum_\lambda r_\lambda(N)s_\lambda([A])s_\lambda([M_1]), \qquad (1\text{-}15)$$

$$\tau_{\tilde{r}}(N,[B],[M_2]) = \sum_\lambda \tilde{r}_\lambda(N)s_\lambda([B])s_\lambda([M_2]). \qquad (1\text{-}16)$$

This doubly externally coupled two-matrix model τ-function can also be expressed in a finite $N \times N$ determinantal form (eq. (4-37), Proposition 4.5).

This approach can also be extended to more general 2D-Toda τ-functions admitting multiple integral representations of the form (4-47). Applying the convolution symmetry (1-6) then gives a new 2D-Toda τ-function expressible either as a multiple integral (eq. (4-48), Proposition 4.6) or as a finite determinant (eq. (4-50), Proposition 4.7).

The key to understanding these constructions, and further results following from them, is the interpretation of the Sato τ-function as a vacuum state expectation value of products of exponentials of bilinear combinations of fermionic creation and annihilation operators [Sato 1981; Jimbo and Miwa 1983; Ueno and Takasaki 1984]. This well-known construction will be summarized in the next section.

2. Fermionic construction of τ-functions

We recall here the approach to the construction of τ-functions for integrable hierarchies of the KP and Toda types due to Sato [Sato 1981; Sato and Sato 1983], the Kyoto school [Date et al. 1981/82; 1983; Jimbo and Miwa 1983; Ueno and Takasaki 1984] and Segal and Wilson [1985].

2.1. Hilbert space Grassmannian and fermionic Fock space.

We begin with the "first quantized" Hilbert space $\mathcal{H}$, which will be identified, as in [Segal and Wilson 1985], with the space of square integrable functions on the unit circle

$$\mathcal{H} = L^2(S^1) = \mathcal{H}_+ + \mathcal{H}_-, \qquad (2\text{-}1)$$

decomposed as the direct sum of the subspaces $\mathcal{H}_+ = \operatorname{span}\{z^i\}_{i\in\mathbb{N}}$ and $\mathcal{H}_- = \operatorname{span}\{z^{-i}\}_{i\in\mathbb{N}^+}$ consisting of functions that admit holomorphic extensions, respectively, to the interior and exterior of the unit circle S^1 in the complex z-plane, with the latter vanishing at $z = \infty$. For consistency with other conventions, the monomial (orthonormal) basis elements $\{e_i\}_{i\in\mathbb{Z}}$ of $\mathcal{H}$ will be denoted by

$$e_i := z^{-i-1}, \quad i \in \mathbb{Z}. \qquad (2\text{-}2)$$

Two infinite abelian groups act on $\mathcal{H}$ by multiplication:

$$\Gamma_+ := \left\{ \gamma_+(t) := e^{\sum_{i=1}^{\infty} t_i z^i} \right\} \quad \text{and} \quad \Gamma_- := \left\{ \gamma_-(t) := e^{\sum_{i=1}^{\infty} t_i z^{-i}} \right\}, \quad (2\text{-}3)$$

where $t := (t_1, t_2, \dots)$ is an infinite sequence of (complex) flow parameters corresponding to the one-parameter subgroups. More generally, we have the general linear group $GL(\mathcal{H})$ consisting of invertible endomorphisms connected to the identity with well defined determinants. (See [Segal and Wilson 1985] for more detailed definitions of this and what follows.)

We consider the Grassmannian $Gr_{\mathcal{H}_+}(\mathcal{H})$ of subspaces $W \subset \mathcal{H}$ that are *commensurable* with $\mathcal{H}_+ \subset \mathcal{H}$ (in the sense of Segal and Wilson, namely that orthogonal projection $\pi_+ : W \to \mathcal{H}_+$ to $\mathcal{H}_+$ is a Fredholm operator while projection $\pi_- : W \to \mathcal{H}_-$ to $\mathcal{H}_-$ is Hilbert–Schmidt). The connected components of $Gr_{\mathcal{H}_+}(\mathcal{H})$, denoted $Gr_{\mathcal{H}_+^N}(\mathcal{H})$, $N \in \mathbb{Z}$, consist of those $W \in Gr_{\mathcal{H}_+}(\mathcal{H})$ for which the Fredholm index of $\pi_+ : W \to \mathcal{H}_+$ is N. These are the $GL(\mathcal{H})$ orbits of the subspaces

$$\mathcal{H}_+^N := z^{-N} \mathcal{H}_+ \subset \mathcal{H}, \quad (2\text{-}4)$$

whose elements are denoted $W_{g,N} = g(\mathcal{H}_+^N) \in Gr_{\mathcal{H}_+^N}(\mathcal{H})$. The solutions to the KP hierarchy are given by the τ-function $\tau_{N,g}(t)$ as defined below, which determines the orbit of $W_{g,N}$ in $Gr_{\mathcal{H}_+^N}(\mathcal{H})$ under Γ_+ through its Plücker coordinates. In the terminology of Segal and Wilson, the index N is called the "virtual dimension" of the elements $W_{g,N} \in Gr_{\mathcal{H}_+^N}(\mathcal{H})$; i.e., their dimension *relative* to the those in the component $Gr_{\mathcal{H}_+^0}(\mathcal{H})$ containing $\mathcal{H}_+$.

The *Fermionic Fock space* is the exterior space $\mathcal{F} := \Lambda \mathcal{H}$ consisting of (a completion of) the span of the semiinfinite wedge products:

$$|\lambda, N\rangle := e_{l_1} \wedge e_{l_2} \wedge \cdots, \quad (2\text{-}5)$$

where $\{l_j\}_{j \in \mathbb{N}^+}$ is a strictly decreasing sequence of integers that saturates, for sufficiently large j, to a descending sequence of consecutive integers. This is equivalent to requiring that there be an associated pair (λ, N) consisting of an integer N and a partition $\lambda = (\lambda_1, \dots, \lambda_{\ell(\lambda)}, 0, 0, \dots)$ of length $\ell(\lambda)$ and weight $|\lambda| = \sum_{i=1}^{\ell(\lambda)} \lambda_i$, where the parts λ_i are a weakly decreasing sequence of nonnegative integers that are positive for $i \le \ell(\lambda)$, and zero for $i > \ell(\lambda)$, such that the sequence $\{l_j\}_{j \in \mathbb{N}^+}$ is given by

$$l_j := \lambda_j - j + N. \quad (2\text{-}6)$$

In particular, for the trivial partition $\lambda = (0)$, we have the "charge N vacuum" vector

$$|0, N\rangle = e_{N-1} \wedge e_{N-2} \wedge \cdots, \quad (2\text{-}7)$$

which will henceforth be denoted $|N\rangle$. The full Fock space $\mathcal{F}$ thus admits a decomposition as an orthogonal direct sum of the subspaces $\mathcal{F}_N$ of states with charge N

$$\mathcal{F} = \bigoplus_{N \in \mathbb{Z}} \mathcal{F}_N. \tag{2-8}$$

Denoting by $\{\tilde{e}^i\}_{i \in \mathbb{Z}}$ the basis for $\mathcal{H}^*$ dual to the monomial basis $\{e_i\}_{i \in \mathbb{Z}}$ for $\mathcal{H}$, we define the Fermi creation and annihilation operators ψ_i and $\psi_i^\dagger$ on an arbitrary vector $v \in \mathcal{F}$ by exterior and interior multiplication, respectively:

$$\psi_i v = e_i \wedge v, \quad \psi_i^\dagger v := i_{\tilde{e}^i} v, \quad v \in \mathcal{F}. \tag{2-9}$$

These satisfy the standard canonical anticommutation relations generating the Clifford algebra on $\mathcal{H} + \mathcal{H}^*$ with respect to the natural corresponding quadratic form

$$[\psi_i, \psi_j]_+ = [\psi_i^\dagger, \psi_j^\dagger]_+ = 0, \quad [\psi_i, \psi_j^\dagger]_+ = \delta_{ij}. \tag{2-10}$$

The basis states $|\lambda, N\rangle$ may be expressed in terms of creation and annihilation operators acting upon the charge N vacuum vector as follows [Harnad and Orlov 2007]

$$|\lambda, N\rangle = (-1)^{\sum_{i=1}^k \beta_i} \prod_{i=1}^k \psi_{N+\alpha_i} \psi_{N-\beta_i-1}^\dagger |N\rangle, \tag{2-11}$$

where $(\alpha_1, \ldots \alpha_k | \beta_1, \ldots, \beta_k)$ is the Frobenius notation (see [Macdonald 1995]) for the partition λ; i.e., α_i is the number of boxes in the corresponding Young diagram to the right of the i-th diagonal element and β_i the number below it.

The Plücker map $\mathfrak{P} : \mathrm{Gr}_{\mathcal{H}_+}(\mathcal{H}) \to \mathbb{P}(\mathcal{F})$ takes the subspace

$$W = \mathrm{span}(w_1, w_2, \ldots) \tag{2-12}$$

into the projectivization of the exterior product of its basis elements:

$$\mathfrak{P} : \mathrm{span}(w_1, w_2, \ldots) \mapsto [w_1 \wedge w_2 \wedge \cdots], \tag{2-13}$$

and may be lifted to a map from the bundle $\mathrm{Fr}_{\mathcal{H}_+}(\mathcal{H})$ of frames on $\mathrm{Gr}_{\mathcal{H}_+}(\mathcal{H})$ to $\mathcal{F}$:

$$\hat{\mathfrak{P}} : \mathrm{Fr}_{\mathcal{H}_+}(\mathcal{H}) \to \mathcal{F}, \quad (w_1, w_2, \ldots) \mapsto w_1 \wedge w_2 \wedge \cdots. \tag{2-14}$$

These interlace the lift of the action of the abelian group $\Gamma_+ \times \mathcal{H} \to \mathcal{H}$ to $\mathrm{Fr}_{\mathcal{H}_+}(\mathcal{H})$ or $\mathrm{Gr}_{\mathcal{H}_+}(\mathcal{H})$ with the following representation of Γ_+ on $\mathcal{F}$ (and its projectivization):

$$\gamma_+(t) : v \mapsto \hat{\gamma}_+(t)v, \quad \hat{\gamma}_+(t) := e^{\sum_{i=1}^\infty t_i H_i}, \quad v \in \mathcal{F}, \tag{2-15}$$

where

$$H_i := \sum_{n \in \mathbb{Z}} \psi_n \psi_{n+i}^\dagger, \quad i \in \mathbb{Z}, \; i \neq 0, \tag{2-16}$$

and $t = (t_1, t_2, \dots)$ is the infinite sequence of flow parameters. Similarly, the Plücker maps $\hat{\mathfrak{P}}$ and $\mathfrak{P}$ interlace the lift of the action of the abelian group $\Gamma_- \times \mathcal{H} \to \mathcal{H}$ to $\mathrm{Fr}_{\mathcal{H}_+}(\mathcal{H})$ or $\mathrm{Gr}_{\mathcal{H}_+}(\mathcal{H})$ with the following representation of Γ_- on $\mathcal{F}$ (and its projectivization):

$$\gamma_-(t) : v \mapsto \hat{\gamma}_-(t)v, \quad \hat{\gamma}_-(t) := e^{\sum_{i=1}^\infty t_i H_{-i}}, \quad v \in \mathcal{F}. \tag{2-17}$$

Remark 2.1. Note that the image under the Plücker map of the virtual dimension N component $\mathrm{Gr}_{\mathcal{H}_+}^N(\mathcal{H})$ of the Grassmannian $\mathrm{Gr}_{\mathcal{H}_+}(\mathcal{H})$ is the $\mathrm{GL}(\mathcal{H})$ orbit of the charged vacuum state $|N\rangle$, consisting of all decomposable elements of $\mathcal{F}_N$.

The KP-Toda τ-function $\tau_g(N, t)$ corresponding to the element $W_{g,N} \in \mathrm{Gr}_{\mathcal{H}_+}(\mathcal{H})$ is given, within a nonzero multiplicative constant, by applying the group elements $\gamma_+(t)$ to $W_{g,N}$, to obtain the Γ_+ orbit

$$\{W_{g,N}(t) := \gamma_+(t)(W_{g,N})\}, \tag{2-18}$$

and taking the linear coordinate (within projectivization) of the image under the Plücker map corresponding to projection along the basis element $|N\rangle$

$$\tau_g(N, t) = \langle N | \hat{\mathfrak{P}}(W_{g,N}(t)) \rangle. \tag{2-19}$$

If the group element $g \in \mathrm{GL}(\mathcal{H})$ is interpreted, relative to the monomial basis $\{e_i\}_{i \in \mathbb{Z}}$, as an infinite matrix exponential $g = e^A$ of an element of the Lie algebra $A \in \mathfrak{gl}(\mathcal{H})$ with matrix elements A_{ij}, then the corresponding representation of $\mathrm{GL}(\mathcal{H})$ on $\mathcal{F}$ is given by

$$\hat{g} := e^{\sum_{i,j \in \mathbb{Z}} A_{ij} : \psi_i \psi_j^\dagger :}, \tag{2-20}$$

where $: :$ denotes normal ordering (i.e., annihilation operators $\psi_j^\dagger$ appearing to the right when $j \geq 0$ and creation operators ψ_i to the right when $i < 0$). This gives the following expression for $\tau_{N,g}(t)$ as a charge N vacuum state expectation value of a product of exponentiated bilinears in the Fermi creation and annihilation operators

$$\tau_g(N, t) = \langle N | \hat{\gamma}_+(t) \hat{g} | N \rangle. \tag{2-21}$$

The equations of the KP hierarchy are then equivalent to the well-known infinite system of Hirota bilinear equations [Jimbo and Miwa 1983; Sato 1981; Sato and Sato 1983] which, in turn, are just the Plücker relations for the decomposable element $\mathfrak{P}(W_{g,N}(t)) \in \mathbb{P}(\mathcal{F}_N)$.

Similarly, we may define a 2-Toda sequence of double KP τ-functions associated to the group element $\hat{g}$

$$\tau_g^{(2)}(N, t, \tilde{t}) = \langle N | \hat{\gamma}_+(t) \hat{g} \hat{\gamma}_-(\tilde{t}) | N \rangle, \tag{2-22}$$

where $\tilde{t} = (\tilde{t}_1, \tilde{t}_2, \dots)$ is a second infinite set of flow parameters. This may similarly be shown to satisfy the Hirota bilinear relations of the 2D-Toda hierarchy.

2.2. Schur function expansions.

Evaluating the matrix elements of $\hat{\gamma}_+(t)$ and $\hat{\gamma}_-(t)$ between the states $|N\rangle$ and $|\lambda, N\rangle$ gives the Schur function

$$\langle N | \hat{\gamma}_+(t) | \lambda, N \rangle = \langle \lambda, N | \hat{\gamma}_-(t) | N \rangle = s_\lambda(t), \tag{2-23}$$

(cf. [Sato 1981; Sato and Sato 1983; Harnad and Orlov 2003; 2006] which is determined through the Jacobi–Trudy formula

$$s_\lambda(t) = \det(h_{\lambda_i - i + j}(t))|_{1 \le i, j \le \ell(\lambda)} \tag{2-24}$$

in terms of the complete symmetric functions $h_i(t)$, defined by

$$e^{\sum_{i=1}^\infty t_i z^i} = \sum_{i=0}^\infty h_i(t) z^i. \tag{2-25}$$

Inserting a sum over a complete set of intermediate states in Equations (2-21), (2-22), we obtain the single and double Schur function expansions

$$\tau_g(N, t) = \sum_\lambda \pi_{N,g}(\lambda) s_\lambda(t), \tag{2-26}$$

$$\tau_g^{(2)}(N, t, \tilde{t}) = \sum_\lambda \sum_\mu B_{N,g}(\lambda, \mu) s_\lambda(t) s_\mu(\tilde{t}). \tag{2-27}$$

Here the sum is over all partitions λ and μ and

$$\pi_{N,g}(\lambda) = \langle \lambda, N | \hat{g} | N \rangle \tag{2-28}$$

is the Plücker coordinate of the image of the element $g(\mathcal{H}_+^N) \in \mathrm{Gr}_{\mathcal{H}_+^N}(\mathcal{H})$ under the Plücker map $\mathfrak{P}$ along the basis direction $|\lambda, N\rangle$ in the charge N sector $\mathcal{F}_N$ of the Fock space. Similarly,

$$B_{N,g}(\lambda, \mu) = \langle \lambda, N | \hat{g} | \mu, N \rangle \tag{2-29}$$

may be viewed as the $|\lambda, N\rangle$ Plücker coordinate of the image of the element $g(w_{\mu,N}) \in \mathrm{Gr}_{\mathcal{H}_+^N}(\mathcal{H})$, where

$$w_{\mu,N} := \mathrm{span}\{e_{\mu_i - i + N}\} \in \mathrm{Gr}_{\mathcal{H}_+^N}(\mathcal{H}). \tag{2-30}$$

In particular, choosing g to be the identity element $\mathbb{1}$, and using Wick's theorem (or equivalently, the Cauchy–Binet identity in semiinfinite form), we obtain [Harnad and Orlov 2003]

$$\tau_{\mathbb{1}}^{(2)}(N, t, \tilde{t}) = \langle N | \hat{\gamma}_+(t)\hat{\gamma}_-(\tilde{t}) | N \rangle = \sum_\lambda s_\lambda(t) s_\lambda(\tilde{t}) = e^{\sum_{i=1}^\infty i t_i \tilde{t}_i}, \quad (2\text{-}31)$$

where the last equality is the Cauchy–Littlewood identity (cf. [Macdonald 1995]).

3. Convolution symmetries

3.1. *Convolution action on $\mathcal{H}$ and $\mathrm{Gr}_{\mathcal{H}_+}(\mathcal{H})$.*

Consider now an infinite sequence of complex numbers $\{T_i\}_{i \in \mathbb{Z}}$, and define

$$\rho_i := e^{T_i}, \quad r_i := \frac{\rho_i}{\rho_{i-1}}, \quad i \in \mathbb{Z}. \quad (3\text{-}1)$$

In the following, we will assume that the series $\sum_{i=1}^\infty T_{-i}$ converges and that

$$\lim_{i \to \infty} |r_i| = r \leq 1 \quad (3\text{-}2)$$

(although, for some purposes, the latter condition may be weakened). It follows that the two series

$$\rho_+(z) = \sum_{i=0}^\infty \rho_i z^i \quad \text{and} \quad \rho_-(z) = \sum_{i=1}^\infty \rho_{-i} z^{-i} \quad (3\text{-}3)$$

are absolutely convergent in the interior and exterior of the unit circle $|z| = 1$, respectively, defining analytic functions $\rho_\pm(z)$ in these regions and that

$$R_\rho := \prod_{i=1}^\infty \rho_{-i} \quad (3\text{-}4)$$

converges to a finite value. If the inequality (3-2) is strict, $\rho_+(z)$ extends to the unit circle, defining a function in $L^2(S^1)$. Henceforth, we denote the pair (ρ_+, ρ_-) by ρ, where the latter can be viewed as a sum $\rho_- + \rho_+$ in the sense of distributional convolutions, as defined below.

If $w \in L^2(S^1)$ has the Fourier series decomposition

$$w(z) = \sum_{i=-\infty}^\infty w_i e_i = \sum_{i=-\infty}^\infty w_i z^{-i-1} = w_+(z) + w_-(z), \quad (3\text{-}5)$$

where

$$w_+(z) := \sum_{i=0}^\infty w_{-i-1} z^i, \quad w_-(z) := \sum_{i=1}^\infty w_i z^{-i-1}, \quad (3\text{-}6)$$

(note the different labelling conventions in (3-3) and (3-6)), we can define a bounded linear map $C_\rho : L^2(S^1) \to L^2(S^1)$ that has the effect of multiplying each Fourier coefficient w_i by the factor ρ_i, and hence each basis element e_i by ρ_i:

$$C_\rho(w)(z) = \sum_{i=-\infty}^{\infty} \rho_i w_i z^{-i-1} = C_\rho(w)_+ + C_\rho(w)_-. \tag{3-7}$$

This can be interpreted as taking a convolution product with the function (or distribution)

$$\tilde{\rho}(z) = \tilde{\rho}_+(z) + \tilde{\rho}_-(z), \tag{3-8}$$

where

$$\tilde{\rho}_+(z) := z^{-1}\rho_-(z^{-1}) = \sum_{i=0}^{\infty} \rho_{-i-1} z^i, \tag{3-9}$$

$$\tilde{\rho}_-(z) := z^{-1}\rho_+(z^{-1}) = \sum_{i=0}^{\infty} \rho_i z^{-i-1}, \tag{3-10}$$

$$C_\rho(w)_+(z) := \lim_{\epsilon \to 0^+} \frac{1}{2\pi i} \oint_{|\zeta|=1-\epsilon} \tilde{\rho}_+(\zeta) w_+(z/\zeta) \zeta^{-1} \, d\zeta, \tag{3-11}$$

$$C_\rho(w)_-(z) := \lim_{\epsilon \to 0^+} \frac{1}{2\pi i} \oint_{|\zeta|=1+\epsilon} \tilde{\rho}_-(\zeta) w_-(z/\zeta) \zeta^{-1} \, d\zeta \tag{3-12}$$

(with the contour integrals taken counterclockwise).

If $\rho_-(z)$ extends analytically to S^1, eq. (3-11) is an ordinary convolution product on the circle (in exponential variables). In the examples detailed below, all but a finite number of the T_{-i} values vanish for $i > 0$, and hence the infinite product (3-4) is really finite, but $\rho_-(z)$ is rational with a pole at $z = 1$ and the convolution product (3-12) may be understood on S^1 only in the sense of distributions.

Remark 3.1. Note that the class of generalized convolution mappings defined by (3-7)–(3-12) only forms a semigroup since, although they may be invertible, their inverse does not generally belong to the same class. It may be extended to a group by dropping the condition (3-2), or restricted to one by requiring $r = 1$, but this will not be needed in the sequel. The linear maps $C_\rho : \mathcal{H} \to \mathcal{H}$ may nevertheless be interpreted as elements of $GL(\mathcal{H})$, and are simply represented in the monomial basis $\{e_i\}$ by the diagonal matrix $\mathrm{diag}\{\rho_i\}$. They thus belong to the abelian subgroup of $GL(\mathcal{H})$ consisting of invertible elements that are diagonal in the monomial basis.

Remark 3.2. Since the Baker–Akhiezer function (1-3), evaluated at all values of the parameters $t = (t_1, t_2, \dots)$, spans the shifted element $z^N(W_{g,N}) \in \mathrm{Gr}_{\mathcal{H}_+^0}(\mathcal{H})$

in the zero virtual dimension component of the Grassmannian, the convolution action (3-11), (3-12), lifted to the Grassmannian, may be obtained by applying its conjugate $z^N \circ C_\rho \circ z^{-N}$ under the shift map

$$z^{-N} : \mathrm{Gr}_{\mathcal{H}^0_+} (\mathcal{H}) \rightarrow \mathrm{Gr}_{\mathcal{H}^N_+} (\mathcal{H}) \tag{3-13}$$

to $\Psi_N(z, t)$. But note that, at fixed values of the flow parameters t, this does not equal the value of the Baker–Akhiezer function corresponding to the transformed τ-function as defined below; only the subspaces of $\mathcal{H}$ that they span, varying over the t values, will coincide. This fact will not be used explicitly in the following, but it underlies the geometrical meaning of generalized convolutions as symmetries of KP-Toda and 2D-Toda hierarchies.

3.2. Convolution action on Fock space.

We now consider the action $\hat{C} \times \mathcal{F} \rightarrow \mathcal{F}$ of the abelian subgroup of $\mathrm{GL}(\mathcal{H})$ consisting of diagonal elements in the monomial basis, and associate an element $\hat{C}_\rho \in \hat{C}$ to each sequence $\{\rho_i\}_{i \in \mathbb{Z}}$ defined as above, such that the Plücker map $\hat{\mathfrak{P}}$ intertwines the $\hat{C}_\rho$ action with that of C_ρ, lifted to the bundle $\mathrm{Fr}_{\mathcal{H}_+} (\mathcal{H})$ of frames over $\mathrm{Gr}_{\mathcal{H}_+} (\mathcal{H})$, and is equivariant with respect to group multiplication in $\hat{C}$.

To do this, we first introduce the abelian algebra generated by the operators

$$K_i := :\psi_i \psi_i^{\dagger}: = \begin{cases} \psi_i \psi_i^{\dagger} & \text{if } i \geq 0, \\ -\psi_i^{\dagger} \psi_i & \text{if } i < 0, \end{cases} \tag{3-14}$$

$$[K_i, K_j] = 0, \quad i, j \in \mathbb{Z}. \tag{3-15}$$

For $\{\rho_i = e^{T_i}\}_{i \in \mathbb{Z}}$ as above, define the operator

$$\hat{C}_\rho := e^{\sum_{i=-\infty}^{\infty} T_i K_i}. \tag{3-16}$$

Definition 3.1. For each pair (λ, N), where $N \in \mathbb{Z}$, and λ is a partition which, expressed in Frobenius notation, is $(\alpha_1 \cdots \alpha_k \mid \beta_1 \cdots \beta_k)$, let

$$r_\lambda(N) := c_r(N) \prod_{(i,j) \in \lambda} r_{N-i+j} = c_r(N) \left(\prod_{i=1}^{k} \frac{\rho_{N+\alpha_i}}{\rho_{N-\beta_i-1}} \right), \tag{3-17}$$

$$c_r(N) := \begin{cases} \prod_{i=0}^{N-1} \rho_i & \text{if } N > 0, \\ 1 & \text{if } N = 0, \\ \prod_{i=N}^{-1} \rho_i^{-1} & \text{if } N < 0. \end{cases} \tag{3-18}$$

Here the inclusion $(i, j) \in \lambda$ is understood to mean that the matrix location (i, j) corresponds to a box within the Young diagram of the partition λ; that is, $1 \le i \le \ell(\lambda)$, $1 \le j \le \lambda_i$. The second equality in (3-17) follows from the definition (3-1).

It follows that $\hat{C}_\rho$ acts diagonally in the basis $\{|\lambda, N\rangle\}$, with eigenvalues $r_\lambda(N)$.

Lemma 3.1.
$$\hat{C}_\rho|\lambda, N\rangle = r_\lambda(N)|\lambda, N\rangle. \tag{3-18}$$

Proof. Since the Fock space basis element $|\lambda, N\rangle$ is an infinite wedge product

$$|\lambda, N\rangle = e_{l_1} \wedge e_{l_2} \wedge \cdots = (-1)^{\sum_{i=1}^{k} \beta_i} \prod_{i=1}^{k} \psi_{N+\alpha_i} \psi^\dagger_{N-\beta_i-1}|N\rangle, \tag{3-20}$$

$$l_j := \lambda_j - j + N, \quad j \in \mathbb{N}^+, \tag{3-21}$$

it follows from the definition (2-9) and the normal ordering in (3-14) that the effect of the action of $e^{T_i K_i}$ on $|\lambda, N\rangle$ is to introduce a multiplicative factor ρ_i if $i \ge 0$ and e_i is present in the wedge product (3-20) or ρ_i^{-1} if $i < 0$ and it is absent, and otherwise no factor. Therefore

$$\hat{C}_\rho|\lambda, N\rangle = \hat{C}_\rho(-1)^{\sum_{i=1}^{k} \beta_i} \prod_{i=1}^{k} \psi_{N+\alpha_i} \psi^\dagger_{N-\beta_i-1}|N\rangle$$

$$= \frac{\prod_{i=1}^{\infty} \rho_{N-i}}{\prod_{i=1}^{\infty} \rho_{-i}} \left(\prod_{i=1}^{k} \frac{\rho_{N+\alpha_i}}{\rho_{N-\beta_i-1}} \right)|\lambda, N\rangle$$

$$= c_r(N) \left(\prod_{i=1}^{k} \frac{\rho_{N+\alpha_i}}{\rho_{N-\beta_i-1}} \right)|\lambda, N\rangle$$

$$= r_\lambda(N)|\lambda, N\rangle. \tag{3-22}$$

$\square$

Now let

$$W = \mathrm{span}\{w_i(z) \in L^2(S^1)\}_{i \in \mathbb{N}^+} \in \mathrm{Gr}_{\mathcal{H}_+}(\mathcal{H}) \tag{3-23}$$

and view $\{w_i\}_{i \in \mathbb{N}^+}$ as a frame for W.

Lemma 3.2. *The Plücker map $\hat{\mathfrak{P}}$ intertwines the convolution action (3-7) and the $\hat{C}$-action on $\mathcal{F}$*

$$\hat{\mathfrak{P}}(\{C_\rho(w_i)\}_{i \in \mathbb{N}^+}) = R_\rho \hat{C}_\rho(\hat{\mathfrak{P}}\{w_i\}_{i \in \mathbb{N}^+}), \tag{3-24}$$

with multiplicative factor $R_\rho := \prod_{i=1}^{\infty} \rho_{-i}$.

Proof. Applying C_ρ to each element $w_i \in L^2(S^1)$ defining the frame for $W \in$ $\mathrm{Gr}_{\mathcal{H}_+}(\mathcal{H})$ just multiplies its Fourier coefficients by the factors ρ_j as in (3-7). It follows that the basis element $|\lambda, N\rangle$ is multiplied by the product of the factors ρ_{l_j} corresponding to the terms e_{l_j} it contains, as in (3-18). Equation (3-24) then follows from the definition of the Plücker map $\widehat{\mathfrak{P}}$ and linearity. $\qquad\square$

Example 3.1. Choose

$$\rho_+(z) = e^z = \sum_{i=0}^{\infty} \frac{z^i}{i!}, \qquad |z| \le 1 \tag{3-25}$$

$$\rho_-(z) = \frac{1}{z-1} = \sum_{i=1}^{\infty} z^{-i}, \quad |z| > 1, \tag{3-26}$$

so

$$\rho_i = \begin{cases} 1/i! & \text{if } i \ge 1, \\ 1 & \text{if } i \le 0, \end{cases} \tag{3-27}$$

$$r_i = \begin{cases} 1/i & \text{if } i \ge 1 \\ 1 & \text{if } i \le 0, \end{cases} \tag{3-28}$$

$$r_\lambda(N) = \frac{1}{\left(\prod\limits_{i=1}^{N-1} i!\right)(N)_\lambda} \quad \text{if } \ell(\lambda) \le N, \tag{3-29}$$
$$\tag{3-30}$$

where

$$(N)_\lambda := \prod_{i=1}^{\ell(\lambda)} \prod_{j=1}^{\lambda_i} (N - i + j) \tag{3-31}$$

is the extended Pochhammer symbol.

Example 3.2. Choose

$$\rho_+(z) = \frac{1}{(1-\zeta z)^a} = \sum_{i=0}^{\infty} (a)_i \frac{(\zeta z)^i}{i!}, \quad |\zeta| < 1, |z| \le 1, \tag{3-32}$$

and $\rho_-(z)$ again as in (3-26), so

$$\rho_i = \begin{cases} (a)_i \zeta^i / i! & \text{if } i \ge 1, \\ 1 & \text{if } i \le 0, \end{cases} \tag{3-33}$$

$$r_i = \begin{cases} (a-1+i)\zeta/i & \text{if } i \ge 1, \\ 1 & \text{if } i \le 0, \end{cases} \tag{3-34}$$

$$r_\lambda(N) = \left(\prod_{i=0}^{N-1} \frac{(a)_i}{i!}\right) \frac{\zeta^{|\lambda| + \frac{1}{2}N(N-1)}(a-1+N)_\lambda}{(N)_\lambda} \quad \text{if } \ell(\lambda) \le N. \tag{3-35}$$

3.3. *Convolutions and Schur function expansions of τ-functions.*

We now consider the KP-Toda tau function

$$\tau_{C_\rho g}(N, t) = \langle N | \hat{\gamma}_+(t) \hat{C}_\rho \hat{g} | N \rangle, \qquad (3\text{-}36)$$

obtained by replacing the group element g in (2-21) by $C_\rho g$. Such a τ-function, obtained from τ_g by applying a convolution symmetry will be denoted

$$\tau_{C_\rho g} =: \tilde{C}_\rho(\tau_g). \qquad (3\text{-}37)$$

Introducing a second pair $(\tilde{\rho}_+, \tilde{\rho}_-)$, defined as in (3-3), with the Fourier coefficients ρ_i replaced by $\tilde{\rho}_i$, we also consider the 2-Toda tau function

$$\tau^{(2)}_{C_\rho g C_{\tilde{\rho}}}(N, t, \tilde{t}*) = \langle N | \hat{\gamma}_+(t) \hat{C}_\rho \hat{g} \hat{C}_{\tilde{\rho}} \hat{\gamma}_-(\tilde{t}) | N \rangle, \qquad (3\text{-}38)$$

obtained by replacing the group element g in (2-22) by $C_\rho g C_{\tilde{\rho}}$, and denote this transformed 2-Toda τ-function

$$\tau^{(2)}_{C_\rho \hat{g} C_{\tilde{\rho}}} =: \tilde{C}^{(2)}_{(\rho, \tilde{\rho})}(\tau^{(2)}_g). \qquad (3\text{-}39)$$

Inserting sums over complete sets of intermediate orthonormal basis states in (3-36) and (3-38), and defining $\tilde{r}_\lambda(N)$ as in (3-17), with the factors ρ_i replaced by $\tilde{\rho}_i$, we obtain the following form for the Schur function expansions (2-26), (2-27).

Proposition 3.1. *The effect of the convolution actions* (3-37), (3-39) *is to multiply the coefficients in the Schur function expansions of* $\tau_{C_\rho g}(N, t)$ *and* $\tau^{(2)}_{C_\rho \hat{g} C_{\tilde{\rho}}}(N, t, \tilde{t})$ *by the diagonal factors* $r_\lambda(N)$ *and* $\tilde{r}_\mu(N)$:

$$\tau_{C_\rho g}(N, t) = \sum_\lambda r_\lambda(N) \pi_{N,g}(\lambda) s_\lambda(t), \qquad (3\text{-}40)$$

$$\tau^{(2)}_{C_\rho g C_{\tilde{\rho}}}(N, t, \tilde{t}) = \sum_\lambda \sum_\mu r_\lambda(N) B_{N,g}(\lambda, \mu) \tilde{r}_\mu(N) s_\lambda(t) s_\mu(\tilde{t}). \qquad (3\text{-}41)$$

The Plücker coordinates for the modified Grassmannian elements $C_\rho g(\mathcal{H}^N_+)$ *and* $C_\rho g C_{\tilde{\rho}}(w_{\mu,N})$ *are thus*

$$\pi_{N, C_\rho g}(\lambda) = r_\lambda(N) \pi_{N,g}(\lambda), \qquad (3\text{-}42)$$

$$B_{N, C_\rho g C_{\tilde{\rho}}}(\lambda, \mu) = r_\lambda(N) B_{N,g}(\lambda, \mu) \tilde{r}_\mu(N). \qquad (3\text{-}43)$$

Proof. This follows immediately from the diagonal form (3-18) of the $\hat{C}$ action in the orthonormal basis $\{|\lambda, N\rangle\}$, substituted into the expansions (2-26), (2-27), using the definitions (2-28) and (2-29) of the Plücker coordinates $\pi_{N, C_\rho g}(\lambda)$ and $B_{N, C_\rho g C_{\tilde{\rho}}}(\lambda, \mu)$. □

In particular, setting $g = C_{\tilde{\rho}} = 1$, in (3-41) we obtain

$$\tau_{C_\rho}^{(2)}(N, t, \tilde{t}) = \sum_\lambda r_\lambda(N) s_\lambda(t) s_\lambda(\tilde{t}) =: \tau_r(N, t, \tilde{t}), \qquad (3\text{-}44)$$

where $\tau_r(N, t, \tilde{t})$ is defined by the second equality. Such τ-functions have been studied as generalizations of hypergeometric functions in [Orlov and Scherbin 2001; Orlov 2006]. (See also [Harnad and Orlov 2003; 2006], where the notation differs slightly due to the presence of the normalization factor $c_r(N)$ in the definition (3-17) of $r_\lambda(N)$.)

In the following, the infinite sequence of parameters $t = (t_1, t_2, \dots)$ will often be chosen as the trace invariants of some square matrix M. The sequence so formed will be denoted

$$t = [M] = \left\{ \frac{1}{i} \operatorname{tr}(M^i) \right\} \Big|_{i \in \mathbb{N}^+}, \quad [M]_i := \frac{1}{i} \operatorname{tr}(M^i). \qquad (3\text{-}45)$$

If t and $\tilde{t}$ in (3-44) are replaced by $[A]$ and $[B]$, respectively, where A and B are a pair of diagonal matrices

$$A = \operatorname{diag}(a_1, \dots, a_N), \quad B = \operatorname{diag}(b_1, \dots, b_N), \qquad (3\text{-}46)$$

with distinct eigenvalues, and

$$\Delta(A) := \prod_{1 \le i < j}^n (a_i - a_j), \quad \Delta(B) := \prod_{1 \le i < j}^n (b_i - b_j) \qquad (3\text{-}47)$$

denote the Vandermonde determinants in the variables $\{a_i\}$ and $\{b_i\}$, we obtain a simple $N \times N$ determinantal expression for $\tau_r(N, [A], [B])$ (cf. [Harnad and Orlov 2006; Orlov 2004]).

Lemma 3.3. *Choosing $\rho_-(z)$ as in (3-26) (i.e., $\rho_{-i} = 1$ for $i < 1$), we have*

$$\tau_r(N, [A], [B]) = \sum_{\ell(\lambda) \le N} r_\lambda(N) s_\lambda([A]) s_\lambda([B]) \qquad (3\text{-}48)$$

$$= \frac{\det(\rho_+(a_i b_j)|_{1 \le i, j \le N}}{\Delta(A)\Delta(B)}. \qquad (3\text{-}49)$$

Remark 3.3. Although various proofs of this result may be found elsewhere (see [Harnad and Orlov 2006], for example), we provide a detailed version here, based on the Cauchy–Binet identity in semiinfinite form, since it involves some useful further relations. An equivalent way is to use the fermionic form of Wick's theorem, which is really just the Cauchy–Binet identity expressed in terms of fermionic operators and matrix elements.

Proof of Lemma 3.3. The Cauchy–Binet identity in semiinfinite form may be expressed by considering two N-dimensional framed subspaces span$\{F_i\}_{1 \le i \le N}$ and span$\{G_i\}_{1 \le i \le N}$ of the complex Euclidean vector space $\ell^2(\mathbb{N}) = \text{span}\{e_i\}_{i \in \mathbb{N}}$, identified with $\mathscr{H}_+ \subset \mathscr{H} = L^2(S^1)$, by choosing the monomials $\{z^i\}_{i \in \mathbb{N}}$ as orthonormal basis. The vectors F_i and G_j are thus identified with elements $F_i(z), G_j(z) \in \mathscr{H}_+$ defined by

$$F_i(z) := \sum_{j=0}^{\infty} F_{ji} z^j, \quad G_i(z) := \sum_{j=0}^{\infty} G_{ji} z^j. \qquad (3\text{-}50)$$

(Note that, to avoid needless use of negative indices, we are not using the same labelling conventions here for the basis elements $\{e_i\}$ as in (2-2).) The complex inner product $(\,,\,)$ is defined by integration

$$(F, G) := \frac{1}{2\pi i} \oint_{z \in S^1} F(z) G(z^{-1}) \frac{dz}{z}. \qquad (3\text{-}51)$$

The Cauchy–Binet identity can then be expressed as

$$\det(F_i, G_j)|_{1 \le i,j \le N} = \sum_{\ell(\lambda) \le N} \det(F_{\lambda_i - i + N, j}) \det(G_{\lambda_i - i + N, j}), \qquad (3\text{-}52)$$

where

$$F_i = \sum_{j \in \mathbb{Z}} F_{ji} e_j, \quad G_i = \sum_{j \in \mathbb{Z}} G_{ji} e_j, \qquad (3\text{-}53)$$

and the sum is over all partitions λ of length $\ell(\lambda) \le N$, completed so that the $N \times N$ submatrices $F_{\lambda_i - i + N, j}$ and $G_{\lambda_i - i + N, j}$ are defined by setting $\lambda_i = 0$ for $i > \ell(\lambda)$. Since all expressions in the sum will be polynomials in the parameters (a_i, b_i) there is no loss of generality in assuming that these lie within the unit disc. We define

$$F_i(z) := \rho_+(a_i z), \quad G_i(z) := (1 - b_i z)^{-1}, \qquad (3\text{-}54)$$

and hence

$$F_{ij} = \rho_i(a_j), \quad G_{ij} = (b_j)^i. \qquad (3\text{-}55)$$

From the character formula

$$s_\lambda([A]) = \frac{\det(a_i^{\lambda_j - j + N})}{\Delta(A)}, \quad s_\lambda([B]) = \frac{\det(b_i^{\lambda_j - j + N})}{\Delta(B)}, \qquad (3\text{-}56)$$

it follows that the determinant factors on the right side of (3-52) are

$$
\det(F_{\lambda_i-i+N,j}) = \det(a_j^{\lambda_i-i+N}\rho_{\lambda_i-i+N})
$$
$$
= \left(\prod_{i=1}^{N}\rho_{\lambda_i-i+N}\right)s_\lambda([A])\Delta([A]), \qquad (3\text{-}57)
$$

$$
\det(G_{\lambda_i-i+N,j}) = \det(b_j^{\lambda_i-i+N}) = s_\lambda([B])\Delta(B). \qquad (3\text{-}58)
$$

From the definitions (3-17) and (3-18), it follows that

$$
\left(\prod_{i=1}^{N}\rho_{\lambda_i-i+N}\right) = r_\lambda(N), \qquad (3\text{-}59)
$$

so the right side of the Cauchy–Binet identity (3-52) is just the right side of (3-48) multiplied by $\Delta([A])\Delta([B])$. On the other hand, from (3-51), the left side of (3-52) is

$$
\det(F_i, G_j) = \det\left(\frac{1}{2\pi i}\oint_{z\in S^1}\frac{\rho_+(a_i z)}{z-b_j}\frac{dz}{z}\right) = \det(\rho_+(a_i b_j)), \qquad (3\text{-}60)
$$

which is just the expression (3-49) multiplied by $\Delta([A])\Delta([B])$. $\square$

Remark 3.4. Note that, for the case of Example 3.1, (3-49) becomes the key identity (cf. [Harnad and Orlov 2006; Zinn-Justin 2002])

$$
\sum_{\ell(\lambda)\le N}\frac{1}{(N)_\lambda}s_\lambda([A])s_\lambda([B]) = \left(\prod_{k=1}^{N-1}k!\right)\frac{\det(e^{a_i b_j})|_{1\le i,j\le N}}{\Delta(A)\Delta(B)}, \qquad (3\text{-}61)
$$

which, together with the character integral [Macdonald 1995]

$$
d_{\lambda,N}\int_{U\in U(N)}d\mu_H(U)s_\lambda([AUXU^\dagger]) = s_\lambda([A])s_\lambda([X]), \qquad (3\text{-}62)
$$

(where $d\mu_H(U)$ is the Haar measure on $U(N)$), implies the Harish-Chandra–Itzykson–Zuber (HCIZ) integral [Itzykson and Zuber 1980]

$$
\int_{U\in U(N)}d\mu_H(U)e^{\operatorname{tr}(AUXU^\dagger)} = \left(\prod_{k=1}^{N-1}k!\right)\frac{\det(e^{a_i x_j})}{\Delta(A)\Delta(X)}. \qquad (3\text{-}63)
$$

Remark 3.5. The condition that the eigenvalues $\{a_i\}$ and $\{b_i\}$ of A and B be distinct can be eliminated simply by taking limits in which some or all of these are made to coincide. In the resulting determinantal formulae, like (3-49), and those appearing in subsequent sections, in which a Vandermonde determinant $\Delta(A)$ or $\Delta(B)$ appears in the denominator, the only modification is that the terms in the numerator determinants depending on the a_i's and b_i's are replaced by their derivatives with respect to these parameters, taken to the same degree as the degeneracy of their values, while the denominator Vandermonde determinants are correspondingly replaced by their lower dimensional analogs. This will not be further developed here, but will be considered elsewhere, in connection with correlation kernels for externally coupled matrix models. All formulae below in which no Vandermonde determinant factors $\Delta(A)$ or $\Delta(B)$ appear in the denominator remain valid in the case of degenerate eigenvalues.

4. Applications to matrix models

We now consider $N \times N$ matrix Hermitian integrals that are τ-functions, and show how the application of convolution symmetries leads to new matrix models of the externally coupled type. In the following, let $d\mu(M)$, be a measure on the space of $N \times N$ Hermitian matrices $M \in \mathbb{H}^{N \times N}$ that is invariant under conjugation by unitary matrices, and such that the reduced measure, projected to the space of eigenvalues by integration over the group $U(N)$, is a product of N identical measures $d\mu_0$ on $\mathbb{R}$, times the Jacobian factor $\Delta^2(X)$,

$$\int_{U \in U(N)} d\mu(UXU^\dagger) = \prod_{a=1}^{N} d\mu_0(x_a)\Delta^2(X), \tag{4-1}$$

where $X = \mathrm{diag}(x_1, \ldots, x_N)$.

4.1. Convolution symmetries, externally coupled Hermitian matrix models and τ-functions as finite determinants.

It is well known that Hermitian matrix integrals of the form

$$Z_N(t) = \int_{M \in \mathbb{H}^{N \times N}} d\mu(M) e^{\mathrm{tr} \sum_{i=1}^{\infty} t_i M^i} \tag{4-2}$$

$$= \prod_{a=1}^{N} \int_{\mathbb{R}} d\mu_0(x_a) e^{\sum_{i=1}^{\infty} t_i x_a^i} \Delta^2(X) \tag{4-3}$$

are KP-Toda τ-functions [Kharchev et al. 1991]. The Schur function expansion is

$$Z_N(t) = \sum_{\ell(\lambda) \leq N} \pi_{N,d\mu}(\lambda)s_\lambda(t), \tag{4-4}$$

where the coefficients $\pi_{N,d\mu}(\lambda)$ are expressible as determinants in terms of the matrix of moments [Harnad and Orlov 2002; 2003; 2006]

$$\pi_{N,d\mu}(\lambda) = \prod_{a=1}^{N} \left(\int_{\mathbb{R}} d\mu_0(x_a) \right) \Delta^2(X)s_\lambda([X]) \tag{4-5}$$

$$= (-1)^{\frac{1}{2}N(N-1)} N! \det(\mathcal{M}_{\lambda_i+N-i,j-1})|_{1 \leq i,j \leq N}, \tag{4-6}$$

$$\mathcal{M}_{ij} := \int_{\mathbb{R}} d\mu_0(x) x^{i+j}. \tag{4-7}$$

Now consider the externally coupled matrix model integral (cf. [Brézin and Hikami 1996; Wang 2009; Zinn-Justin 1998; 2002])

$$Z_{N,\text{ext}}(A) := \int_{M \in \mathbb{H}^{N \times N}} d\mu(M) e^{\text{tr}(AM)}, \tag{4-8}$$

where $A \in \mathbb{H}^{N \times N}$ is a fixed $N \times N$ Hermitian matrix. This can be obtained by simply applying a convolution symmetry transformation of the type given in Example 3.1 to the τ-function defined by the matrix integral (4-3).

Proposition 4.1. *Applying the convolution symmetry $\tilde{C}_\rho$ to the τ-function $Z_N(t)$, where $\rho_+(z)$ and $\rho_-(z)$ are defined as in (3-25), (3-26), and choosing the KP flow parameters as $t = [A]$ gives, within a multiplicative constant, the externally coupled matrix integral (4-8)*

$$\tilde{C}_\rho(Z_N)([A]) = \left(\prod_{i=1}^{N-1} i! \right)^{-1} Z_{N,\text{ext}}(A). \tag{4-9}$$

Proof. Substituting the expansion [Harnad and Orlov 2006]

$$e^{\text{tr } AM} = \sum_{\ell(\lambda) \leq N} \frac{d_{\lambda,N}}{(N)_\lambda} s_\lambda([AM]) \tag{4-10}$$

into (4-8), where

$$d_{\lambda,N} = s_\lambda(\mathbb{1}_N) \tag{4-11}$$

is the dimension of the irreducible $\text{GL}(N)$ tensor representation of symmetry type λ, and expressing M in diagonalized form as

$$M = UXU^\dagger, \tag{4-12}$$

where $U \in U(N)$ and $X = \text{diag}(x_1, \dots x_N)$, gives

$$Z_{N,\text{ext}}(A) = \sum_{\ell(\lambda) \leq N} \int_{U \in U(N)} d\mu_H(U) \prod_{a=1}^{N} \int_{\mathbb{R}} d\mu_0(x_a) e^{\sum_{i=1}^{\infty} t_i x_a^i}$$

$$\times \Delta^2(X) \frac{d_{\lambda,N}}{(N)_\lambda} s_\lambda([AUXU^\dagger]). \tag{4-13}$$

Evaluating the character integral (3-62) and using (4-5), it follows that

$$Z_{N,\text{ext}}(A) = \sum_{\ell(\lambda)\leq N} \prod_{a=1}^{N} \int_{\mathbb{R}} d\mu_0(x_a) e^{\sum_{i=1}^{\infty} t_i x_a^i} \Delta^2(X) \frac{1}{(N)_\lambda} s_\lambda([A]) s_\lambda([X])$$

$$= \sum_{\ell(\lambda)\leq N} \frac{1}{(N)_\lambda} \pi_{N,d\mu}(\lambda) s_\lambda([A])$$

$$= \sum_{\ell(\lambda)\leq N} \left(\prod_{i=1}^{N-1} i! \right) r_\lambda(N) s_\lambda([A])$$

$$= \left(\prod_{i=1}^{N-1} i! \right) \tilde{C}_\rho(Z_N)\big|_{t=[A]}, \tag{4-14}$$

where the third line follows from the expression (3-30) for $r_\lambda(N)$ in Example 3.1 and the last from Proposition 4.1, 3.1. □

More generally, given an arbitrary function $\rho_+(z)$, analytic on the interior of S^1 and choosing $\rho_-(z)$ as in (3-26), we may define a new externally coupled matrix integral

$$Z_{N,\rho}(A) := \int_{M\in\mathbb{H}^{N\times N}} d\mu(M)\, \tau_r(N, [AM]), \tag{4-15}$$

in which $e^{\text{tr}\, AM}$ is replaced by

$$\tau_r(N, [M]) := \tau_r(N, [\mathbb{1}_N], [M]) = \sum_{\ell(\lambda)\leq N} d_{\lambda,N} r_\lambda(N) s_\lambda([M]). \tag{4-16}$$

Then by the same calculation as above, it follows that $Z_{N,\rho}(A)$ is again just the τ-function obtained by applying the convolution symmetry $\tilde{C}_\rho$ to Z_N, evaluated at the parameter values $t = [A]$.

Proposition 4.2. *Applying the convolution symmetry $\tilde{C}_\rho$ to Z_N gives*

$$\tilde{C}_\rho(Z_N)([A]) = Z_{N,\rho}(A). \tag{4-17}$$

In particular, if we take (ρ_+, ρ_-) as in Example 3.2 above, we obtain (cf. [Harnad and Orlov 2006])

$$Z_{N,\rho}(A) =$$

$$\left(\prod_{i=0}^{N-1} \frac{(a)_i}{i!} \right) \zeta^{\frac{1}{2}N(N-1)} \int_{M\in\mathbb{H}^{N\times N}} d\mu(M) \det(1 - \zeta AM)^{-a-N+1}, \tag{4-18}$$

showing that this also is a KP-Toda τ-function evaluated at parameter values $t = [A]$.

Returning to the general case, a finite determinantal formula for $Z_{N,\rho}(A)$ is given by the following.

Proposition 4.3.

$$Z_{N,\rho}(A) = \frac{(-1)^{\frac{1}{2}N(N-1)}N!}{\Delta(A)} \det(G_{ij}(\rho, A))\big|_{1 \le i,j \le N}, \qquad (4\text{-}19)$$

where

$$G_{ij}(\rho, A) := \int_{\mathbb{R}} d\mu_0(x) x^{i-1} \rho_+(a_j x). \qquad (4\text{-}20)$$

Proof. Applying the character integral identity (3-62) to (4-15) gives

$$Z_{N,\rho}(A) = \int_{M \in \mathbb{H}^{N \times N}} d\mu(M) \sum_{\ell(\lambda) \le N} r_\lambda(N) s_\lambda([A]) s_\lambda([M]) \qquad (4\text{-}21)$$

$$= \frac{1}{\Delta(A)} \int d\mu_0(X) \Delta(X) \det(\rho_+(a_i x_j))\big|_{1 \le i,j \le N} \qquad (4\text{-}22)$$

$$= \frac{(-1)^{\frac{1}{2}N(N-1)}N!}{\Delta(A)} \det(G_{ij}(\rho, A))\big|_{1 \le i,j \le N}, \qquad (4\text{-}23)$$

with $G_{ij}(\rho, A)$ defined by (4-20). Here, the integration over the $U(N)$ group has been performed and Lemma 3.3 has been used in (4-22). Equation (4-23) follows from (4-22) by applying the Andréief identity [Andréief 1886] in the form

$$\left(\prod_{m=1}^{N} \int d\mu_0(x_m)\right) \det(\phi_i(x_j)) \det(\psi_k(x_l))\big|_{\substack{1 \le i,j \le N \\ 1 \le k,l \le N}}$$

$$= N! \det\left(\int \phi_i(x) \psi_j(x)\right)\Big|_{1 \le i,j \le N}, \qquad (4\text{-}24)$$

with

$$\phi_i(x) = x^{N-i}, \quad \psi_j(x) := \rho_+(a_j x), \qquad (4\text{-}25)$$

since

$$\Delta(X) = \det(\phi_i(x_j)). \qquad (4\text{-}26)$$

$\square$

4.2. *Externally coupled two-matrix models.*

We now turn to the case of two-matrix models. For simplicity, we only consider Itzykson–Zuber exponential coupling [1980], although the same double convolution transformations may be applied to all the couplings considered in [Harnad and Orlov 2006]. Using the HCIZ identity (3-63) to evaluate the integrals over the unitary groups $U(N)$, we obtain

$$Z_N^{(2)}(t, \tilde{t}) = \int_{M_1 \in \mathbb{H}^{N \times N}} d\mu(M_1) \int_{M_2 \in \mathbb{H}^{N \times N}} d\tilde{\mu}(M_2) \qquad (4\text{-}27)$$
$$\times e^{\text{tr}(\sum_{i=1}^{\infty}(t_i M_1^i + \tilde{t}_i M_2^i) + M_1 M_2)}$$

$$= \prod_{k=1}^{N} k! \prod_{a=1}^{N} \left(\int_{\mathbb{R}} d\mu_0(x_a) \int_{\mathbb{R}} d\tilde{\mu}_0(y_a) e^{\sum_{i=1}^{\infty}(t_i x_a^i + \tilde{t}_i y_a^i + x_a y_a)} \right)$$
$$\times \Delta(X) \Delta(Y),$$

where $Y = \text{diag}(y_1, \ldots, y_N)$. This is known to be a 2D-Toda τ-function [Adler and van Moerbeke 1999; 2005; Harnad and Orlov 2002; 2003; 2006; Orlov and Shiota 2005], with double Schur function expansion

$$Z_N^{(2)}(t, \tilde{t}) = \sum_{\lambda} \sum_{\mu} B_{N, d\mu, d\tilde{\mu}}(\lambda, \mu) s_\lambda(t) s_\mu(\tilde{t}), \qquad (4\text{-}28)$$

where the coefficients $B_{N, d\mu, d\tilde{\mu}}(\lambda, \mu)$ are $N \times N$ determinants of submatrices in terms of the matrix of bimoments

$$B_{N, d\mu, d\tilde{\mu}}(\lambda, \mu) = \prod_{k=1}^{N} k! \prod_{a=1}^{N} \left(\int_{\mathbb{R}} d\mu_0(x_a) \int_{\mathbb{R}} d\tilde{\mu}_0(y_a) e^{x_a y_a} \right)$$
$$\times \Delta(X) \Delta(Y) s_\lambda([X]) s_\mu([Y])$$
$$= (N!) \prod_{k=1}^{N} k! \det(\mathcal{B}_{\lambda_i - i + N, \mu_j - j + N})|_{1 \le i, j \le N}, \qquad (4\text{-}29)$$

$$\mathcal{B}_{ij} := \int_{\mathbb{R}} d\mu_0(x_a) \int_{\mathbb{R}} d\tilde{\mu}_0(y_a) e^{x_a y_a} x^i y^j. \qquad (4\text{-}30)$$

Now, choosing a pair of elements $(\rho, \tilde{\rho})$, with both ρ_- and $\tilde{\rho}_-$ as in (3-26), we may define a family of externally coupled two-matrix models, by

$$Z_{N, \rho, \tilde{\rho}}^{(2)}(A, B) := \int_{M_1 \in \mathbb{H}^{N \times N}} d\mu(M_1) \int_{M_2 \in \mathbb{H}^{N \times N}} d\tilde{\mu}(M_2)$$
$$\times \tau_r(N, [A], [M_1]) \tau_{\tilde{r}}(N, [B], [M_2]) e^{\text{tr}(M_1 M_2)}, \qquad (4\text{-}31)$$

where A, B are hermitian $N \times N$ matrices. This class may be obtained as the 2D-Toda τ-function resulting from applying the convolution symmetry $\tilde{C}_{\rho, \tilde{\rho}}$ to $Z_N^{(2)}$.

Proposition 4.4. *Applying the convolution symmetry $\tilde{C}_{\rho, \tilde{\rho}}$ to $Z_N^{(2)}$ and evaluating at the parameter values $t = [A]$, $\tilde{t} = [B]$ gives the externally coupled matrix integral (4-31)*

$$\tilde{C}_{\rho, \tilde{\rho}}^{(2)}(Z_N^{(2)})([A], [B])) = Z_{N, \rho, \tilde{\rho}}^{(2)}(A, B). \qquad (4\text{-}32)$$

Proof. Because of the $U(N) \times U(N)$ invariance of the measures $d\mu$ and $d\tilde{\mu}$ in (4-31) and all factors in the integrand, except for the coupling term $e^{\operatorname{tr}(M_1 M_2)}$, we may carry out the two $U(N)$ integrations, using the HCIZ identity (3-63), to obtain a reduced integral over the diagonal matrices $X = \operatorname{diag}(x_1, \ldots, x_N)$, $Y = \operatorname{diag}(y_1, \ldots, y_N)$ of eigenvalues of M_1 and M_2:

$$
Z_{N,\rho,\tilde{\rho}}^{(2)}(A, B)
$$

$$
= \prod_{k=1}^{N} k! \prod_{a=1}^{N} \left(\int_{\mathbb{R}} d\mu_0(x_a) \int_{\mathbb{R}} d\tilde{\mu}_0(y_a) \, e^{x_a y_a} \right) \Delta(X)\Delta(Y) \qquad (4\text{-}33)
$$
$$
\times \tau_{C_\rho}(N, [A], [X]) \tau_{C_{\tilde{\rho}}}(N, [B], [Y])
$$

$$
= \sum_{\ell(\lambda) \leq N} \sum_{\ell(\mu) \leq N} r_\lambda(N) B_{N, d\mu, d\tilde{\mu}}(\lambda, \mu) \tilde{r}_\lambda(N) s_\lambda([A]) s_\mu([B]) \qquad (4\text{-}34)
$$

$$
= \tilde{C}_{\rho,\tilde{\rho}}^{(2)}(Z_N^{(2)})([A], [B]). \qquad (4\text{-}35)
$$

where the second equality follows from (3-44) and the last from Proposition 3.1, (3-41). $\qquad\qquad\square$

Since the dependence on A and B is $U(N) \times U(N)$ conjugation invariant we may choose, without loss of generality, A and B to be diagonal matrices

$$
A = \operatorname{diag}(a_1, \ldots, a_N), \quad B = \operatorname{diag}(b_1, \ldots, b_N). \qquad (4\text{-}36)
$$

We then obtain, as in the one-matrix case, a finite determinantal formula for the 2D-Toda τ-function $Z_{N,\rho,\tilde{\rho}}^{(2)}(A, B)$.

Proposition 4.5.

$$
Z_{N,\rho,\tilde{\rho}}^{(2)}(A, B) = \frac{N! \left(\prod\limits_{k=1}^{N} k! \right)}{\Delta(A)\Delta(B)} \det(G_{ij}(\rho, \tilde{\rho}, A, B)|_{1 \leq i,j \leq N}, \qquad (4\text{-}37)
$$

where

$$
G_{ij}(\rho, \tilde{\rho}, A, B) := \int_{\mathbb{R}} d\mu_0(x) \int_{\mathbb{R}} d\tilde{\mu}_0(y) e^{xy} \rho_+(a_i x) \tilde{\rho}_+(b_j y). \qquad (4\text{-}38)
$$

Proof.

$$
Z_{N,\rho,\tilde{\rho}}^{(2)}(A, B) = \int_{M_1 \in \mathbb{H}^{N \times N}} d\mu(M_1) \int_{M_2 \in \mathbb{H}^{N \times N}} d\tilde{\mu}(M_2) e^{\operatorname{tr}(M_1 M_2)} \qquad (4\text{-}39)
$$

$$
\times \sum_{\ell(\lambda) \leq N} r_\lambda(N) s_\lambda([A]) s_\lambda([M_1]) \sum_{\ell(\mu) \leq N} \tilde{r}_\mu(N) s_\mu([B]) s_\mu([M_2])
$$

$$= \frac{\left(\prod_{k=1}^{N} k!\right)}{\Delta(A)\Delta(B)} \int d\mu(X) \int d\tilde{\mu}(Y) \, e^{\sum_{i=1}^{N} x_i y_i} \tag{4-40}$$

$$\times \det(\rho_+(a_k x_l))\big|_{1 \le k,l \le N} \det(\tilde{\rho}_+(b_m y_n))\big|_{1 \le m,n \le N} \tag{4-41}$$

$$= \frac{N! \left(\prod_{k=1}^{N} k!\right)}{\Delta(A)\Delta(B))} \det(G_{ij}(\rho, \tilde{\rho}, A, B))\big|_{1 \le i,j \le N}. \tag{4-42}$$

In (4-41), we have used the HCIZ identity (3-63), antisymmetry of the determinants in the integrand with respect to permutations in the integration variables $(x_1, \ldots, x_N)$ and $(y_1, \ldots, y_N)$ and Lemma 3.3 twice, while in (4-42), we have used the Andréief identity [1886] in the form

$$\left(\prod_{m=1}^{N} \int d\mu(x_m, y_m)\right) \det(\phi_i(x_j)) \det(\psi_k(y_l))\Big|_{\substack{1 \le i,j \le N \\ 1 \le k,l \le N}} \tag{4-43}$$

$$= N! \det\left(\int d\mu(x, y) \phi_i(x) \psi_j(y)\right)\Big|_{1 \le i,j \le N}. \quad \square$$

As the simplest example of a 2D-Toda τ-function obtained through Propositions 4.4 and 4.5, consider the case when the measures $d\mu_0(x)$ and $d\mu_0(y)$ are both Gaussian, and ρ_+ and $\tilde{\rho}_+$ are both taken as the exponential function.

Example 4.1.

$$d\mu_0(x) = e^{-\sigma x^2} dx, \quad d\mu_0(y) = e^{-\sigma y^2} dy, \quad \rho_+(x) = e^x, \quad \tilde{\rho}_+(y) = e^y. \tag{4-44}$$

Evaluating the Gaussian integrals gives

$$G_{ij} = \frac{2\pi}{\sqrt{1+4\sigma^2}} \exp \frac{\sigma(a_i^2 + b_j^2) - a_i b_j}{4\sigma^2 - 1}, \tag{4-45}$$

and hence

$$Z_{N,\rho}(A) = \frac{(2\pi)^N N! \prod_{k=1}^{N} k!}{(1+4\sigma^2)^{N/2} \Delta(A)\Delta(B)} \exp\left(\frac{\sigma}{4\sigma^2 - 1} \sum_{i=1}^{N} (a_i^2 + b_i^2)\right)$$

$$\times \det\left(\exp \frac{\sigma a_i b_j}{1 - 4\sigma^2}\right). \tag{4-46}$$

The exponential factor on the first line of (4-46) is a linear exponential in terms of the 2KP flow variable t_2 and $\tilde{t}_2$ and hence, through the Sato formula (1-3), produces just a gauge factor multiplying the Baker–Akhiezer function [Segal and Wilson 1985]. Therefore (4-46) is just a rescaled, gauge transformed version of the 2KP τ-function of hypergeometric type appearing in the integrand of the Itzykson–Zuber coupled two-matrix model [Itzykson and Zuber 1980].

4.3. *More general 2D-Toda τ-functions as multiple integrals.*

We may extend the above results to more general 2KP-Toda τ-functions expressed as multiple integrals and finite determinants. To begin with, the following multiple integral

$$\tau_{d\mu}^{(2)}(N, t, \tilde{t}) = \prod_{a=1}^{N}\left(\int_{\Gamma}\int_{\tilde{\Gamma}} d\mu(x_a, y_a)e^{\sum_{i=1}^{\infty}(t_i x_a^i + \tilde{t}_i y_a^i)}\right)\Delta(X)\Delta(Y), \quad (4\text{-}47)$$

where Γ, $\tilde{\Gamma}$ are curves in the complex x- and y-planes and $d\mu(x, y)$ is a measure on $\Gamma \times \tilde{\Gamma}$, is a 2D-Toda τ-function [Harnad and Orlov 2006] for a large class of measures $d\mu_0(x, y)$. Applying a double convolution symmetry $\tilde{C}_{\rho, \tilde{\rho}}$, with ρ_- and $\tilde{\rho}_-$ the same as in (3-26), gives a new 2D-Toda τ-function, also having a multiple integral representation.

Proposition 4.6.

$$\tilde{C}_{\rho, \tilde{\rho}}^{(2)}(\tau_{d\mu}^{(2)})(N, t, \tilde{t})$$

$$= \prod_{a=1}^{N}\left(\int_{\Gamma}\int_{\tilde{\Gamma}} d\mu(x_a, y_a)\right)\Delta(X)\Delta(Y)\tau_r(N, t, [X])\tau_{\tilde{r}}(N, \tilde{t}, [Y]). \quad (4\text{-}48)$$

Proof. This is proved similarly to Proposition 4.4, using the Cauchy–Littlewood identity (2-31) twice in the form

$$\prod_{a=1}^{N} e^{\sum_{i=1}^{\infty}(t_i x_a^i + \tilde{t}_i y_a^i)} = \sum_{\ell(\lambda)\leq N} s_\lambda(t)s_\lambda([X]) \sum_{\ell(\mu)\leq N} s_\mu(\tilde{t})s_\mu([Y]). \quad (4\text{-}49)$$

$\square$

Evaluating at parameter values $t = [A]$ and $\tilde{t} = [B]$ and applying Lemma 3.3 again gives the τ-function of (4-48) in $N \times N$ determinantal form.

Proposition 4.7.

$$\tilde{C}_{\rho, \tilde{\rho}}^{(2)}(\tau_{d\mu}^{(2)})([A], [B]) = \frac{N!}{\Delta(A)\Delta(B)} \det(G_{ij}(\rho, \tilde{\rho}, A, B))|_{1\leq i,j\leq N}, \quad (4\text{-}50)$$

where

$$G_{ij}(\rho, \tilde{\rho}, A, B) := \int_{\Gamma}\int_{\tilde{\Gamma}} d\mu(x, y)\rho_+(a_j x)\tilde{\rho}_+(b_j y). \quad (4\text{-}51)$$

Proof.

$$\tilde{C}_{\rho,\tilde{\rho}}^{(2)}(\tau_{d\mu}^{(2)})([A],[B]) = \frac{1}{\Delta(A)\Delta(B)} \prod_{a=1}^{N} \left(\int_{\Gamma} \int_{\tilde{\Gamma}} d\mu(x_a, y_a) \right)$$

$$\times \det(\rho_+(a_k x_l))\big|_{1 \leq k,l \leq N} \det(\tilde{\rho}_+(b_m y_n))\big|_{1 \leq m,n \leq N}$$

$$= \frac{N!}{\Delta(A)\Delta(B)} \det(G_{ij}(\rho, \tilde{\rho}, A, B))\big|_{1 \leq i,j \leq N}, \qquad (4\text{-}52)$$

where again we have used the Lemma 3.3 twice and the Andréief identity in the form (4-43). □

This therefore provides a new class of 2D-Toda τ-functions expressible in such a finite determinantal form, associated to any pair of curves Γ, $\tilde{\Gamma}$, together with a measure $d\mu$ on their product, and a pair of functions $\rho_+(x)$ and $\tilde{\rho}_+(y)$, such that the integrals in (4-51) are well defined and convergent.

Acknowledgements

The authors would like to thank D. Wang for helpful discussions relating to this work.

References

[Adler and van Moerbeke 1999] M. Adler and P. van Moerbeke, "The spectrum of coupled random matrices", *Ann. of Math.* (2) **149**:3 (1999), 921–976.

[Adler and van Moerbeke 2005] M. Adler and P. van Moerbeke, "Virasoro action on Schur function expansions, skew Young tableaux, and random walks", *Comm. Pure Appl. Math.* **58**:3 (2005), 362–408.

[Andréief 1886] C. Andréief, "Note sur une relation entre les intégrales définines des produits des fonctions", *Mém. Soc. Sci. Phys. Nat. Bordeaux* (3) **2** (1886), 1–14.

[Bettelheim et al. 2007] E. Bettelheim, A. G. Abanov, and P. Wiegmann, "Nonlinear dynamics of quantum systems and soliton theory", *J. Phys. A* **40**:8 (2007), F193–F207.

[Brézin and Hikami 1996] E. Brézin and S. Hikami, "Correlations of nearby levels induced by a random potential", *Nuclear Phys. B* **479**:3 (1996), 697–706.

[Date et al. 1981/82] E. Date, M. Jimbo, M. Kashiwara, and T. Miwa, "Transformation groups for soliton equations, IV: a new hierarchy of soliton equations of KP-type", *Phys. D* **4**:3 (1981/82), 343–365.

[Date et al. 1983] E. Date, M. Kashiwara, M. Jimbo, and T. Miwa, "Transformation groups for soliton equations", pp. 39–119 in *Nonlinear integrable systems — classical theory and quantum theory* (Kyoto, 1981), edited by M. Jimbo and T. Miwa, World Sci. Publishing, Singapore, 1983.

[Harnad and Orlov 2002] J. Harnad and A. Y. Orlov, "Matrix integrals as Borel sums of Schur function expansions", pp. 116–123 in *SPT 2002: Symmetry and perturbation theory* (Cala Gonone), edited by S. Abenda et al., World Sci. Publ., 2002.

[Harnad and Orlov 2003] J. Harnad and A. Y. Orlov, "Scalar products of symmetric functions and matrix integrals", *Teoret. Mat. Fiz.* **137**:3 (2003), 375–392. In Russian; translated in *Theoret. Math. Phys.* **137**:3 (2003), 1676-1690.

[Harnad and Orlov 2006] J. Harnad and A. Y. Orlov, "Fermionic construction of partition functions for two-matrix models and perturbative Schur function expansions", *J. Phys. A* **39**:28 (2006), 8783–8809.

[Harnad and Orlov 2007] J. Harnad and A. Y. Orlov, "Fermionic construction of tau functions and random processes", *Phys. D* **235**:1-2 (2007), 168–206.

[Itzykson and Zuber 1980] C. Itzykson and J. B. Zuber, "The planar approximation, II", *J. Math. Phys.* **21**:3 (1980), 411–421.

[Jimbo and Miwa 1983] M. Jimbo and T. Miwa, "Solitons and infinite-dimensional Lie algebras", *Publ. Res. Inst. Math. Sci.* **19**:3 (1983), 943–1001.

[Kharchev et al. 1991] S. Kharchev, A. Marshakov, A. Mironov, A. Orlov, and A. Zabrodin, "Matrix models among integrable theories: forced hierarchies and operator formalism", *Nuclear Phys. B* **366**:3 (1991), 569–601.

[Kontsevich 1992] M. Kontsevich, "Intersection theory on the moduli space of curves and the matrix Airy function", *Comm. Math. Phys.* **147**:1 (1992), 1–23.

[Macdonald 1995] I. G. Macdonald, *Symmetric functions and Hall polynomials*, 2nd ed., Oxford University Press, New York, 1995.

[Okounkov 2000] A. Okounkov, "Toda equations for Hurwitz numbers", *Math. Res. Lett.* **7**:4 (2000), 447–453.

[Okounkov and Pandharipande 2006] A. Okounkov and R. Pandharipande, "Gromov–Witten theory, Hurwitz theory, and completed cycles", *Ann. of Math.* (2) **163**:2 (2006), 517–560.

[Orlov 2004] A. Y. Orlov, "New solvable matrix integrals", *Internat. J. Modern Phys. A* **19** (2004), 276–293.

[Orlov 2006] A. Y. Orlov, "Hypergeometric functions as infinite-soliton tau functions", *Teoret. Mat. Fiz.* **146**:2 (2006), 222–250. In Russian; translated in *Theor. Math. Phys.* **146**, 183-206 (2006).

[Orlov and Scherbin 2001] A. Y. Orlov and D. M. Scherbin, "Multivariate hypergeometric functions as τ-functions of Toda lattice and Kadomtsev–Petviashvili equation", *Phys. D* **152/153** (2001), 51–65.

[Orlov and Shiota 2005] A. Y. Orlov and T. Shiota, "Schur function expansion for normal matrix model and associated discrete matrix models", *Phys. Lett. A* **343**:5 (2005), 384–396.

[Sato 1981] M. Sato, "Soliton equations as dynamical systems on infinite dimensional Grassmann manifolds", *RIMS Kokyuroku* **439** (1981), 30–46.

[Sato and Sato 1983] M. Sato and Y. Sato, "Soliton equations as dynamical systems on infinite-dimensional Grassmann manifold", pp. 259–271 in *Nonlinear partial differential equations in applied science* (Tokyo, 1982), edited by H. Fujita et al., North-Holland Math. Stud. **81**, North-Holland, Amsterdam, 1983.

[Segal and Wilson 1985] G. Segal and G. Wilson, "Loop groups and equations of KdV type", *Inst. Hautes Études Sci. Publ. Math.* **61** (1985), 5–65.

[Silverstein and Bai 1995] J. W. Silverstein and Z. D. Bai, "On the empirical distribution of eigenvalues of a class of large-dimensional random matrices", *J. Multivariate Anal.* **54**:2 (1995), 175–192.

[Takasaki 1984] K. Takasaki, "Initial value problem for the Toda lattice hierarchy", pp. 139–163 in *Group representations and systems of differential equations* (Tokyo, 1982), edited by K. Okamoto, Adv. Stud. Pure Math. **4**, North-Holland, Amsterdam, 1984.

[Ueno and Takasaki 1984] K. Ueno and K. Takasaki, "Toda lattice hierarchy", pp. 1–95 in *Group representations and systems of differential equations* (Tokyo, 1982), edited by K. Okamoto, Adv. Stud. Pure Math. **4**, North-Holland, Amsterdam, 1984.

[Wang 2009] D. Wang, "Random matrices with external source and KP τ functions", *J. Math. Phys.* **50**:7 (2009), 073506, 10.

[Zinn-Justin 1998] P. Zinn-Justin, "Universality of correlation functions of Hermitian random matrices in an external field", *Comm. Math. Phys.* **194**:3 (1998), 631–650.

[Zinn-Justin 2002] P. Zinn-Justin, "HCIZ integral and 2D Toda lattice hierarchy", *Nuclear Phys. B* **634**:3 (2002), 417–432.

harnad@crm.umontreal.ca

Centre de recherches mathématiques,
Université de Montréal,
C. P. 6128, Succ. centre-ville, Montréal H3C 3J7, Canada

Department of Mathematics and Statistics,
Concordia University, 1455 de Maisonneuve Boulevard W.,
Montreal, Quebec, Canada H3G 1M8

orlovs55@mail.ru

Nonlinear Wave Processes Laboratory, Oceanlogy Institute,
36 Nakhimovskii Prospect, Moscow, 117851, Russia

Random Matrices
MSRI Publications
Volume 65, 2014

Universality limits via "old style" analysis

DORON S. LUBINSKY

Techniques from "old style" orthogonal polynomials have turned out to be useful in establishing universality limits for fairly general measures. We survey some of these.

1. Introduction

We focus on the classical setting of random Hermitian matrices: consider a probability distribution $P^{(n)}$ on the space of n by n Hermitian matrices $M = (m_{ij})_{1 \leq i, j \leq n}$:

$$P^{(n)}(M) = cw(M)dM$$
$$= cw(M) \left(\prod_{j=1}^{n} dm_{jj} \right) \left(\prod_{j<k} d(\operatorname{Re} m_{jk}) \, d(\operatorname{Im} m_{jk}) \right).$$

Here w is some nonnegative function defined on Hermitian matrices, and c is a normalizing constant. The most important case is

$$w(M) = \exp(-2n \operatorname{tr} Q(M)),$$

for appropriate functions Q. In particular, the choice $Q(M) = M^2$, leads to the Gaussian unitary ensemble (apart from scaling) that was considered by Wigner, in the context of scattering theory for heavy nuclei. When expressed in spectral form, that is as a probability density function on the eigenvalues $x_1 \leq x_2 \leq \cdots \leq x_n$ of M, it takes the form

$$P^{(n)}(x_1, x_2, \ldots, x_n) = c \left(\prod_{j=1}^{n} w(x_j) \right) \left(\prod_{i<j} (x_i - x_j)^2 \right). \qquad (1\text{-}1)$$

See [Deift 1999, p. 102 ff.]. Again, c is a normalizing constant. Note that w now can be any nonnegative measurable function.

Research supported by NSF grant DMS1001182 and US-Israel BSF grant 2008399.

MSC2010: 15B52, 60B20, 60F99, 42C05, 33C50.

Keywords: orthogonal polynomials, random matrices, unitary ensembles, correlation functions, Christoffel functions.

In most applications, we want to let $n \to \infty$, and obviously the n-fold density complicates issues. So we often integrate out most variables, forming marginal distributions. One particularly important quantity is the m-point correlation function [Deift 1999, p. 112]:

$$\tilde{R}_m(x_1, x_2, \ldots, x_m)$$
$$= \frac{n!}{(n-m)!} \int \cdots \int P^{(n)}(x_1, x_2, \ldots, x_n) \, dx_{m+1} \, dx_{m+2} \ldots dx_n.$$

Here typically, we fix m, and study $\tilde{R}_m$ as $n \to \infty$. $\tilde{R}_m$ is useful in examining spacing of eigenvalues, and counting the expected number of eigenvalues in some set. For example, if B is a measurable subset of $\mathbb{R}$,

$$\int_B \cdots \int_B \tilde{R}_m(x_1, x_2, \ldots, x_m) \, dx_1 \, dx_2 \ldots dx_m$$

counts the expected number of m-tuples $(x_1, x_2, \ldots, x_m)$ of eigenvalues with each $x_j \in B$.

The *universality limit in the bulk* asserts that for fixed $m \geq 2$, and ξ in the "bulk of the spectrum" (where w above "lives") and real $a_1, a_2, \ldots, a_m$, we have

$$\lim_{n \to \infty} \frac{1}{(n\omega(\xi))^m} \tilde{R}_m\left(\xi + \frac{a_1}{n\omega(\xi)}, \xi + \frac{a_2}{n\omega(\xi)}, \ldots, \xi + \frac{a_m}{n\omega(\xi)}\right)$$
$$= \det(\mathbb{S}(a_i - a_j))_{1 \leq i, j \leq m}. \quad (1\text{-}2)$$

Here $\mathbb{S}$ is the *sine* or *sinc* kernel, given by

$$\mathbb{S}(t) = \frac{\sin \pi t}{\pi t}, \quad t \neq 0, \quad (1\text{-}3)$$

and $\mathbb{S}(0) = 1$. What is ω? It is basically an *equilibrium density function*, and we'll discuss this further later. It is appropriate to call the limit (1-2) universal, as it does not depend on ξ, nor on the weight function w.

One of the principal goals has been to establish the universality limit under more and more general conditions, and in this pursuit, orthogonal polynomials have turned out to be a useful tool. Throughout this paper, let μ be a finite positive Borel measure with compact support J and infinitely many points in the support. Define orthonormal polynomials

$$p_n(x) = \gamma_n x^n + \cdots, \quad \gamma_n > 0,$$

$n = 0, 1, 2, \ldots$, satisfying the orthonormality conditions

$$\int_J p_j \, p_k \, d\mu = \delta_{jk}.$$

We may think of w in (1-1) as μ'. The n-th reproducing kernel for μ is

$$K_n(\mu, x, y) = \sum_{k=0}^{n-1} p_k(x) p_k(y),$$

and the normalized kernel is

$$\tilde{K}_n(\mu, x, y) = \mu'(x)^{1/2} \mu'(y)^{1/2} K_n(\mu, x, y).$$

K_n satisfies the very useful extremal property [Freud 1971; Nevai 1979; 1986; Simon 2011]

$$K_n(\mu, \xi, \xi) = \inf_{\deg(P) \le n-1} \frac{P^2(\xi)}{\int P^2 \, d\mu}. \tag{1-4}$$

When $w = \mu'$, there are the remarkable formulae for the probability distribution $P^{(n)}$ [Deift 1999, p.112]:

$$P^{(n)}(x_1, x_2, \ldots, x_n) = \frac{1}{n!} \det(\tilde{K}_n(\mu, x_i, x_j))_{1 \le i, j \le n}, \tag{1-5}$$

and the m-point correlation function:

$$\tilde{R}_m(x_1, x_2, \ldots, x_m) = \det(\tilde{K}_n(\mu, x_i, x_j))_{1 \le i, j \le m}. \tag{1-6}$$

Sometimes we shall find it easier to exclude the measure from the variables $x_1, x_2, \ldots, x_m$, that is we consider the "stripped" m-point correlation function,

$$R_m(x_1, x_2, \ldots, x_m) = \det(K_n(\mu, x_i, x_j))_{1 \le i, j \le m}. \tag{1-7}$$

Because $\tilde{R}_m$ is the determinant of a fixed size m by m matrix, we see that (1-2) reduces to

$$\lim_{n \to \infty} \frac{\tilde{K}_n\left(\mu, \xi + \dfrac{a}{n\omega(\xi)}, \xi + \dfrac{b}{n\omega(\xi)}\right)}{n\omega(\xi)} = \mathbb{S}(a - b), \tag{1-8}$$

for real a, b.

Let us now turn to the choice of ω. As above, suppose that μ has compact support J. Then, ν_J is the probability measure that minimizes the energy integral

$$\iint \log \frac{1}{|x - y|} \, d\nu(x) \, d\nu(y),$$

taken over all probability measures ν on J. It is called the *equilibrium measure* for the set J. It is absolutely continuous in any subinterval of J. Throughout this paper, we set $\omega(x) = \nu_J'(x), x \in J^0$, where J^0 is the interior of J. We call

ω the *equilibrium density* of J and note that $\omega > 0$ in J^0. For example, when $J = [-1, 1]$,

$$\omega(x) = \frac{1}{\pi \sqrt{1 - x^2}}, \quad x \in (-1, 1).$$

Of course, the primary interest in random matrix theory is for varying measures, where at the n-th stage, $\mu'(x) = e^{-2nQ(x)}$, and there ω is an equilibrium density associated with the external field Q.

In some formulations for measures with fixed support, it is easier to prove the limit

$$\lim_{n \to \infty} \frac{\tilde{K}_n\left(\mu, \xi + \dfrac{a}{\tilde{K}_n(\mu, \xi, \xi)}, \xi + \dfrac{b}{\tilde{K}_n(\mu, \xi, \xi)}\right)}{\tilde{K}_n(\mu, \xi, \xi)} = \mathbb{S}(a - b), \qquad (1\text{-}9)$$

and this is consistent with (1-8), since under quite general conditions,

$$\lim_{n \to \infty} \frac{1}{n} \tilde{K}_n(\mu, \xi, \xi) = \lim_{n \to \infty} \frac{1}{n} \mu'(\xi) K_n(\mu, \xi, \xi) = \omega(\xi).$$

The most obvious approach to proving (1-2) is to use the Christoffel–Darboux formula,

$$K_n(\mu, u, v) = \frac{\gamma_{n-1}}{\gamma_n} \frac{p_n(u) p_{n-1}(v) - p_{n-1}(u) p_n(v)}{u - v}, \quad u \neq v, \qquad (1\text{-}10)$$

and to substitute in asymptotics for p_n as $n \to \infty$. This is what effectively was done for the classical weights. Of course there are many approaches, and we cannot survey them here. We simply note that it was the Riemann–Hilbert approach that allowed dramatic breakthroughs, and refer to other papers in this proceedings, and the books [Baik et al. 2007; Baik et al. 2008; Bleher and Its 2001; Deift 1999; Deift and Gioev 2009; Forrester 2010; Mehta 2004].

In terms of "old style" orthogonal polynomials, it was Eli Levin [Levin and Lubinsky 2009] who realized that relatively weak pointwise asymptotics, such as

$$p_n(\cos \theta) = \cos n\theta + o(1), \quad n \to \infty,$$

combined with a Markov–Bernstein inequality, are sufficient for universality. However, it has since been realized that much less suffices.

In subsequent sections, we outline some approaches from classical orthogonal polynomials and complex analysis. In Section 2, it is a comparison method. In Section 3, it is a method based on the theory of entire functions of exponential type. In Section 4, we discuss a recent extremal property. This survey has a narrow focus, and we omit many important contributions and topics.

2. A comparison method

The philosophy behind the comparison method is that a lot of quantities in orthogonal polynomials have a strong local component, and a weak global one. Perhaps the primary example of this is the Christoffel function $\lambda_n(\mu, x)$, or its reciprocal, the reproducing kernel along the diagonal $K_n(\mu, x, x)$. The global component in its asymptotic is determined by the equilibrium density ω of the support of μ, often accompanied by the hypothesis of regularity: we say that a compactly supported measure μ is regular (in the sense of Stahl, Totik, Ullmann) if the leading coefficients $\{\gamma_n\}$ of the orthonormal polynomials satisfy

$$\lim_{n\to\infty} \gamma_n^{1/n} = \frac{1}{\text{cap(supp}[\mu])}.$$

Here cap denotes the logarithmic capacity of the support of μ (see [Ransford 1995; Saff and Totik 1997; Stahl and Totik 1992] for definitions). A simple sufficient criterion for regularity is that of Erdős–Turán: if supp$[\mu]$ consists of finitely many intervals, and $\mu' > 0$ a.e. in each of those intervals, then μ is regular. There are more general criteria in [Stahl and Totik 1992]. Note that pure jump measures and pure singularly continuous measures can be regular.

The archetypal asymptotic for K_n is due to Maté, Nevai, and Totik [Máté et al. 1991] for $[-1, 1]$, and for general support, due to Totik [2000a]:

Theorem 2.1. *Let μ have compact support J and be regular. Let ω be the equilibrium density of J.*

(a) *For a.e. $x \in J^0$, we have*

$$\liminf_{n\to\infty} \frac{1}{n} K_n(\mu, x, x) \geq \frac{\omega(x)}{\mu'(x)}.$$

(b) *If in addition, I is a subinterval of J satisfying*

$$\int_I \log \mu' > -\infty, \tag{2-1}$$

then for a.e. $x \in I$,

$$\lim_{n\to\infty} \frac{1}{n} K_n(\mu, x, x) = \frac{\omega(x)}{\mu'(x)}. \tag{2-2}$$

Why is this local in flavor? Well if two measures μ and ν have the same support, and they are equal when restricted to the interval I, then $K_n(\mu, x, x)$ and $K_n(\nu, x, x)$ have the same asymptotic in I. In fact, more is possible: using fast decreasing polynomials, and the extremal property (1-4), one can prove that the ratio $K_n(\mu, x, x)/K_n(\nu, x, x)$ has limit 1 under much weaker conditions than in (b).

What relevance does this have to universality limits? The answer lies in the following inequality: if $\mu \leq \nu$, then for all real x, y,

$$\frac{|K_n(\mu, x, y) - K_n(\nu, x, y)|}{K_n(\mu, x, x)} \lesseqgtr \left(\frac{K_n(\mu, y, y)}{K_n(\mu, x, x)}\right)^{\frac{1}{2}} \left[1 - \frac{K_n(\nu, x, x)}{K_n(\mu, x, x)}\right]^{\frac{1}{2}}. \quad (2\text{-}3)$$

In particular, if x and y vary with n, and as $n \to \infty$, $\frac{K_n(\nu,x,x)}{K_n(\mu,x,x)}$ has limit 1, while $\frac{K_n(\mu,y,y)}{K_n(\mu,x,x)}$ remains bounded, then $K_n(\mu, x, y)$ and $K_n(\nu, x, y)$ have the same asymptotic. This inequality is easily proven by using the reproducing kernel properties of K_n, and the extremal property (1-4). It enables us to use universality limits for a larger "nice" measure ν to obtain the same for a "not so nice" measure μ, which is locally the same as ν. Thus [Lubinsky 2009a, Theorem 1.1, pp. 916–917]:

Theorem 2.2. *Let μ have support $[-1, 1]$ and be regular. Let $\xi \in (-1, 1)$ and assume μ is absolutely continuous in an open set containing ξ. Assume moreover, that μ' is positive and continuous at ξ. Then uniformly for a, b in compact subsets of the real line, we have*

$$\lim_{n \to \infty} \frac{\tilde{K}_n\left(\xi + \frac{a\pi\sqrt{1-\xi^2}}{n}, \xi + \frac{b\pi\sqrt{1-\xi^2}}{n}\right)}{\tilde{K}_n(\xi, \xi)} = \mathbb{S}(a - b).$$

Weaker integral forms of this limit were also established in [Lubinsky 2009a], when continuity of μ' was replaced by upper and lower bounds. However, the real potential of the inequality (2-3) was soon explored by Findley [2008], Simon [2008b] and Totik [2009]. It was Findley who replaced continuity of μ' by the Szegő condition on $[-1, 1]$. Totik used the method of "polynomial pullbacks", which is based on the observation that if P is a polynomial, then $P^{[-1]}[-1, 1]$ consists of finitely many intervals. This allows one to pass from asymptotics for $[-1, 1]$ to finitely many intervals. In turn, one can use the latter to approximate arbitrary compact sets. Barry Simon used instead Jost functions. Here is Totik's result:

Theorem 2.3. *Let μ have compact support J and be regular. Let I be a subinterval of J in which the local Szegő condition (2-1) holds. Then, for a.e. $x \in I$, and all real a, b,*

$$\lim_{n \to \infty} \frac{\tilde{K}_n\left(\mu, \xi + \frac{a}{n\omega(\xi)}, \xi + \frac{b}{n\omega(\xi)}\right)}{\tilde{K}_n(\mu, \xi, \xi)} = \frac{\sin \pi(a - b)}{\pi(a - b)}.$$

Totik actually showed that the asymptotic holds at any given ξ which is a Lebesgue point of both measure μ, and its local Szegő function. The comparison approach has also been applied to universality on the unit circle [Levin and Lubinsky 2007], to exponential weights [Levin and Lubinsky 2009], at the hard edge of the spectrum [Lubinsky 2008b], to Bergman polynomials [Lubinsky 2010], and in a generalized setting [Lubinsky 2008a].

3. A normal families approach

One pitfall of the comparison inequality, is that it needs a "starting" measure for which universality is known. For general supports, there is no such measure, unless one assumes regularity — which is a global restriction, albeit a weak one. In [Lubinsky 2008d], a method was introduced, that avoids this. It uses basic tools of complex analysis and complex approximation, such as normal families, together with some of the theory of entire functions, and reproducing kernels.

Perhaps the most fundamental idea in this approach is the notion that since K_n is a reproducing kernel for polynomials of degree $\leq n-1$, any scaled asymptotic limit of it must also be a reproducing kernel for a suitable space. It turns out that the correct limit setting is Paley–Wiener space. For given $\sigma > 0$, this is the Hilbert space of entire functions g of exponential type at most $\sigma > 0$, (so that given $\varepsilon > 0$, $|g(z)| = O(e^{(\sigma+\varepsilon)|z|})$, for large $|z|$), whose restriction to the real line is in $L_2(\mathbb{R})$, with the usual $L_2(\mathbb{R})$ inner product. Here the sinc kernel is the reproducing kernel [Stenger 1993, p. 95]:

$$g(x) = \int_{-\infty}^{\infty} g(t) \frac{\sin \sigma (x-t)}{\pi(x-t)} \, dt, \quad x \in \mathbb{R}. \tag{3-1}$$

It is not a trivial exercise to rigorously prove that reproducing kernels for polynomials turn into the reproducing kernel for Paley–Wiener space.

Assume that μ has compact support and that μ' is bounded above and below by positive constants in some open interval O containing the closed interval I. Then it is well known that for some $C_1, C_2 > 0$,

$$C_1 \leq \frac{1}{n} K_n(\mu, x, x) \leq C_2, \tag{3-2}$$

in any proper open subset O_1 of O. Indeed, this follows by comparing λ_n below to the Christoffel function of the weight 1 on a suitable subinterval of O, and comparing it above to a suitable dominating measure. Cauchy–Schwarz inequality's then gives

$$\frac{1}{n}|K_n(\mu, \xi, t)| \leq C \text{ for } \xi, \quad t \in O_1. \tag{3-3}$$

We can extend this estimate into the complex plane, by adapting Bernstein's inequality,

$$|P(z)| \leq |z + \sqrt{z^2 - 1}|^n \|P\|_{L_\infty[-1,1]},$$

which is valid for polynomials of degree $\leq n$ and all complex z. The branch of $\sqrt{\ }$ is taken so that $\sqrt{z^2 - 1} > 0$ for $z \in (1, \infty)$. This leads to

$$\left| \frac{1}{n} K_n\left(\xi + \frac{a}{n}, \xi + \frac{b}{n}\right) \right| \leq C_1 e^{C_2(|\operatorname{Im} a| + |\operatorname{Im} b|)}.$$

Here C_1 and C_2 are independent of n, a and b. In view of (3-2), the same is true of $\{f_n(a, b)\}_{n=1}^{\infty}$, where

$$f_n(a, b) = \frac{K_n\left(\xi + \dfrac{a}{\tilde{K}_n(\xi, \xi)}, \xi + \dfrac{b}{\tilde{K}_n(\xi, \xi)}\right)}{K_n(\xi, \xi)}.$$

Thus, given $A > 0$, we have for $n \geq n_0(A)$ and $|a|, |b| \leq A$, that

$$|f_n(a, b)| \leq C_1 e^{C_2(|\operatorname{Im} a| + |\operatorname{Im} b|)}. \tag{3-4}$$

We emphasize that C_1 and C_2 are independent of n, A, a and b.

Let $f(a, b)$ be the limit of some subsequence $\{f_n(\cdot, \cdot)\}_{n \in \mathcal{S}}$ of $\{f_n(\cdot, \cdot)\}_{n=1}^{\infty}$. It is an entire function in a, b, but (3-4) shows even more: namely that for all complex a, b,

$$|f(a, b)| \leq C_1 e^{C_2(|\operatorname{Im} a| + |\operatorname{Im} b|)}. \tag{3-5}$$

So f is bounded for $a, b \in \mathbb{R}$, and is an entire function of exponential type in each variable. Our goal is to show that

$$f(a, b) = \frac{\sin \pi(a - b)}{\pi(a - b)}. \tag{3-6}$$

So we study the properties of f. The main tool is to take elementary properties of the reproducing kernel K_n, such as properties of its zeros, and then after scaling and taking limits, to analyze the zeros of f, and related quantities. At the end, armed with a range of properties, one proves that these characterize the sinc kernel, and (3-6) follows.

The first result of this type was given in [Lubinsky 2008d]:

Theorem 3.1. *Let μ have compact support J. Let I be compact, and μ be absolutely continuous in an open set containing I. Assume that μ' is positive and continuous at each point of I. The following are equivalent:*

(I) *Uniformly for $\xi \in I$ and a in compact subsets of the real line,*

$$\lim_{n \to \infty} \frac{K_n\left(\xi + \frac{a}{n}, \xi + \frac{a}{n}\right)}{K_n(\xi, \xi)} = 1. \tag{3-7}$$

(II) *Uniformly for $\xi \in I$ and a, b in compact subsets of the complex plane, we have*

$$\lim_{n \to \infty} \frac{K_n\left(\xi + \frac{a}{\widetilde{K}_n(\xi, \xi)}, \xi + \frac{b}{\widetilde{K}_n(\xi, \xi)}\right)}{K_n(\xi, \xi)} = \frac{\sin \pi (a - b)}{\pi (a - b)}. \qquad (3\text{-}8)$$

One can weaken the condition of continuity of μ' to upper and lower bounds and then require ξ to be a Lebesgue point of μ, that is, we assume only

$$\lim_{h, k \to 0+} \frac{\mu([\xi - h, \xi + k])}{k + h} = \mu'(\xi).$$

The clear advantage of the theorem is that there is no global restriction on μ. The downside is that we still have to establish the ratio asymptotic (3-7) for the Christoffel functions/ reproducing kernels, and to date, these have only been established in the stronger form (2-2).

Nevertheless, the method itself has far more promise than the comparison inequality. For varying exponential weights (the "natural" setting for universality limits), it yielded [Levin and Lubinsky 2008b] universality very generally in the bulk, see below. It has also been used at the hard edge of the spectrum in [Lubinsky 2008c], at the soft edge of the spectrum [Levin and Lubinsky 2011], and to Cantor sets with positive measure by Avila, Last and Simon [Avila et al. 2010], as well as for orthogonal rational functions [Deckers and Lubinsky 2012]. Totik has observed that it yields an easier path to his Theorem 2.3 [Totik 2011].

With much more effort, and in particular a new uniqueness theorem for the sinc kernel, this set of methods also yields *universality in measure*, for arbitrary measures μ with compact support [Lubinsky 2012a]:

Theorem 3.2. *Let μ have compact support. Let $\varepsilon > 0$ and $r > 0$. The (linear Lebesgue) measure of the set of ξ satisfying $\mu'(\xi) > 0$ and*

$$\sup_{|u|, |v| \le r} \left| \frac{K_n\left(\xi + \frac{u}{\widetilde{K}_n(\xi, \xi)}, \xi + \frac{v}{\widetilde{K}_n(\xi, \xi)}\right)}{K_n(\xi, \xi)} - \frac{\sin \pi (u - v)}{\pi (u - v)} \right| \ge \varepsilon$$

tends to 0 as $n \to \infty$.

(In the supremum, u, v are complex variables.) Because convergence in measure implies convergence a.e. of subsequences, one obtains pointwise a.e. universality for subsequences, without any local or global assumptions on μ.

Another development involves pointwise universality in the mean [Lubinsky 2012b], under some local conditions. Like all the results of the section, the essential feature is the lack of global regularity assumptions:

Theorem 3.3. *Let μ have compact support. Assume that I is an open interval in which for some $C > 0$, $\mu' \geq C$ a.e. in I. Let $\xi \in I$ be a Lebesgue point of μ. Then, for each $r > 0$,*

$$\lim_{m \to \infty} \frac{1}{m} \sum_{n=1}^{m} \sup_{|u|,|v| \leq r} \left| \frac{K_n\left(\xi + \frac{u}{\tilde{K}_n(\xi,\xi)}, \xi + \frac{v}{\tilde{K}_n(\xi,\xi)}\right)}{K_n(\xi,\xi)} - \frac{\sin \pi(u-v)}{\pi(u-v)} \right| = 0.$$

In particular, this holds for a.e. $\xi \in I$.

Pointwise universality at a given point ξ seems to usually require at least something like μ' being continuous at ξ, or ξ being a Lebesgue point of μ. Indeed, when μ' has a jump discontinuity, the universality limit is different from the sine kernel [Foulquié Moreno et al. 2011], and involves de Branges spaces [Lubinsky 2009b]. It is noteworthy, though, that pure singularly continuous measures can exhibit sine kernel behavior [Breuer 2011].

From a mainstream random matrix point of view, the most impressive application of the normal families method is to exponential weights $W(x) = \exp(-Q(x))$, defined on a closed set Σ on the real line. If Σ is unbounded, we assume that

$$\lim_{|x| \to \infty, x \in \Sigma} W(x)|x| = 0. \tag{3-9}$$

Associated with Σ and Q, we may consider the extremal problem

$$\inf_{\nu}\left(\iint \log \frac{1}{|x-t|} \, d\nu(x) \, d\nu(t) + 2 \int Q \, d\nu \right),$$

where the inf is taken over all positive Borel measures ν with support in Σ and $\nu(\Sigma) = 1$. The inf is attained by a unique equilibrium measure ν_Q, characterized by the following conditions: let

$$V^{\nu_Q}(z) = \int \log \frac{1}{|z-t|} \, d\nu_Q(t)$$

denote the potential for ν_Q. Then

$$V^{\nu_Q} + Q \geq F_Q \text{ on } \Sigma;$$
$$V^{\nu_Q} + Q = F_Q \text{ in } \text{supp}[\nu_Q].$$

Here the number F_Q is a constant. Using asymptotics for Christoffel functions obtained in [Totik 2000b], Eli Levin and I proved this:

Theorem 3.4 [Levin and Lubinsky 2008b, Theorem 1.1, p. 747]. *Let $W = e^{-Q}$ be a continuous nonnegative function on the set Σ, which is assumed to consist* *of at most finitely many intervals. If Σ is unbounded, we assume also* (3-9). *Let h be a bounded positive continuous function on Σ, and for $n \geq 1$, let*

$$d\mu_n(x) = (hW^{2n})(x)\,dx. \tag{3-10}$$

Moreover, let $\tilde{K}_n$ denote the normalized n-th reproducing kernel for μ_n.

Let I be a closed interval lying in the interior of supp$[\nu_Q]$. *Assume that ν_Q is absolutely continuous in a neighborhood of I, and that ν_Q' and Q' are continuous in that neighborhood, while $\nu_Q' > 0$ there. Then uniformly for $\xi \in I$, and a, b in compact subsets of the real line, we have* (1-9).

In particular, when Q' satisfies a Lipschitz condition of some positive order in a neighborhood of I, then [Saff and Totik 1997, p. 216] ν_Q' is continuous there, and hence we obtain universality except near zeros of ν_Q'. Note too that when Q is convex in Σ, or $xQ'(x)$ is increasing there, then the support of ν_Q consists of at most finitely many intervals, with at most one interval per component of Σ [Saff and Totik 1997, p. 199].

4. A variational principle

The methods above intrinsically involve asymptotics for a single reproducing kernel, from which one can pass to the asymptotic for the general m-point correlation function. Remarkably (see [Lubinsky 2013]), there is a variational principle for the m-point correlation function R_m, for arbitrary measures μ, that generalizes the extremal property (1-4) of reproducing kernels, and allows one to investigate general m.

Its formulation involves $\mathcal{AL}_n^m$, the alternating polynomials of degree at most n in m variables. We say that $P \in \mathcal{AL}_n^m$ if

$$P(x_1, x_2, \ldots, x_m) = \sum_{0 \leq j_1, j_2, \ldots, j_m \leq n} c_{j_1 j_2 \ldots j_m} x_1^{j_1} x_2^{j_2} \ldots x_m^{j_m}, \tag{4-1}$$

so that P is a polynomial of degree $\leq n$ in each of its m variables, and in addition is *alternating*, so that for every pair (i, j) with $1 \leq i < j \leq m$,

$$P(x_1, \ldots, x_i, \ldots, x_j, \ldots, x_m) = -P(x_1, \ldots, x_j, \ldots, x_i, \ldots, x_m). \tag{4-2}$$

Thus swapping variables changes the sign.

Observe that if R_i is a univariate polynomial of degree $\leq n$ for each $i = 1, 2, \ldots, m$, then $P(t_1, t_2, \ldots, t_m) = \det[R_i(t_j)]_{1 \leq i,j \leq m} \in \mathcal{AL}_n^m$. Given a fixed m, we shall use the notation

$$\underline{x} = (x_1, x_2, \ldots, x_m), \quad \underline{t} = (t_1, t_2, \ldots, t_m),$$

while $\mu^{\times m}$ denotes the m-fold Cartesian product of μ, so that

$$d\mu^{\times m}(\underline{t}) = d\mu(t_1)\, d\mu(t_2) \ldots d\mu(t_m).$$

Theorem 4.1 [Lubinsky 2013].

$$\det[K_n(\mu, x_i, x_j)]_{1 \le i,j \le m} = m! \sup_{P \in \mathcal{AP}^m_{n-1}} \frac{(P(\underline{x}))^2}{\int (P(\underline{t}))^2 \, d\mu^{\times m}(\underline{t})}. \qquad (4\text{-}3)$$

The supremum is attained for

$$P(\underline{t}) = \det[K_n(\mu, x_i, t_j)]_{1 \le i,j \le m}.$$

Here is an immediate consequence:

Corollary 4.2. $R^n_m(x_1, x_2, \ldots, x_m)$ *is a monotone decreasing function of μ, and a monotone increasing function of n.*

The proof of Theorem 4.1 is based on multivariate orthogonal polynomials built from μ. Given $m \ge 1$, and nonnegative integers $j_1, j_2, \ldots, j_m$, define

$$T_{j_1, j_2, \ldots, j_m}(x_1, x_2, \ldots, x_m) = \det(p_{j_i}(x_k))_{1 \le i,k \le m}.$$

It is easily see that if $0 \le j_1 < j_2 < \cdots < j_m$ and $0 \le k_1 < k_2 < \cdots < k_m$, then

$$\int T_{j_1, j_2, \ldots, j_m}(\underline{t}) T_{k_1, k_2, \ldots, k_m}(\underline{t}) d\mu^{\times m}(\underline{t}) = m! \delta_{j_1 k_1} \delta_{j_2 k_2} \cdots \delta_{j_m k_m}.$$

Define an associated reproducing kernel,

$$K^m_n(\underline{x}, \underline{t}) = \frac{1}{m!} \sum_{1 \le j_1 < j_2 < \cdots < j_m \le n} T_{j_1, j_2, \ldots, j_m}(\underline{x}) T_{j_1, j_2, \ldots, j_m}(\underline{t}).$$

Theorem 4.1 follows easily from the reproducing kernel relation

$$P(\underline{x}) = \int P(\underline{t}) K^m_n(\underline{x}, \underline{t}) \, d\mu^{\times m}(\underline{t}), \quad P \in \mathcal{AP}^m_{n-1}, \ \underline{x} \in \mathbb{R}^n,$$

and the Cauchy–Schwarz inequality.

Just as the extremal property (1-4) for $K_n(\mu, x, x)$ is the main idea in proving Theorem 2.1, so we can use Theorem 4.1 to prove [Lubinsky 2013, Theorem 2.1]:

Theorem 4.3. *Let μ have compact support J. Let $m \ge 1$.*

(a) *For Lebesgue a.e. $(x_1, x_2, \ldots, x_m) \in (J^0)^m$,*

$$\liminf_{n \to \infty} \frac{1}{n^m} \det[K_n(\mu, x_i, x_j)]_{1 \le i,j \le m} \ge \prod_{j=1}^{m} \frac{\omega(x_j)}{\mu'(x_j)}.$$

The right-hand side is interpreted as ∞ if any $\mu'(x_j) = 0$.

(b) *Suppose that I is a compact subinterval of J, for which (2-1) holds. Then for Lebesgue a.e. $(x_1, x_2, \ldots, x_m) \in I^m$,*

$$\limsup_{m \to \infty} \frac{1}{n^m} \det[K_n(\mu, x_i, x_j)]_{1 \le i, j \le m} \le \prod_{j=1}^{m} \frac{\omega_\mu(x_j)}{\mu'(x_j)},$$

where, if ω_L denotes the equilibrium density for the compact set L,

$$\omega_\mu(x) = \inf\{\omega_L(x) : L \subset J \text{ is compact}, \mu_{|L} \text{ is regular}, x \in L\}.$$

A more impressive consequence is pointwise, almost everywhere, one-sided universality, without any local or global restrictions on μ [Lubinsky 2013, Theorem 2.2]:

Theorem 4.4. *Let μ have compact support J. Let $m \ge 1$.*

(a) *For a.e. $x \in J^0 \cap \{\mu' > 0\}$, and for all real $a_1, a_2, \ldots, a_m$,*

$$\liminf_{n \to \infty} \left(\frac{\mu'(x)}{n\omega(x)}\right)^m R_m^n\left(x + \frac{a_1}{n\omega(x)}, \ldots, x + \frac{a_m}{n\omega(x)}\right) \ge \det(\mathbb{S}(a_i - a_j))_{1 \le i, j \le m}.$$

(b) *Suppose that I is a compact subinterval of J, for which (2-1) holds. Then for a.e. $x \in I$, and for all real $a_1, a_2, \ldots, a_m$,*

$$\limsup_{n \to \infty} \left(\frac{\mu'(x)}{n\omega_\mu(x)}\right)^m R_m^n\left(x + \frac{a_1}{n\omega_\mu(x)}, \ldots, x + \frac{a_m}{n\omega_\mu(x)}\right) \le \det(\mathbb{S}(a_i - a_j))_{1 \le i, j \le m}.$$

In closing, we note that the study of universality limits has greatly enriched the asymptotics of orthogonal polynomials. A prime example of this is asymptotics for spacing of zeros [Levin and Lubinsky 2008a; 2010; Simon 2005a; 2005b; 2008a; 2011].

Acknowledgement

I thank the organizers of the conference for the invitation, and especially Percy Deift, for his assistance at the conference.

References

[Avila et al. 2010] A. Avila, Y. Last, and B. Simon, "Bulk universality and clock spacing of zeros for ergodic Jacobi matrices with absolutely continuous spectrum", *Anal. PDE* **3**:1 (2010), 81–108.

[Baik et al. 2007] J. Baik, T. Kriecherbauer, K. T.-R. McLaughlin, and P. D. Miller, *Discrete orthogonal polynomials: asymptotics and applications*, Annals of Mathematics Studies **164**, Princeton University Press, Princeton, NJ, 2007.

[Baik et al. 2008] J. Baik, T. Kriecherbauer, L.-C. Li, K. D. T.-R. McLaughlin, and C. Tomei (editors), *Integrable systems and random matrices: in honor of Percy Deift* (New York University, May 22–26, 2006), Contemporary Mathematics **458**, Amer. Math. Soc., Providence, RI, 2008.

[Bleher and Its 2001] P. Bleher and A. Its (editors), *Random matrix models and their applications*, Math. Sci. Res. Inst. Pub. **40**, Cambridge University Press, Cambridge, 2001.

[Breuer 2011] J. Breuer, "Sine kernel asymptotics for a class of singular measures", *J. Approx. Theory* **163**:10 (2011), 1478–1491.

[Deckers and Lubinsky 2012] K. Deckers and D. S. Lubinsky, "Christoffel functions and universality limits for orthogonal rational functions", *Anal. Appl. (Singap.)* **10**:3 (2012), 271–294.

[Deift 1999] P. A. Deift, *Orthogonal polynomials and random matrices: a Riemann–Hilbert approach*, Courant Lecture Notes in Mathematics **3**, Courant Inst. of Math. Sci., New York, 1999.

[Deift and Gioev 2009] P. Deift and D. Gioev, *Random matrix theory: invariant ensembles and universality*, Courant Lecture Notes in Mathematics **18**, Courant Inst. Math. Sci., New York, 2009.

[Findley 2008] E. Findley, "Universality for locally Szegő measures", *J. Approx. Theory* **155**:2 (2008), 136–154.

[Forrester 2010] P. J. Forrester, *Log-gases and random matrices*, London Mathematical Society Monographs Series **34**, Princeton University Press, 2010.

[Foulquié Moreno et al. 2011] A. Foulquié Moreno, A. Martínez-Finkelshtein, and V. L. Sousa, "Asymptotics of orthogonal polynomials for a weight with a jump on $[-1, 1]$", *Constr. Approx.* **33**:2 (2011), 219–263.

[Freud 1971] G. Freud, *Orthogonal polynomials*, Pergamon and Akadémiai Kiadó, Budapest, 1971.

[Levin and Lubinsky 2007] E. Levin and D. S. Lubinsky, "Universality limits involving orthogonal polynomials on the unit circle", *Comput. Methods Funct. Theory* **7**:2 (2007), 543–561.

[Levin and Lubinsky 2008a] E. Levin and D. S. Lubinsky, "Applications of universality limits to zeros and reproducing kernels of orthogonal polynomials", *J. Approx. Theory* **150**:1 (2008), 69–95.

[Levin and Lubinsky 2008b] E. Levin and D. S. Lubinsky, "Universality limits in the bulk for varying measures", *Adv. Math.* **219**:3 (2008), 743–779.

[Levin and Lubinsky 2009] E. Levin and D. S. Lubinsky, "Universality limits for exponential weights", *Constr. Approx.* **29**:2 (2009), 247–275.

[Levin and Lubinsky 2010] E. Levin and D. S. Lubinsky, "Some equivalent formulations of universality limits in the bulk", pp. 177–188 in *Recent trends in orthogonal polynomials and approximation theory*, edited by F. M. Jorge Arvesú and A. Martínez-Finkelshtein, Contemp. Math. **507**, Amer. Math. Soc., Providence, RI, 2010.

[Levin and Lubinsky 2011] E. Levin and D. S. Lubinsky, "Universality limits at the soft edge of the spectrum via classical complex analysis", *Int. Math. Res. Not.* **2011**:13 (2011), 3006–3070.

[Lubinsky 2008a] D. S. Lubinsky, "Mutually regular measures have similar universality limits", pp. 256–269 in *Approximation theory, XII* (San Antonio, 2007), edited by M. Neamtu and L. L. Schumaker, Nashboro Press, Brentwood, TN, 2008.

[Lubinsky 2008b] D. S. Lubinsky, "A new approach to universality limits at the edge of the spectrum", pp. 281–290 in *Integrable systems and random matrices: in honor of Percy Deift*, edited by J. Baik et al., Contemp. Math. **458**, Amer. Math. Soc., Providence, RI, 2008.

[Lubinsky 2008c] D. S. Lubinsky, "Universality limits at the hard edge of the spectrum for measures with compact support", *Int. Math. Res. Not.* **2008** (2008), Art. ID rnn 099, 39.

[Lubinsky 2008d] D. S. Lubinsky, "Universality limits in the bulk for arbitrary measures on compact sets", *J. Anal. Math.* **106** (2008), 373–394.

[Lubinsky 2009a] D. S. Lubinsky, "A new approach to universality limits involving orthogonal polynomials", *Ann. of Math.* (2) **170**:2 (2009), 915–939.

[Lubinsky 2009b] D. S. Lubinsky, "Universality limits for random matrices and de Branges spaces of entire functions", *J. Funct. Anal.* **256**:11 (2009), 3688–3729.

[Lubinsky 2010] D. S. Lubinsky, "Universality type limits for Bergman orthogonal polynomials", *Comput. Methods Funct. Theory* **10**:1 (2010), 135–154.

[Lubinsky 2012a] D. S. Lubinsky, "Bulk universality holds in measure for compactly supported measures", *J. Anal. Math.* **116** (2012), 219–253.

[Lubinsky 2012b] D. S. Lubinsky, "Bulk universality holds pointwise in the mean for compactly supported measures", *Michigan Math. J.* **61**:3 (2012), 631–649.

[Lubinsky 2013] D. S. Lubinsky, "A variational principle for correlation functions for unitary ensembles, with applications", *Anal. PDE* **6**:1 (2013), 109–130.

[Máté et al. 1991] A. Máté, P. Nevai, and V. Totik, "Szegő's extremum problem on the unit circle", *Ann. of Math.* (2) **134**:2 (1991), 433–453.

[Mehta 2004] M. L. Mehta, *Random matrices*, 3rd ed., Pure and Applied Mathematics (Amsterdam) **142**, Elsevier/Academic Press, Amsterdam, 2004.

[Nevai 1979] P. G. Nevai, *Orthogonal polynomials*, Mem. Amer. Math. Soc. **213**, Amer. Math. Soc., Providence, RI, 1979.

[Nevai 1986] P. Nevai, "Géza Freud, orthogonal polynomials and Christoffel functions: a case study", *J. Approx. Theory* **48**:1 (1986), 167.

[Ransford 1995] T. Ransford, *Potential theory in the complex plane*, London Mathematical Society Student Texts **28**, Cambridge University Press, Cambridge, 1995.

[Saff and Totik 1997] E. B. Saff and V. Totik, *Logarithmic potentials with external fields*, Grundlehren der Math. Wissenschaften **316**, Springer, Berlin, 1997.

[Simon 2005a] B. Simon, *Orthogonal polynomials on the unit circle, I: classical theory*, Amer. Math. Soc. Coll. Pub. **54**, Amer. Math. Soc., Providence, RI, 2005.

[Simon 2005b] B. Simon, *Orthogonal polynomials on the unit circle, II: spectral theory*, Amer. Math. Soc. Coll. Pub. **54**, Ame. Math. Soc., Providence, RI, 2005.

[Simon 2008a] B. Simon, "The Christoffel–Darboux kernel", pp. 295–335 in *Perspectives in partial differential equations, harmonic analysis and applications*, edited by D. Mitrea and M. Mitrea, Proc. Sympos. Pure Math. **79**, Amer. Math. Soc., Providence, RI, 2008.

[Simon 2008b] B. Simon, "Two extensions of Lubinsky's universality theorem", *J. Anal. Math.* **105** (2008), 345–362.

[Simon 2011] B. Simon, *Szegő's theorem and its descendants: spectral theory for L^2 perturbations of orthogonal polynomials*, Princeton University Press, 2011.

[Stahl and Totik 1992] H. Stahl and V. Totik, *General orthogonal polynomials*, Encyclopedia of Mathematics and its Applications **43**, Cambridge University Press, Cambridge, 1992.

[Stenger 1993] F. Stenger, *Numerical methods based on sinc and analytic functions*, Springer Series in Computational Mathematics **20**, Springer, New York, 1993.

[Totik 2000a] V. Totik, "Asymptotics for Christoffel functions for general measures on the real line", *J. Anal. Math.* **81** (2000), 283–303.

[Totik 2000b] V. Totik, "Asymptotics for Christoffel functions with varying weights", *Adv. in Appl. Math.* **25**:4 (2000), 322–351.

[Totik 2009] V. Totik, "Universality and fine zero spacing on general sets", *Ark. Mat.* **47**:2 (2009), 361–391.

[Totik 2011] V. Totik, "Universality under Szegő's condition", 2011.

lubinsky@math.gatech.edu   *School of Mathematics, Georgia Institute of Technology, 686 Cherry Street, Atlanta, GA 30332-0160, United States*

Random Matrices
MSRI Publications
Volume 65, 2014

Fluctuations and large deviations of some perturbed random matrices

MYLÈNE MAÏDA

We review joint results with Benaych-Georges and Guionnet (*Electron. J. Probab.* **16**:60 (2011), 1621–1662 and *Prob. Theory Rel. Fields* **154**:3–4 (2012), 703–751) about fluctuations and large deviations of some spiked models, putting them in the perspective of various works of the last years on extreme eigenvalues of finite-rank deformations of random matrices.

1. Introduction

General statement of the problem. The following algebraic problem is very classical: *Let A and B be two Hermitian matrices of the same size. Assume we know the spectrum of each of them. What can be said about the spectrum of their sum A + B?* The problem was posed by Weyl [1912]. He gave a series of necessary conditions, known as Weyl's interlacing inequalities: if $\lambda_1(A) \geq \cdots \geq \lambda_n(A)$, $\lambda_1(B) \geq \cdots \geq \lambda_n(B)$ and $\lambda_1(A+B) \geq \cdots \geq \lambda_n(A+B)$ are the spectra of A, B and $A+B$, then

$$\lambda_{i+j}(A+B) \leq \lambda_{i+1}(A) + \lambda_{j+1}(B),$$

whenever $0 \leq i, j, i+j < n$. These inequalities have been very fruitful in various fields.

After that, it took a long time to get necessary and sufficient conditions. Horn in the sixties formulated the right conjecture, but the final answer was only given in the late nineties in a series a papers [Klyachko 1998; Helmke and Rosenthal 1995; Knutson and Tao 2001].

If we now look at the problem asymptotically, namely when the size of both matrices goes to infinity, an important breakthrough was made by free probability theory with the notion of asymptotic freeness. This property can be roughly stated as follows: *If A and B are large-dimensional and in generic position relative to one another, the limiting spectrum of their sum depends only on their respective spectra and is given by the free convolution of the two spectra.* We won't go

This work was supported by the Agence Nationale de la Recherche, grant ANR-08-BLAN-0311-03.

further into free probability theory, but we refer the reader to [Emery et al. 2000] for general background on free probability.

In this review paper, we will address the problem of the asymptotic spectrum of the sum of two Hermitian matrices in a particular framework: in the case when one of the matrix is of finite rank, fixed and independent of the size of the matrices. The problem can be stated as follows: take your favorite ensemble of matrices. You know very well the global and local behavior of the spectrum (asymptotic spectral measure, convergence and fluctuations of extreme eigenvalues etc). Add to your random matrix a finite-rank perturbation. How is the spectrum affected by this perturbation?

By Weyl's interlacing inequalities, it is not hard to check that the global behavior is not changed at the macroscopic level. Only extreme eigenvalues can be substantially affected. This review paper discusses almost sure limits, fluctuations and large deviations for these extreme eigenvalues in various models of this type. There exist also a few results on eigenvectors that we won't review here; see [Benaych-Georges and Nadakuditi 2011], for example.

The BBP transition. Before giving a more panoramic view of the literature on these models, we will describe in detail the first set of rigorous results in this direction. This seminal work is due to Baik, Ben Arous and Péché [Baik et al. 2005], and the phenomenon we will describe is therefore often called the BBP transition.

They considered the following model: let G_n be an $n \times m$ matrix whose column vectors are centered Gaussian with covariance matrix Σ and let $S_n = (1/m)G_n G_n^*$ be the corresponding sample covariance matrix. Assume $n/m \to c \in (0, 1)$ and Σ is a perturbation of the identity in the sense that it has at most r eigenvalues different from 1. Let $\ell_1 \geq \ell_2 \geq \cdots \geq \ell_r \geq 1, \ldots, 1$ be the eigenvalues of Σ.

Let us first recall the unperturbed case. If Σ is the identity matrix, the model is known as the Laguerre unitary ensemble (LUE). The following results are now classical:

0. The limiting spectral measure is Marchenko–Pastur, being supported in $[(1 - \sqrt{c})^2, (1 + \sqrt{c})^2]$.

1. For any k fixed, the k largest eigenvalues converge to the edge of the bulk, $(1 + \sqrt{c})^2$.

2. They have fluctuations in the scale $n^{-2/3}$ with Gaussian unitary ensemble (GUE) Tracy–Widom laws.

As mentioned above, it is easy to check that the global regime won't be affected: property 0 remains valid in every regime and we won't repeat it. The interesting thing is the behavior of the largest eigenvalues.

The results of [Baik et al. 2005] are as follows:

- If $\ell_1 < 1 + \sqrt{c}$, properties (1) and (2) remain unchanged: we are in the subcritical regime.
- If $\ell_1 = \cdots = \ell_k > 1 + \sqrt{c}$, we are in the supercritical regime.
  (1) The largest eigenvalue converges outside the bulk to $\ell_1 + c\ell_1/(\ell_1 - 1)$.
  (2) Its fluctuations are in the scale $n^{-1/2}$ with the law of the largest eigenvalue of a matrix from the GUE of size $k \times k$.

In the sequel, we will refer to this phenomenon as the BBP transition.

For completeness, we mention that [Baik et al. 2005] also treats the critical case, when there exists k such that $\ell_1 = \cdots = \ell_k = 1 + \sqrt{c}$, but this critical behavior is hard to generalize; it seems that only a case-by-case analysis is pertinent for critical parameters and we won't dwell on this behavior.

Our last remark about this model is that the perturbation is multiplicative, whereas we are more interested in additive perturbations. But in fact an additive perturbation gives the same kind of behavior: for example, Péché [2006] showed the same kind of transition for a matrix from the GUE perturbed (additively) by a matrix of finite rank.

A quick review of the literature about fluctuations (see also Section 5). Since the appearance of [Baik et al. 2005], there has been quite a lot of work on fluctuations of extreme eigenvalues of different variant of those spiked models.

A large part of the literature has been devoted to applications, more specifically statistical applications of those spiked models. The seminal work in this direction is probably [Johnstone 2001], dealing with applications to principal component analysis (see also [El Karoui 2005]). Indeed, in many papers, the finite-rank matrix is seen as the signal (with a fixed number of significant parameters) and the unperturbed random matrix as the noise. The general question addressed is to know whether the observation of the eigenvalues of "signal plus noise" can give access to the parameters of interest. The results on fluctuations that we will expound below allow us to construct statistical tests on the parameters. All these applied results are a subject in themselves and we won't review them here; we will stick to theoretical results on those spiked models.

As pointed out above, Baik et al. [2005] and Péché [2006] dealt with models with Gaussian entries (perturbed LUE and GUE). There quickly followed attempts to extend it beyond the Gaussian case.

The first results were in the direction of a generalization of the BBP transition. Féral and Péché [2007] showed, by using combinatorics of moments techniques, that the model $W_n + A_n$, where W_n is a Wigner matrix with independent, identically distributed (iid) complex entries with a law having sub-Gaussian moments and A_n is a rank-one perturbation such that $(A_n)_{ij} = \theta/n$ for all i, j, exhibits all the features of the BBP transition. Bai and Yao [2008] studied in more detail the fluctuations of the eigenvalues converging out of the bulk for quite general spiked models and showed that these fluctuations, as in the BBP case, were in the scale $n^{-1/2}$ and governed by small GUE (or GOE in the case of real entries) matrices with sizes the multiplicities of the supercritical eigenvalues of the perturbation.

Then, in 2009, there appeared what we can call the first non-BBP features in such models. Capitaine, Donati-Martin and Féral [Capitaine et al. 2009] showed, among other results, that the fluctuations of the eigenvalues converging out of the bulk are not universal. If again, W_n is a Wigner matrix with complex iid entries with a nice symmetric law μ and A_n has rank one but this time is of the form $A_n = \text{diag}(\theta, 0, \ldots, 0)$, with θ large enough, then the fluctuations of the largest eigenvalue are not Gaussian anymore, the law being rather the convolution of a Gaussian measure together with the law μ.

In [Capitaine et al. 2009], the reason for this nonuniversality remained a bit mysterious, but in [Capitaine et al. 2012] the same authors showed that the crucial feature is whether the perturbation has delocalized eigenvectors (as in [Féral and Péché 2007]), in which case the BBP transition occurs, or localized eigenvectors (as in [Capitaine et al. 2009]), in which case the fluctuations may depend on the law of the entries of the unperturbed matrix.

We emphasize that in the sequel, we are going to work only in the framework of perturbations with delocalized eigenvectors. The main object of this review paper is to present the results of two joint papers of the author with Benaych-Georges and Guionnet [Benaych-Georges et al. 2011; 2012].

2. Fluctuations of extreme eigenvalues of spiked models

Presentation of the deterministic version of the models. As pointed out in the introduction, many versions of the spiked model have been studied in the literature. Let us detail the precise models we have been studying. We first present and detail the case when the unperturbed part is deterministic and we will then explain how the results can be easily generalized to the usual ensembles of matrices (Wigner, Wishart, etc.)

Let X_n be deterministic self-adjoint with eigenvalues $\lambda_1^n \geq \cdots \geq \lambda_n^n$.

We make the following hypothesis on the spectrum of X_n: as n goes to infinity,

$$\text{(H1)} \qquad \frac{1}{n} \sum_{i=1}^{n} \delta_{\lambda_i^n} \to \mu_X, \quad \lambda_1^n \to a, \quad \lambda_n^n \to b,$$

with μ_X a compactly supported probability measure and a and b are respectively the left and right edges of support of μ_X.

We then add to X_n a finite-rank perturbation R_n and consider the perturbed matrix $\widetilde{X}_n = X_n + R_n$. The eigenvalues of $\widetilde{X}_n$ will be denoted by $\widetilde{\lambda}_1^n \geq \cdots \geq \widetilde{\lambda}_n^n$. The finite-rank perturbation R_n will have the following form:

$$R_n = \sum_{j=1}^{r} \theta_i G_i^n (G_i^n)^*,$$

with

$$\theta_1 \geq \cdots \geq \theta_{r_0} > 0 > \theta_{r_0+1} \geq \cdots \geq \theta_r$$

fixed and independent of n, and the G_i^n such that $\sqrt{n}G_i^n$ are vectors with iid entries with law v satisfying a log-Sobolev inequality. This latter hypothesis is technical; it allows us to use some concentration properties for the quantities we will be interested in.

One can also consider R_n of the form

$$R_n = \sum_{j=1}^{r} \theta_i U_i^n (U_i^n)^*,$$

where the U_i^n are obtained from the vectors $\sqrt{n}G_i^n$ by a Gram–Schmidt orthonormalization procedure.

In particular, we stress that in our model the eigenvectors of the perturbation are delocalized.

Almost sure convergence of extreme eigenvalues. Before getting to our results on fluctuations themselves, let us first look at the convergence of those extreme eigenvalues.

We set

$$G_{\mu_X}(z) := \int \frac{1}{z-x} d\mu_X(x),$$

and define

$$\frac{1}{\underline{\theta}} = \lim_{z \to a^-} G_{\mu_X}(z), \qquad \frac{1}{\overline{\theta}} = \lim_{z \to b^+} G_{\mu_X}(z),$$

$$\rho_\theta := \begin{cases} G_{\mu_X}^{-1}(1/\theta) & \text{if } \theta \in (-\infty, \underline{\theta}) \cup (\overline{\theta}, +\infty), \\ a & \text{if } \theta \in [\underline{\theta}, 0), \\ b & \text{if } \theta \in (0, \overline{\theta}]. \end{cases}$$

Theorem 2.1 [Benaych-Georges and Nadakuditi 2011]. *Let $r_0 \in \{0, \ldots, r\}$ be such that*

$$\theta_1 \geq \cdots \geq \theta_{r_0} > 0 > \theta_{r_0+1} \geq \cdots \geq \theta_r.$$

The largest eigenvalues have the following behavior:

$$\tilde{\lambda}_i^n \xrightarrow{\text{a.s.}} \rho_{\theta_i} \quad \text{for all } i \in \{1, \dots, r_0\},$$

$$\tilde{\lambda}_i^n \xrightarrow{\text{a.s.}} b \quad \text{for } i > r_0.$$

Similarly, for the smallest eigenvalues:

$$\tilde{\lambda}_{n-r+i}^n \xrightarrow{\text{a.s.}} \rho_{\theta_i} \quad \text{for all } i \in \{r_0 + 1, \dots, r\},$$

$$\tilde{\lambda}_{n-i}^n \xrightarrow{\text{a.s.}} a \quad \text{for all } i \geq r - r_0.$$

In the sequel we will state only the part of the results concerning the largest eigenvalues, the part concerning the smallest eigenvalues being very similar.

Before moving to the fluctuations, we emphasize that our model exhibits the first feature of the BBP transition: if the perturbation is small, extreme eigenvalues stick to the bulk; if it is strong enough, they converge out of the bulk.

Gaussian fluctuations outside the bulk. The second feature of the BBP transition is the fact that the fluctuations of the eigenvalues converging outside the bulk are in the scale $n^{-1/2}$ and are "of Gaussian type" — in fact, governed by a small matrix from the Gaussian unitary or orthogonal ensemble (GUE/GOE).

Under the additional hypothesis that

(H2) $$\frac{1}{n} \sum_{i=1}^{n} \delta_{\lambda_i^n} \to \mu_X \text{ converges at least as fast as } 1/\sqrt{n},$$

our model exhibits this second feature. More precisely, we have the following result. Let $\alpha_1 > \cdots > \alpha_q > 0$ be the different values of the θ_i such that $\rho_{\theta_i} > b$.

For each j, let I_j be the set of indices i such that $\theta_i = \alpha_j$. Set $k_j = |I_j|$.

Theorem 2.2. *Under hypotheses* (H1) *and* (H2), *if the fourth cumulant* $\kappa_4(\nu)$ *of the law* ν *is zero, the random vector*

$$\left(\gamma_i := \sqrt{n}(\tilde{\lambda}_i^n - \rho_{\theta_i}), i \in I_j\right)_{1 \leq j \leq q}$$

converges in law to the eigenvalues of $(c_j M_j)_{1 \leq j \leq q}$ *with independent matrices* $M_j \in \text{GUE} / \text{GOE}$ *of size* $k_j \times k_j$ *and* c_j *is an explicit constant depending only on* μ_X *and* α_j.

We have a similar result if $\kappa_4(\nu)$ is not zero, only the limiting law will be a bit different. We refer the reader to [Benaych-Georges et al. 2011] for details.

Nonuniversality of the fluctuations near the bulk. In the BBP transition mentioned in the introduction, it turned out that the fluctuations of extreme eigenvalues that sticks to the bulk at the level of almost sure convergence exhibited the same fluctuations as the unperturbed model (in the LUE case, it happened to be governed by Tracy–Widom laws).

In our model, we also addressed the question of the so-called "sticking eigen-values". Their study happened to be much more delicate than the study of the fluctuations of eigenvalues outside the bulk.

As mentioned in the introduction, we won't address the problem of critical parameters. Our result can be roughly stated as follows:

Theorem 2.3. *Under additional hypotheses on X_n, if none of the θ_i is critical, with overwhelming probability, the eigenvalues of $\widetilde{X}_n$ converging to a or b remain at distance at most $n^{-1+\epsilon}$ of the extreme eigenvalues of X_n, for some $\epsilon > 0$.*

We therefore say that the fluctuations of the eigenvalues near the bulk are nonuniversal, in the sense that they follow the fluctuations of the eigenvalues of X_n, that could be in any scale and according to any probability law.

Before giving a more precise statement of the hypotheses and the theorem, one can give a rough explanation of the phenomenon: for fixed values of the θ_i, we have a repulsion phenomenon from the eigenvalues (ev) of X_n at the edge.

- If the repulsion is very strong, the extreme ev of $\widetilde{X}_n$ converge away from the bulk.
- If the repulsion is milder, the extreme ev of $\widetilde{X}_n$ stick to the edge of the bulk.
- If the repulsion is even milder, the extreme ev of $\widetilde{X}_n$ stick to the extreme ev of X_n even at the level of fluctuations.

For the repulsion to be very mild, we need the spacings of the eigenvalues of X_n at the edge to stay small, in the following sense:

(H3)$[p, \alpha]$ There exists a sequence m_n of positive integers tending to infinity with $m_n = O(n^\alpha)$, and constants $\eta_2 > 0$ and $\eta_4 > 0$ such that for any $\delta > 0$ and n large enough,

$$\sum_{i=m_n+1}^{n} \frac{1}{(\lambda_p^n - \lambda_i^n)^2} \leq n^{2-\eta_2},$$

$$\frac{1}{n} \sum_{i=m_n+1}^{n} \frac{1}{\lambda_p^n - \lambda_i^n} \geq \frac{1}{\theta} - \delta,$$

$$\sum_{i=m_n+1}^{n} \frac{1}{(\lambda_p^n - \lambda_i^n)^4} \leq n^{4-\eta_4}$$

The fact that the eigenvalues of the unperturbed matrix are sufficiently spread at the edges to insure the above hypothesis allows the eigenvalues of the perturbed matrix to be very close to them, as stated in the following theorem.

Theorem 2.4. *Let I_b be the set of indices corresponding to the eigenvalues $\widetilde{\lambda}_i^n$ converging to the upper bound of the support of μ_X. Suppose (H1) and (H3)$[r, \alpha]$*

hold. Then for any $\alpha' > \alpha$, we have, for all $i \in I_b$,

$$\min_{1 \le k \le i+r-r_0} |\tilde{\lambda}_i^n - \lambda_k^n| \le n^{-1+\alpha'},$$

with overwhelming probability.[1]

Moreover, in the case where the perturbation has rank one, we can locate exactly in the neighborhood of which eigenvalues of the unperturbed matrix the eigenvalues of the perturbed matrix lie. We won't review this particular case.

Applications to classical models. In comparison with the models presented in the introduction, the model we have chosen till then that is (deterministic + finite-rank random perturbation with delocalized eigenvectors) can seem a bit disappointing. In fact, we can easily extend our theorem to models where the unperturbed matrix is random and therefore generalize some of the results presented in the introduction.

If (X_n) is a sequence of random matrices, we say that it satisfies an hypothesis (H) in probability if the probability that (X_n) satisfies (H) goes to one as n goes to infinity.

To extend our result, we will use the following, which is easy to show.

Theorem 2.5. *Let (X_n) be a sequence of random matrices independent of the column vectors G_i^n or U_i^n of the perturbation.*

(1) *If (H1) holds in probability, Theorem 2.1 holds.*

(2) *If $\kappa_4(\nu) = 0$ and Hypotheses (H1) and (H2) hold in probability, Theorem 2.2 holds.*

(3) *If none of the θ_i is critical and Hypotheses (H1) and (H3) hold in probability, Theorem 2.4 holds "with probability converging to one" instead of "with overwhelming probability".*

This result allows us to treat a lot of classical models. For each of these models, it will be enough to check that the different hypotheses hold in probability. This will allow us to generalize some of the results of [Péché 2006; Féral and Péché 2007; Capitaine et al. 2009]. We detail hereafter some of these generalizations.

Let X_n be one of the following models:

(a) X_n is a Wigner matrix with iid entries (up to symmetry) with zero mean, variance one and finite fourth moment.

(b) X_n is a Wishart matrix of the form $X_n = G_n G_n^* / m$, with G_n a $n \times m$ matrix with (real or complex) iid entries with zero mean, variance one and finite fourth moment, with $n/m \to c \in (0, 1)$.

---

[1] We say that a sequence of events $(E_n)_{n \in \mathbb{N}}$ holds with overwhelming probability if there exists $C, \eta > 0$ such that for n large enough, $\mathbb{P}(E_n) \ge 1 - Ce^{-n^\eta}$.

Then, for the deformed model $\widetilde{X}_n$:

- Almost sure convergence of extreme eigenvalues is governed by Theorem 2.1.
- The eigenvalues converging out of the bulk have Gaussian fluctuations (see Theorem 2.2, where c_j can be computed explicitly).
- If the entries of X_n in the Wigner case and G_n in the Wishart case have a subexponential decay,
  - for a one dimensional perturbation: if the perturbation is supercritical, the largest eigenvalue converges outside and has Gaussian fluctuations, the p-th largest has the $p-1$-th Tracy–Widom law, if it is subcritical, the p-th largest has the p-th Tracy–Widom law;
  - for a multidimensional perturbation: the sticking eigenvalues of $\widetilde{X}_n$ are at distance negligible with respect to $n^{-2/3}$ to the extreme eigenvalues of X_n.

We also get the same kind of results for perturbed Coulomb gases, that is, when the joint law of eigenvalues of the unperturbed part is of the form

$$dP_n(\lambda_1, \ldots, \lambda_n) = (1/Z_n) |\Delta(\lambda)|^\beta e^{-n\beta \sum_{i=1}^n V(\lambda_i)} \prod_{i=1}^n d\lambda_i,$$

with V a strictly convex polynomial potential.

3. Large deviations of extreme eigenvalues of spiked models

Introduction. As the spectrum of a matrix is a very complicated function of the entries, usual large deviations theorems, mainly based on independence do not easily apply. There have been only few works dealing with large deviations in the context of random matrices. The first breakthrough in this direction, which played an important role in the development of the theory, appeared in [Ben Arous and Guionnet 1997], which showed a full large deviation principle (LDP) for the empirical spectral law of Gaussian Wigner matrices or more generally in models where the joint law of the eigenvalues is given by a Coulomb gas distribution, as introduced above (see also [Anderson et al. 2010]). Recently, Adrien Hardy [2012] gave some extension of the result of Ben Arous and Guionnet in the case when the potential is weakly confining. For the empirical spectral law of the sum of a Gaussian Wigner matrix and a deterministic self-adjoint matrix, the LDP is also known thanks to [Guionnet and Zeitouni 2002]. We can also mention [Chatterjee and Varadhan 2012] and [Bordenave and Caputo 2012] on Wigner matrices, and [Hardy and Kuijlaars 2013] on noncentered Wishart matrices.

If we now look at the large deviations for extreme eigenvalues, a little bit more is known. The first result in this direction concerns the largest eigenvalue of a matrix from the GOE or GUE and was shown by Ben Arous, Dembo and Guionnet [Ben Arous et al. 2001] (see also [Anderson et al. 2010] for generalizations). Wishart matrices of the form XX^*, with X a $n \times m$ matrix with iid, not necessarily Gaussian, entries, have been studied in [Fey et al. 2008] in the case when the ratio m/n goes to zero.

Hereafter, we study the problem of the large deviations of extreme eigenvalues in spiked models that are of the same form as the models introduced at the start of Section 2. The only previous result in this direction was established in [Maïda 2007] for a rank-one perturbation of a GOE matrix.

Before going into our results, we stress that among the results mentioned above, all of them, except [Chatterjee and Varadhan 2012; Fey et al. 2008], dealt with Gaussian entries or with cases when the joint law of eigenvalues was explicitly known.

Large deviation principle: the statement. In the paper [Benaych-Georges et al. 2012], we consider the same models (iid and orthonormalized) as introduced at the start of Section 2. The perturbation is more or less the same, except, instead of taking $\sqrt{n}G_i^n$ vectors with iid entries with law ν satisfying a log-Sobolev inequality, we assume that $G = (g_1, \ldots, g_r)$ is a random vector satisfying $E\left(e^{\alpha \sum |g_i|^2}\right) < \infty$ for some $\alpha > 0$ and $(\sqrt{n}G_1^n, \ldots, \sqrt{n}G_r^n)$ are random vectors whose entries are independent copies of G; again $(U_1^n, \ldots, U_r^n)$ are obtained from $(\sqrt{n}G_1^n, \ldots, \sqrt{n}G_r^n)$ by a Gram–Schmidt orthonormalization procedure.

Theorem 3.1. *If X_n satisfies (H1), the law of the r_0 largest eigenvalues of $\widetilde{X}_n$ satisfies a LDP in the scale n with a good rate function L. It has a unique minimizer towards which almost sure convergence holds.*

For the reader that are not familiar with large deviations, we recall that this means that for any open set $O \subset \mathbb{R}^{r_0}$,

$$\liminf_{n \to \infty} \frac{1}{n} \log P\left((\tilde{\lambda}_1, \ldots, \tilde{\lambda}_{r_0}) \in O\right) \geq -\inf_O L,$$

and for any closed set $F \subset \mathbb{R}^{r_0}$,

$$\limsup_{n \to \infty} \frac{1}{n} \log P\left((\tilde{\lambda}_1, \ldots, \tilde{\lambda}_{r_0}) \in F\right) \leq -\inf_F L.$$

In particular, in the simplest case when $X_n = 0$, and for the iid model, we are back to the following: if G_n are $n \times r$ matrices whose rows are i.i.d. copies of G and $\Theta = \text{diag}(\theta_1, \ldots, \theta_r)$, we study the deviations of the eigenvalues of $W_n = (1/n)G_n^*\Theta G_n$ (see [Fey et al. 2008] where they treat the case $\Theta = \text{Id}$).

Before going any further, we want to mention an important generalization of our theorem that will be crucial in the next application: provided the law of $\frac{G}{\sqrt{n}}$ satisfies a LDP, one can relax the hypothesis (H1) in the sense that we do not need to assume that the extreme eigenvalues of X_n converge to the edges (respectively a and b) of the limiting measure μ_X. We can allow a finite number of eigenvalues, that we call outliers, to have their limit outside the support of μ_X.

Application: LDP for perturbed Coulomb gases. As for the study of fluctuations, the deterministic model (i.e., when the unperturbed part is deterministic) may seem a bit artificial. The real interest is in X_n random. But there isn't much hope to get a LDP for the extreme eigenvalues of the perturbed model if we don't even know the deviations of extreme eigenvalues of the original model. As pointed out in the introduction of this section, there are only a few models for which we know a LDP for extreme eigenvalues.

Here, we consider the case when X_n is random with a law with density proportional to $e^{-ntrV(X)}$. We work with the orthonormalized model. More precisely, we assume the U_i^n to be a family of orthonormal vectors, either deterministic or independent of X_n.

Theorem 3.2. *Under appropriate assumptions on V, for any fixed k, the law of the k largest eigenvalues of $\widetilde{X}_n$ satisfies a large deviation principle with a good rate function.*

The strategy of the proof goes as follows: as we have enough information on the deviations of the eigenvalues of X_n, as only a finite number can deviate, we first condition on these deviations so that conditionally to the positions of the eigenvalues of X_n, we can apply the generalization of Theorem 3.1 to the case with outliers.

4. A sketch of the proofs

Without getting too much into the details, we would like to briefly give a few ideas of the proofs of the theorems stated in Sections 2 and 3.

The starting point is a determinant computation: we recall that if $V_n = (V_1^n, \dots V_r^n)$ is the $n \times r$ matrix with column vectors $(G_1^n, \dots G_r^n)$ in the iid model or $(U_1^n, \dots U_r^n)$ in the orthonormalized model then $R_n = V_n \Theta V_n^*$, with $\Theta = \mathrm{diag}(\theta_1, \dots, \theta_r)$. Now, for any z which is not an eigenvalue of X_n we have

$$\det(z - \widetilde{X}_n) = \det(z - X_n - V_n \Theta V_n^*)$$
$$= \det(z - X_n) \det \Theta \det(\Theta^{-1} - V_n^*(z - X_n)^{-1} V_n).$$

Therefore, the eigenvalues of $\widetilde{X}_n$ that are not eigenvalues of X_n satisfy

$$f_n(z) := \det(\Theta^{-1} - V_n^*(z - X_n)^{-1} V_n) = 0.$$

The fact that f_n is the determinant of an $r \times r$, that is fixed size, matrix will considerably ease its study.

From this, the almost sure limits are easy to determine, as one can show, by concentration of measure arguments, that

$$\left(\langle V_i^n, (z - X_n)^{-1} V_j^n \rangle\right)_{1 \le i, j \le r} \simeq \text{diag}\left(G_{\mu_X}(z), \ldots, G_{\mu_X}(z)\right),$$

so that the possible limits are the solutions of $G_{\mu_X}(z) = \theta_i^{-1}$, for $i \in \{1, \ldots r\}$.

The analysis of the fluctuations out of the bulk consists in looking at the fine asymptotics of $\langle V_i^n, (\rho_n - X_n)^{-1} V_j^n \rangle$ around its limits $G_{\mu_X}(\rho_\alpha) = 1/\alpha$ for $\rho_n = \rho_\alpha + x/\sqrt{n}$. The result comes essentially from a precise analysis of the orthonormalization procedure and the use of a central limit theorem for quadratic forms developed in [Bai and Yao 2008].

Then comes the more delicate part which is the analysis of the fluctuations of the eigenvalues sticking to the bulk. The strategy is much more involved as it is hard to distinguish between the eigenvalues of X_n and those of $\widetilde{X}_n$ which are not well separated. The work essentially consists in checking that $f_n(z)$ may vanish only if z is very near to the eigenvalues of X_n.

The starting point to show the LDP is the same as for almost sure convergence and fluctuations, namely the fact that the eigenvalues we are interested in are solutions of $f_n(z) = 0$. Now assume that X_n is diagonal, then $f_n(z)$ is a polynomial function of

$$\langle G_i^n, (z - X_n)^{-1} G_j^n \rangle = \frac{1}{n} \sum_{i=1}^{n} \frac{g_i(k) \overline{g_j(k)}}{z - \lambda_k^n} \quad \text{and} \quad \langle G_i^n, G_j^n \rangle = \frac{1}{n} \sum_{i=1}^{n} g_i(k) \overline{g_j(k)}.$$

By Cramer's theorem or weighted Cramer's theorem and some abstract arguments that are pretty standard in large deviation theory, one can derive LDPs for those two sums and then for f_n.

The eigenvalues being the zeroes of f_n, one could expect to get easily an LDP for them. In fact, they are not continuous functions of f_n in the topology we are dealing with so it will be quite delicate to get this LDP but *in fine*, we will get the expected rate function.

5. A few concluding remarks

There is still a lot of activity around spiked models and as for concluding remarks, we would like to briefly mention the content of a few very recent works that have appeared in the last years or months on the subject.

A first interesting direction was developed by a group of authors and shed a new light on the links between almost sure convergence of the outliers in various spiked models and free probability theory. They successfully used the notion of subordination to characterize the limits of extreme eigenvalues. Capitaine, Donati-Martin, Féral and Février [Capitaine et al. 2011], and then Capitaine

[2013] could approach the problem of spiked models in the case when the perturbation is not of finite rank anymore. Belinschi, Bercovici, Capitaine and Février [Belinschi et al. 2012], address the problem of the outliers of $A + UBU^*$, when A and B are deterministic, U Haar unitary and A has a finite number of outliers but its limiting distribution may not be the Dirac mass at zero.

Baik and Wang [2011; 2013] studied a model in which the law of the matrices is proportional to

$$e^{-n\,\mathrm{Tr}(V(X_n)+A_n X_n)}\, dX_n,$$

with A_n which is of finite rank. In the physics literature (see, e.g., [Brézin and Hikami 1998]), V is usually called the potential and A_n the external field. Although looking very similar to we called perturbed Coulomb gases, this model turns out to be quite different ; in particular it is strongly anisotropic and exhibits some subtle nonuniversal behavior when the potential V is not convex.

In [Pizzo et al. 2013; Renfrew and Soshnikov 2013], the results of [Capitaine et al. 2009] and [Benaych-Georges et al. 2011] on perturbed Wigner matrices have been extended. Using techniques close to those explained in Section 4, the authors could relax the conditions on the moments of the entries of X_n and consider more general forms of finite-rank perturbation, going back to the dichotomy between localized and delocalized eigenvectors.

To conclude we mention that Bloemendal and Virag [2013; 2011] studied the largest eigenvalue of a sample covariance matrix from a spiked population in a model which is the real counterpart of the perturbed LUE studied in [Baik et al. 2005]. In particular, they proved a conjecture from that paper regarding the law of the fluctuations of the outliers when the perturbation is properly scaled around its critical value. These results were generalized to non-Gaussian models by Knowles and Yin [2013; 2014], relying on isotropic local semicircle law for such models.

Acknowledgements

The author thanks MSRI for its hospitality during the program "Random matrix theory, interacting particle systems and integrable systems" and for giving her the opportunity to present this work at the workshop "Connections for women, an introduction to random matrices", which took place in September 2010.

References

[Anderson et al. 2010] G. W. Anderson, A. Guionnet, and O. Zeitouni, *An introduction to random matrices*, Cambridge Studies in Advanced Mathematics **118**, Cambridge University Press, Cambridge, 2010.

[Bai and Yao 2008] Z. Bai and J.-f. Yao, "Central limit theorems for eigenvalues in a spiked population model", *Ann. Inst. Henri Poincaré Probab. Stat.* **44**:3 (2008), 447–474.

[Baik and Wang 2011] J. Baik and D. Wang, "On the largest eigenvalue of a Hermitian random matrix model with spiked external source, I: rank 1 case", *Int. Math. Res. Not.* **2011**:22 (2011), 5164–5240.

[Baik and Wang 2013] J. Baik and D. Wang, "On the largest eigenvalue of a Hermitian random matrix model with spiked external source, II: higher rank cases", *Int. Math. Res. Not.* **2013**:14 (2013), 3304–3370.

[Baik et al. 2005] J. Baik, G. Ben Arous, and S. Péché, "Phase transition of the largest eigenvalue for nonnull complex sample covariance matrices", *Ann. Probab.* **33**:5 (2005), 1643–1697.

[Belinschi et al. 2012] S. Belinschi, H. Bercovici, M. Capitaine, and M. Fevrier, "Outliers in the spectrum of large deformed unitarily invariant models", 2012. arXiv 1207.5443

[Ben Arous and Guionnet 1997] G. Ben Arous and A. Guionnet, "Large deviations for Wigner's law and Voiculescu's non-commutative entropy", *Probab. Theory Related Fields* **108**:4 (1997), 517–542.

[Ben Arous et al. 2001] G. Ben Arous, A. Dembo, and A. Guionnet, "Aging of spherical spin glasses", *Probab. Theory Related Fields* **120**:1 (2001), 1–67.

[Benaych-Georges and Nadakuditi 2011] F. Benaych-Georges and R. R. Nadakuditi, "The eigenvalues and eigenvectors of finite, low rank perturbations of large random matrices", *Adv. Math.* **227**:1 (2011), 494–521.

[Benaych-Georges et al. 2011] F. Benaych-Georges, A. Guionnet, and M. Maida, "Fluctuations of the extreme eigenvalues of finite rank deformations of random matrices", *Electron. J. Probab.* **16** (2011), no. 60, 1621–1662.

[Benaych-Georges et al. 2012] F. Benaych-Georges, A. Guionnet, and M. Maida, "Large deviations of the extreme eigenvalues of random deformations of matrices", *Probab. Theory Related Fields* **154**:3-4 (2012), 703–751.

[Bloemendal and Virág 2011] A. Bloemendal and B. Virág, "Limits of spiked random matrices, II", 2011. arXiv 1109.3704

[Bloemendal and Virág 2013] A. Bloemendal and B. Virág, "Limits of spiked random matrices I", *Probab. Theory Related Fields* **156**:3-4 (2013), 795–825.

[Bordenave and Caputo 2012] C. Bordenave and P. Caputo, "A large deviations principle for Wigner matrices without Gaussian tails", 2012. arXiv 1207.5570

[Brézin and Hikami 1998] E. Brézin and S. Hikami, "Level spacing of random matrices in an external source", *Phys. Rev. E* (3) **58**:6, part A (1998), 7176–7185.

[Capitaine 2013] M. Capitaine, "Additive/multiplicative free subordination property and limiting eigenvectors of spiked additive deformations of Wigner matrices and spiked sample covariance matrices", *J. Theoret. Probab.* **26**:3 (2013), 595–648.

[Capitaine et al. 2009] M. Capitaine, C. Donati-Martin, and D. Féral, "The largest eigenvalues of finite rank deformation of large Wigner matrices: convergence and nonuniversality of the fluctuations", *Ann. Probab.* **37**:1 (2009), 1–47.

[Capitaine et al. 2011] M. Capitaine, C. Donati-Martin, D. Féral, and M. Février, "Free convolution with a semicircular distribution and eigenvalues of spiked deformations of Wigner matrices", *Electron. J. Probab.* **16** (2011), no. 64, 1750–1792.

[Capitaine et al. 2012] M. Capitaine, C. Donati-Martin, and D. Féral, "Central limit theorems for eigenvalues of deformations of Wigner matrices", *Ann. Inst. Henri Poincaré Probab. Stat.* **48**:1 (2012), 107–133.

[Chatterjee and Varadhan 2012] S. Chatterjee and S. R. S. Varadhan, "Large deviations for random matrices", *Commun. Stoch. Anal.* **6**:1 (2012), 1–13.

[El Karoui 2005] N. El Karoui, "Recent results about the largest eigenvalue of random covariance matrices and statistical application", *Acta Phys. Polon. B* **36**:9 (2005), 2681–2697.

[Emery et al. 2000] M. Emery, A. Nemirovski, and D. Voiculescu, *Lectures on probability theory and statistics*, Lecture Notes in Mathematics **1738**, Springer, Berlin, 2000.

[Féral and Péché 2007] D. Féral and S. Péché, "The largest eigenvalue of rank one deformation of large Wigner matrices", *Comm. Math. Phys.* **272**:1 (2007), 185–228.

[Fey et al. 2008] A. Fey, R. van der Hofstad, and M. J. Klok, "Large deviations for eigenvalues of sample covariance matrices, with applications to mobile communication systems", *Adv. in Appl. Probab.* **40**:4 (2008), 1048–1071.

[Guionnet and Zeitouni 2002] A. Guionnet and O. Zeitouni, "Large deviations asymptotics for spherical integrals", *J. Funct. Anal.* **188**:2 (2002), 461–515.

[Hardy 2012] A. Hardy, "A note on large deviations for 2D Coulomb gas with weakly confining potential", *Electron. Commun. Probab.* **17** (2012), no. 19, 12.

[Hardy and Kuijlaars 2013] A. Hardy and A. B. J. Kuijlaars, "Large deviations for a non-centered Wishart matrix", *Random Matrices Theory Appl.* **2**:1 (2013), 1250016, 40.

[Helmke and Rosenthal 1995] U. Helmke and J. Rosenthal, "Eigenvalue inequalities and Schubert calculus", *Math. Nachr.* **171** (1995), 207–225.

[Johnstone 2001] I. M. Johnstone, "On the distribution of the largest eigenvalue in principal components analysis", *Ann. Statist.* **29**:2 (2001), 295–327.

[Klyachko 1998] A. A. Klyachko, "Stable bundles, representation theory and Hermitian operators", *Selecta Math. (N.S.)* **4**:3 (1998), 419–445.

[Knowles and Yin 2013] A. Knowles and J. Yin, "The isotropic semicircle law and deformation of Wigner matrices", *Comm. Pure Appl. Math.* **66**:11 (2013), 1663–1750.

[Knowles and Yin 2014] A. Knowles and J. Yin, "The outliers of a deformed Wigner matrix", *Ann. Probab.* **42**:5 (2014), 1980–2031.

[Knutson and Tao 2001] A. Knutson and T. Tao, "Honeycombs and sums of Hermitian matrices", *Notices Amer. Math. Soc.* **48**:2 (2001), 175–186.

[Maïda 2007] M. Maïda, "Large deviations for the largest eigenvalue of rank one deformations of Gaussian ensembles", *Electron. J. Probab.* **12** (2007), 1131–1150.

[Péché 2006] S. Péché, "The largest eigenvalue of small rank perturbations of Hermitian random matrices", *Probab. Theory Related Fields* **134**:1 (2006), 127–173.

[Pizzo et al. 2013] A. Pizzo, D. Renfrew, and A. Soshnikov, "On finite rank deformations of Wigner matrices", *Ann. Inst. Henri Poincaré Probab. Stat.* **49**:1 (2013), 64–94.

[Renfrew and Soshnikov 2013] D. Renfrew and A. Soshnikov, "On finite rank deformations of Wigner matrices, II: delocalized perturbations", *Random Matrices Theory Appl.* **2**:1 (2013), 1250015, 36.

[Weyl 1912] H. Weyl, "Das asymptotische Verteilungsgesetz der Eigenwerte linearer partieller Differentialgleichungen (mit einer Anwendung auf die Theorie der Hohlraumstrahlung)", *Math. Ann.* **71**:4 (1912), 441–479.

mylene.maida@math.univ-lille1.fr   *Laboratoire Paul Painlevé, Université Lille 1,*
                                     Bâtiment M3, Cité Scientifique,
                                     59655 Villeneuve d'Ascq Cedex,, France

Random Matrices
MSRI Publications
Volume **65**, 2014

Three lectures on free probability

JONATHAN NOVAK

with illustrations by Michael LaCroix

These are notes from a three-lecture mini-course on free probability given at MSRI in the Fall of 2010 and repeated a year later at Harvard. The lectures were aimed at mathematicians and mathematical physicists working in combinatorics, probability, and random matrix theory. The first lecture was a staged rediscovery of free independence from first principles, the second dealt with the additive calculus of free random variables, and the third focused on random matrix models.

Introduction

These are notes from a three-lecture mini-course on free probability given at
MSRI in the Fall of 2010 and repeated a year later at Harvard. The lectures were
aimed at mathematicians and mathematical physicists working in combinatorics,
probability, and random matrix theory. The first lecture was a staged rediscovery
of free independence from first principles, the second dealt with the additive
calculus of free random variables, and the third focused on random matrix
models.

Most of my knowledge of free probability was acquired through informal
conversations with my thesis supervisor, Roland Speicher, and while he is an
expert in the field the same cannot be said for me. These notes reflect my own
limited understanding and are no substitute for complete and rigorous treatments,
such as those of Voiculescu, Dykema and Nica [Voiculescu et al. 1992], Hiai
and Petz [2000], and Nica and Speicher [2006]. In addition to these sources, the
expository articles of Biane [2002], Shlyakhtenko [2005] and Tao [2010] are
very informative.

I would like to thank the organizers of the MSRI semester "Random Matrix
Theory, Interacting Particle Systems and Integrable Systems" for the opportunity
to participate as a postdoctoral fellow. Special thanks are owed to Peter Forrester
for coordinating the corresponding MSRI book series volume in which these
notes appear. I am also grateful to the participants of the Harvard random matrices
seminar for their insightful comments and questions.

I am indebted to Michael LaCroix for making the illustrations which accom-
pany these notes.

1. Lecture one: discovering the free world

1.1. *Counting connected graphs.* Let m_n denote the number of simple, undi-
rected graphs on the vertex set $[n] = \{1, \dots, n\}$. We have $m_n = 2^{\binom{n}{2}}$, since each
pair of vertices is either connected by an edge or not. A more subtle quantity
is the number c_n of connected graphs on $[n]$. The sequence $(c_n)_{n \geq 1}$ is listed

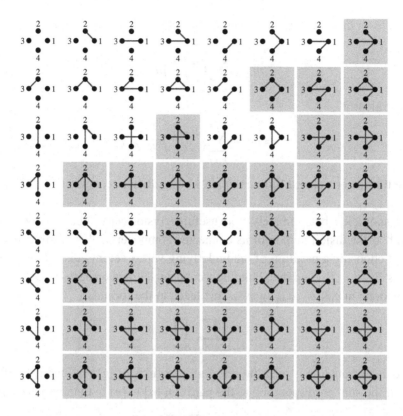

Figure 1. Thirty-eight of sixty-four graphs on four vertices are connected.

as A01187 in Sloane's Online Encyclopedia of Integer Sequences; its first few terms are

$$1, \ 1, \ 4, \ 38, \ 728, \ 26\,704, \ 1\,866\,256, \ \dots.$$

Perhaps surprisingly, there is no closed formula for c_n. However, c_n may be understood in terms of the transparent sequence m_n in several ways, each of which corresponds to a combinatorial decomposition.

First, we may decompose a graph into two disjoint subgraphs: the connected component of a distinguished vertex, say n, and everything else, i.e., the induced subgraph on the remaining vertices. Looking at this the other way around, we may build a graph as follows. From the vertices $1, \dots, n-1$ we can choose k of these in $\binom{n-1}{k}$ ways, and then build an arbitrary graph on these vertices in m_k ways. On the remaining $n-1-k$ vertices together with n, we may build a connected graph in c_{n-k} ways. This construction produces different graphs for different values of k, since the size of the connected component containing the pivot vertex n will be different. Moreover, as k ranges from 1 to $n-1$ we obtain all graphs in this fashion. Thus we have

$$m_n = \sum_{k=0}^{n-1} \binom{n-1}{k} m_k c_{n-k},$$

or equivalently

$$c_n = m_n - \sum_{k=1}^{n-1} \binom{n-1}{k} m_k c_{n-k}.$$

While this is not a closed formula, it allows the efficient computation of c_n given $c_1, \ldots, c_{n-1}$.

A less efficient but ultimately more useful recursion can be obtained by viewing a graph as the disjoint union of its connected components. We construct a graph by first choosing a partition of the underlying vertex set into disjoint nonempty subsets $B_1, \ldots, B_k$, and then building a connected graph on each of these, which can be done in $c_{|B_1|} \ldots c_{|B_k|}$ ways. This leads to the formula

$$m_n = \sum_{\pi \in P(n)} \prod_{B \in \pi} c_{|B|},$$

where the summation is over the set of all partitions of $[n]$. We can split off the term of the sum corresponding to the partition $[n] = [n]$ to obtain the recursion

$$c_n = m_n - \sum_{\substack{\pi \in P(n) \\ b(\pi) \geq 2}} \prod_{B \in \pi} c_{|B|},$$

in which we sum over partitions with at least two blocks.

The above reasoning is applicable much more generally. Suppose that m_n is the number of "structures" which can be built on a set of n labelled points, and that c_n is the number of "connected structures" on these points of the same type. Then the quantities m_n and c_n will satisfy the above (equivalent) relations. This fundamental enumerative link between connected and disconnected structures is ubiquitous in mathematics and the sciences; see [Stanley 1999, Chapter 5]. Prominent examples come from enumerative algebraic geometry [Roth 2009], where connected covers of curves are counted in terms of all covers, and quantum field theory [Etingof 2003], where Feynman diagram sums are reduced to summation over connected terms.

1.2. Cumulants and connectedness.

The relationship between connected and disconnected structures is well-known to probabilists, albeit from a different point of view. In stochastic applications, $m_n = m_n(X) = \mathbb{E}[X^n]$ is the moment sequence of a random variable X, and the quantities $c_n(X)$ defined by either of the equivalent recurrences

$$m_n(X) = \sum_{k=0}^{n-1} \binom{n-1}{k} m_k(X) c_{n-k}(X)$$

and

$$m_n(X) = \sum_{\pi \in P(n)} \prod_{B \in \pi} c_{|B|}(X)$$

are called the cumulants of X. This term was suggested by Harold Hotelling and subsequently popularized by Ronald Fisher and John Wishart in an influential article [1932]. Cumulants were, however, investigated as early as 1889 by the Danish mathematician and astronomer Thorvald Nicolai Thiele, who called them half-invariants. Thiele introduced the cumulant sequence as a transform of the moment sequence defined via the first of the above recurrences, and some years later arrived at the equivalent formulation using the second recurrence. The latter is now called the moment-cumulant formula. Thiele's contributions to statistics and the early theory of cumulants have been detailed by Anders Hald [1981; 2000].

Cumulants are now well-established and frequently encountered in probability and statistics, sufficiently so that the first four have been given names: mean, variance, skewness, and kurtosis.[1] The formulas for mean and variance in terms of moments are simple and familiar:

$$c_1(X) = m_1(X),$$
$$c_2(X) = m_2(X) - m_1(X)^2,$$

whereas the third and fourth cumulants are more involved:

$$c_3(X) = m_3(X) - 3m_2(X)m_1(X) + 2m_1(X)^3,$$
$$c_4(X) = m_4(X) - 4m_3(X)m_1(X) - 3m_2(X)^2 + 12m_2(X)m_1(X)^2 - 6m_1(X)^4.$$

It is not immediately clear why the cumulants of a random variable are of interest. If the distribution of a random variable X is uniquely determined by its moments, then we may think of the moment sequence

$$(m_1(X), m_2(X), \ldots, m_n(X), \ldots)$$

as coordinatizing the distribution of X. Passing from moments to cumulants then amounts to a (polynomial) change of coordinates. Why is this advantageous?

As a motivating example, let us compute the cumulant sequence of the most important random variable, the standard Gaussian X. The distribution of X has

---

[1] In practice, statisticians often define skewness and kurtosis to be the third and fourth cumulants scaled by a power of the variance.

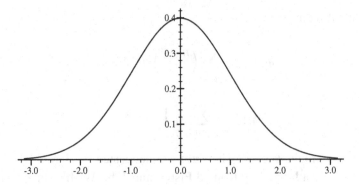

Figure 2. The Gaussian density.

density given by the bell curve

$$\mu_X(dt) = \frac{1}{\sqrt{2\pi}} e^{-t^2/2} dt,$$

depicted in Figure 2.

We will now determine the moments of X. Let z be a complex variable, and define

$$M_X(z) := \int_{\mathbb{R}} e^{tz} \mu_X(dt).$$

Since $e^{-t^2/2}$ decays rapidly as $|t| \to \infty$, $M_X(z)$ is a well-defined entire function of z whose derivatives can be computed by differentiation under the integral sign:

$$M_X'(z) = \int_{\mathbb{R}} t e^{tz} \mu_X(dt), \quad M_X''(z) = \int_{\mathbb{R}} t^2 e^{tz} \mu_X(dt), \quad \dots.$$

In particular, the n-th derivative of $M_X(z)$ at $z = 0$ is

$$M_X^{(n)}(0) = \int_{\mathbb{R}} t^n \mu_X(dt) = m_n(X),$$

so we have the Maclaurin series expansion

$$M_X(z) = \sum_{n=0}^{\infty} m_n(X) \frac{z^n}{n!}.$$

Thus, the integral $M_X(z)$ acts as an exponential generating function for the moments of X. On the other hand, this integral may be explicitly evaluated. Completing the square in the exponent of the integrand we find that

$$M_X(z) = e^{z^2/2} \int_{\mathbb{R}} e^{-(t-z)^2/2} \frac{dt}{\sqrt{2\pi}},$$

whence

$$M_X(z) = e^{z^2/2} = \sum_{k=0}^{\infty} \frac{z^{2k}}{2^k k!}$$

for real z by translation invariance of Lebesgue measure, and hence for all $z \in \mathbb{C}$. We conclude that the odd moments of X vanish, while the even ones are given by the formula

$$m_{2k}(X) = \frac{(2k)!}{2^k k!} = (2k-1) \cdot (2k-3) \cdots 5 \cdot 3 \cdot 1.$$

This is the number of partitions of the set $[2k]$ into blocks of size two, also called "pairings": we have $2k - 1$ choices for the element to be paired with 1, then $2k - 3$ choices for the element to be paired with the smallest remaining unpaired element, etc. Alternatively, we may say that $m_n(X)$ is equal to the number of 1-regular graphs on n labelled vertices. It now follows from the fundamental link between connected and disconnected structures that the cumulant $c_n(X)$ is equal to the number of connected 1-regular graphs. Consequently, the cumulant sequence of a standard Gaussian random variable is simply

$$(0, 1, 0, 0, 0, \dots)$$

That the universality of the Gaussian distribution is reflected in the simplicity of its cumulant sequence signals cumulants as a key concept in probability theory. In Thiele's own words [Hald 2000],

> This remarkable proposition has originally led me to prefer the half-invariants over every other system of symmetrical functions.

This sentiment persists amongst modern-day probabilists. To quote Terry Speed [1983],

> In a sense which it is hard to make precise, all of the important aspects of distributions seem to be simpler functions of cumulants than of anything else, and they are also the natural tools with which transformations of systems of random variables can be studied when exact distribution theory is out of the question.

1.3. Cumulants and independence. The importance of cumulants stems, ultimately, from their relationship with stochastic independence. Suppose that X and Y are a pair of independent random variables whose moment sequences have been given to us by an oracle, and our task is to compute the moments of $X + Y$. Since $\mathbb{E}[X^a Y^b] = \mathbb{E}[X^a] \mathbb{E}[Y^b]$, this can be done using the formula

$$m_n(X + Y) = \sum_{k=0}^{n} \binom{n}{k} m_k(X) m_{n-k}(Y),$$

which is conceptually clear but computationally inefficient because of its dependence on n. For example, if we want to compute $m_{100}(X + Y)$ we must evaluate a sum with 101 terms, each of which is a product of three factors. Computations with independent random variables simplify dramatically if one works with cumulants rather than moments. Indeed, Thiele called cumulants "half-invariants" because

$$X, Y \text{ independent} \implies c_n(X + Y) = c_n(X) + c_n(Y) \ \forall n \geq 1.$$

Thanks to this formula, if the cumulant sequences of X and Y are given, then each cumulant of $X + Y$ can be computed simply by adding two numbers. The mantra to be remembered is that

> *cumulants linearize addition of independent random variables.*

For example, this fact together with the computation we did above yields that the sum of two iid standard Gaussians is a Gaussian of variance two.

In order to precisely understand the relationship between cumulants and independence, we need to extend the relationship between moments and cumulants to a relationship between mixed moments and mixed cumulants. Mixed moments are easy to define: given a set of (not necessarily distinct) random variables $X_1, \ldots, X_n$,

$$m_n(X_1, \ldots, X_n) := \mathbb{E}[X_1 \ldots X_n].$$

It is clear that $m_n(X_1, \ldots, X_n)$ is a symmetric, multilinear function of its arguments. The new notation for mixed moments is related to our old notation for pure moments by

$$m_n(X) = m_n(X, \ldots, X),$$

which we may keep as a useful shorthand.

We now define mixed cumulants recursively in terms of mixed moments using the natural extension of the moment-cumulant formula:

$$m_n(X_1, \ldots, X_n) = \sum_{\pi \in P(n)} \prod_{B \in \pi} c_{|B|}(X_i : i \in B).$$

For example, we have

$$m_2(X_1, X_2) = c_2(X_1, X_2) + c_1(X_1)c_1(X_2),$$

from which we find that the second mixed cumulant of X_1 and X_2 is their covariance,

$$c_2(X_1, X_2) = m_2(X_1, X_2) - m_1(X_1)m_2(X_2).$$

More generally, the recurrence

$$c_n(X_1, \ldots, X_n) = m_n(X_1, \ldots, X_n) - \sum_{\substack{\pi \in P(n) \\ b(\pi) \geq 2}} \prod_{B \in \pi} c_{|B|}(X_i : i \in B)$$

facilitates a straightforward inductive proof that $c_n(X_1, \ldots, X_n)$ is a symmetric, n-linear function of its arguments, which explains Thiele's reference to cumulants as his preferred system of symmetric functions.

The fundamental relationship between cumulants and stochastic independence is the following: X and Y are independent if and only if all their mixed cumulants vanish:

$$c_2(X, Y) = 0,$$
$$c_3(X, X, Y) = c_3(X, Y, Y) = 0,$$
$$c_4(X, X, X, Y) = c_4(X, X, Y, Y) = c_4(X, Y, Y, Y) = 0,$$
$$\vdots$$

The forward direction of this theorem,

$$X, Y \text{ independent} \implies \text{mixed cumulants vanish,}$$

immediately yields Thiele's linearization property, since by multilinearity we have

$$\begin{aligned} c_n(X + Y) &= c_n(X + Y, \ldots, X + Y) \\ &= c_n(X, \ldots, X) + \text{mixed cumulants} + c_n(Y, \ldots, Y) \\ &= c_n(X) + c_n(Y). \end{aligned}$$

Conversely, let X, Y be a pair of random variables whose mixed cumulants vanish. Let us check in a couple of concrete cases that this condition forces X and Y to obey the algebraic identities associated with independent random variables. In the first nontrivial case, $n = 2$, vanishing of mixed cumulants reduces the extended moment-cumulant formula to

$$m_2(X, Y) = c_1(X)c_1(Y) = m_1(X)m_1(Y),$$

which is consistent with the factorization rule $\mathbb{E}[XY] = \mathbb{E}[X]\,\mathbb{E}[Y]$ for independent random variables. Now let us try an $n = 4$ example. We compute $m_4(X, X, Y, Y)$ directly from the extended moment cumulant formula. Referring to Figure 3, we find that vanishing of mixed cumulants implies

$$\begin{aligned} m_4(X, X, Y, Y) = {} & c_2(X, X)c_2(Y, Y) + c_2(X, X)c_1(Y)c_1(Y) \\ & + c_2(Y, Y)c_1(X)c_1(X) + c_1(X)c_1(X)c_1(Y)c_1(Y), \end{aligned}$$

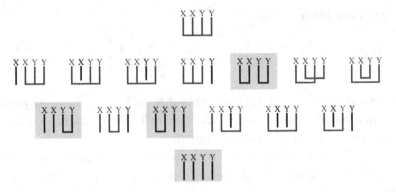

Figure 3. Graphical evaluation of $m_4(X, X, Y, Y)$.

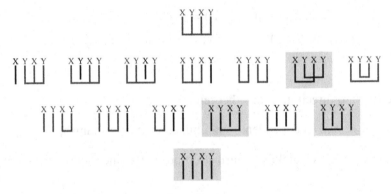

Figure 4. Graphical evaluation of $m_4(X, Y, X, Y)$.

which reduces to the factorization identity $\mathbb{E}[X^2 Y^2] = \mathbb{E}[X^2]\,\mathbb{E}[Y^2]$.

Of course, if we compute $m_4(X, Y, X, Y)$ using the extended moment-cumulant formula we should get the same answer, and indeed this is the case, but it is important to note that the contributions to the sum come from different partitions, as indicated in Figure 4.

1.4. *Central limit theorem by cumulants.* We can use the theory of cumulants presented thus far to prove an elementary version of the central limit theorem. Let $X_1, X_2, X_3 \ldots$ be a sequence of iid random variables, and let X be a standard Gaussian. Suppose that the common distribution of the variables X_i has mean zero, variance one, and finite moments of all orders. Put

$$S_N := \frac{X_1 + \cdots + X_N}{\sqrt{N}}.$$

Then, for each positive integer n,

$$\lim_{N \to \infty} m_n(S_N) = m_n(X).$$

Since moments and cumulants mutually determine one another, in order to prove this CLT it suffices to prove that

$$\lim_{N \to \infty} c_n(S_N) = c_n(X)$$

for each $n \geq 1$. Now, by the multilinearity of c_n and the independence of the X_i, we have

$$\begin{aligned}
c_n(S_N) &= c_n(N^{-1/2}(X_1 + \cdots + X_N)) \\
&= N^{-n/2}(c_n(X_1) + \cdots + c_n(X_N)) \\
&= N^{1-n/2}c_n(X_1),
\end{aligned}$$

where the last line follows from the fact that the X_i are equidistributed. Thus: if $n = 1$,

$$c_1(S_N) = N^{1/2}c_1(X_1) = 0;$$

if $n = 2$,

$$c_2(S_N) = c_2(X_1) = 1;$$

if $n > 2$,

$$c_n(S_N) = N^{\text{negative number}}c_n(X_1).$$

We conclude that

$$\lim_{N \to \infty} c_n(S_N) = \delta_{n2},$$

which we have already identified as the cumulant sequence of a standard Gaussian random variable.

1.5. Geometrically connected graphs. Let us now consider a variation on our original graph-counting question. Given a graph G on the vertex set $[n]$, we may represent its vertices by n distinct points on the unit circle (say, the n-th roots of unity) and its edges by straight line segments joining these points. This is how we represented the set of four-vertex graphs in Figure 1. We will denote this geometric realization of G by $|G|$. The geometric realization of a graph carries extra structure which we may wish to consider. For example, it may happen that $|G|$ is a connected set of points in the plane even if the graph G is not connected in the usual sense of graph theory. Let κ_n denote the number of geometrically connected graphs on $[n]$. This is sequence A136653 in Sloane's database; its first few terms are

$$1, \ 1, \ 4, \ 39, \ 748, \ 27\,162, \ 1\,880\,872, \ \ldots.$$

Since geometric connectivity is a weaker condition than set-theoretic connectivity, κ_n grows faster than c_n; these sequences diverge from one another at $n = 4$, where

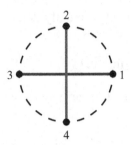

Figure 5. The crosshairs graph.

the unique disconnected but geometrically connected graph is the "crosshairs" graph shown in Figure 5.

Consider now the problem of computing κ_n. As with c_n, we can address this problem by means of a combinatorial decomposition of the set of graphs with n vertices. However, this decomposition must take into account the planar nature of geometric connectivity, which our previous set-theoretic decompositions do not. Consequently, we must formulate a new decomposition.

Given a graph G on $[n]$, let $\pi(G)$ denote the partition of $[n]$ induced by the connected components of G (i and j are in the same block of $\pi(G)$ if and only if they are in the same connected component of G), and let $\pi(|G|)$ denote the partition of $[n]$ induced by the geometrically connected components of $|G|$ (i and j are in the same block of $\pi(|G|)$ if and only if they are in the same geometrically connected component of $|G|$). How are $\pi(G)$ and $\pi(|G|)$ related? To understand this, let us view our geometric graph realizations as living in the hyperbolic plane rather than the Euclidean plane. Thus Figure 1 depicts line systems in the Klein model, in which the plane is an open disc and straight lines are chords of the boundary circle. We could alternatively represent a graph in the Poincaré disc model, where straight lines are arcs of circles orthogonal to the boundary circle, or in the Poincaré half-plane model, where space is an open-half plane and straight lines are arcs of circles orthogonal to the boundary line. The notion of geometric connectedness does not depend on the particular realization chosen. The half-plane model has the useful feature that the geometric realization $|G|$ essentially coincides with the pictorial representation of $\pi(G)$, and we can see clearly that crossings in $|G|$ correspond exactly to crossings in $\pi(G)$. Thus, $\pi(|G|)$ is obtained by fusing together crossing blocks of $\pi(G)$. The resulting partition $\pi(|G|)$ no longer has any crossings — by construction, it is a noncrossing partition; see Figure 6.

We can now obtain a recurrence for κ_n. We construct a graph by first choosing a noncrossing partition of the underlying vertex set into blocks $B_1, \ldots, B_k$ and then building a geometrically connected graph on each block, which can be done

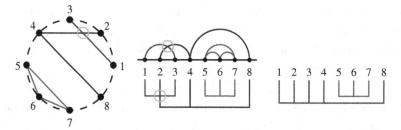

Figure 6. Partition fusion accounts for geometric connectedness.

in $\kappa_{|B_1|} \ldots \kappa_{|B_k|}$ ways. This leads to the formula

$$m_n = \sum_{\pi \in NC(n)} \prod_{B \in \pi} \kappa_{|B|},$$

where the summation is over noncrossing partitions of $[n]$. Just as before, we can split off the term of the sum corresponding to the partition with only one block to obtain the recursion

$$\kappa_n = m_n - \sum_{\substack{\pi \in NC(n) \\ b(\pi) \geq 2}} \prod_{B \in \pi} \kappa_{|B|},$$

in which we sum over noncrossing partitions with at least two blocks.

1.6. Noncrossing cumulants. We have seen above that the usual graph theoretic notion of connectedness manifests itself probabilistically as the cumulant concept. We have also seen that graph theoretic connectedness has an interesting geometric variation, which we called geometric connectedness. This begs the question:

Is there a probabilistic interpretation of geometric connectedness?

Let X be a random variable, with moments $m_n(X)$. Just as the classical cumulants $c_n(X)$ were defined recursively using the relation between all structures and connected structures, we define the noncrossing cumulants of X recursively using the relation between all structures and geometrically connected structures:

$$m_n(X) = \sum_{NC(n)} \prod_{B \in \pi} \kappa_{|B|}(X).$$

We will call this the noncrossing moment-cumulant formula. Since connectedness and geometric connectedness coincide for structures of size $n = 1, 2, 3$, the first three noncrossing cumulants of X are identical to its first three classical cumulants. However, for $n \geq 4$, the noncrossing cumulants become genuinely new statistics of X.

Our first step in investigating these new statistics is to look for a noncrossing analogue of the most important random variable, the standard Gaussian. This should be a random variable whose noncrossing cumulant sequence is

$$0, \ 1, \ 0, \ 0, \ \ldots.$$

If this search leads to something interesting, we may be motivated to further investigate noncrossing probability theory.

From the noncrossing moment-cumulant formula, we find that the moments of the noncrossing Gaussian X are given by

$$m_n(X) = \sum_{\pi \in NC(n)} \prod_{B \in \pi} \delta_{|B|,2} = \sum_{\pi \in NC_2(n)} 1.$$

That is, $m_n(X)$ is equal to the number of partitions in $NC(n)$ all of whose blocks have size 2, i.e., noncrossing pairings of n points. We know that there are no pairings at all on an odd number of points, so the odd moments of X must be zero, which indicates that X likely has a symmetric distribution. The number of pairings on $n = 2k$ points is given by a factorial going down in steps of two, $(2k-1)!! = (2k-1) \cdot (2k-3) \cdots 5 \cdot 3 \cdot 1$, so the number of noncrossing pairings must be smaller than this double factorial.

In order to count noncrossing pairings on $2k$ points, we construct a function f from the set of all pairings on $2k$ points to length $2k$ sequences of ± 1. This function is easy to describe: if $i < j$ constitute a block of π, then the i-th element of $f(\pi)$ is $+1$ and the j-th element of $f(\pi)$ is -1. See Figure 7 for an illustration of this function in the case $k = 3$. By construction, f is a surjection from the set of pairings on $2k$ points onto the set of length $2k$ sequences of ± 1 all of whose partial sums are nonnegative and whose total sum is zero. We leave it to the reader to show that the fibre of f over any such sequence contains exactly one noncrossing pairing, so that f restricts to a bijection from noncrossing pairings onto its image. The image sequences can be neatly enumerated using the Dvoretzky–Motzkin–Raney cyclic shift lemma, as in [Graham et al. 1989, Section 7.5]. They are counted by the Catalan numbers

$$\text{Cat}_k = \frac{1}{k+1} \binom{2k}{k},$$

which are smaller than the double factorials by a factor of $2^k/(k+1)!$. In fact, since $\text{Cat}_k < 4^k$, we can conclude that the distribution of X is compactly supported.

We have discovered that

$$m_n(X) = \begin{cases} 0 & \text{if } n \text{ is odd,} \\ \text{Cat}_{n/2} & \text{if } n \text{ even.} \end{cases}$$

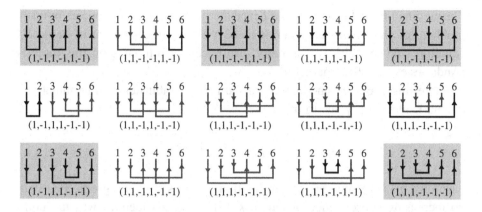

Figure 7. Construction of the function f from pairings to bitstrings.

The Catalan numbers are ubiquitous in enumerative combinatorics (see [Stanley 1999, Exercise 6.19] as well as [Stanley 2013]), and their appearance in this context is the first sign that we are onto something interesting. We are now faced with an inverse problem: we are not trying to calculate the moments of a random variable given its distribution, rather we know that the moment sequence of X is

$$0, \; \mathrm{Cat}_1, \; 0, \; \mathrm{Cat}_2, \; 0, \; \mathrm{Cat}_3, \; 0, \; \ldots.$$

and we would like to write down its distribution μ_X. Equivalently, we are looking for an integral representation of the entire function

$$M_X(z) = \sum_{n=0}^{\infty} \mathrm{Cat}_n \frac{z^{2n}}{(2n)!} = \sum_{n=0}^{\infty} \frac{z^{2n}}{n!(n+1)!}$$

which has the form

$$M_X(z) = \int_{\mathbb{R}} e^{tz} \mu_X(\mathrm{d}t),$$

with μ_X a probability measure on the real line. The solution to this problem can be extracted from the classical theory of Bessel functions.

The modified Bessel function $I_\alpha(z)$ of order α is one of two linearly independent solutions to the modified Bessel equation

$$\left(z^2 \frac{d^2}{dz^2} + z \frac{d}{dz} - (z^2 + \alpha^2) \right) F = 0,$$

the other being the Macdonald function

$$K_\alpha(z) = \frac{\pi}{2} \frac{I_{-\alpha}(z) - I_\alpha(z)}{\sin(\alpha \pi)}.$$

The modified Bessel equation (and hence the functions I_α, K_α) appears in many problems of physics and engineering since it is related to solutions of Laplace's equation with cylindrical symmetry. An excellent reference on this topic is [Andrews et al. 1999, Chapter 4].

Interestingly, Bessel functions also occur in the combinatorics of permutations: a remarkable identity due to Ira Gessel asserts that

$$\det[I_{i-j}(2z)]^k_{i,j=1} = \sum_{n=0}^{\infty} \text{lis}_k(n)\frac{z^{2n}}{(n!)^2},$$

where $\text{lis}_k(n)$ is the number of permutations in the symmetric group $\mathbf{S}(n)$ with no increasing subsequence of length $k + 1$. Gessel's identity was the point of departure in the work of Jinho Baik, Percy Deift and Kurt Johansson who, answering a question posed by Stanislaw Ulam, proved that the limit distribution of the length of the longest increasing subsequence in a uniformly distributed random permutation is given by the ($\beta = 2$) Tracy–Widom distribution. This nonclassical distribution was isolated and studied by Craig Tracy and Harold Widom in a series of works on random matrix theory in the early 1990s where it emerged as the limiting distribution of the top eigenvalue of large random Hermitian matrices. It has a density which may also be described in terms of Bessel functions, albeit indirectly. Consider the ordinary differential equation

$$\frac{d^2}{dx^2}u = 2u^3 + xu$$

for a real function $u = u(x)$, which is known as the Painlevé II equation after the French mathematician (and two-time Prime Minister of France) Paul Painlevé. It is known that this equation has a unique solution, called the Hastings–McLeod solution, with the asymptotics $u(x) \sim - \text{Ai}(x)$ as $x \to \infty$, where

$$\text{Ai}(x) = \frac{1}{\pi}\sqrt{\frac{x}{3}}K_{\frac{1}{3}}(\tfrac{2}{3}x^{3/2})$$

is a scaled specialization of the Macdonald function known as the Airy function. Define the Tracy–Widom distribution function by

$$F(t) = e^{-\int_t^\infty (x-t)u(x)^2 dx},$$

where u is the Hastings–McLeod solution to Painlevé II. The theorem of Baik, Deift and Johansson asserts that

$$\lim_{n\to\infty} \frac{1}{n!}\text{lis}_{2\sqrt{n}+tn^{1/6}}(n) = F(t)$$

for any $t \in \mathbb{R}$. From this one may conclude, for example, that the probability a permutation drawn uniformly at random from the symmetric group $\mathbf{S}(n^2)$ avoids the pattern $1\,2\,\ldots\,2n+1$ converges to $F(0)=0.9694\ldots$. We refer the interested reader to Richard Stanley's survey [2007] for more information on this topic.

Nineteenth century mathematicians knew how to describe the modified Bessel function both as a series,

$$I_\alpha(z) = \sum_{n=0}^{\infty} \frac{(\frac{z}{2})^{2n+\alpha}}{n!\,\Gamma(n+1+\alpha)},$$

and as an integral,

$$I_\alpha(z) = \frac{(\frac{z}{2})^\alpha}{\sqrt{\pi}\,\Gamma(\alpha+\frac{1}{2})} \int_0^\pi e^{(\cos\theta)z}(\sin\theta)^{2\alpha}\,d\theta.$$

From the series representation we find that

$$M_X(z) = \frac{I_1(2z)}{z},$$

and consequently we have the integral representation

$$M_X(z) = \frac{2}{\pi} \int_0^\pi e^{2(\cos\theta)z}\sin^2\theta\,d\theta.$$

This is one step removed from what we want: it tells us that the Catalan numbers are the even moments of the random variable $X = 2\cos(Y)$, where Y is a random variable with distribution

$$\mu_Y(d\theta) = \frac{2}{\pi}\sin^2\theta\,d\theta$$

supported on the interval $[0,\pi]$. However, this is a rather interesting intermediate step since the above measure appears in number theory, where it is called the Sato–Tate distribution; see Figure 8.

The Sato–Tate distribution arises in the arithmetic statistics of elliptic curves. The location of integer points on elliptic curves is a classical topic in number

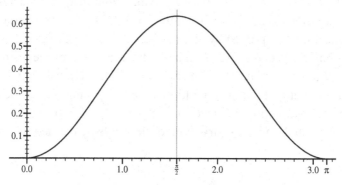

Figure 8. The Sato–Tate density.

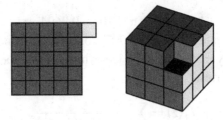

Figure 9. Diophantine perspectives on twenty-six.

theory. For example, Diophantus of Alexandria wrote that the equation

$$y^2 = x^3 - 2$$

has the solution $x = 3$, $y = 5$, and in the 1650s Pierre de Fermat claimed that there are no other positive integer solutions. This is the striking assertion that 26 is the only number one greater than a perfect square and one less than a perfect cube (see Figure 9). That this is indeed the case was proved by Leonhard Euler in 1770, although according to some sources Euler's proof was incomplete and the solution to this problem should be attributed to Axel Thue in 1908.

Modern number theorists study solutions to elliptic Diophantine equations by reducing modulo primes. Given an elliptic curve

$$y^2 = x^3 + ax + b, \quad a, b \in \mathbb{Z},$$

let $\Delta = -16(4a^3 + 27b^2)$ be sixteen times the discriminant of $x^3 + ax + b$, and let S_p be the number of solutions of the congruence

$$y^2 \equiv x^3 + ax + b \qquad \mod p$$

where p is a prime which does not divide Δ. In his 1924 doctoral thesis, Emil Artin conjectured that

$$|S_p - p| \le 2\sqrt{p}$$

for all such good reduction primes. This remarkable inequality states that the number of solutions modulo p is roughly p itself, up to an error of order $\sqrt{p}$. Artin's conjecture was proved by Helmut Hasse in 1933. Around 1960, Mikio Sato and John Tate became interested in the finer question of the distribution of the centred and scaled solution count $(S_p - p)/\sqrt{p}$ for typical elliptic curves E (meaning those without complex multiplication) as p ranges over the infinitely many primes not dividing the discriminant of E. Because of Hasse's theorem, this amounts to studying the distribution of the angle θ_p defined by

$$\frac{S_p - p}{\sqrt{p}} = 2\cos\theta_p$$

in the interval $[0, \pi]$. Define a sequence μ_N^E of empirical probability measures associated to E by

$$\mu_N^E = \frac{1}{\pi(N)} \sum_{p \leq N} \delta_{\theta_p},$$

where $\pi(N)$ is the number of prime numbers less than or equal to N. Sato and Tate conjectured that, for any elliptic curve E without complex multiplication, μ_N^E converges weakly to the Sato–Tate distribution as $N \to \infty$. This is a universality conjecture: it posits that certain limiting behaviour is common to a large class of elliptic curves irrespective of their fine structural details. Major progress on the Sato–Tate conjecture has been made within the last decade; we refer the reader to the surveys of Barry Mazur [2006] and Ram Murty and Kumar Murty [2009] for further information.

The random variable we seek is not the Sato–Tate variable Y, but twice its cosine, $X = 2 \cos Y$. Making the substitution $s = \arccos \theta$ in the integral representation of $M_X(z)$ obtained above, we obtain

$$M_X(z) = \frac{2}{\pi} \int_{-1}^{1} e^{2sz} \sqrt{1 - s^2} \, ds,$$

and further substituting $t = 2s$ this becomes

$$M_X(z) = \frac{1}{2\pi} \int_{-2}^{2} e^{tz} \sqrt{4 - t^2} \, dt.$$

Thus the random variable X with even moments the Catalan numbers and vanishing odd moments

$$\mu_X(dt) = \frac{1}{2\pi} \sqrt{4 - t^2} \, dt,$$

which is both symmetric and compactly supported. This is another famous distribution: it is called the Wigner semicircle distribution after the physicist Eugene Wigner, who considered it in the 1950s in a context ostensibly unrelated to elliptic curves. The density of μ_X is shown in Figure 10 — note that it is not a semicircle, but rather half an ellipse of semi-major axis two and semi-minor axis $1/\pi$.

Wigner was interested in constructing models for the energy levels of complex systems, and hit on the idea that the eigenvalues of large symmetric random matrices provide a good approximation. Wigner considered $N \times N$ symmetric matrices X_N whose entries $X_N(ij)$ are independent random variables, up to the symmetry constraint $X_N(ij) = X_N(ji)$. Random matrices of this form are now known as Wigner matrices, and their study remains a topic of major interest today. Wigner studied the empirical spectral distribution of the eigenvalues of

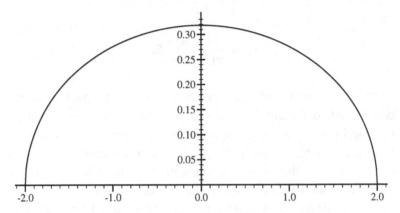

Figure 10. The Wigner semicircle density.

X_N, i.e., the probability measure

$$\mu_N = \frac{1}{N} \sum_{k=1}^{N} \delta_{\lambda_k(N)}$$

which places mass $1/N$ at each eigenvalue of X_N. Note that, unlike in the setting above where we considered the sequence of empirical measures associated to a fixed elliptic curve E, the measure μ_N is a random measure since X_N is a random matrix. Wigner showed that the limiting behaviour of μ_N does not depend on the details of the random variables which make up X_N. In [Wigner 1958], he made the following hypotheses:

(1) Each $X_N(ij)$ has a symmetric distribution.

(2) Each $X_N(ij)$ has finite moments of all orders, each of which is bounded by a constant independent of N, i, j.

(3) The variance of $X_N(ij)$ is $1/N$.

Wigner proved that, under these hypotheses, μ_N converges weakly to the semicircle law which now bears his name. We will see a proof of Wigner's theorem for random matrices with (complex) Gaussian entries in Lecture Three. The universality of the spectral structure of real and complex Wigner matrices holds at a much finer level, and under much weaker hypotheses, both at the edges of the semicircle [Soshnikov 1999] and in the bulk [Erdős et al. 2011; Tao and Vu 2011].

1.7. *Noncrossing independence.* Our quest for the noncrossing Gaussian has brought us into contact with interesting objects (random permutations, elliptic curves, random matrices) and the limit laws which govern them (Tracy–Widom distribution, Sato–Tate distribution, Wigner semicircle distribution). This motivates us to continue developing the rudiments of noncrossing probability theory — perhaps we have hit on a framework within which these objects may be studied.

Our next step is to introduce a notion of noncrossing independence. We know that classical independence is characterized by the vanishing of mixed cumulants. Imitating this, we will define noncrossing independence via the vanishing of mixed noncrossing cumulants. Like classical mixed cumulants, the noncrossing mixed cumulant functionals are defined recursively via the multilinear extension of the noncrossing moment-cumulant formula,

$$m_n(X_1, \ldots, X_n) = \sum_{\pi \in NC(n)} \prod_{B \in \pi} \kappa_{|B|}(X_i : i \in B).$$

The recurrence

$$\kappa_n(X_1, \ldots, X_n) = m_n(X_1, \ldots, X_n) - \sum_{\pi \in NC(n)} \prod_{B \in \pi} \kappa_{|B|}(X_i : i \in B)$$

and induction establish that $\kappa_n(X_1, \ldots, X_n)$ is a symmetric multilinear function of its arguments. Two random variables X, Y are said to be noncrossing independent if their mixed noncrossing cumulants vanish:

$$\kappa_2(X, Y) = 0,$$

$$\kappa_3(X, X, Y) = \kappa_3(X, Y, Y) = 0,$$

$$\kappa_4(X, X, X, Y) = \kappa_4(X, X, Y, Y) = \kappa_4(X, Y, Y, Y) = 0,$$

$$\vdots$$

An almost tautological consequence of this definition is that

$$X, Y \text{ noncrossing independent} \implies \kappa_n(X + Y) = \kappa_n(X) + \kappa_n(Y) \quad \forall n \geq 1.$$

Thus, just as classical cumulants linearize the addition of classically independent random variables,

> *noncrossing cumulants linearize addition*
> *of noncrossing independent random variables.*

We can also note that the semicircular random variable X, whose noncrossing cumulant sequence is $0, 1, 0, 0, \ldots$, plays the role of the standard Gaussian with respect to this new notion of independence. For example, since noncrossing cumulants linearize noncrossing independence, the sum of two noncrossing independent semicircular random variables is a semicircular random variable of variance two. The noncrossing analogue of the central limit theorem asserts that, if $X_1, X_2, \ldots$ is a sequence of noncrossing independent and identically distributed random variables with mean zero and variance one, then the moments of

$$S_N = \frac{X_1 + \cdots + X_N}{\sqrt{N}}$$

converge to the moments of the standard semicircular X as $N \to \infty$. The proof of this fact is identical to the proof of the classical central limit theorem given above, except that classical cumulants are replaced by noncrossing cumulants.

Of course, we don't really know what noncrossing independence means. For example, if X and Y are noncrossing independent, is it true that $\mathbb{E}[XY] = \mathbb{E}[X]\,\mathbb{E}[Y]$? The answer is yes, since classical and noncrossing mixed cumulants agree up to and including order three,

$$c_1(X) = \kappa_1(X), \quad c_2(X, Y) = \kappa_2(X, Y), \quad c_3(X, Y, Z) = \kappa_3(X, Y, Z).$$

But what about higher order mixed moments?

We observed above that, in the classical case, vanishing of mixed cumulants allows us to recover the familiar algebraic identities governing the expectation of independent random variables. We do not have a priori knowledge of the algebraic identities governing the expectation of noncrossing independent random variables, so we must discover them using the vanishing of mixed noncrossing cumulants. Let us see what this implies for the mixed moment $m_4(X, X, Y, Y) = \mathbb{E}[X^2 Y^2]$. Referring to Figure 11 we see that in this case the noncrossing moment-cumulant formula reduces to

$$m_4(X, X, Y, Y) = \kappa_2(X, X)\kappa_2(Y, Y) + \kappa_2(X, X)\kappa_1(Y)\kappa_1(Y)$$
$$+ \kappa_2(Y, Y)\kappa_1(X)\kappa_1(X) + \kappa_1(X)\kappa_1(X)\kappa_1(Y)\kappa_1(Y),$$

which is exactly the formula we obtained for classically independent random variables using the classical moment-cumulant formula.

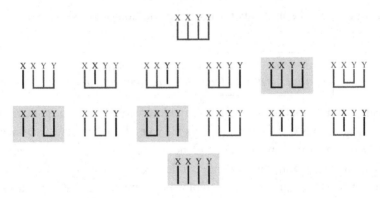

Figure 11. Graphical evaluation of $m_4(X, X, Y, Y)$ using noncrossing cumulants.

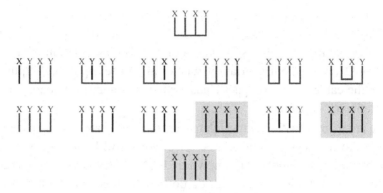

Figure 12. Graphical evaluation of $m_4(X, Y, X, Y)$ using noncrossing cumulants.

However, when we use the noncrossing moment-cumulant formula to evaluate the same mixed moment with its arguments permuted, we instead get

$$m_4(X, Y, X, Y) = \kappa_2(X, X)\kappa_1(Y)\kappa_1(Y)$$
$$+ \kappa_2(Y, Y)\kappa_1(X)\kappa_1(X) + \kappa_1(X)\kappa_1(X)\kappa_1(Y)\kappa_1(Y);$$

see Figure 12. Since $m_4(X, X, Y, Y) = m_4(X, Y, X, Y)$, we are forced to conclude that the two expressions obtained are equal, which in turn forces

$$\kappa_2(X, X)\kappa_2(Y, Y) = 0.$$

Thus, if X, Y are noncrossing independent random variables, at least one of them must have vanishing variance, and consequently must be almost surely constant. The converse is also true — one can show that a (classical or noncrossing) mixed cumulant vanishes if any of its entries are constant random variables. So we have classified pairs of noncrossing independent random variables: they look like $\{X, Y\} = \{\text{arbitrary, constant}\}$. Such pairs of random variables are of no interest from a probabilistic perspective. It would seem that noncrossing probability is a dead end.

1.8. The medium is the message. If Ω is a compact Hausdorff space then the algebra $\mathcal{A}(\Omega)$ of continuous functions $X : \Omega \to \mathbb{C}$ is a commutative C^*-algebra. This means that in addition to its standard algebraic structure (pointwise addition, multiplication and scalar multiplication of functions) $\mathcal{A}(\Omega)$ is equipped with a norm satisfying the Banach algebra axioms and an antilinear involution which is compatible with the norm, $\|X^*X\| = \|X\|^2$. The norm comes from the topology of the source, $\|X\| = \sup_\omega |X(\omega)|$, and the involution comes from the conjugation automorphism of the target, $X^*(\omega) = \overline{X(\omega)}$. Conversely, a famous theorem of Israel Gelfand asserts that any unital commutative C^*-algebra $\mathcal{A}$ can be realized as the algebra of continuous functions on a compact Hausdorff space $\Omega(\mathcal{A})$ in an essentially unique way. In fact, $\Omega(\mathcal{A})$ may be constructed as the set of maximal ideals of $\mathcal{A}$ equipped with a suitable topology. The associations $\Omega \mapsto \mathcal{A}(\Omega)$ and $\mathcal{A} \mapsto \Omega(\mathcal{A})$ are contravariantly functorial and set up a dual equivalence between the category of compact Hausdorff spaces and the category of unital commutative C^*-algebras.

There are many situations in which one encounters a category of spaces dually equivalent to a category of algebras. In a wonderful book [Nestruev 2003], the mathematicians collectively known as Jet Nestruev develop the theory of smooth real manifolds entirely upside-down: the theory is built in the dual algebraic category, whose objects Nestruev terms smooth complete geometric $\mathbb{R}$-algebras, and then exported to the geometric one by a contravariant functor. In many situations, given a category of spaces dually equivalent to a category of algebras it pays to shift our stance and view the algebraic category as primary. In particular, the algebraic point of view is typically easier to generalize. This is the paradigm shift driving Alain Connes' noncommutative geometry programme, and the reader is referred to [Connes 1994] for much more information.

This paradigm shift is precisely what is needed in order to salvage non-crossing probability theory. In probability theory, the notion of space is that of a Kolmogorov triple $(\Omega, \mathcal{F}, P)$ which models the probability to observe a stochastic system in a given state or collection of states. The dual algebraic object associated to a Kolmogorov triple is $L^\infty(\Omega, \mathcal{F}, P)$, the algebra of essentially bounded complex random variables $X : \Omega \to \mathbb{C}$. Just like in the case of continuous functions on a compact Hausdorff space, this algebra has a very special structure: it is a commutative von Neumann algebra equipped with a unital faithful tracial state, $\tau[X] = \int_\Omega X \, dP$. Moreover, there is an analogue of Gelfand's theorem in this setting which says that any commutative von Neumann algebra can be realized as the algebra of bounded complex random variables on a Kolmogorov triple in an essentially unique way. This is the statement that the categories of Kolmogorov triples and commutative von Neumann algebras are dual equivalent.

Noncrossing independence was rendered trivial by the commutativity of random variables. We can rescue it from the abyss by following the lead of noncommutative geometry and dropping commutativity in the dual category: we shift our stance and define a noncommutative probability space to be a pair $(\mathcal{A}, \tau)$ consisting of a possibly noncommutative complex associative unital algebra $\mathcal{A}$ together with a unital linear functional $\tau : \mathcal{A} \to \mathbb{C}$. If we reinstate commutativity and insist that $\mathcal{A}$ is a von Neumann algebra and τ a faithful tracial state, we are looking at essentially bounded random variables on a Kolmogorov triple, but a general noncommutative probability space need not be an avatar of any classical probabilistic entity.

As a nod to the origins of this definition, and in order to foster analogies with classical probability, we refer to the elements of $\mathscr{A}$ as random variables and call τ the expectation functional. This prompts some natural questions. Before this subsection we only discussed real random variables — complex numbers crept in with the abstract nonsense. What is the analogue of the notion of real random variable in a noncommutative probability space? Probabilists characterize random variables in terms of their distributions. Can we assign distributions to random variables living in a noncommutative probability space? Is it possible to give meaning to the phrase "the distribution of a bounded real random variable living in a noncommutative probability space is a compactly supported probability measure on the line"? We will deal with some of these questions at the end of Lecture Two. For now, however, we remain in the purely algebraic framework, where the closest thing to the distribution of a random variable $X \in \mathscr{A}$ is its moment sequence $m_n(X) = \tau[X^n]$. As in [Voiculescu et al. 1992, p. 12]:

> The algebraic context is not used in the pursuit of generality, but rather of transparence.

1.9. A brief history of the free world. Having cast off the yoke of commutativity, we are free — free to explore noncrossing probability in the new framework provided by the noncommutative probability space concept. Noncrossing probability has become *free probability*, and will henceforth be referred to as such. Accordingly, noncrossing cumulants will now be referred to as free cumulants, and noncrossing independence will be termed free independence.

The reader is likely aware that free probability is a flourishing area of contemporary mathematics. This first lecture has been historical fiction, and is essentially an extended version of [Novak and Śniady 2011]. Free probability was not discovered in the context of graph enumeration problems, or by tampering with the cumulant concept, although in retrospect it might have been. Rather, free probability theory was invented by Dan-Virgil Voiculescu in the 1980s in order to address a famous open problem in the theory of von Neumann algebras, the free group factors isomorphism problem. The problem is to determine when the von Neumann algebra of the free group on a generators is isomorphic to the von Neumann algebra of the free group on b generators. It is generally believed that these are isomorphic von Neumann algebras if and only if $a = b$, but this remains an open problem. Free probability theory (and its name) originated in this operator-algebraic context.

Voiculescu's definition of free independence, which was modelled on the free product of groups, is the following: random variables X, Y in a noncommutative probability space $(\mathscr{A}, \tau)$ are said to be freely independent if

$$\tau[f_1(X)g_1(Y)\dots f_k(X)g_k(Y)] = 0$$

whenever $f_1, g_1, \ldots, f_k, g_k$ are polynomials such that

$$\tau[f_1(X)] = \tau[g_1(X)] = \cdots = \tau[f_k(X)] = \tau[g_k(Y)] = 0.$$

This should be compared with the definition of classical independence: random variables X, Y in a noncommutative probability space $(\mathcal{A}, \tau)$ are said to be classically independent if they commute, $XY = YX$, and if

$$\tau[f(X)g(Y)] = 0$$

whenever f and g are polynomials such that $\tau[f(X)] = \tau[g(Y)] = 0$. These two definitions are antithetical: classical independence has commutativity built into it, while free independence becomes trivial if commutativity is imposed. Nevertheless, both notions are accommodated within the noncommutative probability space framework.

The precise statement of equivalence between classical independence and vanishing of mixed cumulants is due to Gian-Carlo Rota [1964]. In the 1990s, knowing both of Voiculescu's new free probability theory and Rota's approach to classical probability theory, Roland Speicher made the beautiful discovery that by excising the lattice of set partitions from Rota's foundations and replacing it with the lattice of noncrossing partitions, much of Voiculescu's theory could be recovered and extended by elementary combinatorial methods. In particular, Speicher showed that free independence is equivalent to the vanishing of mixed free cumulants. The combinatorial approach to free probability is exhaustively applied in [Nica and Speicher 2006], while the original analytic approach of Voiculescu is detailed in [Voiculescu et al. 1992].

2. Lecture two: exploring the free world

Lecture One culminated in the notion of a noncommutative probability space and the realization that this framework supports two types of independence: classical independence and free independence. From here we can proceed in several ways. One option is to prove an abstract result essentially stating that these are the only notions of independence which can occur. This result, due to Speicher, places classical and free independence on equal footing. Another possibility is to present concrete problems of intrinsic interest where free independence naturally appears. We will pursue the second route, and examine problems emerging from the theory of random walks on groups which can be recast as questions about free random variables. In the course of solving these problems we will develop the calculus of free random variables and explore the terrain of the free world.

2.1. *Random walks on the integers.* The prototypical example of a random walk on a group is the simple random walk on $\mathbf{Z}$: a walker initially positioned at zero tosses a fair coin at each tick of the clock — if it lands heads he takes a step of $+1$, if it lands tails he takes a step of -1. A random walk is said to be recurrent if it returns to its initial position with probability one, and transient if not. Is the simple random walk on $\mathbf{Z}$ recurrent or transient?

Let $\alpha(n)$ denote the number of walks which return to zero for the first time after n steps, and let $\phi(n) = 2^{-n}\alpha(n)$ denote the corresponding probability that the first return occurs at time n. Note that $\alpha(0) = \phi(0) = 0$, and define

$$F(z) = \sum_{n=0}^{\infty} \phi(n)z^n.$$

Then

$$F(1) = \sum_{n=0}^{\infty} \phi(n) \le 1$$

is the probability we seek. The radius of convergence of $F(z)$ is at least one, and by Abel's theorem

$$F(1) = \lim_{x \to 1} F(x)$$

as x approaches 1 in the interval $[0, 1)$.

Let $\lambda(n)$ denote the number of length n loops on $\mathbf{Z}$ based at 0, and let $\rho(n) = 2^{-n}\lambda(n)$ be the corresponding probability of return at time n (regardless of whether this is the first return or not). Note that $\lambda(0) = \rho(0) = 1$. We have

$$\lambda(n) = \begin{cases} 0 & \text{if } n \text{ is odd,} \\ \binom{n}{n/2} & \text{if } n \text{ is even.} \end{cases}$$

From Stirling's formula, we see that

$$\rho(2k) \sim \frac{1}{\sqrt{\pi k}}$$

as $k \to \infty$. Thus the radius of convergence of

$$R(z) = \sum_{n=0}^{\infty} \rho(n)z^n$$

is 1.

We can decompose the set of loops of given length according to the number of steps taken to the first return. This produces the equation

$$\lambda(n) = \sum_{k=0}^{n} \alpha(k)\lambda(n - k).$$

Equivalently, since all probabilities are uniform,

$$\rho(n) = \sum_{k=0}^{n} \phi(k)\rho(n-k).$$

Summing on z, this becomes the identity

$$R(z) - 1 = F(z)R(z)$$

in the algebra of holomorphic functions on the open unit disc in $\mathbb{C}$. Since $R(z)$ has nonnegative coefficients, it is nonvanishing for $x \in [0, 1)$ and we can write

$$F(x) = 1 - \frac{1}{R(x)}, \quad 0 \le x < 1.$$

Thus

$$F(1) = \lim_{x \to 1} F(x) = 1 - \frac{1}{\lim_{x \to 1} R(x)}.$$

If $R(1) < \infty$, then by Abel's theorem $\lim_{x \to 1} R(x) = R(1)$ and we obtain $F(1) < 1$. On the other hand, if $R(1) = \infty$, then $\lim_{x \to 1} R(x) = \infty$ and we get $F(1) = 1$. Thus the simple random walk is transient or recurrent according to the convergence or divergence of the series $\sum \rho(n)$. From the Stirling estimate above we find that this sum diverges, so the simple random walk on $\mathbf{Z}$ is recurrent.

2.2. Pólya's theorem. In the category of abelian groups, coproducts are direct sums:

$$\coprod_{i \in I} \mathbf{G}_i = \bigoplus_{i \in I} \mathbf{G}_i.$$

George Pólya [1921] proved that the simple random walk on

$$\mathbf{Z}^d = \underbrace{\mathbf{Z} \oplus \cdots \oplus \mathbf{Z}}_{d}$$

is recurrent for $d = 1, 2$ and transient for $d > 2$. This striking result can be deduced solely from an understanding of the simple random walk on $\mathbf{Z}$.

Let us give a proof of Pólya's theorem. Let $\lambda_d(n)$ denote the number of length n loops on $\mathbf{Z}^d$ based at 0^d. Let $\rho_d(n)$ denote the probability of return to 0^d after n steps,

$$\rho_d(n) = \frac{1}{(2d)^n} \lambda_d(n).$$

As above, the simple random walk on $\mathbf{Z}^d$ is recurrent if the sum $\sum \rho_d(n)$ diverges, and transient otherwise. Form the loop generating function

$$L_d(z) = \sum_{n=0}^{\infty} \lambda_d(n) z^n.$$

We aim to prove that

$$L_d\left(\frac{1}{2d}\right) = \sum_{n=0}^{\infty} \rho_d(n)$$

diverges for $d = 1, 2$ and converges for $d > 2$.

While the ordinary loop generating function is hard to analyze directly, the exponential loop generating function

$$E_d(z) = \sum_{n=0}^{\infty} \lambda_d(n) \frac{z^n}{n!}$$

is quite accessible. Indeed, as in the last subsection we have

$$\lambda_1(n) = \begin{cases} 0 & \text{if } n \text{ is odd,} \\ \binom{n}{n/2} & \text{if } n \text{ is even,} \end{cases}$$

so that

$$E_1(z) = \sum_{k=0}^{\infty} \frac{z^{2k}}{k!\,k!} = I_0(2z)$$

is precisely the modified Bessel function of order zero. Since a loop on $\mathbf{Z}^d$ is just a shuffle of loops on $\mathbf{Z}$, the product formula for exponential generating functions [Stanley 1999] yields

$$E_d(z) = E_1(z)^d = I_0(2z)^d.$$

What we have is the exponential generating function for the loop counts $\lambda_d(n)$, and what we want is the ordinary generating function of this sequence. The integral transform

$$L_f(z) = \int_0^{\infty} f(tz)e^{-t}\,\mathrm{d}t,$$

which looks like the Laplace transform of f but with the z-parameter in the wrong place, converts exponential generating functions into ordinary generating functions. This can be seen by differentiating under the integral sign and using the fact that the moments of the exponential distribution are the factorials,

$$\int_0^{\infty} t^n e^{-t}\,\mathrm{d}t = n!.$$

This trick is constantly used in quantum field theory in connection with Borel summation of divergent series [Etingof 2003]. In particular, we have

$$L_d(z) = \int_0^{\infty} E_d(tz)e^{-t}\,\mathrm{d}t = \int_0^{\infty} I_0(2tz)^d e^{-t}\,\mathrm{d}t.$$

Thus it remains only to show that the integral

$$L_d\left(\frac{1}{2d}\right) = \int_0^\infty I_0\left(\frac{t}{d}\right)^d e^{-t}\,dt$$

is divergent for $d = 1, 2$ and convergent for $d > 2$. This in turn amounts to understanding the asymptotics of $I_0(t/d)$ as $t \to \infty$ along the real line.

We already encountered Bessel functions in Lecture One, and we know that

$$I_0(t/d) = \frac{1}{\pi} \int_0^\pi e^{t\left(\frac{\cos\theta}{d}\right)}\,d\theta.$$

This is an integral of Laplace type,

$$\int_a^b e^{tf(\theta)}\,d\theta,$$

and Laplace integrals localize as $t \to \infty$ with asymptotics given by the classical steepest descent formula (maximum at an endpoint case),

$$\int_a^b e^{tf(\theta)}\,d\theta \sim \sqrt{\frac{\pi}{2t|f''(a)|}}e^{tf(a)}.$$

For our integral, this specializes to

$$I_0(t/d) \sim \sqrt{\frac{1}{2\pi^{3/2}t}}e^{t/d}, \quad t \to \infty,$$

from which it follows that $L_d((2d)^{-1})$ diverges or converges according to the divergence or convergence of the integral

$$\int_1^\infty t^{-d/2}\,dt.$$

This integral diverges for $d = 1, 2$ and converges for $d \geq 3$, which proves Pólya's result. In fact, the probability that the simple random walk on $\mathbf{Z}^3$ returns to its initial position is already less than thirty five percent.

2.3. Kesten's problem. The category of abelian groups is a full subcategory of the category of groups. In the category of groups, coproduct is free product:

$$\coprod_{i \in I} \mathbf{G}_i = *_{i \in I}\mathbf{G}_i.$$

Thus one could equally well ask about the recurrence or transience of the simple random walk on

$$\mathbf{F}_d = \mathbf{Z} * \cdots * \mathbf{Z},$$

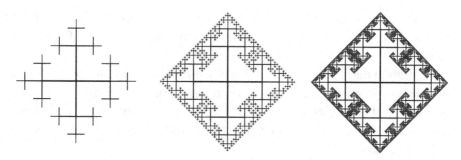

Figure 13. Balls of increasing radius in $\mathbf{F}_2$.

the free group on d generators. Whereas the Cayley graph of the abelian group $\mathbf{Z}^d$ is the $(2d)$-regular hypercubic lattice, the Cayley graph of the free group $\mathbf{F}_d$ is the $(2d)$-regular tree; see Figure 13. What is the free analogue of Pólya's theorem? We will see that the random walk on $\mathbf{F}_d$ can be understood entirely in terms of the random walk on $\mathbf{F}_1 = \mathbf{Z}$, just as in the abelian category. However, the tools we will use are quite different, and the concept of free random variables plays the central role.

The study of random walks on groups was initiated by Harry Kesten in his 1958 Ph.D. thesis, with published results appearing in [Kesten 1959]. A good source of information on this topic, with many pointers to the literature, is Laurent Saloff-Coste's survey article [2001]. Kesten related the behaviour of the simple random walk on a finitely-generated group $\mathbf{G}$ to other properties of $\mathbf{G}$, such as amenability. A countable group is said to be amenable if it admits a finitely additive $\mathbf{G}$-invariant probability measure. The notion of amenability was introduced by John von Neumann in 1929. Finite groups are amenable since they can be equipped with the uniform measure $P(g) = |\mathbf{G}|^{-1}$. For infinite groups the situation is not so clear, and many different characterizations of amenability have been derived. For example, Alain Connes showed that a group is amenable if and only if its von Neumann algebra is hyperfinite. Kesten proved that $\mathbf{G}$ is nonamenable if and only if the probability $\rho_{\mathbf{G}}(n)$ that the simple random walk on $\mathbf{G}$ returns to its starting point at time n decays exponentially in n. We saw above that for $\mathbf{G} = \mathbf{Z}$ the return probability has square root decay, so $\mathbf{Z}$ is amenable. In fact, amenability is preserved by direct sum so all abelian groups are amenable. Is the free group $\mathbf{F}_d$ amenable? Let $\lambda_d(n)$ denote the number of length n loops on $\mathbf{F}_d$ based at id. We will refer to the problem of finding an explicit expression for the loop generating function

$$L_d(z) = 1 + \sum_{n=1}^{\infty} \lambda_d(n) z^n$$

as Kesten's problem. Presumably, if we can obtain an explicit expression for this function then we can read off the asymptotics of $\rho_d(n)$, which is the coefficient of z^n in $L_d(z/2d)$, via the usual methods of singularity analysis of generating functions.

We begin at the beginning: $d = 2$. Let A and B denote the generators of $\mathbf{F}_2$, and let $\mathcal{A} = \mathcal{A}[\mathbf{F}_2]$ be the group algebra consisting of formal $\mathbb{C}$-linear combinations of words in these generators and their inverses, A^{-1} and B^{-1}. The identity element of $\mathcal{A}$ is the empty word, which is identified with id $\in \mathbf{F}_2$. Introduce the expectation functional

$$\tau[X] = \text{coefficient of id in } X$$

for each $X \in \mathcal{A}$. Then $(\mathcal{A}, \tau)$ is a noncommutative probability space. A loop id $\to$ id in $\mathbf{F}_2$ is simply a word in A, A^{-1}, B, B^{-1} which reduces to id. Thus the number of length n loops in $\mathbf{F}_2$ is

$$\lambda_2(n) = m_n(X + Y) = \tau[(X + Y)^n],$$

where $X, Y \in \mathcal{A}$ are the random variables

$$X = A + A^{-1}, \quad Y = B + B^{-1}.$$

We see that the loop generating function for $\mathbf{F}_2$ is precisely the moment generating function for the random variable $X + Y$ in the noncommutative probability space $(\mathcal{A}, \tau)$,

$$L_2(z) = 1 + \sum_{n=1}^{\infty} m_n(X + Y)z^n.$$

We want to compute the moments of the sum $X + Y$ of two noncommutative random variables, and what we know are the moments of its summands:

$$m_n(X) = m_n(Y) = \begin{cases} 0 & \text{if } n \text{ is odd}, \\ \binom{n}{n/2} & \text{if } n \text{ is even} \end{cases}.$$

Now we make the key observation: the random variables X, Y are freely independent. Indeed, suppose that $f_1, g_1, \ldots, f_k, g_k$ are polynomials such that

$$\tau[f_1(X)] = \tau[g_1(Y)] = \cdots = \tau[f_k(X)] = \tau[g_k(Y)] = 0.$$

This means that $f_i(X) = f_i(A + A^{-1})$ is a Laurent polynomial in A with zero constant term, and $g_j(Y) = g_j(B + B^{-1})$ is a Laurent polynomial in B with zero constant term. Since there are no relations between A and B, an alternating product of polynomials of this form cannot produce any occurrences of the empty word, and we have

$$\tau[f_1(X)g_1(Y) \ldots f_k(X)g_k(Y)] = 0.$$

This is precisely Voiculescu's definition of free independence.

We conclude that the problem of computing $\lambda_2(n)$ is a particular case of the problem of computing the moments $m_n(X+Y)$ of the sum of two free random variables given their individual moments, $m_n(X)$ and $m_n(Y)$. This motivates us to solve a fundamental problem in free probability theory:

> *Given a pair of free random variables X and Y, compute the moments of X + Y in terms of the moments of X and the moments of Y.*

We can, in principle, solve this problem using the fact that free cumulants linearize the addition of free random variables, $\kappa_n(X+Y) = \kappa_n(X) + \kappa_n(Y)$. This solution is implemented as the following recursive algorithm.

Input: $\kappa_1(X), \ldots, \kappa_{n-1}(X), \kappa_1(Y), \ldots, \kappa_{n-1}(Y)$.

Step 1: Compute $m_n(X), m_n(Y)$.

Step 2: Compute $\kappa_n(X), \kappa_n(Y)$ using

$$\kappa_n(X) = m_n(X) - \sum_{\substack{\pi \in NC(n) \\ b(\pi) \geq 2}} \prod_{B \in \pi} \kappa_{|\beta|}(X)$$

$$\kappa_n(Y) = m_n(Y) - \sum_{\substack{\pi \in NC(n) \\ b(\pi) \geq 2}} \prod_{B \in \pi} \kappa_{|\beta|}(Y).$$

Step 3: Add:

$$\kappa_n(X+Y) = \kappa_n(X) + \kappa_n(Y).$$

Step 4: Compute $m_n(X+Y)$ using

$$m_n(X+Y) = \kappa_n(X+Y) + \sum_{\substack{\pi \in NC(n) \\ b(\pi \geq 2}} \prod_{B \in \pi} \kappa_{|B|}(X+Y).$$

Output: $m_n(X+Y)$.

This recursive algorithm is conceptually simple but virtually useless as is. In particular, it is not clear how to coax it into computing the loop generating function $L_2(z)$. We need to develop an additive calculus of free random variables which parallels the additive calculus of classically independent random variables.

2.4. *The classical algorithm.* If X, Y are classically independent random variables, we can compute the moments of their sum $X + Y$ using the recursive algorithm above, replacing free cumulants with classical cumulants. But this is not what probabilists do in their daily lives. They have a much better algorithm which uses analytic function theory to efficiently handle the recursive nature of the naive algorithm. The classical algorithm associates to X and Y analytic functions $M_X(z)$ and $M_Y(z)$ which have the property that $M_{X+Y}(z) := M_X(z)M_Y(z)$ encodes the moments of $X + Y$ as its derivatives at $z = 0$. We will give a somewhat roundabout derivation of this algorithm, which is presented in this way specifically to highlight the analogy with Voiculescu's algorithm presented in the next section.

The classical algorithm for summing two random variables is developed in two stages. In the first stage, the relation between the moments and classical cumulants of a random variable is packaged as an identity in the ring of formal power series $\mathbb{C}[\![z]\!]$. Suppose that $(m_n)_{n=1}^\infty$ and $(c_n)_{n=1}^\infty$ are two numerical sequences related by the chain of identities

$$m_n = \sum_{\pi \in P(n)} \prod_{B \in \pi} c_{|B|}, \quad n \geq 1.$$

The π-th term of the sum on the right only depends on the "spectrum" of π, i.e., the integer vector $\Lambda(\pi) = (1^{b_1(\pi)}, 2^{b_2(\pi)}, \ldots, n^{b_n(\pi)})$, where $b_i(\pi)$ is the number of blocks of size i in π. We may view $\Lambda(\pi)$ as the Young diagram with b_i rows of length i. Consequently, we can perform a change of variables to push the summation forward onto a sum over Young diagrams with n boxes provided we can compute the "Jacobian" of the map $\Lambda : P(n) \to Y(n)$ sending π on its spectrum:

$$m_n = \sum_{b_1+2b_2+\cdots+nb_n=n} c_1^{b_1} c_2^{b_2} \ldots c_n^{b_n} |\Lambda^{-1}(1^{b_1}, 2^{b_2}, \ldots, n^{b_n})|.$$

The volume of the fibre of Λ over any given Young diagram can be explicitly computed to be

$$|\Lambda^{-1}(1^{b_1}, 2^{b_2}, \ldots, n^{b_n})| = \frac{n!}{(1!)^{b_1}(2!)^{b_2} \ldots (n!)^{b_n} b_1! b_2! \ldots b_n!},$$

so that we have the chain of identities

$$\frac{m_n}{n!} = \sum_{b_1+2b_2+\cdots+nb_n=n} \frac{(c_1/1!)^{b_1}(c_2/2!)^{b_2} \ldots (c_n/n!)^{b_n}}{b_1! b_2! \ldots b_n!}, \quad n \geq 1.$$

We can bundle these identities as a single relation between power series. Summing on z we obtain

$$1 + \sum_{n=1}^\infty m_n \frac{z^n}{n!} = 1 + \sum_{n=1}^\infty \left(\sum_{b_1+2b_2+\cdots+nb_n=n} \frac{(c_1/1!)^{b_1}(c_2/2!)^{b_2} \ldots (c_n/n!)^{b_n}}{b_1! b_2! \ldots b_n!} \right) z^n$$

$$= 1 + \frac{1}{1!}\left(\sum_{n=1}^\infty c_n \frac{z^n}{n!}\right)^1 + \frac{1}{2!}\left(\sum_{n=1}^\infty c_n \frac{z^n}{n!}\right)^2 + \cdots = \exp\left(\sum_{n=1}^\infty c_n \frac{z^n}{n!}\right).$$

We conclude that the chain of moment-cumulant formulas is equivalent to the single identity $M(z) = e^{C(z)}$ in $\mathbb{C}[[z]]$, where

$$M(z) = 1 + \sum_{n=1}^{\infty} m_n \frac{z^n}{n!}, \quad C(z) = \sum_{n=1}^{\infty} c_n \frac{z^n}{n!}$$

This fact is known in enumerative combinatorics as the exponential formula. In other branches of science it goes by other names, such as the polymer expansion formula or the linked cluster theorem. In the physics literature, the exponential formula is often invoked using colourful phrases such as "connected vacuum bubbles exponentiate" [Samuel 1980]. The exponential formula seems to have been first written down precisely by Adolf Hurwitz [1891].

The exponential formula becomes particularly powerful when combined with complex analysis. Suppose that X, Y are classically independent random variables living in a noncommutative probability space $(\mathcal{A}, \tau)$. Suppose moreover that an oracle has given us probability measures μ_X, μ_Y on the real line which behave like distributions for X, Y insofar as

$$\tau[X^n] = \int_{\mathbb{R}} t^n \mu_X(\mathrm{d}t), \quad \tau[Y^n] = \int_{\mathbb{R}} t^n \mu_Y(\mathrm{d}t), \quad n \geq 1.$$

Let us ask for even more, and insist that μ_X, μ_Y are compactly supported. Then the functions[2]

$$M_X(z) = \int_{\mathbb{R}} e^{tz} \mu_X(\mathrm{d}t), \quad M_Y(z) = \int_{\mathbb{R}} e^{tz} \mu_Y(\mathrm{d}t)$$

are entire, and their derivatives can be computed by differentiation under the integral sign. Consequently, we have the globally convergent power series expansions

$$M_X(z) = 1 + \sum_{n=1}^{\infty} m_n(X) \frac{z^n}{n!},$$

$$M_Y(z) = 1 + \sum_{n=1}^{\infty} m_n(Y) \frac{z^n}{n!}.$$

Since $M_X(0) = M_Y(0) = 1$ and the zeros of holomorphic functions are discrete, we can restrict to a complex domain D containing the origin on which $M_X(z), M_Y(z)$ are nonvanishing. Let Hol(D) denote the algebra of holomorphic functions on D. The following algorithm produces a function $M_{X+Y}(z) \in \mathrm{Hol}(D)$ whose derivatives at $z = 0$ are the moments of $X + Y$.

---

[2]The restriction of M_X to the real axis, $M_X(-x)$, is the two-sided Laplace transform, while the restriction of M_X to the imaginary axis, $M_X(-iy)$, is the Fourier transform.

Input: μ_X and μ_Y.

Step 1: Compute

$$M_X(z) = \int_{\mathbb{R}} e^{tz} \mu_X(dt), \quad M_Y(z) = \int_{\mathbb{R}} e^{tz} \mu_Y(dt).$$

Step 2: Solve

$$M_X(z) = e^{C_X(z)}, \quad M_Y(z) = e^{C_Y(z)}$$

in Hol(D) subject to $C_X(0) = C_Y(0) = 0$.

Step 3: Add:

$$C_{X+Y}(z) := C_X(z) + C_Y(z).$$

Step 4: Exponentiate:

$$M_{X+Y}(z) := e^{C_{X+Y}(z)}.$$

Output: $M_{X+Y}(z)$.

In Step 1, we try to compute the integral transforms $M_X(z)$, $M_Y(z)$ in terms of elementary functions, like e^z, $\log(z)$, $\sin(z)$, $\cos(z)$, $\sinh(z)$, $\cosh(z)$, ... etc, or other classical functions like Bessel functions, Whittaker functions, or anything else that can be looked up in [Andrews et al. 1999]. This is often feasible if the distributions μ_X, μ_Y have known densities, and we saw some examples in Lecture One.

The equations in Step 2 have unique solutions. The required functions $C_X(z)$, $C_Y(z) \in$ Hol(D) are the principal branches of the logarithms of $M_X(z)$ and $M_Y(z)$ on D, and can be represented as contour integrals:

$$C_X(z) = \log M_X(z) = \oint_0^z \frac{M_X'(\zeta)}{M_X(\zeta)} \, d\zeta, \quad C_Y(z) = \log M_Y(z) = \oint_0^z \frac{M_Y'(\zeta)}{M_Y(\zeta)} \, d\zeta$$

for $z \in$ D. Since log has the usual formal properties associated with the logarithm, if Step 1 outputs a reasonably explicit expression then so will Step 2.

Step 2 is the crux of the algorithm. It is performed precisely to change gears from a moment computation to a cumulant computation. Appealing to the exponential formula, we conclude that the holomorphic functions $C_X(z)$, $C_Y(z)$ passed to Step 3 by Step 2 have Maclaurin series

$$C_X(z) = \sum_{n=1}^{\infty} c_n(X) \frac{z^n}{n!}, \quad C_Y(z) = \sum_{n=1}^{\infty} c_n(Y) \frac{z^n}{n!},$$

where $c_n(X)$, $c_n(Y)$ are the cumulants of X and Y. Since cumulants linearize the addition of independent random variables, the new function $C_{X+Y}(z) :=$ $C_X(z)+C_Y(z)$ defined in Step 3 encodes the cumulants of $X+Y$ as its derivatives at $z = 0$.

In Step 4 we define a new function $M_{X+Y}(z) \in \text{Hol}(\text{D})$ by $M_{X+Y}(z) := e^{C_{X+Y}(z)}$. The exponential formula and the moment-cumulant formula now combine in the reverse direction to tell us that the Maclaurin series of $M_{X+Y}(z)$ is

$$M_{X+Y}(z) = 1 + \sum_{n=1}^{\infty} m_n(X+Y)\frac{z^n}{n!}.$$

In summary, assuming that X, Y are classically independent random variables living in a noncommutative probability space $(\mathcal{A}, \tau)$ with affiliated distributions μ_X, μ_Y having nice properties, the classical algorithm takes these distributions as input and outputs a function $M_{X+Y}(z)$ analytic at $z = 0$ whose derivatives are the moments of $X + Y$. It works by combining the exponential formula and the moment-cumulant formula to convert the moment problem into the (linear) cumulant problem, adding, and then converting back to moments. An optional Step 5 is to extract the distribution μ_{X+Y} from $M_{X+Y}(z)$ using the Fourier inversion formula:

$$\mu_{X+Y}([a, b]) = \lim_{T \to \infty} \frac{1}{2\pi} \int_{-T}^{T} \frac{e^{-iat} - e^{-ibt}}{it} M_{X+Y}(it)\,dt.$$

2.5. Voiculescu's algorithm. We wish to develop a free analogue of the classical algorithm. Suppose that X, Y are freely independent random variables living in a noncommutative probability space $(\mathcal{A}, \tau)$ possessing compactly supported real distributions μ_X, μ_Y. The free algorithm should take these distributions as input, build a pair of analytic functions which encode the moments of X and Y respectively, and then convolve these somehow to produce a new analytic function which encodes the moments of $X + Y$. A basic hurdle to be overcome is that, even assuming we know how to construct μ_X and μ_Y, we don't know what to do with them. We could repeat Step 1 of the classical algorithm to obtain analytic functions $M_X(z)$, $M_Y(z)$ whose derivatives at $z = 0$ are the moments of X and Y. If we then perform Step 2 we obtain analytic functions $C_X(z)$, $C_Y(z)$ whose derivatives encode the classical cumulants of X and Y. But classical cumulants do not linearize addition of free random variables.

The classical algorithm is predicated on the existence of a formal power series identity equivalent to the chain of classical moment-cumulant identities. We need a free analogue of this, namely a power series identity equivalent to the chain of numerical identities

$$m_n = \sum_{\pi \in \text{NC}(n)} \prod_{B \in \pi} \kappa_{|B|}, \quad n \geq 1.$$

Proceeding as in the classical case, rewrite this in the form

$$m_n = \sum_{b_1+2b_2+\cdots+nb_n=n} \kappa_1^{b_1}\kappa_2^{b_2}\ldots\kappa_n^{b_n} |\Lambda^{-1}(1^{b_1}, 2^{b_2}, \ldots, n^{b_n}) \cap NC(n)|,$$

where as above $\Lambda : P(n) \to Y(n)$ is the surjection which sends a partition π with b_i blocks of size i to the Young diagram with b_i rows of length i. Now we have to compute the volume of the fibres of Λ intersected with the noncrossing partition lattice. The solution to this enumeration problem is again known in explicit form,

$$|\Lambda^{-1}(1^{m_1}, 2^{m_2}, \ldots, n^{m_n}) \cap NC(n)| = \frac{n!}{(n+1-(b_1+b_2+\cdots+b_n))! \, b_1! b_2! \ldots b_n!}.$$

This formula allows us to obtain the desired power series identity, though the manipulations required are quite involved and require either the use of Lagrange inversion or an understanding of the poset structure of $NC(n)$. In any event, what ultimately comes out of the computation is the fact that two numerical sequences satisfy the chain of free moment-cumulant identities if and only if the ordinary (not exponential) generating functions

$$L(z) = 1 + \sum_{n=1}^{\infty} m_n z^n, \quad K(z) = 1 + \sum_{n=1}^{\infty} \kappa_n z^n$$

solve the equation

$$L(z) = K(zL(z))$$

in the formal power series ring $\mathbb{C}[[z]]$. This is the free analogue of the exponential formula.

As in the classical case, we wish to turn this formal power series encoding into an analytic encoding. Suppose that X, Y admit distributions μ_X, μ_Y supported in the real interval $[-r, r]$. We then have $|m_n(X)|, |m_n(Y)| \le r^n$, so the moment generating functions

$$L_X(z) = 1 + \sum_{n=1}^{\infty} m_n(X) z^n, \quad L_Y(z) = 1 + \sum_{n=1}^{\infty} m_n(Y) z^n,$$

are absolutely convergent in the open disc $D(0, \frac{1}{r})$. One can use the relation between moments and free cumulants to show that the free cumulant generating functions

$$K_X(z) = 1 + \sum_{n=1}^{\infty} \kappa_n(X) z^n, \quad K_Y(z) = 1 + \sum_{n=1}^{\infty} \kappa_n(Y) z^n$$

are also absolutely convergent on a (possibly smaller) neighbourhood of $z = 0$.

However, it turns out that the correct environment for the free algorithm is a neighbourhood of infinity rather than a neighbourhood of zero. This is because what we really want is an integral transform which realizes ordinary generating functions in the same way as the Fourier (or Laplace) transform realizes exponential generating functions. Access to such a transform will allow us to obtain closed forms for generating functions by evaluating integrals, just like in classical probability. Such an object is well-known in analysis, where it goes by the name of the Cauchy (or Stieltjes) transform. The Cauchy transform of a random variable X with real distribution μ_X is

$$G_X(z) = \int_{\mathbb{R}} \frac{1}{z-t} \mu_X(dt).$$

The Cauchy transform is well-defined on the complement of the support of μ_X, and differentiating under the integral sign shows that $G_X(z)$ is holomorphic on its domain of definition. In particular, if μ_X is supported in $[-r, r]$ then $G_X(z)$ admits the convergent Laurent expansion

$$G_X(z) = \frac{1}{z} \sum_{n=0}^{\infty} \frac{\int t^n \mu_X(dt)}{z^n} = \sum_{n=0}^{\infty} \frac{m_n(X)}{z^{n+1}}$$

on $|z| > r$. This is an ordinary generating function for the moments of X with z^{-1} playing the role of the formal variable.

To create an interface between the free moment-cumulant formula and the Cauchy transform, we must rewrite the formal power series identity $L(z) = K(zL(z))$ as an identity in $\mathbb{C}((z)) = \operatorname{Quot} \mathbb{C}[[z]]$, the field of formal Laurent series. Introduce the formal Laurent series

$$G(z) = \frac{1}{z} L\left(\frac{1}{z}\right) = \sum_{n=0}^{\infty} \frac{m_n}{z^{n+1}}.$$

The automorphism $z \mapsto \dfrac{1}{z}$ transforms the noncrossing exponential formula into the identity

$$\frac{K(G(z))}{G(z)} = z.$$

Setting

$$V(z) = \frac{K(z)}{z} = \frac{1}{z} + \sum_{n=0}^{\infty} \kappa_{n+1} z^n,$$

this becomes the identity

$$V(G(z)) = z$$

in $\mathbb{C}((z))$.

We have now associated two analytic functions to X. The first is the Cauchy transform $G_X(z)$, which is defined as an integral transform and admits a convergent Laurent expansion in a neighbourhood of infinity in the z-plane. The second is the Voiculescu transform $V_X(w)$, which is defined by the convergent Laurent series

$$V_X(w) = \frac{1}{w} + \sum_{n=0}^{\infty} \kappa_{n+1} w^n$$

in a neighbourhood of zero in the w-plane. The Voiculescu transform is a meromorphic function with a simple pole of residue one at $w = 0$. The Voiculescu transform less its principal part, $R_X(w) = V_X(w) - \frac{1}{w}$, is an analytic function known as the R-transform of X. From the formal identities $V(G(z)) = z$, $G(V(w)) = w$ and the asymptotics $G_X(z) \sim \frac{1}{z}$ as $|z| \to \infty$ and $V_X(w) \sim \frac{1}{w}$ as $|w| \to 0$, we expect to find a neighbourhood D_∞ of infinity in the z-plane and a neighbourhood D_0 of zero in the w-plane such that $G_X : D_\infty \to D_0$ and $V_X : D_0 \to D_\infty$ are mutually inverse holomorphic bijections. The existence of the required domains hinges on identifying regions where the Cauchy and Voiculescu transforms are injective, and this can be established through a complex-analytic argument; see [Mingo and Speicher $\geq$ 2014, Chapter 4].

With these pieces in place, we can state Voiculescu's algorithm for the addition of free random variables.

Input: μ_X and μ_Y.

Step 1: Compute

$$G_X(z) = \int_{\mathbb{R}} \frac{1}{z-t} \mu_X(dt), \quad G_Y(z) = \int_{\mathbb{R}} \frac{1}{z-t} \mu_Y(dt)$$

Step 2: Solve the first Voiculescu functional equations,

$$(G_X \circ V_X)(w) = w, \quad (G_Y \circ V_Y)(w) = w$$

subject to $V_X(w) \sim \dfrac{1}{w}$ near $w = 0$.

Step 3: Remove principal part:

$$R_X(w) = V_X(w) - \frac{1}{w}, \quad R_Y(w) = V_Y(w) - \frac{1}{w};$$

add:

$$R_{X+Y}(w) := R_X(w) + R_Y(w);$$

restore principal part:

$$V_{X+Y}(w) := R_{X+Y}(w) + \frac{1}{w}.$$

Step 4: Solve the second Voiculescu functional equation,

$$(V_{X+Y} \circ G_{X+Y})(z) = z,$$

subject to $G_{X+Y}(z) \sim \dfrac{1}{z}$ near $z = \infty$.

Output: $G_{X+Y}(z)$.

Voiculescu's algorithm is directly analogous to the classical algorithm presented in the previous section. The analogy can be succinctly summarized by saying that

---
the R-transform is the free analogue of the log of the Fourier transform.

In Step 1, we try to compute the integral transforms $G_X(z)$, $G_Y(z)$ in terms of elementary functions.

Step 2 changes gears from a moment computation to a cumulant computation. Since free cumulants linearize the addition of free random variables, the new function $V_{X+Y}(w) := R_X(w) + R_Y(w) + \frac{1}{w}$ defined in Step 3 encodes the free cumulants of $\kappa_n(X + Y)$ as its Laurent coefficients of nonnegative degree.

In Step 4 we define a new function $G_{X+Y}(z)$ by solving the second Voiculescu functional equation. The free exponential formula and the free moment-cumulant formula combine in the reverse direction to tell us that the Laurent series of $G_{X+Y}(z)$ is

$$G_{X+Y}(z) = \sum_{n=0}^{\infty} \frac{m_n(X+Y)}{z^{n+1}}.$$

An optional fifth step is to extract the distribution μ_{X+Y} from $G_{X+Y}(z)$ using the Stieltjes inversion formula:

$$\mu_{X+Y}(dt) = -\frac{1}{\pi} \lim_{\varepsilon \to 0} \Im G_{X+Y}(t + i\varepsilon).$$

2.6. Solution of Kesten's problem. Our motivation for building up the additive theory of free random variables came from Kesten's problem: explicitly determine the loop generating function of the free group $\mathbf{F}_2$, and more generally of the free group $\mathbf{F}_d$, $d \geq 2$. This amounts to computing the moment generating function

$$L_d(z) = 1 + \sum_{n=1}^{\infty} m_n(S_d)z^d$$

of the sum

$$S_d = X_1 + \cdots + X_d$$

of fid (free identically distributed) random variables with moments

$$\tau[X_i^n] = \begin{cases} 0 & \text{if } n \text{ is odd,} \\ \binom{n}{n/2} & \text{if } n \text{ is even.} \end{cases}$$

Voiculescu's algorithm gives us the means to obtain this generating function provided we can feed it the required input, namely a compactly supported probability measure on $\mathbb{R}$ with moment sequence

$$0, \binom{2}{1}, 0, \binom{4}{2}, 0, \binom{6}{3}, 0, \dots.$$

As we saw above, the exponential generating function of this moment sequence,

$$M(z) = \sum_{k=0}^{\infty} \frac{z^{2k}}{k!\,k!} = I_0(2z),$$

coincides with the modified Bessel function of order zero. From the integral representation

$$I_0(2z) = \frac{1}{\pi} \int_0^{\pi} e^{2(\cos\theta)z}\, d\theta$$

we conclude that a random variable X with odd moments zero and even moments the central binomial coefficients is given by $X = 2\cos(Y)$, where Y has uniform distribution over $[0, \pi]$. Making the same change of variables that we did in Lecture One, we obtain

$$M_X(z) = \frac{1}{\pi} \int_{-2}^{2} e^{tz} \frac{1}{\sqrt{4 - t^2}}\, dt,$$

so that μ_X is supported on $[-2, 2]$ with density

$$\mu_X(dt) = \frac{1}{\pi\sqrt{4 - t^2}}\, dt.$$

This measure is known as the arcsine distribution because its cumulative distribution function is

$$\int_{-2}^{x} \mu_X(dt) = \frac{1}{2} + \frac{\arcsin \frac{x}{2}}{\pi}.$$

So to obtain the loop generating function $L_2(z)$ for the simple random walk on $\mathbf{F}_2$, we should run Voiculescu's algorithm with input $\mu_X = \mu_Y = $ arcsine.

Let us warm up with an easier computation. Suppose that X, Y are not fid arcsine random variables, but rather fid ± 1-Bernoulli random variables:

$$\mu_X = \mu_Y = \tfrac{1}{2}\delta_{-1} + \tfrac{1}{2}\delta_{+1}.$$

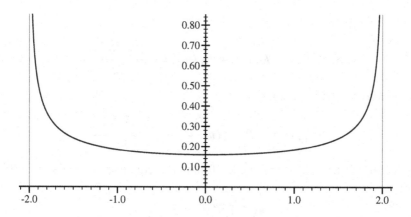

Figure 14. The arcsine density.

We will use Voiculescu's algorithm to obtain the distribution of $X + Y$. If X, Y were classically iid Bernoullis, we would of course obtain the binomial distribution

$$\mu_{X+Y} = \tfrac{1}{4}\delta_{-2} + \tfrac{1}{2}\delta_0 + \tfrac{1}{4}\delta_{+2}$$

giving the distribution of the simple random walk on $\mathbf{Z}$ at time two. The result is quite different in the free case.

Step 1. Obtain the Cauchy transform:

$$G_X(z) = G_Y(z) = \frac{1}{2}\left(\frac{1}{z+1} + \frac{1}{z-1}\right) = \frac{z}{z^2-1} = \sum_{n=0}^{\infty} \frac{1}{z^{2n+1}}.$$

Step 2. Solve the first Voiculescu functional equation. From Step 1, this is

$$w V^2(w) - V(w) - w = 0,$$

which has roots

$$\frac{1+\sqrt{1+4w^2}}{2w} = \frac{1}{w} + w - w^3 + 2w^5 - 5w^7 + \cdots,$$

$$\frac{1-\sqrt{1+4w^2}}{2w} = -w + w^3 - 2w^5 + \cdots.$$

We identify the first of these as the Voiculescu transform $V_X(w) = V_Y(w)$.

Step 3. Compute the R-transform:

$$R_X(w) = R_Y(w) = \frac{1+\sqrt{1+4w^2}}{2w} - \frac{1}{w} = \frac{\sqrt{1+4w^2}-1}{2w},$$

and sum to obtain

$$R_{X+Y}(w) = R_X(w) + R_Y(w) = \frac{\sqrt{1+4w^2}-1}{w}.$$

Now restore the principal part:

$$V_{X+Y}(w) = R_{X+Y}(w) + \frac{1}{w} = \frac{\sqrt{1+4w^2}}{w}.$$

Step 4. Solve the second Voiculescu functional equation. From Step 3, this is the equation

$$\frac{\sqrt{1+4G(z)^2}}{G(z)} = z,$$

which has roots

$$\frac{\pm 1}{\sqrt{z^2-4}} = \frac{\pm 1}{z} + \frac{\pm 2}{z^3} + \frac{\pm 6}{z^5} + \frac{\pm 20}{z^7} + \frac{\pm 70}{z^9} + \frac{\pm 252}{z^{11}} + \cdots.$$

The positive root is identified as $G_{X+Y}(z)$.

Finally, we perform the optional fifth step to recover the distribution μ_{X+Y} whose Cauchy transform is $G_{X+Y}(z)$. This can be done in two ways. First, we could notice that the nonzero Laurent coefficients of G_{X+Y} are the central binomial coefficients $\binom{2k}{k}$, and we just determined that these are the moments of the arcsine distribution. Alternatively we could use Stieltjes inversion:

$$\mu_{X+Y}(dt) = -\frac{1}{\pi} \lim_{\varepsilon \to 0} \frac{1}{\sqrt{(t+i\varepsilon)^2-4}} = -\frac{1}{\pi}\Im\frac{1}{\sqrt{t^2-4}} = \frac{1}{\pi\sqrt{4-t^2}}\delta_{|t|\le 2}.$$

We conclude that the sum of two fid Bernoulli random variables has arcsine distribution. Note the surprising feature that the outcome of a free coin toss has continuous distribution over $[-2, 2]$. More generally, we can say that the sum

$$S_d = X_1 + \cdots + X_{2d}$$

of $2d$ fid ± 1-Bernoulli random variables, i.e., the sum of $2d$ free coin tosses, encodes all information about the simple random walk on $\mathbf{F}_d$ in its moments.

Let us move on to the solution of Kesten's problem for $\mathbf{F}_2$. Here X, Y are fid arcsine random variables.

Step 1. The Cauchy transform $G_X(z) = G_Y(z)$ is the output of our last application of the algorithm, namely

$$G_X(z) = G_Y(z) = \frac{1}{\sqrt{z^2-4}}.$$

Step 2. Solve the first Voiculescu functional equation to obtain

$$V_X(w) = V_Y(w) = \frac{\sqrt{1+4w^2}}{w} = \frac{1}{w} + 2w - 2w^3 + \cdots.$$

Step 3. Switch to R-transforms, add, switch back to get the Voiculescu transform of $X + Y$:

$$V_{X+Y}(w) = \frac{2\sqrt{1+4w^2}-1}{z} = \frac{1}{w} + 4w - 4w^3 + \cdots.$$

Step 4. Solve the second Voiculescu functional equation to obtain

$$G_{X+Y}(z) = \frac{-z+2\sqrt{z^2-12}}{z^2-16} = \frac{1}{z} + \frac{4}{z^3} + \frac{28}{z^5} + \frac{232}{z^7} + \frac{2092}{z^9} + \cdots.$$

We can now calculate the loop generating function for $\mathbf{F}_2$:

$$
\begin{aligned}
L_2(z) &= \frac{1}{z} G_{X+Y}\left(\frac{1}{z}\right) \\
&= \frac{-1+2\sqrt{1-12z^2}}{1-16z^2} = 1 + 4z^2 + 28z^4 + 232z^6 + 2092z^8 + \cdots.
\end{aligned}
$$

More generally, we can run through the above steps for general d to obtain the loop generating function

$$L_d(z) = \frac{-(d-1)+d\sqrt{1-4(2d-1)z^2}}{1-16z^2}$$

for the free group $\mathbf{F}_d$, $d \geq 2$, which in turn leads to the probability generating function

$$L_d\left(\frac{z}{2d}\right) = \frac{-(d-1)+d\sqrt{1-(2d-1)(\frac{z}{d})^2}}{1-4(\frac{z}{d})^2}.$$

Applying standard methods from analytic combinatorics [Flajolet and Sedgewick 2009], this expression leads to the asymptotics

$$\rho_d(n) \sim \text{const}_d \cdot n^{-3/2} \left(\frac{2\sqrt{d}}{d+1}\right)^n$$

for the return probability of the simple random walk on $\mathbf{F}_d$, $d \geq 2$. From this we can conclude that the simple random walk on $\mathbf{F}_d$ is transient for all $d \geq 2$, and indeed that $\mathbf{F}_d$ is nonamenable for all $d \geq 2$.

2.7. Spectral measures and free convolution. Voiculescu's algorithm outputs a function $G_{X+Y}(z)$ which encodes the moments of the sum of two freely independent random variables X and Y. As input, it requires a pair of compactly supported real measures μ_X, μ_Y which act as distributions for X and Y in the sense that

$$\tau[X^n] = \int_{\mathbb{R}} t^n \mu_X(dt), \quad \tau[Y^n] = \int_{\mathbb{R}} t^n \mu_Y(dt).$$

In our applications of Voiculescu's algorithm we were able to find such measures by inspection. Nevertheless, it is of theoretical and psychological importance to determine sufficient conditions guaranteeing the existence of measures with the required properties.

If $X : \Omega \to \mathbb{C}$ is a random variable defined on a Kolmogorov triple $(\Omega, \mathcal{F}, P)$, its distribution μ_X is the pushforward of P by X,

$$\mu_X(B) = (X_* P)(B) = P(X^{-1}(B))$$

for any Borel (or Lebesgue) set $B \subseteq \mathbb{C}$. One has the general change of variables formula

$$\mathbb{E}[f(X)] = \int_{\mathbb{C}} f(z) \mu_X(dz)$$

for any reasonable $f : \mathbb{C} \to \mathbb{C}$. If X is essentially bounded and real-valued, μ_X is compactly supported in $\mathbb{R}$. As a random variable X living in an abstract noncommutative probability space $(\mathcal{A}, \tau)$ is not a function, one must obtain μ_X by some other means.

The existence of distributions is too much to expect within the framework of a noncommtative probability space, which is a purely algebraic object. We need to inject some analytic structure into $(\mathcal{A}, \tau)$. This is achieved by upgrading $\mathcal{A}$ to a $*$-algebra, i.e., a complex algebra equipped with a map $* : \mathcal{A} \to \mathcal{A}$ satisfying

$$(X^*)^* = X, \quad (\alpha X + \beta Y)^* = \overline{\alpha} X^* + \overline{\beta} Y^*, \quad (XY)^* = Y^* X^*.$$

This map, which is an abstraction of complex conjugation, is required to be compatible with the expectation τ in the sense that

$$\tau[X^*] = \overline{\tau[X]}.$$

A noncommutative probability space equipped with this extra structure is called a noncommutative $*$-probability space.

In the framework of a $*$-probability space we can single out a class of random variables analogous to real random variables in classical probability. These are the fixed points of $*$, $X^* = X$. A random variable with this property is called self-adjoint. Self-adjoint random variables have real expected values,

$\tau[X] = \tau[X^*] = \overline{\tau[X]}$, and more generally $\tau[f(X)] \in \mathbb{R}$ for any polynomial f with real coefficients.

The identification of bounded random variables requires one more upgrade. Given a $*$-probability space $(\mathcal{A}, \tau)$, we can introduce a Hermitian form B : $\mathcal{A} \times \mathcal{A} \to \mathbb{C}$ defined by

$$B(X, Y) = \tau[XY^*].$$

If we require that τ has the positivity property $\tau[XX^*] \geq 0$ for all $X \in \mathcal{A}$, then we obtain a seminorm

$$\|X\| = B(X, X)^{1/2}$$

on $\mathcal{A}$, and we can access the Cauchy–Schwarz inequality

$$|B(X, Y)| \leq \|X\| \|Y\|.$$

Once we have Cauchy–Schwarz, we can prove the monotonicity inequalities

$$|\tau[X]| \leq |\tau[X^2]|^{1/2} \leq |\tau[X^4]|^{1/4}$$
$$|\tau[X^3]| \leq |\tau[X^4]|^{1/4} \leq |\tau[X^6]|^{1/6}$$
$$|\tau[X^5]| \leq |\tau[X^6]|^{1/6} \leq |\tau[X^8]|^{1/8}$$

$$\vdots$$

from which the chain of inequalities

$$|\tau[X]| \leq |\tau[X^2]|^{1/2} \leq |\tau[X^4]|^{1/4} \leq |\tau[X^6]|^{1/6} \leq |\tau[X^8]|^{1/8} \leq \cdots$$

can be extracted. From this we conclude that the limit

$$\rho(X) := \lim_{k \to \infty} |\tau[X^{2k}]|^{1/(2k)}$$

exists in $\mathbb{R}_{\geq 0} \cup \{\infty\}$. This limit is called the spectral radius of X. A random variable $X \in \mathcal{A}$ is said to be bounded if its spectral radius is finite, $\rho(X) < \infty$.

In the framework of a noncommutative $*$-probability space $(\mathcal{A}, \tau)$ with nonnegative expectation, bounded self-adjoint random variables play the role of essentially bounded real-valued random variables in classical probability theory. With some work, one may deduce from the Riesz representation theorem that to each bounded self-adjoint X corresponds a unique Borel measure μ_X supported in $[-\rho(X), \rho(X)]$ such that

$$\tau[f(X)] = \int_{\mathbb{R}} f(t) \mu_X(\mathrm{d}t)$$

for all polynomial functions $f : \mathbb{C} \to \mathbb{C}$. The details of this argument, in which a reverse-engineered Cauchy transform plays the key role, are given in Tao's notes [Tao 2010]. The measure μ_X is often called the spectral measure of X, but we will refer to it as the distribution of X. There is also a converse to this result: given any compactly supported measure μ on $\mathbb{R}$, there exists a bounded self-adjoint random variable X living in some noncommutative $*$-probability space $(\mathcal{A}, \tau)$ whose distribution is μ. Consequently, given two compactly supported real probability measures μ, ν we may define a new measure $\mu \boxplus \nu$ as "the distribution of the random variable $X + Y$, where X and Y are freely independent bounded self-adjoint random variables with distributions μ and ν, respectively." Since the sum of two bounded self-adjoint random variables is again bounded self-adjoint, $\mu \boxplus \nu$ is another compactly supported real probability measure. Moreover, $\mu \boxplus \nu$ does not depend on the particular random variables chosen to realize μ and ν. Thus we get a bona fide binary operation $\boxplus$ on the set of compactly supported real measures, which is known as the additive free convolution. For example, we computed above that

$$\text{Bernoulli} \boxplus \text{Bernoulli} = \text{Arcsine}.$$

The additive free convolution of measures is induced by the addition of free random variables. As such, it is the free analogue of the classical convolution of measures induced by the addition of classically independent random variables. Like classical convolution, free convolution can be defined for unbounded measures, but this requires more work [Bercovici and Voiculescu 1993].

2.8. Free Poisson limit theorem. Select positive real numbers λ and α. Consider the measure

$$\mu_N = (1 - \frac{\lambda}{N})\delta_0 + \frac{\lambda}{N}\delta_\alpha$$

which consists of an atom of mass $1 - \frac{\lambda}{N}$ placed at zero and an atom of mass $\frac{\lambda}{N}$ placed at α. For N sufficiently large, μ_N is a probability measure. Its moment sequence is

$$m_n(\mu_N) = \frac{\lambda}{N}\alpha^n, \quad n \geq 1.$$

The N-fold classical convolution of μ_N with itself,

$$\mu_N^{*N} = \underbrace{\mu_N * \cdots * \mu_N}_{N},$$

converges weakly to the Poisson measure of rate λ and jump size α as $N \to \infty$. This is a classical limit theorem in probability known as the Poisson limit theorem, or the law of rare events.

Let us obtain a free analogue of the Poisson limit theorem. This should be a limit law for the iterated free convolution

$$\mu_N^{\boxplus N} = \underbrace{\mu_N \boxplus \cdots \boxplus \mu_N}_{N}.$$

From the free moment-cumulant formula, we obtain the estimate

$$\kappa_n(\mu_N) = m_n(\mu_N) + O\left(\frac{1}{N^2}\right) = \frac{\lambda}{N}\alpha^n + O\left(\frac{1}{N^2}\right).$$

Since free cumulants linearize free convolution, we have

$$\kappa_n(\mu_N^{\boxplus N}) = N\kappa_n(\mu_N) = \lambda\alpha^n + O\left(\frac{1}{N}\right).$$

Thus

$$\lim_{N\to\infty} \kappa_n(\mu_N) = \lambda\alpha^n,$$

and it remains to determine the measure μ with this free cumulant sequence. The Voiculescu transform of μ is

$$V_\mu(w) = \frac{1}{w} + \sum_{n=0}^{\infty} \lambda\alpha^{n+1}w^n = \frac{1}{w} + \frac{\lambda\alpha}{1-\alpha w},$$

so the second Voiculescu functional equation $V_\mu(G_\mu(z)) = z$ yields

$$\frac{1}{G_\mu(z)} + \frac{\lambda\alpha}{1-\alpha G_\mu(z)} = z.$$

This equation has two solutions, and the one which behaves like $1/z$ for $|z| \to \infty$ is the Cauchy transform of μ. We obtain

$$G_\mu(z) = \frac{z + \alpha(1-\lambda) - \sqrt{(z-\alpha(1+\lambda))^2 - 4\lambda\alpha^2}}{2\alpha z}.$$

Applying Stieltjes inversion, we find that the density of μ is given by

$$\mu(dt) = \begin{cases} (1-\lambda)\delta_0 + \lambda m(t)\,dt, & 0 \le \lambda \le 1 \\ m(t)\,dt, & \lambda > 1 \end{cases}$$

where

$$m(t) = \frac{1}{2\pi\alpha t}\sqrt{4\lambda\alpha^2 - (t-\alpha(1+\lambda))^2}.$$

This measure is known as the Marchenko–Pastur distribution after the Ukrainian mathematical physicists Vladimir Marchenko and Leonid Pastur, who discovered it in their study of the asymptotic eigenvalue distribution of a certain class of random matrices.

2.9. *Semicircle flow.* Given $r > 0$, let μ_r be the semicircular measure of radius r:

$$\mu_r(dt) = \frac{2}{\pi r^2}\sqrt{r^2 - t^2}\, dt.$$

Taking $r = 2$ yields the standard semicircular distribution. Let μ be an arbitrary compactly supported probability measure on $\mathbb{R}$. The function

$$f_\mu : \{\text{positive real numbers}\} \to \{\text{compactly supported real measures}\}$$

defined by

$$f_\mu(r) = \mu \boxplus \mu_r$$

is called the semicircle flow. The semicircle flow has very interesting dynamics: in one of his earliest articles on free random variables, Voiculescu [1986] showed that

$$\frac{\partial G(r, z)}{\partial r} + G(r, z)\frac{\partial G(r, z)}{\partial z} = 0,$$

where $G(r, z)$ is the Cauchy transform of $f_\mu(r) = \mu \boxplus \mu_r$. Thus the free analogue of the heat equation is the complex inviscid Burgers equation. For a detailed analysis of the semicircle flow, see [Biane 1997].

3. Lecture three: modelling the free world

Free random variables are of interest for many reasons. First and foremost, Voiculescu's free probability theory is an intrinsically appealing subject worthy of study from a purely esthetic point of view. Adding to this are the many remarkable connections between free probability and other parts of mathematics, including operator algebras, representation theory, and random matrix theory. This lecture is an exposition of Voiculescu's discovery that random matrices provide asymptotic models of free random variables. We follow the treatment of Nica and Speicher [2006].

3.1. *Algebraic model of a free arcsine pair.* In Lecture Two we gave a group-theoretic construction of a pair of free random variables each of which has an arcsine distribution. In this example, the algebra of random variables is the group algebra $\mathscr{A} = \mathscr{A}[\mathbf{F}_2]$ of the free group on two generators A, B, and the expectation τ is the coefficient-of-id functional. We saw that the random variables

$$X = A + A^{-1}, \quad Y = B + B^{-1}$$

are freely independent, and each has an arcsine distribution:

$$\tau[X^n] = \tau[Y^n] = \begin{cases} 0 & \text{if } n \text{ is odd,} \\ \binom{n}{n/2} & \text{if } n \text{ is even.} \end{cases}$$

3.2. *Algebraic model of a free semicircular pair.* We can give a linear-algebraic model of a pair of free random variables each of which has a semicircular distribution. The ingredients in this construction are a complex vector space V and an inner product $B : \mathsf{V} \times \mathsf{V} \to \mathbb{C}$. Our random variables will be endomorphisms of the tensor algebra over V,

$$\mathfrak{F}(\mathsf{V}) = \bigoplus_{n=0}^{\infty} \mathsf{V}^{\otimes n},$$

which physicists and operator algebraists call the full Fock space over V after the Russian physicist Vladimir Fock. We view the zeroth tensor power $\mathsf{V}^{\otimes 0}$ as the line spanned by a distinguished unit vector $\mathsf{v}_{\varnothing}$ called the vacuum vector; $\mathsf{v}_{\varnothing}$ is an abstract vector which is not an element of V. Let $\mathcal{A} = \operatorname{End} \mathfrak{F}(V)$. This is a unital algebra, with unit the identity operator $I : \mathfrak{F}(V) \to \mathfrak{F}(V)$. To make $\mathcal{A}$ into a noncommutative probability space we need an expectation. We get an expectation by lifting the inner product on V to the inner product $\mathfrak{F}(B) : \mathfrak{F}(\mathsf{V}) \times \mathfrak{F}(\mathsf{V}) \to \mathbb{C}$ defined by

$$\mathfrak{F}(B)(\mathsf{v}_1 \otimes \cdots \otimes \mathsf{v}_m, \mathsf{w}_1 \otimes \cdots \otimes \mathsf{w}_n) = \delta_{mn} B(\mathsf{v}_1, \mathsf{w}_1) \ldots B(\mathsf{v}_n, \mathsf{w}_n).$$

Note that this inner product makes $\mathcal{A} = \operatorname{End} \mathfrak{F}(B)$ into a $*$-algebra: for each $X \in \mathcal{A}$, X^* is that linear operator for which the equation

$$\mathfrak{F}(B)(X\mathsf{s}, \mathsf{t}) = \mathfrak{F}(B)(\mathsf{s}, X^*\mathsf{t})$$

holds true for every pair of tensors $\mathsf{s}, \mathsf{t} \in \mathfrak{F}(\mathsf{V})$. The expectation on $\mathcal{A}$ is the linear functional $\tau : \mathcal{A} \to \mathbb{C}$ defined by

$$\tau[X] = \mathfrak{F}(B)(X\mathsf{v}_{\varnothing}, \mathsf{v}_{\varnothing}).$$

This functional is called vacuum expectation. It is unital because

$$\tau[I] = \mathfrak{F}(B)(I\mathsf{v}_{\varnothing}, \mathsf{v}_{\varnothing}) = B(\mathsf{v}_{\varnothing}, \mathsf{v}_{\varnothing}) = 1.$$

Thus $(\mathcal{A}, \tau)$ is a noncommutative $*$-probability space.

To construct a semicircular element in $(\mathcal{A}, \tau)$, notice that to every nonzero vector $\mathsf{v} \in \mathsf{V}$ is naturally associated a pair of linear operators $R_{\mathsf{v}}, L_{\mathsf{v}} : \mathfrak{F}(V) \to \mathfrak{F}(V)$ whose action on decomposable tensors is defined by tensoring:

$$R_{\mathsf{v}}(\mathsf{v}_{\varnothing}) = \mathsf{v},$$
$$R_{\mathsf{v}}(\mathsf{v}_1 \otimes \cdots \otimes \mathsf{v}_n) = \mathsf{v} \otimes \mathsf{v}_1 \otimes \cdots \otimes \mathsf{v}_n, \qquad n \geq 1,$$

and insertion-contraction:

$$L_v(v_\varnothing) = 0,$$
$$L_v(v_1) = B(v_1, v)v_\varnothing,$$
$$L_v(v_1 \otimes v_2 \otimes \cdots \otimes v_n) = B(v_1, v)v_2 \otimes \cdots \otimes v_n, \quad n \geq 2.$$

Since R_v maps $V^{\otimes n} \to V^{\otimes n+1}$ for each $n \geq 0$, it is called the raising (or creation) operator associated to v. Since L_v maps $V^{\otimes n} \to V^{\otimes n-1}$ for each $n \geq 1$ and kills the vacuum, it is called the lowering (or annihilation) operator associated to v. We have $R_v^* = L_v$, and also

$$L_v R_w = B(w, v)I$$

for any vectors $v, w \in V$.

Let $v \in V$ be a unit vector, $B(v, v) = 1$, and consider the self-adjoint random variable

$$X_v = L_v + R_v.$$

We claim that X_v has a semicircular distribution:

$$m_n(X_v) = \tau[X_v^n] = \begin{cases} 0 & \text{if } n \text{ is odd}, \\ \text{Cat}_{n/2} & \text{if } n \text{ is even}. \end{cases}$$

To see this, we write the expansion

$$\tau[X_v^n] = \tau[(X_v + Y_v)^n] = \sum_{W \in \{L_v, R_v\}^n} \tau[W],$$

where the summation is over all words of length n in the operators L_v, R_v. Only a very small fraction of these words have nonzero vacuum expectation. Using the relation $L_v R_v = I$ to remove occurrences of the substring $L_v R_v$, we see that any such word can be placed in normally ordered form

$$W = \underbrace{R_v \ldots R_v}_{a} \underbrace{L_v \ldots L_v}_{b}$$

with $a + b \leq n$. Since the lowering operator kills the vacuum vector, the vacuum expectation of W can only be nonzero if $b = 0$. On the other hand, since $V^{\otimes a}$ is $\mathfrak{F}(B)$-orthogonal to $V^{\otimes 0}$ for $a > 0$, we must also have $a = 0$ to obtain a nonzero contribution. Thus the only words which contribute to the above sum are those whose normally ordered form is that of the identity operator. If we replace each occurrence of L_v in W with a $+1$ and each occurrence of R_v in W with a -1, the condition that W reduces to I becomes the condition that the corresponding bitstring has total sum zero and all partial sums nonnegative. There are no such bitstrings for n odd, and as we saw in Lecture One when n is even the required bitstrings are counted by the Catalan number $\text{Cat}_{n/2}$.

Now let V_1 and V_2 be B-orthogonal vector subspaces of V, each of dimension at least one, and choose unit vectors $x \in V_1, y \in V_2$. According to the above construction, the random variables

$$X = L_x + R_x, \quad Y = L_y + R_y$$

are semicircular. In fact, they are freely independent. To prove this, we must demonstrate that

$$\tau[f_1(X)g_1(Y) \ldots f_k(X)g_k(Y)] = 0$$

whenever $f_1, g_1, \ldots, f_k, g_k$ are polynomials such that

$$\tau[f_1(X)] = \tau[g_1(Y)] = \cdots = \tau[f_k(X)] = \tau[g_k(Y)] = 0.$$

This hypothesis means that $f_i(X) = f_i(L_x + R_x)$ is a polynomial in L_x, R_x none of whose terms are words which normally order to I, and similarly $g_j(Y) = g_j(L_y + R_y)$ is a polynomial in L_y, R_y none of whose terms are words which normally order to I. Consequently, the alternating product

$$f_1(X)g_1(Y) \ldots f_k(X)g_k(Y)$$

is a polynomial in the operators L_x, R_x, L_y, R_y whose terms are words W of the form

$$W_x^1 W_y^1 \ldots W_x^k W_y^k,$$

with W_x^i a word in L_x, R_x which does not normally order to I and W_y^j a word in L_y, R_y which does not normally order to I. Thus the only way that W can have a nonzero vacuum expectation is if we can use the relations $L_x R_y = B(y, x)I$ and $L_y R_x = B(x, y)I$ to normally order W as

$$B(x, y)^m B(y, x)^n I,$$

with m, n nonnegative integers at least one of which is positive. But, since x, y are B-orthogonal, this is the zero element of $\mathcal{A}$, which has vacuum expectation zero.

3.3. Algebraic versus asymptotic models.

We have constructed algebraic models for a free arcsine pair and a free semicircular pair. Perhaps these should be called examples rather than models, since the term model connotes some degree of imprecision or ambiguity and algebra is a subject which allows neither.

Suppose that X, Y are free random variables living in an abstract noncommutative probability space $(\mathcal{A}, \tau)$. An approximate model for this pair will consist of a sequence $(\mathcal{A}_N, \tau_N)$ of concrete or canonical noncommutative probability spaces together with a sequence of pairs X_N, Y_N of random variables from these spaces such that X_N models X and Y_N models Y, i.e.,

$$\tau[f(X)] = \lim_{N \to \infty} \tau[f(X_N)], \quad \tau[g(Y)] = \lim_{N \to \infty} \tau[g(Y_N)]$$

for any polynomials f, g, and such that free independence holds in the large N limit, i.e.,

$$\lim_{N \to \infty} \tau[f_1(X_N)g_1(Y_N) \ldots f_k(X_N)g_k(Y_N)] = 0$$

whenever $f_1, g_1, \ldots, f_k, g_k$ are polynomials such that

$$\lim_{N \to \infty} \tau_N[f_1(X_N)] = \lim_{N \to \infty} \tau_N[g_1(Y_N)] = \cdots = \lim_{N \to \infty} \tau_N[f_k(X_N)]$$
$$= \lim_{N \to \infty} \tau_N[g_k(Y_N)] = 0.$$

The question of which noncommutative probability spaces are considered concrete or canonical, and could therefore serve as potential models, is subjective and determined by individual experience. Three examples of concrete noncommutative probability spaces are:

Group probability spaces: $(\mathscr{A}, \tau)$ consists of the group algebra $\mathscr{A} = \mathscr{A}[\mathbf{G}]$ of a group $\mathbf{G}$, and τ is the coefficient-of-identity expectation. This noncommutative probability space is commutative if and only if $\mathbf{G}$ is abelian.

Classical probability spaces: $(\mathscr{A}, \tau)$ consists of the algebra of complex random variables $\mathscr{A} = L^{\infty-}(\Omega, \mathscr{F}, P) = \bigcap_{p=1}^{\infty} L^p(\Omega, \mathscr{F}, P)$ defined on a Kolmogorov triple which have finite absolute moments of all orders, and τ is the classical expectation $\tau[X] = \mathbb{E}[X]$. Classical probability spaces are always commutative.

Matrix probability spaces: $(\mathscr{A}, \tau)$ consists of the algebra $\mathscr{A} = \mathcal{M}_N(\mathbb{C})$ of $N \times N$ complex matrices $X = [X(ij)]$, and expectation is the normalized trace:

$$\tau[X] = \mathrm{tr}_N[X] = \frac{X(11) + \cdots + X(NN)}{N}.$$

This noncommutative probability space is commutative if and only if $N = 1$.

The first class of model noncommutative probability spaces, group probability spaces, is algebraic and we are trying to move away from algebraic examples. The second model class, classical probability spaces, has genuine randomness but is commutative. The third model class, matrix probability spaces, has a parameter N that can be pushed to infinity but has no randomness. By combining classical probability spaces and matrix probability spaces we arrive at a class of model noncommutative probability spaces which incorporate both randomness and a parameter which can be made large. Thus we are led to consider random matrices.

The space of $N \times N$ complex random matrices is the noncommutative probability space $(\mathscr{A}_N, \tau_N) = (L^{\infty-}(\Omega, \mathscr{F}, P) \otimes \mathcal{M}_N(\mathbb{C}), \mathbb{E} \otimes \mathrm{tr}_N)$. A random variable X_N in this space may be viewed as an $N \times N$ matrix whose entries $X_N(ij)$ belong to the algebra $L^{\infty-}(\Omega, \mathscr{F}, P)$. The expectation $\tau_N[X_N]$ is the expected value of the normalized trace:

$$\tau_N[X_N] = (\mathbb{E} \otimes \mathrm{tr}_N)[X_N] = \mathbb{E}\left[\frac{X_N(11) + \cdots + X_N(NN)}{N}\right].$$

We have already seen indications of a connection between free probability and random matrices. The fact that Wigner's semicircle law assumes the role of the Gaussian distribution in free probability signals a connection between these subjects. Another example is the occurrence of the Marchenko–Pastur distribution in the free version of the Poisson limit theorem — this distribution is well-known in random matrix theory in connection with the asymptotic eigenvalue distribution of Wishart matrices. In Lecture One, we were led to free independence when we tried to solve a counting problem associated to graphs drawn in the plane. The use of random matrices to enumerate planar graphs has been a subject of much interest in mathematical physics since the seminal work of Edouard Brézin, Claude Itzykson, Giorgio Parisi and Jean-Bernard Zuber [Brézin et al. 1978], which built on insights of Gerardus 't Hooft. Then, when we examined the dynamics of the semicircle flow, we found that the free analogue of the heat equation is the complex Burgers equation. This partial differential equation actually appeared in [Voiculescu 1986] before it emerged in random matrix theory [Matytsin 1994] and the discrete analogue of random matrix theory, the dimer model [Kenyon and Okounkov 2007].

In the remainder of these notes, we will model a pair of free random variables X, Y living in an abstract noncommutative probability space using sequences X_N, Y_N of random matrices living in random matrix space. This is first carried out in the important special case where X, Y are semicircular random variables, then adapted to allow Y to have arbitrary distribution while X remains semicircular, and finally relaxed to allow X, Y to have arbitrary specified distributions. The random matrix models of free random variables which we describe below were used by Voiculescu in order to resolve several previously intractable problems in the theory of von Neumann algebras; see [Mingo and Speicher $\geq$ 2014; Voiculescu et al. 1992] for more information. Random matrix models which approximate free random variables in a stronger sense than that described here were subsequently used by Uffe Haagerup and Steen Thorbjørnsen [2005] to resolve another operator algebras conjecture, this time concerning the Ext-invariant of the reduced C^*-algebra of $\mathbf{F}_2$. An important feature of the connection between free probability and random matrices is that it can sometimes be inverted to obtain information about random matrices using the free calculus. For each of the three matrix models constructed we give an example of this type.

3.4. *Random matrix model of a free semicircular pair.* In this subsection we construct a random matrix model for a free semicircular pair X, Y.

In Lecture One, we briefly discussed Wigner matrices. A real Wigner matrix is a symmetric matrix whose entries are centred real random variables which are independent up to the symmetry constraint. A complex Wigner matrix is a Hermitian matrix whose entries are centred complex random variables which are independent up to the complex symmetry constraint. Our matrix model for a free semicircular pair will be built out of complex Wigner matrices of a very special type: they will be GUE random matrices.

To construct a GUE random matrix X_N, we start with a Ginibre matrix Z_N. Let $(\Omega, \mathcal{F}, P)$ be a Kolmogorov triple. The N^2 matrix elements $Z_N(ij) \in L^{\infty-}(\Omega, \mathcal{F}, P)$ of a Ginibre matrix are iid complex Gaussian random variables of mean zero and variance $1/N$. Thus Z_N is a random variable in the noncommutative probability space $(\mathcal{A}_N, \tau_N) = (L^{\infty-}(\Omega, \mathcal{F}, P) \otimes \mathcal{M}_N(\mathbb{C}), \mathbb{E} \otimes \mathrm{tr}_N)$. The symmetrized random matrix $X_N = \frac{1}{2}(Z_N + Z_N^*)$ is again a member of random matrix space. The joint distribution of the eigenvalues of X_N can be explicitly computed, and is given by

$$P(\lambda_N(1) \in I_1, \ldots, \lambda_N(N) \in I_N) \propto \int_{I_1} \cdots \int_{I_N} e^{-N^2 \mathcal{H}(\lambda_1, \ldots, \lambda_N)} \, d\lambda_1 \ldots d\lambda_N$$

for any intervals $I_1, \ldots, I_N \subseteq \mathbb{R}$, where $\mathcal{H}$ is the log-gas Hamiltonian [Forrester 2010]

$$\mathcal{H}(\lambda_1, \ldots, \lambda_N) = \frac{1}{N} \sum_{i=1}^{N} \frac{\lambda_i^2}{2} - \frac{1}{N^2} \sum_{1 \le i \ne j \le N} \log |\lambda_i - \lambda_j|.$$

The random point process on the real line driven by this Hamiltonian is known as the Gaussian Unitary Ensemble, and X_N is termed a GUE random matrix. GUE random matrices sit at the nexus of the two principal strains of complex random matrix theory: they are simultaneously Hermitian Wigner matrices and unitarily invariant matrices. The latter condition means that the distribution of a GUE matrix in the space of $N \times N$ Hermitian matrices is invariant under conjugation by unitary matrices. The spectral statistics of a GUE random matrix can be computed in gory detail from knowledge of the joint distribution of eigenvalues, and virtually any question can be answered. The universality programme in random matrix theory seeks to show that, in the limit $N \to \infty$ and under mild hypotheses, Hermitian Wigner matrices as well as unitarily invariant Hermitian matrices exhibit the same spectral statistics as GUE matrices.

Given the central role of the GUE in random matrix theory, it is fitting that our matrix model for a free semicircular pair is built from a pair of independent GUE matrices. The first step in proving this is to show that a single GUE matrix X_N in random matrix space $(\mathscr{A}_N, \tau_N)$ is an asymptotic model for a single semicircular random variable X living in an abstract noncommutative probability space $(\mathscr{A}, \tau)$. In other words, we need to prove that

$$\lim_{N \to \infty} \tau_N[X_N^n] = \lim_{N \to \infty} (\mathbb{E} \otimes \mathrm{tr}_N)[X_N^n] = \begin{cases} 0 & \text{if } n \text{ is odd,} \\ \mathrm{Cat}_{n/2} & \text{if } n \text{ is even.} \end{cases}$$

In order to establish this, we will not need access to the eigenvalues of X_N. Rather, we work with the correlation functions of its entries.

Let $X_N = [X_N(ij)]$ be a GUE random matrix. Mixed moments of the random variables $X_N(ij)$, i.e., expectations of the form

$$\mathbb{E}\left[\prod_{k=1}^{n} X_N(i(k)j(k)) \right]$$

where i, j are functions $[n] \to [N]$, are called correlation functions. All correlations may be computed in terms of pair correlations (i.e., covariances)

$$\mathbb{E}[X_N(ij)\overline{X_N(kl)}] = \mathbb{E}[X_N(ij)X_N(lk)] = \frac{\delta_{ik}\delta_{jl}}{N}$$

using a convenient combinatorial formula known as Wick's formula. This formula, named for the Italian physicist Gian-Carlo Wick, is yet another manifestation of the moment-cumulant/exponential formulas. It asserts that

$$\mathbb{E}\left[\prod_{k=1}^{n} X_N(i(k)j(k))) \right] = \sum_{\pi \in P_2(n)} \prod_{\{r,s\} \in \pi} \mathbb{E}[X_N(i(r)j(r))X_N(i(s)j(s))]$$

for any integer $n \geq 1$ and functions $i, j : [n] \to [N]$. The sum on the right hand side is taken over all pair partitions of $[n]$, and the product is over the blocks of π. For example,

$$\mathbb{E}[X_N(i(1)j(1))X_N(i(2)j(2))X_N(i(3)j(3))] = 0$$

since there are no pairings on three points, whereas

$$\mathbb{E}[X_N(i(1)j(1))X_N(i(2)j(2))X_N(i(3)j(3))X_N(i(4)j(4))]$$
$$= \mathbb{E}[X_N(i(1)j(1))X_N(i(2)j(2))]\,\mathbb{E}[X_N(i(3)j(3))X_N(i(4)j(4))]$$
$$+ \mathbb{E}[X_N(i(1)j(1))X_N(i(3)j(3))]\,\mathbb{E}[X_N(i(2)j(2))X_N(i(4)j(4))]$$
$$+ \mathbb{E}[X_N(i(1)j(1))X_N(i(4)j(4))]\,\mathbb{E}[X_N(i(2)j(2))X_N(i(3)j(3))],$$

corresponding to the three pair partitions

$$\{1, 2\} \sqcup \{3, 4\}, \quad \{1, 3\} \sqcup \{2, 4\}, \quad \{1, 4\} \sqcup \{2, 3\}$$

of [4]. The Wick formula is a special feature of Gaussian random variables which, ultimately, is a consequence of the moment formula

$$\mathbb{E}[X^n] = \sum_{\pi \in P_2(n)} 1$$

for a single standard real Gaussian X which we proved in Lecture One. A proof of the Wick formula may be found in Alexandre Zvonkin's article [1997].

We now compute the moments of the trace of a GUE matrix X_N using the Wick formula, and then take the $N \to \infty$ limit. We have

$$\tau_N[X_N^n] = \frac{1}{N} \sum_{i:[n] \to [N]} \mathbb{E}[X_N(i(1)i(2))X_N(i(2)i(3)))\ldots X_N(i(n)i(1))]$$

$$= \frac{1}{N} \sum_{i:[n] \to [N]} \mathbb{E}\left[\prod_{k=1}^{n} X_N(i(k)i\gamma(k))\right],$$

where $\gamma = (1\ 2\ \ldots\ n)$ is the full forward cycle in the symmetric group $\mathbf{S}(n)$. Let us apply the Wick formula to each term of this sum, and then use the covariance structure of the matrix elements. We obtain

$$\mathbb{E}\left[\prod_{k=1}^{n} X_N(i(k)i\gamma(k))\right] = \sum_{\pi \in P_2(n)} \prod_{\{r,s\} \in \pi} \mathbb{E}[X_N(i(r)i\gamma(r))X_N(i(s)i\gamma(s))]$$

$$= N^{-n/2} \sum_{\pi \in P_2(n)} \prod_{\{r,s\} \in \pi} \delta_{i(r)i\gamma(s)}\delta_{i(s)i\gamma(r)}.$$

Now, any pair partition of $[n]$ can be viewed as a product of disjoint two-cycles in $\mathbf{S}(n)$. For example, the three pair partitions of [4] enumerated above may be viewed as the fixed-point-free involutions

$$(1\ 2)(3\ 4), \quad (1\ 3)(2\ 4), \quad (1\ 4)(2\ 3)$$

in $\mathbf{S}(4)$. This is a useful shift in perspective because partitions are inert combinatorial objects whereas permutations are functions which act on points. Our computation above may thus be rewritten as

$$\mathbb{E}\left[\prod_{k=1}^{n} X_N(i(k)i\gamma(k))\right] = N^{-n/2} \sum_{\pi \in P_2(n)} \prod_{k=1}^{n} \delta_{i(k)i\gamma\pi(k)}.$$

Putting this all together and changing order of summation, we obtain

$$\tau_N[X_N^n] = N^{1-n/2} \sum_{i:[n]\to[N]} \sum_{\pi\in P_2(n)} \prod_{k=1}^{n} \delta_{i(k)i\gamma\pi(k)}$$

$$= N^{1-n/2} \sum_{\pi\in P_2(n)} \sum_{i:[n]\to[N]} \prod_{k=1}^{n} \delta_{i(k)i\gamma\pi(k)},$$

from which we see that the internal sum is nonzero if and only if the function $i : [n] \to [N]$ is constant on the cycles of the permutation $\gamma\pi \in S(n)$. In order to build such a function, we must specify one of N possible values to be taken on each cycle. We thus obtain

$$\tau_N[X_N^n] = \sum_{\pi\in P_2(n)} N^{c(\gamma\pi)-1-n/2},$$

where $c(\sigma)$ denotes the number of cycles in the disjoint cycle decomposition of a permutation $\sigma \in S(n)$. For example, when $n = 3$ we have $\tau_n[X_N^3] = 0$ since there are no fixed-point-free involutions in $S(3)$. In order to compute $\tau_N[X_N^4]$, we first compute the product of γ with all fixed-point-free involutions in $S(4)$,

$$(1\ 2\ 3\ 4)(1\ 2)(3\ 4) = (1\ 3)(2)(4)$$
$$(1\ 2\ 3\ 4)(1\ 3)(2\ 4) = (1\ 4\ 3\ 2)$$
$$(1\ 2\ 3\ 4)(1\ 4)(2\ 3) = (2\ 4)(1)(3),$$

and from this we obtain

$$\tau_N[X_N^4] = 2 + \frac{1}{N^2}.$$

More generally, $\tau_N[X_N^n] = 0$ whenever n is odd since there are no pairings on an odd number of points. When $n = 2k$ is even the product $\gamma\pi$ has the form

$$\gamma\pi = (1\ 2\ \ldots\ 2k)(s_1\ t_1)(s_2\ t_2)\ldots(s_k\ t_k).$$

In this product, each transposition factor $(s_i\ t_i)$ acts either as a "cut" or as a "join", meaning that it may either cut a cycle of $(1\ 2\ \ldots\ 2k)(s_1\ t_1)\ldots(s_{i-1}\ t_{i-1})$ in two, or join two disjoint cycles together into one. More geometrically, we can view the product $\gamma\pi$ as a walk of length k on the (right) Cayley graph of $S(2k)$; this walk is nonbacktracking and each step taken augments the distance from the identity permutation by ± 1 (see Figure 15).

A cut (step towards the identity) occurs when s_i and t_i reside on the same cycle in the disjoint cycle decomposition of $(1\ 2\ \ldots\ 2k)(s_1\ t_1)\ldots(s_{i-1}\ t_{i-1})$, while a join (step away from the identity) occurs when s_i and t_i are on different cycles. In general, the number of cycles in the product will be

$$c(\gamma\pi) = 1 + \#\text{cuts} - \#\text{joins},$$

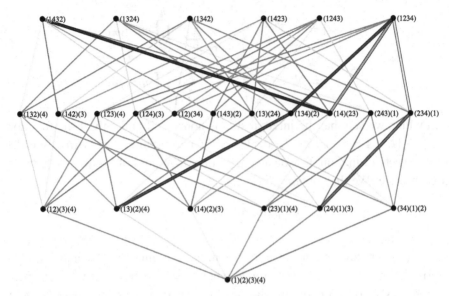

Figure 15. Walks corresponding to the products $\gamma\pi$ in $\mathbf{S}(4)$.

so $c(\gamma\pi)$ is maximal at $c(\gamma\pi) = 1+k$ when it is acted on by a sequence of k cut transpositions. In this case we get a contribution of $N^{1+k-1-k} = N^0$ to $\tau[X_N^n]$. In fact, we always have

$$\#\text{cuts} - \#\text{joins} = k - 2g$$

for some nonnegative integer g, leading to a contribution of the form N^{-2g} and resulting in the formula

$$\tau_N[X_N^{2k}] = \sum_{g \geq 0} \frac{\varepsilon_g(2k)}{N^{2g}}$$

where $\varepsilon_g(2k)$ is the number of products $\gamma\pi$ of the long cycle with a fixed-point-free involution in $\mathbf{S}(2k)$ which terminate at a point of the sphere $\partial B(\mathrm{id}, 2k - 1 - 2g)$. We are only interested in the first term of this expansion, $\varepsilon_0(2k)$, which counts fixed-point-free involutions in $\mathbf{S}(2k)$ entirely composed of cuts. It is not difficult to see that $(s_1\ t_1)\ldots(s_k\ t_k)$ is a sequence of cuts for γ if and only if it corresponds to a noncrossing pair partition of $[2k]$, and as we know the number of these is Cat_k.

We have now shown that

$$\lim_{N \to \infty} \tau_N[X_N^n] = \lim_{N \to \infty} (\mathbb{E} \otimes \mathrm{tr}_N)[X_N^n] = \begin{cases} 0 & \text{if } n \text{ is odd,} \\ \mathrm{Cat}_{n/2} & \text{if } n \text{ is even.} \end{cases}$$

for a GUE matrix X_N. This establishes that X_N is an asymptotic random matrix model of a single semicircular random variable X. It remains to use this fact to construct a sequence of pairs of random matrices which model a pair X, Y of freely independent semicircular random variables.

What should we be looking for? Let $X^{(1)}$, $X^{(2)}$ be a pair of free semicircular random variables. Let $e : [n] \to [2]$ be a function, and apply the free moment-cumulant formula to the corresponding mixed moment:

$$\tau[X^{(e(1))}\dots X^{(e(n))}] = \sum_{\pi \in NC(n)} \prod_{B \in \pi} \kappa_{|B|}(X^{(e(i))} : i \in B)$$

$$= \sum_{\pi \in NC_2(n)} \prod_{\{r,s\} \in \pi} \delta_{e(r)e(s)}.$$

This reduction occurs because $X^{(1)}$, $X^{(2)}$ are free, so that all mixed free cumulants in these variables vanish. Moreover, these variables are semicircular so only order two pure cumulants survive. We can think of the function e as a bicolouring of $[n]$. The formula for mixed moments of a semicircular pair then becomes

$$\tau[X^{(e(1))}\dots X^{(e(n))}] = \sum_{\pi \in NC_2^{(e)}(n)} 1,$$

where $\pi \in NC_2^{(e)}(n)$ is the set of noncrossing pair partitions of $[n]$ which pair elements of the same colour. This is very much like the Wick formula for Gaussian expectations, but with Gaussians replaced by semicirculars and summation restricted to noncrossing pairings. We need to realize this structure in the combinatorics of GUE random matrices.

This construction goes as follows. Let $Z_N^{(e)}(ij)$, $1 \le e \le 2$, $1 \le i, j \le N$ be a collection of $2N^2$ iid centred complex Gaussian random variables of variance $1/N$. Form the corresponding Ginibre matrices $Z_N^{(1)} = [Z_N^{(1)}(ij)]$, $Z_N^{(2)} = [Z_N^{(2)}(ij)]$ and GUE matrices $X_N^{(1)} = \frac{1}{2}(Z_N^{(1)} + (Z_N^{(1)})^*)$, $X_N^{(2)} = \frac{1}{2}(Z_N^{(2)} + (Z_N^{(2)})^*)$. The resulting covariance structure of matrix elements is

$$\mathbb{E}[X_N^{(p)}(ij)\overline{X_N^{(q)}(kl)}] = \mathbb{E}[X_N^{(p)}(ij)X_N^{(q)}(lk)] = \frac{\delta_{ik}\delta_{jl}\delta_{pq}}{N}.$$

We can prove that $X_N^{(1)}$, $X_N^{(2)}$ are asymptotically free by showing that

$$\lim_{N \to \infty} \tau_N[X_N^{(e(1))}\dots X_N^{(e(n))}] = |NC_2^{(e)}(n)|,$$

and this can in turn be proved using the Wick formula and the above covariance structure. Computations almost exactly like those appearing in the one-matrix case lead to the formula

$$\tau_N[X_N^{(e(1))}\dots X_N^{(e(n))}] = \sum_{\pi \in P_2^{(e)}(n)} N^{c(\gamma\pi)-1-n/2},$$

with the summation being taken over the set $P_2^{(e)}(n)$ of pairings on $[n]$ which respect the colouring $e : [n] \to [2]$. Arguing as above, each such pairing makes a contribution of the form N^{-2g} for some $g \geq 0$, and those which make contributions on the leading order N^0 correspond to sequences of cut transpositions for the full forward cycle π, which we know come from noncrossing pairings. So in the limit $N \to \infty$ this expectation converges to $|NC_2^{(e)}(n)|$, as required.

3.5. *Random matrix model of a free pair with one semicircle.*

In the previous subsection we modelled a free pair of semicircular random variables X, Y living in an abstract noncommutative probability space $(\mathscr{A}, \tau)$ using a sequence of independent GUE random matrices X_N, Y_N living in random matrix space $(\mathscr{A}_N, \tau_N)$.

It is reasonable to wonder whether we have not overlooked the possibility of modelling X, Y in a simpler way, namely using deterministic matrices. Indeed, we have

$$\tau[X^n] = \int_{\mathbb{R}} t^n \mu_X(dt)$$

with

$$\mu_X(dt) = \frac{1}{2\pi} \sqrt{4 - t^2}\, dt$$

the Wigner semicircle measure, and this fact leads to a deterministic matrix model for X. For each $N \geq 1$, define the N-th classical locations $L_N(1) < L_N(2) < \cdots < L_N(N)$ of μ_X implicitly by

$$\int_{-2}^{L_N(i)} \mu_X(dt) = \frac{i}{N}.$$

That is, we start at $t = -2$ and integrate along the semicircle until a mass of i/N is achieved, at which time we mark off the corresponding location $L_N(i)$ on the t-axis. The measure μ_N which places mass $1/N$ at each of the N-th classical locations converges weakly to μ_X as $N \to \infty$. Consequently, the diagonal matrix X_N with entries $X_N(ij) = \delta_{ij} L_N(i)$ is a random variable in deterministic matrix space $(\mathscr{M}_N(\mathbb{C}), \mathrm{tr}_N)$ which models X,

$$\lim_{N \to \infty} \mathrm{tr}_N[X_N^n] = \tau[X^n].$$

Since X and Y are equidistributed, putting $Y_N := X_N$ we have that X_N models X and Y_N models Y. However, X_N and Y_N are not asymptotically free. Indeed, asymptotic freeness of X_N and Y_N would imply that

$$\lim_{N \to \infty} \mathrm{tr}_N[X_N Y_N] = \lim_{N \to \infty} \mathrm{tr}_N[X_N] \lim_{N \to \infty} \mathrm{tr}_N[X_N] = 0,$$

but instead we have

$$\mathrm{tr}_N[X_N Y_N] = \frac{L_N(1)^2 + \cdots + L_N(N)^2}{N},$$

the mean squared classical locations of the Wigner measure, which is strictly positive and increasing in N. Thus while X_N and Y_N model X and Y respectively, they cannot model the free relation between them. However, this does not preclude the possibility that a pair of free random variables can be modelled by one random and one deterministic matrix.

Let X and Y be a pair of free random variables with X semicircular, and Y of arbitrary distribution. Let X_N be a sequence of GUE matrices modelling X, and suppose that Y_N is a sequence of deterministic matrices modelling Y,

$$\lim_{N \to \infty} \mathrm{tr}_N[Y_N^n] = \tau[Y^n].$$

X_N lives in random matrix space $(\mathcal{A}_N, \tau_N) = (L^{\infty-}(\Omega, \mathcal{F}, P) \otimes \mathcal{M}_N(\mathbb{C}), \mathbb{E} \otimes \mathrm{tr}_N)$ while Y_N lives in deterministic matrix space $(\mathcal{M}_N(\mathbb{C}), \mathrm{tr}_N)$, so a priori it is meaningless to speak of the potential asymptotic free independence of X_N and Y_N. However, we may think of a deterministic matrix as a random matrix whose entries are constant random variables in $L^{\infty-}(\Omega, \mathcal{F}, P)$. This corresponds to an embedding of deterministic matrix space in random matrix space satisfying $\tau_N|_{\mathcal{M}_N(\mathbb{C})} = (\mathbb{E} \otimes \mathrm{tr}_N)|_{\mathcal{M}_N(\mathbb{C})} = \mathrm{tr}_N$. From this point of view, Y_N is a random matrix model of Y and we can consider the possibility that $X_N, Y_N \in \mathcal{A}_N$ are asymptotically free with respect to τ_N. We now show that this is indeed the case.

As in the previous subsection, we proceed by identifying the combinatorial structure governing the target pair X, Y and then looking for this same structure in the $N \to \infty$ asymptotics of X_N, Y_N. Our target is a pair of free random variables with X semicircular and Y arbitrary. Understanding their joint distribution means understanding the collection of mixed moments

$$\tau[X^{p(1)} Y^{q(1)} \ldots X^{p(n)} Y^{q(n)}],$$

with $n \geq 1$ and $p, q : [n] \to \{0, 1, 2, \ldots\}$. This amounts to understanding mixed moments of the form

$$\tau[X Y^{q(1)} \ldots X Y^{q(n)}],$$

since we can artificially insert copies of $Y^0 = 1_{\mathcal{A}}$ to break up powers of X greater than one. We can expand this expectation using the free moment-cumulant formula and simplify the resulting expression using the fact that mixed cumulants in free random variables vanish. Further simplification results from the fact that, since X is semicircular, its only nonvanishing pure cumulant is $\kappa_2(X) = 1$. This leads to a formula for $\tau[X Y^{q(1)} \ldots X Y^{q(n)}]$ which is straightforward but whose statement requires some notions which we have not covered (in particular, the complement of a noncrossing partition; see [Nica and Speicher 2006]). However, in the case where τ is a tracial expectation, meaning that $\tau[AB] = \tau[BA]$, the formula in question can be stated more simply as

$$\tau[XY^{q(1)} \dots XY^{q(n)}] = \sum_{\pi \in \mathrm{NC}_2(n)} \tau_{\pi\gamma}[Y^{q(1)}, \dots, Y^{q(n)}].$$

Here, as in the last subsection, we think of a pair partition $\pi \in P_2(n)$ as a product of disjoint two-cycles in the symmetric group $S(n)$, and γ is the full forward cycle $(1\ 2\ \dots\ n)$. Given a permutation $\sigma \in S(n)$, the expression $\tau_\sigma[A_1, \dots, A_N]$ is defined to be the product of τ extended over the cycles of σ. For example,

$$\tau_{(1\ 6\ 2)(4\ 5)(3)}[A_1, A_2, A_3, A_4, A_5, A_6] = \tau[A_1 A_6 A_2]\tau[A_4 A_5]\tau[A_3].$$

This definition is kosher since τ is tracial. We now have our proof strategy: we will prove that X_N, Y_N are asymptotically free by showing that

$$\lim_{N\to\infty} \tau_N[X_N Y_N^{q(1)} \dots X_N Y_N^{q(n)}] = \sum_{\pi \in \mathrm{NC}_2(n)} \tau_{\pi\gamma}[Y^{q(1)}, \dots, Y^{q(n)}].$$

The computation proceeds much as in the last section — we expand everything in sight and apply the Wick formula. We have

$$\tau_N[X_N Y_N^{q(1)} \dots X_N Y_N^{q(n)}]$$
$$= \frac{1}{N} \sum_a \mathbb{E}[X_N(a(1)a(2))Y_N^{q(1)}(a(2)a(3)) \cdots$$
$$\times X_N(a(2n-1)a(2n))Y_N^{q(n)}(a(2n)a(1))],$$

the summation being over all functions $a : [2n] \to [N]$. Let us reparametrize each term of the sum with $i, j : [n] \to [N]$ defined by

$$(a(1), a(2), \dots, a(2n-1), a(2n)) = (i(1), j(1), \dots, i(n), j(n)).$$

Our computation so far becomes

$$\tau_N[X_N Y_N^{q(1)} \dots X_N Y_N^{q(n)}]$$
$$= \frac{1}{N} \sum_{i,j} \mathbb{E}\left[\prod_{k=1}^n X_N(i(k)j(k))\right] \prod_{k=1}^n Y_N^{q(k)}(j(k)i\gamma(k)).$$

Applying the Wick formula, the calculation evolves as follows:

$$\tau_N[X_N Y_N^{q(1)} \ldots X_N Y_N^{q(n)}]$$

$$= \frac{1}{N} \sum_{i,j} \sum_{\pi \in P_2(n)} \prod_{\{r,s\} \in \pi} \mathbb{E}[X_N(i(r)j(r))X_N(i(s)j(s))] \prod_{k=1}^{n} Y_N^{q(k)}(j(k)i\gamma(k))$$

$$= N^{-1-n/2} \sum_{i,j} \sum_{\pi \in P_2(n)} \prod_{k=1}^{n} \delta_{i(k)j\pi(k)} Y_N^{q(k)}(j(k)i\gamma(k))$$

$$= N^{-1-n/2} \sum_{\pi \in P_2(n)} \sum_{j} \prod_{k=1}^{n} Y_N^{q(k)}(j(k)j\pi\gamma(k))$$

$$= N^{-1-n/2} \sum_{\pi \in P_2(n)} \mathrm{Tr}_{\pi\gamma}[Y_N^{q(1)}, \ldots, Y_N^{q(n)}]$$

$$= \sum_{\pi \in P_2(n)} N^{c(\pi\gamma)-1-n/2} \mathrm{tr}_{\pi\gamma}[Y_N^{q(1)}, \ldots, Y_N^{q(n)}].$$

As in the previous subsection, the dominant contributions to this sum are of order N^0 and come from those pair partitions $\pi \in P_2(n)$ for which $c(\pi\gamma)$ is maximal, and these are the noncrossing pairings. Hence we obtain

$$\lim_{N \to \infty} \tau_N[X_N Y_N^{q(1)} \ldots X_N Y_N^{q(n)}] = \sum_{\pi \in NC_2(n)} \tau_{\pi\gamma}[Y^{q(1)}, \ldots, Y^{q(n)}],$$

as required.

3.6. Random matrix model of an arbitrary free pair.

In the last section we saw that a pair of free random variables can be modelled by one random and one deterministic matrix provided that at least one of the target variables is semicircular. In this case, the semicircular target is modelled by a sequence of GUE random matrices.

In this section we show that any pair of free random variables can be modelled by one random and one deterministic matrix, provided each target variable can be individually modelled by a sequence of deterministic matrices. The idea is to randomly rotate one of the deterministic matrix models so as to create the free relation.

Let X, Y be a pair of free random variables living in an abstract noncommutative probability space $(\mathscr{A}, \tau)$. We make no assumption on their moments. What we assume is the existence of a pair of deterministic matrix models

$$\tau[X^n] = \lim_{N \to \infty} \mathrm{tr}_N[X_N^n], \quad \tau[Y^n] = \lim_{N \to \infty} \mathrm{tr}_N[Y_N^n].$$

If X, Y happen to have distributions μ_X, μ_Y which are compactly supported probability measures on $\mathbb{R}$, then such models can always be constructed. In particular, this will be the case if X, Y are bounded self-adjoint random variables living in a $*$-probability space.

As in the previous subsection, we view X_N, Y_N as random matrices with constant entries so that they reside in random matrix space $(\mathscr{A}_N, \tau_N)$, with the $\mathbb{E}$ part of $\tau_N = \mathbb{E} \otimes \mathrm{tr}_N$ acting trivially. As we saw above, there is no guarantee that X_N, Y_N are asymptotically free. On the other hand, we also saw that special pairs of free random variables can be modelled by one random and one deterministic matrix. Therefore it is reasonable to hope that making X_N genuinely random might lead to asymptotic freeness. We have to randomize X_N in such a way that its moments will be preserved. This can be achieved via conjugation by a unitary random matrix $U_N \in \mathscr{A}_N$,

$$X_N \mapsto U_N X_N U_N^*.$$

The deterministic matrix X_N and its randomized version $U_N X_N U_N^*$ have the same moments since

$$\begin{aligned}
\tau_N[(U_N X_N U_N^*)^n] &= (\mathbb{E} \otimes \mathrm{tr}_N)[(U_N X_N U_N^*)^n] \\
&= (\mathbb{E} \otimes \mathrm{tr}_N)[U_N X_N^n U_N^*] \\
&= (\mathbb{E} \otimes \mathrm{tr}_N)[U_N^* U_N X_N^n] \\
&= (\mathbb{E} \otimes \mathrm{tr}_N)[X_N^n] \\
&= \tau_N[X_N^n].
\end{aligned}$$

Consequently, the sequence $U_N X_N U_N^*$ is a random matrix model for X.

We aim to prove that $U_N X_N U_N^*$ and Y_N are asymptotically free. Since we are making no assumptions on the limiting variables X, Y, we cannot verify this by looking for special structure in the limiting mixed moments of $U_N X_N U_N^*$ and Y_N, as we did above. Instead, we must verify asymptotic freeness directly, using the definition:

$$\lim_{N \to \infty} \tau_N[f_1(U_N X_N U_N^*) g_1(Y_N) \dots f_n(U_N X_N U_N^*) g_n(Y_N)] = 0$$

whenever $f_1, g_1, \dots, f_n, g_n$ are polynomials such that

$$\lim_{N \to \infty} \tau_N[f_1(U_N X_N U_N^*)] = \lim_{N \to \infty} \tau_n[g_1(Y_N)] = \cdots$$
$$= \lim_{N \to \infty} \tau_N[f_n(U_N X_N U_N^*)] = \lim_{N \to \infty} \tau_n[g_n(Y_N)] = 0.$$

Though the brute force verification of this criterion may seem an impossible task, we will see that it can be accomplished for a well-chosen sequence of unitary random matrices U_N. Let us advance as far as possible before specifying U_N precisely.

As an initial reduction, note the identity

$$\tau_N[f_1(U_N X_N U_N^*)g_1(Y_N)\dots f_n(U_N X_N U_N^*)g_n(Y_N)]$$
$$= \tau_N[U_N f_1(X_N)U_N^* g_1(Y_N)\dots U_N f_n(X_N)U_N^* g_n(Y_N)].$$

Since the f_i and g_j are polynomials and τ_N is linear, the right hand side of this equation may be expanded as a sum of monomial expectations,

$$\tau_N[U_N f_1(X_N)U_N^* g_1(Y_N)\dots U_N f_n(X_N)U_N^* g_n(Y_N)]$$
$$= \sum_{p,q} c(pq)\tau_N[U_N X_N^{p(1)} U_N^* Y_N^{q(1)} \dots U_N X_N^{p(n)} U_N^* Y_N^{q(n)}]$$

weighted by some scalar coefficients $c(pq)$, the sum being over functions $p : [n] \to \{0,\dots,\max \deg f_i\}$, $q : [n] \to \{0,\dots,\max \deg g_j\}$. Each monomial expectation can in turn be expanded as

$$\tau_N[U_N X_N^{p(1)} U_N^* Y_N^{q(1)} \dots U_N X_N^{p(n)} U_N^* Y_N^{q(n)}]$$
$$= \frac{1}{N}\sum_a \mathbb{E}\big[U_N(a(1)a(2))X_N^{p(1)}(a(2)a(3))\dots$$
$$\times U_N^*(a(4n-1)a(4n))Y_N^{q(n)}(a(4n)a(1))\big]$$
$$= \frac{1}{N}\sum_a \mathbb{E}\big[U_N(a(1)a(2))X_N^{p(1)}(a(2)a(3))\dots$$
$$\times \bar{U}_N(a(4n)a(4n-1))Y_N^{q(n)}(a(4n)a(1))\big].$$

Let us reparametrize the summation index $a : [4n] \to [N]$ by a quadruple of functions $i, j, i', j' : [n] \to [N]$ according to

$$(a(1), a(2), a(3), a(4), \dots, a(4n-3), a(4n-2), a(4n-1), a(4n))$$
$$= (i(1), j(1), j'(1), i'(1), \dots, i(n), j(n), j'(n), i'(n)).$$

Our monomial expectations then take the more streamlined form

$$\tau_N[U_N X_N^{p(1)} U_N^* Y_N^{q(1)} \dots U_N X_N^{p(n)} U_N^* Y_N^{q(n)}]$$
$$= \frac{1}{N}\sum_{i,j,i',j'} \mathbb{E}\Big[\prod_{k=1}^n U_N(i(k)j(k))\bar{U}_N(i'(k)j'(k))\Big]$$
$$\times \prod_{k=1}^n X_N^{p(k)}(j(k)j'(k))Y_N^{q(k)}(i'(k)i\gamma(k)),$$

where as always $\gamma = (1\ 2\ \dots\ n)$ is the full forward cycle in the symmetric group $S(n)$. In order to go any further with this calculation, we must deal with the correlation functions

$$\mathbb{E}\left[\prod_{k=1}^{n} U_N(i(k)j(k))\overline{U}_N(i'(k)j'(k))\right].$$

of the matrix elements of U_N. We would like to have an analogue of the Wick formula which will enable us to address these correlation functions. A formula of this type is known for random matrices sampled from the Haar probability measure on the unitary group $\mathbf{U}(N)$.

Haar-distributed unitary matrices are the second most important class of random matrices after GUE matrices. Like GUE matrices, they can be constructively obtained from Ginibre matrices. Let $\tilde{Z}_N = \sqrt{N}Z_N$ be an $N \times N$ random matrix whose entries $\tilde{Z}_N(ij)$ are iid complex Gaussian random variables of mean zero and variance one. This is a renormalized version of the Ginibre matrix which we previously used to construct a GUE random matrix. The Ginibre matrix $\tilde{Z}_N$ is almost surely nonsingular. Applying the Gram–Schmidt orthonormalization procedure to the columns of $\tilde{Z}_N$, we obtain a random unitary matrix U_N whose distribution in the unitary group $\mathbf{U}(N)$ is given by the Haar probability measure. The entries $U_N(ij)$ are bounded random variables, so U_N is a noncommutative random variable living in random matrix space $(\mathcal{A}_N, \tau_N)$. The eigenvalues $\lambda_N(1) = e^{i\theta_N(1)}, \ldots, \lambda_N(N) = e^{i\theta_N(N)}$, $0 \le \theta_N(1) \le \cdots \le \theta_N(N) \le 2\pi$ of U_N form a random point process on the unit circle with joint distribution

$$P(\theta_N(1) \in I_1, \ldots, \theta_N(N) \in I_N) \propto \int_{I_1} \cdots \int_{I_N} e^{-N^2\mathcal{H}(\theta_1,\ldots,\theta_N)}\, d\theta_1 \ldots d\theta_N$$

for any intervals $I_1, \ldots, I_N \subseteq [0, 2\pi]$, where $\mathcal{H}$ is the log-gas Hamiltonian [Forrester 2010]

$$\mathcal{H}(\theta_1, \ldots, \theta_N) = -\frac{1}{N^2} \sum_{1 \le i \ne j \le N} \log |e^{i\theta_i} - e^{i\theta_j}|.$$

The random point process on the unit circle driven by this Hamiltonian is known as the Circular Unitary Ensemble, and U_N is termed a CUE random matrix. As with GUE random matrices, almost any question about the spectrum of CUE random matrices can be answered using this explicit formula; see, for example, [Diaconis 2003] for a survey of many interesting results.

We are not interested in the eigenvalues of CUE matrices, but rather in the correlation functions of their matrix elements. These can be handled using a Wick-type formula known as the Weingarten formula, after the American physicist Donald H. Weingarten.[3] Like the Wick formula, the Weingarten formula is a combinatorial rule which reduces the computation of general correlation

---

[3]Further information on Weingarten and his colleagues in the first Fermilab theory group may be found at http://bama.ua.edu/~lclavell/Weston/.

functions to the computation of a special class of correlations. Unfortunately, the Weingarten formula is more complicated than the Wick formula. It reads

$$\mathbb{E}\left[\prod_{k=1}^{n} U_N(i(k)j(k))\bar{U}_N(i'(k)j'(k))\right]$$

$$= \sum_{\rho,\sigma\in\mathbf{S}(n)} \delta_{i\sigma,i'}\delta_{j\rho,j'}\mathbb{E}\left[\prod_{k=1}^{n} U_N(kk)\bar{U}_N(k\rho^{-1}\sigma(k))\right].$$

Note that his formula only makes sense when $N \geq n$, and instead of a sum over fixed-point-free involutions we are faced with a double sum over all of $\mathbf{S}(n)$. Worse still, the Weingarten formula does not reduce our problem to the computation of pair correlators, but only to the computation of arbitrary permutation correlators

$$\mathbb{E}\left[\prod_{k=1}^{n} U_N(kk)\bar{U}_N(k\pi(k))\right], \quad \pi \in \mathbf{S}(n),$$

and these have a rather complicated structure. Their computation is the subject of a large literature both in physics and mathematics, a unified treatment of which may be found in [Collins et al. ≥ 2014]. We delay dealing with these averages for the moment and press on in our calculation.

We return to the expression

$$\tau_N[U_N X_N^{p(1)} U_N^* Y_N^{q(1)} \dots U_N X_N^{p(n)} U_N^* Y_N^{q(n)}]$$

$$= \frac{1}{N} \sum_{i,j,i',j'} \mathbb{E}\left[\prod_{k=1}^{n} U_N(i(k)j(k))\bar{U}_N(i'(k)j'(k))\right]$$

$$\times \prod_{k=1}^{n} X_N^{p(k)}(j(k)j'(k))Y_N^{q(k)}(i'(k)i\gamma(k)),$$

and apply the Weingarten formula. The calculation evolves as follows:

$$\tau_N[U_N X_N^{p(1)} U_N^* Y_N^{q(1)} \dots U_N X_N^{p(n)} U_N^* Y_N^{q(n)}]$$

$$= \frac{1}{N} \sum_{i,j,i',j'} \sum_{\rho,\sigma\in\mathbf{S}(n)} \delta_{i\sigma,i'}\delta_{j\rho,j'}\mathbb{E}\left[\prod_{k=1}^{n} U_N(kk)\bar{U}_N(k\rho^{-1}\sigma(k))\right]$$

$$\times \prod_{k=1}^{n} X_N^{p(k)}(j(k)j'(k))Y_N^{q(k)}(i'(k)i\gamma(k))$$

$$= \frac{1}{N} \sum_{\rho,\sigma\in\mathbf{S}(n)} \mathbb{E}\left[\prod_{k=1}^{n} U_N(kk)\bar{U}_N(k\rho^{-1}\sigma(k))\right]$$

$$\times \sum_{i',j}\prod_{k=1}^{n} X_N^{p(k)}(j(k)j\rho(k))Y_N^{q(k)}(i'(k)i\sigma^{-1}\gamma(k))$$

$$= \frac{1}{N} \sum_{\rho,\sigma \in S(n)} \mathbb{E}\left[\prod_{k=1}^{n} U_N(kk)\bar{U}_N(k\rho^{-1}\sigma(k))\right]$$
$$\times \mathrm{Tr}_\rho(X_N^{p(1)}, \ldots, X_N^{p(n)}) \, \mathrm{Tr}_{\sigma^{-1}\gamma}(Y_N^{p(1)}, \ldots, Y_N^{p(n)})$$

$$= \sum_{\rho,\sigma \in S(n)} \mathbb{E}\left[\prod_{k=1}^{n} U_N(kk)\bar{U}_N(k\rho^{-1}\sigma(k))\right] N^{c(\rho)+c(\sigma^{-1}\gamma)-1} \, \mathrm{tr}_\rho(X_N^{p(1)}, \ldots, X_N^{p(n)})$$
$$\times \mathrm{tr}_{\sigma^{-1}\gamma}(Y_N^{p(1)}, \ldots, Y_N^{p(n)}).$$

At this point we are forced to deal with the permutation correlators

$$\mathbb{E}\left[\prod_{k=1}^{n} U_N(kk)\bar{U}_N(k\pi(k))\right].$$

Perhaps the most appealing presentation of these expectations is as a power series in N^{-1}. It may be shown [Novak 2010] that

$$\mathbb{E}\left[\prod_{k=1}^{n} U_N(kk)\bar{U}_N(k\pi(k))\right] = \frac{1}{N^n}\sum_{r=0}^{\infty}(-1)^r \frac{c_{n,r}(\pi)}{N^r},$$

for any $\pi \in S(n)$, where the coefficient $c_{n,r}(\pi)$ equals the number of factorizations

$$\pi = (s_1 \ t_1) \ldots (s_r \ t_r)$$

of π into r transpositions $(s_i \ t_i) \in S(n)$, $s_i < t_i$, which have the property that

$$t_1 \leq \cdots \leq t_r.$$

This series is absolutely convergent for $N \geq n$, but divergent for $N < n$. This will not trouble us since we are looking for $N \to \infty$ asymptotics with n fixed. Indeed, let $|\pi| = n - c(\pi)$ denote the distance from the identity permutation to π in the Cayley graph of $S(n)$. Then, since any permutation is either even or odd, we have

$$\mathbb{E}\left[\prod_{k=1}^{n} U_N(kk)\bar{U}_N(k\pi(k))\right] = \frac{1}{N^n}\sum_{r=0}^{\infty}(-1)^r \frac{c_{n,r}(\pi)}{N^r}$$
$$= \frac{(-1)^{|\pi|}}{N^{n+|\pi|}}\sum_{g=0}^{\infty} \frac{c_{n,|\pi|+2g}(\pi)}{N^{2g}}$$
$$= \frac{a(\pi)}{N^{n+|\pi|}} + O\left(\frac{1}{N^{n+|\pi|+2}}\right),$$

where $a(\pi) = (-1)^{|\pi|} c_{n,|\pi|}(\pi)$ is the leading asymptotics. We may now continue our calculation:

$$\tau_N[U_N X_N^{p(1)} U_N^* Y_N^{q(1)} \ldots$$

$$\tau_N[U_N X_N^{p(1)} U_N^* Y_N^{q(1)} \ldots U_N X_N^{p(n)} U_N^* Y_N^{q(n)}]$$

$$= \sum_{\rho,\sigma \in S(n)} \left(\frac{a(\rho^{-1}\sigma))}{N^{n+|\rho^{-1}\sigma|}} + O\left(\frac{1}{N^{n+|\rho^{-1}\sigma|+2}} \right) \right) N^{c(\rho)+c(\sigma^{-1}\gamma)-1}$$

$$\times \mathrm{tr}_\rho(X_N^{p(1)}, \ldots, X_N^{p(n)}) \, \mathrm{tr}_{\sigma^{-1}\gamma}(Y_N^{p(1)}, \ldots, Y_N^{p(n)})$$

$$= \sum_{\rho,\sigma \in S(n)} \left(a(\rho^{-1}\sigma) + O\left(\frac{1}{N^2} \right) \right) N^{|\gamma|-|\rho|-|\rho^{-1}\sigma|-|\sigma^{-1}\gamma|}$$

$$\times \mathrm{tr}_\rho(X_N^{p(1)}, \ldots, X_N^{p(n)}) \, \mathrm{tr}_{\sigma^{-1}\gamma}(Y_N^{p(1)}, \ldots, Y_N^{p(n)}).$$

Putting everything together, we have shown that

$$\tau_N[U_N f_1(X_N) U_N^* g_1(Y_N) \ldots U_N f_n(X_N) U_N^* g_n(Y_N)]$$

$$= \sum_{\rho,\sigma \in S(n)} \left(a(\rho^{-1}\sigma) + O\left(\frac{1}{N^2} \right) \right) N^{|\gamma|-|\rho|-|\rho^{-1}\sigma|-|\sigma^{-1}\gamma|}$$

$$\times \mathrm{tr}_\rho(f_1(X_N), \ldots, f_n(X_N)) \, \mathrm{tr}_{\sigma^{-1}\gamma}(g_1(Y_N), \ldots, g_n(Y_N)),$$

and it remains to show that the $N \to \infty$ limit of this complicated expression is zero. To this end, consider the order $|\gamma| - |\rho| - |\rho^{-1}\sigma| - |\sigma^{-1}\gamma|$ of the ρ, σ term in this sum. The positive part, $|\gamma| = n - 1$, is simply the length of any geodesic joining the identity permutation to γ in the Cayley graph of $S(n)$. The negative part, $-|\rho| - |\rho^{-1}\sigma| - |\sigma^{-1}\gamma|$, is the length of a walk from the identity to γ made up of three legs: a geodesic from id to ρ, followed by a geodesic from ρ to σ, followed by a geodesic from σ to γ. Thus the order of the ρ, σ term is at most N^0, and this occurs precisely when ρ and σ lie on a geodesic from id to γ; see Figure 16. Thus

$$\lim_{N \to \infty} \tau_N[U_N f_1(X_N) U_N^* g_1(Y_N) \ldots U_N f_n(X_N) U_N^* g_n(Y_N)]$$

$$= \sum_{|\rho|+|\rho^{-1}\sigma|+|\sigma^{-1}\gamma|=|\gamma|} a(\rho^{-1}\sigma) \tau_\rho(f_1(X), \ldots, f_n(X)) \tau_{\sigma^{-1}\gamma}(g_1(Y), \ldots, g_n(Y)).$$

Since

$$\tau[f_1(X)] = \tau[g_1(Y)] = \cdots = \tau[f_n(X)] = \tau[g_n(Y)] = 0,$$

in order to show that the sum on the right has all terms equal to zero it suffices to show that the condition $|\rho| + |\rho^{-1}\sigma| + |\sigma^{-1}\gamma| = |\gamma|$ forces either ρ or $\sigma^{-1}\gamma$ to have a fixed point. This is because τ_ρ and $\tau_{\sigma^{-1}\gamma}$ are products determined by the cycle structure of the indexing permutation. Since ρ, σ lie on a geodesic id $\to \gamma$, we have $|\rho| + |\sigma^{-1}\gamma| \le |\gamma| = n - 1$, so that one of ρ or $\sigma^{-1}\gamma$ is a product of at most $(n-1)/2$ transpositions. In the extremal case, all of these transpositions

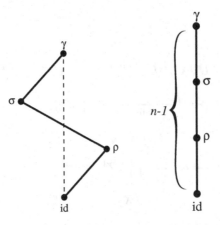

Figure 16. Only geodesic paths survive in the large N limit.

are joins, leading to a permutation consisting of an $(n-1)$-cycle and a fixed point.

3.7. GUE+GUE. Imagine that we had been enumeratively lazy in our construction of the GUE matrix model of a free semicircular pair, and had only shown that two iid GUE matrices $X_N^{(1)}, X_N^{(2)}$ are asymptotically free without determining their individual limiting distributions. We could then appeal to the free central limit theorem to obtain that the limit distribution of the random matrix

$$S_N = \frac{X_N^{(1)} + \cdots + X_N^{(n)}}{\sqrt{N}},$$

where the $X_N^{(i)}$ are iid GUE samples, is standard semicircular. On the other hand, since the matrix elements of the $X_N^{(i)}$ are independent Guassians whose variances add, we see that the rescaled sum S_N is itself an $N \times N$ GUE random matrix for each finite N. Thus we recover Wigner's semicircle law (for GUE matrices) from the free central limit theorem.

3.8. GUE+deterministic. Let X_N be an $N \times N$ GUE random matrix. Let Y_N be an $N \times N$ deterministic Hermitian matrix whose spectral measure ν_N converges weakly to a compactly supported probability measure ν. Let σ be the limit distribution of the random matrix $X_N + Y_N$. Since X_N, Y_N are asymptotically free, we have

$$\sigma = \mu \boxplus \nu,$$

where μ is the Wigner semicircle.

3.9. *Randomly rotated + diagonal.* Consider the $2N \times 2N$ diagonal matrix

$$D_{2N} = \begin{bmatrix} 1 & & & & \\ & -1 & & & \\ & & \ddots & & \\ & & & 1 & \\ & & & & -1 \end{bmatrix}$$

whose diagonal entries are the first $2N$ terms of an alternating sequence of ± 1, all other entries being zero. Let U_{2N} be a $2N \times 2N$ CUE random matrix, and consider the random Hermitian matrix

$$A_{2N} = U_{2N} D_{2N} U_{2N}^* + D_{2N}.$$

Let μ_{2N} denote the spectral measure of A_{2N}. We claim that μ_{2N} converges weakly to the arcsine distribution

$$\mu(dt) = \frac{1}{\pi \sqrt{4 - t^2}} \, dt, \quad t \in [-2, 2],$$

as $N \to \infty$.

Proof: Set $X_{2N} = U_{2N} D_{2N} U_{2N}^*$ and $Y_{2N} = D_{2N}$. Then X_N, Y_N is a random matrix model for a pair of free random variables X, Y each of which has the ± 1-Bernoulli distribution

$$\frac{1}{2}\delta_{-1} + \frac{1}{2}\delta_{+1}.$$

Thus the limit distribution of their sum is

$$\text{Bernoulli} \boxplus \text{Bernoulli} = \text{Arcsine}.$$

References

[Andrews et al. 1999] G. E. Andrews, R. Askey, and R. Roy, *Special functions*, Encyclopedia of Mathematics and its Applications **71**, Cambridge University Press, Cambridge, 1999.

[Bercovici and Voiculescu 1993] H. Bercovici and D. Voiculescu, "Free convolution of measures with unbounded support", *Indiana Univ. Math. J.* **42**:3 (1993), 733–773.

[Biane 1997] P. Biane, "On the free convolution with a semi-circular distribution", *Indiana Univ. Math. J.* **46**:3 (1997), 705–718.

[Biane 2002] P. Biane, "Free probability and combinatorics", pp. 765–774 in *Proceedings of the International Congress of Mathematicians, II* (Beijing, 2002), edited by T. Li, Higher Ed. Press, Beijing, 2002.

[Brézin et al. 1978] E. Brézin, C. Itzykson, G. Parisi, and J. B. Zuber, "Planar diagrams", *Comm. Math. Phys.* **59**:1 (1978), 35–51.

[Collins et al. ≥ 2014] B. Collins, S. Matsumoto, and J. Novak, *An invitation to Weingarten calculus*, In preparation.

[Connes 1994] A. Connes, *Noncommutative geometry*, Academic Press, San Diego, CA, 1994.

[Diaconis 2003] P. Diaconis, "Patterns in eigenvalues: the 70th Josiah Willard Gibbs lecture", *Bull. Amer. Math. Soc. (N.S.)* **40**:2 (2003), 155–178.

[Erdős et al. 2011] L. Erdős, B. Schlein, and H.-T. Yau, "Universality of random matrices and local relaxation flow", *Invent. Math.* **185**:1 (2011), 75–119.

[Etingof 2003] P. Etingof, "Mathematical ideas and notions of quantum field theory", 2003, http://math.mit.edu/~etingof/lect.ps. lecture notes.

[Fisher and Wishart 1932] R. A. Fisher and J. Wishart, "The derivation of the pattern formulae of two–way partitions from those of simpler patterns", *Proc. London Math. Soc.* **S2-33**:1 (1932), 195.

[Flajolet and Sedgewick 2009] P. Flajolet and R. Sedgewick, *Analytic combinatorics*, Cambridge Univ. Press, 2009.

[Forrester 2010] P. J. Forrester, *Log-gases and random matrices*, London Mathematical Society Monographs Series **34**, Princeton University Press, Princeton, NJ, 2010.

[Graham et al. 1989] R. L. Graham, D. E. Knuth, and O. Patashnik, *Concrete mathematics: a foundation for computer science*, Addison-Wesley Publishing Company, Advanced Book Program, Reading, MA, 1989.

[Haagerup and Thorbjørnsen 2005] U. Haagerup and S. Thorbjørnsen, "A new application of random matrices: $\mathrm{Ext}(C^*_{\mathrm{red}}(F_2))$ is not a group", *Ann. of Math. (2)* **162**:2 (2005), 711–775.

[Hald 1981] A. Hald, "T. N. Thiele's contributions to statistics", *Internat. Statist. Rev.* **49**:1 (1981), 1–20 (one plate).

[Hald 2000] A. Hald, "The early history of cumulants and the Gram–Charlier series", *International Statistical Review* **68**:2 (2000), 137–153.

[Hiai and Petz 2000] F. Hiai and D. Petz, *The semicircle law, free random variables, and entropy*, Amer. Math. Soc., Providence, RI, 2000.

[Hurwitz 1891] A. Hurwitz, "Über Riemann'sche Flächen mit gegebenen Verzweigungspunkten", *Mathematische Annalen* **39** (1891), 1–66.

[Kenyon and Okounkov 2007] R. Kenyon and A. Okounkov, "Limit shapes and the complex Burgers equation", *Acta Math.* **199**:2 (2007), 263–302.

[Kesten 1959] H. Kesten, "Symmetric random walks on groups", *Trans. Amer. Math. Soc.* **92** (1959), 336–354.

[Matytsin 1994] A. Matytsin, "On the large-N limit of the Itzykson–Zuber integral", *Nuclear Phys. B* **411**:2-3 (1994), 805–820.

[Mazur 2006] B. Mazur, "Controlling our errors", *Nature* **443** (2006), 38–39.

[Mingo and Speicher ≥ 2014] J. A. Mingo and R. Speicher, *Free probability and random matrices*, Fields Institute Monographs, Amer. Math. Soc., Providence, RI. To appear.

[Murty and Murty 2009] M. R. Murty and V. K. Murty, "The Sato–Tate conjecture and generalizations", pp. 639–646 in *Current trends in science: platinum jubilee special*, edited by N. Mukunda, Indian Academy of Sciences, Bangalore, 2009.

[Nestruev 2003] J. Nestruev, *Smooth manifolds and observables*, Graduate Texts in Mathematics **220**, Springer, New York, 2003.

[Nica and Speicher 2006] A. Nica and R. Speicher, *Lectures on the combinatorics of free probability*, London Mathematical Society Lecture Note Series **335**, Cambridge University Press, Cambridge, 2006.

[Novak 2010] J. I. Novak, "Jucys–Murphy elements and the unitary Weingarten function", pp. 231–235 in *Noncommutative harmonic analysis with applications to probability, II*, edited by M. Bożejko et al., Banach Center Publ. **89**, Polish Acad. Sci. Inst. Math., Warsaw, 2010.

[Novak and Śniady 2011] J. Novak and P. Śniady, "What is... a free cumulant?", *Notices Amer. Math. Soc.* **58**:2 (2011), 300–301.

[Pólya 1921] G. Pólya, "Über eine Aufgabe der Wahrscheinlichkeitsrechnung betreffend die Irrfahrt im Straßennetz", *Math. Ann.* **84**:1-2 (1921), 149–160.

[Rota 1964] G.-C. Rota, "On the foundations of combinatorial theory, I: theory of Möbius functions", *Z. Wahrscheinlichkeitstheorie und Verw. Gebiete* **2** (1964), 340–368.

[Roth 2009] M. Roth, "Counting covers of an elliptic curve", 2009, http://www.mast.queensu.ca/~mikeroth/notes/covers.pdf.

[Saloff-Coste 2001] L. Saloff-Coste, "Probability on groups: random walks and invariant diffusions", *Notices Amer. Math. Soc.* **48**:9 (2001), 968–977.

[Samuel 1980] S. Samuel, "$U(N)$ integrals, $1/N$, and the De Wit–'t Hooft anomalies", *J. Math. Phys.* **21**:12 (1980), 2695–2703.

[Shlyakhtenko 2005] D. Shlyakhtenko, "Notes on free probability theory", preprint, 2005. arXiv 0504063

[Soshnikov 1999] A. Soshnikov, "Universality at the edge of the spectrum in Wigner random matrices", *Comm. Math. Phys.* **207**:3 (1999), 697–733.

[Speed 1983] T. P. Speed, "Cumulants and partition lattices", *Austral. J. Statist.* **25**:2 (1983), 378–388.

[Stanley 1999] R. P. Stanley, *Enumerative combinatorics*, vol. 2, Cambridge Studies in Advanced Mathematics **62**, Cambridge University Press, Cambridge, 1999.

[Stanley 2007] R. P. Stanley, "Increasing and decreasing subsequences and their variants", pp. 545–579 in *International Congress of Mathematicians, I*, edited by J. L. V. Marta Sanz-Solé, Javier Soria and J. Verdera, Eur. Math. Soc., Zürich, 2007.

[Stanley 2013] R. P. Stanley, "Catalan addendum", 2013, http://www-math.mit.edu/~rstan/ec/catadd.pdf.

[Tao 2010] T. Tao, "254A, notes 5: free probability", 2010, http://terrytao.wordpress.com/2010/02/10/245a-notes-5-free-probability.

[Tao and Vu 2011] T. Tao and V. Vu, "Random matrices: universality of local eigenvalue statistics", *Acta Math.* **206**:1 (2011), 127–204.

[Voiculescu 1986] D. Voiculescu, "Addition of certain noncommuting random variables", *J. Funct. Anal.* **66**:3 (1986), 323–346.

[Voiculescu et al. 1992] D. V. Voiculescu, K. J. Dykema, and A. Nica, *Free random variables*, CRM Monograph Series 1, American Mathematical Society, Providence, RI, 1992.

[Wigner 1958] E. P. Wigner, "On the distribution of the roots of certain symmetric matrices", *Ann. of Math.* (2) **67** (1958), 325–327.

[Zvonkin 1997] A. Zvonkin, "Matrix integrals and map enumeration: an accessible introduction", *Math. Comput. Modelling* **26**:8-10 (1997), 281–304.

jnovak@math.mit.edu   *Department of Mathematics, Massachusetts Institute of Technology, 77 Massachusetts Avenue, Cambridge, 02139-4307, United States*

Random Matrices
MSRI Publications
Volume **65**, 2014

Whittaker functions and related stochastic processes

NEIL O'CONNELL

We review some recent results on connections between Brownian motion, Whittaker functions, random matrices and representation theory.

1. Harish-Chandra formula, Duistermaat–Heckman measure and Gelfand–Tsetlin patterns

Define $J_\lambda(x) = h(\lambda)^{-1} \det(e^{\lambda_i x_j})$, where $h(\lambda) = \prod_{i<j}(\lambda_i - \lambda_j)$. For each x, $J_\lambda(x)$ is an analytic function of λ; in particular,

$$J_0(x) = \left(\prod_{j=1}^{n-1} j! \right)^{-1} h(x).$$

The functions $J_\lambda(x)$ play a central role in random matrix theory. For example, if Λ and X are Hermitian matrices with eigenvalues given by λ and x, respectively, then

$$\int_{U(n)} e^{\operatorname{tr} \Lambda U X U^*} dU = \frac{J_\lambda(x)}{J_0(x)}, \tag{1}$$

where the integral is with respect to normalised Haar measure on the unitary group. This is known as the Harish-Chandra, or Itzykson–Zuber, formula.

Let $\beta = (\beta_t, \ t \geq 0)$ be a standard Brownian motion in $\mathbb{R}^n$ with drift λ. Denote by $\mathbb{P}_x$ the law of β started at x and by $\mathbb{E}_x$ the corresponding expectation. Set

$$\Omega = \{x \in \mathbb{R}^n : x_1 > x_2 > \cdots > x_n\}, \quad T = \inf\{t > 0 : \beta_t \notin \Omega\}.$$

For $\lambda, x \in \mathbb{R}^n$, write $\lambda(x) = \sum_i \lambda_i x_i$.

Proposition 1. *For* $x, \lambda \in \Omega$, $J_\lambda(x) = h(\lambda)^{-1} e^{\lambda(x)} \mathbb{P}_x(T = \infty)$.

Proof. This is well known; see for example [Biane et al. 2005]. The function $u(x) = \mathbb{P}_x(T = \infty)$, $x \in \Omega$, satisfies $\frac{1}{2}\Delta u + \lambda \cdot \nabla u = 0$, vanishes on the boundary of Ω and $\lim_{x \to \infty} u(x) = 1$. Here we write $x \to \infty$ to mean $x_i - x_{i+1} \to \infty$ for $i = 1, \ldots, n-1$. Hence $v(x) = e^{\lambda(x)} u(x)$ satisfies $\Delta v = \sum_i \lambda_i^2 v$, vanishes on

MSC2010: primary 60J65, 82B23; secondary 06B15.

the boundary of Ω and $\lim_{x\to\infty} e^{-\lambda(x)} v(x) = 1$. The function $\det(e^{\lambda_i x_j})$ also has these properties, so by uniqueness, $v(x) = \det(e^{\lambda_i x_j})$, as required. $\qquad\square$

The Harish-Chandra formula has the following interpretation. Pick U at random according to the normalised Haar measure on $U(n)$ and let $\mu^x(dy)$ denote the law of the diagonal of the random matrix UXU^*. Then the integral becomes

$$\int_{U(n)} e^{\operatorname{tr} \Lambda UXU^*} dU = \int_{\mathbb{R}^n} e^{\lambda(y)} \mu^x(dy).$$

Setting $m^x(dy) = J_0(x)\mu^x(dy)$, we obtain

$$\int_{\mathbb{R}^n} e^{\lambda(y)} m^x(dy) = J_\lambda(x).$$

The measure m^x is known as the Duistermaat–Heckman measure associated with the point $x \in \Omega$. It has the following properties, which are well-known. The symmetric group S_n acts naturally on $\mathbb{R}^n$ by permuting coordinates. The support of the measure m^x is the convex hull of the set of images of x under the action of S_n. It has a piecewise polynomial density. This comes from the fact, which we will now explain, that the Duistermaat–Heckman measure is the push-forward via an affine map of the Lebesgue measure on a higher dimensional polytope known as the Gelfand–Tsetlin polytope.

Let $x \in \Omega$ and denote by $GT(x)$ the polytope of Gelfand–Tsetlin patterns with bottom row equal to x:

$$GT(x) = \{P_{k,j}, \ 1 \le j \le k \le n : P_{k,j+1} \le P_{k-1,j} \le P_{k,j}, \ 1 \le j < k \le n, \ P_{n,\cdot} = x\}.$$

Define the *type* of a pattern P to be the vector

$$\text{type } P = \left(P_{1,1}, \ P_{2,1} + P_{2,2} - P_{1,1}, \ \ldots, \ \sum_{j=1}^{n} P_{n,j} - \sum_{j=1}^{n-1} P_{n-1,j} \right). \tag{2}$$

Consider the map from $U(n)$ to $GT(x)$ defined by $U \mapsto P$ where: for each $1 \le k \le n$, $P_{k,\cdot}$ is the vector of eigenvalues of the k-th principal minor of UXU^*. It is well-known (see, for example, [Baryshnikov 2001] or [Alexeev and Brion 2004, Section 5.6] for a more general statement) that the push-forward of Haar measure under this map is the standard Euclidean measure on the polytope $GT(x)$. Moreover, the diagonal of the matrix UXU^* is equal to the type of the pattern P. From this we obtain another integral representation for the function J_λ as

$$J_\lambda(x) = \int_{GT(x)} e^{\lambda \cdot \text{type } P} \, dP. \tag{3}$$

2. Whittaker functions

Set $H = \Delta - 2\sum_{i=1}^{n-1} e^{-\alpha_i(x)}$, where $\alpha_i = e_i - e_{i+1}$, $i = 1, \ldots, n-1$. Write $H = H^{(n)}$ for the moment; we will drop the superscript again later, whenever it is unnecessary. For convenience we define $H^{(1)} = d^2/dx^2$ and $\psi_\lambda^{(1)}(x) = e^{\lambda x}$. Following [Gerasimov et al. 2006], for $n \geq 2$ and $\theta \in \mathbb{C}$, define a kernel on $\mathbb{R}^n \times \mathbb{R}^{n-1}$ by

$$Q_\theta^{(n)}(x, y) = \exp\left(\theta\left(\sum_{i=1}^n x_i - \sum_{i=1}^{n-1} y_i\right) - \sum_{i=1}^{n-1}(e^{y_i - x_i} + e^{x_{i+1} - y_i})\right).$$

Denote the corresponding integral operator by $\mathcal{Q}_\theta^{(n)}$, defined on a suitable class of functions by

$$\mathcal{Q}_\theta^{(n)} f(x) = \int_{\mathbb{R}^{n-1}} Q_\theta^{(n)}(x, y) f(y)\, dy.$$

The Whittaker functions $\psi_\lambda^{(n)}$, $\lambda \in \mathbb{C}^n$ are defined recursively by

$$\psi_{\lambda_1, \ldots, \lambda_n}^{(n)} = \mathcal{Q}_{\lambda_n}^{(n)} \psi_{\lambda_1, \ldots, \lambda_{n-1}}^{(n-1)}. \tag{4}$$

As observed in [Gerasimov et al. 2006], the following intertwining relation holds:

$$(H^{(n)} - \theta^2) \circ \mathcal{Q}_\theta^{(n)} = \mathcal{Q}_\theta^{(n)} \circ H^{(n-1)}. \tag{5}$$

This follows from the identity $(H_x^{(n)} - \theta^2) Q_\theta^{(n)}(x, y) = H_y^{(n-1)} Q_\theta^{(n)}(x, y)$, which is readily verified. Combining (4) with the intertwining relation (5) yields the eigenvalue equation:

$$H^{(n)} \psi_\lambda^{(n)} = \left(\sum_{i=1}^n \lambda_i^2\right) \psi_\lambda^{(n)}. \tag{6}$$

Let us now drop the superscripts and write $H = H^{(n)}$, $\psi_\lambda = \psi_\lambda^{(n)}$. Iterating (4) gives the following integral formula, due to Givental [1997] (see also [Joe and Kim 2003; Gerasimov et al. 2006]):

$$\psi_\lambda(x) = \int_{\Gamma(x)} e^{\mathscr{F}_\lambda(T)} \prod_{k=1}^{n-1} \prod_{i=1}^k dT_{k,i}, \tag{7}$$

where $\Gamma(x)$ denotes the set of real triangular arrays $(T_{k,i},\ 1 \leq i \leq k \leq n)$ with $T_{n,i} = x_i$, $1 \leq i \leq n$, and

$$\mathscr{F}_\lambda(T) = \sum_{k=1}^n \lambda_k\left(\sum_{i=1}^k T_{k,i} - \sum_{i=1}^{k-1} T_{k-1,i}\right) - \sum_{k=1}^{n-1} \sum_{i=1}^k (e^{T_{k,i} - T_{k+1,i}} + e^{T_{k+1,i+1} - T_{k,i}}).$$

Now, it is shown in [Baudoin and O'Connell 2011] that, for each $\lambda \in \Omega$, the equation $Hf = \sum_i \lambda_i^2 f$ has a unique solution $f = f_\lambda$ such that $e^{-\lambda(x)} f_\lambda(x)$ is bounded and $\lim_{x \to +\infty} e^{-\lambda(x)} f_\lambda(x) = 1$, where we write $x \to +\infty$ to mean $\alpha_i(x) = x_i - x_{i+1} \to +\infty$ for each i. Moreover, by Feynman–Kac,

$$f_\lambda(x) = e^{\lambda(x)} \mathbb{E}_x \exp\left(-\sum_{i=1}^{n-1} \int_0^\infty e^{-\alpha_i(\beta_s)}\, ds\right), \tag{8}$$

where β_s is a Brownian motion in $\mathbb{R}^n$ with drift λ as in the previous section. The relation between the functions f_λ and the Whittaker functions ψ_λ is thus determined by the following proposition.

Proposition 2. *For $\lambda \in \Omega$,*

$$\lim_{x \to +\infty} e^{-\lambda(x)} \psi_\lambda(x) = \prod_{i<j} \Gamma(\lambda_i - \lambda_j). \tag{9}$$

Proof. We prove this by induction on n using the recursion (4). Write $\psi_\lambda = \psi_\lambda^{(n)}$ as before, setting $\psi_\lambda^{(1)}(x) = e^{\lambda x}$. Then $e^{-\lambda(x)} \psi_\lambda^{(1)}(x) = 1$ and, for $n \geq 2$,

$$e^{-\lambda(x)} \psi_\lambda^{(n)}(x) = \int_{\mathbb{R}^{n-1}} \exp\left(-\sum_{i=1}^n \lambda_i x_i + \lambda_n\left(\sum_{i=1}^n x_i - \sum_{i=1}^{n-1} y_i\right) - \sum_{i=1}^{n-1}\left(e^{y_i - x_i} + e^{x_{i+1} - y_i}\right)\right)$$

$$\times \psi_{\lambda_1,\ldots,\lambda_{n-1}}^{(n-1)}(y_1, \ldots, y_{n-1})\, dy_1 \ldots dy_{n-1}$$

$$= \int_{\mathbb{R}^{n-1}} e^{\sum_{i=1}^{n-1}(\lambda_i - \lambda_n) y_i} \exp\left(-\sum_{i=1}^{n-1} e^{y_i} - \sum_{i=1}^{n-1} e^{x_{i+1} - x_i - y_i}\right)$$

$$\times e^{-\sum_{i=1}^{n-1} \lambda_i (x_i + y_i)} \psi_{\lambda_1,\ldots,\lambda_{n-1}}^{(n-1)}(x_1 + y_1, \ldots, x_{n-1} + y_{n-1})\, dy_1 \ldots dy_{n-1}.$$

By induction, we immediately conclude that, for each n, if $x, \lambda \in \Omega$ then $e^{-\lambda(x)} \psi_\lambda^{(n)}(x) \leq \prod_{i<j} \Gamma(\lambda_i - \lambda_j)$. Here we are using

$$\int_{\mathbb{R}^{n-1}} e^{\sum_{i=1}^{n-1}(\lambda_i - \lambda_n) y_i} \exp\left(-\sum_{i=1}^{n-1} e^{y_i} - \sum_{i=1}^{n-1} e^{x_{i+1} - x_i - y_i}\right) dy_1 \ldots dy_{n-1}$$

$$\leq \int_{\mathbb{R}^{n-1}} e^{\sum_{i=1}^{n-1}(\lambda_i - \lambda_n) y_i} \exp\left(-\sum_{i=1}^{n-1} e^{y_i}\right) dy_1 \ldots dy_{n-1} = \prod_{i=1}^{n-1} \Gamma(\lambda_i - \lambda_n).$$

It follows, again by induction and using the dominated convergence theorem, that (9) holds for $\lambda \in \Omega$. $\qquad\square$

Corollary 3. *For $\lambda \in \Omega$,*

$$\psi_\lambda(x) = \prod_{i<j} \Gamma(\lambda_i - \lambda_j) e^{\lambda(x)} \mathbb{E}_x \exp\left(-\sum_{i=1}^{n-1} \int_0^\infty e^{-\alpha_i(\beta_s)}\, ds\right). \tag{10}$$

Corollary 4. *For* $x, \lambda \in \Omega$,

$$J_\lambda(x) = \lim_{\beta \to \infty} \beta^{-n(n-1)/2} \psi_{\lambda/\beta}(\beta x).$$

Proof. By Proposition 1, the statement is equivalent to

$$\lim_{\beta \to \infty} \beta^{-n(n-1)/2} \psi_{\lambda/\beta}(\beta x) = h(\lambda)^{-1} e^{\lambda(x)} \mathbb{P}_x(T = \infty).$$

This follows directly from (10) by Brownian rescaling.   □

As shown in [Baudoin and O'Connell 2011], the function ψ_λ, which can be defined by (10), is a class-one Whittaker function, as defined by Jacquet [2004] and Hashizume [1982]. In the notation of [Baudoin and O'Connell 2011] we are taking $\Pi = \{\alpha_i/2, \ i = 1, \ldots, n-1\}$, $m(2\alpha) = 0$, $|\eta_\alpha|^2 = 1$ and $\psi_\nu(x) = 2^q k_\nu(x)$ where $q = n(n-1)/2$. In [Gerasimov et al. 2008], the relationship between Givental integral formula and a recursive integral formula due to Stade [1990] based on Jacquet's definition (see also [Ishii and Stade 2007]) is described.

Givental's integral formula (7) has a very similar structure to the formula (3). Indeed, if we define the type of an array $(T_{k,i}, \ 1 \le i \le k \le n)$ to be the vector

$$\text{type } T = \left(T_{1,1}, T_{2,1} + T_{2,2} - T_{1,1}, \ldots, \sum_{j=1}^{n} T_{n,j} - \sum_{j=1}^{n-1} T_{n-1,j} \right),$$

and a measure

$$g(dT) = \prod_{k=1}^{n-1} \prod_{i=1}^{k} e^{-e^{T_{k,i} - T_{k+1,i}}} e^{-e^{T_{k+1,i+1} - T_{k,i}}} \, dT_{k,i} = e^{\mathscr{F}_0(T)} \prod_{k=1}^{n-1} \prod_{i=1}^{k} dT_{k,i},$$

then

$$\psi_\lambda(x) = \int_{\Gamma(x)} e^{\lambda \cdot \text{type } T} g(dT).$$

On the other hand, if we replace the functions $e^{-e^{x-y}}$ in the reference measure g by the indicator functions $1_{x < y}$ to get a new reference measure

$$g_0(dT) = \prod_{k=1}^{n-1} \prod_{i=1}^{k} 1_{T_{k,i} < T_{k+1,i}} 1_{T_{k+1,i+1} < T_{k,i}},$$

then (3) can be written as

$$J_\lambda(x) = \int_{\Gamma(x)} e^{\lambda \cdot \text{type } T} g_0(dT).$$

We note the following. If $\lambda \in \iota\mathbb{R}^n$ then $\overline{\psi_\lambda(x)} = \psi_{-\lambda}(x)$; if $\lambda \in \iota\mathbb{R}^n$ and $\nu \in \mathbb{R}^n$, then $|\psi_{\lambda+\nu}(x)| \le \psi_\nu(x)$. For each $x \in \mathbb{R}^n$, $\psi_\lambda(x)$ is an entire, symmetric function of $\lambda \in \mathbb{C}^n$ [Gerasimov et al. 2008; Hashizume 1982; Kharchev and Lebedev

1999]. There is a Plancherel theorem [Wallach 1992; Arnold and Novikov 1994; Gerasimov et al. 2008; Kharchev and Lebedev 1999] which states that the integral transform

$$\hat{f}(\lambda) = \int_{\mathbb{R}^n} f(x)\psi_\lambda(x)\,dx \tag{11}$$

is an isometry from $L_2(\mathbb{R}^n, dx)$ onto $L_2^{\mathrm{sym}}(\iota\mathbb{R}^n, s_n(\lambda)\,d\lambda)$, where L_2^{sym} is the space of L_2 functions which are symmetric in their variables, $\iota = \sqrt{-1}$ and $s_n(\lambda)\,d\lambda$ is the *Sklyanin measure* defined by

$$s_n(\lambda) = \frac{1}{(2\pi\iota)^n n!} \prod_{j\neq k} \Gamma(\lambda_j - \lambda_k)^{-1}. \tag{12}$$

For x, $\mu \in \mathbb{R}^n$, denote by σ_μ^x the probability measure on the set of real triangular arrays $(T_{k,i})_{1\leq i\leq k\leq n}$ defined by

$$\int f\,d\sigma_\mu^x = \psi_\mu(x)^{-1} \int_{\Gamma(x)} f(T)e^{\mathscr{F}_\mu(T)} \prod_{k=1}^{n-1}\prod_{i=1}^{k} dT_{k,i}.$$

Define a probability measure γ_μ^x by

$$\int_{\mathbb{R}^n} e^{\lambda\cdot y}\gamma_\mu^x(dy) = \frac{\psi_{\mu+\lambda}(x)}{\psi_\mu(x)}, \quad \lambda \in \mathbb{C}^n.$$

The probability measure $\gamma^x = \gamma_0^x$ is the analogue of the (normalised) Duistermaat–Heckman measure in this setting. The integral operator K with kernel

$$K(x, dy) = \psi_0(x)\gamma^x(dy)$$

satisfies the intertwining relation $HK = K\Delta$. We can write

$$K(x, dy) = k(x, y)\rho_x(dy),$$

where k is a smooth kernel from $\mathbb{R}^n$ to $\mathbb{R}_x^n = \{y \in \mathbb{R}^n : \sum_i y_i = \sum_i x_i\}$ and ρ_x denotes the Euclidean measure on $\mathbb{R}_x^n$. For $n = 2$,

$$k(x, y) = \exp(-e^{x_2-y_1} - e^{y_1-x_1})$$

and, for $n = 3$,

$$k(x, y) = \psi_0^{(2)}(a, b) = 2K_0(2e^{(b-a)/2})$$

where

$$e^{-a} = e^{x_3-y_1-y_2} + e^{-x_1}, \quad e^b = e^{y_1} + e^{y_2} + e^{y_1+y_2-x_2} + e^{x_2},$$

and K_0 denotes the Macdonald function with index 0.

3. Interpretation of γ^x in terms of Brownian motion

A reduced decomposition of an element $w \in S_n$ is a minimal expression of w as a product of adjacent transpositions, that is, $w = s_{i_1} \ldots s_{i_r}$, where s_i denotes the transposition $(i, i+1)$. We will also refer to the word $\boldsymbol{i} = i_1 i_2 \ldots i_r$ as a reduced decomposition. By definition, any reduced decomposition has the same length $l(w)$, defined to be the length of w. There is a unique longest element in S_n, namely the permutation

$$w_0 = \begin{pmatrix} 1 & 2 & \cdots & n \\ n & n-1 & \cdots & 1 \end{pmatrix}.$$

Its length is $n(n-1)/2$, as can be seen by taking the reduced decomposition

$$\boldsymbol{i} = 1\ 21\ 321\ \ldots\ n\,n{-}1\ldots 21.$$

The symmetric group acts on $\mathbb{R}^n$ by permutation of coordinates, and as such is an example of a finite reflection group. It is generated by the hyperplane reflections $s_i = s_{\alpha_i}$, $i = 1, \ldots, n-1$, defined for $x \in \mathbb{R}^n$ by

$$s_i x = x - \alpha_i(x)\alpha_i,$$

where $\alpha_i = e_i - e_{i+1}$. Note that s_i corresponds to the adjacent transposition $(i, i+1)$.

For a continuous path $\eta : (0, \infty) \to \mathbb{R}^n$, define $T_i = T_{\alpha_i}$ by

$$T_i \eta(t) = \eta(t) + \left(\log \int_0^t e^{-\alpha_i(\eta(s))}\, ds \right) \alpha_i, \quad t > 0.$$

Let $w = s_{i_1} \cdots s_{i_r}$ be a reduced decomposition. Then

$$T_w := T_{i_r} \cdots T_{i_1}$$

depends only on w, not on the chosen decomposition [Biane et al. 2005].

We now introduce a probability measure $\mathbb{P}$ under which η is a Brownian motion in $\mathbb{R}^n$ with a drift μ and $\eta(0) = 0$. In this setting, a very special role is played by the transform $T^{(n)} = T_{w_0}$. In the following we use the fact that this is well-defined for each n. Write $\eta = (\eta^1, \ldots, \eta^n)$. For each $k \leq n$, set

$$(T_{k,1}, \ldots, T_{k,k}) = T^{(k)}(\eta^1, \ldots, \eta^k).$$

The evolution of the triangular array $T_{k,j}$, $1 \leq j \leq k \leq n$, is given recursively as follows: $dT_{1,1} = d\eta^1$ and, for $k \geq 2$,

$$dT_{k,1} = dT_{k-1,1} + e^{T_{k,2}-T_{k-1,1}} \, dt$$

$$dT_{k,2} = dT_{k-1,2} + \left(e^{T_{k,3}-T_{k-1,2}} - e^{T_{k,2}-T_{k-1,1}} \right) dt$$

$$\vdots$$

$$dT_{k,k-1} = dT_{k-1,k-1} + \left(e^{T_{k,k}-T_{k-1,k-1}} - e^{T_{k,k-1}-T_{k-1,k-2}} \right) dt$$

$$dT_{k,k} = d\eta^k - e^{T_{k,k}-T_{k-1,k-1}} \, dt. \tag{13}$$

The process, which is clearly Markov, contains a number of projections which are also Markov. For example, setting $\xi_k = T_{k,k}$, we have, for $k \leq n$,

$$d\xi_k = d\eta^k - e^{\xi_k-\xi_{k-1}} \, dt.$$

This defines a simple interacting particle system on the real line, which has very nice properties. For example, in the coordinates $\sum_i \xi_i$ and $\xi_i - \xi_{i+1}$ $1 \leq i \leq n-1$, it has a product form invariant measure, that is, a product measure which is invariant.

A remarkable fact is that each row in the pattern $T_{k,j}$ is a Markov process with respect to its own filtration. This gives an interpretation of the measures γ_μ^x and σ_μ^x defined in the previous section.

Theorem 5 [O'Connell 2012]. $T_{w_0}\eta(t)$, $t > 0$, *is a diffusion process in* $\mathbb{R}^n$ *with infinitesimal generator*

$$\mathcal{L}_\mu = \frac{1}{2}\psi_\mu^{-1}\left(H - \sum_{i=1}^n \mu_i^2 \right)\psi_\mu = \tfrac{1}{2}\Delta + \nabla \log \psi_\mu \cdot \nabla.$$

For $t > 0$, *the conditional law of* $\{T_{k,j}(t), \ 1 \leq j \leq k \leq n\}$, *given*

$$\{T_{w_0}\eta(s), \ s \leq t; \ T_{w_0}\eta(t) = x\},$$

is σ_μ^x, *and the conditional law of* $\eta(t)$, *given the same, is* γ_μ^x. *The law of* $T_{w_0}\eta(t)$ *is given by*

$$\nu_t^\mu(dx) = e^{-\sum_i \mu_i^2 t/2}\psi_\mu(x)\theta_t(x)\,dx,$$

where

$$\theta_t(x) = \int_{\iota\mathbb{R}^n} \psi_{-\lambda}(x)e^{\sum_i \lambda_i^2 t/2}s_n(\lambda)\,d\lambda. \tag{14}$$

In the case $n = 2$, this is equivalent to a theorem of Matsumoto and Yor [1999].

Write $\mathcal{L} = \mathcal{L}_0$ and $\nu_t = \nu_t^0$. The diffusion with generator $\mathcal{L}$ is the analogue of Dyson's Brownian motion in this setting and the measures ν_t and θ_t (the latter requires normalisation) are the analogues of the Gaussian unitary and Gaussian orthogonal ensembles, respectively. The diffusion with generator $\mathscr{L}_\mu$ was introduced in [Baudoin and O'Connell 2011]. When $\mu \in \bar{\Omega}$, it can be interpreted as a Brownian motion in $\mathbb{R}^n$ killed according to the potential $\sum_i e^{x_{i+1} - x_i}$ and then conditioned to survive forever [Katori 2011; 2012]. The path-transformation T_{w_0} is closely related to the geometric (lifting of the) RSK correspondence introduced by A. N. Kirillov [2001] and studied further by Noumi and Yamada [2004]. A discrete-time version of the above theorem, which works directly in the setting of the geometric RSK correspondence, is given in [Corwin et al. 2014]. In the discrete-time setting the Whittaker functions continue to play a central role. See also [Borodin and Corwin 2014; Borodin et al. 2013; Chhaibi 2012; Gorsky et al. 2012; O'Connell et al. 2014; O'Connell and Warren 2011] for further related developments.

4. Application to random polymers

The following model was introduced in [O'Connell and Yor 2001]. The environment is given by a sequence $B_1, B_2, \ldots$ independent standard 1-dimensional Brownian motions. For up/right paths $\phi \equiv \{0 < t_1 < \cdots < t_{N-1} < t\}$ (as shown in Figure 1), define

$$E(\phi) = B_1(t_1) + B_2(t_2) - B_2(t_1) + \cdots + B_N(t) - B_N(t_{N-1}),$$

$$P(d\phi) = Z_t^n(\beta)^{-1} e^{\beta E(\phi)} d\phi, \quad Z_t^n(\beta) = \int e^{\beta E(\phi)} d\phi.$$

Set $X_1^n(t) = \log Z_t^n$ and, for $k = 2, \ldots, n$,

$$X_1^n(t) + \cdots + X_k^n(t) = \log \int e^{E(\phi_1) + \cdots + E(\phi_k)} d\phi_1 \ldots d\phi_k,$$

where the integral is over nonintersecting paths $\phi_1, \ldots, \phi_k$ from $(0, 1), \ldots, (0, k)$ to $(t, n - k + 1), \ldots, (t, n)$.

Let $\eta = (B_n, \ldots, B_1)$. Then $X = T_{w_0} \eta$ and the following holds.

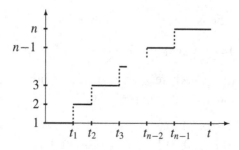

Figure 1. An up/right path $\phi \equiv \{0 < t_1 < \ldots < t_{n-1} < t\}$.

Theorem 6 [O'Connell 2012]. *The process $X(t), t > 0$ is a diffusion in $\mathbb{R}^n$ with infinitesimal generator $\mathcal{L}$. The distribution of $X(t)$ is given by v_t. For $s > 0$,*

$$Ee^{-sZ_t^n} = \int s^{-\sum \lambda_i} \prod_i \Gamma(\lambda_i)^n e^{\frac{1}{2}\sum_i \lambda_i^2 t} s_n(\lambda)\, d\lambda,$$

where the integral is along (upwards) vertical lines with $\mathfrak{R}\lambda_i > 0$ for all i.

The free energy for this model is given by [O'Connell and Yor 2001; Moriarty and O'Connell 2007]

$$\lim_{n\to\infty} \frac{1}{n} \log Z_n^n = \inf_{t>0}[t - \Psi(t)],$$

almost surely, where $\Psi(z) = \Gamma'(z)/\Gamma(z)$. The conjectured KPZ scaling behaviour for the fluctuations of $\log Z_n^n$ was (essentially) established by Seppäläinen and Valkó [2010]; more recently, Borodin, Corwin and Ferrari have proved the full KPZ universality conjecture for this model, namely that $\log Z_n^n$, suitably centred and rescaled, converges in law to the Tracy–Widom F_2 distribution of random matrix theory [Borodin and Corwin 2014; Borodin et al. 2014]. See also [Spohn 2014].

5. Reduced double Bruhat cells and their parametrisations

The Weyl group associated with $GL(n)$ is the symmetric group S_n. Each element $w \in S_n$ has a representative $\bar{w} \in GL(n)$ defined as follows. Denote the standard generators for $\mathfrak{gl}_n$ by h_i, e_i and f_i. For example, for $n = 3$,

$$h_1 = \begin{pmatrix} 1 & 0 & 0 \\ 0 & 0 & 0 \\ 0 & 0 & 0 \end{pmatrix}, \quad h_2 = \begin{pmatrix} 0 & 0 & 0 \\ 0 & 1 & 0 \\ 0 & 0 & 0 \end{pmatrix}, \quad h_3 = \begin{pmatrix} 0 & 0 & 0 \\ 0 & 0 & 0 \\ 0 & 0 & 1 \end{pmatrix},$$

$$e_1 = \begin{pmatrix} 0 & 1 & 0 \\ 0 & 0 & 0 \\ 0 & 0 & 0 \end{pmatrix}, \quad e_2 = \begin{pmatrix} 0 & 0 & 0 \\ 0 & 0 & 1 \\ 0 & 0 & 0 \end{pmatrix}, \quad f_1 = \begin{pmatrix} 0 & 0 & 0 \\ 1 & 0 & 0 \\ 0 & 0 & 0 \end{pmatrix}, \quad f_2 = \begin{pmatrix} 0 & 0 & 0 \\ 0 & 0 & 0 \\ 0 & 1 & 0 \end{pmatrix}.$$

For adjacent transpositions $s_i = (i, i+1)$, define

$$\bar{s}_i = \exp(-e_i)\exp(f_i)\exp(-e_i) = (I - e_i)(I + f_i)(I - e_i).$$

In other words, $\bar{s}_i = \varphi_i\begin{pmatrix} 0 & -1 \\ 1 & 0 \end{pmatrix}$, where φ_i is the natural embedding of SL(2) into $GL(n)$ given by h_i, e_i and f_i. For example, when $n = 3$,

$$\bar{s}_1 = \begin{pmatrix} 0 & -1 & 0 \\ 1 & 0 & 0 \\ 0 & 0 & 1 \end{pmatrix}, \quad \bar{s}_2 = \begin{pmatrix} 1 & 0 & 0 \\ 0 & 0 & -1 \\ 0 & 1 & 0 \end{pmatrix}.$$

Now let $w = s_{i_1} \ldots s_{i_r}$ be a reduced decomposition and define $\bar{w} = \bar{s}_{i_1} \ldots \bar{s}_{i_r}$. Note that $\overline{uv} = \bar{u}\bar{v}$ whenever $l(uv) = l(u) + l(v)$. For $n = 2$, $w_0 = s_1$ and

$$\bar{w}_0 = \bar{s}_1 = \begin{pmatrix} 0 & -1 \\ 1 & 0 \end{pmatrix}.$$

For $n = 3$, $w_0 = s_1 s_2 s_1 = s_2 s_1 s_2$ is represented by

$$\bar{w}_0 = \bar{s}_1 \bar{s}_2 \bar{s}_1 = \bar{s}_2 \bar{s}_1 \bar{s}_2 = \begin{pmatrix} 0 & 0 & 1 \\ 0 & -1 & 0 \\ 1 & 0 & 0 \end{pmatrix}.$$

Denote the upper (respectively lower) triangular matrices in $GL(n)$ by B and B_-, and the upper (respectively lower) uni-triangular matrices in $GL(n)$ by N and N_-. The group $GL(n)$ has two Bruhat decompositions

$$GL(n) = \bigcup_{u \in S_n} B\bar{u}B = \bigcup_{v \in S_n} B_-\bar{v}B_-.$$

The double Bruhat cells $G^{u,v}$ are defined, for $u, v \in S_n$, by

$$G^{u,v} = B\bar{u}B \cap B_-\bar{v}B_-.$$

The reduced double Bruhat cells $L^{u,v}$ are defined by

$$L^{u,v} = N\bar{u}N \cap B_-\bar{v}B_-.$$

We also define the opposite reduced double Bruhat cells $M^{u,v}$ by

$$M^{u,v} = B\bar{u}B \cap N_-\bar{v}N_-.$$

The reduced double Bruhat cell $L^{w,e}$ (where e denotes the identity in S_n) admits the following parametrisations, one for each reduced decomposition of w. See [Lusztig 1994; Berenstein and Zelevinsky 2001; Berenstein et al. 1996; Fomin and Zelevinsky 1999]. Set

$$Y_i(u) = \varphi_i \begin{pmatrix} u & 0 \\ 1 & u^{-1} \end{pmatrix}, \quad i = 1, \ldots, n-1.$$

Then, for any reduced decomposition $i = i_1 \ldots i_r$ of w, the map

$$(u_1, \ldots, u_r) \mapsto Y_{i_1}(u_1) \cdots Y_{i_r}(u_r)$$

defines a bijection between $\mathbb{C}^r_{\neq 0}$ and $L^{w,e}$. This bijection has the property that the totally positive part $L^{u,v}_{>0}$ of $L^{u,v}$ corresponds precisely to the subset $\mathbb{R}^r_{>0}$ of $\mathbb{C}^r_{\neq 0}$. There are explicit transition maps which relate the parameters $(u_1, \ldots, u_r)$ corresponding to different reduced decompositions of w.

In the case $n = 3$, the two representations of an element in $L^{w_0,e}$ corresponding to the words 121 and 212, denoting the corresponding parameters by (u_1, u_2, u_3) and (u_1', u_2', u_3'), respectively, are given by

$$
\begin{pmatrix} u_1 u_3 & 0 & 0 \\ u_3 + u_2/u_1 & u_2/u_1 u_3 & 0 \\ 1 & 1/u_3 & 1/u_2 \end{pmatrix} = \begin{pmatrix} u_2' & 0 & 0 \\ u_1' & u_1' u_3'/u_2' & 0 \\ 1 & u_3'/u_2' + 1/u_1' & 1/u_1' u_3' \end{pmatrix}.
$$

The transition maps are given by

$$
u_1' = u_3 + u_2/u_1, \quad u_2' = u_1 u_3, \quad u_3' = u_1 u_2/(u_2 + u_1 u_3). \tag{15}
$$

Their is a similar parametrisation for $M^{w,e}$, due to Lusztig [1994]. For $i = 1, \ldots, n-1$, set $X_i(v) = I + v f_i$. Take any reduced decomposition $i = i_1 \ldots i_r$ for w. Then the map

$$
(v_1, \ldots, v_r) \mapsto X_{i_1}(v_1) \cdots X_{i_r}(v_r)
$$

defines a bijection between $\mathbb{C}_{\neq 0}^r$ and $M^{w,e}$. This bijection also has the property that the totally positive part $M_{>0}^{u,v}$ of $M^{u,v}$ corresponds precisely to the subset $\mathbb{R}_{>0}^r$ of $\mathbb{C}_{\neq 0}^r$.

In the case $n = 3$, the two representations of an element in $M^{w_0,e}$ corresponding to the words 121 and 212, denoting the corresponding parameters by (v_1, v_2, v_3) and (v_1', v_2', v_3'), respectively, are given by

$$
\begin{pmatrix} 1 & 0 & 0 \\ v_1 + v_3 & 1 & 0 \\ v_2 v_3 & v_2 & 1 \end{pmatrix} = \begin{pmatrix} 1 & 0 & 0 \\ v_2' & 1 & 0 \\ v_1' v_2' & v_1' + v_3' & 1 \end{pmatrix},
$$

with transition maps

$$
v_1' = \frac{v_2 v_3}{v_1 + v_3}, \quad v_2' = v_1 + v_3, \quad v_3' = \frac{v_1 v_2}{v_1 + v_3}. \tag{16}
$$

We conclude this section with a simple lemma. Let $b \in G^{e,w}$ and write $b = an$ where $a = \operatorname{diag}(a_1, \ldots, a_n)$, say, and $n \in N$. Then, for any $w \in S_n$, $b\bar{w}$ has a Gauss (or LDU) decomposition $b\bar{w} = [b\bar{w}]_-[b\bar{w}]_0[b\bar{w}]_+$ and $n\bar{w}$ has a Gauss decomposition $n\bar{w} = [n\bar{w}]_-[n\bar{w}]_0[n\bar{w}]_+$ [Fomin and Zelevinsky 1999]. Moreover, $[n\bar{w}]_{-0} = [n\bar{w}]_-[n\bar{w}]_0 \in L^{w,e}$ and $[b\bar{w}]_- \in M^{w,e}$. Let $i = i_1 \ldots i_r$ be a reduced decomposition for w. Then we can write

$$
[n\bar{w}]_{-0} = Y_{i_1}(u_1) \ldots Y_{i_r}(u_r), \quad [b\bar{w}]_- = X_{i_1}(v_1) \ldots X_{i_r}(v_r).
$$

Define $Z_i(u) = \varphi_i \left(\begin{smallmatrix} u & 0 \\ 0 & u^{-1} \end{smallmatrix} \right)$. Set $a^0 = a$ and, for $1 \le k \le r$, $a^k = a^{k-1} Z_{i_k}(u_k)$. Write $a^k = \operatorname{diag}(a_1^k, \ldots, a_n^k)$.

Lemma 7. *The following relations holds:* $[b\bar{w}]_0 = a^r$ *and, for* $k = 1, \ldots, r$,
$v_k = u_k^{-1} a_{i_k+1}^{k-1} / a_{i_k}^{k-1}$.

Proof. Note that $a[n\bar{w}]_{-0} = [b\bar{w}]_{-0} = [b\bar{w}]_- [b\bar{w}]_0$, hence

$$a Y_{i_1}(u_1) \ldots Y_{i_r}(u_r) = X_{i_1}(v_1) \ldots X_{i_r}(v_r)[b\bar{w}]_0.$$

The result follows by repeated application of the identity $a Y_i(u) = X_i(v)a'$, where $a' = a Z_i(u)$ and $v = u^{-1} a_{i+1}/a_i$. $\qquad\square$

6. An evolution on upper triangular matrices

As shown in [Biane et al. 2005], the path-transformations $T_w \eta$ can also be represented in terms of an evolution on the upper triangular matrices in $GL(n, \mathbb{R})$. Let $w = s_{i_1} \cdots s_{i_r}$ be a reduced decomposition and $\eta : (0, \infty) \to \mathbb{R}^n$ a continuous path. Set $\eta_0 = \eta$ and, for $k \leq r$,

$$\eta_k = T_{i_k} \ldots T_{i_1} \eta \quad x_k(t) = \log \int_0^t e^{-\alpha_{i_k}(\eta_{k-1}(s))} ds. \tag{17}$$

Then $\eta_r = T_w \eta$ and, for each $k \leq r$, $\eta_k = \eta + \sum_{j=1}^k x_j \alpha_{i_j}$.

Write $\eta(t) = (\eta_t^1, \ldots, \eta_t^n)$. Define a path $b(t)$ taking values in B by

$$b_{ij}(t) = e^{\eta^i(t)} \int_{0 < s_{j-1} < s_{j-2} < \cdots < s_i < t} \exp\left(-\sum_{k=i}^{j-1} \alpha_k(\eta(s_k))\right) ds_i \cdots ds_{j-1}.$$

If η is smooth, the b satisfies the ordinary differential equation

$$db = \left(\sum_{i=1}^n h_i \, d\eta^i + \sum_{i=1}^{n-1} e_i \, dt\right) b, \quad b(0) = I.$$

If η is a Brownian path (as in the next section) then b satisfies the equation interpreted as a Stratonovich SDE.

When $n = 2$,

$$db = \begin{pmatrix} d\eta^1 & dt \\ 0 & d\eta^2 \end{pmatrix} b, \quad b(t) = \begin{pmatrix} e^{\eta_t^1} & \int_0^t e^{\eta_s^2 - \eta_s^1 + \eta_t^1} \, ds \\ 0 & e^{\eta_t^2} \end{pmatrix}.$$

When $n = 3$,

$$db = \begin{pmatrix} d\eta^1 & dt & 0 \\ 0 & d\eta^2 & dt \\ 0 & 0 & d\eta^3 \end{pmatrix} b,$$

and the solution is given by

$$b(t) = \begin{pmatrix} e^{\eta_t^1} & \int_0^t e^{\eta_s^2 - \eta_s^1 + \eta_t^1}\, ds & \iint_{0<r<s<t} e^{\eta_r^3 - \eta_r^2 + \eta_s^2 - \eta_s^1 + \eta_t^1}\, dr\, ds \\ 0 & e^{\eta_t^2} & \int_0^t e^{\eta_s^3 - \eta_s^2 + \eta_t^2}\, ds \\ 0 & 0 & e^{\eta_t^3} \end{pmatrix}.$$

Write $b = an$, where $a = \mathrm{diag}(e^{\eta^1}, \ldots, e^{\eta^n})$ and $n \in N$. Set $u_k = e^{x_k}$ and $v_k = e^{-x_k - \alpha_k(\eta_{k-1})}$.

Theorem 8 [Biane et al. 2005; 2009]. *For each $t > 0$, $b(t)\bar{w}$ has a Gauss decomposition $b\bar{w} = [b\bar{w}]_-[b\bar{w}]_0[b\bar{w}]_+$, with $[b\bar{w}]_0 = \exp(T_w\eta(t))$. Moreover, $[n\bar{w}]_{-0} = Y_{i_1}(u_1) \cdots Y_{i_r}(u_r) \in L_{>0}^{w,e}$.*

By Lemma 7, we also have $[b\bar{w}]_- = X_{i_1}(v_1) \cdots X_{i_r}(v_r) \in M_{>0}^{w,e}$.

6.1. *The case $n = 2$.* From the definitions: $\alpha_1 = e_1 - e_2$, $w_0 = s_1 = s_{e_1 - e_2}$,

$$u := u_1 = e^{x_1} = \int_0^t e^{-\eta_s^1 + \eta_s^2}\, ds, \qquad v := v_1 = e^{-\eta^1 + \eta^2} u^{-1},$$

$$e^{T_{w_0}\eta} = (e^{\eta^1} u, e^{\eta^2} u^{-1}) = \left(\int_0^t e^{\eta_s^2 + \eta_t^1 - \eta_s^1}\, ds, \int_0^t e^{-(\eta_s^1 + \eta_t^2 - \eta_s^2)} \right),$$

$$b = \begin{pmatrix} e^{\eta^1} & \int_0^t e^{\eta_s^2 - \eta_s^1 + \eta_t^1}\, ds \\ 0 & e^{\eta^2} \end{pmatrix} = \begin{pmatrix} e^{\eta^1} & e^{\eta^1} u \\ 0 & e^{\eta^2} \end{pmatrix} = \begin{pmatrix} e^{\eta^1} & 0 \\ 0 & e^{\eta^2} \end{pmatrix} \begin{pmatrix} 1 & u \\ 0 & 1 \end{pmatrix} = an.$$

Taking $\bar{w}_0 = \begin{pmatrix} 0 & -1 \\ 1 & 0 \end{pmatrix}$, we see that

$$b\bar{w}_0 = \begin{pmatrix} e^{\eta^1} u & -e^{\eta^1} \\ e^{\eta^2} & 0 \end{pmatrix} = \begin{pmatrix} 1 & 0 \\ v & 1 \end{pmatrix} \begin{pmatrix} e^{\eta^1} u & 0 \\ 0 & e^{\eta^2} u^{-1} \end{pmatrix} \begin{pmatrix} 1 & -u^{-1} \\ 0 & 1 \end{pmatrix}.$$

and

$$n\bar{w}_0 = \begin{pmatrix} 1 & u \\ 0 & 1 \end{pmatrix} \begin{pmatrix} 0 & -1 \\ 1 & 0 \end{pmatrix} = \begin{pmatrix} u & -1 \\ 1 & 0 \end{pmatrix} = \begin{pmatrix} u & 0 \\ 1 & u^{-1} \end{pmatrix} \begin{pmatrix} 1 & -u^{-1} \\ 0 & 1 \end{pmatrix}$$

Hence

$$[b\bar{w}_0]_0 = e^{T_{w_0}\eta}, \qquad [b\bar{w}_0]_- = \begin{pmatrix} 1 & 0 \\ v & 1 \end{pmatrix} = X_1(v), \qquad [n\bar{w}_0]_{-0} = \begin{pmatrix} u & 0 \\ 1 & u^{-1} \end{pmatrix} = Y_1(u),$$

as claimed.

6.2. *The case $n = 3$.* From the definitions:

$$\alpha_1 = e_1 - e_2, \qquad \alpha_2 = e_2 - e_3, \qquad w_0 = s_1 s_2 s_1 = s_2 s_1 s_2.$$

For the reduced decomposition $w_0 = s_1 s_2 s_1$, we have

$$u_1 = e^{x_1} = \int_0^t e^{-\eta^1(s) + \eta^2(s)}\, ds, \qquad e^{\eta_1} = (e^{\eta^1} u_1, e^{\eta^2}/u_1, e^{\eta^3}),$$

$$u_2 = e^{x_2} = \int_0^t e^{-\eta_1^2(s)+\eta_1^3(s)} \, ds, \quad e^{\eta_2} = (e^{\eta^1} u_1, \, e^{\eta^2} u_2/u_1, \, e^{\eta^3}/u_2),$$

$$u_3 = e^{x_3} = \int_0^t e^{-\eta_2^1(s)+\eta_2^2(s)} \, ds, \quad e^{\eta_3} = e^{T_{w_0}\eta} = (e^{\eta^1} u_1 u_3, \, e^{\eta^2} u_2/u_1 u_3, \, e^{\eta^3}/u_2),$$

$$v_1 = e^{-\eta^1+\eta^2}/u_1, \quad v_2 = e^{-\eta^2+\eta^3}\frac{u_1}{u_2}, \quad v_3 = e^{-\eta^1+\eta^2}\frac{u_2}{u_1^2 u_3},$$

$$b = \begin{pmatrix} e^{\eta^1} & \int_0^t e^{\eta_s^2-\eta_s^1+\eta_t^1} \, ds & \iint_{0<r<s<t} e^{\eta_r^3-\eta_r^2+\eta_s^2-\eta_s^1+\eta_t^1} \, dr \, ds \\ 0 & e^{\eta^2} & \int_0^t e^{\eta_s^3-\eta_s^2+\eta_t^2} \, ds \\ 0 & 0 & e^{\eta^3} \end{pmatrix}$$

$$= \begin{pmatrix} e^{\eta^1} & 0 & 0 \\ 0 & e^{\eta^2} & 0 \\ 0 & 0 & e^{\eta^3} \end{pmatrix} \begin{pmatrix} 1 & u_1 & u_1 u_3 \\ 0 & 1 & u_3 + u_2/u_1 \\ 0 & 0 & 1 \end{pmatrix} = an.$$

The identity

$$\int_0^t e^{\eta_s^3-\eta_s^2+\eta_t^2} \, ds = u_3 + u_2/u_1$$

follows from (15). Now,

$$\bar{w}_0 = \begin{pmatrix} 0 & 0 & 1 \\ 0 & -1 & 0 \\ 1 & 0 & 0 \end{pmatrix},$$

so we have

$$b\bar{w}_0 = \begin{pmatrix} e^{\eta^1} u_1 u_3 & -e^{\eta^1} u_1 & e^{\eta^1} \\ e^{\eta^2}(u_3+u_2/u_1) & -e^{\eta^2} & 0 \\ e^{\eta^3} & 0 & 0 \end{pmatrix}$$

$$= \begin{pmatrix} 1 & 0 & 0 \\ v_1+v_3 & 1 & 0 \\ v_2 v_3 & v_2 & 1 \end{pmatrix} \begin{pmatrix} e^{\eta^1} u_1 u_3 & 0 & 0 \\ 0 & e^{\eta^2} u_2/u_1 u_3 & 0 \\ 0 & 0 & e^{\eta^3}/u_2 \end{pmatrix} \begin{pmatrix} 1 & -1/u_3 & 1/u_1 u_3 \\ 0 & 1 & -u_3/u_2-1/u_1 \\ 0 & 0 & 1 \end{pmatrix}$$

and

$$n\bar{w}_0 = \begin{pmatrix} u_1 u_3 & -u_1 & 1 \\ u_3+u_2/u_1 & -1 & 0 \\ 1 & 0 & 0 \end{pmatrix}$$

$$= \begin{pmatrix} u_1 u_3 & 0 & 0 \\ u_3+u_2/u_1 & u_2/u_1 u_3 & 0 \\ 1 & 1/u_3 & 1/u_2 \end{pmatrix} \begin{pmatrix} 1 & -1/u_3 & 1/u_1 u_3 \\ 0 & 1 & -u_3/u_2-1/u_1 \\ 0 & 0 & 1 \end{pmatrix}.$$

Thus $[b\bar{w}_0]_0 = e^{T_{w_0}\eta}$, and, as claimed,

$$[b\bar{w}_0]_- = \begin{pmatrix} 1 & 0 & 0 \\ v_1+v_3 & 1 & 0 \\ v_2v_3 & v_2 & 1 \end{pmatrix} = X_1(v_1)X_2(v_2)X_3(v_3),$$

$$[n\bar{w}_0]_{-0} = \begin{pmatrix} u_1u_3 & 0 & 0 \\ u_3+u_2/u_1 & u_2/u_1u_3 & 0 \\ 1 & 1/u_3 & 1/u_2 \end{pmatrix} = Y_1(u_1)Y_2(u_2)Y_3(u_3).$$

6.3. Evolution of the Lusztig parameters. As before, we introduce a probability measure $\mathbb{P}$ under which η is a Brownian motion in $\mathbb{R}^n$ with a drift μ and $\eta(0) = 0$. For each $k \leq n$, set

$$(T_{k,1}, \ldots, T_{k,k}) = T^{(k)}(\eta^1, \ldots, \eta^k).$$

Note that this is given in terms of the principal minors $b^{(k)}$, $k \leq n$, of b by $T^{(k)}(\eta^1, \ldots, \eta^k) = \log[b^{(k)} \bar{w}_0^{(k)}]_0$, where $w_0^{(k)}$ denotes the longest element in S_k. The evolution of the triangular array $T_{k,j}$, $1 \leq j \leq k \leq n$, is given by (13). As remarked earlier, this process contains a number of projections which are also Markov. In particular, setting $\xi_k = T_{k,k}$, we have, for $k \leq n$,

$$d\xi_k = d\eta^k - e^{\xi_k - \xi_{k-1}} dt.$$

This defines a simple interacting particle system on the real line which, in the coordinates $\sum_i \xi_i$ and $\xi_i - \xi_{i+1}$, $1 \leq i \leq n-1$, has a product form invariant measure. There is an extension of this process, involving the Lusztig parameters, which is also Markov and, moreover, also has a product form invariant measure. Let $v_1, \ldots, v_q$ be the Lusztig parameters corresponding to a reduced decomposition $w_0 = s_{i_1} \ldots s_{i_q}$, that is,

$$[b\bar{w}_0]_- = X_{i_1}(v_1) \cdots X_{i_q}(v_q).$$

Set $y_k = -\log v_k$. The evolution of y_k, $1 \leq k \leq q$, is given by

$$dy_k = d\alpha_{i_k}(\eta_{k-1}) + e^{-y_k} dt,$$

where $\eta_k = T_{i_k} \ldots T_{i_1} \eta$. Setting $x_k = y_k - \alpha_{i_k}(\eta_{k-1})$, note that $dx_k = e^{-y_k} dt$ and $\eta_k = \eta + \sum_{j=1}^k x_j \alpha_j$. Hence,

$$dy_k = d\alpha_{i_k}(\eta) + \sum_{j=1}^{k-1} \alpha_{i_k}(\alpha_{i_j}) e^{-y_j} dt + e^{-y_k} dt. \tag{18}$$

Let $\beta_1 = \alpha_{i_1}$ and, for $2 \leq k \leq q$, $\beta_k = s_{i_1} \ldots s_{i_{k-1}} \alpha_{i_k}$. Set $\theta_k = -\beta_k(\mu)$. If $\mu \in w_0\Omega = -\Omega$, then $\theta_k > 0$ for all k and the diffusion has stationary distribution given by the product measure

$$\pi = \bigotimes_{k=1}^{q} \Gamma(\theta_k)^{-1} g_{\theta_k},$$

where $g_\theta(dx) = \exp(-\theta x - e^{-x}) \, dx$. This can be seen as a consequence of the following fact, which is the analogue in this setting of the output theorem for the $M/M/1$ queue [O'Connell and Yor 2001]. Let x_t be a standard one-dimensional Brownian motion with negative drift $-\theta$, and consider the one-dimensional diffusion

$$dy = \sqrt{2} dx + e^{-y} \, dt.$$

This has a unique invariant distribution $\Gamma(\theta)^{-1} g_\theta$. If we start this diffusion in equilibrium and define $\tilde{x}_t = x_t + 2(y_0 - y_t)$, then $\tilde{x}$ has the same law as x and, moreover, $\tilde{x}_s$, $s \le t$, is independent of y_u, $u \ge t$, for all t. It follows that the measure π is invariant. For an analytic proof of this fact, see [O'Connell and Ortmann 2012]. See also [Biane et al. 2009, Proposition 5.9], where the equivalent property is proved in the "zero-temperature" setting.

If we choose the reduced decomposition $i = 1\ 21\ 321\ n{-}1n{-}2\ldots21$, and define, for $m \le n - 1$ and $1 \le i \le n - m$, $q_{m,i} = T_{i+m-1,i} - T_{i+m,i+1}$, then

$$(y_1, y_2, \ldots, y_q) = (q_{1,1}, q_{1,2}, \ldots, q_{1,n}, q_{2,1}, \ldots, q_{2,n-1}, \ldots, q_{n-1,1}).$$

Note that $q_{1,i} = \xi_i - \xi_{i+1}$, for $1 \le i \le n - 1$. In these coordinates, the evolution is given by

$$dq_{m,i} = d\alpha_i(\eta) + e^{-q_{m,i}} \, dt + \sum_{l=1}^{m-1} (2e^{-q_{l,i}} - e^{-q_{l,i+1}} - e^{-q_{l,i-1}}) \, dt,$$

with the conventions that the empty sum is zero and $q_{l,0} = +\infty$. Setting $\theta_{m,i} = \mu_{m+i} - \mu_m$, an invariant measure for this diffusion is given by the product measure $\bigotimes_{m,i} g_{\theta_{m,i}}$. The dynamics of this process can be viewed as a network, as follows. Consider the dynamics

$$dQ = d(A - S) + e^{-Q} \, dt, \quad dD = dA - dQ, \quad dT = dS + dQ.$$

We think of A, S as the input and D, T as the output, and represent this system graphically as follows:

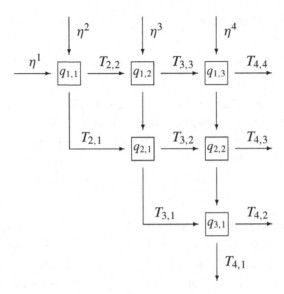

Figure 2. Graphical representation of the evolution of Lusztig parameters.

Then the evolution of the $q_{m,i}$ can be represented as in Figure 2. To see directly from this picture the product-form invariant measure, note that, if A and S are independent standard one-dimensional Brownian motions with respective drifts λ and σ, with $\lambda < \sigma$, then the diffusion Q has invariant distribution $\Gamma(\theta)^{-1} g_\theta$, where $\theta = \sigma - \lambda$. Moreover, if we start this diffusion in equilibrium, then $D_t = A_t + Q_0 - Q_t$ and $T_t = S_t - Q_0 + Q_t$ are independent standard one-dimensional Brownian motions with respective drifts λ and σ, and for each $t > 0$, (D_s, T_s), $s \leq t$, is independent of Q_u, $u \geq t$. The analogue of this fact in the setting of Poisson queueing networks is the cornerstone of classical queueing theory. It is called the output, or Burke, theorem. Finally, we remark that the dynamics indicated by Figure 2 is the analogue, in this setting, of the dynamical interpretation given in [O'Connell 2003] of the RSK correspondence as a kind of "queueing network".

7. From the Feynman–Kac formula to Givental's integral formula

The fact that the evolution equation (18) for the Lusztig parameters has a product form invariant measure sheds some light on the relation between the Feynman–Kac formula (10) and the integral formula of Givental. It follows from this that, for any given reduced decomposition of w_0, the random variables

$$\int_0^\infty e^{-\alpha_i(\beta_s)} \, ds, \quad i = 1, \ldots, n-1$$

can be expressed, via the transition maps, as rational functions of a collection of

$q = n(n-1)/2$ independent Gamma-distributed random variables with respective parameters θ_k, $k \leq q$, defined as above with $\beta = -\eta$. Note that β is a Brownian motion with drift $\lambda = -\mu \in \Omega$. Since the sets $\{\theta_k, k \leq q\}$ and $\{\lambda_i - \lambda_j, i < j\}$ are the same, this allows (10) to be written as a q-dimensional integral

$$\psi_\lambda(x) = \prod_{i<j} \Gamma(\lambda_i - \lambda_j) e^{\lambda(x)} \mathbb{E}_x \exp\left(-\sum_{i=1}^{n-1} \int_0^\infty e^{-\alpha_i(\beta_s)} \, ds\right)$$

$$= e^{\lambda(x)} \int_{\mathbb{R}_+^q} e^{-\sum_{i=1}^{n-1} e^{-\alpha_i(x)} r_i(v_1,\dots,v_q)} \prod_{i=1}^q v_i^{\theta_i - 1} e^{-v_i} \, dv_i. \qquad (19)$$

For example, when $n = 3$ and $i = 121$, we have

$$\theta_1 = \lambda_1 - \lambda_2, \quad \theta_2 = \lambda_1 - \lambda_3, \quad \theta_3 = \lambda_2 - \lambda_3,$$

and, using (16),

$$r_1(v_1, v_2, v_3) = \frac{1}{v_1}, \quad r_2(v_1, v_2, v_3) = \frac{1}{v_1'} = \frac{v_1 + v_3}{v_2 v_3}.$$

In this case, the integral formula (19) becomes

$$\psi_\lambda(x) = e^{\lambda_1 x_1 + \lambda_2 x_2 + \lambda_3 x_3} \int_{\mathbb{R}_+^3} v_1^{\lambda_1 - \lambda_2 - 1} v_2^{\lambda_1 - \lambda_3 - 1} v_3^{\lambda_2 - \lambda_3 - 1}$$

$$\times \exp\left(-v_1 - v_2 - v_3 - e^{-x_1 + x_2} \frac{1}{v_1} - e^{-x_2 + x_3} \frac{v_1 + v_3}{v_2 v_3}\right) dv_1 \, dv_2 \, dv_3.$$

Under the change of variables

$$v_1 = e^{T_{32} - T_{21}}, \quad v_2 = e^{T_{33} - T_{22}}, \quad v_3 = e^{T_{22} - T_{11}},$$

where $T = (T_{ki}, 1 \leq i \leq k \leq 3)$ is an array with $(T_{31}, T_{32}, T_{33}) = (x_1, x_2, x_3)$, this integral becomes

$$\psi_\lambda(x) = \int_{\mathbb{R}^3} e^{\lambda_1 (T_{31} + T_{32} + T_{33} - T_{21} - T_{11}) + \lambda_2 (T_{21} + T_{22} - T_{11}) + \lambda_3 T_{11}}$$

$$\times \exp\left(-e^{T_{32} - T_{21}} - e^{T_{33} - T_{22}} - e^{T_{22} - T_{11}} - e^{T_{21} - T_{31}} - e^{T_{11} - T_{22}} - e^{T_{22} - T_{32}}\right) dT_{11} \, dT_{21} \, dT_{22}.$$

Since $\Psi_\lambda(x)$ is a symmetric function of λ we see that this agrees with Givental's integral formula (7). We note that this is reminiscent of the derivation of Givental's formula given in [Gerasimov et al. 2008] (see also [Gerasimov et al. 2006; 1997]).

8. Fundamental Whittaker functions

The eigenvalue equation (6) also has series solutions known as *fundamental Whittaker functions*. Define a collection of analytic functions $a_{n,m}(v)$, $n \geq 2$, $m \in (\mathbb{Z}_+)^{n-1}$, $v \in \mathbb{C}^n$ recursively by

$$a_{2,m}(v) = \frac{1}{m! \, \Gamma(v_1 - v_2 + m + 1)},$$

and for $n > 2$,

$$a_{n,m}(v) = \sum_k a_{n-1,k}(\mu) \prod_{i=1}^{n-1} \frac{1}{(m_i - k_i)!} \frac{1}{\Gamma(v_i - v_n + m_i - k_{i-1})},$$

where $\mu_i = v_i + v_n/(n-1)$, $i \leq n-1$, and the sum is over $k \in (\mathbb{Z}_+)^{n-2}$ satisfying $k_i \leq m_i$, $1 \leq i \leq n-2$, with the convention that $k_0 = k_{n-1} = 0$. Then for each n, $a_{n,m}(v)$ satisfies the recursion

$$\left[\sum_{i=1}^{n-1} m_i^2 - \sum_{i=1}^{n-2} m_i m_{i+1} + \sum_{i=1}^{n-1} (v_i - v_{i+1}) m_i \right] a_{n,m}(v) = \sum_{i=1}^{n-1} a_{n,m-e_i}(v)$$

[Ishii and Stade 2007, Theorem 15], with the convention that $a_{n,m} = 0$ for $m \notin (\mathbb{Z}_+)^{n-1}$, and $a_{n,0}(v) = \prod_{i<j} \Gamma(v_i - v_j + 1)^{-1}$. Writing $m' = \sum_{i=1}^{n-1} m_i (e_i - e_{i+1})$, the series

$$m_v(x) = \sum_m a_{n,m}(v) e^{-(m'+v,x)} \tag{20}$$

is a fundamental Whittaker function as defined by Hashizume [1982], and satisfies the eigenvalue equation (6). We adopt a slightly different normalisation than the ones used in [Hashizume 1982] or [Ishii and Stade 2007]. Note that, for each $x \in \mathbb{R}^n$, $m_v(x)$ is an analytic function of v. Moreover:

Proposition 9. $\qquad \psi_v(x) = \prod_{i<j} \frac{\pi}{\sin \pi(v_i - v_j)} \sum_{w \in S_n} (-1)^w m_{-wv}(x).$

Proof. This comes from [Baudoin and O'Connell 2011]. In the notation of that paper we are taking $\Pi = \{\alpha_i/2 : i = 1, \ldots, n-1\}$, $m(2\alpha) = 0$, $|\eta_\alpha|^2 = 1$ and $\psi_v(x) = 2^q k_v(x)$, where $q = n(n-1)/2$. $\qquad \square$

Now consider the function $\theta_t(x)$ defined by (14). Note that we can write

$$s_n(\lambda) = \frac{1}{(2\pi \iota)^n n!} h(\lambda) \prod_{i>j} \frac{\sin \pi(\lambda_i - \lambda_j)}{\pi}.$$

Corollary 10. $\qquad \theta_t(x) = \frac{1}{(2\pi \iota)^n} \int_{\iota \mathbb{R}^n} m_\lambda(x) h(\lambda) e^{\sum_i \lambda_i^2 t/2} d\lambda.$

9. Relativistic Toda and q-deformed Whittaker functions

The algebraic structure underlying Theorem 5 is an intertwining relation between certain differential operators associated with the open quantum Toda chain with n particles. This structure should carry over to the setting of Ruijsenaars' relativistic Toda difference operators and q-deformed Whittaker functions [Ruijsenaars 1990; 1999; Etingof 1999; Gerasimov et al. 2010]. A recent (related, but different) development along these lines is given in [Borodin and Corwin 2014]. We will describe here the q-analogue of Theorem 5 in the rank one case, which corresponds to $n = 2$.

In the case $n = 2$, the Whittaker function is given by

$$\psi_\lambda(x) = 2\exp\left(\tfrac{1}{2}(\lambda_1 + \lambda_2)(x_1 + x_2)\right) K_{\lambda_1 - \lambda_2}(2e^{(x_2 - x_1)/2}),$$

where $K_\nu(z)$ is the Macdonald function. In this case, Theorem 5 is equivalent to the following theorem of Matsumoto and Yor [1999].

Theorem 11. (1) *Let* $(B_t^{(\mu)}, t \geq 0)$ *be a Brownian motion with drift* μ, *and define*

$$Z_t^{(\mu)} = \int_0^t e^{2B_s^{(\mu)} - B_t^{(\mu)}} ds.$$

Then $\log Z^{(\mu)}$ *is a diffusion process with infinitesimal generator*

$$\frac{1}{2}\frac{d^2}{dx^2} + \left(\frac{d}{dx}\log K_\mu(e^{-x})\right)\frac{d}{dx}.$$

(2) *The conditional law of* $B_t^{(\mu)}$, *given* $\{Z_s^{(\mu)}, s \leq t : Z_t^{(\mu)} = z\}$, *is given by the generalised inverse Gaussian distribution*

$$\tfrac{1}{2}K_\mu(1/z)^{-1}e^{\mu x}\exp(-\cosh(x)/z)\,dx.$$

Let $0 \leq q < 1$. Denote the q-Pochhammer symbol by

$$(q)_n = (q; q)_n = (1 - q)\cdots(1 - q^n),$$

with the conventions that $(q)_0 = 1$ and $(0)_n = 1$. In what follows we also adopt the convention that $0^0 = 1$.

For $\lambda \in \mathbb{C}$ and $z \geq 0$, define

$$\psi_\lambda(z) = \sum_{y=0}^z \frac{q^{\lambda(2y - z)}}{(q)_y(q)_{z-y}}.$$

This is a q-deformed Whittaker function associated with $\mathfrak{sl}_2$ [Gerasimov et al. 2010]. It satisfies the difference equation

$$(1 - q^{z+1})\psi_\lambda(z + 1) + \psi_\lambda(z - 1) = (q^\lambda + q^{-\lambda})\psi_\lambda(z)$$

where we set $\psi_\lambda(-1) = 0$, and is related to the q-Hermite polynomials by

$$(q)_z \psi_\lambda(z) = H_z\left(\frac{q^\lambda + q^{-\lambda}}{2}\,\middle|\,q\right).$$

Fix $0 \le q < 1$, $0 \le p \le 1$, and let $(Y_n, Z_n)_{n\ge0}$ be a Markov chain with state space $\{(y, z) \in \mathbb{Z}^2 : z \ge y \ge 0\}$ and transition probabilities given by

$$\Pi((y, z), (y+1, z+1)) = p, \quad \Pi((y, z), (y, z+1)) = (1-p)q^y,$$
$$\Pi((y, z), (y-1, z-1)) = (1-p)(1-q^y).$$

Note that Y is itself a Markov chain with transition probabilities

$$P(y, y+1) = p, \quad P(y, y) = (1-p)q^y, \quad P(y, y-1) = (1-p)(1-q^y),$$

and $X = 2Y - Z$ is a simple random walk on the integers which increases by one with probability p and decreases by one with probability $1 - p$. Choose $v \in \mathbb{R}$ such that $p = q^v/(q^v + q^{-v})$.

Theorem 12. *Let $Y_0 = Z_0 = 0$. The process $(Z_n, n \ge 0)$ is a Markov chain with transition probabilities*

$$Q(z, z+1) = \frac{1 - q^{z+1}}{q^v + q^{-v}}\frac{\psi_v(z+1)}{\psi_v(z)}, \quad Q(z, z-1) = \frac{1}{q^v + q^{-v}}\frac{\psi_v(z-1)}{\psi_v(z)}.$$

Moreover, for each $n \ge 0$, the conditional distribution of Y_n, given $\sigma\{Z_m, m \le n\}$ and $Z_n = z$, is given by

$$\pi_z(y) = \psi_v(z)^{-1}\frac{q^{v(2y-z)}}{(q)_y(q)_{z-y}}, \quad y = 0, 1, \ldots, z.$$

The proof is straightforward using the theory of Markov functions, by which it suffices to check that the transition operators Π and Q satisfy the intertwining relation $QK = K\Pi$ where

$$K(z, (y, z')) = \frac{\delta_{z,z'}q^{v(2y-z)}}{\psi_v(z)(q)_y(q)_{z-y}}.$$

This intertwining relation is readily verified. When $q = 0$ and $v = 0$, $\psi_v(z) = z$ and the above theorem can be interpreted as the discrete version of Pitman's $2M - X$ theorem, which states that if X_n is a simple symmetric random walk and $M_n = \max_{m \le n} X_m$, then $2M - X$ is a Markov chain with transition probabilities $Q(z, z+1) = (z+1)/2z$, $Q(z, z-1) = (z-1)/2z$. When $q \to 1$, it should rescale to Theorem 11.

The analogue of the output/Burke theorem in the setting of Theorem 12 is the following. If $p < 1/2$, then the Markov chain Y has a stationary distribution. If Y_0 is chosen according to this distribution and $Z_0 = 0$, the process $(Z_n, n \ge 0)$

is a simple random walk on the integers which increases by one with probability p and decreases by one with probability $1 - p$.

Acknowledgements

Many thanks to Nikos Zygouras and an anonymous referee for a careful reading of the manuscript and for valuable suggestions, which lead to a much improved version. Thanks also to Alexei Borodin for helpful discussions regarding Proposition 9.

This review started out as draft lecture notes for the summer school "Random matrix theory and applications" held at the Indian Institute of Science, Bangalore, in January, 2012. Unfortunately, I didn't make it to Bangalore (for bureaucratic reasons) but I am very grateful to the organisers for all their efforts. Research supported in part by EPSRC grant EP/I014829/1.

References

[Alexeev and Brion 2004] V. Alexeev and M. Brion, "Toric degenerations of spherical varieties", *Selecta Math. (N.S.)* **10**:4 (2004), 453–478.

[Arnold and Novikov 1994] V. I. Arnold and S. P. Novikov (editors), *Dynamical systems, VII*, Encyclopaedia of Mathematical Sciences **16**, Springer, Berlin, 1994.

[Baryshnikov 2001] Y. Baryshnikov, "GUEs and queues", *Probab. Theory Related Fields* **119**:2 (2001), 256–274.

[Baudoin and O'Connell 2011] F. Baudoin and N. O'Connell, "Exponential functionals of Brownian motion and class-one Whittaker functions", *Ann. Inst. Henri Poincaré Probab. Stat.* **47**:4 (2011), 1096–1120.

[Berenstein and Zelevinsky 2001] A. Berenstein and A. Zelevinsky, "Tensor product multiplicities, canonical bases and totally positive varieties", *Invent. Math.* **143**:1 (2001), 77–128.

[Berenstein et al. 1996] A. Berenstein, S. Fomin, and A. Zelevinsky, "Parametrizations of canonical bases and totally positive matrices", *Adv. Math.* **122**:1 (1996), 49–149.

[Biane et al. 2005] P. Biane, P. Bougerol, and N. O'Connell, "Littelmann paths and Brownian paths", *Duke Math. J.* **130**:1 (2005), 127–167.

[Biane et al. 2009] P. Biane, P. Bougerol, and N. O'Connell, "Continuous crystal and Duistermaat–Heckman measure for Coxeter groups", *Adv. Math.* **221**:5 (2009), 1522–1583.

[Borodin and Corwin 2014] A. Borodin and I. Corwin, "Macdonald processes", *Probab. Theory Related Fields* **158**:1-2 (2014), 225–400.

[Borodin et al. 2013] A. Borodin, I. Corwin, and D. Remenik, "Log-gamma polymer free energy fluctuations via a Fredholm determinant identity", *Comm. Math. Phys.* **324**:1 (2013), 215–232.

[Borodin et al. 2014] A. Borodin, I. Corwin, and P. Ferrari, "Free energy fluctuations for directed polymers in random media in $1 + 1$ dimension", *Comm. Pure Appl. Math.* **67**:7 (2014), 1129–1214.

[Chhaibi 2012] R. Chhaibi, *Modèle de Littelmann pour cristaux géométriques, fonctions de Whittaker sur des groupes de Lie et mouvement brownien*, Ph.D. thesis, Université Paris VI - Pierre et Marie Curie, 2012, http://tel.archives-ouvertes.fr/docs/00/78/20/28/PDF/These-CHHAIBI.pdf.

[Corwin et al. 2014] I. Corwin, N. O'Connell, T. Seppäläinen, and N. Zygouras, "Tropical combinatorics and Whittaker functions", *Duke Math. J.* **163** (2014), 513–563.

[Etingof 1999] P. Etingof, "Whittaker functions on quantum groups and q-deformed Toda operators", pp. 9–25 in *Differential topology, infinite-dimensional Lie algebras, and applications*, edited by A. Astashkevich and S. Tabachnikov, Amer. Math. Soc. Transl. Ser. 2 **194**, Amer. Math. Soc., Providence, RI, 1999.

[Fomin and Zelevinsky 1999] S. Fomin and A. Zelevinsky, "Double Bruhat cells and total positivity", *J. Amer. Math. Soc.* **12**:2 (1999), 335–380.

[Gerasimov et al. 1997] A. Gerasimov, S. Kharchev, A. Morozov, M. Olshanetsky, A. Marshakov, and A. Mironov, "Liouville type models in the group theory framework, I: finite-dimensional algebras", *Internat. J. Modern Phys. A* **12**:14 (1997), 2523–2583.

[Gerasimov et al. 2006] A. Gerasimov, S. Kharchev, D. Lebedev, and S. Oblezin, "On a Gauss–Givental representation of quantum Toda chain wave function", *Int. Math. Res. Not.* **2006** (2006), Art. ID 96489, 23.

[Gerasimov et al. 2008] A. Gerasimov, D. Lebedev, and S. Oblezin, "Baxter operator and Archimedean Hecke algebra", *Comm. Math. Phys.* **284**:3 (2008), 867–896.

[Gerasimov et al. 2010] A. Gerasimov, D. Lebedev, and S. Oblezin, "On q-deformed $\mathfrak{gl}_{l+1}$-Whittaker function I", *Comm. Math. Phys.* **294**:1 (2010), 97–119.

[Givental 1997] A. Givental, "Stationary phase integrals, quantum Toda lattices, flag manifolds and the mirror conjecture", pp. 103–115 in *Topics in singularity theory*, edited by A. B. Sossinsky, Amer. Math. Soc. Transl. Ser. 2 **180**, Amer. Math. Soc., Providence, RI, 1997.

[Gorsky et al. 2012] A. Gorsky, S. Nechaev, R. Santachiara, and G. Schehr, "Random ballistic growth and diffusion in symmetric spaces", *Nuclear Phys. B* **862**:1 (2012), 167–192.

[Hashizume 1982] M. Hashizume, "Whittaker functions on semisimple Lie groups", *Hiroshima Math. J.* **12**:2 (1982), 259–293.

[Ishii and Stade 2007] T. Ishii and E. Stade, "New formulas for Whittaker functions on GL$(n, \mathbb{R})$", *J. Funct. Anal.* **244**:1 (2007), 289–314.

[Jacquet 2004] H. Jacquet, "Integral representation of Whittaker functions", pp. 373–419 in *Contributions to automorphic forms, geometry, and number theory*, edited by H. Hida et al., Johns Hopkins Univ. Press, Baltimore, MD, 2004.

[Joe and Kim 2003] D. Joe and B. Kim, "Equivariant mirrors and the Virasoro conjecture for flag manifolds", *Int. Math. Res. Not.* **2003**:15 (2003), 859–882.

[Katori 2011] M. Katori, "O'Connell's process as a vicious Brownian motion", *Phys. Rev. E* **84** (2011), 061144/1–11.

[Katori 2012] M. Katori, "Survival probability of mutually killing Brownian motions and the O'Connell process", *J. Stat. Phys.* **147**:1 (2012), 206–223.

[Kharchev and Lebedev 1999] S. Kharchev and D. Lebedev, "Integral representation for the eigenfunctions of a quantum periodic Toda chain", *Lett. Math. Phys.* **50**:1 (1999), 53–77.

[Kirillov 2001] A. N. Kirillov, "Introduction to tropical combinatorics", pp. 82–150 in *Physics and combinatorics, 2000 (Nagoya)*, edited by A. N. Kirillov and N. Liskova, World Sci. Publ., 2001.

[Lusztig 1994] G. Lusztig, *Introduction to quantum groups*, Birkhäuser, New York, 1994.

[Matsumoto and Yor 1999] H. Matsumoto and M. Yor, "A version of Pitman's $2M - X$ theorem for geometric Brownian motions", *C. R. Acad. Sci. Paris Sér. I Math.* **328**:11 (1999), 1067–1074.

[Moriarty and O'Connell 2007] J. Moriarty and N. O'Connell, "On the free energy of a directed polymer in a Brownian environment", *Markov Process. Related Fields* **13**:2 (2007), 251–266.

[Noumi and Yamada 2004] M. Noumi and Y. Yamada, "Tropical Robinson–Schensted–Knuth correspondence and birational Weyl group actions", pp. 371–442 in *Representation theory of algebraic groups and quantum groups*, edited by T. Shoji et al., Adv. Stud. Pure Math. **40**, Math. Soc. Japan, Tokyo, 2004.

[O'Connell 2003] N. O'Connell, "A path-transformation for random walks and the Robinson–Schensted correspondence", *Trans. Amer. Math. Soc.* **355**:9 (2003), 3669–3697.

[O'Connell 2012] N. O'Connell, "Directed polymers and the quantum Toda lattice", *Ann. Probab.* **40**:2 (2012), 437–458.

[O'Connell and Ortmann 2012] N. O'Connell and J. Ortmann, "Product-form invariant measures for Brownian motion with drift satisfying a skew-symmetry type condition", 2012. arXiv 1201. 5586

[O'Connell and Warren 2011] N. O'Connell and J. Warren, "A multi-layer extension of the stochastic heat equation", 2011. arXiv 1104.3509

[O'Connell and Yor 2001] N. O'Connell and M. Yor, "Brownian analogues of Burke's theorem", *Stochastic Process. Appl.* **96**:2 (2001), 285–304.

[O'Connell et al. 2014] N. O'Connell, T. Seppäläinen, and N. Zygouras, "Geometric RSK correspondence, Whittaker functions and symmetrized random polymers", *Inventiones Math.* **197** (2014), 361–416.

[Ruijsenaars 1990] S. N. M. Ruijsenaars, "Relativistic Toda systems", *Comm. Math. Phys.* **133**:2 (1990), 217–247.

[Ruijsenaars 1999] S. N. M. Ruijsenaars, "Systems of Calogero–Moser type", pp. 251–352 in *Particles and fields* (Banff, AB, 1994), edited by G. Semenoff and L. Vinet, Springer, New York, 1999.

[Seppäläinen and Valkó 2010] T. Seppäläinen and B. Valkó, "Bounds for scaling exponents for a 1 + 1 dimensional directed polymer in a Brownian environment", *ALEA Lat. Am. J. Probab. Math. Stat.* **7** (2010), 451–476.

[Spohn 2014] H. Spohn, "KPZ scaling theory and the semidiscrete directed polymer model", pp. 483–493 in *Random matrix theory, interacting particle systems, and integrable systems*, Mathematical Sciences Research Institute Publications **65**, Cambridge University Press, New York, 2014.

[Stade 1990] E. Stade, "On explicit integral formulas for GL$(n, \mathbb{R})$-Whittaker functions", *Duke Math. J.* **60**:2 (1990), 313–362.

[Wallach 1992] N. R. Wallach, *Real reductive groups, II*, Pure and Applied Mathematics **132**, Academic Press, Boston, 1992.

n.m.o-connell@warwick.ac.uk   *Mathematics Institute, University of Warwick,*
Zeeman Building, Coventry CV4 7AL, United Kingdom

Random Matrices
MSRI Publications
Volume 65, 2014

How long does it take
to compute the eigenvalues
of a random symmetric matrix?

CHRISTIAN W. PFRANG, PERCY DEIFT AND GOVIND MENON

We present the results of an empirical study of the performance of the QR
algorithm (with and without shifts) and the Toda algorithm on random symmetric matrices. The random matrices are chosen from six ensembles, four
of which lie in the Wigner class. For all three algorithms, we observe a form
of universality for the deflation time statistics for random matrices within the
Wigner class. For these ensembles, the empirical distribution of a normalized
deflation time is found to collapse onto a curve that depends only on the
algorithm, but not on the matrix size or deflation tolerance provided the
matrix size is large enough. For the QR algorithm with the Wilkinson shift,
the observed universality is even stronger and includes certain non-Wigner
ensembles. Our experiments also provide a quantitative statistical picture of
the accelerated convergence with shifts.

1. Introduction

We present the results of a statistical study of the performance of the QR and
Toda eigenvalue algorithms on random symmetric matrices. Our work is mainly
inspired by progress in quantifying the "probability of difficulty" and "typical
behavior" for several numerical algorithms [Demmel 1988; Goldstine and von
Neumann 1951]. This approach has led to a deeper understanding of the efficacy
of fundamental numerical algorithms such as Gaussian elimination and the
simplex method [Rudelson and Vershynin 2008; Sankar et al. 2006; Smale
1983; Tao and Vu 2010]. It has also stimulated new ideas in random matrix
theory [Dumitriu and Edelman 2002; Edelman 1988; Edelman and Sutton 2007].
Testing eigenvalue algorithms with random input continues this effort. In related

The work of Pfrang and Menon was supported in part by NSF grant DMS 07-48482. The work of
Deift was supported in part by NSF grant DMS 10-01886. Deift also acknowledges support by a
grant from The Simonyi Fund at the Institute for Advanced Study in Princeton.
MSC2010: 65F15, 65Y20, 82B44, 60B20.
Keywords: symmetric eigenvalue problem, QR algorithm, Toda algorithm, matrix sign algorithm,
random matrix theory.

work [Pfrang 2011], we have also studied the performance of a version of the matrix sign algorithm. However, these results are of a different character, and apart from some theoretical observations, we do not present any experimental results for this algorithm (see [Pfrang 2011] for more information). Our study is empirical — a study of the eigenvalue problem from the viewpoint of complexity theory is presented in [Armentano 2014].

1.1. Algorithms and ensembles.

1.1. *Algorithms and ensembles*. It is natural to study the QR algorithm because of its elegance and fundamental practical importance. But in fact all the algorithms we study are linked by a common framework. In each case, an initial matrix L_0 is diagonalized via a sequence of isospectral iterates L_m. The gist of the framework is that the L_m correspond exactly to the flow of a completely integrable Hamiltonian system evaluated at integer times. The Hamiltonian for these flows is of the form tr $G(L)$ where G is a real-valued function defined on an interval. Different choices of G generate different algorithms: $G(x) = x(\log x - 1)$ yields unshifted QR, $G(x) = x^2/2$ yields Toda, and $G(x) = |x|$ yields the matrix sign algorithm. As noted above, we will not present any numerical experiments on the matrix sign algorithm (but see Section 2). We note that the practical implementation of the QR algorithm requires an efficient shifting strategy. Our work includes a study of the QR algorithm with the Wilkinson shift as discussed below.

Initial matrices are drawn from six ensembles that arise in random matrix theory. These are listed below in Section 2.4. For many random matrix ensembles, as the size of the matrix grows, the density of eigenvalues and suitably rescaled fluctuations have limiting distributions that may be computed explicitly. Four of the ensembles we study consist of random matrices with independent entries subject to the constraint of symmetry. The law of these entries is chosen so that these ensembles have the Wigner semicircle law as limiting spectral density. We say that these ensembles are in the *Wigner class*. Numerical experiments with these ensembles are contrasted with two ensembles that do not belong to the Wigner class.

1.2. *Deflation and QR with the Wilkinson shift*. In evaluating these algorithms we focus on the statistics of deflation. Given a real, symmetric, $n \times n$ matrix L and an integer k between 1 and n, we write

$$L = \begin{pmatrix} L_{11} & L_{12} \\ L_{12}^T & L_{22} \end{pmatrix}, \quad \tilde{L} = \begin{pmatrix} L_{11} & 0 \\ 0 & L_{22} \end{pmatrix}, \tag{1}$$

where L_{11} is a $k \times k$ block. Let λ_j and $\tilde{\lambda}_j$, $j = 1, \ldots, n$, denote the eigenvalues of L and $\tilde{L}$. For a fixed tolerance $\epsilon > 0$ we say that L is deflated to $\tilde{L}$ when the off-diagonal block L_{12} is so small that $\max_j |\lambda_j - \tilde{\lambda}_j| < \epsilon$. The *deflation time* is the number of iterations m before L_m can be deflated by a tolerance $\epsilon > 0$ at some index k. The *deflation index* is this value of k. Since the iterative eigenvalue algorithms correspond to Hamiltonian flows, there is also a natural notion of deflation time for the Hamiltonian flows (see equations (17) and (18) below).

Let us now explain why deflation serves as a useful measure of the time required to compute the eigenvalues of a matrix. The cost of practical computation requires an analysis of algorithms, hardware and software. In our study, we only focus on the algorithm, and "time" is taken to mean the number of iterations required for convergence. In our experiments we have observed that the QR and Toda algorithms deflate a matrix at the upper-left or lower-right corner with high probability. The deflation index for the shifted QR algorithm is $n - 1$ with overwhelming probability. The deflation index for unshifted QR is also typically $n - 1$ (see Figures 19 and 20). As a consequence, the deflation time is typically the same as the time taken to compute an eigenvalue. We then expect that the time taken to compute all eigenvalues with these algorithms is determined by n deflations. By contrast, we find that the matrix sign algorithm typically deflates a matrix in the middle and does not immediately yield any eigenvalues. Instead, these are obtained after a divide-and-conquer procedure that consists of approximately $\log_2 n$ deflations. Thus, for all these algorithms a finite sequence of deflation times determines the number of iterations necessary to compute eigenvalues. We must note however, that we do not track all deflations in our experiments, only the first. This restriction is necessary to keep the datasets manageable as n increases. A more extensive study that tracks all deflation times for these algorithms will certainly yield further interesting information. Finally, as we show in Section 2.6 below, the notion of deflation time is also of theoretical value since it is the starting point for an analysis of the expected number of iterations for eigenvalue algorithms that is similar in spirit to [Smale 1983].

The convergence of the QR algorithm is greatly accelerated by shifts. We will only consider the Wilkinson shift, i.e., the shift is the eigenvalue of the 2×2 lower diagonal corner of the matrix that is closer to L_{nn}. The QR algorithm on tridiagonal matrices is cubically convergent with this choice of shift (this is generically true [Wilkinson 1968]; see also [Leite et al. 2010] for a more careful analysis). As noted above, the unshifted QR algorithm deflates at index $n - 1$ with very high probability. Since the Wilkinson shift utilizes the lower 2×2 block of the matrix, the number of the iterations required for shifted QR, as opposed to unshifted QR, to deflate is far smaller. While such acceleration of convergence is well-known, some features of our experiments still come as a surprise. For example, a striking feature of Figures 1 and 2 is that the number of iterations required to deflate a random matrix with the QR algorithm (shifted and unshifted) is almost independent of n for matrices as large as 190×190.

1.3. _Universality._ Our main empirical findings concern universal fluctuations in the deflation time distribution for the QR algorithm (shifted and unshifted) and the Toda algorithm for ensembles in the Wigner class. We sample the deflation time for a range of matrix size and deflation tolerance combinations and normalize these empirical distributions to mean zero and variance one. The resulting histograms have the same general shape and in particular, the same tails on the right side (see in particular Figures 4, 7 and 10). In other words, _the fluctuations in deflation time are universal._ For the Toda and unshifted QR algorithm, the observed limiting fluctuations for Wigner and non-Wigner ensembles are distinct (see Figures 6 and 9). In addition, we find that the universal distributions for Wigner ensembles have exponential tails for unshifted QR and Gaussian tails for Toda (Figures 6 and 12). Universality of the tails is quantified with a statistical methodology developed in [Clauset et al. 2009]. Quite remarkably, for the (Wilkinson) shifted QR algorithm, the observed universality is stronger: to a good approximation _all_ tested ensembles show the same limiting distribution (see Figure 9).

The origin of such universality is not clear. We do not understand fully if our results are connected with the now familiar universality theorems of random matrix theory such as those that describe fluctuations in the bulk and at the edge of the spectrum for the Wigner ensembles [Erdős and Yau 2012; Mehta 2004; Tracy and Widom 1994]. Unlike these universality theorems, where the mean and variance are known theoretically, in our work the mean and variance of the deflation time are computed empirically and we have not yet been able to determine analytically how these depend on n. It does appear however that the mean deflation time is linearly proportional to $\log \epsilon$ (see Figures 13 and 15).

More broadly, our experiments are suggestive of a wider class of questions concerning universality of fluctuations for computations in numerical linear algebra. For example, in similar experiments to be reported elsewhere, one of the authors (P.D.) and Sheehan Olver have studied the solution x to the linear equation $Ax = b$ empirically, when A is a random positive symmetric matrix and b is a random vector. They compute the solution using the conjugate gradient method and observe universal fluctuations in the number of iterations required for convergence, independent of the choice of ensemble for A and b.

We now discuss the algorithms and ensembles in greater detail. This is followed by a description of the results in Section 3. The implementation of the algorithms is discussed briefly in Section 4.

2. Algorithms, ensembles and deflation statistics

2.1. *Notation.* We denote the space of real, symmetric $n \times n$ matrices by $\text{Symm}(n)$ and the space of real $n \times m$ matrices by $\mathbb{R}^{n \times m}$. Matrices in $\text{Symm}(n)$ are denoted L or M and the iterates of an eigenvalue algorithm are denoted L_m or M_m, $m = 0, 1, 2, \ldots$. We use Q to denote an orthogonal matrix and R an upper triangular matrix with positive diagonal entries, typically with reference to a QR factorization. We use $\sigma(L)$ to denote the spectrum of L. The spectral decomposition of $L \in \text{Symm}(n)$ is written

$$L = U \Lambda U^T,$$

where $\Lambda = \text{diag}(\lambda_1, \lambda_2, \ldots, \lambda_n)$ the matrix of eigenvalues and U denotes the orthogonal matrix of eigenvectors of L. We also use the following standard notation. The $n \times n$ identity matrix is I; the standard basis in $\mathbb{R}^n$ is $(e_1, \ldots, e_n)$; the unit sphere in $\mathbb{R}^{n+1}$ and its positive orthant are S^n and S^n_+, respectively; the symmetric group of order n is S_n.

When $L \in \text{Symm}(n)$ is tridiagonal, we let $(a_1, \ldots, a_n)$ be its diagonal entries and $(b_1, \ldots, b_{n-1})$ its off-diagonal entries. A Jacobi matrix is a tridiagonal matrix with $b_i > 0$ for all i. The space of Jacobi matrices is denoted $\text{Jac}(n)$. We use $u = U^T e_1$ to denote the first row of the matrix of eigenvectors. When $L \in \text{Jac}(n)$, $\sigma(L)$ is simple and we may assume that $\lambda_1 > \lambda_2 > \cdots > \lambda_n$. Moreover, all the components u_i are nonzero and we may assume that $u_i > 0$ (this is also generically true for $L \in \text{Symm}(n)$). Let $\mathcal{M}$ denote the manifold $\{(\Lambda, u) \in \mathbb{R}^n \times S^n_+ | \lambda_1 > \lambda_2 > \cdots > \lambda_n\}$. It is a basic result in the spectral and inverse spectral theory of Jacobi matrices that the matrix L can be reconstructed if Λ and u are given. More precisely, the spectral map $\mathcal{S} : \text{Jac}(n) \to \mathcal{M}$ defined by $L \mapsto (\Lambda, u)$ is a diffeomorphism [Deift et al. 1983, Theorem 2].

We use the following standard notation for probabilistic notions. The phrase independent and identically distributed is abbreviated to iid. A normal random variable with mean μ and variance σ^2 is denoted $\mathcal{N}(\mu, \sigma^2)$; a Bernoulli random variable that is ± 1 with probability $1/2$ is denoted $\mathcal{B}$; a random variable with the χ-distribution with parameter k is denoted χ_k. The notation $X \sim Y$ means that X has the same law as Y.

2.2. *The QR algorithm and Hamiltonian eigenvalue algorithms.* We assume the reader is familiar with the QR algorithm (excellent textbook presentations are [Demmel 1997; Golub and Van Loan 1996; Trefethen and Bau 1997]). In the unshifted QR algorithm the iterates M_m are generated through QR factorizations and matrix multiplication in the reverse order:

$$Q_m R_m = M_m, \quad M_{m+1} = R_m Q_m, \quad m = 0, 1, 2, \ldots. \tag{2}$$

The shifted QR algorithm relies on a shift μ_m at each step, and the modified steps

$$Q_m R_m = M_m - \mu_m I, \quad M_{m+1} = R_m Q_m + \mu_m I, \quad m = 0, 1, 2, \ldots. \quad (3)$$

Typical shifts, such as the Wilkinson shift, are constructed from the lower 2×2 block of M_m [Golub and Van Loan 1996, p. 418].

In the early 80s it was discovered that the QR algorithm is intimately connected with integrable Hamiltonian systems [Symes 1980; 1981/82; Deift et al. 1986; 1983; Nanda 1985]. We summarize these results below. An expanded presentation of these connections may be found in [Deift et al. 1996; 1993; Pfrang 2011]. A different exposition that explains these ideas in a fashion "intrinsic" to numerical linear algebra is [Watkins 1984].

Assume G is a piecewise smooth real-valued function defined on an interval, and set $g = G'$. If g is defined on $\sigma(L)$, we define $g(L) := U g(\Lambda) U^T$. Let M_- denote the strictly lower triangular part of the square matrix M, and $\mathrm{pr}_{\mathfrak{k}} M := M_-^T - M_-$, the projection of M onto skew-symmetric matrices. We then consider the ordinary differential equation

$$\dot{L} = [\mathrm{pr}_{\mathfrak{k}} g(L), L]. \quad (4)$$

Equation (4) defines a completely integrable Hamiltonian flow on the space of (generic) symmetric matrices with Hamiltonian $H(L) = \mathrm{tr}\, G(L)$ and symplectic structure detailed in [Deift et al. 1986]. This flow is connected to the unshifted QR algorithm as follows.

Theorem 1. *Let g be a real-valued function defined on $\sigma(L_0)$. Then*

(a) *The solution to Equation (4) with initial condition L_0 is an isospectral deformation*

$$L(t) = Q(t)^T L_0 Q(t), \quad (5)$$

where the orthogonal matrix $Q(t)$ is given by the unique QR factorization

$$e^{tg(L_0)} = Q(t) R(t), \quad t \geq 0, \quad (6)$$

that has $Q(0) = I$ and depends smoothly on t.

(b) *At integer times $m = 0, 1, 2 \ldots$, the solution $L(m)$ satisfies*

$$e^{g(L(m))} = M_m, \quad (7)$$

where M_m is the m-th step of the QR algorithm applied to the initial matrix $M_0 = e^{g(L_0)}$.

(c) *Assume that the spectrum $\sigma(L_0)$ is simple and that g is injective on $\sigma(L_0)$. Then $L_\infty = \lim_{t\to\infty} L(t)$ is a diagonal matrix consisting of the eigenvalues of L_0.*

The case of tridiagonal matrices is of practical and theoretical importance. When L_0 is tridiagonal, so is $L(t)$, and the flow can be linearized using the spectral map $\mathcal{S}$ for Jacobi matrices.

Theorem 2. *Assume $L_0 \in \mathrm{Jac}(n)$. Then the solution $L(t)$ to (4) is an isospectral deformation $L(t) = U(t)^T \Lambda U(t)$ and the evolution of $u(t) = U(t)^T e_1$ and $L(t)$ is given explicitly by*

$$u(t) = \frac{e^{tg(\Lambda)} u_0}{\|e^{tg(\Lambda)} u_0\|}, \qquad L(t) = \mathcal{S}^{-1}(\lambda, u(t)). \tag{8}$$

Assume g is injective on $\sigma(L_0)$. Then

$$\lim_{t\to\infty} L(t) = \mathrm{diag}(\lambda_{\sigma_1}, \ldots, \lambda_{\sigma_n}), \tag{9}$$

where $\sigma \in S_n$ is the permutation such that $g(\lambda_{\sigma_1}) > \cdots > g(\lambda_{\sigma_n})$.

Theorem 1 and Theorem 2 may be used to develop numerical schemes. The main observation is that each choice of a Hamiltonian $H(L) = \mathrm{tr}\, G(L)$ corresponds to a choice of an algorithm. In particular, we have:

(1) the *unshifted QR algorithm:* $g(x) = \log x$, $G(x) = x(\log x - 1)$ and $H_{\mathrm{QR}}(L) = \mathrm{tr}\,[L \log L - L]$ [Nanda 1985];

(2) the *Toda algorithm:* $g(x) = x$, $G(x) = x^2/2$ and $H_{\mathrm{Toda}}(L) = \frac{1}{2}\,\mathrm{tr}\, L^2$ (in this case, (4) describes the evolution of the Toda lattice [Moser and Zehnder 2005]);

(3) the *matrix sign algorithm:* $g(x) = \mathrm{sign}\, x$, $G(x) = |x|$ and $H_{\mathrm{sign}}(L) = \mathrm{tr}\,|L|$.

Of course, each step of the shifted QR algorithm, $L \mapsto L - \mu I$, is Hamiltonian, with Hamiltonian $H_{\mathrm{QR,shift}}(L) = H_{\mathrm{QR}}(L - \mu I)$. While every function G defines a Hamiltonian not all choices are equally relevant. Since our goal is to find the spectral decomposition of L_0, we must assume that U and Λ are unknown. But then how are we to compute the matrix-valued functions $g(L)$ or $e^{g(L)}$ efficiently? The choices $g(x) = \log x$ and $g(x) = x$ are special since these give $e^{g(L)} = L$ and $g(L) = L$, respectively. The first choice gives the QR algorithm (strictly speaking a branch of the logarithm must be chosen so that (4) is well-defined, but this does not affect the QR algorithm because of (7)). For the second choice $g(x) = x$, the vector field (4) is faster to compute than the matrix exponential $e^{L(m)}$ and it is natural to use an ordinary differential equation solver for (4) to diagonalize L. This is the essence of the Toda algorithm.

Our final choice $g(x) = \text{sign}(x)$ requires further comment since the observation that the matrix sign algorithm is Hamiltonian seems to us to be new. Assume zero is not an eigenvalue of L_0 and let $\Sigma_\pm$ denote the eigenspaces of L_0 corresponding to positive and negative eigenvalues, respectively. Consider matrices $Q_\pm$ whose columns form an orthonormal basis for $\Sigma_\pm$, respectively. Then the matrices $P_+ = Q_+ Q_+^T$ and $P_- = Q_- Q_-^T$ are orthogonal projections onto $\Sigma_\pm$, respectively, and we find $\text{sign}(L_0) = P_+ - P_-$ and $(I \pm \text{sign}(L_0))/2 = P_\pm$. It is immediate that

$$e^{t \, \text{sign}(L_0)} = e^t P_+ - e^{-t} P_-, \quad \text{and} \quad \lim_{t \to \infty} e^{-t} e^{t \, \text{sign}(L_0)} = P_+. \tag{10}$$

The projection P_+ has a rank-revealing QR factorization $P_+ = U_\infty R_\infty \Pi$ [Higham 2008, Chapter 2.5]. The matrix sign algorithm rests on the fact that with U_∞ as above, $U_\infty^T U_\infty = I$, and the matrix

$$\tilde{L} = U_\infty^T L_0 U_\infty \tag{11}$$

is block-diagonal as in (1), where L_{11} is $k \times k$ with $k = \dim(\Sigma_+)$. Clearly, $\sigma(\tilde{L}) = \sigma(L_0)$.

Thus, the procedure to deflate a matrix using the matrix sign algorithm is:

(1) Given L_0, compute $\text{sign}(L_0)$ and hence $P_+ = (I + \text{sign}(L_0))/2$.

(2) Compute U_∞ using a rank-revealing QR decomposition of P_+.

(3) Compute $\tilde{L} = U_\infty^T L_0 U_\infty$.

We note that $\text{sign}(L_0)$ can be computed efficiently using a scaled Newton iteration and inverse-free modifications of this procedure [Bai et al. 1997; Higham 2008; Malyshev 1993]. The complete spectral decomposition of L_0 may be determined in a sequence of deflation steps. At each stage, the number of iterations required to deflate the matrix depends on the number of iterations required to compute $\text{sign}(L_0)$.

From the dynamical point of view, let $L(t)$ denote the solution to (4) with $g(L) = \text{sign}(L)$. Then it may be shown that for generic initial data $\Pi = I$ and $\lim_{t \to \infty} L(t) = \tilde{L}$ where $\tilde{L} = U_\infty^T L_0 U_\infty$ is the block-diagonal matrix obtained above by the matrix sign algorithm. While this dynamical interpretation of the matrix sign algorithm is of theoretical interest, it is not clear how to implement the algorithm numerically in an effective manner.

We have not tested the performance of the matrix sign algorithm with random input in full generality. Instead, we have tested the deflation behavior of this algorithm in a more restricted setting by first precomputing $\text{sign}(L_0)$ and then using Theorem 2. These results are not presented in this paper: the interested reader is referred to [Pfrang 2011].

2.3. Deflation criterion. Consider a symmetric matrix A with eigenvalues

$$\lambda_1 \geq \cdots \geq \lambda_n,$$

a symmetric matrix B, a positive number ϵ and the perturbed matrix $A + \epsilon B$ with eigenvalues

$$\lambda_1(\epsilon) \geq \cdots \geq \lambda_n(\epsilon).$$

Standard perturbation theory [Demmel 1997, Theorem 5.1] implies

$$|\lambda_i - \lambda_i(\epsilon)| \leq \epsilon \|B\|_2. \tag{12}$$

When deflating Jacobi matrices the perturbation matrix is of the form

$$B = \begin{pmatrix} 0 & E_{1k}^T \\ E_{1k} & 0 \end{pmatrix}, \tag{13}$$

where the only nonzero entry in $E_{1k} \in \mathbb{R}^{(n-k) \times k}$ is a one in the upper right corner. Clearly, $\|B\|_2 = 1$ in this case. For the deflation of full symmetric matrices, the perturbation matrix has the structure

$$B = \begin{pmatrix} 0 & B_{21}^T \\ B_{21} & 0 \end{pmatrix}, \tag{14}$$

where again $B_{21} \in \mathbb{R}^{(n-k) \times k}$, but now all entries of B satisfy $|b_{ij}| \leq 1$. In this case, one may show that $\|B\|_2 \leq \sqrt{k(n-k)}$.

We now define the deflation criterion. If L is a Jacobi matrix define

$$\hat{\epsilon}_k = b_k. \tag{15}$$

If $L = (l_{ij})$ is a full symmetric matrix, set

$$\hat{\epsilon}_k = \sqrt{k(n-k)} \max_{\substack{k < i \leq n \\ 1 \leq j \leq k}} |l_{ij}|. \tag{16}$$

Assume L_m is a sequence of iterates (Jacobi or full symmetric) obtained through an iterative eigenvalue algorithm. For a given tolerance $\epsilon > 0$ and initial matrix L_0 we define the *deflation time*

$$\tau_{n,\epsilon}(L_0) = \min\{m \mid \hat{\epsilon}_k(L_m) < \epsilon \text{ for some } 1 \leq k \leq n-1\}. \tag{17}$$

For calculations based on the Hamiltonian flow (4) it is more natural to consider the real valued deflation time

$$\tau_{n,\epsilon}(L_0) = \inf\{t > 0 \mid \hat{\epsilon}_k(L(t)) < \epsilon \text{ for some } 1 \leq k \leq n-1\}. \tag{18}$$

The location where the matrix deflates is called the *deflation index*:

$$\iota_{n,\epsilon}(L_0) = \arg \min_{1 \le k \le n-1} \hat{\epsilon}_k(L_{\tau_{n,\epsilon}(L_0)}). \tag{19}$$

There is an important difference between deflation and the asymptotic convergence guaranteed by Theorem 1. While Theorem 1 may be used to compute asymptotic rates of convergence as $t \to \infty$ [Deift et al. 1983, Theorem 3], in practice the rate of convergence is determined by deflation and transients play an important role. We illustrate this with a simple example.

Fix $\lambda_1 > \lambda_2 > 0$, let $\Lambda = \text{diag}(\lambda_1, \lambda_2)$ and consider the QR flow on Symm(2) with the initial matrix

$$L_0 = Q_0 \Lambda Q_0^T, \quad Q_0 = \begin{pmatrix} \cos \theta_0 & \sin \theta_0 \\ \sin \theta_0 & -\cos \theta_0 \end{pmatrix}. \tag{20}$$

According to Theorem 2, $\lim_{t\to\infty} L(t) = \Lambda$ for every θ_0. However, if $\theta_0 \approx \pi/2$, L_0 is a small perturbation of $\text{diag}(\lambda_2, \lambda_1)$, and in practice, the algorithm would immediately deflate and return L_0. But according to Theorem 2, $L(t)$ must evolve so that the initial diagonal terms "turn around" and are presented in the correct order $\text{diag}(\lambda_1, \lambda_2)$ as $t \to \infty$ (see (9)). More generally, consider $\Lambda = (\lambda_1, \ldots, \lambda_n)$ with $\lambda_1 > \lambda_2 > \cdots > \lambda_n > 0$. Each permutation $\sigma \in S_n$ yields a distinct fixed point $\Lambda_\sigma = (\lambda_{\sigma_1}, \ldots, \lambda_{\sigma_n})$ for the QR and Toda algorithms. In a numerical calculation, an initial condition close to Λ_σ is immediately deflated. Alternatively, iterates may pass close to one of the permutations Λ_σ and again deflation occurs at finite times. However, only the equilibrium $(\lambda_1, \ldots, \lambda_n)$ attracts generic initial conditions [Deift et al. 1983]. Thus the notion of convergence as $t \to \infty$ and deflation are completely distinct.

2.4. Ensembles. We now introduce the six ensembles of random matrices that we will analyze. For general introductions on random matrices see [Deift 1999; Edelman and Rao 2005; Mehta 2004]. The simplest way to construct an ensemble of random matrices is to choose entries independently subject only to the constraint of symmetry. Such ensembles are called *Wigner ensembles*. We also say that an ensemble lies in the *Wigner class* if the limiting spectral distribution for this ensemble is the Wigner semicircle law (described below). We consider four Wigner ensembles in the Wigner class:

1. the Gaussian orthogonal ensemble (GOE) (independent entries where we have $M_{ii} \sim \sqrt{2}\mathcal{N}(0, 1)$, $M_{ij} \sim \mathcal{N}(0, 1)$, $i > j$);

2. the Gaussian Wigner ensemble (GWE) (iid $M_{ij} \sim \mathcal{N}(0, 1)$, $i \ge j$);

3. the Bernoulli ensemble (iid $M_{ij} \sim \mathcal{B}$, $i \ge j$);

4. the Hermite-1 ensemble on Jacobi matrices (iid $a_k \sim \mathcal{N}(0, 1)$, $k = 1, \ldots, n$ and independent $b_k \sim \chi_k$, $k = 0, \ldots, n - 1$).

Items 1–3 are ensembles of full symmetric matrices. The distinction between 1 and 2 is that the variance of the diagonal and off-diagonal entries of matrices in GOE is different to ensure orthogonal invariance (see [Mehta 2004]). Hermite-1 is an ensemble of Jacobi matrices obtained by applying the Householder tridiagonalization procedure to the GOE ensemble. It is a remarkable fact that the entries remain independent under tridiagonalization (this is not true when matrices from ensembles (2) and (3) are tridiagonalized).

A choice of an ensemble of random, symmetric matrices is a choice of a probability measure on the space of symmetric matrices. When the matrix entries are independent this measure is a product measure. For example, the measure corresponding to GOE has density

$$P_{\text{GOE}}(M) = 2^{2n/2}(2\pi)^{-n(n+1)/4}e^{-\frac{1}{4}\operatorname{tr}(M^2)}. \tag{21}$$

For all these ensembles, while the matrix entries are independent, the eigenvalues are not. The joint density of eigenvalues for GOE and Hermite-1 may be computed explicitly and is given by the determinantal formula [Mehta 2004, Chapter 3]

$$f_1(\Lambda) = \frac{1}{Z_n}|\Delta_n(\lambda)|e^{-\frac{|\lambda|^2}{2}}, \qquad \Delta_n(\lambda) = \prod_{i<j}(\lambda_i - \lambda_j). \tag{22}$$

The normalization constant Z_n may be computed explicitly. By contrast, while the analogues of (21) for ensembles 2 and 3 are clear, there is no explicit analogue for (22).

The ensembles 1–4 are in the Wigner class, i.e., for each of these ensembles

$$\lim_{n \to \infty} \frac{1}{n}\#\{\lambda_i \in \sqrt{n}(a, b)\} = \int_a^b \nu(x)\, dx, \tag{23}$$

where $\nu(x)$ denotes the density of the *Wigner semicircle law*

$$\nu(x) = \frac{1}{2\pi}\sqrt{4 - x^2}\, \mathbb{1}_{|x| \le 2}. \tag{24}$$

We will contrast our results on these ensembles with two ensembles of Jacobi matrices that are not in the Wigner class. These are:

5. the uniform doubly stochastic Jacobi ensemble (UDSJ);

6. the Jacobi uniform ensemble (JUE).

Doubly stochastic Jacobi matrices of dimension $n \times n$ form a compact polytope in $\mathbb{R}^{n-1}$ which can be equipped with its uniform measure [Diaconis and Wood

2010]. This is the UDSJ ensemble. We can approximately sample from this ensemble using a Gibbs sampler.

JUE is defined using the spectral map $\mathcal{S}$ for $\mathrm{Jac}(n)$. Since we may describe Jacobi matrices by their spectral data (Λ, u), a probability measure on the spectral data pulls back under $\mathcal{S}^{-1}$ to a probability measure on $\mathrm{Jac}(n)$. For JUE, we replace (22) with eigenvalues chosen independently and uniformly on an interval and u distributed uniformly on the orthant S_+^{n-1}. In our numerical simulations we assume the eigenvalues are uniformly distributed on $[-2\sqrt{n}, 2\sqrt{n}]$ because this interval corresponds to the support of the semicircle law and allows a comparison between JUE and ensembles in the Wigner class. A particularly important aspect of JUE is that the eigenvalues do not repel one another. This strongly affects the statistics of $\tau_{n,\epsilon}$ as shown below (for unshifted QR and Toda, but not for shifted QR!).

2.5. The normalized deflation time.

We have now defined the algorithms, ensembles and deflation criterion. For a given algorithm and ensemble, $\tau_{n,\epsilon}(L)$ and $\iota_{n,\epsilon}(L)$ are random variables that depends on the random initial matrix L and $\epsilon > 0$. We explore the empirical distributions of $\tau_{n,\epsilon}$ and $\iota_{n,\epsilon}$ in simulations. Our main empirical finding is that for each algorithm these empirical distributions collapse into a universal distribution for the Wigner ensembles 1–4. Let $\mu_{n,\epsilon}$ and $\sigma_{n,\epsilon}^2$ denote the empirically determined mean and variance of $\tau_{n,\epsilon}(L)$ for a particular algorithm and ensemble.

Our simulations suggest that the normalized deflation time

$$T_{n,\epsilon} = \frac{\tau_{n,\epsilon} - \mu_{n,\epsilon}}{\sigma_{n,\epsilon}} \tag{25}$$

converges in distribution as $n \to \infty$ and $\epsilon \to 0$ and that the limit is the same for ensembles in the Wigner class (see Figures 4 and 10). Both $\mu_{n,\epsilon}$ and $\sigma_{n,\epsilon}$ are computed empirically. Our numerical calculations also suggest that $\mu_{n,\epsilon} \sim C |\log \epsilon|$ for all ensembles in the Wigner class (see Figures 13 and 15). As already noted above, a surprising outcome of our simulations is that universality for shifted QR is more encompassing, and actually holds for all six ensembles 1–6.

In order to prove convergence in distribution of $T_{n,\epsilon}$ it is first necessary to estimate the mean and variance of τ. We present below a calculation of $\mu_{2,\epsilon}$ that illustrates the subtle role of eigenvalue repulsion.

2.6. The scaling of the expected deflation time.

In this section we estimate the expected deflation time of the Toda flow on $\mathrm{Symm}(2)$. We show that

$$\mu_{2,\epsilon,\mathrm{GOE}} \sim C |\log \epsilon|, \quad \text{but} \quad \mu_{2,\epsilon,\mathrm{JUE}} \sim C |\log \epsilon|^2, \quad \epsilon \to 0. \tag{26}$$

The interval of support for the JUE density is chosen here to be $[-1, 1]$. This choice only affects the prefactor C, not the term $|\log \epsilon|^2$.

In order to establish these asymptotics, we first determine the deflation time τ_ϵ as a function of the initial condition (for brevity we write τ_ϵ for $\tau_{2,\epsilon}$ since $n = 2$ is fixed). Since $M(t) \in \text{Symm}(2)$ we may write $M = U(t)\Lambda U(t)^T$, where $\Lambda = \text{diag}(\lambda_1, \lambda_2)$, $\lambda_1 > \lambda_2$, and

$$U(t) = \begin{pmatrix} \cos \theta(t) & \sin \theta(t) \\ \sin \theta(t) & -\cos \theta(t) \end{pmatrix}. \tag{27}$$

Note that $m_{12} > 0$ corresponds to $\theta \in (0, \pi/2)$. We use Theorem 2 to obtain

$$m_{12}(t) = (\lambda_1 - \lambda_2) \cos \theta(t) \sin \theta(t) = (\lambda_1 - \lambda_2) \cdot \frac{e^{t(\lambda_2-\lambda_1)} \cdot \tan \theta_0}{1 + e^{2t(\lambda_2-\lambda_1)} \tan^2 \theta_0}. \tag{28}$$

Here $\theta_0 = \theta(0)$. Now we set $m_{12}(\tau_\epsilon) = \epsilon$ and solve to find

$$(\lambda_1 - \lambda_2)\tau_\epsilon = \begin{cases} 0 & m_{12}(0) \le \epsilon, \\ \log \tan \theta_0 - \log\left[\dfrac{\lambda_1 - \lambda_2}{2\epsilon} - \sqrt{\dfrac{(\lambda_1 - \lambda_2)^2}{4\epsilon^2} - 1}\right] & m_{12}(0) > \epsilon. \end{cases} \tag{29}$$

The asymptotics of τ_ϵ are easily determined. We have

$$(\lambda_1 - \lambda_2)\tau_\epsilon \sim -\log \epsilon + \log \tan \theta_0 + \log(\lambda_1 - \lambda_2), \quad \epsilon \to 0. \tag{30}$$

In order to compute the mean deflation time for GOE and JUE we first change to spectral variables. As noted above, the spectral map $\mathscr{S}$ is a diffeomorphism between the set of 2×2 symmetric matrices with $m_{12} > 0$ and the set $\{\lambda_1 > \lambda_2\} \times (0, \pi/2)$. The Jacobian of this transformation is $\lambda_1 - \lambda_2$, so that

$$dm_{11}\, dm_{22}\, dm_{12} = (\lambda_1 - \lambda_2)\, d\lambda_1\, d\lambda_2\, d\theta. \tag{31}$$

The mean deflation time for GOE is then given by

$$\mu_{2,\epsilon,\text{GOE}} = \frac{1}{Z_1} \int_{-\infty}^{\infty} \int_{-\infty}^{\lambda_1} \int_0^{\pi/2} \tau_\epsilon(\lambda_1, \lambda_2, \theta) e^{-(\lambda_1^2+\lambda_2^2)/4} (\lambda_1 - \lambda_2)\, d\lambda_1\, d\lambda_2\, d\theta. \tag{32}$$

For JUE, the eigenvalues are chosen uniformly from $[-1, 1]$ and we find

$$\mu_{2,\epsilon,\text{JUE}} = \frac{1}{Z_2} \int_{-1}^{1} \int_{-1}^{\lambda_1} \int_0^{\pi/2} \tau_\epsilon(\lambda_1, \lambda_2, \theta)\, d\lambda_2\, d\lambda_2\, d\theta. \tag{33}$$

Here Z_1 and Z_2 are normalizing constants for these probability densities.

The asymptotic behavior of (30), combined with (32) and (33), suggests the following leading order behavior as $\epsilon \to 0$:

$$\mu_{2,\epsilon,\text{GOE}} \sim \frac{|\log \epsilon|}{Z_1} \int_{-\infty}^{\infty} \int_{-\infty}^{\lambda_1} \int_0^{\pi/2} e^{-(\lambda_1^2 + \lambda_2^2)/4} \, d\lambda_1 \, d\lambda_2 \, d\theta \sim C_1 |\log \epsilon|,$$

$$\mu_{2,\epsilon,\text{JUE}} \sim \frac{|\log \epsilon|}{Z_2} \int_{-1}^{1} \int_{-1}^{\lambda_1} \int_0^{\pi/2} \frac{1}{\lambda_1 - \lambda_2} \mathbb{1}_{m_{12} > \epsilon} \, d\lambda_1 \, d\lambda_2 \, d\theta \sim C_2 |\log \epsilon|^2.$$

Here C_i denote constants that may be computed explicitly. The second integral is divergent without the cut-off $\mathbb{1}_{m_{12} > \epsilon}$: the cut-off gives rise to an additional factor of $|\log \epsilon|$. With more effort, these formal estimates may be made rigorous.

The analogous calculations for $M(t) \in \text{Jac}(n)$, $n > 2$ are quite subtle. For Jacobi matrices deflation occurs when $M(t)$ approaches the boundary $\partial \text{Jac}(n)$ of $\text{Jac}(n)$ (see for example [Deift et al. 1983, Figs. 6 and 7]). A theoretical analysis of such deflations, which we have not carried out yet, is a significant challenge as it requires a detailed understanding of the geometry of both the flow and the initial probability distribution in the vicinity of $\partial \text{Jac}(n)$ in high dimensions. For this reason, we are reduced to using the empirical mean $\mu_{n,\epsilon}$ and variance $\sigma_{n,\epsilon}^2$ to define the normalized deflation time in (25).

3. Results

We generated a large number (typically 5000–10,000) of samples of the deflation time and the deflation index for each choice of the following parameters:

1. an eigenvalue algorithm (QR without shift, QR with shift, Toda);

2. a random matrix ensemble;

3. matrix size n (typically ranging from 10, 30, ..., 190);

4. tolerance ϵ (typically 10^{-k}, $k = 2, 4, 6, 8$).

We present a representative sample of our main results. Further statistical tests, figures and tables that amplify our conclusions may be found in [Pfrang 2011].

3.1. Unscaled deflation time statistics for GOE.
We first present deflation time statistics for $\tau_{n,\epsilon}$ for a fixed ensemble (GOE) for both the QR (shifted and unshifted) and Toda algorithms. The statistics of $\tau_{n,\epsilon}$ for the unshifted QR algorithm are shown in Figure 1. Similar statistics for the QR algorithm with Wilkinson shift and the Toda algorithm are shown in Figures 2 and 3, respectively. These figures reflect the typical dependence of these algorithms on n and ϵ for ensembles 1–6. Similar statistics for other ensembles may be found in [Pfrang 2011, Chapter 7]. In all cases, we observe that the histograms for the QR algorithm are relatively insensitive to n and shift to the right as ϵ decreases. The

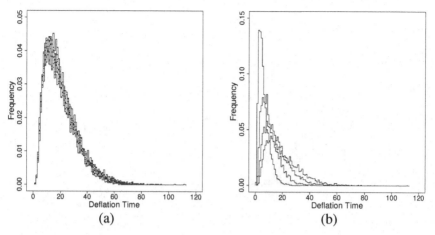

Figure 1. QR algorithm applied to GOE. (a) Histogram for empirical frequency $\tau_{n,\epsilon}$ as n ranges from $10, 30, \ldots, 190$ for a fixed deflation tolerance $\epsilon = 10^{-8}$. The curves (10 of them, plotted one on top of another) do not depend significantly on n. (b) Histogram for empirical frequency of $\tau_{n,\epsilon}$ when $\epsilon = 10^{-k}$, $k = 2, 4, 6, 8$ for fixed matrix size $n = 190$. Curves move to the right as ϵ decreases.

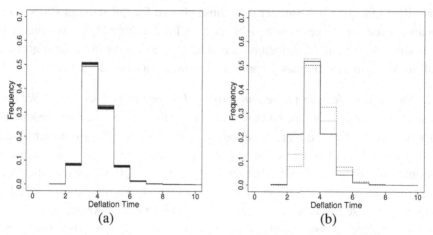

Figure 2. Shifted QR algorithm applied to GOE. (a) Histogram for empirical frequency $\tau_{n,\epsilon}$ as n ranges from $10, 30, \ldots, 190$ for a fixed deflation tolerance $\epsilon = 10^{-12}$. In the case of the unshifted QR algorithm, curves are insensitive to n, though the tail becomes more pronounced for larger n. (b) Histogram for empirical frequency $\tau_{n,\epsilon}$ when $\epsilon = 10^{-k}$, $k = 8, 10, 12$ for fixed matrix size $n = 190$. Curves move to the right as ϵ decreases.

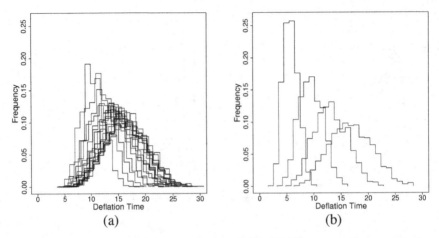

Figure 3. Toda algorithm applied to GOE. (a) Histogram for empirical frequency of $\tau_{n,\epsilon}$ as n ranges from $10, 30, \ldots, 190$ for a fixed deflation tolerance $\epsilon = 10^{-8}$. Curves drift to the right as n increases. (b) Histogram for empirical frequency $\tau_{n,\epsilon}$ when $\epsilon = 10^{-k}$, $k = 2, 4, 6, 8$ for fixed matrix size $n = 190$. Curves move to the right as ϵ decreases.

effect of the Wilkinson shift is to sharply reduce the number of iterations required (note the different scale of the abscissa in Figures 1 and 2). The values of ϵ for shifted QR are much smaller than those chosen for QR without shifts. This choice is necessary to generate a viable data set for the shifted QR algorithm with sufficient variation in the deflation time. The histograms for the Toda algorithm shift to the right as n increases and ϵ decreases, as discussed below.

3.2. _Normalized deflation time and universality for the Wigner class._ We now present results that show the collapse of all data onto universal curves depending only on the algorithm under the rescaling (25). The statistics of the empirical mean $\mu_{n,\epsilon}$ and standard deviation $\sigma_{n,\epsilon}$ are discussed a little later. The empirical distribution of the normalized deflation time $T_{n,\epsilon}$ for the QR algorithm with initial data from the Wigner ensembles is shown in Figure 4. All the data contained in Figure 1 collapse onto the single curve seen in Figure 4(a). Analogous data for the other Wigner class ensembles 2–4 collapse onto the _same universal curve_. The normalized deflation time distributions for UDSJ and JUE are shown in Figure 5. While we again observe a collapse of the data, it is not onto the curve of Figure 4(a). This contrast is amplified in the comparison of the tails of the normalized deflation time (see Figure 6). QQ plots that directly compare the histograms of these distributions may be found in [Pfrang 2011].

The most obvious difference between the behavior of the unshifted and shifted QR algorithm is that the spread in the deflation time for the shifted QR algorithm is

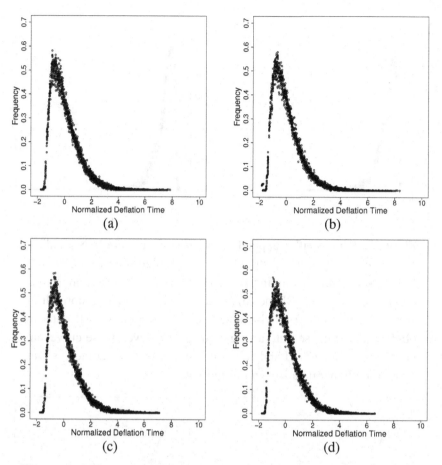

Figure 4. Universal deflation time statistics for QR algorithm applied to Wigner class. Empirical deflation time normalized as in (25) for $\epsilon = 10^{-k}$, $k = 2, 4, 6, 8$ and n ranging from $10, 30, \ldots, 190$. Random matrix ensembles are (a) GOE; (b) Hermite-1; (c) GWE; and (d) Bernoulli, with (a)–(d) obtained by rescaling data of 10×4 fixed-n and fixed-ϵ histograms and plotting them together. All these data collapse onto one universal curve. Plotting all 160 histograms together in Figure 6 further demonstrates universality of the deflation algorithm.

much narrower. However, this does not seem to affect our general conclusion that there is universality for each Hamiltonian eigenvalue algorithm. The normalized deflation time distribution for shifted QR is shown in Figures 7 and 8. Moreover, for shifted QR, the deflation times vary far less with the choice of underlying ensemble than the unshifted QR algorithm. In particular, we see a strong similarity for all ensembles in Figure 9. This behavior is in contrast with that of unshifted QR, shown in Figure 6.

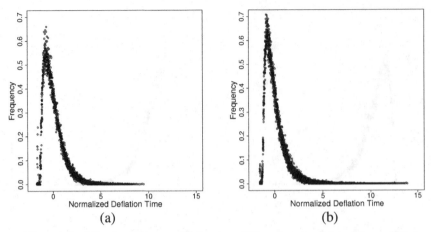

Figure 5. QR algorithm applied to non-Wigner ensembles. Normalized empirical deflation time distributions for QR algorithm with $\epsilon = 10^{-k}$, $k = 2, 4, 6, 8$ and n ranging from $10, 30, \ldots, 190$. Random matrix ensembles are (a) UDSJ and (b) JUE. Each figure contains normalized empirical data of 40 fixed-n and fixed-ϵ histograms. All data are observed to collapse onto a single curve. However, these curves are not the same for UDSJ and JUE, and neither of these coincides with curve for Wigner data shown in Figure 4.

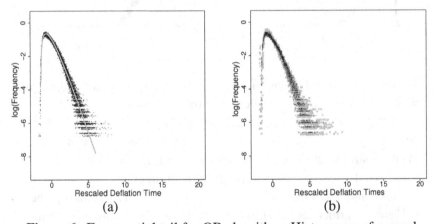

Figure 6. Exponential tail for QR algorithm. Histograms of normalized deflation time for QR algorithms on a logarithmic scale. (a) Wigner data: Empirical normalized deflation time distributions from all 160 histograms of Wigner class initial data (black dots) are compared with a gamma distribution with parameters $k = 2$ and $\theta = 1$ shifted to mean zero (gray line). (b) non-Wigner data: Empirical normalized deflation time distributions from 40 GOE histograms (black dots) contrasted with data from 40 UDSJ histograms (gray squares).

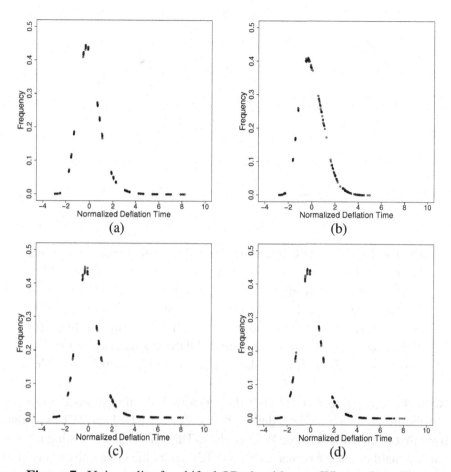

Figure 7. Universality for shifted QR algorithm on Wigner class. Empirical deflation time for QR algorithm with Wilkinson shift normalized as in (25) with $\epsilon = 10^{-k}$, $k = 8, 10, 12$ and n ranging from $10, 30, \ldots, 190$. Note that ϵ is significantly smaller than for unshifted QR algorithm. Ensembles are (a) GOE; (b) Hermite-1; (c) GWE; and (d) Bernoulli. Figures (a)–(d) are obtained by collapsing data as in Figure 4. Peak of the TE1 ensemble is lower, and tail shorter, than those for other three ensembles.

Finally, we have also observed universality for the Toda algorithm. The empirical distribution of the normalized deflation time for the Wigner ensembles is shown in Figure 10. Again, all the data contained in Figure 3 collapse onto the single curve seen in Figure 10(a). Further, analogous data for the other Wigner ensembles 2–4 collapse onto the same curve. The data for UDSJ and JUE collapse under normalization, but not onto the same distribution (see Figures 5 and 12).

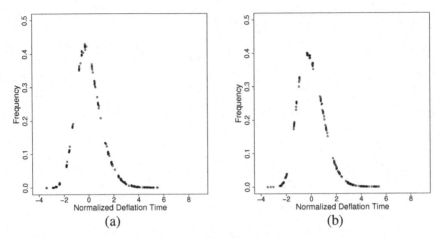

Figure 8. Shifted QR algorithm applied to non-Wigner ensembles. Normalized empirical deflation time distributions for QR algorithm with Wilkinson shift for $\epsilon = 10^{-k}$, $k = 8, 10, 12$ and n ranging from $10, 30, \ldots, 190$. Random matrix ensembles are (a) UDSJ and (b) JUE. Note that results for these ensembles seem very similar to those for Wigner class data shown in Figure 7. UDSJ is similar to full matrix ensembles, while JUE is similar to TE1, also a tridiagonal ensemble.

Remark 1. We note that for both the QR and Toda algorithms the limiting distribution of the normalized deflation time $T_{n,\epsilon}$ for UDSJ and JUE is distinct from that of ensembles in the Wigner class. This raises the interesting issue in random matrix theory whether UDSJ and JUE are in the same universality class as Wigner ensembles and invariant ensembles. As JUE does not have eigenvalue repulsion built in, this is unlikely to be the case.

3.3. *Universal tails for deflation times.* We used a hypothesis testing approach to quantify the statement that the rescaled deflation time has an exponential tail for QR and a Gaussian tail for Toda. Our approach is modeled on the methodology of [Clauset et al. 2009]. Given deflation time data D we perform maximum likelihood estimation of parameters for distribution families conditioned on observing only values above a cutoff value $x_{\min}(D)$ and use a semiparametric approach to compute p-values for these parameters. Based on D and our parameter estimate, we compute resampled data sets and a modified Kolmogorov–Smirnov statistic measuring the distance between the empirical distribution function and the ones resulting from our maximum likelihood estimates. The semiparametric p-value is given as the proportion of instances that the resampled data sets yield larger modified KS statistics than the original. If this p-value is large we accept the hypothesis that the original data set has in fact the proposed decay in the right tail.

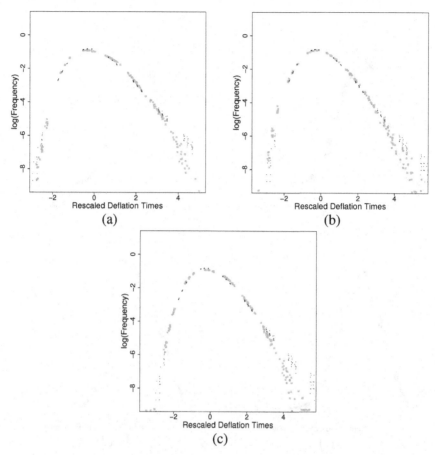

Figure 9. Comparison of ensembles for shifted QR. Histograms of
normalized deflation time for shifted QR algorithms on a logarithmic
scale. In (a)–(c), GOE (black dots) is contrasted with data from a
second ensemble (gray dots). (a) GOE and TE1: Empirical normalized
deflation time distributions from 40 GOE histograms (black dots) contrasted with data from 40 TE1 histograms (gray squares); (b) GOE and
UDSJ; (c) GOE and JUE.

We applied this approach with the Gaussian, Exponential, Weibull and Gamma
families. We found that the exponential tails fit the QR runtime data especially
well for small values of the deflation tolerance. The fit of the Toda runtime data
to Gaussian tails is very compelling across most experimental regimes. Direct
pictorial comparisons of the normalized Toda runtimes with the standard normal
as well as normalized QR runtimes with normalized Gamma distributions are
shown in Figure 6. Further details of the statistical tests may be found in [Pfrang
2011].

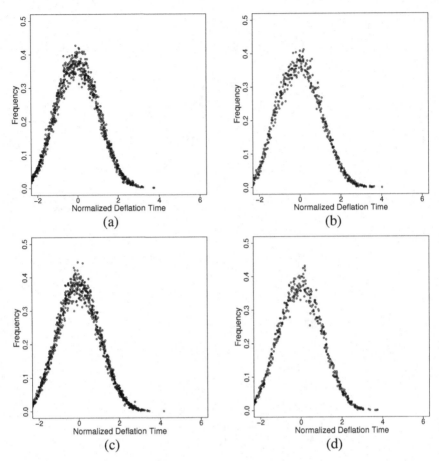

Figure 10. Universal deflation time statistics for Toda algorithm applied to Wigner class. Empirical deflation time normalized as in (25) for $\epsilon = 10^{-k}$, $k = 2, 4, 6, 8$ and n ranging from $10, 30, \ldots, 190$. Random matrix ensembles are (a) GOE; (b) Hermite-1; (c) GWE; and (d) Bernoulli, with (a)–(d) obtained by rescaling data of 40 fixed-n and fixed-ϵ histograms and plotting them together. All these data collapse onto one universal curve. Universality is amplified in Figure 12.

3.4. The dependence of $\mu_{n,\epsilon}$ and $\sigma_{n,\epsilon}$ on n and ϵ. We used linear regression to express $\mu_{n,\epsilon}$ and $\sigma_{n,\epsilon}$ as functions of $\log \epsilon$ and n. Only the best fits are reported here. The data for the QR algorithm was matched very well by

$$\mu_{n,\epsilon} \approx a_0 + a_1 n + a_2 \log \epsilon, \qquad (34)$$

$$\sigma_{n,\epsilon} \approx b_0 + b_1 n + b_2 \log \epsilon. \qquad (35)$$

This regression is compared visually with the numerical data in Figures 13 and 14.

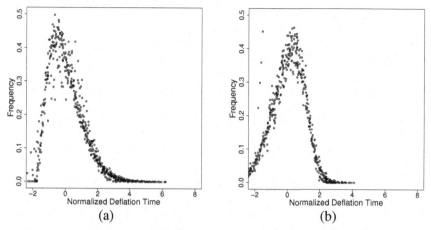

Figure 11. Toda algorithm applied to non-Wigner ensembles. Normalized empirical deflation time distributions for Toda algorithm with $\epsilon = 10^{-k}$, $k = 2, 4, 6, 8$ and n ranging from $10, 30, \ldots, 190$. Random matrix ensembles are (a) UDSJ and (b) JUE; each contains normalized empirical data of 40 fixed-n and fixed-ϵ histograms. All data are observed to collapse onto a single curve. However, curves are not the same for UDSJ and JUE and neither of these coincides with Wigner data curve shown in Figure 10.

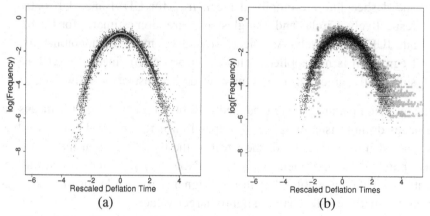

Figure 12. Gaussian tail for Toda algorithm. Histograms of normalized deflation time for QR algorithms on a logarithmic scale. (a) Wigner data: Empirical normalized deflation time distributions from all 160 histograms of Wigner class initial data (black dots) compared with standard normal distribution (gray line). (b) non-Wigner data: Empirical normalized deflation time distributions from 40 GOE histograms (black dots) contrasted with data from 40 UDSJ histograms (gray squares).

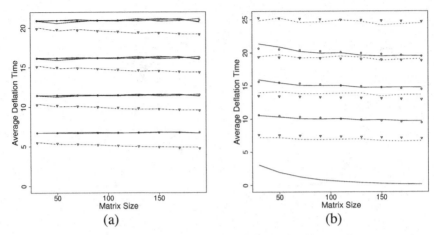

Figure 13. Mean deflation time $\mu_{n,\epsilon}$ for QR algorithm. Empirical average of deflation time for $\epsilon = 10^{-k}$, $k = 2, 4, 6, 8$ and n in the range $10, 30, \ldots, 190$. (a) Wigner class initial data: Full lines are empirical mean $\mu_{n,\epsilon}$ for GOE, GWE and Bernoulli ensembles. Note that they seem to align well with one another. Circles are values obtained from the regression estimate (34) with parameters listed in Table 1. Dashed line and triangles represent empirical data and regression, respectively, for Hermite-1 ensemble. (b) JUE and UDSJ initial data: Full line and dashed line are empirical mean $\mu_{n,\epsilon}$ for UDSJ and JUE data, respectively. Circles and triangles are regression estimates for UDSJ and JUE, respectively. As ϵ decreases, curves move up monotonically. Regression is not applied to lowest curve in (b) since $\epsilon = 0.01$ is sufficiently large that several matrices deflate instantaneously.

The regression parameters are tabulated in Tables 1 and 2. Since the means and variances do not visually appear to depend on n for ensembles 1–3 we have also included the p-values for the t-test of the hypothesis that the coefficient corresponding to the dimension is zero. Note that $\mu_{n,\epsilon}$ and $\sigma_{n,\epsilon}$ are almost identical for the ensembles 1–3 in the Wigner class, while for the Hermite-1 initial data both statistics have a slightly larger value.

Ensemble	a_0	a_1	a_2	p-value a_1
GOE, GWE, Bernoulli	1.96824	0.0004690	-1.0263649	0.0095
Hermite-1	.802338	$-.004554$	-1.042907	$< 2 \cdot 10^{-16}$
UDSJ	0.7648330	-0.0072921	-1.0916354	$4.16 \cdot 10^{-8}$
JUE	1.844126	-0.003467	-1.276037	0.0256

Table 1. Regression parameters for $\mu_{n,\epsilon}$ for the unshifted QR algorithm.

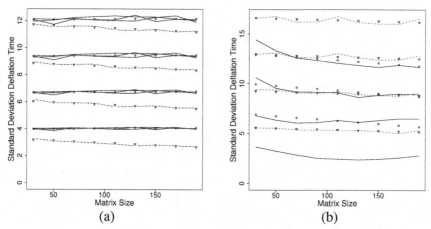

Figure 14. Standard deviation $\sigma_{n,\epsilon}$ of deflation time for QR algorithm. (a) Ensembles in Wigner class; (b) JUE and UDSJ. Legend as in Figure 13 with regression parameters from Table 2.

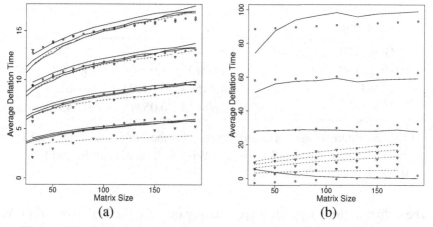

Figure 15. Mean deflation time $\mu_{n,\epsilon}$ for Toda algorithm. Mean deflation time distributions of Toda algorithm for initial data described in Figure 13. (a) Wigner class initial data; (b) JUE and UDSJ initial data. Legend as in Figure 13 and regression parameters as in Table 3.

Ensemble	b_0	b_1	b_2	p-value b_1
GOE, GWE, Bernoulli	1.2799509	0.0005311	-0.5854859	0.0118
Hermite-1	0.442622	$-.003329$	$-.617517$	$8 \cdot 10^{-15}$
UDSJ	1.066713	-0.007584	-0.658920	0.000353
JUE	2.0044243	-0.0026034	-0.7961700	0.000185

Table 2. Regression parameters for $\sigma_{n,\epsilon}$ for the unshifted QR algorithm.

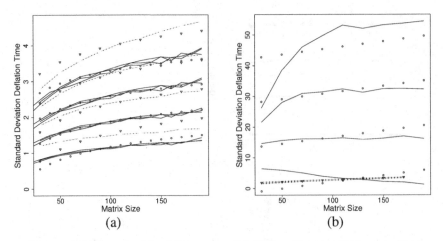

Figure 16. Standard deviation $\sigma_{n,\epsilon}$ of deflation time for Toda algorithm.
(a) Ensembles in the Wigner class; (b) JUE and UDSJ. Legend as in
Figure 13 and regression parameters as in Table 4.

Ensemble	a_0	a_1	a_2
GOE, GWE, Bernoulli	−6.0669	1.2888	−0.7302
Hermite-1	−7.0273	1.6795	−0.7708
UDSJ	−34.01514	0.02984	−6.60133
JUE	−2.78614	0.05318	−0.74315

Table 3. Regression parameters for $\mu_{n,\epsilon}$ for the Toda algorithm. UDSJ
and JUE are fit to (34)–(35) and the Wigner class ensembles are fit to
(36)–(37).

The deflation time depends more strongly on n for the Toda algorithm. We
explored several regressions but our results for Toda are more ambiguous than
for QR. We found that the non-Wigner ensembles (UDSJ and JUE) could be fit
with an expression of the form (34)–(35). However, the Wigner class ensembles
were better suited to the regression

$$\mu_{n,\epsilon} \approx a_0 + a_1 \log n + a_2 \log \epsilon \qquad (36)$$

$$\sigma_{n,\epsilon} \approx b_0 + b_1 \log n + b_2 \log \epsilon \qquad (37)$$

The results of this regression are presented in Figures 15–16 and Tables 3–4.

3.5. *Deflation index statistics and the effect of the Wilkinson shift.* The remarkable acceleration of QR by shifting is of course well known. Our experiments
provide a quantitative statistical picture for the efficacy of the shift. Figure 17

Ensemble	b_0	b_1	b_2
GOE, Gaussian Wigner, Bernoulli	-1.6532	0.3347	-0.1569
Hermite-1	-2.1233	0.6324	-0.1727
UDSJ	-16.46367	0.04845	-3.10561
JUE	0.97525	0.04068	0.01451

Table 4. Regression parameters for $\sigma_{n,\epsilon}$ for the Toda algorithm. UDSJ and JUE are fit to (34)–(35) and the Wigner class ensembles are fit to (36)–(37).

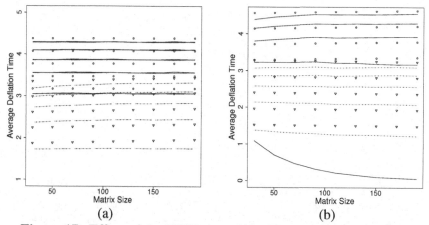

(a)   (b)

Figure 17. Effect of the Wilkinson shift. Mean deflation time $\mu_{n,\epsilon}$ for QR algorithm with Wilkinson shift. (a) Wigner class ensembles; (b) JUE and UDSJ. Empirical data are generated for $\epsilon = 10^{-2}, \ldots, 10^{-8}$ and $n = 20, \ldots, 190$. Empirical data and a regression of the form (34) are presented in same line-styles as in Figure 13. Observe that $\mu_{n,\epsilon}$ is almost independent of n and that curves move upwards as ϵ decreases, as in Figure 13, but the scale of the ordinate is different. Regression is not applied to lowest curve in (b) since $\epsilon = 0.01$ is sufficiently large that several matrices deflate instantaneously.

shows that the deflation time is sharply reduced by the Wilkinson shift. Figure 18 shows that the standard deviation of the deflation time is also sharply reduced by the shift. Deflation takes only a few iterations independent of the size of the matrix. This is in sharp contrast with the unshifted QR algorithm.

An explanation for the speed-up lies in the statistics of the deflation index shown in Figure 19. We find that the unshifted QR algorithm deflates at the bottom right corner of the matrix with high probability. Since the Wilkinson shift uses only the 2×2 lower-right block of the matrix, small off-diagonal terms

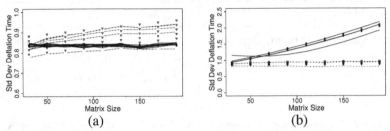

Figure 18. Standard deviation of deflation time with Wilkinson shift. (a) Wigner class ensembles; (b) JUE and UDSJ. Line styles are as in Figure 17 with a regression of the form (35).

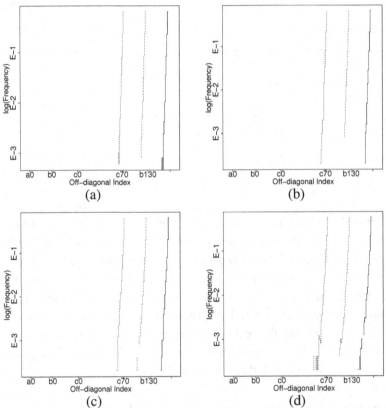

Figure 19. Empirical distributions of deflation index $\iota_{n,\epsilon}$ for unshifted QR algorithm. Figures show histograms of frequency with which deflation occurs at a given off-diagonal index. To aid visibility, distribution is centered so that peaks do not overlap. Off-diagonal index takes values between 0 and $n - 2$. Here "a", "b" and "c" refer to ensembles with $n = 190$, 130 and 70, respectively. Ensembles shown are (a) Hermite-1; (b) GOE; (c) UDSJ; and (d) JUE.

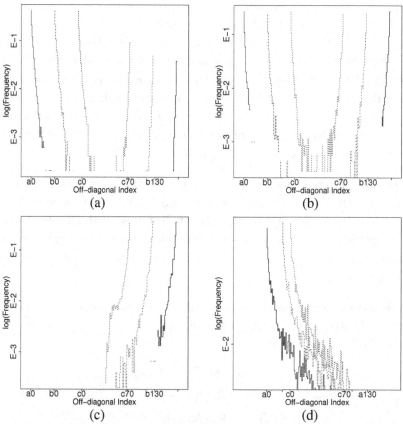

Figure 20. Empirical distributions of the deflation index $\iota_{n,\epsilon}$ for the Toda algorithm. Figures show histograms of frequency with which deflation occurs at a given off-diagonal index. Ensembles are as in Figure 19.

in this block accelerate the unshifted algorithm greatly. In contrast with the QR algorithm, the Toda algorithm deflates at both the upper-left and lower-right corner of the matrix (Figure 20). Note though that deflation is still predominantly at the corners of the matrix. Similar statistics for other ensembles may be found in [Pfrang 2011].

4. Methods and implementation

The algorithms were implemented in Python and run on a computing cluster using the module mpi4py. For numerical computations we relied on the scipy module except in the case of the RKPW spectral reconstruction procedure [Gragg and Harrod 1984] which was implemented in C. Our simulation strategy was to generate a number N of samples for each (ϵ, n)-pair of tolerances given by $\epsilon \in \{10^{-k} : k = 2, 4, 6, 8\}$ and matrix dimensions $n \in \{10, 30, \ldots, 190\}$. One initial matrix sample of size $n_i \times n_i$ is used to generate deflation time and deflation index samples for all pairs $(\tilde{\epsilon}, n_i)$, where $\tilde{\epsilon}$ is in our list of tolerances. To do this we advance the matrix using the algorithm under consideration until we undercut each of the tolerances in the list and save the corresponding statistics along the way. Typically for each (ϵ, n)-combination we generate between 1000 and 5000 samples. In the following we present a short summary of the implementation strategies chosen for the individual algorithms.

4.1. *QR algorithm.* Our simulation code uses the QR decomposition and matrix multiplication methods provided by scipy for the case of full symmetric matrices. For Jacobi matrices we implemented the efficient (unshifted) QR step presented for example in [Golub and Van Loan 1996]. We augment these implementations to include the Wilkinson shift by subtracting (adding) the shift value before (after) the QR step, respectively.

4.2. *Toda algorithm.* Both Jacobi and full symmetric matrices are treated similarly for this algorithm. The implementation uses the QR representation (8) to generate Toda steps T_n as follows:

$$M_k = \exp(T_k) = Q_k R_k, \tag{38}$$

$$M_{k+1} = R_k Q_k, \tag{39}$$

$$T_{k+1} = \log(M_{k+1}). \tag{40}$$

Our implementation uses scipy routines for the matrix exponentials and matrix logarithms. Note that in general the matrix exponential of a Jacobi matrix is full symmetric. scipy is also used for the QR decomposition and standard matrix multiplication routines for the reverse order multiplication.

Note that we do not use an ordinary differential equation solver to solve (4) and diagonalize the matrix as proposed in [Deift et al. 1983]. This is because our goal here is not to develop a competitive numerical scheme, but to compute reliable statistics of the deflation time for different algorithms. The above numerical scheme based on QR factorization was validated against both an ordinary differential equation solver based method and the use of the explicit solution (8) with the RKPW implementation of the inverse spectral map.

5. Acknowledgments

The numerical results presented here are part of the first author's Ph.D dissertation at Brown University [Pfrang 2011]. The help of the support staff at the Center for Computation and Visualization at Brown University is gratefully acknowledged.

We also thank Jim Demmel, Luen-Chau Li, Irina Nenciu and Nick Trefethen for their interest in this study. Finally, we thank the anonymous referee for several valuable comments and suggestions regarding our work.

References

[Armentano 2014] D. Armentano, "Complexity of path-following methods for the eigenvalue problem", *Found. Comput. Math.* **14**:2 (2014), 185–236.

[Bai et al. 1997] Z. Bai, J. Demmel, and M. Gu, "An inverse free parallel spectral divide and conquer algorithm for nonsymmetric eigenproblems", *Numer. Math.* **76**:3 (1997), 279–308.

[Clauset et al. 2009] A. Clauset, C. R. Shalizi, and M. E. J. Newman, "Power-law distributions in empirical data", *SIAM Rev.* **51**:4 (2009), 661–703.

[Deift 1999] P. A. Deift, *Orthogonal polynomials and random matrices: a Riemann–Hilbert approach*, Courant Lecture Notes in Mathematics **3**, Courant Institute of Mathematical Sciences, New York, 1999.

[Deift et al. 1983] P. Deift, T. Nanda, and C. Tomei, "Ordinary differential equations and the symmetric eigenvalue problem", *SIAM J. Numer. Anal.* **20**:1 (1983), 1–22.

[Deift et al. 1986] P. Deift, L.-C. Li, T. Nanda, and C. Tomei, "The Toda flow on a generic orbit is integrable", *Comm. Pure Appl. Math.* **39**:2 (1986), 183–232.

[Deift et al. 1993] P. Deift, L.-C. Li, and C. Tomei, "Symplectic aspects of some eigenvalue algorithms", pp. 511–536 in *Important developments in soliton theory*, edited by A. S. Fokas and V. E. Zakharov, Springer, Berlin, 1993.

[Deift et al. 1996] P. Deift, C. D. Levermore, and C. E. Wayne (editors), *Dynamical systems and probabilistic methods in partial differential equations: Proceedings of the 1994 AMS-SIAM Summer Seminar* (Berkeley, California, June 20–July 1, 1994), Lectures in Applied Mathematics **31**, American Mathematical Society, Providence, RI, 1996.

[Demmel 1988] J. W. Demmel, "The probability that a numerical analysis problem is difficult", *Math. Comp.* **50**:182 (1988), 449–480.

[Demmel 1997] J. W. Demmel, *Applied numerical linear algebra*, Society for Industrial and Applied Mathematics (SIAM), Philadelphia, PA, 1997.

[Diaconis and Wood 2010] P. Diaconis and P. Wood, "Random doubly stochastic Jacobi matrices", 2010. Unpublished manuscript.

[Dumitriu and Edelman 2002] I. Dumitriu and A. Edelman, "Matrix models for beta ensembles", *J. Math. Phys.* **43**:11 (2002), 5830–5847.

[Edelman 1988] A. Edelman, "Eigenvalues and condition numbers of random matrices", *SIAM J. Matrix Anal. Appl.* **9**:4 (1988), 543–560.

[Edelman and Rao 2005] A. Edelman and N. R. Rao, "Random matrix theory", *Acta Numer.* **14** (2005), 233–297.

[Edelman and Sutton 2007] A. Edelman and B. D. Sutton, "From random matrices to stochastic operators", *J. Stat. Phys.* **127**:6 (2007), 1121–1165.

[Erdős and Yau 2012] L. Erdős and H.-T. Yau, "Universality of local spectral statistics of random matrices", *Bull. Amer. Math. Soc.* (*N.S.*) **49**:3 (2012), 377–414.

[Goldstine and von Neumann 1951] H. H. Goldstine and J. von Neumann, "Numerical inverting of matrices of high order, II", *Proc. Amer. Math. Soc.* **2** (1951), 188–202.

[Golub and Van Loan 1996] G. H. Golub and C. F. Van Loan, *Matrix computations*, 3rd ed., Johns Hopkins University Press, Baltimore, MD, 1996.

[Gragg and Harrod 1984] W. B. Gragg and W. J. Harrod, "The numerically stable reconstruction of Jacobi matrices from spectral data", *Numer. Math.* **44**:3 (1984), 317–335.

[Higham 2008] N. J. Higham, *Functions of matrices: theory and computation*, Society for Industrial and Applied Mathematics (SIAM), Philadelphia, PA, 2008.

[Leite et al. 2010] R. S. Leite, N. C. Saldanha, and C. Tomei, "The asymptotics of Wilkinson's shift: loss of cubic convergence", *Found. Comput. Math.* **10**:1 (2010), 15–36.

[Malyshev 1993] A. N. Malyshev, "Parallel algorithm for solving some spectral problems of linear algebra", *Linear Algebra Appl.* **188/189** (1993), 489–520.

[Mehta 2004] M. L. Mehta, *Random matrices*, 3rd ed., Pure and Applied Mathematics (Amsterdam) **142**, Elsevier/Academic Press, Amsterdam, 2004.

[Moser and Zehnder 2005] J. Moser and E. J. Zehnder, *Notes on dynamical systems*, Courant Lecture Notes in Mathematics **12**, Courant Institute of Mathematical Sciences, New York, 2005.

[Nanda 1985] T. Nanda, "Differential equations and the QR algorithm", *SIAM J. Numer. Anal.* **22**:2 (1985), 310–321.

[Pfrang 2011] C. W. Pfrang, *Diagonalizing random matrices with integrable systems*, Ph.D. thesis, Brown University, 2011, https://repository.library.brown.edu/studio/item/bdr:11289/.

[Rudelson and Vershynin 2008] M. Rudelson and R. Vershynin, "The Littlewood–Offord problem and invertibility of random matrices", *Adv. Math.* **218**:2 (2008), 600–633.

[Sankar et al. 2006] A. Sankar, D. A. Spielman, and S.-H. Teng, "Smoothed analysis of the condition numbers and growth factors of matrices", *SIAM J. Matrix Anal. Appl.* **28**:2 (2006), 446–476.

[Smale 1983] S. Smale, "On the average number of steps of the simplex method of linear programming", *Math. Programming* **27**:3 (1983), 241–262.

[Symes 1980] W. W. Symes, "Hamiltonian group actions and integrable systems", *Phys. D* **1**:4 (1980), 339–374.

[Symes 1981/82] W. W. Symes, "The QR algorithm and scattering for the finite nonperiodic Toda lattice", *Phys. D* **4**:2 (1981/82), 275–280.

[Tao and Vu 2010] T. Tao and V. Vu, "Random matrices: the distribution of the smallest singular values", *Geom. Funct. Anal.* **20**:1 (2010), 260–297.

[Tracy and Widom 1994] C. A. Tracy and H. Widom, "Level-spacing distributions and the Airy kernel", *Comm. Math. Phys.* **159**:1 (1994), 151–174.

[Trefethen and Bau 1997] L. N. Trefethen and D. Bau, III, *Numerical linear algebra*, Society for Industrial and Applied Mathematics (SIAM), Philadelphia, PA, 1997.

[Watkins 1984] D. S. Watkins, "Isospectral flows", *SIAM Rev.* **26**:3 (1984), 379–391.

[Wilkinson 1968] J. H. Wilkinson, "Global convergence of tridiagonal QR algorithm with origin shifts", *Linear Algebra and Appl.* **1** (1968), 409–420.

christian.w.pfrang@jpmorgan.com

J. P. Morgan Securities, 383 Madison Avenue,
New York, New York 10033, United States

deift@cims.nyu.edu

Courant Institute of Mathematical Sciences,
New York University, 251 Mercer Street,
New York, New York 10012, United States

menon@dam.brown.edu

Division of Applied Mathematics, Brown University,
182 George Street, Providence, 02912, United States

Random Matrices
MSRI Publications
Volume **65**, 2014

Exact solutions of the Kardar–Parisi–Zhang equation and weak universality for directed random polymers

JEREMY QUASTEL

We survey recent results of convergence to random matrix distributions of directed random polymer free energy fluctuations in the intermediate disorder regime. These are obtained by passing through the exact formulas for fluctuations of KPZ at finite time.

1. Directed random polymers

Directed random polymers were introduced in the mid eighties as models of defect lines in media with impurities (see [Kardar 2007] for a review). They became popular in physics because besides their applicability as models and inherent interest, they are a case where the replica methods developed for the more difficult spin glass models give consistent answers. We will be interested in the $1 + 1$ dimensional case. We are given a random environment $\xi(i, j)$ of independent identically distributed real random variables for i, j in $\mathbb{Z}_+ \times \mathbb{Z}$. Given the environment, the energy of an n-step nearest neighbour walk $\boldsymbol{x} = (x_1, \ldots, x_n)$ is

$$H_n^\xi(\boldsymbol{x}) = \sum_{i=1}^n \xi(i, x_i). \tag{1}$$

The polymer measure on such walks starting at 0 at time 0 and ending at x at time n is then defined by

$$P_{n,x}^{\beta,\xi}(\boldsymbol{x}) = \frac{1}{Z^{\beta,\xi}(n, x)} e^{-\beta H_n^\xi(\boldsymbol{x})} P(\boldsymbol{x}). \tag{2}$$

The parameter $\beta > 0$, which measures how much the path prefers to travel through areas of low energy, is called the inverse temperature. P is the uniform probability measure on such walks, and $Z(n, x)$ is the partition function

$$Z^{\beta,\xi}(n, x) = \sum_{\boldsymbol{x}} e^{-\beta H_n^\xi(\boldsymbol{x})} P(\boldsymbol{x}). \tag{3}$$

This is the point-to-point free energy. If we do not specify the endpoint, we get the point-to-line free energy, which we denote by $Z^{\beta,\xi}(n)$.

What happens is that for large n that path is localized about a path which is special for that n; it has lateral fluctuations of size $n^{2/3}$. In terms of the free energy, the key conjecture is that its fluctuations are of size $n^{1/3}$ and given by Tracy–Widom distributions. More precisely, as long as $E[\xi(i,j)_-^5] < \infty$, it is expected [Biroli et al. 2007] that there exist c and σ depending on $\beta > 0$ and the distribution of the environment such that

$$\frac{\log Z^{\beta,\xi}(n,x) - cn}{\sigma n^{1/3}} \Rightarrow F_{\text{GUE}}. \tag{4}$$

Here F_{GUE} is the Tracy–Widom limiting distribution of the largest eigenvalue from the Gaussian orthogonal ensemble. For the point-to-line free energy $Z_{\beta,\xi}(n)$, the analogous statement is conjectured to hold except that now the asymptotic fluctuations are governed by the GOE Tracy–Widom distribution.

Remarkably, not a single case was known. The conjecture is extrapolation from the $\beta = \infty$ case with exponential or geometric distribution, where exact calculations are possible [Johansson 2000].

2. Continuum random polymer

The (point-to-point) continuum random polymer is the probability measure $P_{T,x}^{\beta,\xi}$ on continuous functions $x(t)$ on $[0,T]$ with $x(0) = 0$ and $x(T) = x$ and formal density

$$\exp\left(-\beta \int_0^T \xi(t,x(t))\,dt - \frac{1}{2}\int_0^T |\dot{x}(t)|^2\,dt\right), \tag{5}$$

where $\xi(t,x)$, $t \geq 0$, $x \in \mathbb{R}$ is space-time white noise.[1] One can think of it as an elastic band in a random energy background.

One can also think of the continuum random polymer as having a density

$$\exp\left(-\beta \int_0^T \xi(t,x(t))\,dt\right) \tag{6}$$

with respect to the Brownian bridge. Neither prescription makes mathematical sense; however, if one smooths out the noise, so that it does make sense, and removes the smoothing, there is a limiting measure on continuous functions $C[0,T]$ which we call $P_{T,x}^{\beta,\xi}$. Of course, the measure depends on the background randomness ξ just as in the discrete case. So it is a random probability measure on $C[0,T]$. In fact, it is a Markov process, and one can define it directly as

---

[1]That is, the distribution valued Gaussian variable such that for smooth functions φ of compact support in $\mathbb{R}_+ \times \mathbb{R}$, $\langle \varphi, \xi \rangle := \int_{\mathbb{R}_+ \times \mathbb{R}} \varphi(t,x)\xi(t,x)\,dx\,dt$ are mean zero Gaussian with covariance $E[\langle \varphi_1, \xi \rangle \langle \varphi_1, \xi \rangle] = \langle \varphi_1, \varphi_2 \rangle$.

follows. Let $z(s, x, t, y)$ denote the solution of the stochastic heat equation after time $s \geq 0$ starting with a delta function at x,

$$\partial_t z = \tfrac{1}{2} \partial_y^2 z - \beta \xi z, \qquad t > s, \ y \in \mathbb{R}, \tag{7}$$

$$z(s, x, s, y) = \delta_x(y). \tag{8}$$

It is important that they are all using the same noise ξ. Note that the stochastic heat equation is well-posed [Walsh 1986]. The solutions look locally like exponential Brownian motion in space. They are Hölder $\tfrac{1}{2} - \delta$ for any $\delta > 0$ in x and $\tfrac{1}{4} - \delta$ for any $\delta > 0$ in t. In fact, exponential Brownian motion $e^{B(x)}$ is invariant up to multiplicative constants, that is, if one starts (7) with $e^{B(x)}$ where $B(x)$ is a two-sided Brownian motion, then there is a (random) $C(t)$ so that $C(t)z(t, x)$ is another exponential of two sided Brownian motion [Bertini and Giacomin 1997].

$P_{T,x}^{\beta,\xi}$ is then defined to be the probability measure on continuous functions $x(t)$ on $[0, T]$ with $x(0) = 0$ and $x(T) = x$ and finite dimensional distributions

$$P_{T,x}^{\beta,\xi}(x(t_1) \in dx_1, \ldots, x(t_n) \in dx_n)$$
$$= \frac{z(0, 0, t_1, x_1) z(t_1, x_1, t_2, x_2) \cdots z(t_{n-1}, x_{n-1}, t_n, x_n) z(t_n, x_n, T, x)}{z(0, 0, T, x)} dx_1 \cdots dx_n$$

for $0 < t_1 < t_2 < \cdots < t_n < T$.

One can check these are a.s. a consistent family of finite dimensional distributions. It is basically because of the Chapman–Kolmogorov equation

$$\int_{\mathbb{R}} z(s, x, \tau, u) z(\tau, u, t, y) \, du = z(s, x, t, y), \tag{9}$$

which is a consequence of the linearity of the stochastic heat equation.

We can also define the joint measure $\mathbb{P}_{T,x}^{\beta} = P_{T,x}^{\beta,\xi} \otimes Q(\xi)$ where Q is the distribution of the ξ, that is, the probability measure of the white noise.

Theorem 2.1 [Alberts et al. 2014]. (i) *The measures $P_{T,x}^{\beta,\xi}$ and $\mathbb{P}_{T,x}^{\beta}$ are well-defined (the former, almost surely).*

(ii) *$P_{T,x}^{\beta,\xi}$ is a Markov process supported on Hölder continuous functions of exponent $\tfrac{1}{2} - \delta$ for any $\delta > 0$, for Q almost every ξ.*

(iii) *Let $t_k^n = k2^{-n}$. Then with $\mathbb{P}_{T,x}^{\beta}$ probability one, we have that for all $0 \leq t \leq 1$,*

$$\sum_{k=1}^{\lfloor 2^n t \rfloor} (x(t_k^n) - x(t_{k-1}^n))^2 \to t \tag{10}$$

as $n \to \infty$; i.e., the quadratic variation exists, and is the same as $\mathbb{P}_{T,x}^{0}$ (Brownian bridge).

(iv) *$P_{T,x}^{\beta,\xi}$ is singular with respect to $P_{T,x}^{0}$ (Brownian bridge) for almost every ξ.*

So the continuum random polymer looks locally like, but is singular with respect to, Brownian motion. One can also define the point-to-line continuum random polymer $\mathbb{P}_T^\beta$, in the same way as in the discrete case. For large T, one expects $\mathrm{Var}_{\mathbb{P}_T^\beta}(x(T)) \sim T^{4/3}$ in the point-to-line case or $\mathrm{Var}_{\mathbb{P}_{T,0}^\beta}(x(T/2)) \sim T^{4/3}$ in the point-to-point case. Here the variance is over the random background as well as $P_{T,x}^{\beta,\xi}$. The conditional variance given ξ should be much smaller.

3. Connection with KPZ

In the previous section we saw that if $z(t, x)$ is the solution of (7) with initial data (8) then $h(t, x) = -\beta^{-1} \log z(t, x)$ can be thought of as the free energy of the point-to-point continuum random polymer[2]. It is also the *Hopf–Cole solution of the Kardar–Parisi–Zhang equation*

$$\partial_t h = -\tfrac{1}{2}\beta^{-1}(\partial_x h)^2 + \tfrac{1}{2}\partial_x^2 h + \xi. \tag{11}$$

The equation was introduced by Kardar, Parisi and Zhang [1986], and has become the canonical model for random interface growth in physics. Formally, it is equivalent to the stochastic Burgers equation

$$\partial_t u = -\tfrac{1}{2}\beta^{-1}\partial_x u^2 + \tfrac{1}{2}\partial_x^2 u + \partial_x \xi, \tag{12}$$

which, if things were nice, would be satisfied by $u = \partial_x h$. Since $\log z(t, x)$ looks locally like Brownian motion, (11) is not well-posed (see [Hairer 2013] for recent progress on this question). If ξ were smooth, then the Hopf–Cole transformation takes (7) to (11). For white noise ξ, we take $h(t, x) = -\beta^{-1} \log z(t, x)$ with $z(t, x)$ a solution of (7) to be the *definition* of the solution of (11). It is known that these are the solutions one obtains if one smooths the noise, solves the equation, and takes a limit as the smoothing is removed.[3] They are also the solutions obtained as the limit of discrete models in the weakly asymmetric limit.

To understand the weakly asymmetric limit we consider how the KPZ equation (11) rescales. Let

$$h_\epsilon(t, x) = \epsilon^a h(\epsilon^{-z} t, \epsilon^{-1} x). \tag{13}$$

Recall the white noise has the distributional scale invariance

$$\xi(t, x) \stackrel{\text{dist}}{=} \epsilon^{(z+1)/2} \xi(\epsilon^z t, \epsilon^1 x). \tag{14}$$

Hence, setting $\beta = 1$ for clarity,

$$\partial_t h_\epsilon = -\tfrac{1}{2}\epsilon^{2-z-a}(\partial_x h_\epsilon)^2 + \tfrac{1}{2}\epsilon^{2-z}\partial_x^2 h_\epsilon + \epsilon^{a-\frac{1}{2}z+\frac{1}{2}}\xi. \tag{15}$$

---

[2]Because of the conditioning it is perhaps more appropriate to call $h(t, x) - \frac{x^2}{2t} - \log\sqrt{2\pi t}$ the free energy.

[3]After subtraction of a diverging constant.

Because the paths of h are locally Brownian in x we are forced to take $a = \frac{1}{2}$ to see nontrivial limiting behaviour. This forces us to take

$$z = \tfrac{3}{2}. \tag{16}$$

The nontrivial limiting behaviour of models in the KPZ universality class are all obtained in this scale.

On the other hand, if we started with KPZ with a weak asymmetry

$$\partial_t h = -\tfrac{1}{2}\epsilon^{1/2}(\partial_x h)^2 + \tfrac{1}{2}\partial_x^2 h + \xi, \tag{17}$$

then a diffusive scaling,

$$h_\epsilon(t, x) = \epsilon^{1/2} h(\epsilon^{-2}t, \epsilon^{-1}x), \tag{18}$$

would bring us back to the standard KPZ equation (11). In this way, KPZ and the continuum random polymer can be obtained from discrete models having an adjustable asymmetry.

4. Invariance principle for directed random polymers

Consider the distribution of the rescaled polymer path

$$x_\epsilon(t) := \epsilon x_{\lfloor \epsilon^{-2}t \rfloor}, \qquad 0 \le t \le T \tag{19}$$

under the measure $P^{\epsilon^{1/2}\beta,\xi}_{\epsilon^{-2}T,0}$ from (2). Note that the asymmetry here is the temperature which has been scaled into a crossover regime near zero. Again we have a joint measure on paths and noise which we call $\mathbb{P}^{\beta,\epsilon}_{T,0}$.

Theorem 4.1 [Alberts et al. 2014]. *Assume that $E[\xi] = 0$ and $E[\xi_-^8] < \infty$. Then the $\mathbb{P}^{\beta,\epsilon}_{T,0}$, $\epsilon > 0$ are a tight family and the limiting measure is the continuum random polymer $\mathbb{P}^{2^{1/2}\beta}_{T,0}$. In particular, for the free energy (3),*

$$\log Z^{\epsilon^{1/2}\beta,\xi}(\lfloor \epsilon^{-2}t \rfloor, 0) - \epsilon^{-2}\hat{\lambda}(\epsilon^{1/2}\beta)t + \tfrac{1}{2}\log(\epsilon^{-2}/4) \to \log z_{2^{1/2}\beta}(t, x), \tag{20}$$

where z_β is the solution of the stochastic heat equation $\partial_t z = \tfrac{1}{2}\partial_y^2 z - \beta\xi z$ with initial data $z_\beta(0, x) = \delta_0(x)$, and

$$\hat{\lambda}(\beta) = \tfrac{1}{2}E[\xi^2]\beta^2 + \tfrac{1}{3!}E[\xi^3]\beta^3 + \tfrac{1}{4!}E[\xi^4]\beta^4 \tag{21}$$

are the first four terms in the expansion of the log-moment generating function of the random variables $\xi = \xi(i, j)$.

The condition $E[\xi_-^8] < \infty$ is not optimal. One expects it to be true if $E[\xi_-^6] < \infty$ and false otherwise. The reason is that there are $\mathbb{O}(\epsilon^{-3})$ sites at play in the heat cone. Each $-\xi$ should not be larger than $\epsilon^{-1/2}$ or else it becomes an attractive point for the polymer. By Chebyshev inequality $P(\xi < -\epsilon^{-1/2}) = o(\epsilon^3)$, if $E[\xi_-^6] < \infty$, so we do not observe any such attractive points in the heat cone. The reason for the condition $E[\xi_-^8] < \infty$ is that there are no more than $\mathcal{O}(\epsilon^{-4})$ sites in all, so the argument becomes easier because we don't have to do tight estimates at the edge of the heat cone. Similar arguments, together with the conjectured localization of the polymer path lead to the conjectured $E[\xi_-^5] < \infty$ condition for the strong noise limit (4).

5. Asymmetric simple exclusion

The asymmetric simple exclusion process is a Markov process whose state space consists of particle configurations on $\mathbb{Z}$ with at most one particle per site. Each particle attempts to walk as a continuous time simple random walk on $\mathbb{Z}$, independently of the other particles, attempting jumps to the left as a Poisson process with rate q and to the right as a Poisson process with rate $p = 1 - q$. However, the jumps only take place if the target site is unoccupied. Because of the continuous time one does not have to face the issue of possible ties. The process can be thought of as a height function $h^{\mathrm{ASEP}}(t, x)$ given as

$$
h^{\mathrm{ASEP}}(t, x) = \begin{cases} 2N(t) + \sum_{0 < y \le x} \hat{\eta}(t, y), & x > 0, \\ 2N(t), & x = 0, \\ 2N(t) - \sum_{0 < y \le x} \hat{\eta}(t, y), & x < 0, \end{cases} \tag{22}
$$

where $N(t)$ records the net number of particles to cross from site 1 to site 0 in time t and where $\hat{\eta}(t, x)$ equals 1 if there is a particle at x at time t and -1 otherwise. The state space is now random walk paths in x, and the special definition with the $N(t)$ means that the entire dynamics for the height function is that local maxima become local minima at rate q and local minima become local maxima at rate p, independently for different nearest neighbour pairs.

The special case in which the initial data has all sites to the right of the origin occupied and all sites to the left unoccupied is called the *corner growth model*.

Theorem 5.1 (Tracy–Widom ASEP formula [Tracy and Widom 2009]). *Consider the corner growth model with $q > p$ such that $q + p = 1$. Let $\gamma = q - p$ and $\tau = p/q$. For $m = \lfloor \frac{1}{2}(s + x) \rfloor$, $t \ge 0$ and $x \in \mathbb{Z}$,*

$$
P(h_\gamma(t, x) \ge s) = \int_{S_{\tau^+}} \frac{d\mu}{\mu} \prod_{k=0}^{\infty} (1 - \mu \tau^k) \det(I + \mu J_{t,m,x,\mu})_{L^2(\Gamma_\eta)}, \tag{23}
$$

where S_{τ^+} is a positively oriented circle centred at zero of radius strictly between τ and 1, and where the kernel of the determinant is given by

$$
J_{t,m,x,\mu}(\eta, \eta') = \int_{\Gamma_\zeta} \exp\{\Psi_{t,m,x}(\zeta) - \Psi_{t,m,x}(\eta')\} \frac{f(\mu, \zeta/\eta')}{\eta'(\zeta - \eta)} d\zeta. \tag{24}
$$

Here η and η' are on Γ_η, a circle centred at zero of radius strictly between τ and 1; the ζ integral is on Γ_ζ, a circle centred at zero of radius strictly between 1 and τ^{-1}; and

$$f(\mu, z) = \sum_{k=-\infty}^{\infty} \frac{\tau^k}{1 - \tau^k \mu} z^k,$$

$$\Psi_{t,m,x}(\zeta) = \Lambda_{t,m,x}(\zeta) - \Lambda_{t,m,x}(\xi), \tag{25}$$

$$\Lambda_{t,m,x}(\zeta) = -x \log(1 - \zeta) + \frac{t\zeta}{1 - \zeta} + m \log \zeta.$$

6. Weakly asymmetric limit and KPZ crossover formula

It has been known since [Bertini and Giacomin 1997] that KPZ can be obtained as the weakly asymmetric limit of simple exclusion. In our case, we need this for the corner growth initial conditions which is not covered by their results.

Theorem 6.1 [Amir et al. 2011]. *For the corner growth model,*

$$\epsilon^{1/2} h^{\mathrm{ASEP}}_{q-p=\epsilon^{1/2}}(\epsilon^{-2}t, \epsilon^{-1}x) - \tfrac{1}{2}\epsilon^{-3/2} - \tfrac{1}{8}\epsilon^{-1/2} - \log(\tfrac{1}{2}\epsilon^{-1/2}) \to -\log z(t, x), \tag{26}$$

where $z(t, x)$ is the solution of the stochastic heat equation (7) with $z(0, x) = \delta_0(x)$ and $\beta = 1$.

The expression $h(t, x) = -\log z(t, x)$ is called the *narrow wedge solution of KPZ* and governs growth models with curved initial data.

In February 2010, Amir, Corwin and the author [Amir et al. 2011] and Sasamoto and Spohn [2010] independently studied the limit (26) of (24) by steepest descent. The methods were basically the same, however Amir et al. supply a mathematical proof, while Sasamoto and Spohn used physical arguments at various points. This gives the following exact formula for KPZ. Consider the solution of the stochastic heat equation with $z(0, x) = \delta_0(x)$ and $\beta = 1$ and define $\mathcal{A}_t$ by

$$z(t, x) = \frac{1}{\sqrt{2\pi t}} \exp\left(-\frac{x^2}{2t} + \frac{t}{24} + t^{1/3}\mathcal{A}_t(t^{-2/3}x)\right). \tag{27}$$

It is not hard to check that for each t, $\mathcal{A}_t(x)$ is stationary in x. It is called the *crossover Airy$_2$ process*.

Theorem 6.2 [Amir et al. 2011; Sasamoto and Spohn 2010].

$$P(t^{1/3}\mathcal{A}_t(x) \le s) = \int_{-\infty}^{\infty} e^{-e^{s-a}} \det(I - K_t) \operatorname{Tr}((I - K_t)^{-1} \operatorname{Proj}_{\mathrm{Ai}})_{L^2(t^{-1/3}a, \infty)} \, da,$$

where

$$K_t(x, y) = \int_{-\infty}^{\infty} \frac{1}{1 - e^{-t^{1/3}s}} \mathrm{Ai}(x + s)\mathrm{Ai}(y + s) \, ds. \tag{28}$$

In particular,

$$P(\mathscr{A}_t(x) \le s) \overset{t \to \infty}{\longrightarrow} F_{\mathrm{GUE}}(s). \tag{29}$$

From (20) and (29) we obtain:

Corollary 6.3 (weak universality for directed random polymers in $1 + 1$ dimensions). *Assume that ω are i.i.d. with $E[\xi_-^8] < \infty$. Then as $n \to \infty$ followed by $\beta \to \infty$,*

$$\frac{\log Z^{n^{-1/4}\beta, \xi}(n, 0) - n\hat{\lambda}(\beta n^{-1/4}) + \log \sqrt{\pi n/2} + 2\beta^4/3}{2\beta^{4/3}} \overset{(d)}{\longrightarrow} F_{\mathrm{GUE}}.$$

Here $\hat{\lambda}$ is defined in (21). As explained before the corollary is expected to be true exactly under the condition $E[\xi_-^6] < \infty$ which we hope to achieve in future work.

References

[Alberts et al. 2014] T. Alberts, K. Khanin, and J. Quastel, "The continuum directed random polymer", *J. Stat. Phys.* **154**:1-2 (2014), 305–326.

[Amir et al. 2011] G. Amir, I. Corwin, and J. Quastel, "Probability distribution of the free energy of the continuum directed random polymer in $1 + 1$ dimensions", *Comm. Pure Appl. Math.* **64**:4 (2011), 466–537.

[Bertini and Giacomin 1997] L. Bertini and G. Giacomin, "Stochastic Burgers and KPZ equations from particle systems", *Comm. Math. Phys.* **183**:3 (1997), 571–607.

[Biroli et al. 2007] G. Biroli, J.-P. Bouchaud, and M. Potters, "Extreme value problems in random matrix theory and other disordered systems", *J. Stat. Mech. Theory Exp.* **2007**:7 (2007), P07019.

[Hairer 2013] M. Hairer, "Solving the KPZ equation", *Ann. of Math.* (2) **178**:2 (2013), 559–664.

[Johansson 2000] K. Johansson, "Shape fluctuations and random matrices", *Comm. Math. Phys.* **209**:2 (2000), 437–476.

[Kardar 2007] M. Kardar, *Statistical physics of fields*, Cambridge University Press, Cambridge, 2007.

[Kardar et al. 1986] M. Kardar, G. Parisi, and Y.-C. Zhang, "Dynamic scaling of growing interfaces", *Phys. Rev. Lett.* **56** (1986), 889–892.

[Sasamoto and Spohn 2010] T. Sasamoto and H. Spohn, "Exact height distributions for the KPZ equation with narrow wedge initial condition", *Nuclear Phys. B* **834**:3 (2010), 523–542.

[Tracy and Widom 2009] C. A. Tracy and H. Widom, "Asymptotics in ASEP with step initial condition", *Comm. Math. Phys.* **290**:1 (2009), 129–154.

[Walsh 1986] J. B. Walsh, "An introduction to stochastic partial differential equations", pp. 265–439 in *École d'été de probabilités de Saint-Flour, XIV–1984*, edited by P. L. Hennequin, Lecture Notes in Math. **1180**, Springer, Berlin, 1986.

quastel@math.toronto.edu   *Departments of Mathematics and Statistics,*
                                  University of Toronto, 40 Saint George Street,
                                  Toronto, ON M5S 1L2, Canada

Random Matrices
MSRI Publications
Volume **65**, 2014

Replica analysis of the one-dimensional KPZ equation

TOMOHIRO SASAMOTO

In the last few years several exact solutions have been obtained for the one-dimensional KPZ equation, which describes the dynamics of growing interfaces. In particular the computations based on replica method have allowed to study fine fluctuation properties of the interface for various initial conditions including the narrow wedge, flat and stationary cases. In addition, an interesting aspect of the replica analysis of the KPZ equation is that the calculations are not only exact but also "almost rigorous". In this article we give a short review of this development.

1. Introduction

The one-dimensional Kardar–Parisi–Zhang (KPZ) equation,

$$\frac{\partial h(x,t)}{\partial t} = \frac{\lambda}{2}\left(\frac{\partial h(x,t)}{\partial x}\right)^2 + \nu\frac{\partial^2 h(x,t)}{\partial x^2} + \eta(x,t), \qquad (1)$$

is a well known prototypical equation which describes a growing interface [Kardar et al. 1986; Barabási and Stanley 1995]. Here $h(x,t)$ represents the height of the surface at position $x \in \mathbb{R}$ and time $t \geq 0$. The first term represents a nonlinearity effect and the second term describes a smoothing mechanism. The parameters λ and ν measure the strengths of these effects. The last term $\eta(x,t)$ indicates the existence of randomness in our description of surface growth. For the standard KPZ equation it is taken to be the Gaussian white noise with covariance,

$$\langle \eta(x,t)\eta(x',t')\rangle = D\delta(x-x')\delta(t-t'). \qquad (2)$$

Here and in the remainder of the article, $\langle \cdots \rangle$ indicates an average with respect to the randomness η.

The KPZ equation (1) is a nonlinear stochastic partial differential equation (SPDE), which is difficult to handle in general. But fortunately the KPZ equation has a nice integrable structure which has allowed detailed studies of its properties. In particular the one-point height distribution has been computed explicitly for three different initial conditions: the narrow wedge $h(x,0) = -|x|/\delta, \delta \to 0$

[Sasamoto and Spohn 2010a; Amir et al. 2011], the flat $h(x, 0) = 0, x \in \mathbb{R}$ [Calabrese and Le Doussal 2011], and the stationary, $h(x, 0) = B(x)$ [Imamura and Sasamoto 2012], cases. Here $B(x), x \in \mathbb{R}$, represents the (two-sided) one-dimensional Brownian motion with $B(0) = 0$. From the narrow wedge initial condition the surface grows to a shape of parabola macroscopically, which is representative of a curved surface. The flat case has been the most typical initial condition for Monte Carlo simulation studies, whereas the stationary case is regarded as one of the most important situation from the point of view of nonequilibrium statistical mechanics. The macroscopic shape from the flat and the BM cases are both flat but the fluctuations are different.

Historically the narrow wedge case was first "solved" by using a fact that the KPZ equation can be regarded as a certain weakly asymmetric limit of the asymmetric simple exclusion process (ASEP) [Sasamoto and Spohn 2010a; 2010b; 2010c; Amir et al. 2011]. Soon after the same result was rederived by using a replica method, which subsequently allowed the analysis of the other two cases as well [Calabrese and Le Doussal 2011; Imamura and Sasamoto 2012]. So the replica method has the advantage of being suited for various generalizations but it also has a disadvantage related to the analytic continuation about the replica number. In this article we explain and discuss a few aspects of the application of the replica method to the KPZ equation.

2. Cole–Hopf transformation

By a set of scalings of space, time and height,

$$x \to \alpha^2 x, \quad t \to 2\nu\alpha^4 t, \quad h \to \frac{\lambda}{2\nu}h,$$

with $\alpha = (2\nu)^{-3/2}\lambda D^{1/2}$, we can and will do set $\nu = \frac{1}{2}, \lambda = D = 1$. Applying the Cole–Hopf transformation,

$$Z(x, t) = e^{h(x,t)}, \tag{3}$$

(1) is linearized as

$$\frac{\partial Z(x, t)}{\partial t} = \frac{1}{2}\frac{\partial^2 Z(x, t)}{\partial x^2} + \eta(x, t)Z(x, t). \tag{4}$$

Without the last term η, the second term is absent and this is simply the diffusion equation which can be solved easily by Fourier analysis. For the KPZ equation, however, there remains the second term which has η as a multiplicative factor.

One should regard the noise to be the cylindrical Brownian motion. Then (4) written in the form of a stochastic differential equation is well defined and one can *define* the solution of the KPZ equation to be $h(x,t) = \log Z(x,t)$. This is called the Cole–Hopf solution.

A merit of this transformation is that this equation can be regarded as the imaginary time Schrödinger equation for a single particle under a random potential η. In other words this can be regarded as a statistical mechanical problem of a directed polymer in a random potential where Z has the meaning of its partition function. In particular one can write down the Feynman–Kac formula for this quantity [Bertini and Cancrini 1995],

$$Z(x,t) = \mathbb{E}_x \left(\exp \left[\int_0^t \eta(b(s), t - s) \, ds \right] Z(b(t), 0) \right), \tag{5}$$

where $\mathbb{E}_x$ represents the averaging over the standard Brownian motion $b(s)$, $0 < s < t$ with $b(0) = x$. The information of the initial condition is given by specifying $Z(x, t = 0)$ in this formula. We take $Z(x, t = 0) = \delta(x)$ for the narrow wedge case, $Z(x = 0, t) = 1$ for the flat case, and

$$Z(x, t = 0) = e^{B(x)} \tag{6}$$

for the stationary case.

3. Replica method

Originally we were interested in the distribution of the height $h(x,t)$ of the solution for the KPZ equation. After the Cole–Hopf transformation in the previous section, it is equivalent to the distribution of the logarithm of the partition function, $\log Z(x,t)$, of the directed polymer. But considering this quantity directly seems very difficult. Within the replica method, we instead compute N-th replica partition function $\langle Z^N(x,t) \rangle$ and try to retrieve the information about $\log Z$ from it. Z^N means that one is considering N copies of directed polymer systems with the same randomness and hence the name "replica".

The replica method is widely used when studying systems with randomness. For example, in spin glass theory, one considers a Hamiltonian like

$$H = \sum_{\langle ij \rangle} J_{ij} s_i s_j. \tag{7}$$

Here i is a site on a d-dimensional hypercube, the summation is taken over all nearest neighbor pairs of sites and $s_i = \pm 1$ is an Ising spin at the site i. The coupling constant $J_{i,j}$ is taken to be random, e.g., Bernoulli distributed independently for all $\langle ij \rangle$s. This is the Ising model with randomness. For low enough temperature in $d \geq 3$, there appears the spin glass phase in which the spin is frozen randomly [Nishimori 2001]. The quantity of main interest is the averaged free energy $\langle \log Z \rangle$, where $Z = \sum_{s_i = \pm 1} e^{-H}$ is the partition function. To study this one often resorts to the following identity,

$$\langle \log Z \rangle = \lim_{n \to 0} \frac{\langle Z^n \rangle - 1}{n}. \tag{8}$$

This is somehow true if n is taken to be a complex number. By writing $\langle Z^n \rangle = \langle e^{n \log Z} \rangle$, for a pure imaginary n one can consider $\langle Z^n \rangle$ as a characteristic function of $\log Z$ which exists in general. Furthermore when $\langle |\log Z| \rangle$ exists, $\langle \log Z \rangle$ is given by its first derivative:

$$\langle \log Z \rangle = \left. \frac{\partial \langle Z^n \rangle}{\partial n} \right|_{n=0}. \tag{9}$$

In many cases, however, one can compute the replica partition function only for integer $n = 1, 2, \ldots$. Then using the identity (8) implies that one is assuming the analytic continuation with respect to n. There is a theorem due to Carlson for this kind of situation but unfortunately in many cases of physical interest the assumption of the theorem does not hold and hence the application of (8) is not justified in general [Tanaka 2007]. But one can still try to utilize (8) which might give the correct answer. In fact this has been accepted as a very powerful techniques to study systems with randomness but when using it one always has to be careful about the pitfalls. The computation of the averaged free energy $\langle \log Z \rangle$ using this procedure is called the replica trick.

For the case of the KPZ equation, we are interested not only in the average but also in the full distribution of $\log Z$. We compute their generating function $G_t(s)$ of $\langle Z^N(x, t) \rangle$ defined as

$$G_t(s) = \sum_{N=0}^{\infty} \frac{(-e^{-\gamma_t s})^N}{N!} \langle Z^N(x, t) \rangle e^{N \frac{\gamma_t^3}{12}}, \tag{10}$$

with $\gamma_t = (t/2)^{1/3}$. Formally one can recover the probability density by inverting the generating function. Of course there is a problem of the uniqueness as discussed above. But this implies the possibility that one can recover the distribution of $\log Z$ by way of the computations of moments. In fact for the KPZ equation with narrow wedge initial condition, one can check that the correct distribution is obtained in this way [Calabrese et al. 2010; Dotsenko 2010]. It gives us a strong motivation to study the KPZ equation with other initial conditions by using the replica method.

4. Replica Bethe ansatz for KPZ equation: δ-Bose gas

Using the Feynman path integral representation of Z and remembering that the noise η is Gaussian, one can take the average with respect to the random potential

η for the replica partition function. As a result it is written as

$$\langle Z^N(x,t)\rangle = \langle x|e^{-H_N t}|\Phi\rangle. \tag{11}$$

More details about this procedure can be found in [Imamura and Sasamoto 2011b]. Here H_N is a nonrandom Hamiltonian of N particles, $\langle x|$ represents the state with all N particles being at the position x and the $|\Phi\rangle$ the initial state. For the KPZ equation, the Hamiltonian H_N turns out to be that of the delta-function Bose gas (δ-Bose gas) with attractive interaction [Kardar 1987]:

$$H_N = -\frac{1}{2}\sum_{j=1}^{N}\frac{\partial^2}{\partial x_j^2} - \frac{1}{2}\sum_{j\neq k}^{N}\delta(x_j - x_k). \tag{12}$$

The eigenvalues and eigenfunctions can be constructed by using the Bethe ansatz [Lieb and Liniger 1963; McGuire 1964; Dotsenko 2010; Calabrese et al. 2010]. In particular the ground state is a bound state whose wave function is

$$\langle x_1,\ldots,x_N \mid \Psi_z\rangle = Ce^{-\sum_{i,j=1}^{N}|x_i-x_j|}, \tag{13}$$

where C is a normalization constant and the corresponding energy is given by $E = -\frac{1}{24}(N^3 - N)$. Kardar [1987] argued that the N^3 term is responsible for the KPZ exponent $\frac{1}{3}$.

Fortunately, for the δ-Bose gas, one can give a description of all the eigenfunctions and eigenvalues. So at least formally one can expand $\langle Z^N(x,t)\rangle$ in terms of the eigenstates of the Hamiltonian and the corresponding eigenvalues. It had been anticipated for a long time that this might lead to more detailed information on h beyond the scaling exponent (see, for instance, [Dotsenko 2001]), but performing the summation over excited states is very involved and it was only very recently that this program was performed successfully to give an expression for the height distribution.

Now we give a description of the eigenstate and its eigenvalues. Let $|\Psi_z\rangle$ and E_z be the eigenstate and its eigenvalue of H_N:

$$H_N|\Psi_z\rangle = E_z|\Psi_z\rangle. \tag{14}$$

By the Bethe ansatz, they are given as follows. For a set of quasimomenta z_js, the eigenfunction is given by

$$\langle x_1,\ldots,x_N \mid \Psi_z\rangle = C_z \sum_{P\in S_N} \operatorname{sgn} P \prod_{1\leq j<k\leq N}\left(z_{P(j)}-z_{P(k)}+i\,\operatorname{sgn}(x_j-x_k)\right)$$
$$\times \exp\left(i\sum_{l=1}^{N}z_{P(l)}x_l\right), \tag{15}$$

where S_N is the set of permutations with N elements and C_z is the normalization constant,

$$C_z = \left(\frac{\prod_{\alpha=1}^{M} n_\alpha}{N!} \prod_{1 \le j < k \le N} \frac{1}{|z_j - z_k - i|^2} \right)^{1/2}, \tag{16}$$

taken to be a positive real number. The corresponding eigenvalue is simply given by $E_z = \frac{1}{2} \sum_{j=1}^{N} z_j^2$. For the δ-Bose gas with attractive interaction, the quasimomenta z_j $(1 \le j \le N)$ are in general complex numbers. They are divided into M groups $(1 \le M \le N)$ and the α-th group consists of n_α quasimomenta which share the common real part q_α. With this notation, the eigenvalue E_z is given by [Dotsenko 2010]

$$E_z = \frac{1}{2} \sum_{\alpha=1}^{M} n_\alpha q_\alpha^2 - \frac{1}{24} \sum_{\alpha=1}^{M} (n_\alpha^3 - n_\alpha). \tag{17}$$

Note that for $N = M$ and $q_\alpha = 0$, $1 \le \alpha \le N$, this gives the ground state energy $-\frac{1}{24}(N^3 - N)$ mentioned above.

We expand the replica partition function $\langle Z^N(x,t) \rangle$ (11) by the eigenstates as

$$\langle Z^N(x,t) \rangle = \sum_z e^{-E_z t} \langle x \mid \Psi_z \rangle \langle \Psi_z \mid \Phi \rangle. \tag{18}$$

For the case of the narrow wedge initial condition, $|\Phi\rangle$ is simply $|0\rangle$ and hence one only needs to take the summation over z. But in general we would write as

$$\langle Z^N(x,t) \rangle = \int_{-\infty}^{\infty} dy_1 \cdots \int_{-\infty}^{\infty} dy_N \langle x | e^{-H_N t} | y_1, \ldots, y_N \rangle \langle y_1, \ldots, y_N \mid \Phi \rangle. \tag{19}$$

Expanding the propagator $\langle x | e^{-H_N t} | y_1, \ldots, y_N \rangle$ by the Bethe eigenstates of the δ-Bose gas (15), we have

$$\langle Z^N(x,t) \rangle = \sum_{M=1}^{N} \frac{N!}{M!} \prod_{j=1}^{N} \int_{-\infty}^{\infty} dy_j \left(\int_{-\infty}^{\infty} \prod_{\alpha=1}^{M} \frac{dq_\alpha}{2\pi} \sum_{n_\alpha=1}^{\infty} \right) \delta_{\sum_{\beta=1}^{M} n_\beta, N}$$
$$\times e^{-E_z t} \langle x \mid \Psi_z \rangle \langle \Psi_z \mid y_1, \ldots, y_N \rangle \langle y_1, \ldots, y_N \mid \Phi \rangle. \tag{20}$$

Here we want to perform the integrations over y_j $(1 \le j \le N)$ and write

$$\langle Z^N(x,t) \rangle = \sum_{M=1}^{N} \frac{N!}{M!} \left(\int_{-\infty}^{\infty} \prod_{\alpha=1}^{M} \frac{dq_\alpha}{2\pi} \sum_{n_\alpha=1}^{\infty} \right) \delta_{\sum_{\beta=1}^{M} n_\beta, N}$$
$$\times e^{-E_z t} \langle x \mid \Psi_z \rangle \prod_{j=1}^{N} \int_{-\infty}^{\infty} dy_j \langle \Psi_z \mid y_1, \ldots, y_N \rangle \langle y_1, \ldots, y_N \mid \Phi \rangle. \tag{21}$$

At this point this is not allowed in general because in some cases like the stationary situation the integrations over q_α, $(1 \leq \alpha \leq M)$ must be performed before those over y_j, $(1 \leq j \leq N)$. But here with this remark in mind we will write the last factor as $\langle \Psi_z | \Phi \rangle$ and then

$$\langle Z^N(x,t) \rangle = \sum_{M=1}^{N} \frac{N!}{M!} \left(\int_{-\infty}^{\infty} \prod_{\alpha=1}^{M} \frac{dq_\alpha}{2\pi} \sum_{n_\alpha=1}^{\infty} \right) \delta_{\sum_{\beta=1}^{M} n_\beta, N} e^{-E_z t}$$
$$\times \langle x | \Psi_z \rangle \langle \Psi_z | \Phi \rangle. \quad (22)$$

5. Narrow wedge

For the narrow wedge case, $|\Phi\rangle = |0\rangle$. The wave function (15) with $x_l = x$ can be simplified by applying a combinatorial identity:

$$\sum_{P \in S_N} \text{sgn} \, P \prod_{1 \leq j < k \leq N} \left(w_{P(j)} - w_{P(k)} + if(j,k) \right) = N! \prod_{1 \leq j < k \leq N} (w_j - w_k). \quad (23)$$

This holds for any complex variables w_j $(1 \leq j \leq N)$ and $f(j,k)$ and was derived as Lemma 1 in [Prolhac and Spohn 2011a]. We find

$$\langle x | \Psi_z \rangle \langle \Psi_z | \Phi \rangle = N! \prod_{\alpha=1}^{M} \frac{(n_\alpha!)^2}{n_\alpha} \prod_{1 \leq j < k \leq N} \frac{|z_j - z_k|^2}{|z_j - z_k - i|^2} \prod_{l=1}^{N} e^{iz_l x}$$

$$= N! \prod_{\alpha < \beta}^{M} \frac{|q_\alpha - q_\beta - \frac{i}{2}(n_\alpha - n_\beta)|^2}{|q_\alpha - q_\beta - \frac{i}{2}(n_\alpha + n_\beta)|^2} \prod_{\alpha=1}^{M} \frac{e^{in_\alpha q_\alpha x}}{n_\alpha}$$

$$= 2^M N! \prod_{\alpha=1}^{M} \left(\int_0^\infty d\omega_\alpha \right)$$

$$\times \det \left(e^{in_j q_j x - n_j (\omega_j + \omega_k) - 2iq_j(\omega_j - \omega_k)} \right)_{j,k=1}^{M}, \quad (24)$$

where in the last equality we use the Cauchy determinantal formula.

Using this we get the expression for the generating function. Taking $x = 0$ for simplicity, we have

$$G_t(s) = \sum_{M=0}^{\infty} \frac{(-1)^M}{M!} \prod_{k=1}^{M} \int_0^\infty d\omega_k \, \det(A_{j,k})_{j,k=1}^{M}, \quad (25)$$

where

$$A_{j,k} = \sum_{n=1}^{\infty} (-1)^{n-1} \int_{\mathbb{R}} \frac{dq}{\pi} e^{-n(\omega_j + \omega_k) - 2iq(\omega_j - \omega_k) - \gamma_t^3 n q^2 + \frac{\gamma_t^3}{12} n^3 - \gamma_t n s}.$$

Here notice that the summation over n is divergent because of a factor $e^{\gamma_t^3 n^3/12}$. This is a serious difficulty of the replica analysis of the KPZ equation. But for the moment let us proceed by using a formula,

$$e^{n^3} = \int_{\mathbb{R}} \mathrm{Ai}(y)e^{ny}\, dy, \tag{26}$$

where Ai is the standard Airy function. This linearizes n^3 in the exponent and then one can take the geometric series to arrive at an expression,

$$G_t(s) = \det(1 - P_0 K_{t,s} P_0), \tag{27}$$

where det means the Fredholm determinant, P_0 is the projection to $[0, \infty)$ and the kernel is

$$K_{t,s}(\xi_j, \xi_k) = \int_{-\infty}^{\infty} dy\, \mathrm{Ai}(\xi_j + y)\mathrm{Ai}_\Gamma(\xi_k + y)\frac{e^{\gamma_t y}}{e^{\gamma_t y} + e^{\gamma_t s}}. \tag{28}$$

By inverting this one gets an expression for the height distribution, which agrees with the expression which has been derived by unarguably correct methods [Sasamoto and Spohn 2010a; Amir et al. 2011]. This may be a posteriori evidence that the replica method is useful for studying the KPZ equation. In addition, this can be considered as a "singular" limit of a rigorous analysis for a discrete model; see the remark in 7.2. The computations are involved and one awaits full details of their derivations.

6. Flat and stationary case

6.1. _Flat._ The state $|\Phi\rangle$ corresponding to the flat initial condition is constant. Hence one has to perform the y integration in (21) which is already not easy. Calabrese and Le Doussal [2011] found a formula for this case using the idea of studying a half-infinite system at first.

6.2. _Stationary._ In [Imamura and Sasamoto 2012], in order to take the average over the BM initial condition, we employed the strategy that we first consider a generalized initial condition,

$$h(x, 0) = \begin{cases} B_{-,v_-}(-x) := \tilde{B}(-x) + v_-x, & x < 0, \\ B_{+,v_+}(x) := B(x) - v_+x, & x > 0, \end{cases} \tag{29}$$

where $B(x)$, $\tilde{B}(x)$ are independent standard BMs and $v_\pm$ are the strengths of the drifts. The point is that once this generalized case is solved, one can study the stationary case by taking the $v_\pm \to 0$ limit.

Because the Brownian motion is a Gaussian process, one can perform the average over the initial distribution (29) and the dependence of $|\Phi\rangle$ on $x_1, \ldots, x_N$

can be explicitly calculated. For the region where

$$x_1 < \cdots < x_l < 0 < x_{l+1} < \cdots < x_N, \quad 1 \le l \le N,$$

one finds

$$\langle x_1, \ldots, x_N \mid \Phi \rangle$$
$$= e^{v_- \sum_{j=1}^l x_j - v_+ \sum_{j=l+1}^N x_j} \prod_{j=1}^l e^{\frac{1}{2}(2l-2j+1)x_j} \prod_{j=1}^{N-l} e^{\frac{1}{2}(N-l-2j+1)x_{l+j}}. \quad (30)$$

At this point one has to put the conditions $v_\pm > 0$ to have the wave function decaying at infinity. Since we are considering a bosonic system, this should be symmetrized with respect to $x_1, \ldots, x_N$. Using this together with another combinatorial identity, one can compute $\langle \Psi_z \mid \Phi \rangle$ as

$$\langle \Psi_z \mid \Phi \rangle = N! \, C_z \frac{\prod_{m=1}^N (v_+ + v_- - m) \prod_{1 \le j < k \le N} (z_j^* - z_k^*)}{\prod_{m=1}^N (-i z_m^* + v_- - 1/2)(-i z_m^* - v_+ + 1/2)}. \quad (31)$$

Then we can proceed in a fairy similar way as for the narrow wedge case. After some computation, we get an expression for the generating function,

$$G_t(s) = \sum_{N=0}^\infty \prod_{l=1}^N (v_+ + v_- - l) \sum_{M=1}^N \frac{(-e^{-\gamma_t s})^N}{M!} \prod_{\alpha=1}^M \left(\int_0^\infty d\omega_\alpha \sum_{n_\alpha=1}^\infty \right) \delta_{\sum_{\beta=1}^M n_\beta, N}$$

$$\times \det \left(\int_{\mathbb{R}-ic} \frac{dq}{\pi} \frac{e^{-\gamma_t^3 n_j q^2 + \frac{\gamma_t^3}{12} n_j^3 - n_j (\omega_j + \omega_k) - 2iq(\omega_j - \omega_k)}}{\prod_{r=1}^{n_j} (-iq + v_- + \frac{1}{2}(n_j - 2r))(iq + v_+ + \frac{1}{2}(n_j - 2r))} \right)_{j,k=1}^M, \quad (32)$$

with c taken large enough. A big difference from the narrow wedge [Sasamoto and Spohn 2010a; 2010b; 2010c; Amir et al. 2011] and the half BM initial condition [Imamura and Sasamoto 2011a] is that this generating function itself is not a Fredholm determinant because of the existence of the factor $\prod_{l=1}^N (v_+ + v_- - l)$. But this difficulty can be overcome by considering a further generalization of the initial condition in which the initial overall height is distributed as the inverse gamma distribution. After some computation, we obtain the height distribution for the initial condition (29) given by

$$F_{v_\pm, t}(s) = \frac{\Gamma(v_+ + v_-)}{\Gamma(v_+ + v_- + \gamma_t^{-1} d/ds)} \left[1 - \int_{-\infty}^\infty du \, e^{-e^{\gamma_t (s-u)}} v_{v_\pm, t}(u) \right]. \quad (33)$$

Here $v_{v_\pm, t}(u)$ is expressed as a difference of two Fredholm determinants,

$$v_{v_\pm, t}(u) = \det(1 - P_u (B_t^\Gamma - P_{\mathrm{Ai}}^\Gamma) P_u) - \det(1 - P_u B_t^\Gamma P_u), \quad (34)$$

where P_s represents the projection onto (s, ∞),

$$P_{\mathrm{Ai}}^{\Gamma}(\xi_1, \xi_2) = \mathrm{Ai}_{\Gamma}^{\Gamma}\left(\xi_1, \frac{1}{\gamma_t}, v_-, v_+\right) \mathrm{Ai}_{\Gamma}^{\Gamma}\left(\xi_2, \frac{1}{\gamma_t}, v_+, v_-\right), \qquad (35)$$

$$B_t^{\Gamma}(\xi_1, \xi_2) = \int_{-\infty}^{\infty} dy \frac{1}{1 - e^{-\gamma_t y}} \mathrm{Ai}_{\Gamma}^{\Gamma}\left(\xi_1 + y, \frac{1}{\gamma_t}, v_-, v_+\right)$$
$$\times \mathrm{Ai}_{\Gamma}^{\Gamma}\left(\xi_2 + y, \frac{1}{\gamma_t}, v_+, v_-\right), \qquad (36)$$

and

$$\mathrm{Ai}_{\Gamma}^{\Gamma}(a, b, c, d) = \frac{1}{2\pi} \int_{\Gamma_{i\frac{d}{b}}} dz e^{iza + i\frac{z^3}{3}} \frac{\Gamma(ibz + d)}{\Gamma(-ibz + c)}, \qquad (37)$$

where Γ_{z_p} represents the contour from $-\infty$ to ∞ and, along the way, passing below the pole at $z = id/b$. Note the similarity of our formulas with the narrow wedge case.

Once this generalized case is solved, it is not difficult to find a formula for the height distribution for the stationary situation. Furthermore by a simple generalization one can also study the stationary two point correlation function. For more details see [Imamura and Sasamoto 2012; 2013].

7. A few remarks

7.1. *Multipoint distribution.* Prolhac and Spohn [2011a; 2011b] applied the replica analysis to study the distribution at more than one point. Unfortunately a summation which appears in the computation seems impossible to perform. But they showed that if one introduces an "factorization approximation", one can proceed further and that in the scaling limit it tends to the Airy process which is the expected limiting process.

7.2. *Replica analysis for discretized models.* Recently, Borodin and Corwin [2014] introduced a new discrete model called the q-TASEP. In a certain limit this model reduces to the KPZ equation. On the other hand, for this model, the series become convergent and the replica computation can be made rigorous. In this sense, one could say that the replica method for the KPZ equation is "almost rigorous".

Acknowledgments

The author thanks the organizers and participants of the MSRI workshop "Random Matrix Theory, Interacting Particle Systems and Integrable Systems" for giving him the opportunities to discuss with many people on the subject. In particular he acknowledges A. Borodin, I. Corwin, P. L. Ferrari, T. Imamura, S. Prolhac, J. Quastel and H. Spohn for useful discussions. The work of Sasamoto is supported by KAKENHI (22740054).

References

[Amir et al. 2011] G. Amir, I. Corwin, and J. Quastel, "Probability distribution of the free energy of the continuum directed random polymer in 1 + 1 dimensions", *Comm. Pure Appl. Math.* **64**:4 (2011), 466–537.

[Barabási and Stanley 1995] A.-L. Barabási and H. E. Stanley, *Fractal concepts in surface growth*, Cambridge University Press, Cambridge, 1995.

[Bertini and Cancrini 1995] L. Bertini and N. Cancrini, "The stochastic heat equation: Feynman–Kac formula and intermittence", *J. Statist. Phys.* **78**:5-6 (1995), 1377–1401.

[Borodin and Corwin 2014] A. Borodin and I. Corwin, "Macdonald processes", *Probab. Theory Related Fields* **158**:1-2 (2014), 225–400.

[Calabrese and Le Doussal 2011] P. Calabrese and P. Le Doussal, "An exact solution for the KPZ equation with flat initial conditions", *Phys.Rev.Lett.* **106** (2011), 250603.

[Calabrese et al. 2010] P. Calabrese, P. L. Doussal, and A. Rosso, "Free-energy distribution of the directed polymer at high temperature", *Euro. Phys. Lett.* **90** (2010), 200002.

[Dotsenko 2001] V. Dotsenko, *Introduction to the replica theory of disordered statistical systems*, Collection Aléa-Saclay: Monographs and Texts in Statistical Physics, Cambridge University Press, Cambridge, 2001.

[Dotsenko 2010] V. Dotsenko, "Bethe ansatz derivation of the Tracy–Widom distribution for one-dimensional directed polymers", *EPL (Europhysics Letters)* **90**:2 (2010), 20003.

[Imamura and Sasamoto 2011a] T. Imamura and T. Sasamoto, "Current moments of 1D ASEP by duality", *J. Stat. Phys.* **142**:5 (2011), 919–930.

[Imamura and Sasamoto 2011b] T. Imamura and T. Sasamoto, "Replica approach to the KPZ equation with the half Brownian motion initial condition", *J. Phys. A* **44**:38 (2011), 385001.

[Imamura and Sasamoto 2012] T. Imamura and T. Sasamoto, "Exact solution for the stationary Kardar–Parisi–Zhang Equation", *Phys. Rev. Lett.* **108** (2012), 190603.

[Imamura and Sasamoto 2013] T. Imamura and T. Sasamoto, "Stationary correlations for the 1D KPZ equation", *J. Stat. Phys.* **150**:5 (2013), 908–939.

[Kardar 1987] M. Kardar, "Replica Bethe ansatz studies of two-dimensional interfaces with quenched random impurities", *Nuclear Phys. B* **290**:4 (1987), 582–602.

[Kardar et al. 1986] M. Kardar, G. Parisi, and Y.-C. Zhang, "Dynamic scaling of growing interfaces", *Phys. Rev. Lett.* **56** (1986), 889–892.

[Lieb and Liniger 1963] E. H. Lieb and W. Liniger, "Exact analysis of an interacting Bose gas, I: the general solution and the ground state", *Phys. Rev.* (2) **130** (1963), 1605–1616.

[McGuire 1964] J. B. McGuire, "Study of exactly soluble one-dimensional N-body problems", *J. Mathematical Phys.* **5** (1964), 622–636.

[Nishimori 2001] H. Nishimori, *Statistical physics of spin glasses and information processing*, International Series of Monographs on Physics **111**, Oxford University Press, New York, 2001.

[Prolhac and Spohn 2011a] S. Prolhac and H. Spohn, "Two-point generating function of the free energy for a directed polymer in a random medium", *J. Stat. Mech. Theory Exp.* 1 (2011), P01031.

[Prolhac and Spohn 2011b] S. Prolhac and H. Spohn, "The one-dimensional KPZ equation and the Airy process", *J. Stat. Mech. Theory Exp.* 3 (2011), P03020.

[Sasamoto and Spohn 2010a] T. Sasamoto and H. Spohn, "The crossover regime for the weakly asymmetric simple exclusion process", *J. Stat. Phys.* **140**:2 (2010), 209–231.

[Sasamoto and Spohn 2010b] T. Sasamoto and H. Spohn, "Exact height distributions for the KPZ equation with narrow wedge initial condition", *Nuclear Phys. B* **834**:3 (2010), 523–542.

[Sasamoto and Spohn 2010c] T. Sasamoto and H. Spohn, "One-Dimensional Kardar–Parisi–Zhang Equation: An Exact Solution and its Universality", *Phys. Rev. Lett.* **104** (2010), 230602.

[Tanaka 2007] T. Tanaka, "Moment problem in replica method", *Interdiscip. Inform. Sci.* **13**:1 (2007), 17–23.

sasamoto@phys.titech.ac.jp   *Department of Physics, Tokyo Institute of Technology, 2-12-1 Ookayama, Meguro-ku, Tokyo 152-8550, Japan*

Random Matrices
MSRI Publications
Volume **65**, 2014

Asymptotic expansions for β matrix models and their applications to the universality conjecture

MARIYA SHCHERBINA

We consider β matrix models with real analytic potentials for both one-cut and multi-cut regimes. We discuss recent results on the asymptotic expansion of the correlators and partition functions and their applications to the studies of random matrices.

1. Introduction

We consider the probability measure on $\mathbb{R}^n$ of the form

$$
p_{n,\beta}(\lambda_1, \dots, \lambda_n) = Q_{n,\beta}^{-1}[V] \prod_{i=1}^{n} e^{-n\beta V(\lambda_i)/2} \prod_{1 \le i < j \le n} |\lambda_i - \lambda_j|^\beta \tag{1-1}
$$

$$
=: Q_{n,\beta}^{-1}[V] e^{\beta H(\lambda_1, \dots, \lambda_n)/2},
$$

$$
\mathbf{E}_{n,\beta}\{(\dots)\} = \int (\dots) p_{n,\beta}(\lambda_1, \dots, \lambda_n) \, d\bar{\lambda}, \tag{1-2}
$$

where the function H, which we call the Hamiltonian to stress the analogy with statistical mechanics, and the normalizing constant $Q_{n,\beta}[V]$ (partition function) have the form

$$
H(\lambda_1, \dots, \lambda_n) = -n \sum_{i=1}^{n} V(\lambda_i) + \sum_{i \ne j} \log |\lambda_i - \lambda_j|
$$

$$
= \int e^{\beta H(\lambda_1, \dots, \lambda_n)/2} d\bar{\lambda}. \tag{1-3}
$$

The function V (called the potential) is a real-valued Hölder function satisfying the condition

$$
V(\lambda) \ge 2(1 + \epsilon) \log(1 + |\lambda|). \tag{1-4}
$$

We will study the asymptotic behavior (for large n) of $Q_{n,\beta}[V]$ and the marginal densities of (1-1) (correlation functions)

$$p_{l,\beta}^{(n)}(\lambda_1, \ldots, \lambda_l) = \int_{\mathbb{R}^{n-l}} p_{n,\beta}(\lambda_1, \ldots \lambda_l, \lambda_{l+1}, \ldots, \lambda_n) \, d\lambda_{l+1} \ldots d\lambda_n. \quad (1\text{-}5)$$

The distribution (1-1) can be considered for any $\beta > 0$, but the cases $\beta = 1, 2, 4$ are especially important, since they correspond to the eigenvalue distribution of real symmetric, hermitian, and symplectic matrix models respectively.

Since the papers [Boutet de Monvel et al. 1995; Johansson 1998] it is known that if V is a Hölder function, then

$$n^{-2} \log Q_{n,\beta}[V] = \frac{\beta}{2}\mathscr{E}[V] + O(\log n/n),$$

where

$$\mathscr{E}[V] = -\min_{m \in \mathcal{M}_1} \left\{ L[dm, dm] + \int V(\lambda) m(d\lambda) \right\} = \mathscr{E}_V(m^*), \quad (1\text{-}6)$$

and the minimizing measure m^* (called the equilibrium measure) has a compact support $\sigma := \operatorname{supp} m^*$. Here and below we denote

$$L[dm, dm] = \int \log |\lambda - \mu|^{-1} dm(\lambda) \, dm(\mu),$$

$$L[f](\lambda) = \int \log |\lambda - \mu|^{-1} f(\mu) \, d\mu, \quad L[f, g] = (L[f], g), \quad (1\text{-}7)$$

where $(.\,,.)$ is a standard inner product of $L^2[\mathbb{R}]$.

Moreover, it was proved in [Boutet de Monvel et al. 1995] that for any h whose first derivative is bounded in σ_ε (ε-neighborhood of σ) we have

$$\left| \int h(\lambda) \left(p_{1,\beta}^{(n)} d\lambda - dm(\lambda) \right) \right| \leq C \|h'\|_\infty (\log n/n)^{1/2}. \quad (1\text{-}8)$$

Here and below $\|\varphi\|_\infty = \sup_{\lambda \in \sigma_\varepsilon} |\varphi(\lambda)|$.

If V' is a Hölder function, then the equilibrium measure m^* has a density ρ (equilibrium density). The support σ and the density ρ are uniquely defined by the conditions:

$$v(\lambda) := 2 \int \log |\mu - \lambda| \rho(\mu) \, d\mu - V(\lambda) = \sup v(\lambda) := v^*, \quad \lambda \in \sigma,$$

$$v(\lambda) \leq \sup v(\lambda), \quad \lambda \notin \sigma, \quad \sigma = \operatorname{supp}\{\rho\}. \quad (1\text{-}9)$$

Without loss of generality we will assume below that $\sigma \subset (-1, 1)$ and $v^* = 0$.

In this paper we discuss the asymptotic expansion of the partition function $Q_{n,\beta}[V]$ and of the Stieltjes transforms of the marginal densities. Problems of this kind appear in many fields of mathematics, including the statistical mechanics of log-gases, combinatorics (graphical enumeration), and the theory of orthogonal polynomials (see [Ercolani and McLaughlin 2003] for the detailed and interesting discussion on the motivation of the problem). Here we are going to discuss with more details the applications of the problems to studies of the eigenvalue distribution of random matrices.

The first important problem of the eigenvalue distribution is the behavior of the random variables called the linear eigenvalue statistics, which correspond to the smooth test function h

$$\mathcal{N}_n[h] = \sum_{i=1}^{n} h(\lambda_i). \tag{1-10}$$

The result of (1-8) gives us the main term of the expectation of $\mathbf{E}_{n,\beta}\{\mathcal{N}_n[h]\}$. It was also proved in [Boutet de Monvel et al. 1995] that the variance of $\mathcal{N}_n[h]$ tends to zero, as $n \to \infty$. But the behavior of the fluctuations of $\mathcal{N}_n[h]$ was studied only in the case of one-cut potentials (see [Johansson 1998]). Even the bound for $\mathbf{Var}_{n,\beta}\{\mathcal{N}_n[h]\}$ in the multi-cut regime was known only for $\beta = 2$. Thus the behavior of the characteristic functional, corresponding to the linear eigenvalue statistics (1-10) of the test function h

$$Z_{n,\beta}[h] = \mathbf{E}_{n,\beta}\{e^{\mathcal{N}_n[h] - \mathbf{E}_{n,\beta}\{\mathcal{N}_n[h]\}}\} = \frac{Q_{n,\beta}[V - \frac{2}{\beta}(h - \mathbf{E}_{n,\beta}\{n^{-1}\mathcal{N}_n[h]\})]}{Q_{n,\beta}[V]} \tag{1-11}$$

is one of the questions of primary interest in the random matrix theory. It is evident from the right-hand side of (1-11) that since $Z_{n,\beta}[h]$ is a ratio of two partition functions, to study the behavior of $Z_{n,\beta}[h]$, it suffices to find the coefficients of the expansion of $\log Q_{n,\beta}[V]$ up to the order $o(1)$.

The other very important question of the theory of random matrices is so-called the universality conjecture for the local eigenvalue statistics. According to this conjecture, for example, for the bulk of the spectrum, the behavior of the scaled correlation functions of (1-5)

$$p_{k,\beta}^{(n)}(\lambda_0 + x_1/(n\rho(\lambda_0)), \ldots, \lambda_0 + x_k/(n\rho(\lambda_0)))$$

in the limit $n \to \infty$ is universal, that is, do not depend on V and depends only on β. For $\beta = 2$ this problem is very well studied now. It is well known (see, e.g., [Mehta 1991]) that for $\beta = 2$ all correlation functions of (1-5) can be expressed in the terms of the reproducing kernel of the system of polynomials orthogonal with a varying weight $e^{-n\beta V}$. The orthogonal polynomial machinery, in particular, the Christoffel–Darboux formula and Christoffel function simplify considerably the studies of marginal densities (1-5). This allows to study the local eigenvalue statistics in many different cases: bulk of the spectrum, edges of the spectrum, special points, etc. (see [Pastur and Shcherbina 1997; 2008; 2011; Deift et al. 1999; Bleher and Its 2003; Claeys and Kuijlaars 2006; Levin and Lubinsky 2008; McLaughlin and Miller 2008; Shcherbina 2011]).

For $\beta = 1, 4$ the situation is more complicated. It was shown in [Tracy and Widom 1998] that all correlation functions can be expressed in terms of some 2×2-matrix kernels. But the representation is less convenient than that in the case $\beta = 2$. It makes difficult the problems, which for $\beta = 2$ are just simple exercises. For example, the bound for the variance of linear eigenvalue statistics for $\beta = 2$ is a trivial corollary of the Christoffel–Darboux formula for any σ, while for $\beta = 1, 4$, as it was mentioned above, in the multi-cut regime the bound was not known till the recent time. As for the universality conjecture, there were a number of papers with improving results, first for monomial $V = \lambda^{2m} + o(1)$, (see [Stojanovic 2000; Deift and Gioev 2007b; 2007a; Deift et al. 2007]) proving the bulk and edge universality for $\beta = 1, 4$, then for arbitrary real analytic one-cut potential (see [Shcherbina 2009b; 2009a]). But combining interesting observations of the papers [Widom 1999; Stojanovic 2000], we conclude that to prove the bulk universality for $\beta = 1, 4$, it is enough to control $\log Q_{n,\beta}[V]$ up to the $O(1)$ terms. This was done first for the one-cut case in [Kriecherbauer and Shcherbina 2010] and then in the multi-cut case in [Shcherbina 2011] (see Section 3 for a more detailed discussion of the universality proof).

Let us mention now the most important results on the expansion of $\log Q_{n,\beta}[V]$ and the correlators. The CLT for linear eigenvalue statistics in the one-cut regime for any β and polynomial V was proved in [Johansson 1998]. The expansion for the first and the second correlators for $\beta = 2$ and one-cut real analytic V and $\beta = 2$ was proved in [Albeverio et al. 2001]. The expansion of $\log Q_{n,\beta}[V]$ for a one-cut polynomial V and $\beta = 2$ was obtained in [Ercolani and McLaughlin 2003]. The formal expansion for any β and polynomial V were obtained in the physical papers [Chekhov and Eynard 2006; Eynard 2009]. The CLT for real analytic multi-cut V and special $h = V$ for $\beta = 2$ was obtained in [Pastur 2006]. The control of $\log Q_{n,\beta}[V]$ up to $O(1)$ for one-cut real analytic V and multi-cut real analytic V was performed in [Kriecherbauer and Shcherbina 2010] and [Shcherbina 2011], respectively. The expansion of the partition function and all the correlators for the one-cut real analytic V and any β was constructed in [Borot and Guionnet 2013]. And the CLT for linear eigenvalue statistics in the multi-cut regime for any β and polynomial V was proved recently in [Shcherbina 2013].

The paper is organized as follows. In Section 2 we discuss the CLT and the expansion of the partition function and correlators in the one-cut regime, obtained in [Johansson 1998; Kriecherbauer and Shcherbina 2010; Borot and Guionnet 2013]. In Section 3 we discuss the applications of the results on the control of $\log Q_{n,\beta}[V]$ up to $O(1)$ in the multi-cut case to the proof of the bulk universality for $\beta = 1, 4$ in the multi-cut case, following [Kriecherbauer and Shcherbina 2010; Shcherbina 2011], and in Section 4 we discuss the results of [Shcherbina 2013] on the CLT in the multi-cut case.

Throughout the paper we assume the following conditions on the potential V:

C1. *V is a Hölder function satisfying (1-4), which is analytic in some open domain of $\mathbf{D} \subset \mathbb{C}$ containing the support σ of the corresponding equilibrium measure, and*

$$\sigma = \bigcup_{\alpha=1}^{q} \sigma_\alpha, \qquad \sigma_\alpha = [a_\alpha, b_\alpha]; \tag{1-12}$$

C2. *The equilibrium density ρ can be represented in the form*

$$\rho(\lambda) = \frac{1}{2\pi} P(\lambda) \Im X^{1/2}(\lambda + i0), \quad \inf_{\lambda \in \sigma} |P(\lambda)| > 0, \tag{1-13}$$

where

$$X(z) = \prod_{\alpha=1}^{q} (z - a_\alpha)(z - b_\alpha), \tag{1-14}$$

and we choose a branch of $X^{1/2}(z)$ such that $X^{1/2}(z) \sim z^q$, as $z \to +\infty$. Moreover, the function v defined by (1-9) attains its maximum only if λ belongs to σ.

Remark. It is known (see, e.g., [Albeverio et al. 2001]) that for analytic V the equilibrium density ρ always has the form (1-13)–(1-14). The function P in (1-13) is analytic and can be represented in the form

$$P(z) = \frac{1}{2\pi i} \oint_{\mathscr{L}} \frac{V'(z) - V'(\zeta)}{(z - \zeta) X^{1/2}(\zeta)} \, d\zeta. \tag{1-15}$$

Hence condition C2 means that ρ has no zeros in the internal points of σ and behaves like square root near the edge points. This behavior of V is usually called generic (see [Kuijlaars and McLaughlin 2000] for the results explaining the term).

2. Asymptotic expansion and CLT for β matrix models in the one-cut regime

The one-cut case is the simplest version of the possible spectrum of the β-models. As it is clear from the physical papers [Chekhov and Eynard 2006; Eynard 2009], it is the only case when it is expected that fluctuations of eigenvalue statistics are asymptotically Gaussian and the asymptotic expansions of $\log Q_{n,\beta}[V]$ does not contain some kind of θ-function. Hence, almost all known before results on the expansions of $\log Q_{n,\beta}[V]$ and the correlators (see the definition in (2-4) below) were obtained for the one-cut potentials V. One of the first results in this direction is the CLT for linear eigenvalue statistics, which was proved by Johansson [1998] and improved in [Kriecherbauer and Shcherbina 2010; Shcherbina 2013].

Theorem 1. *Let V satisfy condition C1–C2 and $\sigma = \mathrm{supp}\, \rho = [a, b]$. Then for any real-valued h with $\|h^{(4)}\|_\infty$, $\|h'\|_\infty \leq \log n$ the characteristic functional $Z_{n,\beta}[h]$ of (1-11) has the form*

$$Z_{n,\beta}[h] = \exp\left\{ \frac{\beta}{2}\left(\left(\frac{2}{\beta} - 1 \right)(h, v) + \tfrac{1}{4}(\overline{D}h, h) \right) \right\}$$

$$\cdot \left(1 + n^{-1} O\big(\|h'\|_\infty^3 + \|h^{(4)}\|_\infty^3 \big) \right), \quad (2\text{-}1)$$

where the operator $\overline{D}_\sigma$ is defined as

$$\overline{D}_\sigma = \frac{1}{2}(D_\sigma + D_\sigma^*), \quad D_\sigma h(\lambda) = \frac{X^{-1/2}(\lambda)}{\pi^2} \int_\sigma \frac{h'(\mu) X^{1/2}(\mu)\, d\mu}{(\lambda - \mu)}, \quad (2\text{-}2)$$

and the nonpositive measure v has the form

$$(v, h) := \tfrac{1}{4}(h(b) + h(a)) - \frac{1}{2\pi} \int_\sigma \frac{h(\lambda)\, d\lambda}{X^{1/2}(\lambda)} + \tfrac{1}{2}(D_\sigma \log P, h), \quad (2\text{-}3)$$

with P defined by (1-15) and $X^{1/2}(\lambda) := \Im X^{1/2}(\lambda + i0)$ with X of (1-14).

The method of the proof proposed in [Johansson 1998] was based on the analysis of the first loop equation (see (2-10) below) combined with a priory bound (1-8), obtained in [Boutet de Monvel et al. 1995]. But it was used essentially in the proof that V is a polynomial. Then in [Kriecherbauer and Shcherbina 2010] the method of [Johansson 1998] was generalized to any one-cut analytic potential. Moreover, $\log Q_{n,\beta}[V]$ was found up to $O(1)$ term by using the idea of the interpolation between the Gaussian potential and the arbitrary one-cut potential. The last step in the construction of the asymptotic expansion in n^{-1} was done recently in [Borot and Guionnet 2013]. The authors studied the asymptotic expansion of all correlators, defined as

$$w_k(z_1, \ldots, z_k) := \frac{1}{n} \frac{\partial^k}{\partial t_1 \ldots \partial t_k} \log Q_{n,\beta}\left[V - \frac{2}{\beta n} \sum_{j=1}^{k} t_k \phi_{z_k} \right]\bigg|_{t_1 = \cdots = t_k = 0}, \quad (2\text{-}4)$$

where

$$\phi_z(\lambda) = \frac{1}{z - \lambda}.$$

One can easily see that then, for example, w_1 is the Stieltjes transform of the first marginal density (1-5) and

$$w_1(z) = n^{-1}\mathbf{E}_{n,\beta}\{\mathcal{N}_n[\phi_z]\} = (\phi_z, p_{1,\beta}^{(n)}),$$

$$w_2(z_1, z_2) = n^{-1}\mathbf{Cov}_{n,\beta}\{\mathcal{N}_n[\phi_{z_1}], \mathcal{N}_n[\phi_{z_2}]\}. \tag{2-5}$$

The main result of [Borot and Guionnet 2013] is the following theorem.

Theorem 2. *Under conditions C1–C2, any correlator (2-4) admits an asymptotic expansion of any order m, which means that*

$$w_k(z_I) = \sum_{j=k-1}^{m} n^{-j} w_k^{(j)}(z_I) + O(n^{-m-1}), \tag{2-6}$$

where the bound is uniform in $z_1, \ldots, z_k$ varying in any compact K of the upper half-plane.

Moreover, $\log Q_{n,\beta}$ also admits the asymptotic expansion in n of any order m:

$$\log(Q_{n,\beta}/n!) = \frac{\beta n^2}{2}\mathscr{E}[V] + F_\beta(n) + n\left(\frac{\beta}{2}-1\right)((\log\rho, \rho) - 1 - \log 2\pi)$$

$$+ \sum_{j=0}^{m} n^{-j} q^{(j)}[\rho] + O(n^{-m-1}), \tag{2-7}$$

where the coefficients of the expansion $q^{(j)}[\rho]$ are defined in terms of the integrals with the Stieltjes transform of the equilibrium density ρ, $F_\beta(n)$ collects the term which appears in the Gaussian case

$$F_\beta(n) = \log(Q_{n,\beta}^*/n!) + \frac{3\beta n^2}{8},$$

and $Q_{n,\beta}^*$ is the partition function of the Gaussian case, that is, corresponds to $V(\lambda) = \frac{1}{2}\lambda^2$.

Remark. By the Selberg formula (see, e.g., [Forrester 2010]), we have

$$Q_{n,\beta}^*/n! = \left(\frac{n\beta}{2}\right)^{-\beta n^2/4 - n(1-\beta/2)/2} (2\pi)^{n/2} \prod_{j=1}^{n} \frac{\Gamma(\beta j/2)}{\Gamma(\beta/2)}. \tag{2-8}$$

Moreover, it is known (see [Forrester 2010]) that

$$F_\beta(n) = n\left(\frac{\beta}{2}-1\right)\left(\log\frac{n\beta}{2} - \frac{1}{2}\right) + n\log\frac{\sqrt{2\pi}}{\Gamma(\beta/2)} - c_\beta \log n + c_\beta^{(1)} + o(1), \tag{2-9}$$

where

$$c_\beta = \frac{\beta}{24} - \frac{1}{4} + \frac{1}{6\beta},$$

and $c_\beta^{(1)}$ is some constant, depending only on β (for $\beta = 2$, $c_\beta^{(1)} = \zeta'(1)$).

Sketch of the proof. As was mentioned above, the proof (given in [Borot and Guionnet 2013]) is a nice combination of the methods and results of [Johansson 1998; Kriecherbauer and Shcherbina 2010] with the analysis of the loop equations given in the physical papers [Chekhov and Eynard 2006; Eynard 2009]. The first loop equation is well known and used in many papers [Pastur and Shcherbina 1997; Johansson 1998; Kriecherbauer and Shcherbina 2010]:

$$w_1^2(z) - V'(z)w_1(z) + \frac{1}{2\pi i} \oint_L \frac{V'(z) - V'(\zeta)}{z - \zeta} w_1(\zeta)\, d\zeta$$

$$= \frac{1}{n}\left(\frac{2}{\beta} - 1\right)\partial_z w_1(z) - \frac{1}{n}w_2(z, z). \quad (2\text{-}10)$$

Here and below the contours L, L' (and so on) in $\mathbf{D}$ encircle the ε-neighborhood of the spectrum Σ, but do not contain z and zeros of P of (1-15). The other loop equations can be obtained from the first one by differentiating as in (2-4):

$$(2w_1(z) - V'(z))w_{k+1}(z, z_I) + \frac{1}{2\pi i}\oint \frac{V'(z) - V'(\zeta)}{z - \zeta} w_{k+1}(\zeta, z_I)\, d\zeta$$

$$= F_{k+1}\big(z; \{w_j\}_{j=2}^{k+2}\big),$$

where

$$F_{k+1}\big(z; \{w_j\}_{j=2}^{k+2}\big)$$

$$:= \frac{1}{n}\left(\frac{2}{\beta} - 1\right)\partial_z w_{k+1}(z, z_I) - \sum_{\substack{J \subset I \\ |J| \neq 0, k}} w_{|J|+1}(z, z_J) w_{k+1-|J|}(z, z_{I \setminus J})$$

$$- \frac{2}{\beta}\sum_{j=1}^{k}\partial_{z_j} \frac{w_k(z, z_{I \setminus \{j\}}) - w_k(z_I)}{z - z_j} - \frac{1}{n}w_{k+2}(z, z, z_I).$$

It was proved in [Johansson 1998; Kriecherbauer and Shcherbina 2010] that

$$w_1(z) = g(z) + n^{-1}w_1^{(1)}(z) + O(n^{-2}),$$
$$g(z) = \tfrac{1}{2}\big(V'(z) - P(z)X^{1/2}(z)\big). \quad (2\text{-}11)$$

Substituting this expression in (2-10) and multiplying the result by n, we obtain an equation with respect to $w_1^{(1)}$, which (combined with equations for $\{w_k\}_{k \geq 2}$ above) gives us the system of equations:

$$\mathcal{K}w_1^{(1)}(z) = \left(1 - \frac{2}{\beta}\right)\partial_z\left(g(z) + \frac{1}{n}w_1^{(1)}(z)\right) - w_2(z, z) - \frac{1}{n}\big(w_1^{(1)}(z)\big)^2$$

$$=: F_1\big(z; w_1^{(1)}, w_2\big),$$

$$\mathcal{K}w_{k+1}(z, z_I) = F_{k+1}\big(z; \{w_j\}_{j=2}^{k+2}\big) - \frac{2}{n}w_1^{(1)}(z)w_{k+1}(z, z_I), \quad (2\text{-}12)$$

where the linear operator $\mathcal{H} : \text{Hol} [\mathbf{D} \setminus \sigma] \to \text{Hol} [\mathbf{D} \setminus \sigma]$ is defined as

$$\mathcal{H} f(z) = -P(z) X^{1/2}(z) f(z) + \frac{1}{2\pi i} \oint \frac{V'(z) - V'(\zeta)}{z - \zeta} f(\zeta) \, d\zeta.$$

Consider also the operator $\mathcal{H}^{(-1)} : \text{Hol} [\mathbf{D} \setminus \sigma] \to \text{Hol} [\mathbf{D} \setminus \sigma]$ of the form

$$\mathcal{H}^{(-1)} f(z) := \frac{1}{2\pi i X^{1/2}(z)} \oint_{\mathcal{L}} \frac{f(\zeta) \, d\zeta}{P(\zeta)(z - \zeta)}. \tag{2-13}$$

Till now we have not used that we have a one-cut potential. The loop equations can be written in the multi-cut case as well as in the one-cut and the operators $\mathcal{H}$ and $\mathcal{H}^{(-1)}$ can be constructed by the same formulas, if we use P and X of (1-15) and (1-14). It is straightforward to check that if we apply to the both parts of (2-12) the operator (2-13), then in the multi-cut case (when $X^{1/2}(z) \sim z^q$) we obtain

$$w_{k+1}(z, z_I) + \frac{p_{k+1}(z; z_I)}{X^{1/2}(z)} =$$
$$\mathcal{H}^{(-1)} F_{k+1}\big(z; \{w_j\}_{j=2}^{k+2}\big) - \frac{2}{n} \mathcal{H}^{(-1)} w_1^{(1)}(z) w_{k+1}(z, z_I),$$

where $p_{k+1}(z; z_I)$ is a polynomial with respect to z of degree $q - 2$, whose coefficients are the linear combinations of the first $q - 1$ coefficient in the asymptotic expansion of $w_{k+1}(z, z_I)$ with respect to z^{-j}. The main technical obstacle to study the multi-cut case by the method described in this section is that for $q \neq 1$ we do not know these coefficients, while in the case $q = 1$ (one-cut case) it is easy to see that $p_{k+1}(z; z_I) = 0$ for all $k \geq 0$ and we obtain the system of equations

$$w_{k+1}(z, z_I) = \mathcal{H}^{(-1)} F_{k+1}\big(z; \{w_j\}_{j=2}^{k+2}\big) - \frac{2}{n} \mathcal{H}^{(-1)} w_1^{(1)}(z) w_{k+1}(z, z_I). \tag{2-14}$$

The key technical point in the analysis of the last equations is a priory estimate

$$|w_{k+1}(z, z_I)| \leq n^{-1} C(z, z_I), \quad k \geq 1. \tag{2-15}$$

It can be derived from the bound proven in [Johansson 1998] (see also [Kriecherbauer and Shcherbina 2010]). Let $\mathcal{L}$ be any contour enclosed σ. Then there is a constant C_L such that for any real analytic function φ,

$$\mathbf{E}_{n,\beta}\big\{\exp\{\mathring{\mathcal{N}}_n[\varphi] / (C_L \sup_{\zeta \in \mathcal{L}} |\varphi(\zeta)|)\}\big\} \leq 6$$
$$\Rightarrow \mathbf{E}_{n,\beta}\{|\mathring{\mathcal{N}}_n[\varphi]|^p\} \leq C_p (C_L \sup_{\zeta \in \mathcal{L}} |\varphi(\zeta)|)^p,$$

where $\mathring{\mathcal{N}}_n[\varphi] = \mathcal{N}_n[\varphi] - \mathbf{E}_{n,\beta}\{\mathcal{N}_n[\varphi]\}$. The last bound implies (2-15). With this bound in hands it is easy to see that (2-14) has "triangle" form: the right-hand side of the equation for w_{k+1} contains $w_2, \ldots, w_k$, the derivative of $n^{-1} w_{k+1}$,

$n^{-1}w_1^{(1)}w_{k+1}$, and $n^{-1}w_{k+2}$. Hence it can be solved in each order in n^{-1} starting from the first equation and going down step by step. This leads to the assertion (2-6).

To derive the assertion (2-7) from (2-6), we use the idea of [Kriecherbauer and Shcherbina 2010] of interpolation between the Gaussian (quadratic) potential with the same support $\sigma = [a, b]$ and the potential V. Consider the functions $V^{(0)}$ and V_t of the form

$$V^{(0)}(\lambda) = 2(\lambda - c)^2/d, \quad c = \tfrac{1}{2}(a+b), \ d = b - a,$$
$$V_t(\lambda) = tV(\lambda) + (1 - t)V^{(0)}(\lambda). \tag{2-16}$$

Let $Q_{n,\beta}(t) := Q_{n,\beta}[V_t]$ be defined by (1-3) with V replaced by V_t. Then, evidently, $Q_{n,\beta}(1) = Q_{n,\beta}[V]$, and $Q_{n,\beta}(0) = Q_{n,\beta}[V^{(0)}]$. Hence

$$\frac{1}{n^2}\log Q_{n,\beta}(1) - \frac{1}{n^2}\log Q_{n,\beta}(0) = \frac{1}{n^2}\int_0^1 dt\, \frac{d}{dt}\log Q_{n,\beta}(t)$$
$$= -\frac{\beta}{2\pi i}\int_0^1 dt \oint_L dz(V(z) - V^{(0)}(z))w_1(z; t), \tag{2-17}$$

where $w_1(z; t)$ is defined by (2-4) for V_t. Using (1-9), one can check that for the distribution (1-1) with V replaced by V_t the equilibrium density ρ_t has the form

$$\rho_t(\lambda) = t\rho(\lambda) + (1 - t)\rho^{(0)}(\lambda), \quad \rho^{(0)}(\lambda) = \frac{2X^{1/2}(\lambda)}{\pi d^2}, \tag{2-18}$$

with X of (1-14). Hence, substituting (2-11) for V_t into (2-17), we get

$$\log Q_{n,\beta}[V] = \log Q_{n,\beta}[V^{(0)}] - n^2\frac{\beta}{2}\mathscr{E}[V^{(0)}] + n^2\frac{\beta}{2}\mathscr{E}[V]$$
$$+ \frac{\beta n}{2}\frac{1}{(2\pi i)}\int_0^1 dt \oint_{\mathscr{L}} (V(z) - V^{(0)}(z))w_1^{(1)}(z; t)\, dz,$$

Then we use the expression for $w_1^{(1)}(z; t)$ which follow from the first equations of (2-12). After some transformations we arrive at (2-7). $\qquad\square$

3. Bulk universality for orthogonal and symplectic ensembles

As it was mentioned in Introduction one of the most important applications of the asymptotic expansion of $\log Q_{n,\beta}[V]$ is the proof of the universality of the local regime in the case of $\beta = 1, 4$ (for real symmetric and symplectic matrix models). Throughout this section we will assume that V is a polynomial of degree $2m$, satisfying condition C2, and n is even, but the result can be generalized on V, satisfying conditions C1–C2. According to the results of [Tracy and Widom

1998], the matrix kernels for $\beta = 1, 4$ can be expressed in terms of the scalar kernels

$$S_{n,1}(\lambda, \mu) = -\sum_{j,k=0}^{n-1} \psi_j^{(n)}(\lambda)(M_n^{(n)})_{jk}^{-1}(\epsilon \psi_k^{(n)})(\mu), \tag{3-1}$$

$$S_{n/2,4}(\lambda, \mu) = -\sum_{j,k=0}^{n-1} (\psi_j^{(n)})'(\lambda)(D_n^{(n)})_{jk}^{-1} \psi_k^{(n)}(\mu); \tag{3-2}$$

here $\epsilon(\lambda) = \frac{1}{2}\,\mathrm{sgn}(\lambda)$ (sgn denoting the standard signum function),

$$(\epsilon f)(\lambda) := \int_{\mathbb{R}} \epsilon(\lambda - \mu) f(\mu)\, d\mu,$$

and $D_n^{(n)}$ and $M_n^{(n)}$ are the top left corner $n \times n$ blocks of the semiinfinite matrices that correspond to the differentiation operator and to some integration operator, respectively:

$$D_\infty^{(n)} := \big((\psi_j^{(n)})', \psi_k^{(n)}\big)_{j,k \geq 0}, \quad D_n^{(n)} = \{D_{jk}^{(n)}\}_{j,k=0}^{n-1},$$

$$M_\infty^{(n)} := \big(\epsilon \psi_j^{(n)}, \psi_k^{(n)}\big)_{j,k \geq 0}, \quad M_n^{(n)} = \{M_{jk}^{(n)}\}_{j,k=0}^{n-1}. \tag{3-3}$$

Both matrices $D_\infty^{(n)}$ and $M_\infty^{(n)}$ are skew-symmetric, and since $\epsilon(\psi_j^{(n)})' = \psi_j^{(n)}$, we have for any $j, l \geq 0$ that

$$\delta_{jl} = \big(\epsilon(\psi_j^{(n)})', \psi_l\big) = \sum_{k=0}^{\infty} (D_\infty^{(n)})_{jk}(M_\infty^{(n)})_{kl} \iff D_\infty^{(n)} M_\infty^{(n)} = 1 = M_\infty^{(n)} D_\infty^{(n)}.$$

It was observed in [Widom 1999] that if V is a rational function, in particular, a polynomial of degree $2m$, then the kernels $S_{n,1}, S_{n,4}$ can be written as

$$S_{n,1}(\lambda, \mu) = K_{n,2}(\lambda, \mu) + n \sum_{j,k=-(2m-1)}^{2m-1} F_{jk}^{(1)} \psi_{n+j}^{(n)}(\lambda)\epsilon\psi_{n+k}^{(n)}(\mu),$$

$$\tag{3-4}$$

$$S_{n/2,4}(\lambda, \mu) = K_{n,2}(\lambda, \mu) + n \sum_{j,k=-(2m-1)}^{2m-1} F_{jk}^{(4)} \psi_{n+j}^{(n)}(\lambda)\epsilon\psi_{n+k}^{(n)}(\mu),$$

where $F_{jk}^{(1)}$, $F_{jk}^{(4)}$ can be expressed in terms of the matrix T_n^{-1}, where T_n is the $(2m-1) \times (2m-1)$ block in the bottom right corner of $D_n^{(n)} M_n^{(n)}$:

$$(T_n)_{jk} := (D_n^{(n)} M_n^{(n)})_{n-2m+j,n-2m+k}, \quad 1 \leq j, k \leq 2m-1. \tag{3-5}$$

The main technical obstacle to study the kernels $S_{n,1}, S_{n,4}$ is the problem to prove that $(T_n^{-1})_{jk}$ are bounded uniformly in n. Till the recent time this technical problem was solved only in a few cases. In [Deift and Gioev 2007b; 2007a] the case $V(\lambda) = \lambda^{2m}(1 + o(1))$ (in our notations) was studied and the problem of invertibility of T_n was solved by computing the entries of T_n explicitly. Similar method was used in [Deift et al. 2007] to prove bulk and edge universality (including the case of the hard edge) for the Laguerre type ensembles with monomial V. In [Stojanovic 2000] the problem of invertibility of T_n was solved also by computing the entries of T_n for V being an even quartic polynomial. In [Shcherbina 2009b; Shcherbina 2009a] similar problem was solved without explicit computation of the entries of T_n. It was shown that for any real analytic V with one interval support of the equilibrium density $(M_n^{(n)})^{-1}$ is uniformly bounded in the operator norm.

But there is also a possibility to prove that T_n is invertible with another technique. As a by product of the calculation in [Tracy and Widom 1998] one also obtains relations between the partition functions $Q_{n,\beta}$ and the determinants of $M_n^{(n)}$ and $D_n^{(n)}$:

$$\det M_n^{(n)} = \left(\frac{Q_{n,1} \Gamma_n}{n!\, 2^{n/2}} \right)^2, \quad \det D_n^{(n)} = \left(\frac{Q_{n/2,4} \Gamma_n}{(n/2)!\, 2^{n/2}} \right)^2,$$

where

$$\Gamma_n := \prod_{j=0}^{n-1} \gamma_j^{(n)},$$

$\gamma_j^{(n)}$ being the leading coefficient of the j-th orthogonal polynomial $p_j^{(n)}(\lambda)$. It is also known (see [Mehta 1991]) that $Q_{n,2} = n!/\Gamma_n^2$. Since $d_\infty^{(n)} M_\infty^{(n)} = 1$ and since $(D_\infty^{(n)})_{j,k} = n\,\text{sign}(j - k) V'(J^{(n)})_{jk}$ implies that

$$(D_\infty^{(n)})_{j,k} = 0 \quad \text{if } |j - k| \geq 2m,$$
$$|(D_\infty^{(n)})_{j,k}| \leq nC \quad \text{if } |j - n| \leq nc, \tag{3-6}$$

we have $D_n^{(n)} M_n^{(n)} = 1 + \Delta_n$ with Δ_n being zero except for the bottom $2m - 1$ rows, and we arrive at this formula, first observed in [Stojanovic 2000]:

$$\det(T_n) = \det(D_n^{(n)} M_n^{(n)}) = \left(\frac{Q_{n,1} Q_{n/2,4}}{Q_{n,2}(n/2)!\, 2^n} \right)^2. \tag{3-7}$$

Hence to control $\det(T_n)$, it suffices to control $\log Q_{n,\beta}$ for $\beta = 1, 2, 4$ up to the order $O(1)$. One can easily see that for a one-cut case the control can be done by using Theorem 1. But, as it was mentioned in the previous section, the method used there does not work for the multi-cut case (see discussion after Equation (2-13)).

In [Shcherbina 2011] the problem to control $Q_{n,\beta}[V]$ for $\beta = 1, 2, 4$ is solved in a little bit different way. It is proved that for any analytical potential V with

q-interval support σ of the equilibrium density $Q_{n,\beta}[V]$ up to the order $O(1)$ can be factorized to a product of some one-cut partition functions with appropriate "effective potentials" $V_\alpha^{(a)}$ defined in terms of σ, V and ρ.

Set

$$\mu_\alpha = \int_{\sigma_\alpha} \rho(\lambda)\, d\lambda, \quad n_\alpha^* := [n\mu_\alpha] + d_\alpha, \tag{3-8}$$

where $[x]$ means an integer part of x, and $d_\alpha = 0, \pm 1, \pm 2$ are chosen in a way which makes n_α^* even (recall that n is even) and

$$\sum n_\alpha^* = n. \tag{3-9}$$

Note, that the choice of d_α is not unique, but a different choice differs only by $O(1)$ and leads to the same expression for $Q_{n,\beta}[V]$ in (3-12) which is up to $O(1)$.

Introduce the "effective potentials"

$$V_\alpha^{(a)}(\lambda) = \mathbf{1}_{\sigma_{\alpha,\varepsilon}}(\lambda)\left(V(\lambda) - 2\int_{\sigma\setminus\sigma_\alpha} \log|\lambda - \mu|\rho(\mu)\, d\mu\right), \tag{3-10}$$

and denote by Σ^* the "cross energy"

$$\Sigma^* := \sum_{\alpha \neq \alpha'} \int_{\sigma_\alpha} d\lambda \int_{\sigma_{\alpha'}} d\mu \log|\lambda - \mu|\rho(\lambda)\rho(\mu). \tag{3-11}$$

Theorem 3. *Let V be a polynomial of degree $2m$ satisfying condition C2 and n be even. Then the matrices $F^{(1)}$ and $F^{(4)}$ in (3-4) are bounded in the operator norm uniformly in n. Moreover, the logarithm of the partition function $Q_{n,\beta}[V]$ can be obtained up to $O(1)$ term from the representation*

$$\log(Q_{n,\beta}[V]/n!) = \sum_{\alpha=1}^{q} \log(Q_{n_\alpha^*,\beta}[V_\alpha^{(a)}]/n_\alpha^*!) - \frac{\beta n^2}{2}\Sigma^* + O(1), \tag{3-12}$$

where $V_\alpha^{(a)}$ and Σ^* are defined in (3-10) and (3-11).

As it was mentioned above, Theorem 3 together with some asymptotic results for orthogonal polynomials of [Deift et al. 1999] proves the universality conjecture for local eigenvalue statistics of the matrix models (1-1).

Theorem 4. *Let V be a polynomial of degree $2m$ satisfying condition C2. Then we have for (even) $n \to \infty$, $\lambda_0 \in \mathbb{R}$ with $\rho(\lambda_0) > 0$, and for $\beta \in \{1,4\}$ that*

$$(n\rho(\lambda_0))^{-1} S_{n,1}\big(\lambda_0 + \xi/n\rho(\lambda_0), \lambda_0 + \eta/n\rho(\lambda_0)\big) = \frac{\sin\pi(\xi - \eta)}{\pi(\xi - \eta)} + O(n^{-1/2}),$$

$$(n\rho(\lambda_0))^{-1} S_{n/2,4}\big(\lambda_0 + \xi/n\rho(\lambda_0), \lambda_0 + \eta/n\rho(\lambda_0)\big) = \frac{\sin\pi(\xi - \eta)}{\pi(\xi - \eta)} + O(n^{-1/2}).$$

the error bound is uniform for bounded ξ, η *and for* λ_0 *contained in some compact subset of* $\bigcup_{\alpha=1}^{q}(a_\alpha, b_\alpha)$.

It is an immediate consequence of Theorem 4 and of the formulas which express the correlation functions in terms of $S_{n,1}$ or $S_{n,4}$ (see [Tracy and Widom 1998]) that the corresponding rescaled l-point correlation functions

$$p_{l,1}^{(n)}(\lambda_0 + \xi_1/n\rho(\lambda_0), \dots, \lambda_0 + \xi_l/n\rho(\lambda_0)),$$

$$p_{l,4}^{(n/2)}(\lambda_0 + \xi_1/n\rho(\lambda_0), \dots, \lambda_0 + \xi_l/n\rho(\lambda_0))$$

converge for n (even) $\to \infty$ to some limit that depends on $\beta = 1, 4$ but not on the choice of V.

Sketch the proof of Theorem 3. Set

$$\sigma_\varepsilon = \bigcup_{\alpha=1}^{q} \sigma_{\alpha,\varepsilon}, \quad \sigma_{\alpha,\varepsilon} = [a_\alpha - \varepsilon, b_\alpha + \varepsilon], \quad \text{dist}\{\sigma_{\alpha,\varepsilon}, \sigma_{\alpha',\varepsilon}\} > \delta > 0, \quad \alpha \neq \alpha'. \quad (3\text{-}13)$$

First of all we replace the integration domain in the definition of $Q_{n,\beta}[V]$ and $p_{k,\beta}^{(n)}$ from $\mathbb{R}$ to σ_ε. Then, according to [Pastur and Shcherbina 2008], $Q_{n,\beta}[V]$ and $p_{k,\beta}^{(n)}$ will be changed by $(1 + O(e^{-nc}))$ factor.

To understand how the potentials $V_\alpha^{(a)}$ of (3-10) appear, let us represent $H(\bar{\lambda})$ as

$$-n\sum_{i=1}^{n} V(\lambda_i) + \sum_{\substack{i \neq j \\ \alpha,\alpha'=1}}^{q} \chi_\alpha(\lambda_i)\chi_{\alpha'}(\lambda_j) \log |\lambda_i - \lambda_j|$$

$$= -n\sum_{i=1}^{n} V(\lambda_i) + \sum_{i \neq j}\sum_{\alpha=1}^{q} \chi_\alpha(\lambda_i)\chi_\alpha(\lambda_j) \log |\lambda_i - \lambda_j|$$

$$+ 2n\sum_{j=1}^{n} \chi_\alpha(\lambda_i) \sum_{\alpha' \neq \alpha} \int \log |\lambda_i - \mu|\chi_{\alpha'}(\mu)\rho(\mu)\, d\mu - n^2\Sigma^*$$

$$+ \sum_{\substack{i,j=1,\dots,n \\ \alpha \neq \alpha'}} \int d\lambda\, d\mu \log |\lambda - \mu|\chi_\alpha(\lambda)\chi_{\alpha'}(\mu)(\delta_{\lambda_i}(\lambda) - \rho(\lambda))(\delta_{\lambda_j}(\mu) - \rho(\mu))$$

$$= H_a(\bar{\lambda}) + \Delta H(\bar{\lambda}), \quad (3\text{-}14)$$

where χ_α is the indicator function of the interval $\sigma_{\alpha,\varepsilon}$, $\delta_{\lambda_i}(\lambda) = \delta(\lambda - \lambda_i)$ is a delta-function, the "cross energy" Σ^* is defined in (3-11), and we introduce

$$H_a(\lambda_1 \dots \lambda_n)$$

$$= -n\sum_{\alpha=1}^{q}\sum_{i=1}^{n} V_\alpha^{(a)}(\lambda_i) + \sum_{i \neq j} \log |\lambda_i - \lambda_j|\left(\sum_{\alpha=1}^{q} \chi_\alpha(\lambda_i)\chi_\alpha(\lambda_j)\right) - n^2\Sigma^*,$$

in which the "effective potential" $V_\alpha^{(a)}(\lambda)$ is defined by (3-10).

Consider

$$Q_{n,\beta}^{(a)}[V] = \int_{\sigma_\varepsilon^n} e^{\beta H_a(\lambda_1,\ldots,\lambda_n)} d\lambda_1 \ldots d\lambda_n.$$

By the Jensen inequality

$$\frac{\beta}{2}\langle\Delta H\rangle_{H_a} \le \log Q_{n,\beta}[V] - \log Q_{n,\beta}^{(a)}[V] \le \frac{\beta}{2}\langle\Delta H\rangle_H.$$

Then it was shown that both the right-hand side and the left-hand side of this inequality are $O(1)$ and

$$\log Q_{n,\beta}^{(a)}[V] = \sum_{\alpha=1}^{q} \log Q_{n_\alpha^*,\beta}[(n/n_\alpha^*)V^{(a)}] - n^2\Sigma^* + O(1),$$

where the n_α^*, $\alpha = 1, \ldots, q$, are chosen to satisfy (3-8) and (3-9). We do not give more details here, because the result follows from (4-10) in the next section. $\square$

4. CLT for β-model in the multi-cut regime

The idea of using some factorization of $Q_{n,\beta}[V]$ into a product of one-cut partition functions for the effective potentials $V_\alpha^{(a)}$ was used in [Shcherbina 2013] to prove the CLT for linear eigenvalue statistics (1-10). In order to formulate corresponding result we need some extra definitions. Consider the Hilbert space

$$\mathcal{H} = \bigoplus_{\alpha=1}^{q} L^2[\sigma_\alpha] \tag{4-1}$$

with the standard inner product $(.,.)$. Define the operator $\mathcal{L}$ (cf. (1-7)) by

$$\mathcal{L}f = 1_\sigma L[f], \quad \mathcal{L}_\alpha f := 1_{\sigma_\alpha} L[f 1_{\sigma_\alpha}], \tag{4-2}$$

the block diagonal operators

$$\bar{D} := \bigoplus_{\alpha=1}^{q} \bar{D}_\alpha, \quad \widehat{\mathcal{L}} := \bigoplus_{\alpha=1}^{q} \widehat{\mathcal{L}}_\alpha, \tag{4-3}$$

where $\bar{D}_\alpha$ is defined by (2-2) for σ_α. Moreover, denote

$$\widetilde{\mathcal{L}} := \mathcal{L} - \widehat{\mathcal{L}}, \quad \mathcal{G} := (1 + \bar{D}\widetilde{\mathcal{L}})^{-1}. \tag{4-4}$$

An important role below belongs to a positive definite matrix $\mathcal{Q} = \{\mathcal{Q}_{\alpha\alpha'}\}_{\alpha,\alpha'=1}^{q}$ of the form

$$\mathcal{Q}_{\alpha\alpha'} = (\mathcal{L}\psi^{(\alpha)}, \psi^{(\alpha')}), \tag{4-5}$$

where $\psi^{(\alpha)}(\lambda) = p_\alpha(\lambda) X^{-1}(\lambda)$ (p_α is a polynomial of degree $q-1$) is the unique solution of the system of equations

$$(\mathcal{L}\psi^{(\alpha)})_{\alpha'} = \delta_{\alpha\alpha'}, \quad \alpha' = 1, \ldots, q. \tag{4-6}$$

One can easily see that the function $\Psi_\alpha(z) = \int \log|z - \lambda| \psi^{(\alpha)}(\lambda) \, d\lambda$ is the harmonic measure of σ_α with respect to $\mathbb{C} \setminus \sigma$. Denote also

$$I[h] = (I_1[h], \ldots, I_q[h]), \quad I_\alpha[h] := \sum_{\alpha'} \mathcal{Q}_{\alpha\alpha'}^{-1}(h, \psi^{(\alpha')}). \tag{4-7}$$

The main result of [Shcherbina 2013] is this:

Theorem 5. *Let the potential V satisfy conditions C1–C2, and let $\|h^{(4)}\|_\infty < \infty$. Then*

$$Z_{n,\beta}[h] = \exp\left\{\frac{\beta}{8}(\mathcal{G}\bar{D}h, h) + \left(\frac{\beta}{2} - 1\right)(\mathcal{G}v, h)\right\} \frac{\Theta(\bar{I}[h]; \{n\bar{\mu}\})}{\Theta(0; \{n\bar{\mu}\})} (1 + O(n^{-\delta})), \tag{4-8}$$

where $\delta > 0$ and

$$\Theta(I[h]; \{n\bar{\mu}\}) :=$$

$$\sum_{n_1 + \cdots + n_q = n_0} \exp\left\{-\frac{\beta}{2}(\mathcal{Q}^{-1}\Delta\bar{n}, \Delta\bar{n}) + \frac{\beta}{2}(\Delta\bar{n}, I[h]) + \left(\frac{\beta}{2} - 1\right)(\Delta\bar{n}, I[\log\bar{\rho}])\right\},$$

$$(\{n\bar{\mu}\})_\alpha = \{n\mu_\alpha\}, \quad (\Delta\bar{n})_\alpha = n_\alpha - \{n\mu_\alpha\},$$

$$(\log\bar{\rho})_\alpha = \log\rho_\alpha, \quad n_0 = \sum_{\alpha=1}^q \{n\mu_\alpha\}, \tag{4-9}$$

with a positive definite matrix $\mathcal{Q}$ of (4-5) and $I[h]$ defined by (4-7).

For $h = 0$ we have

$$Q_{n,\beta}[V] = \mathcal{Z}_{n,\beta} \frac{\exp\left\{\frac{2}{\beta}\left(\frac{\beta}{2} - 1\right)^2(\widetilde{\mathcal{L}}\mathcal{G}v, v)\right\}}{\det^{1/2}(1 - \bar{D}\widetilde{\mathcal{L}})} \Theta(0; \{n\bar{\mu}\})(1 + O(n^{-\kappa})),$$

$$Q_{n,\beta}[V] = \exp\left\{\frac{n^2\beta}{2}\mathcal{E}[V] + F_\beta(n) + n\left(\frac{\beta}{2} - 1\right)\left((\log\rho, \rho) - 1 - \log 2\pi\right)\right.$$

$$\left. -c_\beta(q-1)\log n + \sum_{\alpha=1}^q (q_\beta^{(0)}[\mu_\alpha^{-1}\rho_\alpha] - c_\beta\log\mu_\alpha)\right\}, \tag{4-10}$$

where μ_α, ρ_α are defined in (3-8), $q_\beta^{(0)}[\rho]$ is defined in (2-7), $F_\beta(n)$ and c_β are defined in (2-9) and det means the Fredholm determinant of $\bar{D}\widetilde{\mathcal{L}}$ on σ.

Let us remark that since the kernel of $\mathcal{L}$ is an analytic function, it is easy to prove that $\bar{D}\widetilde{\mathcal{L}}$ is a trace class operator. Moreover, it is proven in [Shcherbina 2013] that $\|\bar{D}\widetilde{\mathcal{L}}\| < 1$. Hence $\det^{1/2}(1 + \bar{D}\widetilde{\mathcal{L}})^{-1}$ is well defined.

Note also that since $\widetilde{\mathcal{L}} = 0$ in the one-cut case, the formulas of Theorem 5 for $q = 1$ coincide with that of Theorem 1.

According to Theorem 5, the fluctuations of the linear eigenvalue statistics $\mathcal{N}_n[h]$ are not Gaussian in the multi-cut regime, because the exponent of the generating functional $Z_{n,\beta}[h]$ is not quadratic with respect to h. We obtain that $Z_{n,\beta}[h]$ contains some quasiperiodic Θ-function in which the quadratic form $\mathfrak{D}^q_{\alpha,\alpha'=1}$ is determined by the geometrical structure of σ. The fluctuations of $\mathcal{N}_n[h]$ become Gaussian if and only if all parameters $I_\alpha[h] = 0$ (see (4-7)). Similar results for $\beta = 2$ were predicted in [Pastur 2006] on the basis of the analysis of the asymptotics of orthogonal polynomials obtained in [Deift et al. 1999]. One more interesting observation is that the operator $\widetilde{\mathcal{L}}\mathcal{G}$ which appears in the place of the "variance" differ from $\mathcal{L}^{-1}$ (see (4-2) for the definition of $\mathcal{L}$) only by the final rank perturbation. This perturbation provides, in particular, that $\widetilde{\mathcal{L}}\mathcal{G}f = 0$, if $f(\lambda) = \text{const}, \lambda \in \sigma$.

Sketch of the proof of Theorem 5. Let $\bar{n} := (n_1, \ldots, n_q)$ and set

$$|\bar{n}| := \sum_{\alpha=1}^q n_\alpha, \quad \mathbf{1}_{\bar{n}}(\bar{\lambda}) := \prod_{j=1}^{n_1} \mathbf{1}_{\sigma_{1,\varepsilon}}(\lambda_j) \cdots \prod_{j=|\bar{n}|-n_q+1}^n \mathbf{1}_{\sigma_{q,\varepsilon}}(\lambda_j). \quad (4\text{-}11)$$

It is evident that

$$Q_{n,\beta}[V]/n! = \sum_{|\bar{n}|=n} \frac{\int \mathbf{1}_{\bar{n}}(\bar{\lambda}) e^{\beta H(\bar{\lambda})/2}}{n_1! \ldots n_q!} = \sum_{|\bar{n}|=n} \frac{\int \mathbf{1}_{\bar{n}}(\bar{\lambda}) e^{\beta (H_a(\bar{\lambda}) + \Delta H(\bar{\lambda}))/2}}{n_1! \ldots n_q!}. \quad (4\text{-}12)$$

Since $\log|\lambda - \mu|$ for $\lambda \in \sigma_{\alpha,\varepsilon}$, $\mu \in \sigma_{\alpha',\varepsilon}$, $\alpha \neq \alpha'$ is an analytic function, the expansion of it in the Fourier series with respect to some appropriate basis (e.g., Chebyshev polynomials $\{p_k^{(\alpha)}(\lambda)\}$, $\{p_k^{(\alpha')}(\mu)\}$) will converge exponentially fast. Hence, if we choose $M = [\log^2 n]$, then

$$\log|\lambda - \mu| = \sum_{k,m=1}^M L_{k,m}^{(\alpha,\alpha')} p_k^{(\alpha)}(\lambda) p_m^{(\alpha')}(\mu) + O(e^{-c\log^2 n}),$$

$$\lambda \in \sigma_{\alpha,\varepsilon}, \; \mu \in \sigma_{\alpha',\varepsilon}, \; \alpha \neq \alpha'. \quad (4\text{-}13)$$

Thus,

$$\Delta H(\bar{\lambda}) \mathbf{1}_{\bar{n}}(\bar{\lambda})$$

$$= \mathbf{1}_{\bar{n}}(\bar{\lambda}) \sum_{\substack{j,j'=1 \\ \alpha \neq \alpha'}}^n \sum_{k,m=1}^M L_{k,m}^{(\alpha,\alpha')} \left(p_k^{(\alpha)}(\lambda_j) - \frac{n}{n_\alpha} c_k^{(\alpha)} \right) \left(p_m^{(\alpha')}(\lambda_{j'}) - \frac{n}{n_{\alpha'}} c_k^{(\alpha)} \right) + O(e^{-c\log^2 n})$$

with

$$c_k^{(\alpha)} := (p_k^{(\alpha)}, \rho 1_{\sigma_\alpha}).$$

Then we represent the matrix

$$\widetilde{\mathcal{L}}^{(M)} := \big\{ L_{k,m}^{(\alpha,\alpha')} \big\}_{k,m=1,\dots m, \alpha, \alpha'=1,\dots q},$$

which consists of q^2 blocks $M \times M$, as a difference of two positive block matrix of the same dimensionality

$$\widetilde{\mathcal{L}}^{(M)} = \hat{\mathcal{A}}^{(M)} - \mathcal{A}^{(M)}, \quad \mathcal{A}^{(M)}, \hat{\mathcal{A}}^{(M)} > 0$$

and apply the Hubbard–Stratonovich transformation to $e^{\beta \Delta H(\bar\lambda)/2} 1_{\bar n}(\bar\lambda)$:

$$e^{\beta(\hat{\mathcal{A}}^{(M)}\bar x, \bar x)/2} = \Big(\frac{\beta}{2\pi}\Big)^{(Mq)^2/2} \int_{\mathbb{R}^{(Mq)^2}} d\bar u^{(1)} e^{\beta((\hat{\mathcal{A}}^{(M)})^{1/2}\bar x, \bar u^{(1)})/2 - \beta(\bar u^{(1)}, \bar u^{(1)})/8},$$

$$e^{-\beta(\mathcal{A}^{(M)}\bar x, \bar x)/2} = \Big(\frac{\beta}{2\pi}\Big)^{(Mq)^2/2} \int_{\mathbb{R}^{(Mq)^2}} d\bar u^{(2)} e^{i\beta((\mathcal{A}^{(M)})^{1/2}\bar x, \bar u^{(2)})/2 - \beta(\bar u^{(2)}, \bar u^{(2)})/8}.$$

$$(4\text{-}14)$$

We obtain the linear with respect to $p_k^{(\alpha)}(\lambda_j)$ expression $\tilde h(\bar u_1, \bar u_2)$ in the exponent. Then apply Theorem 1 to

$$V = \mu_\alpha^{-1} V_\alpha^{(a)}, \quad h = \frac{\beta}{2}(\mu_\alpha^{-1} - n/n_\alpha)V_\alpha^{(a)} + \tilde h(u_1, \bar u_2).$$

This gives a quadratic form with respect to $(\bar u^{(1)}, \bar u^{(2)})$ in the exponent, and the quasi periodic quadratic form in the exponent appears due to the coefficient in front of $V_\alpha^{(a)}$. Then we integrate with respect to $(\bar u^{(1)}, \bar u^{(2)})$. After some transformations we obtain the assertion of Theorem 5. □

In principle, this way could be used to construct the asymptotic expansion of $Q_{n,\beta}[V]$ with respect to n^{-1}, because after the Hubbard–Stratonovich transformation we can apply the result Theorem 2. But there is a problem that in this case we have to apply (2-7) to a non real perturbation of V (see (4-14)). There is a way to extend the bounds obtained for th with a real t to a non real t, but for $|t| \le |\log n|^{1/2}$. It is enough for the CLT but not enough for the construction of the expansion.

Acknowledgements

The author would like to thank MSRI and the organizers of the semester "Random Matrix Theory and its Applications", where this work was started.

References

[Albeverio et al. 2001] S. Albeverio, L. Pastur, and M. Shcherbina, "On the $1/n$ expansion for some unitary invariant ensembles of random matrices", *Comm. Math. Phys.* **224**:1 (2001), 271–305.

[Bleher and Its 2003] P. Bleher and A. Its, "Double scaling limit in the random matrix model: the Riemann–Hilbert approach", *Comm. Pure Appl. Math.* **56**:4 (2003), 433–516.

[Borot and Guionnet 2013] G. Borot and A. Guionnet, "Asymptotic expansion of β matrix models in the one-cut regime", *Comm. Math. Phys.* **317**:2 (2013), 447–483.

[Chekhov and Eynard 2006] L. Chekhov and B. Eynard, "Matrix eigenvalue model: Feynman graph technique for all genera", *J. High Energy Phys.* **2006**:12 (2006), 026.

[Claeys and Kuijlaars 2006] T. Claeys and A. B. J. Kuijlaars, "Universality of the double scaling limit in random matrix models", *Comm. Pure Appl. Math.* **59**:11 (2006), 1573–1603.

[Deift and Gioev 2007a] P. Deift and D. Gioev, "Universality at the edge of the spectrum for unitary, orthogonal, and symplectic ensembles of random matrices", *Comm. Pure Appl. Math.* **60**:6 (2007), 867–910.

[Deift and Gioev 2007b] P. Deift and D. Gioev, "Universality in random matrix theory for orthogonal and symplectic ensembles", *Int. Math. Res. Pap.* **2007**:2 (2007), Art ID rpm004.

[Deift et al. 1999] P. Deift, T. Kriecherbauer, K. T.-R. McLaughlin, S. Venakides, and X. Zhou, "Uniform asymptotics for polynomials orthogonal with respect to varying exponential weights and applications to universality questions in random matrix theory", *Comm. Pure Appl. Math.* **52**:11 (1999), 1335–1425.

[Deift et al. 2007] P. Deift, D. Gioev, T. Kriecherbauer, and M. Vanlessen, "Universality for orthogonal and symplectic Laguerre-type ensembles", *J. Stat. Phys.* **129**:5-6 (2007), 949–1053.

[Ercolani and McLaughlin 2003] N. M. Ercolani and K. D. T.-R. McLaughlin, "Asymptotics of the partition function for random matrices via Riemann–Hilbert techniques and applications to graphical enumeration", *Int. Math. Res. Not.* **2003**:14 (2003), 755–820.

[Eynard 2009] B. Eynard, "Large N expansion of convergent matrix integrals, holomorphic anomalies, and background independence", *J. High Energy Phys.* **2009**:3 (2009), 003.

[Forrester 2010] P. J. Forrester, *Log-gases and random matrices*, London Mathematical Society Monographs Series **34**, Princeton University Press, Princeton, NJ, 2010.

[Johansson 1998] K. Johansson, "On fluctuations of eigenvalues of random Hermitian matrices", *Duke Math. J.* **91**:1 (1998), 151–204.

[Kriecherbauer and Shcherbina 2010] T. Kriecherbauer and M. Shcherbina, "Fluctuations of eigenvalues of matrix models and their applications", 2010. arXiv 1003.6121

[Kuijlaars and McLaughlin 2000] A. B. J. Kuijlaars and K. T.-R. McLaughlin, "Generic behavior of the density of states in random matrix theory and equilibrium problems in the presence of real analytic external fields", *Comm. Pure Appl. Math.* **53**:6 (2000), 736–785.

[Levin and Lubinsky 2008] E. Levin and D. S. Lubinsky, "Universality limits in the bulk for varying measures", *Adv. Math.* **219**:3 (2008), 743–779.

[McLaughlin and Miller 2008] K. T.-R. McLaughlin and P. D. Miller, "The $\bar{\partial}$ steepest descent method for orthogonal polynomials on the real line with varying weights", *Int. Math. Res. Not.* **2008** (2008), Art. ID rnn 075.

[Mehta 1991] M. L. Mehta, *Random matrices*, 2nd ed., Academic Press, Boston, MA, 1991.

[Boutet de Monvel et al. 1995] A. Boutet de Monvel, L. Pastur, and M. Shcherbina, "On the statistical mechanics approach in the random matrix theory: integrated density of states", *J. Statist. Phys.* **79**:3-4 (1995), 585–611.

[Pastur 2006] L. Pastur, "Limiting laws of linear eigenvalue statistics for Hermitian matrix models", *J. Math. Phys.* **47**:10 (2006), 103303.

[Pastur and Shcherbina 1997] L. Pastur and M. Shcherbina, "Universality of the local eigenvalue statistics for a class of unitary invariant random matrix ensembles", *J. Statist. Phys.* **86**:1-2 (1997), 109–147.

[Pastur and Shcherbina 2008] L. Pastur and M. Shcherbina, "Bulk universality and related properties of Hermitian matrix models", *J. Stat. Phys.* **130**:2 (2008), 205–250.

[Pastur and Shcherbina 2011] L. Pastur and M. Shcherbina, *Eigenvalue distribution of large random matrices*, Mathematical Surveys and Monographs **171**, American Mathematical Society, Providence, RI, 2011.

[Shcherbina 2009a] M. Shcherbina, "Edge universality for orthogonal ensembles of random matrices", *J. Stat. Phys.* **136**:1 (2009), 35–50.

[Shcherbina 2009b] M. Shcherbina, "On universality for orthogonal ensembles of random matrices", *Comm. Math. Phys.* **285**:3 (2009), 957–974.

[Shcherbina 2011] M. Shcherbina, "Orthogonal and symplectic matrix models: universality and other properties", *Comm. Math. Phys.* **307**:3 (2011), 761–790.

[Shcherbina 2013] M. Shcherbina, "Fluctuations of linear eigenvalue statistics of β matrix models in the multi-cut regime", *J. Stat. Phys.* **151**:6 (2013), 1004–1034.

[Stojanovic 2000] A. Stojanovic, "Universality in orthogonal and symplectic invariant matrix models with quartic potential", *Math. Phys. Anal. Geom.* **3**:4 (2000), 339–373.

[Tracy and Widom 1998] C. A. Tracy and H. Widom, "Correlation functions, cluster functions, and spacing distributions for random matrices", *J. Statist. Phys.* **92**:5-6 (1998), 809–835.

[Widom 1999] H. Widom, "On the relation between orthogonal, symplectic and unitary matrix ensembles", *J. Statist. Phys.* **94**:3-4 (1999), 347–363.

shcherbi@ilt.kharkov.ua

Institute for Low Temperature Physics, National Academy of Sciences of Ukraine, Kharkov, 61103, Ukraine

Department of Mathematics, V. N. Karazin Kharkov National University, Kharkov, 61077, Ukraine

Random Matrices
MSRI Publications
Volume **65**, 2014

KPZ scaling theory and the semidiscrete directed polymer model

HERBERT SPOHN

We explain how the claims of KPZ scaling theory are confirmed by a recent proof of Borodin and Corwin on the asymptotics of the semidiscrete directed polymer.

1. Introduction

The Kardar–Parisi–Zhang (KPZ) equation [1986] is a stochastic partial differential equation modeling surface growth and, more generally, the motion of an interface bordering a stable against a metastable phase. Scaling theory is an educated guess on the nonuniversal coefficients in the asymptotics for models in the KPZ universality class. Scaling theory has been developed in a landmark contribution by Krug, Meakin and Halpin-Healy [Krug et al. 1992]. The purpose of our article is to explain how to apply scaling theory to the semidiscrete directed polymer. This model has been discussed in depth at the 2010 random matrix workshop at MSRI and, so-to-speak as a spin-off, Borodin and Corwin [2014] developed the beautiful theory of Macdonald processes, which provides the tools for an asymptotic analysis of the semidiscrete directed polymer. As we will establish, scaling theory is consistent with the results of Borodin and Corwin, thereby providing a highly nonobvious control check.

To place the issue in focus, let me start with a simple example. Assume as given the stationary sequence X_j, $j \in \mathbb{Z}$, of mean zero random variables and let us consider the partial sums

$$S_n = \sum_{j=1}^{n} X_j. \tag{1-1}$$

As well studied, it is fairly common that $S_n/\sqrt{n}$ converges to a Gaussian as $n \to \infty$, i.e.,

$$\lim_{n \to \infty} \mathbb{P}\left(S_n \leq \sqrt{D}\sqrt{n}s\right) = F_{\mathrm{G}}(s), \tag{1-2}$$

where F_{G} is the distribution function of a unit Gaussian random variable. Here F_{G} is the universal object, while the coefficient $D > 0$ depends on the law $\mathbb{P}$ and is in this sense model dependent, resp. nonuniversal. However, using stationarity, D is readily guessed as

$$D = \sum_{j=-\infty}^{\infty} \mathbb{E}(X_0 X_j). \tag{1-3}$$

The KPZ class deals with strongly dependent random variables, for which partial sums are of size $n^{1/3}$ rather than $n^{1/2}$. F_G is to be substituted by the GUE Tracy–Widom distribution function, F_{GUE}, which first appeared in the context of the largest eigenvalue of a GUE random matrix [Forrester 1993; Tracy and Widom 1994]. F_{GUE} is defined through a Fredholm determinant as

$$F_{GUE}(s) = \det(1 - P_s K_{Ai} P_s). \tag{1-4}$$

Here K_{Ai} is the Airy kernel,

$$K_{Ai}(x, y) = \int_0^{\infty} d\lambda \, \text{Ai}(x + \lambda) \, \text{Ai}(y + \lambda), \tag{1-5}$$

with Ai the Airy function, and P_x projects onto the interval $[x, \infty)$. For each x, $P_x K_{Ai} P_x$ is a trace class operator in $L^2(\mathbb{R})$, hence (1-4) is well-defined. To determine the scale coefficient is less obvious than in the example above, but will be explained in due course. Let me stress that scaling theory is crucial for the proper statistical analysis of either physics [Takeuchi and Sano 2010; Takeuchi et al. 2011] or computer experiments [Alves et al. 2011]. Without this input, the comparison with theoretical results would be considerably less reliable.

Our paper is divided into two parts. We first explain scaling theory in the context of a specific class of growth models. In the second part the theory is applied to the semidiscrete directed polymer model. The convergence to the GUE Tracy–Widom distribution is established in [Borodin and Corwin 2014], of course including an expression for the nonuniversal scale coefficient. Our goal is to explain, how this coefficient can be determined independently, not using the import from the proof in [Borodin and Corwin 2014].

2. Scaling theory for the single-step model

In the single step growth model the moving surface is described by the graph of the height function $h(t) : \mathbb{Z} \to \mathbb{Z}$ with the constraint

$$|h(j + 1, t) - h(j, t)| = 1, \tag{2-1}$$

hence the name. The random deposition/evaporation events are modeled by a Markov jump process constrained to satisfy (2-1). The allowed local moves are then transitions from $h(j, t)$ to $h(j, t) \pm 2$. The dynamics should be invariant under a shift in the h-direction. Hence the rates for the deposition/evaporation events are allowed to depend only on the local slopes. It is then convenient to switch to height differences

$$\eta_j(t) = h(j+1, t) - h(j, t), \quad \eta_j(t) = \pm 1. \tag{2-2}$$

A single growth step at the bond $(j, j+1)$ is given by

$$\eta \to \eta^{j,j+1}, \tag{2-3}$$

where $\eta^{j,j+1}$ is the configuration with the slopes at j and $j+1$ interchanged. If the corresponding rates are denoted by $c_{j,j+1}(\eta)$, depending on the local neighborhood of $(j, j+1)$, the Markov generator reads

$$Lf(\eta) = \sum_{j \in \mathbb{Z}} c_{j,j+1}(\eta)(f(\eta^{j,j+1}) - f(\eta)). \tag{2-4}$$

The slope field η is locally conserved, in the sense that the sum $\sum_{j=a}^{b} \eta(j, t)$ changes only through the fluxes at the two boundaries a, b. To keep things concretely, in the following we consider only the wedge initial condition

$$h(j, 0) = |j|. \tag{2-5}$$

Scaling theory is based on the following.

Assumption. *The spatially ergodic and time stationary measures of the slope process $\eta(t)$ are precisely labeled by the average density*

$$\rho = \lim_{a \to \infty} \frac{1}{2a+1} \sum_{|j| \le a} \eta_j, \tag{2-6}$$

with $|\rho| \le 1$.

Our assumption was formulated more than 30 years ago. Except for special cases, it remains open even today; see [Liggett 1999] for more details. The stationary measures from the assumption are denoted by μ_ρ, as a probability measure on $\{-1, 1\}^{\mathbb{Z}}$.

Given μ_ρ one defines two natural quantities:

- the average steady state current:

$$j(\rho) = \mu_\rho(c_{0,1}(\eta)(\eta_0 - \eta_1)); \tag{2-7}$$

- the integrated covariance of the conserved slope field:

$$A(\rho) = \sum_{j \in \mathbb{Z}} (\mu_\rho(\eta_0 \eta_j) - \mu_\rho(\eta_0)^2). \tag{2-8}$$

We first notice that for long times there is a law of large numbers stating that

$$h(j, t) \simeq t\phi(j/t), \qquad (2\text{-}9)$$

for large j, t with a deterministic profile function ϕ. In fact ϕ is the Legendre transform of j in the sense that

$$\phi(y) = \sup_{|\rho| \leq 1} (y\rho - j(\rho)). \qquad (2\text{-}10)$$

The argument is based on the hydrodynamic limit for nonreversible lattice gases [Spohn 1991], which asserts that on the macroscopic scale the density $\rho(x, t)$ of the conserved field satisfies

$$\frac{\partial}{\partial t} \rho(x, t) + \frac{\partial}{\partial x} j(\rho(x, t)) = 0, \qquad (2\text{-}11)$$

with initial condition

$$\rho(x, 0) = \begin{cases} 1 & \text{if } x \geq 0, \\ -1 & \text{if } x < 0. \end{cases} \qquad (2\text{-}12)$$

The entropy solution to (2-11), (2-12) is indeed given by (2-9), (2-10).

The average current can be fairly arbitrary, except for the linear behavior near $\rho = \pm 1$. On the other hand ϕ is convex up with $\phi(x) = |x|$ for $|x| \geq x_c$. At points, where ϕ is either linear or cusp-like, the fluctuations may be different from the generic case and have to be discussed separately. We set

$$\lambda(\rho) = -j''(\rho). \qquad (2\text{-}13)$$

Conjecture (KPZ class). *Let y be such that ϕ is twice differentiable at y with $\phi''(y) \neq 0$ and set $\rho = \phi'(y)$, $|\rho| < 1$. If $A(\rho) < \infty$ and $\lambda(\rho) \neq 0$, then*

$$\lim_{t \to \infty} \mathbb{P}\big(h(\lfloor yt \rfloor, t) - t\phi(y) \leq -\big(-\tfrac{1}{2}\lambda A^2\big)^{1/3} t^{1/3} s\big) = F_{\text{GUE}}(s). \qquad (2\text{-}14)$$

We use $\lfloor \cdot \rfloor$ to denote the integer part. Since $\phi''(y) > 0$, we have $\lambda(\rho) < 0$ by Legendre transform.

On the scale $\big(-\tfrac{1}{2}\lambda A^2 t\big)^{1/3}$ the height fluctuations are governed by the Tracy–Widom distribution. λA^2 is the model dependent coefficient which, at least in principle, can be computed once μ_ρ is available. $\lambda \equiv 0$ for reversible slope dynamics, since $j = 0$. In that case the fluctuations are of scale $t^{1/4}$ and Gaussian; see [Spohn 1991, Part II, Chapter 3] for a discussion. For nonreversible slope dynamics one can still arrange for $\lambda \equiv 0$. If j is nonzero, there could be isolated points at which λ vanishes. At a cubic inflection point of j the height fluctuations are expected to be of order $(t(\log t)^{1/2})^{1/2}$ [Paczuski et al. 1992; Derrida et al. 1991].

In our set-up the conjecture has been proved by Tracy and Widom [2009] for the PASEP. In this case the exchange $+-$ to $-+$ occurs with rate p and the exchange $-+$ to $+-$ with rate $1-p$, $0 \le p < \frac{1}{2}$. Then μ_ρ is a Bernoulli measure, hence $A(\rho) = 1 - \rho^2$, $j(\rho) = \frac{1}{2}(2p - 1)(1 - \rho^2)$, and

$$-\tfrac{1}{2}\lambda A^2 = \tfrac{1}{2}(1 - 2p)(1 - \rho^2)^2. \tag{2-15}$$

The profile function is

$$\phi(y) = \begin{cases} \frac{1}{2}(1 - 2p)(1 + (y/(1 - 2p))^2) & \text{for } |y| \le 1 - 2p, \\ |y| & \text{for } |y| \ge 1 - 2p. \end{cases}$$

For the totally asymmetric case, $p = 0$, the limit (2-14) has been established before by Johansson [2000]. The PushTASEP falls also under our scheme with a proof by Borodin and Ferrari [2008]. Scaling theory is further confirmed for growth models different from single-step, to mention the discrete time TASEP [Johansson 2000], the polynuclear growth model [Prähofer and Spohn 2000], and the KPZ equation [Amir et al. 2011; Sasamoto and Spohn 2010]. For a discussion of the scaling theory in the context of TASEP we refer to [Prähofer and Spohn 2002].

The theory of Macdonald process [Borodin and Corwin 2014] has brought a q-deformed version of the TASEP in focus. For us here it is a further example for which the nonuniversal constants can be computed. For the rate in (2-4) we set

$$c_{j,j+1}(\eta) = \tfrac{1}{4}(1 - \eta_j)(1 + \eta_{j+1})g((n_{j+1}^-(\eta)), \tag{2-16}$$

where n_j^- is the number of consecutive $-$ slopes to the left of site j. We have $g(0) = 0$, $g(j) > 0$ for $j > 0$, and g increases at most linearly. The q-TASEP is the special case where $g(j) = 1 - q^j$, $0 \le q < 1$, with the TASEP recovered in the limit $q \to 0$. The slope system maps onto the totally asymmetric zero range process for length of consecutive gaps between the $+$ slope, denoted by Y_j, $j \in \mathbb{Z}$, $Y_j = 0, 1, \ldots$. In the stationary measure the Y_j are i.i.d. and

$$\mathbb{P}(Y_0 = k) = \begin{cases} Z(\alpha)^{-1} & \text{if } k = 0, \\ Z(\alpha)^{-1} \left(\prod_{j=1}^k g(j) \right)^{-1} \alpha^k & \text{if } k = 1, 2, \ldots, \end{cases} \tag{2-17}$$

where

$$Z(\alpha) = 1 + \sum_{k=1}^\infty \left(\prod_{j=1}^k g(j) \right)^{-1} \alpha^k, \tag{2-18}$$

and $\alpha > 0$ such that $Z(\alpha) < \infty$. The translation invariant, time stationary measures for the $\eta(t)$-process with rates (2-16) are stationary renewal processes on $\mathbb{Z}$ with renewal distribution (2-17), (2-18).

The coefficient A can be computed from

$$\lim_{N\to\infty} \frac{1}{N} \log \left\langle \exp\left[\lambda \sum_{j=1}^{N} \eta_j\right] \right\rangle_\alpha = r(\lambda), \qquad (2\text{-}19)$$

where the average is over the stationary renewal process with parameter α. Then

$$\rho = r'(\lambda) \quad \text{and} \quad A = r''(\lambda) \quad \text{at } \lambda = 0.$$

The rate function r is implicitly determined by

$$r(\lambda) = -\lambda - \log z(\lambda), \quad \frac{1}{Z(\alpha)} z(\lambda) Z(\alpha z(\lambda)) e^{2\lambda} = 1. \qquad (2\text{-}20)$$

A and ρ are computed by successive differentiations. The result is best expressed through $G(\alpha) = \log Z(\alpha)$. Then

$$\tfrac{1}{2}(1+\rho) = (1+\alpha G')^{-1}, \qquad (2\text{-}21)$$

$$A = 4(1+\alpha G')^{-3} \alpha (\alpha G')' = -\alpha(1+\rho)\frac{d\rho}{d\alpha}. \qquad (2\text{-}22)$$

The average current is given by

$$j(\rho) = -2\langle c_{0,1}\rangle_\alpha, \quad j = -\alpha(1+\rho). \qquad (2\text{-}23)$$

One can use (2-23) together with (2-21) to work out λ. But no particularly illuminating formula results for the combination λA^2.

In addition to (2-14) there is a second scale, which will play no role here, but should be mentioned. Instead of the height statistics at the single point $\lfloor yt \rfloor$ one could consider, for example, the joint distribution of $h(j_1, t)$, $h(j_2, t)$, referred to as transverse correlations. The transverse scale tells us at which separation $|j_1 - j_2|$ there are nontrivial correlations in the limit $t \to \infty$. KPZ scaling theory asserts that this scale is

$$(2\lambda^2 A t^2)^{1/3}. \qquad (2\text{-}24)$$

The factor 2 comes from the requirement that the limit joint distribution is the two-point distribution of the Airy process. Corresponding predictions hold for the multipoint statistics. Also when considering the two-point function of the stationary $\eta(t)$ process, $\mathbb{E}(\eta_0(0)\eta_j(t)) - (\mathbb{E}(\eta_0(0)))^2$, up to a shift linear in t, j has to vary on the scale of (2-24) to have a nontrivial limit as $t \to \infty$. The scale (2-24) is confirmed for the PNG [Prähofer and Spohn 2004], TASEP [Ben Arous and Corwin 2011], and PushTASEP [Borodin and Ferrari 2008] two-point function in case of step initial conditions, for the stationary TASEP [Ferrari and Spohn 2006] and stationary KPZ equation [Imamura and Sasamoto 2012], and for TASEP and PNG [Borodin et al. 2007; 2008] in case of flat initial conditions.

3. The semidiscrete directed polymer model

Our starting point is a very particular discretization of the stochastic heat equation
as

$$dZ_j = Z_{j-1} dt + Z_j db_j. \tag{3-1}$$

Here $j \in \mathbb{Z}$, $t \geq 0$, and $\{b_j(t), j \in \mathbb{Z}\}$ is a collection of independent standard
Brownian motions. The analogue of the wedge initial condition is

$$Z_j(0) = \delta_{j,0}. \tag{3-2}$$

Hence $Z_j(t) = 0$ for $j < 0$ and

$$dZ_j = Z_{j-1} dt + Z_j db_j, \quad j = 1, 2, \dots,$$
$$dZ_0 = Z_0 db_0. \tag{3-3}$$

Let us introduce the totally asymmetric random walk, $w(t)$, on $\mathbb{Z}$ moving with
rate 1 to the right. Denoting by $\mathbb{E}_0$ expectation for $w(t)$ with $w(0) = 0$, one can
represent

$$Z_j(t) = \mathbb{E}_0 \left(\exp\left[\int_0^t db_{w(s)}(s) \right] \delta_{w(t),j} \right) e^t. \tag{3-4}$$

$Z_j(t)$ is the random partition function of the directed polymer $w(t)$, length t,
endpoints 0 and j, in the random potential $db_j(s)/ds$. This model was first
introduced by O'Connell and Yor [2001]; see also [Moriarty and O'Connell
2007; O'Connell 2012]. In the zero temperature limit one would maximize over
the term in the exponential at fixed $\{b_j(s)\}$ and fixed end points; see [Glynn and
Whitt 1991] for an early study. The statistics of the maximizer is closely related
to GUE and Dyson's Brownian motion [Baryshnikov 2001].

The height corresponds to the random free energy and we set

$$h_j(t) = \log Z_j(t), \quad j \geq 0, \ t > 0. \tag{3-5}$$

Then h_j is the solution to

$$dh_j = e^{h_{j-1} - h_j} dt + db_j, \tag{3-6}$$

and the slope $u_j = h_{j+1} - h_j$ is governed by

$$du_j = (e^{-u_j} - e^{-u_{j-1}}) dt + db_{j+1} - db_j, \quad j = 1, 2, \dots, \tag{3-7}$$
$$du_0 = e^{-u_0} + db_1 - db_0. \tag{3-8}$$

The slope $u_j(t)$ is locally conserved.

Somewhat unexpectedly, one can still find the stationary and translation
invariant measures for the interacting diffusions (3-7) on the lattice $\mathbb{Z}$. They are

labeled by a parameter $r > 0$ and are of product form. The single site measure is a double exponential of the form

$$\mu_r(dx) = \Gamma(r)^{-1} e^{-e^{-x}} e^{-rx} \, dx, \quad r > 0. \tag{3-9}$$

Averages with respect to μ_r are denoted by $\langle \cdot \rangle_r$. The parameters of scaling theory are now easily computed. We find

$$\rho = \langle u_0 \rangle_r = -\psi(r), \tag{3-10}$$

with $\psi = \Gamma'/\Gamma$ the digamma function on $\mathbb{R}^+$. Note that $\psi' > 0$, $\psi'' < 0$, and ρ ranges over $\mathbb{R}$. From (3-7) the random current is $-e^{-u_j} dt - db_{j+1}$ and hence the average current is

$$j = -\langle e^{-u_0} \rangle_r = -r. \tag{3-11}$$

Finally

$$A(r) = \langle u_0^2 \rangle_r - \langle u_0 \rangle_r^2 = \psi'(r). \tag{3-12}$$

Since the initial conditions force ϕ to be convex down, the signs from Section 2 are reversed. In particular, the sup in (2-10) is replaced by the inf and

$$\phi(y) = \inf_{\rho \in \mathbb{R}} (-y\rho - j(-\rho)), \quad y \geq 0. \tag{3-13}$$

Then $\phi(0) = 0$, $\phi'' < 0$, and ϕ has a single strictly positive maximum before dropping to $-\infty$ as $y \to \infty$. Thus $t\phi(y/t)$ reproduces the singular initial conditions for (3-6) as $t \to 0$. Moriarty and O'Connell [2007] prove that

$$\lim_{N \to \infty} \frac{1}{N} h_N(\kappa N) = f(\kappa), \tag{3-14}$$

with

$$f(\kappa) = \inf_{s \geq 0} (\kappa s - \psi(s)). \tag{3-15}$$

Scaling theory claims that

$$\lim_{t \to \infty} \frac{1}{t} h_{\lfloor yt \rfloor}(t) = \phi(y), \quad y > 0. \tag{3-16}$$

Hence, using that $\psi' > 0$ and $y\kappa = 1$, we have

$$\phi(y) = \frac{1}{\kappa} f(\kappa) = \inf_{s \geq 0} (s - y\psi(s)) = \inf_{\tilde{s} \in \mathbb{R}} (\psi^{-1}(\tilde{s}) - y\tilde{s}), \tag{3-17}$$

in agreement with (3-13).

We turn to the asymptotic analysis of the height fluctuations:

Theorem [Borodin and Corwin 2014, Theorem 5.12]. *There exists a κ^* such that, for $0 < \kappa^* < \kappa$,*

$$\lim_{n \to \infty} \mathbb{P}\big(h_n(\kappa n) - nf(\kappa) \leq (-2f''(\kappa))^{-1/3} n^{1/3} s\big) = F_{\text{GUE}}(s). \tag{3-18}$$

According to (3-11), $\lambda = -j'' > 0$. Hence in (2-14) $-\lambda$ is replaced by λ and the sign to the right of $\leq$ is $+$. To see whether with these changes scaling theory is confirmed, we start from

$$h_{\lfloor yt \rfloor}(t) = t\phi(y) + \left(\tfrac{1}{2}\lambda A^2 t\right)^{1/3} \xi_{\text{TW}}, \tag{3-19}$$

with ξ_{TW} a GUE Tracy–Widom distributed random variable, hence

$$h_n(\kappa n) = \kappa n \phi(\kappa^{-1}) + \left(\tfrac{1}{2}\lambda A^2 \kappa n\right)^{1/3} \xi_{\text{TW}}. \tag{3-20}$$

Now $\rho = -\psi(r(\rho))$ is differentiated as

$$1 = -\psi' r', \quad 0 = \psi''(r')^2 + \psi' r''. \tag{3-21}$$

Since $-\lambda = \mathrm{j}''(\rho) = -r''(\rho)$ and $A(r) = \psi'(r)$, one has

$$\lambda A^2 = \psi'' r'. \tag{3-22}$$

Since $y = \mathrm{j}'(\rho) = -r'(\rho)$ and $y\kappa = 1$, we conclude

$$\lambda A^2 \kappa = -\psi'', \tag{3-23}$$

and, since f is the Legendre transform of ψ,

$$\lambda A^2 \kappa = -\frac{1}{f''}, \tag{3-24}$$

in agreement with (3-18).

Conclusion

KPZ scaling theory makes a prediction on the nonuniversal coefficients for models in the KPZ class and has been confirmed for PASEP, discrete TASEP, and PNG. We add to this list the semidiscrete directed polymer. The corresponding stochastic "particle" model is a system of diffusions, $u_j(t)$, with nearest neighbor interactions such that the sums $\sum_j u_j(t)$ are locally conserved. This model has a flavor rather distinct from driven lattice gases. Still the long time asymptotics in all models is the Tracy–Widom statistics.

References

[Alves et al. 2011] S. G. Alves, T. J. Oliveira, and S. Ferreira, "Universal fluctuations in radial growth models belonging to the KPZ universality class", *Europhys. Lett.* **96** (2011), 48003.

[Amir et al. 2011] G. Amir, I. Corwin, and J. Quastel, "Probability distribution of the free energy of the continuum directed random polymer in $1 + 1$ dimensions", *Comm. Pure Appl. Math.* **64**:4 (2011), 466–537.

[Baryshnikov 2001] Y. Baryshnikov, "GUEs and queues", *Probab. Theory Related Fields* **119**:2 (2001), 256–274.

[Ben Arous and Corwin 2011] G. Ben Arous and I. Corwin, "Current fluctuations for TASEP: a proof of the Prähofer–Spohn conjecture", *Ann. Probab.* **39**:1 (2011), 104–138.

[Borodin and Corwin 2014] A. Borodin and I. Corwin, "Macdonald processes", *Probab. Theory Related Fields* **158**:1-2 (2014), 225–400.

[Borodin and Ferrari 2008] A. Borodin and P. L. Ferrari, "Large time asymptotics of growth models on space-like paths, I: PushASEP", *Electron. J. Probab.* **13** (2008), no. 50, 1380–1418.

[Borodin et al. 2007] A. Borodin, P. L. Ferrari, M. Prähofer, and T. Sasamoto, "Fluctuation properties of the TASEP with periodic initial configuration", *J. Stat. Phys.* **129**:5-6 (2007), 1055–1080.

[Borodin et al. 2008] A. Borodin, P. L. Ferrari, and T. Sasamoto, "Large time asymptotics of growth models on space-like paths, II: PNG and parallel TASEP", *Comm. Math. Phys.* **283**:2 (2008), 417–449.

[Derrida et al. 1991] B. Derrida, J. L. Lebowitz, E. R. Speer, and H. Spohn, "Fluctuations of a stationary nonequilibrium interface", *Phys. Rev. Lett.* **67**:2 (1991), 165–168.

[Ferrari and Spohn 2006] P. L. Ferrari and H. Spohn, "Scaling limit for the space-time covariance of the stationary totally asymmetric simple exclusion process", *Comm. Math. Phys.* **265**:1 (2006), 1–44.

[Forrester 1993] P. J. Forrester, "The spectrum edge of random matrix ensembles", *Nuclear Phys. B* **402**:3 (1993), 709–728.

[Glynn and Whitt 1991] P. W. Glynn and W. Whitt, "Departures from many queues in series", *Ann. Appl. Probab.* **1**:4 (1991), 546–572.

[Imamura and Sasamoto 2012] T. Imamura and T. Sasamoto, "Exact Solution for the Stationary Kardar-Parisi-Zhang Equation", *Phys. Rev. Lett.* **108** (2012), 190603.

[Johansson 2000] K. Johansson, "Shape fluctuations and random matrices", *Comm. Math. Phys.* **209**:2 (2000), 437–476.

[Kardar et al. 1986] M. Kardar, G. Parisi, and Y.-C. Zhang, "Dynamic scaling of growing interfaces", *Phys. Rev. Lett.* **56** (1986), 889–892.

[Krug et al. 1992] J. Krug, P. Meakin, and T. Halpin-Healy, "Amplitude universality for driven interfaces and directed polymers in random media", *Phys. Rev. A* **45** (1992), 638–653.

[Liggett 1999] T. M. Liggett, *Stochastic interacting systems: contact, voter and exclusion processes*, Grundlehren der Mathematischen Wissenschaften **324**, Springer, Berlin, 1999.

[Moriarty and O'Connell 2007] J. Moriarty and N. O'Connell, "On the free energy of a directed polymer in a Brownian environment", *Markov Process. Related Fields* **13**:2 (2007), 251–266.

[O'Connell 2012] N. O'Connell, "Directed polymers and the quantum Toda lattice", *Ann. Probab.* **40**:2 (2012), 437–458.

[O'Connell and Yor 2001] N. O'Connell and M. Yor, "Brownian analogues of Burke's theorem", *Stochastic Process. Appl.* **96**:2 (2001), 285–304.

[Paczuski et al. 1992] M. Paczuski, M. Barma, S. N. Majumdar, and T. Hwa, "Fluctuations of a nonequililbrium interface", *Phys. Rev. Lett.* **69** (1992), 2735–2735.

[Prähofer and Spohn 2000] M. Prähofer and H. Spohn, "Universal distributions for growth processes in $1 + 1$ dimensions and random matrices", *Phys. Rev. Lett.* **84** (2000), 4882–4885.

[Prähofer and Spohn 2002] M. Prähofer and H. Spohn, "Current fluctuations for the totally asymmetric simple exclusion process", pp. 185–204 in *In and out of equilibrium* (Mambucaba, 2000), edited by V. Sidoravicius, Progr. Probab. **51**, Birkhäuser, Boston, 2002.

[Prähofer and Spohn 2004] M. Prähofer and H. Spohn, "Exact scaling functions for one-dimensional stationary KPZ growth", *J. Statist. Phys.* **115**:1-2 (2004), 255–279.

[Sasamoto and Spohn 2010] T. Sasamoto and H. Spohn, "Exact height distributions for the KPZ equation with narrow wedge initial condition", *Nuclear Phys. B* **834**:3 (2010), 523–542.

[Spohn 1991] H. Spohn, *Large scale dynamics of interacting particles*, Springer, Heidelberg, 1991.

[Takeuchi and Sano 2010] K. A. Takeuchi and M. Sano, "Universal fluctuations of growing interfaces: evidence in turbulent liquid crystals", *Phys. Rev. Lett.* **104** (2010), 230601.

[Takeuchi et al. 2011] K. A. Takeuchi, M. Sano, T. Sasamoto, and H. Spohn, "Growing interfaces uncover universal fluctuations behind scale invariance", *Scientific Reports* **1** (2011), 34.

[Tracy and Widom 1994] C. A. Tracy and H. Widom, "Level-spacing distributions and the Airy kernel", *Comm. Math. Phys.* **159**:1 (1994), 151–174.

[Tracy and Widom 2009] C. A. Tracy and H. Widom, "Asymptotics in ASEP with step initial condition", *Comm. Math. Phys.* **290**:1 (2009), 129–154.

spohn@ma.tum.de   *Department of Mathematics and Physics, Technische Universität München, D-85747 Garching, Germany*

Random Matrices
MSRI Publications
Volume 65, 2014

Experimental realization of Tracy–Widom distributions and beyond: KPZ interfaces in turbulent liquid crystal

KAZUMASA A. TAKEUCHI

Analytical studies have shown the Tracy–Widom distributions and the Airy processes in the asymptotics of a few growth models in the Kardar–Parisi–Zhang (KPZ) universality class. Here the author shows evidence that these mathematical objects arise even in a real experiment: more specifically, in growing interfaces of turbulent liquid crystal. The present article is devoted to overviewing the current status of this experimental approach to the KPZ class, which directly concerns random matrix theory and related fields of mathematical physics. In particular, the author summarizes those statistical properties which were derived rigorously for simple solvable models and realized here experimentally, and those which were evidenced in the experiment and remain to be explained by further mathematical or theoretical studies.

1. Introduction

The first decade of the 21st century and a couple of preceding years have been marked by a series of remarkable analytical developments, which have revealed profound and rigorous connections among random matrix theory, combinatorial problems, and the physical problem of the fluctuating interface growth ([Baik and Rains 2001; Kriecherbauer and Krug 2010; Sasamoto and Spohn 2010a; Corwin 2012] and references therein). Their primary conclusions in terms of the interface growth problem, first pointed out by Johansson [2000] for TASEP and by Prähofer and Spohn [2000] for the PNG model, are the following [Kriecherbauer and Krug 2010; Sasamoto and Spohn 2010a; Corwin 2012]: (i) The distribution function and the spatial correlation function were obtained rigorously for the asymptotic interface fluctuations. (ii) The results depend on the global shape of the interfaces, or on the initial condition. For the two prototypical cases of the curved and flat growing interfaces, the distribution function is given by the Tracy–Widom distribution [Tracy and Widom 1994; 1996] for GUE and GOE, respectively, and the spatial two-point correlation function by the covariance of the Airy$_2$ and Airy$_1$ process [Prähofer and Spohn 2002; Sasamoto 2005], respectively. (iii) All or some of these conclusions were reached for a number of models, namely the TASEP [Johansson 2000; Borodin et al. 2007; 2008; Sasamoto 2005] and the PASEP [Tracy and Widom 2009], the PNG model [Prähofer and Spohn 2000; 2002; Borodin et al. 2008], and the KPZ equation [Sasamoto and Spohn 2010c; 2010b; Amir et al. 2011; Calabrese et al. 2010; Dotsenko 2010; Calabrese and Le Doussal 2011; Prolhac and Spohn 2011]. They are believed to be universal characteristics of the KPZ class, which is the basic universality class for describing scale-invariant growth of interfaces due to local interactions [Kardar et al. 1986].

This geometry-dependent universality of the KPZ class and the nontrivial connection to random matrix theory were recently made visible by a real experiment [Takeuchi and Sano 2010; 2012; Takeuchi et al. 2011]. Using the electrically driven convection of nematic liquid crystal [de Gennes and Prost 1995], the author and his coworker generated expanding domains of turbulence amidst another turbulent state which is only metastable (Figure 1) and measured the fluctuations of their growing interfaces. Although the growth mechanism in this experiment is far more complicated than the solvable mathematical models — it is realized by proliferation and random transport of topological defects due to local turbulent flow in the electroconvection, such microscopic difference is scaled out in the macroscopic dynamics according to the universality hypothesis. The author then indeed found the aforementioned statistical properties of the KPZ class emerge in the scaling limit [Takeuchi and Sano 2010; 2012; Takeuchi et al. 2011].

The present contribution is devoted to overviewing the current status of this experimental investigation. It summarizes, one by one, those statistical properties which were derived rigorously for the solvable models and confirmed here experimentally (Section 2) and those which were evidenced in the experiment and remain to be explained by mathematical or theoretical studies (Section 3). Note however that, because of the space constraint, the present survey does *not* cover all the experimental results obtained so far; for the complete description of the experimental system and the results, the readers are referred to the recent article by the author and the coworker [Takeuchi and Sano 2012].

2. Experimental realization of analytically solved properties

Scaling exponents. The generated growing interfaces become rougher and rougher as time elapses. In other words, the local height $h(x, t)$ measured along the average growth direction (see Figure 1) is fluctuating in both space and time and the fluctuations grow with time. The roughness can be quantified by, for example, the height-difference correlation function $C_h(l, t) \equiv \langle [h(x + l, t) - h(x, t)]^2 \rangle$

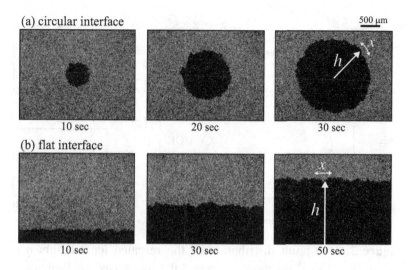

Figure 1. Growing turbulent domain (black) in the liquid-crystal convection, bordered by a circular (a) and flat (b) interface. Indicated below each image is the elapsed time for this growth process. Movies are also available as supplementary information of [Takeuchi et al. 2011]. Such interfaces were generated about a thousand times to evaluate all the statistical properties presented in this article.

with the ensemble average $\langle \cdots \rangle$. It then turned out to obey the following power law called the Family–Vicsek scaling [Family and Vicsek 1985]:

$$C_h(l, t)^{1/2} \sim t^\beta F_h(lt^{-1/z}) \sim \begin{cases} l^\alpha & \text{for } l \ll l_*, \\ t^\beta & \text{for } l \gg l_*, \end{cases} \tag{1}$$

with a scaling function F_h, a crossover length scale $l_* \sim t^{1/z}$, and the KPZ characteristic exponents $\alpha = 1/2$, $\beta = 1/3$, and $z \equiv \alpha/\beta = 3/2$ for $1 + 1$ dimensions [Kardar et al. 1986]. The same set of the exponents was found for both circular and flat interfaces. This implies that the one-point fluctuations of the local height h can be described, for large t, as

$$h \simeq v_\infty t + (\Gamma t)^{1/3} \chi, \tag{2}$$

with two constant parameters v_∞ and Γ and with a random variable χ that captures the fluctuations of the growing interfaces. The two parameters are related to those of the KPZ equation, $\partial_t h = \nu \partial_x^2 h + (\lambda/2)(\partial_x h)^2 + \sqrt{D}\xi$ with white noise ξ, by $v_\infty = \lambda$ and $\Gamma = A^2 \lambda/2$ with $A \equiv D/2\nu$ [Takeuchi and Sano 2012].

Distribution function. The random variable χ in (2) turned out to be identical to the variable $\chi_2 \equiv \chi_{GUE}$ obeying the GUE Tracy–Widom distribution for the

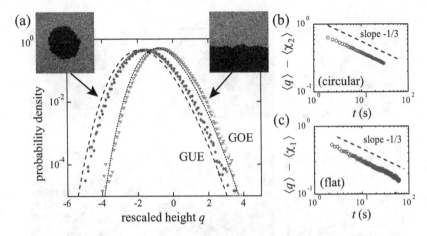

Figure 2. One-point distribution of the rescaled local height $q \equiv (h - v_\infty t)/(\Gamma t)^{1/3}$ for the circular and flat interfaces. (a) Probability density of q measured at different times, $t = 10$ s and 30 s for the circular interfaces (solid symbols) and $t = 20$ s and 60 s for the flat ones (open symbols), from right to left. The dashed and dotted curves show the GUE and GOE Tracy–Widom distributions, respectively (with the factor $2^{-2/3}$ for the latter). (b,c) Finite-time correction in the mean. The figures are reprinted from [Takeuchi and Sano 2012] with adaptations, with kind permission from Springer Science+Business Media.

circular interfaces, and to $\chi_1 \equiv 2^{-2/3}\chi_{GOE}$ with χ_{GOE} being the GOE Tracy–Widom random variable for the flat interfaces, in the limit $t \to \infty$. This is in agreement with the analytical results for all the above-mentioned solvable models [Kriecherbauer and Krug 2010; Sasamoto and Spohn 2010a; Corwin 2012]. It was shown by plotting histograms of the rescaled height $q \equiv (h - v_\infty t)/(\Gamma t)^{1/3} \simeq \chi$ [Figure 2(a)] using the experimentally measured values of the parameters v_∞ and Γ. The slight horizontal shifts visible in Figure 2(a) are due to finite-time correction in the mean $\langle q \rangle$, which decays by a power law $\langle q \rangle - \langle \chi_i \rangle \sim t^{-1/3}$ [Figure 2(b,c)] with $i = 1$ (flat) or 2 (circular). These finite-time corrections will be revisited in Section 3.

Spatial correlation function. In the solvable case, it is analytically proved that the two-point spatial correlation function

$$C_s(l; t) \equiv \langle h(x+l, t)h(x, t) \rangle - \langle h(x+l, t) \rangle \langle h(x, t) \rangle \tag{3}$$

is given by the covariance of the Airy$_1$ process $\mathcal{A}_1(u)$ for the flat interfaces and the Airy$_2$ process $\mathcal{A}_2(u)$ for the curved ones [Kriecherbauer and Krug 2010;

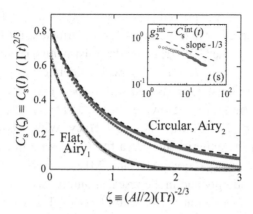

Figure 3. Two-point spatial correlation function $C_s(l;t)$ in the rescaled units. The symbols are the experimental data at $t = 10$ s and 30 s for the circular case and $t = 20$ s and 60 s for the flat one (from bottom to top for each pair). The dashed and dashed-dotted curves indicate the covariance of the Airy$_2$ and Airy$_1$ processes, respectively. The inset shows the finite-time correction for the circular case, expressed in terms of the integral $C_s^{int}(t) \equiv \int_0^\infty C_s'(\zeta;t)\,d\zeta$ and $g_2^{int} \equiv \int_0^\infty g_2(\zeta)\,d\zeta$. The figure is reprinted from [Takeuchi and Sano 2012] with adaptations, with kind permission from Springer Science+Business Media.

Sasamoto and Spohn 2010a; Corwin 2012], through the single expression

$$C_s(l;t) \simeq (\Gamma t)^{2/3} g_i\left(\tfrac{1}{2} A l (\Gamma t)^{-2/3}\right). \tag{4}$$

Here, $g_i(\zeta) \equiv \langle \mathscr{A}_i(u+\zeta)\mathscr{A}_i(u)\rangle - \langle \mathscr{A}_i(u)\rangle^2$ and the Airy processes are normalized to have the same variance as χ_i, $\langle \mathscr{A}_i^2(u)\rangle_c = \langle \chi_i^2 \rangle_c$. The experimental data at large times also indicate (4) for both flat and circular interfaces (Figure 3), providing information on finite-time corrections as well.

Extreme-value statistics. Analytical studies of the curved PNG interface, or the related mathematical problems of the vicious walker and the directed polymer, have also successfully solved the asymptotic distribution for the maximal height H_{max} and, very recently, its position X_{max} in this model [Johansson 2003; Moreno Flores et al. 2013; Schehr 2012]. In the rescaled units, the two extremal quantities are described as

$$q_{max}^{(h)} \equiv (H_{max} - v_\infty t)/(\Gamma t)^{1/3} \rightarrow \max(\mathscr{A}_2(u) - u^2),$$
$$X_{max}' \equiv (A X_{max}/2)/(\Gamma t)^{2/3} \rightarrow \arg\max(\mathscr{A}_2(u) - u^2)$$

in the limit $t \to \infty$. It was proved [Johansson 2003; Moreno Flores et al. 2013; Schehr 2012] in particular that the asymptotic distribution for the rescaled maximal height is given by the GOE Tracy–Widom distribution (with the factor $2^{-2/3}$); in other words, $H_{\max} \simeq v_\infty t + (\Gamma t)^{1/3} \chi$ with χ identical to χ_1. Experimentally, this maximal height must be measured with respect to a fictitious flat substrate that includes the origin of the growing cluster, and hence $H_{\max} = \max(h \sin \phi)$ with the azimuth ϕ. In this way the GOE Tracy–Widom distribution was indeed identified in the experimental data for $H_{\max}$, with finite-time correction $\langle q_{\max}^{(h)} \rangle - \langle \chi_1 \rangle \sim t^{-1/3}$ [Takeuchi and Sano 2012]. The position $X_{\max}$ was also measured correspondingly and in the rescaled unit it was shown to approach the asymptotic analytical solution in [Moreno Flores et al. 2013; Schehr 2012] with increasing time [Takeuchi and Sano 2012].

3. Experimental fact for analytically unsolved properties

Finite-time corrections. The experimental data were obviously obtained at finite times and therefore allow studying the finite-time corrections from the analytical expressions derived in the asymptotic limit. For the one-point distribution, the corrections in the nth-order cumulants were found to be $\langle q^n \rangle_c - \langle \chi_i^n \rangle_c \sim \mathbb{O}(t^{-n/3})$ up to $n = 4$ for both flat and circular cases, except that the second- and fourth-order cumulants for the circular interfaces were too small to identify any systematic variation in time [Takeuchi and Sano 2012]. Although one can show the same exponents $\mathbb{O}(t^{-n/3})$ up to $n = 4$ for the curved exact solution of the KPZ equation [Sasamoto and Spohn 2010c; 2010b; Takeuchi and Sano 2012], for the TASEP, PASEP, and PNG model, only the corrections in the n-th-order *moments* were evaluated: $\mathbb{O}(t^{-1/3})$ for $n = 1$ and $\mathbb{O}(t^{-2/3})$ for $n \geq 2$ [Ferrari and Frings 2011; Baik and Jenkins 2013]. It would be useful to show if the corrections in the *cumulants* are in the order of $\mathbb{O}(t^{-n/3})$ or $\mathbb{O}(t^{-2/3})$ for these solvable models. From the numerical side, circular interfaces of an off-lattice Eden model showed corrections in the order of $\mathbb{O}(t^{-2/3})$ for both first- and second-order cumulants within the time window of the simulation [Takeuchi 2012]. Although one cannot exclude the possibility of crossover to $\mathbb{O}(t^{-1/3})$ for the first-order cumulant, one could also speculate that this leading term is somehow absent in the off-lattice Eden model because of some sort of symmetry. To add, in contrast to the exponents, the coefficients for these finite-time corrections are understood to be model-dependent. Indeed, that for the first-order cumulant, or the mean, $\langle q \rangle - \langle \chi_i \rangle$, is negative for the curved solution of the KPZ equation [Sasamoto and Spohn 2010c; 2010b] but positive for the TASEP [Ferrari and Frings 2011; Baik and Jenkins 2013], the simulation of the off-lattice Eden model [Takeuchi 2012], and the liquid-crystal experiment [Takeuchi and Sano 2012]. See Figure 2(b,c).

Similar data analysis was performed for the maximal height H_{max} and found, for the mean, the same exponent as for the one-point distribution:

$$\langle q_{max}^{(h)} \rangle - \langle \chi_1 \rangle \sim \mathcal{O}(t^{-1/3})$$

for the experiment [Takeuchi and Sano 2012] and

$$\langle q_{max}^{(h)} \rangle - \langle \chi_1 \rangle \sim \mathcal{O}(t^{-2/3})$$

for the off-lattice Eden simulation [Takeuchi 2012]. Concerning the distribution of the position X_{max}, opposite signs of the corrections were found for the second- and fourth-order cumulants between the experiment and the numerically solved PNG droplet [Takeuchi and Sano 2012]. These finite-time distributions of the extremal quantities remain inaccessible by analytic means.

The finite-time corrections were also measured experimentally for the spatial correlation function $C_s(l; t)$ [Takeuchi and Sano 2012]. The corrections were quantified in terms of the integral of the rescaled correlation function, $C_s^{int}(t) \equiv \int_0^\infty C_s'(\zeta; t) d\zeta$ with $C_s'(\zeta; t) \equiv C_s(l; t)/(\Gamma t)^{2/3}$ and $\zeta \equiv (Al/2)(\Gamma t)^{-2/3}$. For the circular interfaces, it was shown to approach the value of the Airy$_2$ covariance $g_2^{int} \equiv \int_0^\infty g_2(\zeta) d\zeta$ as $g_2^{int} - C_s^{int} \sim \mathcal{O}(t^{-1/3})$ [Takeuchi and Sano 2012] (Figure 3 inset). The same exponent was also found numerically in the circular interfaces of the off-lattice Eden model [Takeuchi 2012], though the way the function $C_s'(\zeta; t)$ approaches $g_2(\zeta)$ appears to be different. It is therefore important to have analytical solutions for the spatial correlation function at finite times, which are not yet obtained in a controlled manner in any solvable models.

Spatial persistence probability. Although it is considered that the spatial profile of the growing interfaces itself is given by the corresponding Airy process, to the knowledge of the author, statistical quantities other than the two-point correlation function have not been explicitly calculated in the analytical studies. In other words, measuring such quantities on the spatial correlation of the interfaces can also shed light on the temporal correlation of the Airy processes, as well as that for the largest eigenvalue in Dyson's Brownian motion for GUE random matrices, which is equivalent to the Airy$_2$ process [Johansson 2003].

This strategy was also pursued in the liquid-crystal experiment [Takeuchi and Sano 2012], in which the persistence property of the height fluctuation $\delta h(x, t) \equiv h(x, t) - \langle h \rangle$ was measured. The spatial persistence probability $P_\pm^{(s)}(l; t)$ is defined as the probability that a positive $(+)$ or negative $(-)$ fluctuation continues over length l in a spatial profile of the interfaces at time t. The experimental data then indicated, within the experimental accuracy, exponential decay

$$P_\pm^{(s)}(l; t) \sim e^{-\kappa_\pm^{(s)} \zeta}, \tag{5}$$

with $\zeta \equiv (Al/2)(\Gamma t)^{-2/3}$ for both flat and circular interfaces [Takeuchi and Sano 2012]. The decay coefficients $\kappa_{\pm}^{(s)}$ were however different between the two cases:

$$\begin{cases} \kappa_+^{(s)} = 1.9(3) \\ \kappa_-^{(s)} = 2.0(3) \end{cases} \text{(flat)} \quad \text{and} \quad \begin{cases} \kappa_+^{(s)} = 1.07(8) \\ \kappa_-^{(s)} = 0.87(6) \end{cases} \text{(circular)}, \quad (6)$$

where the numbers in the parentheses indicate the range of error expected in the last digit of the estimates. The exponential decay (5) in the spatial persistence probability was also identified numerically for the circular interfaces of the off-lattice Eden model, which gave $\kappa_+^{(s)} = 0.90(2)$ and $\kappa_-^{(s)} = 0.89(4)$ [Takeuchi and Sano 2012]. Since a similar set of the coefficients was numerically found in the temporal persistence probability of the GUE Dyson Brownian motion [Takeuchi and Sano 2012], namely $\kappa_+^{(s)} = 0.90(8)$ and $\kappa_-^{(s)} = 0.90(6)$, the author considers that the experimental value of $\kappa_+^{(s)}$ for the circular interfaces is somewhat affected by finite-time effect and/or experimental error. To resolve this issue, it is important to derive a theoretical expression for this persistence probability, whether rigorously or approximatively, and to provide a direct numerical evaluation with the aid of, e.g., Bornemann's method [2010] to estimate the Fredholm determinant numerically.

Temporal correlation. In contrast to the spatial correlation of the interfaces which can be dealt with in terms of the Airy processes, their temporal correlation remains inaccessible in analytical studies. Given that it is also expected to be universal in the scaling limit, explicit information provided by experimental and numerical studies may hint at the form of the solution that should be reached, if reachable, on the temporal correlation of solvable growth models.

The two-point temporal correlation function

$$C_t(t, t_0) \equiv \langle h(x, t)h(x, t_0)\rangle - \langle h(x, t)\rangle\langle h(x, t_0)\rangle \quad (7)$$

was experimentally measured along the characteristic lines, or in the vertical and radial direction for the flat and circular interfaces, respectively, and turned out to be very different between the two cases [Takeuchi and Sano 2012]. For the flat case, it is governed by the scaling form $C_t(t, t_0) \simeq (\Gamma^2 t_0 t)^{1/3} F_t(t/t_0)$ with a scaling function $F_t(t/t_0) \sim (t/t_0)^{-\bar{\lambda}}$ and $\bar{\lambda} = 1$. In contrast, for the circular case, the raw correlation function $C_t(t, t_0)$ does *not* decay to zero, presumably even in the limit $t \to \infty$. The author found that the experimental curve for $C_t(t, t_0)$ at each t_0 is proportional to the functional form obtained by Singha after rough theoretical approximations [Singha 2005]:

$$\frac{C_t(t, t_0)}{C_t(t_0, t_0)} \approx c(t_0) F_{\text{Singha}}(t/t_0; b(t_0)), \quad (t \neq t_0), \tag{8}$$

$$F_{\text{Singha}}(\tau; b) \equiv \frac{e^{b(1-1/\sqrt{\tau})}\Gamma(2/3, b(1 - 1/\sqrt{\tau}))}{\Gamma(2/3)}, \tag{9}$$

with the upper incomplete Gamma function $\Gamma(s, x)$, the Gamma function $\Gamma(s)$, and unknown parameters $b(t_0)$ and $c(t_0)$ which turned out to depend on t_0 [Takeuchi and Sano 2012]. This functional form also indicates

$$\lim_{t \to \infty} C_t(t, t_0) > 0,$$

as suggested by the experimental data. The ever-lasting temporal correlation in the circular case formally implies $\bar{\lambda} = 1/3$, in contrast with $\bar{\lambda} = 1$ for the flat case. This supports Kallabis and Krug's conjecture [1999] that the autocorrelation exponent $\bar{\lambda}$ derived for the linear growth equations:

$$\bar{\lambda} = \begin{cases} \beta + d/z & \text{(flat)}, \\ \beta & \text{(circular)}, \end{cases} \tag{10}$$

where d is the spatial dimension, also applies to the KPZ universality class. This conjecture, as well as the interesting functional form for the temporal correlation of the circular interfaces, need to be explained on the basis of more refined, hopefully rigorous, theoretical arguments.

The temporal correlation was also characterized in terms of the persistence probability. Along the characteristic lines, the temporal persistence probability $P_{\pm}(t, t_0)$ is defined as the joint probability that the interface fluctuation $\delta h(x, t)$ is positive $(+)$ or negative $(-)$ at time t_0 and maintains the same sign until time t. Experimentally, it was found to decay algebraically

$$P_{\pm}(t, t_0) \sim (t/t_0)^{-\theta_{\pm}} \tag{11}$$

with different sets of the exponents $\theta_{\pm}$ for the flat and circular interfaces [Takeuchi and Sano 2012]:

$$\begin{cases} \theta_+ = 1.35(5) \\ \theta_- = 1.85(10) \end{cases} \text{(flat)} \quad \text{and} \quad \begin{cases} \theta_+ = 0.81(2) \\ \theta_- = 0.80(2) \end{cases} \text{(circular).} \tag{12}$$

It is interesting to note that θ_+ and θ_- are asymmetric in the flat case, which had also been reported in numerical work [Kallabis and Krug 1999] and associated with the nonlinearity in the KPZ equation, whereas this asymmetry is somehow canceled for the circular interfaces. This latter statement was also confirmed by the simulation of the off-lattice Eden model, which gave $\theta_+ = 0.81(3)$ and $\theta_- = 0.77(4)$ [Takeuchi 2012]. Theoretical accounts should hopefully be made on how the asymmetry $\theta_+ \neq \theta_-$, which is present in the flat case, is canceled for the circular interfaces.

4. Concluding remarks

We have briefly overviewed the main experimental results obtained for the growing interfaces of the liquid-crystal turbulence [Takeuchi and Sano 2010; 2012; Takeuchi et al. 2011]. On the one hand, this experiment provides an interesting situation where the deep and beautiful mathematical concepts developed in random matrix theory and other domains of mathematical physics arise in a real phenomenon (Section 2). It would be remarkable that we can directly look at the Tracy–Widom distributions and the Airy processes by our eyes, or more precisely by a microscope, all the more because there is no random matrix which explicitly arises in this problem. On the other hand, and more importantly for future developments, such an experimental study allows us to access statistical properties that remain unsolved in the rigorous analytical treatments (Section 3). The author believes that providing proof or theoretical accounts for those unsolved statistical properties in any solvable model will further advance our understanding on the KPZ universality class, as well as in the wide variety of related mathematical fields in this context. Finally, the author would like to refer the interested readers to the article [Takeuchi and Sano 2012], in which one can find much more complete descriptions on the experiment and the results.

Note added in proof

After submission of this article, Ferrari and Frings [2013] derived analytic expressions for the persistence probability of negative fluctuations for the Airy$_1$ and Airy$_2$ processes (corresponding to the spatial persistence probability $P_-^{(s)}$ for the interfaces). They numerically evaluated the decay coefficient $\kappa_-^{(s)}$ for the Airy$_1$ process and found it in agreement with the experimental value. However, in my viewpoint, two problems remain open:

(1) The persistence probability of positive fluctuations remains to be solved.

(2) The present article reported $\kappa_-^{(s)} \approx \kappa_+^{(s)}$ with the sign of the fluctuations defined with respect to the mean value $\langle h \rangle$, while Ferrari and Frings showed that $\kappa_-^{(s)}$ depends continuously on the reference value c used to define the sign: specifically, $\kappa_-^{(s)}$ decreases with increasing c. For $\kappa_+^{(s)}$, one naturally expects that it increases with c. Thus, it is not known whether and why $\kappa_-^{(s)}$ and $\kappa_+^{(s)}$ take the same value at $c = \langle h \rangle$, or they just happen to be close.

Likewise, after submission of the article, Alves et al. [2013] reported results of extensive simulations of the off-lattice Eden model and showed that the peculiar finite-time correction of $\mathcal{O}(t^{-2/3})$ for the mean height of this model is replaced by the usual scaling $\mathcal{O}(t^{-1/3})$ at larger times. Other open problems on finite-time corrections mentioned in Section 3 remain unsolved to the knowledge of the author.

Acknowledgements

The author is indebted to T. Sasamoto for his suggestion on the content of this contribution, whereas any problem in it is obviously attributed to the author. The author wishes to thank M. Prähofer for providing him with the theoretical curves for the Tracy–Widom distributions (used in Figure 2) and F. Bornemann for those of the covariance of the Airy processes (Figure 3), evaluated numerically by his accurate algorithm [Bornemann 2010]. This work is supported in part by Grant for Basic Science Research Projects from The Sumitomo Foundation.

References

[Alves et al. 2013] S. G. Alves, T. J. Oliveira, and S. C. Ferreira, "Non-universal parameters, corrections and universality in Kardar–Parisi–Zhang growth", *J. Stat. Mech.* **05** (2013), P05007.

[Amir et al. 2011] G. Amir, I. Corwin, and J. Quastel, "Probability distribution of the free energy of the continuum directed random polymer in $1 + 1$ dimensions", *Comm. Pure Appl. Math.* **64**:4 (2011), 466–537.

[Baik and Jenkins 2013] J. Baik and R. Jenkins, "Limiting distribution of maximal crossing and nesting of Poissonized random matchings", *Ann. Probab.* **41**:6 (2013), 4359–4406.

[Baik and Rains 2001] J. Baik and E. M. Rains, "Symmetrized random permutations", pp. 1–19 in *Random matrix models and their applications*, edited by P. Bleher and A. Its, Math. Sci. Res. Inst. Publ. **40**, Cambridge Univ. Press, 2001.

[Bornemann 2010] F. Bornemann, "On the numerical evaluation of Fredholm determinants", *Math. Comp.* **79**:270 (2010), 871–915.

[Borodin et al. 2007] A. Borodin, P. L. Ferrari, M. Prähofer, and T. Sasamoto, "Fluctuation properties of the TASEP with periodic initial configuration", *J. Stat. Phys.* **129**:5-6 (2007), 1055–1080.

[Borodin et al. 2008] A. Borodin, P. L. Ferrari, and T. Sasamoto, "Large time asymptotics of growth models on space-like paths, II: PNG and parallel TASEP", *Comm. Math. Phys.* **283**:2 (2008), 417–449.

[Calabrese and Le Doussal 2011] P. Calabrese and P. Le Doussal, "Exact solution for the Kardar–Parisi–Zhang equation with flat initial conditions", *Phys. Rev. Lett.* **106** (2011), 250603.

[Calabrese et al. 2010] P. Calabrese, P. Le Doussal, and A. Rosso, "Free-energy distribution of the directed polymer at high temperature", *Europhys. Lett.* **90** (2010), 20002.

[Corwin 2012] I. Corwin, "The Kardar–Parisi–Zhang equation and universality class", *Random Matrices Theory Appl.* **1**:1 (2012), 1130001.

[Dotsenko 2010] V. Dotsenko, "Bethe ansatz derivation of the Tracy–Widom distribution for one-dimensional directed polymers", *Europhys. Lett.* **90** (2010), 20003.

[Family and Vicsek 1985] F. Family and T. Vicsek, "Scaling of the active zone in the Eden process on percolation networks and the ballistic deposition model", *J. Phys. A* **18** (1985), L75–L81.

[Ferrari and Frings 2011] P. L. Ferrari and R. Frings, "Finite time corrections in KPZ growth models", *J. Stat. Phys.* **144**:6 (2011), 1123–1150.

[Ferrari and Frings 2013] P. L. Ferrari and R. Frings, "On the spatial persistence for Airy processes", *J. Stat. Mech. Theory Exp.* 2 (2013), P02001.

[de Gennes and Prost 1995] P. G. de Gennes and J. Prost, *The physics of liquid crystals*, 2nd ed., International Series of Monographs on Physics **83**, Oxford Univ. Press, New York, 1995.

[Johansson 2000] K. Johansson, "Shape fluctuations and random matrices", *Commun. Math. Phys.* **209** (2000), 437–476.

[Johansson 2003] K. Johansson, "Discrete polynuclear growth and determinantal processes", *Comm. Math. Phys.* **242**:1-2 (2003), 277–329.

[Kallabis and Krug 1999] H. Kallabis and J. Krug, "Persistence of Kardar–Parisi–Zhang interfaces", *Europhys. Lett.* **45** (1999), 20.

[Kardar et al. 1986] M. Kardar, G. Parisi, and Y.-C. Zhang, "Dynamic scaling of growing interfaces", *Phys. Rev. Lett.* **56** (Mar 1986), 889–892.

[Kriecherbauer and Krug 2010] T. Kriecherbauer and J. Krug, "A pedestrian's view on interacting particle systems, KPZ universality and random matrices", *J. Phys. A* **43** (2010), 403001.

[Moreno Flores et al. 2013] G. Moreno Flores, J. Quastel, and D. Remenik, "Endpoint distribution of directed polymers in $1 + 1$ dimensions", *Commun. Math. Phys.* **317**:2 (2013), 363–380.

[Prähofer and Spohn 2000] M. Prähofer and H. Spohn, "Universal distributions for growth processes in $1 + 1$ dimensions and random matrices", *Phys. Rev. Lett.* **84** (2000), 4882–4885.

[Prähofer and Spohn 2002] M. Prähofer and H. Spohn, "Scale Invariance of the PNG droplet and the Airy process", *J. Stat. Phys.* **108** (2002), 1071–1106.

[Prolhac and Spohn 2011] S. Prolhac and H. Spohn, "Two-point generating function of the free energy for a directed polymer in a random medium", *J. Stat. Mech. Theory Exp.* 1 (2011), P01031.

[Sasamoto 2005] T. Sasamoto, "Spatial correlations of the 1D KPZ surface on a flat substrate", *J. Phys. A* **38**:33 (2005), L549–L556.

[Sasamoto and Spohn 2010a] T. Sasamoto and H. Spohn, "The $1 + 1$-dimensional Kardar–Parisi–Zhang equation and its universality class", *J. Stat. Mech.* **2010** (2010), P11013.

[Sasamoto and Spohn 2010b] T. Sasamoto and H. Spohn, "Exact height distributions for the KPZ equation with narrow wedge initial condition", *Nucl. Phys. B* **834** (2010), 523–542.

[Sasamoto and Spohn 2010c] T. Sasamoto and H. Spohn, "One-dimensional Kardar-Parisi–Zhang equation: an exact solution and its universality", *Phys. Rev. Lett.* **104** (2010), 230602.

[Schehr 2012] G. Schehr, "Extremes of N vicious walkers for large N: application to the directed polymer and KPZ interfaces", *J. Stat. Phys.* **149**:3 (2012), 385–410.

[Singha 2005] S. B. Singha, "Persistence of surface fluctuations in radially growing surfaces", *J. Stat. Mech.* **2005** (2005), P08006.

[Takeuchi 2012] K. A. Takeuchi, "Statistics of circular interface fluctuations in an off-lattice Eden model", *J. Stat. Mech.* **2012**:05 (2012), P05007.

[Takeuchi and Sano 2010] K. A. Takeuchi and M. Sano, "Universal fluctuations of growing interfaces: evidence in turbulent liquid crystals", *Phys. Rev. Lett.* **104** (2010), 230601.

[Takeuchi and Sano 2012] K. A. Takeuchi and M. Sano, "Evidence for geometry-dependent universal fluctuations of the Kardar–Parisi–Zhang interfaces in liquid-crystal turbulence", *J. Stat. Phys.* **147**:5 (2012), 853–890.

[Takeuchi et al. 2011] K. A. Takeuchi, M. Sano, T. Sasamoto, and H. Spohn, "Growing interfaces uncover universal fluctuations behind scale invariance", *Sci. Rep.* **1** (jul 2011), 34.

[Tracy and Widom 1994] C. A. Tracy and H. Widom, "Level-spacing distributions and the Airy kernel", *Comm. Math. Phys.* **159**:1 (1994), 151–174.

[Tracy and Widom 1996] C. A. Tracy and H. Widom, "On orthogonal and symplectic matrix ensembles", *Comm. Math. Phys.* **177**:3 (1996), 727–754.

[Tracy and Widom 2009] C. A. Tracy and H. Widom, "Asymptotics in ASEP with step initial condition", *Comm. Math. Phys.* **290**:1 (2009), 129–154.

kat@kaztake.org   *Department of Physics, University of Tokyo, 7-3-1 Hongo, Bunkyo-ku, Tokyo 113-0033, Japan*

Random Matrices
MSRI Publications
Volume **65**, 2014

Random matrices:
the four-moment theorem
for Wigner ensembles

TERENCE TAO AND VAN VU

We survey some recent progress on rigorously establishing the universality of various spectral statistics of Wigner random matrix ensembles, focussing in particular on the four-moment theorem and its applications.

1. Introduction

This paper surveys the *four-moment theorem* and its applications in understanding the asymptotic spectral properties of random matrix ensembles of Wigner type. Due to limitations of space, this survey will be far from exhaustive; an extended version will appear elsewhere. (See also [Erdős 2011; Guionnet 2011; Schlein 2011] for some recent surveys in this area.)

To simplify the exposition (at the expense of stating the results in maximum generality), we shall restrict attention to a model class of random matrix ensembles, in which we assume somewhat more decay and identical distribution hypotheses than are strictly necessary for the main results.

Definition 1 (Wigner matrices). Let $n \geq 1$ be an integer (which we view as a parameter going off to infinity). An $n \times n$ *Wigner Hermitian matrix* M_n is defined to be a random Hermitian $n \times n$ matrix $M_n = (\xi_{ij})_{1 \leq i, j \leq n}$, in which the ξ_{ij} for $1 \leq i \leq j \leq n$ are jointly independent with $\xi_{ji} = \overline{\xi_{ij}}$ (in particular, the ξ_{ii} are real-valued). For $1 \leq i < j \leq n$, we require that the ξ_{ij} have mean zero and variance one, while for $1 \leq i = j \leq n$ we require that the ξ_{ij} have mean zero and variance σ^2 for some $\sigma^2 > 0$ independent of i, j, n. To simplify some of the statements of the results here, we will also assume that the $\xi_{ij} \equiv \xi$ are identically distributed for $i < j$, and the $\xi_{ii} \equiv \xi'$ are also identically distributed for $i = j$, and furthermore that the real and imaginary parts of ξ are independent. We refer to the distributions $\mathrm{Re}\,\xi$, $\mathrm{Im}\,\xi$, and ξ' as the *atom distributions* of M_n.

Tao is supported by a grant from the MacArthur Foundation, by NSF grant DMS-0649473, and by the NSF Waterman award. Vu is supported by research grants DMS-0901216 and AFOSAR-FA-9550-09-1-0167.

MSC2000: 15A52.

We say that the Wigner matrix ensemble *obeys Condition* C0 if we have the exponential decay condition

$$P(|\xi_{ij}| \geq t^C) \leq e^{-t}$$

for all $1 \leq i, j \leq n$ and $t \geq C'$, and some constants C, C' (independent of i, j, n).

We refer to the matrix $W_n := (1/\sqrt{n})M_n$ as the *coarse-scale normalised Wigner Hermitian matrix*, and $A_n := \sqrt{n}M_n$ as the *fine-scale normalised Wigner Hermitian matrix*.

Example 2 (invariant ensembles). An important special case of a Wigner Hermitian matrix M_n is the *gaussian unitary ensemble* (GUE), in which $\xi_{ij} \equiv N(0, 1)_{\mathbb{C}}$ are complex gaussians with mean zero and variance one for $i \neq j$, and $\xi_{ii} \equiv N(0, 1)_{\mathbb{R}}$ are real gaussians with mean zero and variance one for $1 \leq i \leq n$ (thus $\sigma^2 = 1$ in this case). Another important special case is the *gaussian orthogonal ensemble* (GOE), in which $\xi_{ij} \equiv N(0, 1)_{\mathbb{R}}$ are real gaussians with mean zero and variance one for $i \neq j$, and $\xi_{ii} \equiv N(0, 2)_{\mathbb{R}}$ are real gaussians with mean zero and variance 2 for $1 \leq i \leq n$ (thus $\sigma^2 = 2$ in this case). These ensembles obey Condition C0. These ensembles are invariant with respect to conjugation by unitary and orthogonal matrices respectively.

Given an $n \times n$ Hermitian matrix A, we will denote its n eigenvalues in increasing order as

$$\lambda_1(A) \leq \cdots \leq \lambda_n(A),$$

and write $\lambda(A) := (\lambda_1(A), \ldots, \lambda_n(A))$. We also let $u_1(A), \ldots, u_n(A) \in \mathbb{C}^n$ be an orthonormal basis of eigenvectors of A with $Au_i(A) = \lambda_i(A)u_i(A)$.

We also introduce the *eigenvalue counting function*

$$N_I(A) := \left|\{1 \leq i \leq n : \lambda_i(A) \in I\}\right| \tag{1}$$

for any interval $I \subset \mathbb{R}$. We will be interested in both the *coarse-scale* eigenvalue counting function $N_I(W_n)$ and the *fine-scale* eigenvalue counting function $N_I(A_n)$.

2. The local semicircular law

The most fundamental result about the spectrum of Wigner matrices is the *Wigner semicircular law*. We state here a powerful local version of this law, due to Erdős, Schlein, and Yau [Erdős et al. 2009a; 2009b; 2010d] (see also [Erdős et al. 2012a; 2012c; 2012d; 2013] for further refinements). Denote by ρ_{sc} the semicircle density function with support on $[-2, 2]$:

$$\rho_{sc}(x) := \frac{1}{2\pi}(4 - x^2)_+^{1/2}. \tag{2}$$

Theorem 3 (local semicircle law). *Let M_n be a Wigner matrix obeying Condition C0, let $\varepsilon > 0$, and let $I \subset \mathbb{R}$ be an interval of length $|I| \geq n^{-1+\varepsilon}$. Then with overwhelming probability,[1] one has[2]*

$$N_I(W_n) = n \int_I \rho_{sc}(x)\, dx + o(n|I|). \tag{3}$$

Proof. See, for example, [Tao and Vu 2010, Theorem 1.10]. For the most precise estimates currently known of this type (and with the weakest decay hypotheses on the entries), see [Erdős et al. 2013]. The proofs are based on the Stieltjes transform method; see [Bai and Silverstein 2006] for an exposition of this method. $\square$

A variant of Theorem 3, established subsequently[3] in [Erdős et al. 2012d], is the extremely useful *eigenvalue rigidity property*

$$\lambda_i(W_n) = \lambda_i^{cl}(W_n) + O_\varepsilon(n^{-1+\varepsilon}), \tag{4}$$

valid with overwhelming probability in the bulk range $\delta n \leq i \leq (1 - \delta)n$ for any fixed $\delta > 0$ (and assuming Condition C0). This result is key in some of the strongest applications of the theory. Here the *classical location* $\lambda_i^{cl}(W_n)$ of the i-th eigenvalue is the element of $[-2, 2]$ defined by the formula

$$\int_{-2}^{\lambda_i^{cl}(W_n)} \rho_{sc}(y)\, dy = \frac{i}{n}.$$

Roughly speaking, results such as Theorem 3 and (4) control the spectrum of W_n at scales $n^{-1+\varepsilon}$ and above. However, they break down at the fine scale n^{-1}; indeed, for intervals I of length $|I| = O(1/n)$, one has $n \int_I \rho_{sc}(x)\, dx = O(1)$, while $N_I(W_n)$ is clearly a natural number, so that one can no longer expect an asymptotic of the form (3). Nevertheless, local semicircle laws are an essential part of the fine-scale theory. One particularly useful consequence of these laws is that of *eigenvector delocalisation*, first established in [Erdős et al. 2010d]:

Corollary 4 (eigenvector delocalisation). *Let M_n be a Wigner matrix obeying Condition C0, and let $\varepsilon > 0$. Then with overwhelming probability, one has $u_i(W_n)^* e_j = O(n^{-1/2+\varepsilon})$ for all $1 \leq i, j \leq n$, where the $e_1, \ldots, e_n$ are the standard basis of $\mathbb{C}^n$.*

---

[1] By this, we mean that the event occurs with probability $1 - O_A(n^{-A})$ for each $A > 0$.

[2] We use the asymptotic notation $o(X)$ to denote any quantity that goes to zero as $n \to \infty$ when divided by X, and $O(X)$ to denote any quantity bounded in magnitude by CX, where C is a constant independent of n.

[3] The result in [Erdős et al. 2012d] actually proves a more precise result that also gives sharp results in the edge of the spectrum, though due to the sparser nature of the $\lambda_i^{cl}(W_n)$ in that case, the error term $O_\varepsilon(n^{-1+\varepsilon})$ must be enlarged.

Note from Pythagoras' theorem that $\sum_{j=1}^{n} |u_i(M_n)^* e_j|^2 = \|u_i(M_n)\|^2 = 1$; thus Corollary 4 asserts, roughly speaking, that the coefficients of each eigenvector are as spread out (or *delocalised*) as possible.

Corollary 4 can be established in a number of ways. One particularly slick approach proceeds via control of the resolvent (or Green's function) $(W_n - zI)^{-1}$, taking advantage of the identity

$$\mathrm{Im}((W_n - zI)^{-1})_{jj} = \sum_{i=1}^{n} \frac{\eta}{(\lambda_i(W_n) - E)^2 + \eta^2} |u_i(W_n)^* e_j|^2$$

for $z = E + \sqrt{-1}\eta$; it turns out that the machinery used to prove Theorem 3 also can be used to control the resolvent. See for instance [Erdős 2011] for details of this approach.

3. GUE and gauss divisible ensembles

We now turn to the question of the fine-scale behaviour of eigenvalues of Wigner matrices, starting with the model case of GUE. Here, it is convenient to work with the fine-scale normalisation $A_n := \sqrt{n} M_n$. For simplicity we will restrict attention to the bulk region of the spectrum, which in the fine-scale normalisation corresponds to eigenvalues $\lambda_i(A_n)$ of A_n that are near nu for some fixed $-2 < u < 2$ independent of n.

A basic object of study are the *k-point correlation functions*

$$R_n^{(k)} = R_n^{(k)}(A_n) : \mathbb{R}^k \to \mathbb{R}^+,$$

defined via duality to be the unique symmetric function (or measure) for which one has

$$\int_{\mathbb{R}^k} F(x_1, \ldots, x_k) R_n^{(k)}(x_1, \ldots, x_k) \, dx_1 \ldots dx_k$$

$$= k! \sum_{1 \le i_1 < \cdots < i_k} \mathbf{E} F(\lambda_{i_1}(A_n), \ldots, \lambda_{i_k}(A_n)) \quad (5)$$

for all symmetric continuous compactly supported functions $F : \mathbb{R}^k \to \mathbb{R}$. Alternatively, one can write

$$R_n^{(k)}(x_1, \ldots, x_k) = \frac{n!}{(n-k)!} \int_{\mathbb{R}^{n-k}} \rho_n(x_1, \ldots, x_n) \, dx_{k+1} \ldots dx_n,$$

where $\rho_n := (1/n!) R_n^{(n)}$ is the symmetrised joint probability distribution of all n eigenvalues of A_n.

From the semicircular law, we expect that at the energy level nu for some $-2 < u < 2$, the eigenvalues of A_n will be spaced with average spacing $1/\rho_{sc}(u)$.

It is thus natural to consider the *normalised k-point correlation function* $\rho_{n,u}^{(k)} = \rho_{n,u}^{(k)}(A_n) : \mathbb{R}^k \to \mathbb{R}^+$, defined by the formula

$$\rho_{n,u}^{(k)}(x_1, \ldots, x_k) := R_n^{(k)}\left(nu + \frac{x_1}{\rho_{sc}(u)}, \ldots, nu + \frac{x_k}{\rho_{sc}(u)}\right). \tag{6}$$

It has been generally believed (and in many cases explicitly conjectured; see e.g., [Mehta 1967, p. 9]) that the asymptotic statistics for the quantities mentioned above are *universal*, in the sense that the limiting laws do not depend on the distribution of the atom variables (assuming of course that they have been normalised as stated in Definition 1). This phenomenon was motivated by examples of similarly universal laws in physics, such as the laws of thermodynamics or of critical percolation; see, for example, [Mehta 1967; Deift 1999; Deift 2007] for further discussion.

It is clear that if one is able to prove the universality of a limiting law, then it suffices to compute this law for one specific model in order to describe the asymptotic behaviour for all other models. A natural choice for the specific model is GUE, as for this model, many limiting laws can be computed directly thanks to the availability of an explicit formula for the joint distribution of the eigenvalues, as well as the useful identities of determinantal processes. For instance, one has *Ginibre's formula*

$$\rho_n(x_1, \ldots, x_n) = \frac{1}{(2\pi n)^{n/2}} e^{-|x|^2/2n} \prod_{1 \le i < j \le n} |x_i - x_j|^2, \tag{7}$$

for the joint eigenvalue distribution, as can be verified from a standard calculation; see [Ginibre 1965]. From this formula, the theory of determinantal processes, and asymptotics for Hermite polynomials, one can then obtain the limiting law

$$\lim_{n \to \infty} \rho_{n,u}^{(k)}(x_1, \ldots, x_k) = \rho_{Sine}^{(k)}(x_1, \ldots, x_k) \tag{8}$$

locally uniformly in $x_1, \ldots, x_k$ where

$$\rho_{Sine}^{(k)}(x_1, \ldots, x_k) := \det(K_{Sine}(x_i, x_j))_{1 \le i, j \le k}$$

and K_{Sine} is the *Dyson sine kernel*

$$K_{Sine}(x, y) := \frac{\sin(\pi(x - y))}{\pi(x - y)}$$

(with the usual convention that $\frac{\sin x}{x}$ equals 1 at the origin); see [Ginibre 1965; Mehta 1967].

Using a general central limit theorem for determinantal processes due to Costin and Leibowitz [1995] and Soshnikov [2002], one can then give a limiting law for

$N_I(A_n)$ in the case of the macroscopic intervals $I = [nu, +\infty)$. More precisely, one has the central limit theorem

$$\frac{N_{[nu,+\infty)}(A_n) - n \int_u^\infty \rho_{sc}(y)\, dy}{\sqrt{\frac{1}{2\pi^2} \log n}} \to N(0,1)_{\mathbb{R}}$$

in the sense of probability distributions, for any $-2 < u < 2$; see [Gustavsson 2005]. By using the counting functions $N_{[nu,+\infty)}$ to solve for the location of individual eigenvalues $\lambda_i(A_n)$, one can then conclude the central limit theorem

$$\frac{\lambda_i(A_n) - \lambda_i^{cl}(A_n)}{\sqrt{\log n/2\pi/\rho_{sc}(u)}} \to N(0,1)_{\mathbb{R}} \qquad (9)$$

whenever $\lambda_i^{cl}(A_n) := n\lambda_i^{cl}(W_n)$ is equal to $n(u+o(1))$ for some fixed $-2 < u < 2$; see [Gustavsson 2005].

Much of the above analysis extends to many other classes of invariant ensembles (such as GOE), for which the joint eigenvalue distribution has a form similar to (7); see [Deift 1999] for further discussion. Another important extension of the above results is to the *gauss divisible* ensembles, which are Wigner matrices M_t of the form

$$M_n^t = e^{-t/2} M_n^0 + (1 - e^{-t})^{1/2} G_n,$$

where G_n is a GUE matrix independent of M_n^0. In particular, the random matrix M_n^t is distributed as M_n^0 for $t = 0$ and then continuously deforms towards the GUE distribution as $t \to +\infty$. By using explicit formulae for the eigenvalue distribution of a gauss divisible matrix, Johansson [2001] was able[4] to extend the asymptotic (8) for the k-point correlation function from GUE to the more general class of gauss divisible matrices with fixed parameter $t > 0$ (independent of n).

It is of interest to extend this analysis to as small a value of t as possible, since if one could set $t = 0$ then one would obtain universality for all Wigner ensembles. By optimising Johansson's method (and taking advantage of the local semicircle law), Erdős, Péché, Ramirez, Schlein, and Yau [Erdős et al. 2010a] was able to extend the universality of (8) (interpreted in a suitably weak convergence topology, such as vague convergence) to gauss divisible ensembles for t as small as $n^{-1+\varepsilon}$ for any fixed $\varepsilon > 0$.

An important alternate approach to these results was developed by Erdős et al. [2010c; 2011; 2012b], based on a stability analysis of the Dyson Brownian motion [Dyson 1970] governing the evolution of the eigenvalues of a matrix

---

[4]Some additional technical hypotheses were assumed in [Johansson 2001], namely that the diagonal variance σ^2 was equal to 1, that the real and imaginary parts of each entry of M_n' were independent, and that the matrix entries had bounded C_0-th moment for some $C_0 > 6$.

Ornstein–Uhlenbeck process. We refer to [Erdős 2011] for a discussion of this method. Among other things, this argument reproves a weaker version of the result in [Erdős et al. 2010a] mentioned earlier, in which one obtained universality for the asymptotic (8) after an additional averaging in the energy parameter u. However, the method was simpler and more flexible than that in [Erdős et al. 2010a], as it did not rely on explicit identities, and has since been extended to many other types of ensembles, including the real symmetric analogue of gauss divisible ensembles in which the role of GUE is replaced instead by GOE.

4. The four-moment theorem

The results discussed above for invariant or gauss divisible ensembles can be extended to more general Wigner ensembles via a powerful swapping method known as the *Lindeberg exchange strategy*, introduced in Lindeberg's classic proof [1922] of the central limit theorem, and first applied to Wigner ensembles in [Chatterjee 2006]. This method can be used to control expressions such as $\mathbf{E}F(M_n) - F(M'_n)$, where M_n, M'_n are two (independent) Wigner matrices. If one can obtain bounds such as

$$\mathbf{E}F(M_n) - \mathbf{E}F(\tilde{M}_n) = o(1/n)$$

when $\tilde{M}_n$ is formed from M_n by replacing[5] one of the diagonal entries ξ_{ii} of M_n by the corresponding entry ξ'_{ii} of M'_n, and bounds such as

$$\mathbf{E}F(M_n) - \mathbf{E}F(\tilde{M}_n) = o(1/n^2)$$

when $\tilde{M}_n$ is formed from M_n by replacing one of the off-diagonal entries ξ_{ij} of M_n with the corresponding entry ξ'_{ij} of M'_n (and also replacing $\xi_{ji} = \overline{\xi_{ij}}$ with $\xi'_{ji} = \overline{\xi'_{ij}}$, to preserve the Hermitian property), then on summing an appropriate telescoping series, one would be able to conclude asymptotic agreement of the statistics $\mathbf{E}F(M_n)$ and $\mathbf{E}F(M'_n)$:

$$\mathbf{E}F(M_n) - \mathbf{E}F(M'_n) = o(1). \tag{10}$$

The *four-moment theorem* asserts, roughly speaking, that we can obtain conclusions of the form (10) for suitable statistics F as long as M_n, M'_n match to fourth order. More precisely, we have

---

[5]Technically, the matrices $\tilde{M}_n$ formed by such a swapping procedure are not Wigner matrices as defined in Definition 1, because the diagonal or upper-triangular entries are no longer identically distributed. However, all of the relevant estimates for Wigner matrices can be extended to the nonidentically distributed case at the cost of making the notation slightly more complicated. As this is a relatively minor issue, we will not discuss it further here.

Definition 5 (matching moments). Let $k \geq 1$. Two complex random variables ξ, ξ' are said to *match to order k* if one has $\mathbf{E}\,\mathrm{Re}(\xi)^a\,\mathrm{Im}(\xi)^b = \mathbf{E}\,\mathrm{Re}(\xi')^a\,\mathrm{Im}(\xi')^b$ whenever $a, b \geq 0$ are integers such that $a + b \leq k$.

In the model case when the real and imaginary parts of ξ or of ξ' are independent, the matching moment condition simplifies to the assertion that $\mathbf{E}\,\mathrm{Re}(\xi)^a = \mathbf{E}\,\mathrm{Re}(\xi')^a$ and $\mathbf{E}\,\mathrm{Im}(\xi)^b = \mathbf{E}\,\mathrm{Im}(\xi')^b$ for all $0 \leq a, b \leq k$.

Theorem 6 (four-moment theorem). *Let $c_0 > 0$ be a sufficiently small constant. Let $M_n = (\xi_{ij})_{1 \leq i,j \leq n}$ and $M'_n = (\xi'_{ij})_{1 \leq i,j \leq n}$ be two Wigner matrices obeying Condition C0. Assume furthermore that for any $1 \leq i < j \leq n$, ξ_{ij} and ξ'_{ij} match to order 4 and for any $1 \leq i \leq n$, ξ_{ii} and ξ'_{ii} match to order 2. Set $A_n := \sqrt{n} M_n$ and $A'_n := \sqrt{n} M'_n$, let $1 \leq k \leq n^{c_0}$ be an integer, and let $G : \mathbb{R}^k \to \mathbb{R}$ be a smooth function obeying the derivative bounds*

$$|\nabla^j G(x)| \leq n^{c_0} \tag{11}$$

for all $0 \leq j \leq 5$ and $x \in \mathbb{R}^k$. Then for any $1 \leq i_1 < i_2 \cdots < i_k \leq n$, and for n sufficiently large we have

$$\left| \mathbf{E}\big(G(\lambda_{i_1}(A_n), \ldots, \lambda_{i_k}(A_n))\big) - \mathbf{E}\big(G(\lambda_{i_1}(A'_n), \ldots, \lambda_{i_k}(A'_n))\big) \right| \leq n^{-c_0}. \tag{12}$$

A preliminary version of Theorem 6 was first established by the authors in [Tao and Vu 2011b], in the case[6] of bulk eigenvalues (thus $\delta n \leq i_1, \ldots, i_k \leq (1 - \delta)n$ for some absolute constant $\delta > 0$). In [Tao and Vu 2010], the restriction to the bulk was removed; and in [Tao and Vu 2012a], Condition C0 was relaxed to a finite moment condition. We will discuss the proof of this theorem in Section 5. There is strong evidence that the condition of four matching moments is necessary to obtain the conclusion (12); see [Tao and Vu 2011a].

A key technical result used in the proof of the four-moment theorem, which is also of independent interest, is the *gap theorem*:

Theorem 7 (gap theorem). *Let M_n be a Wigner matrix obeying Condition C0. Then for every $c_0 > 0$ there exists a $c_1 > 0$ (depending only on c_0) such that*

$$\mathbf{P}(|\lambda_{i+1}(A_n) - \lambda_i(A_n)| \leq n^{-c_0}) \ll n^{-c_1}$$

for all $1 \leq i < n$.

For reasons of space we will not discuss the proof of this theorem here, but refer the reader to [Tao and Vu 2011b; 2012a]. Among other things, the gap theorem tells us that eigenvalues of a Wigner matrix are usually simple. Closely related *level repulsion* estimates were established (under an additional

---

[6]In the paper, k was held fixed, but an inspection of the argument reveals that it extends without difficulty to the case when k is as large as n^{c_0}, for c_0 small enough.

smoothness hypothesis on the atom distributions) in [Erdős et al. 2010d].

Another variant of the four-moment theorem was subsequently introduced in [Erdős et al. 2012c], in which the eigenvalues $\lambda_{i_j}(A_n)$ appearing in Theorem 6 were replaced by the components of the resolvent (or Green's function) $(W_n - z)^{-1}$, but with slightly different technical hypotheses on the matrices M_n, M'_n; see [Erdős et al. 2012c] for full details. As the resolvent-based quantities are averaged statistics that sum over many eigenvalues, they are far less sensitive to the eigenvalue repulsion phenomenon than the individual eigenvalues, and as such the version of the four-moment theorem for Green's function has a somewhat simpler proof (based on resolvent expansions rather than the Hadamard variation formulae and Taylor expansion). Conversely, though, to use the four-moment theorem for Green's function to control individual eigenvalues, while possible, requires a significant amount of additional argument; see [Knowles and Yin 2013]. Finally, we remark that the four-moment theorem has also been extended to cover eigenvectors as well as eigenvalues; see [Tao and Vu 2012b; Knowles and Yin 2013] for details.

5. Sketch of proof of the four-moment theorem

In this section we discuss the proof of Theorem 6, following the arguments that originated in [Tao and Vu 2011b] and refined in [Tao and Vu 2012a].

In addition to Theorem 7, a key ingredient is the following truncated version of the four-moment theorem, in which one removes the event that two consecutive eigenvalues are too close to each other. For technical reasons, we need to introduce quantities

$$Q_i(A_n) := \sum_{j \neq i} \frac{1}{|\lambda_j(A_n) - \lambda_i(A_n)|^2}$$

for $i = 1, \ldots, n$, which is a regularised measure of extent to which $\lambda_i(A_n)$ is close to any other eigenvalue of A_n.

Theorem 8 (truncated four-moment theorem). *Let $c_0 > 0$ be a sufficiently small constant. Let $M_n = (\xi_{ij})_{1 \leq i, j \leq n}$ and $M'_n = (\xi'_{ij})_{1 \leq i, j \leq n}$ be two Wigner matrices obeying Condition C0. Assume furthermore that for any $1 \leq i < j \leq n$, ξ_{ij} and ξ'_{ij} match to order 4 and for any $1 \leq i \leq n$, ξ_{ii} and ξ'_{ii} match to order 2. Set $A_n := \sqrt{n} M_n$ and $A'_n := \sqrt{n} M'_n$, let $1 \leq k \leq n^{c_0}$ be an integer, and let*

$$G = G(\lambda_{i_1}, \ldots, \lambda_{i_k}, Q_{i_1}, \ldots, Q_{i_k})$$

be a smooth function from $\mathbb{R}^k \times \mathbb{R}^k_+$ to $\mathbb{R}$ that is supported in the region

$$Q_{i_1}, \ldots, Q_{i_k} \leq n^{c_0} \tag{13}$$

and obeys the derivative bounds

$$|\nabla^j G(\lambda_{i_1}, \ldots, \lambda_{i_k}, Q_{i_1}, \ldots, Q_{i_k})| \leq n^{c_0} \qquad (14)$$

for all $0 \leq j \leq 5$. *Then*

$$\mathbf{E}\, G(\lambda_{i_1}(A_n), \ldots, \lambda_{i_k}(A_n), Q_{i_1}(A_n), \ldots, Q_{i_k}(A_n))$$
$$= \mathbf{E}\, G(\lambda_{i_1}(A'_n), \ldots, \lambda_{i_k}(A'_n), Q_{i_1}(A'_n), \ldots, Q_{i_k}(A'_n)) + O(n^{-1/2+O(c_0)}). \quad (15)$$

We will discuss the proof of this theorem shortly. Using Theorem 7, one can then deduce Theorem 6 from Theorem 8 by smoothly truncating in the Q variables: see [Tao and Vu 2011b, Section 3.3].

It remains to establish Theorem 8. To simplify the exposition slightly, let us assume that the matrices M_n, M'_n are real symmetric rather than Hermitian.

As indicated in Section 4, the basic idea is to use the Lindeberg exchange strategy. To illustrate the idea, let $\tilde{M}_n$ be the matrix formed from M_n by replacing a single entry ξ_{pq} of M_n with the corresponding entry ξ'_{pq} of M'_n for some $p < q$, with a similar swap also being performed at the ξ_{qp} entry to keep $\tilde{M}_n$ Hermitian. Strictly speaking, $\tilde{M}_n$ is not a Wigner matrix as defined in Definition 1, as the entries are no longer identically distributed, but this will not significantly affect the arguments. (One also needs to perform swaps on the diagonal, but this can be handled in essentially the same manner.)

Set $\tilde{A}_n := \sqrt{n}\tilde{M}_n$ as usual. We will sketch the proof of the claim that

$$\mathbf{E}\, G(\lambda_{i_1}(A_n), \ldots, \lambda_{i_k}(A_n), Q_{i_1}(A_n), \ldots, Q_{i_k}(A_n))$$
$$= \mathbf{E}\, G(\lambda_{i_1}(\tilde{A}_n), \ldots, \lambda_{i_k}(\tilde{A}_n), Q_{i_1}(\tilde{A}_n), \ldots, Q_{i_k}(\tilde{A}_n)) + O(n^{-5/2+O(c_0)});$$

by telescoping together $O(n^2)$ estimates of this sort one can establish (15). (For swaps on the diagonal, one only needs an error term of $O(n^{-3/2+O(c_0)})$, since there are only $O(n)$ swaps to be made here rather than $O(n^2)$. This is ultimately why there are two fewer moment conditions on the diagonal than off it.)

We can write $A_n = A(\xi_{pq})$, $\tilde{A}_n = A(\xi'_{pq})$, where

$$A(t) = A(0) + tA'(0)$$

is a (random) Hermitian matrix depending linearly[7] on a real parameter t, with $A(0)$ being a Wigner matrix with one entry (and its adjoint) zeroed out, and

---

[7]If we were working with Hermitian matrices rather than real symmetric matrices, then one could either swap the real and imaginary parts of the ξ_{ij} separately (exploiting the hypotheses that these parts were independent), or else repeat the above analysis with t now being a complex parameter (or equivalently, two real parameters) rather than a real one. In the latter case, one needs to replace all instances of single variable calculus below (such as Taylor expansion) with double variable calculus, but aside from notational difficulties, it is a routine matter to perform this modification.

$A'(0)$ is the explicit elementary Hermitian matrix

$$A'(0) = e_p e_q^* + e_p^* e_q. \tag{16}$$

We note the crucial fact that the random matrix $A(0)$ is independent of both ξ_{pq} and ξ'_{pq}. Note from Condition C0 that we expect ξ_{pq}, ξ'_{pq} to have size $O(n^{O(c_0)})$ most of the time, so we should (heuristically at least) be able to restrict attention to the regime $t = O(n^{O(c_0)})$. If we then set

$$F(t) := \mathbf{E}\, G\big(\lambda_{i_1}(A(t)), \ldots, \lambda_{i_k}(A(t)), Q_{i_1}(A(t)), \ldots, Q_{i_k}(A(t))\big) \tag{17}$$

then our task is to show that

$$\mathbf{E} F(\xi_{pq}) = \mathbf{E} F(\xi'_{pq}) + O(n^{-5/2+O(c_0)}). \tag{18}$$

Suppose that we have Taylor expansions of the form

$$\lambda_{i_l}(A(t)) = \lambda_{i_l}(A(0)) + \sum_{j=1}^{4} c_{l,j} t^j + O(n^{-5/2+O(c_0)}) \tag{19}$$

for all $t = O(n^{O(c_0)})$ and $l = 1, \ldots, k$, where the Taylor coefficients $c_{l,j}$ have size $c_{l,j} = O(n^{-j/2+O(c_0)})$, and similarly for the quantities $Q_{i_l}(A(t))$. Then by using the hypothesis (14) and further Taylor expansion, we can obtain a Taylor expansion

$$F(t) = F(0) + \sum_{j=1}^{4} f_j t^j + O(n^{-5/2+O(c_0)})$$

for the function $F(t)$ defined in (17), where the Taylor coefficients f_j have size $f_j = O(n^{-j/2+O(c_0)})$. Setting t equal to ξ_{pq} and taking expectations, and noting that the Taylor coefficients f_j depend only on F and $A(0)$ and is thus independent of ξ_{ij}, we conclude that

$$\mathbf{E} F(\xi_{pq}) = \mathbf{E} F(0) + \sum_{j=1}^{4} (\mathbf{E} f_j)(\mathbf{E}\, \xi_{pq}^j) + O(n^{-5/2+O(c_0)}),$$

and similarly for $\mathbf{E} F(\xi'_{pq})$. If ξ_{pq} and ξ'_{pq} have matching moments to fourth order, this gives (18).

It remains to establish (19) (as well as the analogue for $Q_{i_l}(A(t))$, which turns out to be analogous). We abbreviate i_l simply as i. By Taylor's theorem with remainder, it would suffice to show that

$$\frac{d^j}{dt^j} \lambda_i(A(t)) = O(n^{-j/2+O(c_0)}) \tag{20}$$

for $j = 1, \ldots, 5$. As it turns out, this is not quite true as stated, but it becomes true (with overwhelming probability[8]) if one can assume that $Q_i(A(t))$ is bounded by $n^{O(c_0)}$. In principle, one can reduce to this case due to the restriction (13) on the support of G, although there is a technical issue because one will need to establish the bounds (20) for values of t other than ξ_{pq} or $\tilde{\xi}_{pq}$. This difficulty can be overcome by a continuity argument; see [Tao and Vu 2011b]. For the purposes of this informal discussion, we shall ignore this issue and simply assume that we may restrict to the case where

$$Q_i(A(t)) \ll n^{O(c_0)}. \tag{21}$$

In particular, the eigenvalue $\lambda_i(A(t))$ is simple, which ensures that all quantities depend smoothly on t (locally, at least).

To prove (20), one can use the classical *Hadamard variation formulae* for the derivatives of $\lambda_i(A(t))$, which can be derived for instance by repeatedly differentiating the eigenvector equation $A(t)u_i(A(t)) = \lambda_i(A(t))u_i(A(t))$. The formula for the first derivative is

$$\frac{d}{dt}\lambda_i(A(t)) = u_i(A(t))^* A'(0)u_i(A(t)).$$

But recall from eigenvalue delocalisation (Corollary 4) that with overwhelming probability, all coefficients of $u_i(A(t))$ have size $O(n^{-1/2+o(1)})$; given the nature of the matrix (16), we can then obtain (20) in the $j = 1$ case.

Now consider the $j = 2$ case. The second derivative formula reads

$$\frac{d^2}{dt^2}\lambda_i(A(t)) = -2\sum_{j \neq i} \frac{|u_i(A(t))^* A'(0)u_j(A(t))|^2}{\lambda_j(A(t)) - \lambda_i(A(t))}.$$

Using eigenvalue delocalisation as before, we see with overwhelming probability that the numerator is $O(n^{-1+o(1)})$. To deal with the denominator, one has to exploit the hypothesis (21) and the local semicircle law (Theorem 3). Using these tools, one can conclude (20) in the $j = 2$ case with overwhelming probability.

It turns out that one can continue this process for higher values of j, although the formulae for the derivatives for $\lambda_i(A(t))$ (and related quantities, such as $P_i(A(t))$ and $Q_i(A(t))$) become increasingly complicated, being given by a certain recursive formula in j. See [Tao and Vu 2011b] for details.

---

[8]Technically, each value of t has a different exceptional event of very small probability for which the estimates fail. Since there are uncountably many values of t, this could potentially cause a problem when applying the union bound. In practice, though, it turns out that one can restrict t to a discrete set, such as the multiples of n^{-100}, in which case the union bound can be applied without difficulty. See [Tao and Vu 2011b] for details.

6. Distribution of individual eigenvalues

One of the simplest applications of the above machinery is to extend the central limit theorem (9) of Gustavsson [2005] for eigenvalues $\lambda_i(A_n)$ in the bulk from GUE to more general ensembles:

Theorem 9. *The gaussian fluctuation law* (9) *continues to hold for Wigner matrices obeying Condition* C0, *and whose atom distributions match that of GUE to second order on the diagonal and fourth order off the diagonal; thus, one has*

$$\frac{\lambda_i(A_n) - \lambda_i^{\mathrm{cl}}(A_n)}{\sqrt{\log n / 2\pi} / \rho_{\mathrm{sc}}(u)} \to N(0, 1)_{\mathbb{R}}$$

whenever $\lambda_i^{\mathrm{cl}}(A_n) = n(u + o(1))$ *for some fixed* $-2 < u < 2$.

Proof. Let M_n' be drawn from GUE, thus by (9) one already has

$$\frac{\lambda_i(A_n') - \lambda_i^{\mathrm{cl}}(A_n)}{\sqrt{\log n / 2\pi} / \rho_{\mathrm{sc}}(u)} \to N(0, 1)_{\mathbb{R}}$$

(note that $\lambda_i^{\mathrm{cl}}(A_n) = \lambda_i^{\mathrm{cl}}(A_n')$). To conclude the analogous claim for A_n, it suffices to show that

$$\mathbf{P}(\lambda_i(A_n') \in I_-) - n^{-c_0} \le \mathbf{P}(\lambda_i(A_n) \in I) \le \mathbf{P}(\lambda_i(A_n') \in I_+) + n^{-c_0} \qquad (22)$$

for all intervals $I = [a, b]$, and n sufficiently large, where

$$I_+ := [a - n^{-c_0/10}, b + n^{-c_0/10}] \quad \text{and} \quad I_- := [a + n^{-c_0/10}, b - n^{-c_0/10}].$$

We will just prove the second inequality in (22), as the first is very similar. We define a smooth bump function $G : \mathbb{R} \to \mathbb{R}^+$ equal to one on I_- and vanishing outside of I_+. Then we have

$$\mathbf{P}(\lambda_i(A_n) \in I) \le \mathbf{E}\, G(\lambda_i(A_n)) \quad \text{and} \quad \mathbf{E}\, G(\lambda_i(A_n')) \le \mathbf{P}(\lambda_i(A_n') \in I).$$

On the other hand, one can choose G to obey (11). Thus by Theorem 6 we have

$$\left| \mathbf{E}\, G(\lambda_i(A_n)) - \mathbf{E}\, G(\lambda_i(A_n')) \right| \le n^{-c_0},$$

and the second inequality in (22) follows from the triangle inequality. The first inequality is similarly proven using a smooth function that equals 1 on I_- and vanishes outside of I. $\qquad \square$

Remark 10. In [Gustavsson 2005] the asymptotic joint distribution of k distinct eigenvalues $\lambda_{i_1}(M_n), \ldots, \lambda_{i_k}(M_n)$ in the bulk of a GUE matrix M_n was computed (it is a gaussian k-tuple with an explicit covariance matrix). By using the above argument, one can extend that asymptotic for any fixed k to other Wigner matrices, so long as they match GUE to fourth order off the diagonal and to second order on the diagonal.

If one could extend the results in [Gustavsson 2005] to broader ensembles of matrices, such as gauss divisible matrices, the same argument would allow some of the moment matching hypotheses to be dropped, using tools such as Lemma 13.

Remark 11. In [Doering and Eichelsbacher 2011], a moderate deviations property of the distribution of the eigenvalues $\lambda_i(A_n)$ was established first for GUE, and then extended to the same class of matrices considered in Theorem 9 by using the four-moment theorem. An analogue of Theorem 9 for real symmetric matrices (using GOE instead of GUE) was established in [O'Rourke 2010].

There are similar results at the edge of the spectrum, though with several additional technicalities; see [Soshnikov 1999; Ruzmaikina 2006; Khorunzhiy 2012; Tao and Vu 2010; Johansson 2012; Erdős et al. 2012d].

7. The Wigner–Dyson–Mehta conjecture

We now consider the extent to which the asymptotic (8), which asserts that the normalised k-point correlation functions $\rho_{n,u}^{(k)}$ converge to the universal limit $\rho_{\text{Sine}}^{(k)}$, can be extended to more general Wigner ensembles. A long-standing conjecture of Wigner, Dyson, and Mehta (see, e.g., [Mehta 1967]) asserts (informally speaking) that (8) is valid for all fixed k, all Wigner matrices and all fixed energy levels $-2 < u < 2$ in the bulk. However, to make this conjecture precise one has to specify the nature of convergence in (8). For GUE, the convergence is quite strong (in the local uniform sense), but one cannot expect such strong convergence in general, particularly in the case of discrete ensembles in which $\rho_{n,u}^{(k)}$ is a discrete probability distribution (i.e., a linear combination of Dirac masses) and thus is unable to converge uniformly or pointwise to the continuous limiting distribution $\rho_{\text{Sine}}^{(k)}$. We will thus instead settle for the weaker notion of *vague convergence*. More precisely, we say that (8) holds in the vague sense if one has

$$
\int_{\mathbb{R}^k} F(x_1, \ldots, x_k) \rho_{n,u}^{(k)}(x_1, \ldots, x_k) \, dx_1 \ldots dx_k
$$
$$
= \int_{\mathbb{R}^k} F(x_1, \ldots, x_k) \rho_{\text{Sine}}^{(k)}(x_1, \ldots, x_k) \, dx_1 \ldots dx_k \quad (23)
$$

for all continuous, compactly supported functions $F : \mathbb{R}^k \to \mathbb{R}$. By the Stone–Weierstrass theorem we may take F to be a test function (i.e., smooth and compactly supported) without loss of generality.

The Wigner–Dyson–Mehta conjecture is largely resolved in the vague convergence category, in the case of Hermitian Wigner matrices whose variances match those of GUE:

Theorem 12 (Wigner–Dyson–Mehta conjecture for Hermitian matrices in the vague sense). *Let M_n be a Wigner matrix obeying Condition C0 and which matches GUE to second order (i.e., the real and imaginary parts of the off-diagonal entries have variance $1/2$, an the diagonal entries have variance 1), and let $-2 < u < 2$ and $k \geq 1$ be fixed. Then (8) holds in the vague sense.*

This theorem, proven in [Tao and Vu 2011c] (in fact the **C0** condition can be relaxed to a finite moment condition [Tao and Vu 2011c]), builds upon a long sequence of partial results towards the Wigner–Dyson–Mehta conjecture [Johansson 2001; Erdős et al. 2010a; 2010b; 2010c; 2012c; 2012d; Tao and Vu 2011b], which we will summarise (in a slightly nonchronological order) below. An alternate proof of the result (and a strengthening to the case of matrices with variable variance) was then established in [Erdős and Yau 2012]. The analogous problem for matrices matching GOE (i.e., real symmetric matrices whose diagonal entries have variance 2) remains open, unless one performs an additional averaging in the energy parameter.

As recalled in Section 3, the asymptotic (8) for GUE (in the sense of locally uniform convergence, which is far stronger than vague convergence) follows as a consequence of the Gaudin–Mehta formula and the Plancherel–Rotach asymptotics for Hermite polynomials.[9]

The next major breakthrough was by Johansson [2001], who, as discussed previously, established (8) for gauss divisible ensembles at some fixed time parameter $t > 0$ independent of n, obtained (8) in the vague sense (in fact, the slightly stronger convergence of *weak convergence* was established in that paper, in which the function F in (23) was allowed to merely be L^∞ and compactly supported, rather than continuous and compactly supported). The main tool used in [Johansson 2001] was an explicit determinantal formula for the correlation functions in the gauss divisible case, essentially due to Brézin and Hikami [1997].

In Johansson's result, the time parameter $t > 0$ had to be independent of n. It was realised by Erdős, Ramirez, Schlein, and Yau that one could obtain many further cases of the Wigner–Dyson–Mehta conjecture if one could extend Johansson's result to much shorter times t that decayed at a polynomial rate in n. This was first achieved (again in the context of weak convergence) for $t > n^{-3/4+\varepsilon}$ for an arbitrary fixed $\varepsilon > 0$ in [Erdős et al. 2010c], and then to the essentially optimal case $t > n^{-1+\varepsilon}$ (for weak convergence, and (implicitly) in the local L^1 sense as well) in [Erdős et al. 2010a]. By combining this with the method of reverse heat flow discussed in Section 4, the asymptotic (8) (again in

---

[9]Analogous results are known for much wider classes of invariant random matrix ensembles, see, for example, [Deift et al. 1999; Pastur and Shcherbina 1997; Bleher and Its 1999]. However, we will not discuss these results further here, as they do not directly impact on the case of Wigner ensembles.

the sense of weak convergence) was established for all Wigner matrices whose distribution obeyed certain smoothness conditions (e.g., when $k = 2$ one needs a C^6 type condition), and also decayed exponentially. The methods used in [Erdős et al. 2010a] were an extension of those in [Johansson 2001], combined with an approximation argument (the "method of time reversal") that approximated a continuous distribution by a gauss divisible one (with a small value of t); the arguments in [Erdős et al. 2010c] are based instead on an analysis of the Dyson Brownian motion.

By combining the above observation with the moment matching lemma presented below, one immediately concludes Theorem 12 assuming that the off-diagonal atom distributions are supported on at least three points.

Lemma 13 (moment matching lemma). *Let ξ be a real random variable with mean zero, variance one, finite fourth moment, and which is supported on at least three points. Then there exists a gauss divisible, exponentially decaying real random variable ξ' that matches ξ to fourth order.*

For a proof of this lemma, see [Tao and Vu 2011b, Lemma 28]. The requirement of support on at least three points is necessary; indeed, if ξ is supported in just two points a, b, then $\mathbf{E}(\xi - a)^2(\xi - b)^2 = 0$, and so any other distribution that matches ξ to fourth order must also be supported on a, b and thus cannot be gauss divisible.

To remove the requirement that the atom distributions be supported on at least three points, one can observe from the proof of the four-moment theorem that one only needs the moments of M_n and M_n' to *approximately* match to fourth order in order to be able to transfer results on the distribution of spectra of M_n to that of M_n'. In particular, if $t = n^{-1+\varepsilon}$ for some small $\varepsilon > 0$, then the gauss divisible matrix M_n^t associated to M_n at time t is already close enough to matching the first four moments of M_n to apply (a version of) the four-moment theorem. The results of [Erdős et al. 2010a] give the asymptotic (8) for M_n^t, and the eigenvalue rigidity property (4) then allows one to transfer this property to M_n, giving Theorem 12.

Remark 14. The above presentation (drawn from the most recent paper [Tao and Vu 2011c]) is somewhat ahistorical, as the arguments used above emerged from a sequence of papers, which obtained partial results using the best technology available at the time. In [Tao and Vu 2011b], where the first version of the four-moment theorem was introduced, the asymptotic (8) was established under the additional assumptions of Condition C0, and matching the GUE to fourth order; the former hypothesis was due to the weaker form of the four-moment theorem known at the time, and the latter was due to the fact that the eigenvalue rigidity result (4) was not yet established (and was instead deduced from the results of [Gustavsson 2005] combined with the four-moment theorem, thus necessitating the matching moment hypothesis). For related reasons, the paper in [Erdős et al. 2010b] (which first introduced the use of an approximate four-moment theorem) was only able to establish (8) after an additional averaging in the energy parameter u (and with Condition C0). The subsequent progress in [Erdős et al. 2011] via heat flow methods gave an alternate approach to establishing (8), but also required an averaging in the energy and a hypothesis that the atom distributions be supported on at least three points, although the latter condition was then removed in [Erdős et al. 2012d]. In a very recent paper [Erdős et al. 2012a], Condition C0 has been relaxed to finite $(4 + \varepsilon)$-th moment of the entries for any fixed $\varepsilon > 0$, though still at the cost of averaging in the energy parameter. Some generalisations in other directions (e.g., to covariance matrices, or to generalised Wigner ensembles with nonconstant variances) were also established in [Ben Arous and Péché 2005; Tao and Vu 2012a; Erdős et al. 2012a; 2012b; 2012c; 2012d; 2013; Wang 2012].

Remark 15. While Theorem 12 is the "right" result for discrete Wigner ensembles, one expects stronger notions of convergence when one has more smoothness hypotheses on the atom distribution; in particular, one should have local uniform convergence of the correlation functions when the distribution is smooth enough. Some very recent progress in this direction in the $k = 1$ case was obtained by Maltsev and Schlein [2011a; 2011b].

Acknowledgements

We are indebted to the anonymous referee for helpful suggestions and references.

References

[Bai and Silverstein 2006] Z. D. Bai and J. Silverstein, *Spectral analysis of large dimensional random matrices*, Mathematics Monograph Series **2**, Science Press, Beijing, 2006.

[Ben Arous and Péché 2005] G. Ben Arous and S. Péché, "Universality of local eigenvalue statistics for some sample covariance matrices", *Comm. Pure Appl. Math.* **58**:10 (2005), 1316–1357.

[Bleher and Its 1999] P. Bleher and A. Its, "Semiclassical asymptotics of orthogonal polynomials, Riemann–Hilbert problem, and universality in the matrix model", *Ann. of Math.* (2) **150**:1 (1999), 185–266.

[Brézin and Hikami 1997] E. Brézin and S. Hikami, "An extension of level-spacing universality", 1997. arXiv cond-mat/9702213

[Chatterjee 2006] S. Chatterjee, "A generalization of the Lindeberg principle", *Ann. Probab.* **34**:6 (2006), 2061–2076.

[Costin and Lebowitz 1995] O. Costin and J. L. Lebowitz, "Gaussian fluctuation in random matrices", *Phys. Rev. Lett.* **75**:1 (1995), 69–72.

[Deift 1999] P. A. Deift, *Orthogonal polynomials and random matrices: a Riemann–Hilbert approach*, Courant Lecture Notes in Mathematics 3, Courant Institute of Mathematical Sciences, New York, 1999.

[Deift 2007] P. Deift, "Universality for mathematical and physical systems", pp. 125–152 in *International Congress of Mathematicians, I*, edited by M. Sanz-Solé et al., Eur. Math. Soc., Zürich, 2007.

[Deift et al. 1999] P. Deift, T. Kriecherbauer, K. T.-R. McLaughlin, S. Venakides, and X. Zhou, "Uniform asymptotics for polynomials orthogonal with respect to varying exponential weights and applications to universality questions in random matrix theory", *Comm. Pure Appl. Math.* 52:11 (1999), 1335–1425.

[Doering and Eichelsbacher 2011] H. Doering and P. Eichelsbacher, "Moderate deviations for the eigenvalue counting function of Wigner matrices", 2011. arXiv 1104.0221

[Dyson 1970] F. J. Dyson, "Correlations between eigenvalues of a random matrix", *Comm. Math. Phys.* 19 (1970), 235–250.

[Erdős 2011] L. Erdős, "Universality of Wigner random matrices: a survey of recent results", *Uspekhi Mat. Nauk* 66:3(399) (2011), 67–198. In Russian; translated in *Russian Mathematical Surveys*, 66:3 (2011), 507–626. arXiv 1004.0861

[Erdős and Yau 2012] L. Erdős and H.-T. Yau, "A comment on the Wigner–Dyson–Mehta bulk universality conjecture for Wigner matrices", *Electron. J. Probab.* 17 (2012), no. 28, 5.

[Erdős et al. 2009a] L. Erdős, B. Schlein, and H.-T. Yau, "Semicircle law on short scales and delocalization of eigenvectors for Wigner random matrices", *Ann. Probab.* 37:3 (2009), 815–852.

[Erdős et al. 2009b] L. Erdős, B. Schlein, and H.-T. Yau, "Local semicircle law and complete delocalization for Wigner random matrices", *Comm. Math. Phys.* 287:2 (2009), 641–655.

[Erdős et al. 2010a] L. Erdős, S. Péché, J. A. Ramírez, B. Schlein, and H.-T. Yau, "Bulk universality for Wigner matrices", *Comm. Pure Appl. Math.* 63:7 (2010), 895–925.

[Erdős et al. 2010b] L. Erdős, J. Ramírez, B. Schlein, T. Tao, V. Vu, and H.-T. Yau, "Bulk universality for Wigner Hermitian matrices with subexponential decay", *Math. Res. Lett.* 17:4 (2010), 667–674.

[Erdős et al. 2010c] L. Erdős, J. A. Ramírez, B. Schlein, and H.-T. Yau, "Universality of sine-kernel for Wigner matrices with a small Gaussian perturbation", *Electron. J. Probab.* 15 (2010), no. 18, 526–603.

[Erdős et al. 2010d] L. Erdős, B. Schlein, and H.-T. Yau, "Wegner estimate and level repulsion for Wigner random matrices", *Int. Math. Res. Not.* 2010:3 (2010), 436–479.

[Erdős et al. 2011] L. Erdős, B. Schlein, and H.-T. Yau, "Universality of random matrices and local relaxation flow", *Invent. Math.* 185:1 (2011), 75–119.

[Erdős et al. 2012a] L. Erdős, A. Knowles, H.-T. Yau, and J. Yin, "Spectral statistics of Erdős–Rényi Graphs II: Eigenvalue spacing and the extreme eigenvalues", *Comm. Math. Phys.* 314:3 (2012), 587–640.

[Erdős et al. 2012b] L. Erdős, B. Schlein, H.-T. Yau, and J. Yin, "The local relaxation flow approach to universality of the local statistics for random matrices", *Ann. Inst. Henri Poincaré Probab. Stat.* 48:1 (2012), 1–46.

[Erdős et al. 2012c] L. Erdős, H.-T. Yau, and J. Yin, "Bulk universality for generalized Wigner matrices", *Probab. Theory Related Fields* 154:1-2 (2012), 341–407.

[Erdős et al. 2012d] L. Erdős, H.-T. Yau, and J. Yin, "Rigidity of eigenvalues of generalized Wigner matrices", *Adv. Math.* 229:3 (2012), 1435–1515.

[Erdős et al. 2013] L. Erdős, A. Knowles, H.-T. Yau, and J. Yin, "Spectral statistics of Erdős–Rényi graphs, I: local semicircle law", *Ann. Probab.* **41**:3B (2013), 2279–2375.

[Ginibre 1965] J. Ginibre, "Statistical ensembles of complex, quaternion, and real matrices", *J. Mathematical Phys.* **6** (1965), 440–449.

[Guionnet 2011] A. Guionnet, "Grandes matrices aléatoires et théorèmes d'universalité (d'après Erdős, Schlein, Tao, Vu et Yau)", in *Séminaire Bourbaki* 2009/2010 (Exposé 1019), Soc. Math. de France, Paris, 2011.

[Gustavsson 2005] J. Gustavsson, "Gaussian fluctuations of eigenvalues in the GUE", *Ann. Inst. H. Poincaré Probab. Statist.* **41**:2 (2005), 151–178.

[Johansson 2001] K. Johansson, "Universality of the local spacing distribution in certain ensembles of Hermitian Wigner matrices", *Comm. Math. Phys.* **215**:3 (2001), 683–705.

[Johansson 2012] K. Johansson, "Universality for certain Hermitian Wigner matrices under weak moment conditions", *Ann. Inst. Henri Poincaré Probab. Stat.* **48**:1 (2012), 47–79.

[Khorunzhiy 2012] O. Khorunzhiy, "High moments of large Wigner random matrices and asymptotic properties of the spectral norm", *Random Oper. Stoch. Equ.* **20**:1 (2012), 25–68.

[Knowles and Yin 2013] A. Knowles and J. Yin, "Eigenvector distribution of Wigner matrices", *Probab. Theory Related Fields* **155**:3-4 (2013), 543–582.

[Lindeberg 1922] J. W. Lindeberg, "Eine neue Herleitung des Exponentialgesetzes in der Wahrscheinlichkeitsrechnung", *Math. Z.* **15**:1 (1922), 211–225.

[Maltsev and Schlein 2011a] A. Maltsev and B. Schlein, "Average density of states of Hermitian Wigner matrices", *Adv. Math.* **228**:5 (2011), 2797–2836.

[Maltsev and Schlein 2011b] A. Maltsev and B. Schlein, "A Wegner estimate for Wigner matrices", 2011. arXiv 1103.1473

[Mehta 1967] M. L. Mehta, *Random matrices and the statistical theory of energy levels*, Academic Press, New York, 1967.

[O'Rourke 2010] S. O'Rourke, "Gaussian fluctuations of eigenvalues in Wigner random matrices", *J. Stat. Phys.* **138**:6 (2010), 1045–1066.

[Pastur and Shcherbina 1997] L. Pastur and M. Shcherbina, "Universality of the local eigenvalue statistics for a class of unitary invariant random matrix ensembles", *J. Statist. Phys.* **86**:1-2 (1997), 109–147.

[Ruzmaikina 2006] A. Ruzmaikina, "Universality of the edge distribution of eigenvalues of Wigner random matrices with polynomially decaying distributions of entries", *Comm. Math. Phys.* **261**:2 (2006), 277–296.

[Schlein 2011] B. Schlein, "Spectral properties of Wigner matrices", pp. 79–94 in *Mathematical results in quantum physics*, edited by P. Exner, World Sci. Publ., Hackensack, NJ, 2011.

[Soshnikov 1999] A. Soshnikov, "Universality at the edge of the spectrum in Wigner random matrices", *Comm. Math. Phys.* **207**:3 (1999), 697–733.

[Soshnikov 2002] A. Soshnikov, "Gaussian limit for determinantal random point fields", *Ann. Probab.* **30**:1 (2002), 171–187.

[Tao and Vu 2010] T. Tao and V. Vu, "Random matrices: universality of local eigenvalue statistics up to the edge", *Comm. Math. Phys.* **298**:2 (2010), 549–572.

[Tao and Vu 2011a] T. Tao and V. Vu, "Random matrices: localization of the eigenvalues and the necessity of four moments", *Acta Math. Vietnam.* **36**:2 (2011), 431–449.

[Tao and Vu 2011b] T. Tao and V. Vu, "Random matrices: universality of local eigenvalue statistics", *Acta Math.* **206**:1 (2011), 127–204.

[Tao and Vu 2011c] T. Tao and V. Vu, "The Wigner–Dyson–Mehta bulk universality conjecture for Wigner matrices", *Electron. J. Probab.* **16**:77 (2011), 2104–2121.

[Tao and Vu 2012a] T. Tao and V. Vu, "Random covariance matrices: universality of local statistics of eigenvalues", *Ann. Probab.* **40**:3 (2012), 1285–1315.

[Tao and Vu 2012b] T. Tao and V. Vu, "Random matrices: universal properties of eigenvectors", *Random Matrices Theory Appl.* **1**:1 (2012), 1150001, 27.

[Wang 2012] K. Wang, "Random covariance matrices: universality of local statistics of eigenvalues up to the edge", *Random Matrices Theory Appl.* **1**:1 (2012), 1150005, 24.

tao@math.ucla.edu   *Department of Mathematics,*
                                University of California, Los Angeles, 405 Hilgard Avenue,
                                Los Angeles, CA 90095-1555, United States

van.vu@yale.edu   *Department of Mathematics, Yale University,*
                                10 Hillhouse Avenue, New Haven, CT 06511, United States

Printed in the United States
By Bookmasters